PRAISE FOR THE FIRST EDITION

FROM STUDENTS:

"The art is absolutely wonderful! I really like how the book divides topics into plates so the readers can specifically go to their weak areas... It is perfect for a new student in physiology."
  Tracey Mullauer, Culver-Stockton College (MO)

"The rich use of analogies in the text and the terrific illustrations... produce an explication of the concepts which was consistently more memorable and more readily understandable than the A&P textbook I was using. The style is a good mix of analogy, example, and details carefully woven together in a very readable format... The simple, memorable, and often beautiful drawings are one of the book's strongest parts."
  John Pozar, Chemeketa Community College (OR)

"The level is just right—a thorough summary... I appreciate the brevity of this style and prefer the delivery of "just the facts" compared to textbooks that are overly friendly and chatty."
  Jacqueline Tilly, Mesa Community College (AZ)

"The Physiology Coloring Book offers better, more complete explanations than the Princeton *Physiology Coloring Workbook*, which is a bit overly simplistic and contains diagrams that are too abstract. This one definitely comes across as more authoritative and scholarly."
  Peter Kaye, Brookdale Community College (NJ)

"The diagrams are quite effective and the illustrative analogies are amusing, such as the superman acting on an ion pump... This book is dynamic, exciting and fun."
  Meredith Blodget, Middlesex Community-Technical College (CT)

FROM PROFESSORS:

"The Physiology Coloring Book presents topics with an imaginative, interesting and refreshingly innovative style. It focuses attention on major points accurately and efficiently. The book is valuable for undergraduate, graduate, medical, nursing and dental students, as well as for others interested in the basic biological and clinical sciences."
  Thomas Adams, Michigan State University

"Absolutely excellent, concise, clearly written explanations of many basic physical principles that are essential to physiology but that are often times confusing to students... Illustrations are excellent."
  Terry Machen, University of California – Berkeley

"One of the strengths of this text is its amazing detail for such a brief treatment of the various topics. It also sticks with the basics in order to facilitate understanding."
  Jim Herman, Texas A&M University

"The graphics are excellent, providing a vibrant visual framework to help illustrate concepts. At the same time, a well written text adds detail... The art is far superior to the simplistic line art of the Princeton Review series and is much richer in texture."
  Steve Wickler, California State Polytechnic University – Pomona

"I think the book provides a fun and useful study aid for students."
  Mark Nelson, University of Illinois, Urbana-Champaign

"I reviewed a number of coloring books before selecting this book. I selected it because of my familiarity with the author's *Anatomy Coloring Book*. I liked the line drawings, support text, and price...The coloring book, in my course, is an integral part of the learning process."
  Stanley Irvine, College of Eastern Utah

ABOUT THE AUTHORS

WYNN KAPIT

Wynn Kapit is the designer of *The Physiology Coloring Book*. He also designed the immensely popular *Anatomy Coloring Book* and *The Geography Coloring Book*. *The Anatomy Coloring Book* spawned the genre of scientific coloring books and has become a classic with more than 2.5 million copies in print. Mr. Kapit received a B.B.A. and an L.L.B. from the University of Miami and an M.A. from the University of California, Berkeley.

ROBERT I. MACEY

Robert Macey is Professor Emeritus in the Department of Molecular and Cell Biology at the University of California, Berkeley. He was a Professor of Physiology and Chair of the Department of Physiology-Anatomy at Berkeley. Dr. Macey has written extensive research articles and reviews on membrane transport as well as a successful text on human physiology. He received his Ph.D. from the University of Chicago.

ESMAIL MEISAMI

Esmail (Essie) Meisami is a professor in the Department of Molecular and Integrative Physiology at the University of Illinois at Urbana-Champaign. He has authored and edited books on biology, physiology, human growth and development, and developmental neurobiology and has written numerous research papers on sensory systems and hormones in brain development. He received his Ph.D. in Physiology from the University of California, Berkeley.

Other Coloring Books available from Benjamin/Cummings Science:

The Anatomy Coloring Book, Second Edition by Wynn Kapit and Lawrence Elson
The Microbiology Coloring Book, by I. Edward Alcamo and Lawrence Elson

SECOND EDITION

THE PHYSIOLOGY COLORING BOOK

Wynn Kapit

Designer and Illustrator

Robert I. Macey

Professor Emeritus
of Physiology
University of California
at Berkeley

Esmail Meisami

Professor of Physiology
University of Illinois

An imprint of Addison Wesley Longman, Inc.

Originally published by HarperCollins College Publishers

San Francisco • Reading, Massachusetts • New York • Harlow, England
Don Mills, Ontario • Sydney • Mexico City • Madrid • Amsterdam

Publisher: Daryl Fox

Sponsoring Editor: Amy Folsom

Marketing Manager: Lauren Harp

Managing Editor: Wendy Earl

Production Editor: David Novak

Copy Editor: Jill Breedon

Typographer: The TypeStudio,
            Santa Barbara, CA

ISBN 0-321-03663-8

34 17

Benjamin/Cummings Science Publishing
1301 Sansome Street
San Francisco, California 94111

CONTENTS

Preface
Introduction

CELL PHYSIOLOGY

NERVE, MUSCLE & SYNAPSE

CIRCULATION

CONTENTS

RESPIRATION

KIDNEY

DIGESTION

CONTENTS

METABOLIC PHYSIOLOGY

BLOOD AND DEFENSE

REPRODUCTION

In this new edition of the *Physiology Coloring Book*, our methods and aspirations remain substantially unchanged. We render a self-contained, modern synopsis of human physiology. The material begins at the beginning—it is developed from the ground up and is suitable for both college and health professional students, as well as for self-study by educated laypersons. To accomplish our goals within the confines of 161 plates, we utilize the unique pedagogical features made available by the active process of coloring. The result is a non-conventional book providing an alternative or supplement to commonly used texts.

What are these features, and how do they apply to physiology? In the case of anatomy, the virtues of coloring are unmistakable. Classical anatomy is a visual science concerned with well-defined physical structures. Drawing these structures is a time-honored process that works because it cannot be done without personal attention to detail. In many ways, coloring structures is similar to drawing. It develops appreciation for shapes and relative proportions, and, perhaps more importantly, it introduces a kinesthetic sense into the learning process as hand motion is integrated with visual stimuli. Further, the use of color coding enhances awareness of relationships in and simplifies complex drawings in a way that is hardly attainable by other means. In addition, the use of colors to associate names with structures brings scientific jargon to life.

To the extent that physiology depends on structure, the same benefits accrue to it. However, descriptions of static anatomical structures are merely the starting point. Physiology's most distinctive feature is that it deals with dynamic processes. This is reflected in the prevailing and effective use of flow diagrams to describe forces, chemical reactions, flows, steady states, signals, and feedback. Because these concepts are necessarily abstract, they have not been standardized into any universally accepted symbolic representation, and they introduce significant difficulties for beginning students. This book addresses these difficulties in several ways.

In the first place, the liberal use of illustrations and even cartoons adds "flesh and blood" to flow diagrams, allowing students to associate process with locale or other more familiar ideas. In addition, using colors to associate names with structures and processes provides a seamless means to acquire scientific vocabulary. Further, the use of color codes makes it easy to follow common elements (e.g., H^+ ions in acid-base balance) in complex diagrams. But most importantly, the coloring process provides an urgently needed focus for first encounters with complex phenomena. In these instances, beginners commonly quit in despair, while the more experienced will break the problem into smaller, manageable parts and gradually piece the entire problem together. The act of coloring forces the student to confront a complex diagram one part at a time, making the novice feel more

secure in a state of "not knowing" for longer periods so that learning has a greater chance. Finally, the individual choices of colors make the project both personal and fun—a welcome diversion from stereotypical studies in which many long hours are spent going through the motions of soaking up information. We have had fun producing this volume; we hope you will have good times too.

Although the chapters are placed in linear sequence, it is not always necessary to follow them in the order of presentation. Some readers may find the beginning plates too abstract on first encounter; they may profit by starting with one of the organ systems in the latter sections of the book, returning to earlier plates as needed. In any case, it is highly desirable to read (and refer back to) the Introduction on the following pages, which explains a number of codes and symbols that are utilized consistently throughout the book.

In preparing this new edition we have reworked the text and/or figures of virtually every plate, making corrections and pedagogical improvements and introducing recent developments. We have also added nine plates with new material. Our attempts to depict physiological phenomena in semi-literal caricatures inevitably involve compromise; some topics are developed more intensely, at the expense of others. We are interested in readers' opinions on these issues, and we would also appreciate responses that point out any inaccuracies.

We are grateful for the expert critical advice of a number of our colleagues who have reviewed sections of the first edition and recommended changes. We thank Thomas Adams, Michigan State University; Sonya Conway, Northern Illinois University; John Forte, University of California, Berkeley; Jim Herman, Texas A&M University; Matilde Holzwarth, University of Illinois, Urbana-Champaign; Stanley R. Irvine, College of Eastern Utah; John J. Lepri, University of North Carolina, Greensboro; John Lovell, Kent State University; Terry Machen, University of California, Berkeley; Ann Nardulli, University of Illinois, Urbana-Champaign; Mark E. Nelson, University of Illinois, Urbana-Champaign; Shelia L. Taylor, Ozarks Technical Community College; and Steve Wickler, California Polytechnic State University. We were also fortunate to be guided by the perspectives of our own students as well as student reviewers: Meredith Blodget, Suzanne Click, Dorislee Jackson, Peter Kaye, Tracy Mullauer, Tami Platisha, John Pozar, and Jackie Tilley. In addition we thank Jill Breedon, our excellent copy editor, and Gerry Ichikawa of the TypeStudio in Santa Barbara for his fine contribution.

Special thanks to Lauren Kapit, Christa Zvegintzov, and Nooshin Meisami for their patience as well as valuable advice on literary and artistic matters. Finally, we thank Amy Folsom, our sponsoring editor at Benjamin Cummings, for her enthusiastic encouragement and for her gracious, but firm, hand in guiding this revision to completion.

Wynn Kapit
Robert I. Macey
Esmail Meisami

INTRODUCTION

(Just a few simple points to follow)

How Many Colors Needed

You should have at least 10 colored pens or pencils (not crayons). Pencils are more flexible in that they can be lightened or darkened depending on how hard you press. Pens, on the other hand, usually provide brighter colors and your finished work will more closely resemble a printed page.

Whichever medium you choose, the more colors you have, the greater your pleasure in coloring and the better your results will be. If you buy your colors separately (as opposed to a set), stock up on the lighter colors and be sure to include gray and black.

How the Coloring System Works

It is very simple: each of the illustrations has certain parts drawn with dark outlines and marked by small letter labels (A, B, C, etc.). Each of these parts to be colored is also identified by its name printed in outlined letters and followed by the same small label. Color both the name and the outlined part of the illustration to which it refers with the same color. Where a name is just a general heading and does not refer to a specific structure, the letter label is followed by a dash (A-, B-) and only the name is colored.

You should not use the same color for different letter labels on the same page unless the plate calls for more colors than you have, in which case you will have to begin repeating colors already used.

Occasionally, different outlined parts with different names will be sufficiently related to each other that they will all be marked with the same letter label. In such cases the letter label will have different superscripts (A^1, A^2, etc.) and the names and structures will all receive the same color. Please look at the front cover to see examples of labels with superscripts.

If the outlined names or darkly outlined structures are labeled with asterisks ($*$), color them gray. If they are labeled with the "don't color" symbol (-¦-), leave them uncolored.

How the Book is Arranged

The book is divided into sections: Respiration, Digestion, etc. Each section contains a group of plates. Each plate is composed of a text page on the left and an illustration page on the right. The text page introduces the topic of the plate and provides an overview.

A plate usually deals with only one topic, so it shouldn't be necessary for you to refer to preceding or following pages when focusing on that particular topic.

How to Approach Each Plate

Whether to read the text first and then color, or color the illustrations first and then read the text, is something you will probably decide for yourself. You may find that reading, coloring, and rereading is the best approach. But even if you prefer to color first you will still be aided by the presence of captions placed next to each illustration.

Before starting to color, please look over the color notes (CN) placed in the lower right-hand corner of the text page. They are there to tell you whether there is something unusual to take note of, or when to use a special color, or when to follow a certain sequence in coloring.

Where to Begin

You can begin anywhere, but it is recommended that you start at the beginning of whichever section you choose.

If you don't plan to color the entire book and just want to work on certain plates, then by all means color whichever one you wish. Each plate is self-contained and provides enough information to both easily color and comprehend the material.

SYMBOLS USED THROUGHOUT THE BOOK

These symbols are meant to save space and reduce the need for repeated explanations of certain actions, processes, or acting agents in various physiological events.

It isn't necessary to memorize them, as they will quickly become familiar to you as you begin to color the book. They are assembled here as a reference guide in the event you forget what they represent. It may or may not be necessary to color them; that will depend on how they are used and labeled within the illustration.

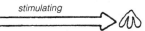

Any structure, substance, or action connected to either of these symbols is either increasing or growing in size.

These symbols represent the opposite activity: a decrease, decline, or reduction in size.

A long, solid arrow indicates a stimulating or activating action.

A long, broken arrow indicates a slowing down or inhibiting action.

A dotted arrow suggests that the object of this action has been inhibited or stopped.

A glucose (sugar) molecule; their combination into glycogen; and the test tube symbols for high or low blood glucose (sugar) levels.

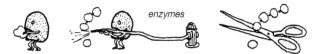

A common representation of a typical nerve cell.

The common fat (triglyceride) molecule and its glycerol and fatty acid components.

Individual amino acid molecules and their linking into proteins.

A gradient symbol representing the movement of substances from high to low concentrations.

Symbols for various enzymes and their activities: devouring, breaking down, or cutting apart molecules.

The powerful ATP molecule, and the energy released as it changes into the weak ADP.

Different transport mechanisms for moving substances across cell membranes.

A symbol for metabolism occurring within a cell.

"Living things consist of one or more cells."
"Each cell can live independently of the rest."
"Cells can arise only from other cells."

These three statements express the "cell doctrine," which implies that those parts of our body that live—that eat, breathe, move about, and reproduce—do so only through the cells that make up about two-thirds of our body weight. If physiology seeks to discover how living things work, it needs to express the explanations in terms of cellular activities.

Cells come in different sizes, shapes, and internal structures. Liver cells differ from brain cells, which differ from blood cells. All cells contain "mini-organs" called *organelles,* each specialized to perform a function. Although the cell portrayed in the plate cannot represent all cells, it does contain the following structures and organelles that commonly occur in most.

Cell membrane—This outer boundary of the cell consists of a thin (4–5 nm) sheet of fatty (lipid) molecules with embedded protein. In addition to structural proteins, some of these proteins provide pathways for transport and regulate flow of materials into and out of the cell. Others serve as receptors for chemical signals coming from other cells. Further, some membrane proteins serve as enzymes, while still others function as antigens that identify the individual self.

Nucleus—The most prominent cellular organelle, the nucleus contains genetic material: genes, DNA, and chromosomes. Information stored in genes is utilized in everyday cell life and reproduction. The nucleus also contains a smaller body, the *nucleolus,* that consists of densely packed chromosome regions together with some protein and some RNA strands. The nucleolus initiates the formation of *ribosomes,* structures that are required for protein synthesis. The nucleus is surrounded by a double membrane that is riddled with pores involved in transporting materials between the nucleus and the rest of the cell.

Cytoplasm/Cytosol—Occupying the space between the nucleus and the plasma membrane, the cytoplasm contains membrane-bound organelles, ribosomes for synthesizing cytoplasmic proteins, and a complex network of filaments and tubules called the *cytoskeleton.* The fluid portion of the cytoplasm in between these structures, the *cytosol,* contains many protein enzymes (catalysts used in cellular chemistry).

Mitochondria—These "power houses" of the cell are the sites where chemical energy contained within nutrients is trapped and stored through the formation of *ATP* molecules. ATP, in turn, serves as an energy "currency" to carry out cellular work, supplying the energy required for movement, secretion, and synthesis of complex structures.

Endoplasmic reticulum—The endoplasmic reticulum (ER) is a network of tubes and flattened sacs, formed by membranes, that is distributed throughout the cytoplasm. Some ER (*rough ER*) has a granular appearance because of attached ribosome particles. These are sites for synthesis of proteins destined for organelles, for cell membrane components, or for secretion to the cell exterior (e.g., hormones). *Smooth ER* lacks attached ribosomes. It is commonly involved in lipid metabolism, but it can also serve in detoxification of drugs and deactivation of steroid hormones. In muscle cells, smooth ER (called sarcoplasmic reticulum) sequesters large amounts of calcium, which are used to trigger muscular contraction.

Golgi apparatus—Sets of smooth membranes forming flattened, fluid-filled sacs that are stacked like pancakes are called the Golgi apparatus, which is involved in modifying, sorting, and packaging proteins for delivery to other organelles or for secretion out of the cell. Numerous membrane-bound vesicles are frequently found around the Golgi apparatus. They probably carry material between the Golgi and other organelles of the cell (e.g., receiving protein-laden vesicles from the rough ER or delivering other vesicles to the plasma membrane).

Endo- and Exocytotic vesicles—These membrane-enclosed vesicles traveling from (and to) the plasma membrane are important carriers for protein delivery into (or out of) the cell. Exocytosis (secretion) involves an actual fusion of the vesicular membrane with the plasma membrane, enabling vesicle contents to be expelled (secreted) outside the cell. In *endocytosis (pinocytosis, phagocytosis)* the reverse occurs: the plasma membrane infolds and engulfs extracellular material; then a membrane-bound vesicle (containing the material and surrounding fluid) buds off and is incorporated into the cell.

Lysosomes—These membrane-bound vesicles contain many enzymes capable of digesting cellular products or damaged organelles, as well as bacteria brought into the cell via endocytosis. The fatal Tay-Sachs disease arises from a congenital lack of lysosomal enzymes that digest components of nerve cells (glycolipids). These accumulate in the cells, causing them to swell and degenerate.

Peroxisomes—These also are membrane-bound vesicles containing digestive enzymes. They break down long-chain fatty acids as well as some toxic substances. Genetic defects in peroxisome membrane transport causes the fatal childhood diseases Zwellinger's syndrome and X-linked adrenoleukodystrophy.

Cytoskeleton—The cytoskeleton consists of arrays of protein filaments that form networks within the cytosol, giving the cell its shape. These filaments also provide a basis for movement of the entire cell as well as internal movements of its component organelles and proteins. The three major types of cytoskeleton filaments are microtubules (25 nm diameter), actin filaments (25 nm—see next plate), and intermediate filaments (10 nm—next plate). Intermediate filaments are strong stable structures which protect cells from mechanical stress. Microtubules undergo frequent changes, growing or retracting by the addition or subtraction of their molecular building blocks (*tubulin*). They usually grow out of organizing centers, e.g., *centrosomes* (which are important during cell division). As they extend outward, they form a system of intracellular tracks that are used to transport vesicles, organelles, and other cellular components to different positions within the cell. This movement is driven by specialized *motor molecules (dynein, kinesin).* By subtle changes in shape, these motor molecules can attach, release, and then reattach to successive positions so that the motor molecule "walks" on the filament. The other end of the motor may be attached to the cargo that is moved. Although actin filaments can form stiff permanent structures, like microtubules they can also grow and retract, and they are involved in a large array of cell movements including cell crawling, phagocytosis, and muscle contraction. A number of different proteins can bind to the filaments. The action of the filament is determined by the specific proteins that are bound. The proteins called myosin, are motor molecules (plate 21).

CN: Use your lightest colors for A and G.
1. Begin in the upper left corner by coloring the title, structural example, and the related structure in the central illustration of the entire cell. Do this for each structure as you work clockwise around the page. Note that the space between membranes of the rough endoplasmic reticulum (is not colored in the example on the right but colored in the central drawing for reasons of identification.

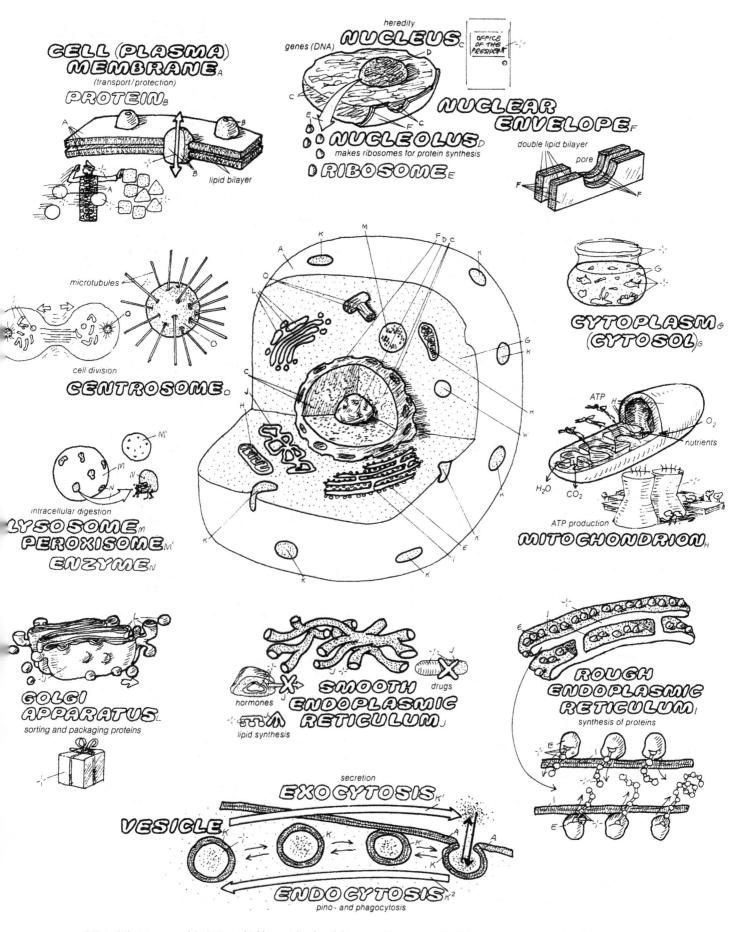

CELL (PLASMA) MEMBRANE A
(transport/protection)
PROTEIN B
lipid bilayer

heredity
NUCLEUS
genes (DNA)
OFFICE OF THE PRESIDENT
NUCLEAR ENVELOPE F
NUCLEOLUS D
makes ribosomes for protein synthesis
RIBOSOME E
double lipid bilayer
pore

microtubules
cell division
CENTROSOME O

intracellular digestion
LYSOSOME M
PEROXISOME M'
ENZYME N

CYTOPLASM (CYTOSOL) G

ATP
O₂
nutrients
H₂O   CO₂
ATP production
MITOCHONDRION H

GOLGI APPARATUS
sorting and packaging proteins

hormones
SMOOTH ENDOPLASMIC RETICULUM J
drugs
lipid synthesis

ROUGH ENDOPLASMIC RETICULUM I
synthesis of proteins

secretion
EXOCYTOSIS K'
VESICLE K
ENDOCYTOSIS K²
pino- and phagocytosis

Cells in different organs of the body are highly specialized, and this specialization is often reflected in structural variations. Although the generalized cell depicted above cannot represent any particular cell, it does contain structures and organelles that commonly occur in most. All cells are bounded by a plasma membrane, a continuous double-layered sheet of lipid molecules containing embedded proteins. Similar membranes form a number of structures within the cell. Essentially, all cells have a membrane-bound nucleus that contains the genetic instructions (genes). By expressing information stored in the genes, the nucleus directs everyday cell life and reproduction. The space between the plasma membrane and the nucleus is called the cytoplasm. Membrane-bound organelles, as well as the filaments and microtubules that compose the cytoskeleton, are suspended within the cytoplasmic fluid, the cytosol.

Although there are many different kinds of human cells, they can be classified into four broad types: (1) muscle cells, specialized for generation of mechanical force and movement; (2) nerve cells, specialized for rapid communication; (3) connecting and supporting tissue cells, including blood and lymph; and (4) epithelial cells for protection, selective secretion, and absorption. This plate focuses on epithelial cells to illustrate how groups of these cells adhere to one another to form tissues and how specialized structures—cell junctions, microvilli, and cilia—support special functions. Other cell types are taken up in more detail in the context of specific organs.

SHEETS OF EPITHELIAL CELLS DIVIDE BODY COMPARTMENTS

Epithelial cells adhere to one another, often forming layered sheets with very little space between cells. They are found at surfaces that cover the body or that line the walls of tubular or hollow structures. Thus, epithelial cells are found in the skin, kidney, glands, and linings of the lungs, gastrointestinal tract, bladder, and blood vessels. Sheets of them often form the boundaries between different body compartments, where they regulate exchange of molecules between compartments. Virtually all substances that enter or leave the body must cross at least one epithelial layer. For example, the small intestine forms a hollow cylinder whose interior lining is populated by several types of epithelial cells. Some secrete digestive enzymes, others absorb nutrients, still others secrete a protective mucus. In each case, the epithelial cells are called upon to transport materials in one direction only: either from blood vessels (embedded within the intestinal walls) to the hollow interior (lumen) of the cylinder in the case of secretion, or from lumen to blood in the case of absorption. Thus, the cell must have a "sense of direction"; it must "know" the difference between the lumen side and the blood side. The cell cannot be completely symmetrical, and its asymmetry in function is reflected in an asymmetrical or polar structure.

Structural asymmetry, revealed by both cell shape and organelle position, is established and maintained by an elaborate cytoskeleton. In addition, there are striking differences in the plasma membranes located at various sides of the cell. We identify three different surfaces of epithelial cells: (1) The apical or mucosal surface faces the outside environment or the lumen of a particular organ. (2) The basal surface is on the opposite side, the side that lies closest to the blood vessels. (3) The lateral sides face neighboring epithelial cells. Each of these membrane surfaces contains different proteins and structures required for normal function.

EPITHELIAL CELLS ADHERE TO & COMMUNICATE WITH THEIR NEIGHBORS

The lateral surfaces of epithelial cells must adhere to one another to maintain their sheetlike structure and to provide tight seals that retard leakage of solute and water between adjacent cells. If substances do move across the epithelial layer, it is generally because they are selectively recognized and transported by the cells themselves. Discrete structures called desmosomes provide a major source of this adhesion. They lie close to, or within, the membrane and bind the cells together where they come in contact. Other specialized contact sites (tight junctions) are used to plug potential leaks; still

others (gap junctions) are used for cell-to-cell communication. Collectively, these contact sites are called cell junctions.

DESMOSOMES PROVIDE STRONG ADHESION

Desmosomes are regions of tight adhesion between cells that give the tissue a structural integrity. At a desmosome, there is a small extracellular space between the two cell membranes that is filled with a fine filamentous material that probably cements the two cells together. There are two types of desmosomes: belt desmosomes (continuous zones of attachment that encircle the cell) and spot desmosomes (attachments to small regions of contact, often compared to "spot welds").

TIGHT JUNCTIONS PREVENT LEAKAGE & PRESERVE POLARITY

Tight junctions form very close contacts between neighboring cells, leaving virtually no space between. These junctions extend around the entire circumference of the cell, providing a tight seal that prevents leakage of fluids and materials. They also preserve the cell's asymmetry by preventing cell membrane–bound proteins from migrating, within the membrane, along the circumference of the cell, from one side to the other.

GAP JUNCTIONS PROVIDE COMMUNICATION

Gap junctions are specialized for communication between adjacent cells. They consist of an array of six cylindrical protein subunits that spans the plasma membrane and reaches out a short distance into the extracellular space. The subunits are bunched together with their long axes parallel to one another in a manner that forms an open space or channel about 1.5 nm wide running the entire length of the array. These channels act as pores that tunnel through the membrane, but the tunnels do not empty into the extracellular space. Instead, each array attaches to a similar array in an adjacent cell, forming a tunnel of double the length with the entrance in one cell and the exit in the adjacent cell. These tunnels are wide enough to allow small solutes and common ions to pass. Thus, the junctions provide for passage of electrical and chemical signals between cells, allowing them to function in unison. Under certain circumstances (e.g., a rise in intracellular Ca^{++}), the central channel closes, isolating the involved cell from others. Gap junctions are particularly important in coordinating heart, smooth muscle, and epithelial cell activities.

MICROVILLI INCREASE CELL SURFACE AREA

Microvilli are small, fingerlike projections found on the apical surface of epithelial cells. They are most abundant in tissues that primarily transport molecules across the epithelial sheet. Microvilli are advantageous because they greatly increase the surface area available for transport (e.g., by a factor of 25 in the intestine). Actin filaments, anchored at their base in the terminal web of fibers and running the entire length of the microvilli, provide support for their upright position.

CILIA PROPEL FLUIDS & PARTICLES ALONG CELL SURFACE

Cilia are very long projections from the apical surface that are involved in transporting material along (i.e., tangential to) the epithelial surface rather than through it. They are abundant in the respiratory tract, oviducts, and uterus. They function by "beating" (i.e., by whiplike movements that mechanically propel fluids and particles on the cellular surface in the direction of a rapid forward stroke). An array of microtubules that runs the length of each cilium mediates these motions.

CN: Use the same color for the plasma membrane (F) as was used on plate 1.
1. Begin with the 3-dimensional drawing of epithelial cells on the right. As you color each structure, complete its corresponding structure in the cross-sectional diagram on the left. The latter contains

additional structures that should be colored as well. Note that A, D, and L all are parts of the plasma membrane (F) but receive different colors.
2. Among the list of titles, notice that the functions of structures H–N are placed in parentheses and are colored gray

APICAL SURFACE:
CILIA,
MICROTUBULE,
BASAL BODY,
MICROVILLI,
MICROFILAMENT (ACTIN)

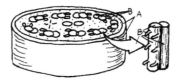

LATERAL SURFACE:
LATERAL PLASMA MEMBRANE,
TIGHT JUNCTION, (IMPERMEABLE)*
DESMOSOME: BELT (CELL-TO-CELL)*
DESMOSOME: SPOT (ADHERENCE)*
GAP JUNCTION, (INTERCELLULAR
COMMUNICATION)*

BASAL SURFACE:
HEMIDESMOSOME (ADHERENCE)*
BASAL PLASMA MEMBRANE,

TERMINAL WEB,
CYTOSKELETON (INTERMEDIATE FILAMENTS)
EXTRACELLULAR
SPACE,
NUTRIENTS &
METABOLITES,

Epithelial cells adhere to one another, often forming layered sheets that cover the body and organs, or line the wells of tubular or hollow structures (skin, kidney, glands, and linings of the lungs, gastrointestinal tract, bladder, and blood vessels). They have 3 distinct surfaces: (1) the apical surface facing the outside environment or the lumen of an organ, (2) the basal surface, on the opposite side, facing blood vessels, and (3) the lateral sides facing neighboring epithelial cells.

THE APICAL SURFACE sometimes contains microvilli and cilia. **Microvilli** increase the area of the apical surface severalfold. Actin **filaments**, anchored in the terminal web and running the length of each microvillus, are believed to support their upright position. **Cilia** are involved in transporting material tangential to the epithelial surface by whiplike movements that propel fluids and particles on the cellular surface in the direction of a rapid toward stroke. These motions are mediated by **microtubules** that run the length of each cilium in a characteristic 9 + 2 array (9 pairs of microtubules forming a ring around a central pair). Each cilium is anchored in a **basal body**. The cilium is bent as the pairs of microtubules slide past one another.

THE LATERAL SURFACE contains 3 types of junctions: (1) **desmosomes**, for adherence of neighboring cells, (2) **tight junctions**, for sealing leaks between cells, and (3) **gap junctions**, which provide open channels between cells for electrical and chemical communication.

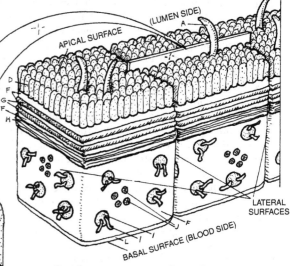

Spot desmosomes cement cells together at discrete locations. Within the cell, they are connected by a profuse network of filaments, a part of the **cytoskeleton** that helps provide mechanical stability. **Belt desmosomes** encircle the entire cell with an intracellular filamentous cement. Within the cell, the belt desmosome has a band of cylindrical actin filaments (shown in cross-section) just adjacent to the inner part of the cell membrane.

THE BASAL SURFACE's **plasma membrane** is attached to the basement membrane (basal lamina), a porous structure containing collagen and glycoproteins that separates the epithelial cells from underlying connective tissues, nerves, and blood vessels. The attachment is strengthened by **hemidesmosomes** (half–spot desmosomes).

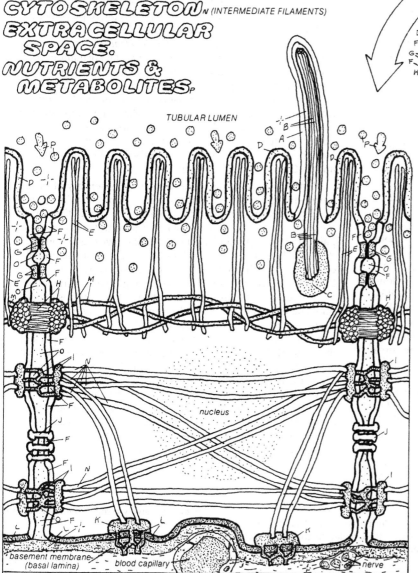

TUBULAR LUMEN

nucleus

basement membrane (basal lamina)　blood capillary　nerve

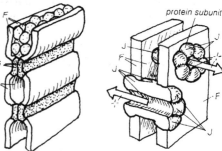

TIGHT JUNCTIONS　　GAP JUNCTIONS

protein subunit

No cell lives forever. With a few exceptions (notably nerve and muscle cells), the cells of your body are not the same ones that were present just a few years ago. "Old" cells apparently wear out, die, and are continually replaced by new ones. On average, intestinal cells live for only 36 hours, white blood cells for 2 days, and red blood cells for 4 months; brain cells may live for 60 years or more. Growth also requires the production of new cells. As cell size increases, cells become less efficient because distances from the plasma membrane to the more central portions of the cell also increase, making the transport of such essentials as O_2 into and CO_2 out of the cell more difficult. These difficulties do not arise because growth occurs primarily by increasing the number of cells rather than increasing the mass of individual cells.

CELL DIVISION

In cell division one parent cell divides into two daughter cells to create new cells. Although some characteristics (e.g., weight) of the daughters may be different from the parent, they are identical in the most important way: they both carry the same fundamental set of genetic instructions that govern their activities and reproduction. This instruction set, *genetic code*, is provided by the detailed structure of DNA (*deoxyribonucleic acid*) molecules that are packaged within the cell nucleus. Replication of these molecules and their distribution to each daughter cell ensures the continuity of cell characteristics with each division. Processes involved in the cell cycle take place in three phases.

1. Interphase: The cell increases in mass — This occurs through the synthesis of a diversity of molecules, including an exact copy of its DNA. That portion of interphase in which DNA synthesis takes place is called the S period; it is preceded and followed by two "gap" periods called G1 and G2 respectively (see illustration). During the S period, the centrosomes also duplicate.

2. Mitosis: DNA is replicated and moved — Following G2, the cell enters mitosis, a stage in which the replicate sets of DNA are bundled off to opposite ends of the cell in preparation for the final stages in which the cell splits in two (follow the diagrams in the plate for details). Mitosis begins when DNA molecules, which had been unwound during interphase, become highly coiled and condense into rod-shaped bodies known as *chromosomes*. At this stage, each chromosome is split longitudinally into two identical halves called chromatids. Each *chromatid* contains a copy of the duplicated DNA along with some protein that provides a scaffold for the long DNA molecules and helps regulate DNA activity. Meanwhile, the nuclear envelope begins to degenerate, and, outside the nucleus, *centrosomes* migrate to opposite ends of the cell to form an elaborate structure of *microtubules* called a *spindle*. Each chromosome, attached to these microtubules, lines up at the cell's equator in such a way that its two chromatids are attached to microtubules leading to opposite ends of the cell. The microtubules then pull on the chromatids, moving a complete set to opposite parts of the cell. Finally, the chromatids at both ends of the cell begin to unwind and become indistinct while a new nuclear envelope forms around each of the two sets of chromatids.

3. Cytokinesis: The cell divides — This is the final stage. Cytoplasm division takes place as a furrow develops, becoming deeper and deeper until the original cell is pinched in two, and the daughter nuclei, formed during mitosis, are enclosed in separate cells. At this point, the daughter cells enter the G1 stage of interphase, completing the cycle.

DNA REPLICATION

If DNA is the heredity material, two important questions arise. First, how is DNA replicated so that it can be passed undiluted from generation to generation? Second, how does DNA carry the information needed for directing cellular activities? Answers to both questions require information about the chemical structure of DNA.

DNA forms a double helix — A DNA molecule contains two extremely long "backbone" chains made of many 5-carbon sugars (*deoxyriboses*) connected, end on end, via a phosphate linkage (i.e.,...sugar-phosphate-sugar-phosphate...). Like the legs of a ladder, these backbone chains run parallel to one another. They are connected at regular intervals by *nitrogenous bases*, which form the "rungs" of the ladder. It takes two bases to span the distance between the legs; the two are connected in the center of the span by weak chemical bonds, *hydrogen bonds*. Finally, the legs of the ladder are twisted into a helical structure, making one complete turn of the helix every ten "rungs" of the ladder. Since each leg of the ladder forms a helix, DNA is a double helix

Base pairs A–T and G–C are complementary — The particular bases that form the rungs and their relative placement within the ladder structure are the key to our problems. Only four different base species form DNA: *adenine* (abbreviated as A), *guanine* (G), *cytosine* (C), and *thymine* (T). Formation of each ladder rung requires two of these, but not any two. The two bases, like pieces in a jig-saw puzzle, must have the proper size and shape and must be able to interlock (form hydrogen bonds) within a given constellation. Examination of DNA structure shows that rungs can be formed by a combination of A with T (A–T) or G with C (G–C), but all other possible combinations, like A–A, A–C, or G–T, will not work. A–T and G–C are called *complementary base pairs*.

Replication requires separation and reassembly of base pairs — Imagine that you and another person each take hold of one leg of the ladder and pull. It will come apart at the seams (i.e., at the center of the rungs where the complementary base pairs are held together by relatively weak hydrogen bonds). You each take one strand (half of the structure) consisting of one long leg with single bases attached and, separately, you both begin to reconstruct the missing half. The missing leg is no problem; it is always the same string of deoxyribose and phosphate. But the bases are also prescribed: to every A on the single strand, you attach a T, to every T an A, to every G a C, and to every C a G. You have reconstructed an exact replica of the original DNA, and so has your partner. There are now two copies in place of the original; precise replication has been accomplished. A similar process takes place within the cell, only here the strands are separated bit by bit, and synthesis of new DNA follows closely behind in the wake of the separation, aided by the action of special enzymes, *DNA polymerases*. A discussion of our second problem, how DNA carries the hereditary material, is taken up in plate 4.

CN: Use a dark color for D.
1. Begin by coloring the cell at the top of the page, and then color the circular diagram of the cell cycle immediately below it.
2. Color the stages of the cell cycle, beginning with interphase near the top left-hand side and continuing progressively through the stages of mitosis and cytokinesis.
3. Color the schematic respresentation of DNA structure replication along the right-hand side. Among the bases note that guanine (I) and cytosine (I') are cross-hatched.

INTERPHASE c'

During interphase: 1. Uncoiled DNA (contained in chromatin) replicates. 2. Following replication, DNA becomes active in directing the RNA and protein synthesis required for cell division. 3. The centrosomes duplicate.

MITOSIS d'

PROPHASE d'

During prophase: 1. The nuclear envelope begins to break down. 2. The two copies of DNA begin to coil and supercoil, forming chromosomes. 3. Centrosomes separate, migrating toward the cell poles (located at the extreme right and left in the illustration). 4. The centrosomes organize microtubules which form the mitotic apparatus.

METAPHASE d'

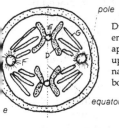

pole

equator

e

During metaphase: 1. The nuclear envelope and the nucleolus disappear. 2. The chromosomes line up around the cell equator (imaginary line connecting the top and bottom in the illustration).

ANAPHASE d'

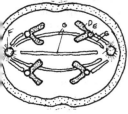

During anaphase: 1. Microtubules attach to proteins (kinetochore) that are bound to a constricted region (centromere) of the chromatid. 2. The sister chromatids are pulled to opposite poles of the cell by microtubules of the spindle.

TELOPHASE d'

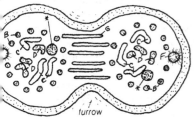

furrow

During telophase: 1. A new nuclear envelope forms around the chromosomes near each of the two poles as two nuclei and two nucleoli begin to appear. 2. The chromosomes uncoil, forming chromatin. 3. The spindle fibers disappear.

CYTOKINESIS A'

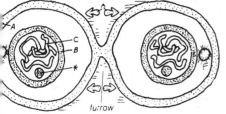

furrow

During cytokinesis, the two daughter cells separate. A furrow (narrowing) forms along the equator and progressively constricts the cell until it separates in two. First signs of a furrow can be seen as early as anaphase.

PLASMA MEMBRANE A
NUCLEAR ENVELOPE B
CHROMATIN C
CHROMOSOME (4-6) D
KINETOCHORE E
CENTROSOME F
SPINDLE FIBERS G

DNA REPLICATION c'

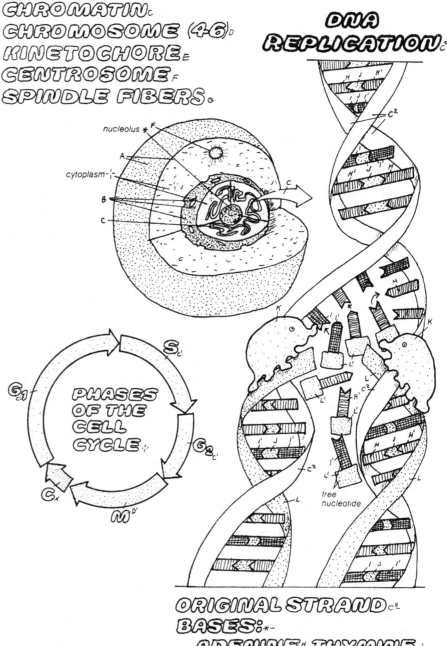

nucleolus * F
cytoplasm i'
free nucleotide

PHASES OF THE CELL CYCLE

ORIGINAL STRAND c²
BASES: *
ADENINE, THYMINE H'
GUANINE, CYTOSINE I'
HYDROGEN BOND J
DNA POLYMERASE K
NEW STRAND L
BACKBONE L'

Replication is the process in which an exact copy of the DNA molecule is made. The double-stranded DNA in the chromatin unfolds, separating at sites where the two strands are attached (hydrogen bonding sites between the complementary bases adenine, thymine, guanine, and cytosine). Each strand consists of a backbone (leg of the ladder) plus attached bases. Aided by enzymes such as DNA polymerase and using the original separated strands as a template, two new strands are synthesized as nucleotide building blocks (molecules containing the base together with the backbone materials, sugar and phosphate) are attached to the template. An exact copy is always obtained because complementary bases are the only ones that are attached. An adenine will attach only to a thymine (and vice versa); a guanine attaches only to a cytosine (and vice versa).

To understand how DNA directs the cell, we begin by observing that such activities as growth, reproduction, secretion, and motility are all derived in the final analysis from chemical reactions. Of the large numbers of products that theoretically could be formed from chemicals used by the cell, only a few are produced within the cell. These products are "selected" by the action of enzymes, catalysts that speed specific reactions. Left by themselves, most of the plausible reactions proceed too slowly to be significant. The presence of a specific enzyme "turns on" a specific reaction simply by speeding it. In this way, enzymes control chemical reactions and cellular activities. But what controls the enzymes? They are made of protein and are synthesized within each cell. It follows that whatever controls protein synthesis controls which enzymes are present and therefore controls the cell. DNA plays its dominant role because it contains detailed plans for each protein that is synthesized. This determines the growth and development of individual cells, of tissues, and of the entire organism.

PROTEINS ARE MADE OF AMINO ACIDS
Proteins are giant molecules constructed by linking large numbers of amino acids, end to end, by special chemical bonds (peptide bonds) so that they form a chain. There are only 20 different kinds of amino acids in proteins, and because proteins often contain hundreds of them, the same kind of amino acid must appear in more than one position along the chain. We can compare amino acids with letters of the alphabet and protein molecules with huge words. Just as the word is determined by the precise sequence of letters, so is the protein (and its properties) determined by the sequence of amino acids along the chain. It follows that if DNA contains the "blueprints" for protein construction it must contain the amino acid sequence of that protein. But how?

EACH AMINO ACID IS CODED BY A THREE-BASE SEQUENCE
DNA (plate 3) is also made of large numbers of building blocks, the *nitrogenous bases*, and the properties of the DNA molecule are determined by the sequential placement of these bases as "rungs" in the ladderlike chain structure. Each DNA is also like a huge word with the bases representing letters of the alphabet. However, although proteins are based on a 20-letter "alphabet" (20 amino acids), DNA has only four bases: *adenine* (A), *guanine* (G), *cytosine* (C), and *thymine* (T). Somehow the sequence of just four different kinds of bases along the DNA ladder provides a code for the placement of 20 different kinds of amino acids in a protein chain. There cannot be a one-to-one correspondence of the letters in the two alphabets, for if each base corresponded to a single amino acid, then DNA would be able to code only for proteins containing at most four different amino acids. Instead, a sequence of three bases is used to code for each amino acid. For example, when the bases C, C, G occur one right after the other in the DNA ladder, it is a code for the amino acid glycine; the sequence A, G, T codes for the amino acid serine. The sequence C, C, G, A, G, T is a signal for part of a protein where serine follows glycine. By using bases three at a time, it is possible to form 64 unique combinations (e.g., AAA, AAG, ... CCA, CTC, ... TTC, ... etc.), far more than necessary to code for 20 amino acids.

MESSENGER AND TRANSFER RNA
How do cells actually translate the code and build proteins? DNA always remains within the nucleus, yet proteins are synthesized in the cytoplasm. A first step is to make a copy of the "blueprints" and transport it into the cytoplasm, a process called *transcription*. The transcript (copy) of this genetic code is a molecule called *messenger ribonucleic acid* (mRNA), which moves to the cytoplasm, where it associates with particles called *ribosomes*, the assembly sites for new proteins. Meanwhile other RNA molecules, tRNA (*transfer ribonucleic acid*), pick up loose amino acids in the cytoplasm that have been activated (energized) in preparation for use. Each tRNA molecule, with a single specific amino acid attached, migrates to the ribosomes, where its amino acid will be utilized at the appropriate position as it detaches from the tRNA and links to the emerging protein chain.

TRANSCRIPTION: mRNA GETS THE "MESSAGE"
Given this scenario, two problems arise. The first is *transcription*: how are DNA blueprints copied onto RNA? The second is *translation*: how is the code utilized so that amino acids are always linked to the protein in the proper sequence? Answers to both questions are based on RNA's close resemblance to DNA. They differ in that (1) they have slightly different sugars (deoxyribose and ribose); (2) RNA is usually single-stranded, containing only one leg of the ladder together with nitrogenous bases forming half "rungs" along its length; and (3) like DNA, RNA contains A, G, and C, but T is replaced by a very similar molecule, *uracil* (U). Thus, RNA is a similar "four-alphabet" molecule with letters A, G, C, and U. All RNA, but in particular mRNA, is formed from DNA in the same way that DNA makes more DNA. The double-stranded DNA "unzips" a bit, and one of the legs serves as a template for RNA construction. As in DNA synthesis, the sequence of bases in RNA is complementary to the sequence in the DNA template that formed it. A piece of DNA with sequence AGATCTTGT, for example, will make a piece of RNA with sequence UCUAGAACA. Each base triple (three letters) in mRNA is called a *codon*. The transcription problem is solved by constructing a strand of RNA, which does not duplicate the base sequence of the original DNA, but rather contains the complementary base sequence as a codon.

TRANSLATION: mRNA AND tRNA INTERACT
tRNA molecules are shaped like a cloverleaf. The stem contains the attachment site for the amino acid, and the loop contains a specific set of three bases (called an *anticodon*), which is the code for the amino acid that will become attached. Because the mRNA codons contain the complementary bases to the DNA and hence to the amino acid code, it follows that mRNA and tRNA have complementary sets of bases and that they will easily form loose H bonds. The tRNA simply lines up along the mRNA sites, as illustrated, so that the amino acids are now in proper sequence and can be linked by peptide bonds. Actually, the ribosome moves along the mRNA strand and, as illustrated, handles only two amino acids at a time. After the peptide bond is formed between the two amino acids, the tRNA that has resided longest on the ribosome detaches, leaving a vacant position for the next tRNA (and amino acid) with the complementary anticodon to attach. In this way, the protein chain grows until the final one or two codons on the mRNA signal the end. Following this translation process, proteins are often modified by folding, shortening, or adding carbohydrates, a process called *postranslational modification*.

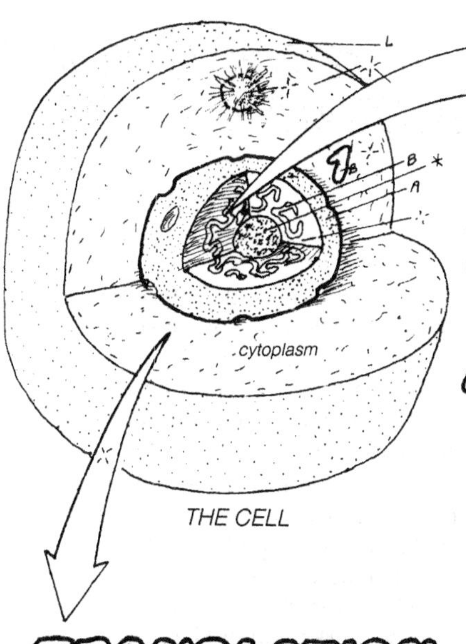

TRANSCRIPTION
DNA ⇒ RNA
(IN THE NUCLEOUS)

THE CELL

cytoplasm

NUCLEAR ENVELOPE A
CHROMATIN B

DNA HELIX:
BACKBONE B' BASE C

hydrogen bond

nuclear pore

RNA SYNTHESIS:*
MESSENGER RNA, (BACKBONE D
CODON TRIPLETS E (BASES) E'
RNA POLYMERASE ENZYME F
TRANSFER RNA (t RNA) G
ANTICODON TRIPLET H

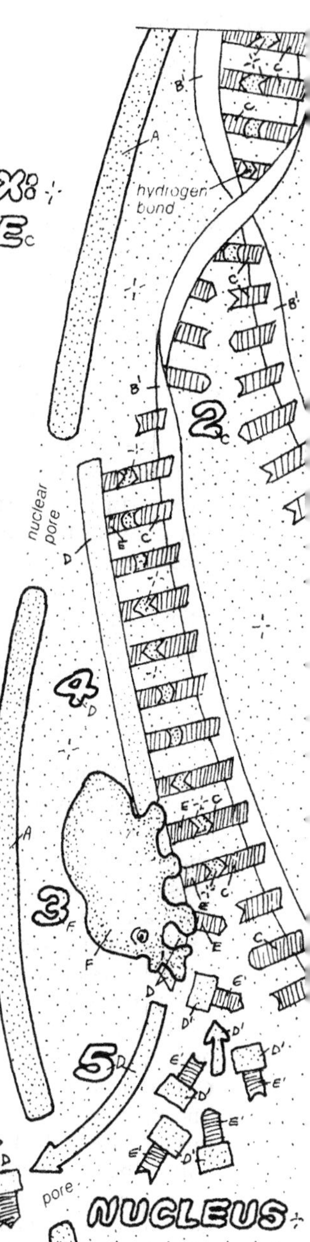

TRANSLATION
RNA ⋯ PROTEIN
(IN THE CYTOPLASM)

When DNA is not engaged in replication, it directs the synthesis of proteins, which carry out various cell functions. To do this, the DNA produces special copies of its codes in the form of functionally distinct types of RNA molecules (TRANSCRIPTION). These RNA molecules then move to the cytoplasm, where they interact to express the codes relayed by the DNA, resulting in the synthesis of specific polypeptide chains/proteins (TRANSLATION). All proteins are chains of 20 different amino acids (AA) arranged in various orders. To make RNA, the DNA in the nucleus (1) partially unfolds (2). Aided by the RNA polymerase enzymes (3), the appropriate RNA molecules (4), such as the messenger (mRNA) (5) and transfer (tRNA) (6), are formed and sent to the cytoplasm, where AAs are supplied (7). Energized by ATP (8), each AA binds with the corresponding tRNA molecule to form tRNA-AA complexes (9). The mRNA (5) carries the codes for the arrangement of AAs in the polypeptide chain. The mRNA interacts with the ribosomes (R) (10). Each R has a small and large subunit. As the R moves along the mRNA (11), the tRNA molecules bind with the R-mRNA complex, according to matching of the codes in the mRNA with those in the tRNA. While in the R, the AA attached to the various tRNA molecules form peptide bonds between themselves, resulting in the formation of a polypeptide (protein) chain. The chain dissociates from the R assembly line (12) when synthesis is completed. In secretory cells, R synthesizes proteins while attached to the endoplasmic reticulum.

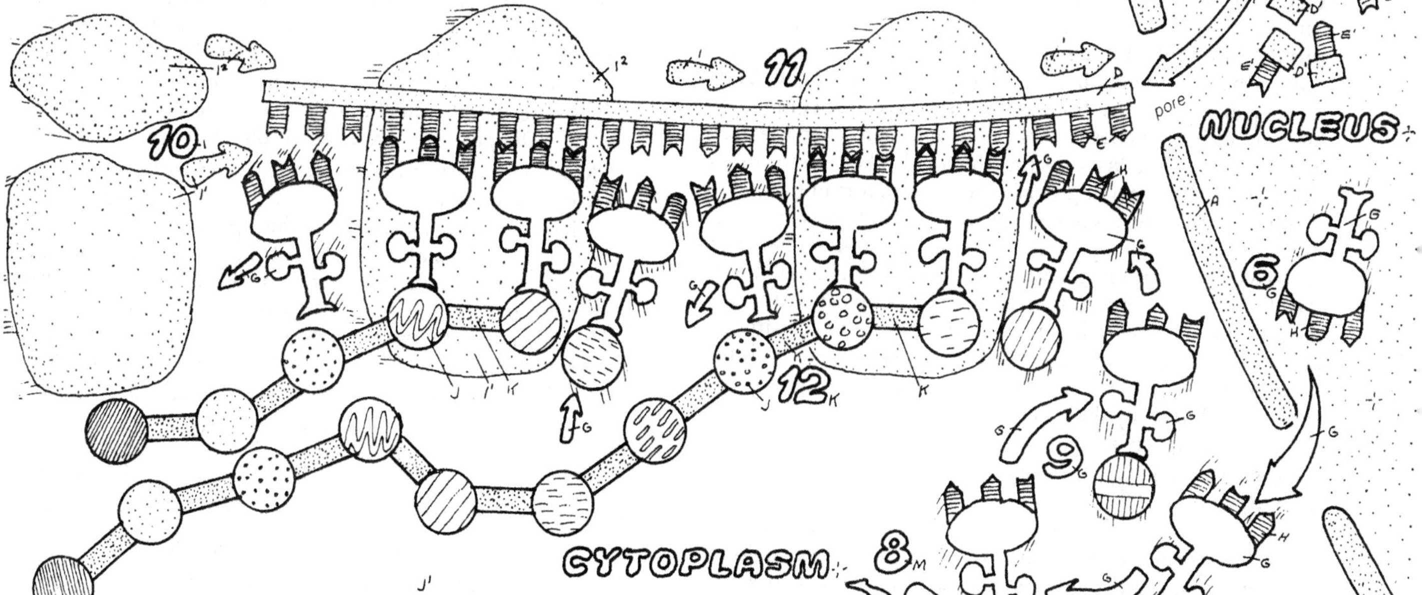

RIBOSOME I
LARGE SUBUNIT I'
SMALL SUBUNIT I²
AMINO ACID J
POLYPEPTIDE
CHAIN J'
PEPTIDE BOND K
CELL MEMBRANE L
ATP M

NUCLEUS

pore

CYTOPLASM

POSTRANSLATIONAL
MODIFICATION
POLYSACCHARIDE N

Often, the completed ("translated") polypeptide requires folding (13) or shortening (14) before it can be fully functional. In some cases, polysaccharides are added (15).

Moving about, pumping blood, producing complex cellular structures, transporting molecules—these and other everyday activities that we normally take for granted all extract a price: they require energy. That energy is supplied by food. On the one hand, we have the machines that do the work (muscles, for example); on the other hand, we have the food as an energy source. Somehow they must be linked; energy has to be extracted from the food and stored in a form that is directly utilizable by the machine. The primary storage form living organisms use is the molecule ATP (adenosine triphosphate).

ATP IS THE CELLULAR ENERGY CURRENCY
ATP contains three phosphate groups joined in tandem. When the terminal phosphate is split off, it becomes ADP (adenosine diphosphate), and considerable energy is released. If the proper machinery is present, most of this energy can be captured and used for work. The ADP is not a simple waste product; it is recycled and utilized to synthesize new ATP.

$$ATP \underset{\text{energy trapped from food*}}{\overset{\text{energy source for work*}}{\rightleftharpoons}} ADP + P + energy*$$

The reaction goes to the right to power cellular machinery for contraction, transport, and synthesis. But if the split phosphate group is simply transferred to the machine, the energy goes with it, and the machine becomes energized. (The molecular part of the machine that receives the phosphate now has a higher energy content, which allows it to enter reactions it otherwise could not have entered.) ATP is the universal energy currency because of its ability to phosphorylate (transfer the phosphate to) cellular machines and boost them into a higher energy state.

GLUCOSE IS BROKEN DOWN IN SMALL STEPS
The reaction goes to the left as carbohydrates, fats, and proteins are broken down by chemical reactions occurring within the cell (metabolism). In this plate, we focus on ATP formation via carbohydrate metabolism. Glucose contains large quantities of energy that can be released when the chemical bonds holding its atoms together are broken. For example, if 1 mole (180 grams) of glucose is oxidized, forming CO_2 and water, 686,000 calories of energy are liberated. We can imagine many different ways to arrive at the same products, but in each case the same energy would be released. The cell must take the glucose apart in small controlled steps and capture most of this energy in the form of ATP before it is dissipated as heat. The cell accomplishes this in part because it contains a number of specific enzymes that speed the reaction along a specific path (i.e., by their presence, they determine the path of "least resistance").

ANAEROBIC (NO O_2) METABOLISM PRODUCES LACTIC ACID AND MINIMAL ATP
Energy release from glucose or from glycogen (the storage form of glucose) always begins with a sequence of reactions called glycolysis that converts glucose into pyruvate with the concomitant production of ATP. Beginning with the 6-carbon glucose, the reaction sequence is primed by investing two

molecules of ATP to phosphorylate the molecule before it is broken into two 3-carbon fragments. These are processed further to yield four new ATP, a net profit of two (4 – 2 [priming ATP] = 2). The entire sequence involves 10 reactions, each catalyzed by a specific enzyme, ending in the production of two molecules of pyruvate (a 3-carbon structure).

The presence of O_2 is not required for any of these steps; although only a small fraction (about 2%) of the available energy in the original glucose has been trapped as ATP, the cell apparently can generate ATP anaerobically (in the absence of air or free oxygen). However, this glycolytic process of breaking down glucose works only if H atoms are stripped off the carbon skeletons and transferred to other molecules called NAD^+.

$$2H \text{ (from carbohydrate)} + NAD^+ \rightarrow NADH + H^+$$

For every glucose, 4 H are transferred to $2 NAD^+$. But the total amount of NAD^+ is very small (it is built from the vitamin niacin), and the reaction will stop if we run out of NAD^+. NADH needs to dump its H somewhere so it can return for more. Normally, O_2 serves as the final resting place for H, and H_2O forms. In the absence of O_2, pyruvate itself serves as a dumping ground for H, and lactic acid forms. NAD^+ circulates, carrying H from high up in the glycolytic scheme to pyruvate and back (see plate).

AEROBIC (O_2) METABOLISM PRODUCES FAR MORE ATP VIA THE RESPIRATORY CHAIN
When O_2 is present, glycolysis proceeds as before, but now the role of NAD^+ (and a similar H carrier, FAD) becomes more apparent. They have succeeded in trapping a good portion of the energy in the original glucose, and the presence of O_2 allows this energy to be utilized to form ATP. Now, instead of using pyruvate, the H carriers transfer their H and energy to the respiratory chain, a system of carriers that reside within the mitochondrial membranes. In turn, the energized membranes of the mitochondria are able to produce 3 ATP for each NADH passed (only 2 ATP if the H donor is FAD).

Moreover, the availability of the respiratory chain allows energy contained in pyruvate to be tapped. Instead of absorbing H and forming lactate, pyruvate splits off a CO_2, and the remaining 2-C (acetate) portion is transferred via acetyl-CoA to the citric acid cycle, where it is further degraded into two molecules of CO_2 (see plate 6). Again H are stripped off the carbon skeletons by the H carriers, which deliver them to the respiratory chain and return for more. The final bookkeeping record for cellular combustion of one molecule of glucose is

glycolysis:	2 ATP + 2 NADH + 0 $FADH_2$	
2 pyruvate → acetyl-CoA:	0 ATP + 2 NADH + 0 $FADH_2$	
2 turns of citric acid cycle:	2 ATP + 6 NADH + 2 $FADH_2$	
Total:	4 ATP + 10 NADH + 2 $FADH_2$	

Total ATP (after cashing in   4   + $(10 \times 3) + (2 \times 2)$ = 38 ATP!
H carriers at resp. chain)

CN: Use red for A and another bright color for B. Use a pale color for D and a light color for M.
1. Begin with the upper panel, tracing the process of oxygen and food, representing high energy, into a body cell's metabolism.
2. Color the anaerobic process on the left, down to the roadblock of O_2.

3. Color the aerobic process. Note that the glycolysis portion (above the dotted line) is a simplified version of the same process that occurs anaerobically, except that lactic acid is not produced. Follow this chart down to the eventual production of 34 ATP. A more detailed explanation of the citric acid cycle and respiration appears on plate 6.

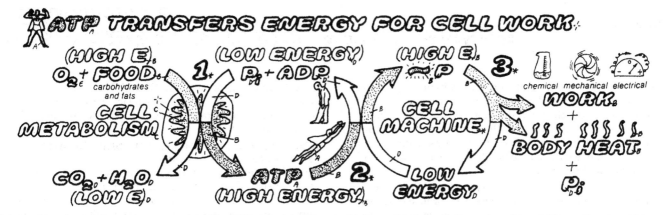

ATP TRANSFERS ENERGY FOR CELL WORK

Cellular machines do work. Some transport materials, others lift weights (muscle cells), and still others build complex molecules and structures out of simple raw materials. Food supplies the energy, but an intermediary substance, ATP (adenosine triphosphate), transfers the energy from food to the cellular machine. When food is burned in a furnace, energy is liberated as heat and light. When the same food is "burned" in the metabolic reactions of the cell (1), a good portion of this energy is trapped through the synthesis of ATP from ADP and inorganic phosphate (P_i). ATP, in turn, can energize the cell machine. It does this (2) by transferring its terminal phosphate ($\sim P$) to it, raising the machine to a higher energy state where it can participate in more reactions and perform cellular work (3).

HOW ATP IS MADE

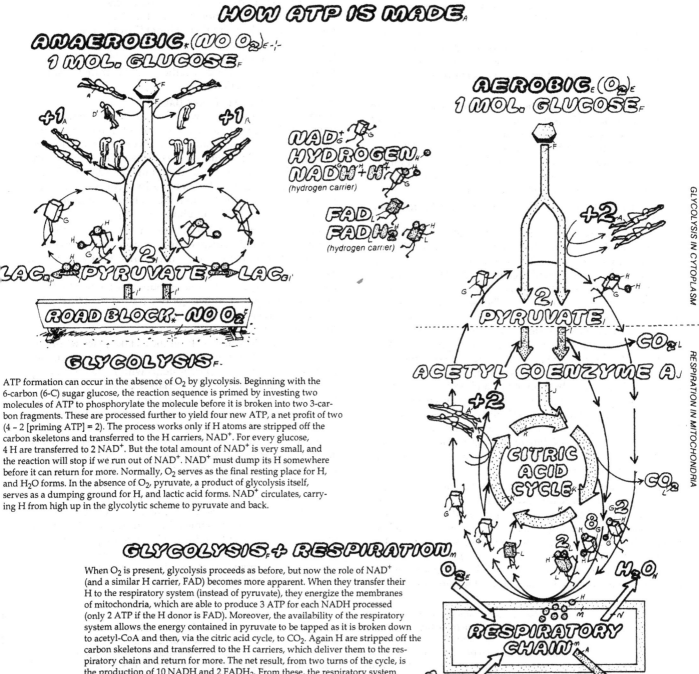

GLYCOLYSIS

ATP formation can occur in the absence of O_2 by glycolysis. Beginning with the 6-carbon (6-C) sugar glucose, the reaction sequence is primed by investing two molecules of ATP to phosphorylate the molecule before it is broken into two 3-carbon fragments. These are processed further to yield four new ATP, a net profit of two (4 – 2 [priming ATP] = 2). The process works only if H atoms are stripped off the carbon skeletons and transferred to the H carriers, NAD^+. For every glucose, 4 H are transferred to 2 NAD^+. But the total amount of NAD^+ is very small, and the reaction will stop if we run out of NAD^+. NAD^+ must dump its H somewhere before it can return for more. Normally, O_2 serves as the final resting place for H, and H_2O forms. In the absence of O_2, pyruvate, a product of glycolysis itself, serves as a dumping ground for H, and lactic acid forms. NAD^+ circulates, carrying H from high up in the glycolytic scheme to pyruvate and back.

GLYCOLYSIS + RESPIRATION

When O_2 is present, glycolysis proceeds as before, but now the role of NAD^+ (and a similar H carrier, FAD) becomes more apparent. When they transfer their H to the respiratory system (instead of pyruvate), they energize the membranes of mitochondria, which are able to produce 3 ATP for each NADH processed (only 2 ATP if the H donor is FAD). Moreover, the availability of the respiratory system allows the energy contained in pyruvate to be tapped as it is broken down to acetyl-CoA and then, via the citric acid cycle, to CO_2. Again H are stripped off the carbon skeletons and transferred to the H carriers, which deliver them to the respiratory chain and return for more. The net result, from two turns of the cycle, is the production of 10 NADH and 2 FADH$_2$. From these, the respiratory system produces $10 \times 3 + (2 \times 2) = 34$ ATP. Add the two ATP produced during glycolysis plus an additional two produced in the citric acid cycle, and we have a net of 38 ATP! Compare this to the two formed when O_2 was absent and the path took the detour to lactate.

Plate 5 focused on the degradation of carbohydrates, in particular glucose, to form ATP. Fats and proteins are also used for these purposes, but the final common pathway is the same—through the citric acid cycle and the respiratory chain, as outlined below. Taking an overview of the processes involved in ATP generation, we can conveniently divide the oxidation of foodstuffs into three stages:

Stage I. Recovery of Glucose, Glycerol, and Fatty and Amino Acids—Large molecules in food are broken down into simpler forms. *Proteins* are broken down into *amino acids, fats* are broken down into *glycerol* and *fatty acids*, and large *carbohydrates* (e.g., starch, glycogen, sucrose) are broken down into simple *6-carbon sugars* similar to glucose.

Stage II. Glucose, Glycerol, Fatty and Amino Acid Metabolism Converges on Acetyl-CoA—These elements play a central role in metabolism. Most of them, including simple sugars, fatty acids, glycerol, and several amino acids, are broken down into the same 2-carbon fragment called *acetate* that attaches to the same pivotal molecule, *coenzyme A* (abbreviated *CoA*), and enters the citric acid cycle as the compound *acetyl-CoA*.

Stage III. The Final Common Path: Citric Acid Cycle and Respiratory Chain—This final stage consists of the citric acid cycle and the respiratory chain (also called the electron transport chain) together with the ensuing synthesis of ATP. From acetyl-CoA onward, the metabolic path is the same for all foodstuffs. This plate focuses on this final common pathway, stage III, which occurs only in the presence of O_2.

ACETYL COA INITIATES THE CITRIC ACID CYCLE PRODUCING 3 NADH, 1 FADH$_2$, 1 ATP
Returning to our example of glucose metabolism, recall that one molecule of glucose produces a net gain of two ATP, two NADH and two pyruvate. In the presence of O_2, pyruvate is not utilized to form lactic acid because the NADH deposits its H on O_2 (via the respiratory chain—see below), freeing pyruvate to enter into further reactions. The two pyruvate move into the mitochondria in preparation for their entrance in the citric acid cycle. A precycle step breaks them into 2-C fragments (acetate), which are attached to a coenzyme A (CoA), forming acetyl-CoA. In the process, energy is recovered by transfer of H to NAD^+, and CO_2 is formed. Acetyl-CoA is also formed during the combustion of fats and proteins, and it plays a key role in feeding the 2-C acetate to the citric acid cycle. The acetate is split from CoA and combines with a 4-C structure, forming the 6-C *citrate* molecule, and the cycle begins. As illustrated, *each turn of the cycle produces 3 NADH, 1 FADH2, and 1 ATP*, with the carbon remains of the original acetate finally discarded as 2 CO_2.

NADH AND FADH$_2$ UNLOAD THEIR H ON RESPIRATORY CHAIN
Like glycolysis, the citric acid cycle will come to grinding halt as soon as all the H carriers NADH and FADH$_2$ are loaded with H. However, H is unloaded into the respiratory chain at the mitochondrial membrane, regenerating NAD^+ and FAD to participate in further metabolism. *The respiratory chain is a system of electron and H carriers* embedded in the inner member of the double membrane that surrounds the mitochondria.

In addition to regenerating NAD^+ and FAD, the respiratory chain also pumps H^+ ions into the space between the two mitochondrial membranes. These ions will be used in the final synthesis of ATP.

RESPIRATORY CHAIN PUMPS H$^+$ INTO SPACE BETWEEN MITOCHONDRIAL MEMBRANES
To understand the respiratory chain, recall that a neutral H atom consists of one electron and one H^+ (i.e., H = H^+ + 1 electron). When NADH arrives at the first carrier of the respiratory chain at the inner face of the inner mitochondrial membrane, it transfers two electrons and one H^+. Another H^+ is picked up from the surrounding solution, and the electrons and H^+ (=2H) are carried from the inner to the outer face. At this point, the H components, H^+ and electrons, part company. The H^+ are deposited in the small space between the two mitochondrial membranes, and the electrons return to the inner face to pick up another pair of H^+ from the surrounding solution. This trip across the inner face is repeated twice for a total of three round-trips. After the third trip, the electrons are picked up by the O_2, and, together with H^+ from the surrounding fluids, form water.

ATP IS FORMED AS H$^+$ LEAKS BACK
At each of the three trips, two H^+ are deposited in the inter-membrane space so that the concentration of H^+ builds up. These H^+s leak back into the matrix of the mitochondria through a special protein complex, called ATP synthase, that forms a channel through the membrane. The energy dissipated by the H^+ as it is driven by its high concentration, as well as electrical force, through these channels is used to synthesize ATP from ADP and phosphate (P_i).

FADH2 differs from NADH; it transfers its 2H to the respiratory chain downstream from the transfer point of NADH, where only two round-trips across the membrane are available. As a result, it transfers only 4H^+ across the mitochondrial membrane, and this accounts for the fact that it provides energy for synthesis of two (rather than three) ATP.

SPECIAL SYSTEMS TRANSPORT SUBSTRATES ACROSS MITOCHONDRIAL MEMBRANE
The citric acid cycle and ATP synthesis take place within the mitochondria; other functions (e.g., glycolysis) occur in the cytoplasm. Special transport systems within the mitochondrial membrane that move materials in and out overcome these restrictions. The transport system for newly synthesized ATP moves ATP out in exchange (countertransport, see plate 9) for ADP, which will be used for further ATP production. Other specialized transport systems are available for pyruvate and the NADH that arises from glycolysis. NADH itself does not cross the membrane; instead, it transfers its H at the outer face to H carriers that transport the H to the inner surfaces. Here the H is picked up by NAD^+ that is trapped inside the mitochondria. It becomes NADH, which now has access to the respiratory chain. In some mitochondria, the H are picked up by the FAD rather than the NAD^+, resulting in some energy loss. In these mitochondria, the net ATP production arising from combustion of one glucose molecule will be 36 rather than 38.

CN: Use the same color for the following titles as was used on the preceding page (note that the letter labels are different, so check carefully): A C, D, E, F, G, I, J, P, Q, R, and S. Use dark colors for B and K.
1. Begin at the top of the page with the entry, from the cytoplasm, of pyruvate into the citric acid cycle. It isn't necessary to color the titles of the various acids in this cycle.
2. Begin the lower panel with the small cutaway drawing of an entire mitochondrion. Color the enlarged rectangular portion, noting the arrows on the left representing the entry of pyruvate (F) and hydrogen ions (D).
3. Color the next enlargement. Note that the inner matrix (M) and intermembrane space (L) are left uncolored. Begin the coloring on the upper left side with the passage of H^+ through the membrane via the electron carrier (P). Note that the electrons that are carried along the path of this system are not shown. Follow the buildup of H^+ in the intermembrane space and its passage back into the matrix.

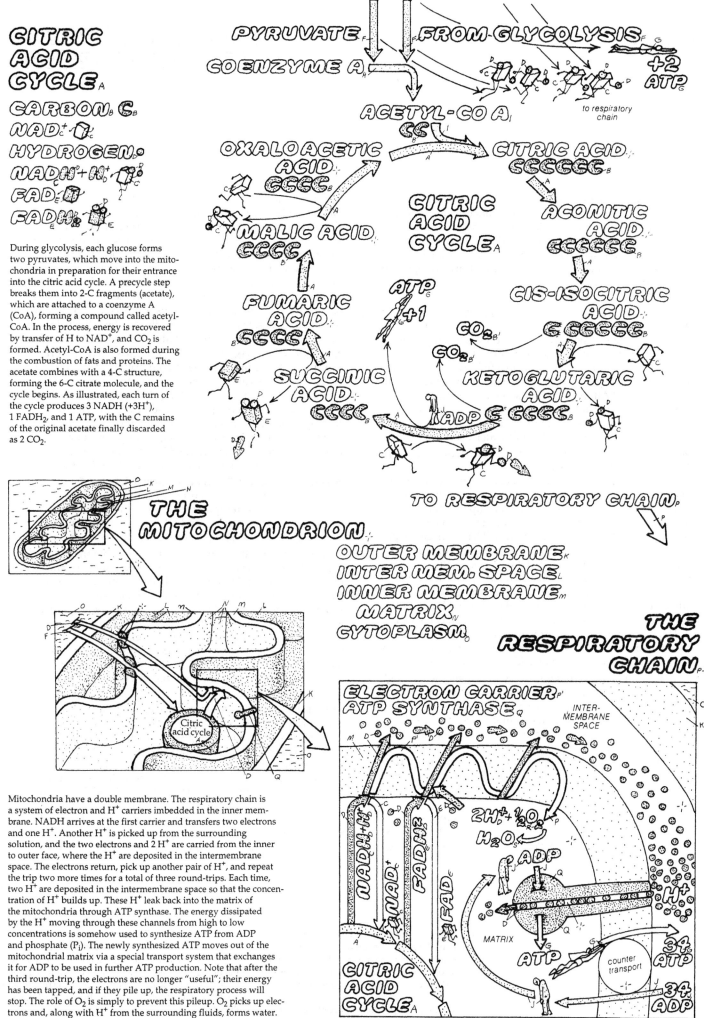

CITRIC ACID CYCLE

CARBON **C**

NAD$^+$

HYDROGEN

NADH + H$^+$

FAD

FADH$_2$

During glycolysis, each glucose forms two pyruvates, which move into the mitochondria in preparation for their entrance into the citric acid cycle. A precycle step breaks them into 2-C fragments (acetate), which are attached to a coenzyme A (CoA), forming a compound called acetyl-CoA. In the process, energy is recovered by transfer of H to NAD$^+$, and CO_2 is formed. Acetyl-CoA is also formed during the combustion of fats and proteins. The acetate combines with a 4-C structure, forming the 6-C citrate molecule, and the cycle begins. As illustrated, each turn of the cycle produces 3 NADH (+3H$^+$), 1 FADH$_2$, and 1 ATP, with the C remains of the original acetate finally discarded as 2 CO_2.

THE MITOCHONDRION

OUTER MEMBRANE
INTER MEM. SPACE
INNER MEMBRANE
MATRIX
CYTOPLASM

THE RESPIRATORY CHAIN

Mitochondria have a double membrane. The respiratory chain is a system of electron and H$^+$ carriers imbedded in the inner membrane. NADH arrives at the first carrier and transfers two electrons and one H$^+$. Another H$^+$ is picked up from the surrounding solution, and the two electrons and 2 H$^+$ are carried from the inner to outer face, where the H$^+$ are deposited in the intermembrane space. The electrons return, pick up another pair of H$^+$, and repeat the trip two more times for a total of three round-trips. Each time, two H$^+$ are deposited in the intermembrane space so that the concentration of H$^+$ builds up. These H$^+$ leak back into the matrix of the mitochondria through ATP synthase. The energy dissipated by the H$^+$ moving through these channels from high to low concentrations is somehow used to synthesize ATP from ADP and phosphate (P$_i$). The newly synthesized ATP moves out of the mitochondrial matrix via a special transport system that exchanges it for ADP to be used in further ATP production. Note that after the third round-trip, the electrons are no longer "useful"; their energy has been tapped, and if they pile up, the respiratory process will stop. The role of O$_2$ is simply to prevent this pileup. O$_2$ picks up electrons and, along with H$^+$ from the surrounding fluids, forms water.

Membranes are ubiquitous! Not only do they define the boundaries of cells and subcellular organelles, they transmit signals across these boundaries, contain cascades of enzymes that are essential for metabolism, and regulate which substances enter and which leave. In many ways, the cell membrane resembles the wall around an ancient city; by regulating traffic in and out, the membrane is a major determinant of cellular economy.

CELL MEMBRANES ARE COMPOSED OF PROTEINS FLOATING IN A FLUID BILAYER OF LIPID

Although different membranes have different compositions, all cell membranes share a common primitive structure. They are composed of proteins floating in a fluid *bilayer* (two layers back to back) of *lipid*. How do these bilayers form? What holds them together? Although membrane molecules interact with one another, the primitive forces holding the membrane together appear to arise from interactions of membrane with water and from interactions of water with water.

WATER MOLECULES INTERACT WITH EACH OTHER AND WITH OTHER POLAR MOLECULES

Water is not a symmetrical molecule; the two hydrogens are at one side of the molecule and the oxygen is at the other. Moreover, the electron clouds that make up the molecule tend to hover closer to oxygen than to hydrogen. As a result, because water bears no net charge, it will have a negative pole near the oxygen and a positive pole near the hydrogens. We classify water as *polar* to contrast it with electrically symmetric molecules like *hydrocarbons*, which are *nonpolar*. In ordinary water, not only is the positive hydrogen end of one water attracted to the negative oxygen end of its neighbor, but it may even form a weak chemical bond (called an H bond) with it. Thus, liquid water has a structure. Although it is not a rigid crystal like ice, it does contain numerous aggregates of up to eight to ten molecules weakly held together by H bonds.

Just as water molecules interact with one another through polar attractions or H bond formation, so do they interact and form microstructures with other polar molecules. However, they do not interact with nonpolar molecules. As a result, nonpolar molecules are excluded from the water phase; they are insoluble. The water molecules, "seeking each other," eject the nonpolar solute much like the ejection of a bar of soap from a clenched fist. The forces arising from this phenomenon are called *hydrophobic forces*, and nonpolar solutes are called *hydrophobic solutes*.

PHOSPHOLIPIDS HAVE POLAR HEADS AND NON-POLAR TAILS

Most membrane lipids are *phospholipids*. Phospholipids have a dual structure. One end, the *"head,"* is *polar (hydrophilic)*, and it is attracted to water. The remaining "tail" *is nonpolar (hydrophobic)*, and it is ejected from water. This incompatibility of the two ends is the key to membrane structure. When phospholipids are mixed with water, forces are set up that tend to incorporate the head within the water and eject the tail. The illustrations show how both of these forces can be accommodated through formation of *micelles or bilayers. Cholesterol* is another common lipid constituent of cell membranes. It fills some of the spaces between the tails of adjacent phospholipids and it moderates the membrane fluidity by influencing the interactions and motions of the tails.

MEMBRANE PROTEINS HAVE POLAR REGIONS EXPOSED TO WATER & NON-POLAR REGIONS IMBEDDED IN THE BILAYER

Similar principles apply to those *proteins* that are part of the membrane. Proteins consist of long chains of amino acids, some of which are more polar and some of which are more hydrophobic. Although proteins do not have an obvious head and tail, they are folded so that their hydrophobic portions are out of the water phase, imbedded in the hydrophobic body of the membrane; their polar portions are anchored in the water.

SPECIFIC MEMBRANE PROTEINS PERFORM SPECIFIC FUNCTIONS

Some are involved in transport of materials into or out of the cell, others serve as receptors for hormones, others catalyze specific chemical reactions, and others act as links between cells or as "anchors" for structural elements inside the cell.

Those proteins that traverse the entire membrane are exposed to both surfaces. They are often involved in transport of materials into and out of the cell. Some appear to form clusters of perhaps two or four molecules in the membrane, and some are believed to fold in ways that confine polar parts to an interior core that runs through the center of the molecule or through the center of a cluster. This would provide a *channel* for small polar molecules, particularly water and ions, to move through the membrane. Some of these channels may contain "gates," "filters," or other devices that regulate traffic. These are taken up in plates that follow.

THE CELL SURFACE HAS A CARBOHYDRATE COAT

Some proteins are confined to a single side of the membrane. Proteins as well as phospholipids exposed to the external surface (facing the outside of the cell) frequently have *carbohydrate* chains attached; they are called *glycoproteins* and *glycolipids* respectively. These external carbohydrates lubricate the cell surface, preventing it from "sticking" and they protect the cell from mechanical and chemical damage. In addition they play an important role in allowing the cell-cell recognition and adhesion. These carbohydrates also have antigenic properties (e.g., they form ABO blood antigens—plate 144).

CN: Use light blue for B and a dark color for H.
1. Begin with the polar portion of the phospholipid molecule, noting the heavy outline dividing it into three areas of color (C, D, and E). These areas make up the head (A) in the symbolic representation to the right.
2. Color the nonpolar portion, and continue with the central cartoon. Then color the two forms that the result can take: micelle or lipid bilayer.
3. Color the material describing the cell membrane, including the small drawing of a cell. Note that the areas of water bordering both sides of the membrane are left uncolored except for the titles describing their location.

THE PHOSPHOLIPID MOLECULE.

POLAR PORTION (HYDROPHILIC).
CHARGED GROUP:*
ALCOHOLS.
PHOSPHATE.
GLYCEROL.

carbon
nitrogen
hydrogen
oxygen
phosphorus

HEAD.

WATER.

NONPOLAR PORTION (HYDROPHOBIC).
FATTY ACID CHAIN.

Mixing phospholipids with water yields structures where polar head groups remain in the water while nonpolar tails are excluded. This can result in different structures: micelle spheres are simplest; their tails avoid contact with water by pointing toward the interior of the sphere where water is absent. Bilayer vesicles are more like cells. Their tails avoid water by contacting one another at their tips so that only head groups are exposed to water at the exterior or interior of the sphere.

TAIL.

SYMBOL

STRUCTURAL
MODEL

MICELLE.*

OR

THE CELL MEMBRANE.

LIPID BILAYER:* POLAR. NONPOLAR.
PROTEIN.
CHANNEL.
CARBOHYDRATE.

Membrane proteins are folded so that their polar parts are exposed to water, but their nonpolar parts are imbedded more deeply into hydrophobic portions of the bilayer. Some proteins traverse the entire membrane; they are often involved in transport of materials into and out of the cell. Other proteins are confined to a single side of the bilayer. Specific proteins imbedded in the membrane perform specific functions. Some are involved in transport, others serve as receptors for hormones, others catalyze specific chemical reactions, and still others act as links between cells.

CELL

sugar
molecule

glycoprotein

LIPID BILAYER.*

EXTRACELLULAR FLUID.

CYTOPLASM OF CELL.

non-polar

polar

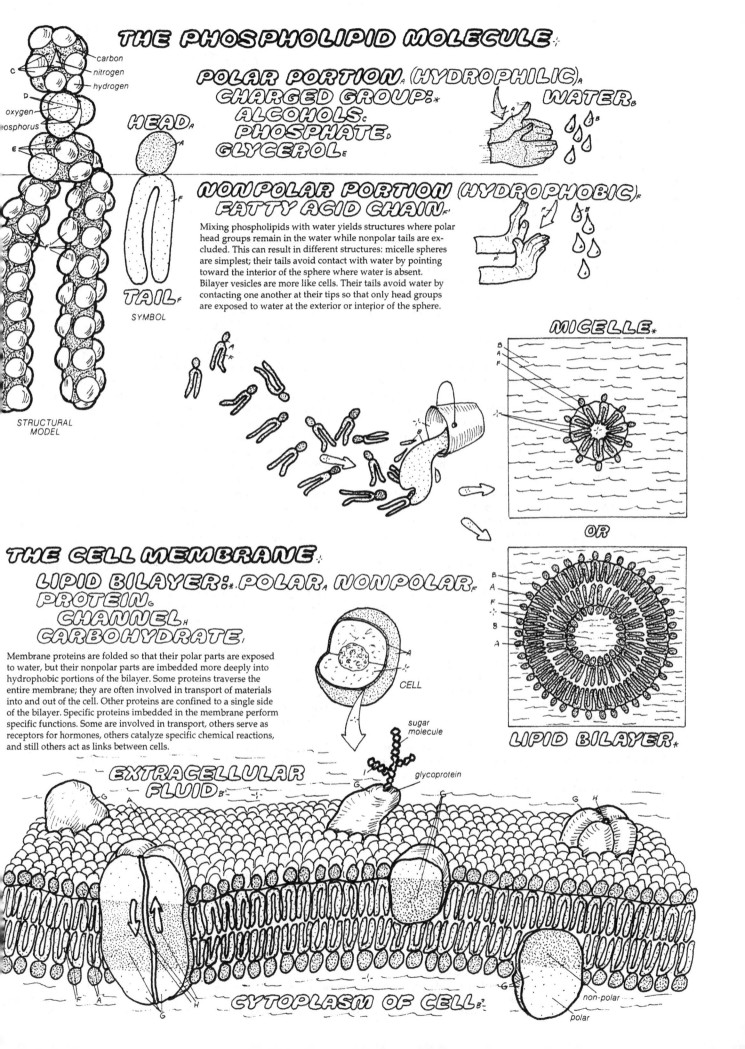

There are several different types of force; balls roll downhill because of gravitational force, and electrons (negative charge) flow toward protons (positive charge) because of electrical forces. Movements through membranes are driven by forces that arise from differences in concentration, pressure, and electrical charge on the two sides of the membrane. The forces created by these differences are called gradients. All else being equal, the larger the gradient (e.g., the larger the difference), the greater the flow through the membrane. *Concentration gradients* give rise to movements called *diffusion, pressure gradients* cause *bulk flow,* and *electrical gradients,* more commonly called *voltage gradients* or *membrane potentials,* drive the *flow of ions (ionic current).* In addition, water flows to the side of the membrane that has the most concentrated solute, a process known as *osmosis.*

DIFFUSION: SOLUTES MOVE DOWN CONCENTRATION GRADIENTS

Whenever there is a difference of concentration of a solute between two regions, the solute tends to move (diffuse) from the highly concentrated region to the less concentrated region. Solutes diffuse down their concentration gradient. This net movement arises because molecules are always moving about at random. At first thought, it seems surprising that an orderly net movement arises from molecular chaos, but the simple example described in the legend to the diffusion figure illustrates how this comes about. Net movement stops when concentrations are equal. Under most circumstances, diffusion of uncharged substances are independent of each other. For example, the diffusion of sugar down its concentration gradient would take place at the same speed even if another diffusing substance, say urea, were present.

The time required for diffusion over small distances, such as the size of a cell, is only a fraction of a second, but larger distances take surprisingly longer times. To diffuse 10 cm takes about 53 days, and it would take years for O_2 to diffuse from your lungs to your toes. But O_2 does not travel from lungs to body tissues by diffusion. Instead, it is transported by bulk flow through the circulating blood. Once the O_2 reaches the tissues, it diffuses the short distance through the wall of the blood vessel (capillary) into the tissue and to any cell in the near vicinity. The process is over in seconds.

BULK FLOW: MASSIVE FLUID MOVEMENT DOWN PRESSURE GRADIENTS

In contrast to diffusion, in bulk flow the whole mass (fluid plus any dissolved solutes) moves. For example, when you push the plunger on the top of a syringe, fluid flows out the needle (by bulk flow). In the diagram, the man on the left is pushing harder than the one on the right, so fluid flows from left to right. We call the "push" *pressure.* (More precisely, the pressure is the force he exerts on each square centimeter of plunger.) Fluid flows down pressure gradients, from high to low pressures, by bulk flow.

The diagram to the right shows how pressure can be measured. A light (ideally weightless) moveable partition is placed on top of the fluid, and mercury is poured on top until the weight of the mercury is just sufficient to stop the flow. At this point, the pressure gradient has fallen to zero; pressures

to the left and right are equal. But the pressure on the right is determined by the weight of the mercury (divided by the area of the partition), and this is determined by the height of the mercury column. We measure pressure in millimeters of mercury (mm Hg); it is the height of the column of mercury required to balance the pressure so that no movement occurs.

OSMOSIS: WATER FLOWS UP OSMOTIC GRADIENTS

The diagram (next to bottom) shows a membrane separating two solutions. The membrane is permeable to water but impermeable to the solute. Water will flow from left to right, from the region where the concentration of dissolved solute is low (actually zero in our example) toward the region where it is high. The flow is called osmosis. The forces that cause osmosis can be measured by the same technique used to measure pressure (see illustration on the right). Use the "weight-less" moveable piston to cover the solution, and pour on mercury until movement stops. The height of the mercury measures the pressure required to stop the osmotic flow; it is called *osmotic pressure.* The osmotic pressure will depend on the solution used on the right-hand side; the more concentrated the solution, the higher the osmotic pressure. As a result, we identify the osmotic pressure with the solution. The osmotic pressure of a solution is simply the pressure that would be measured if the solution were placed on the right-hand side of our device. Note that according to this convention, water flows *up* the osmotic pressure gradient. To a good approximation, all molecules and ions make equal contributions to osmotic pressures.

Osmosis involves bulk flow. Suppose, in our example, that the solute on the right is protein and that we dissolve equal concentrations of sugar on both sides of the membrane. Because proteins are so much larger than sugar, it is easy to find a porous membrane that will allow water and sugar to pass but will restrain the protein. What happens in this case? As before, the water flows from left to right, but now it carries the sugar with it as if the water were being driven by the same type of pressure gradient described in the bulk flow panel. Osmotic flow takes place in bulk; the solvent (water) drags all solutes along with it except those that are restrained by the membrane.

IONIC CURRENT: POSITIVE IONS FLOW DOWN, NEGATIVE IONS FLOW UP VOLTAGE GRADIENTS

The bottom diagram shows ionic movements (current) that arise from tiny differences in electrical charge on the two membrane surfaces. Ions of like sign repel; ions of unlike sign attract. When positive and negative ions are separated, they tend to move back together. Gradients associated with this attraction (or repulsion) are easy to measure. They are called voltage gradients. Positive ions move down voltage gradients; negative ions move up voltage gradients.

ENERGY GRADIENTS

In each of our examples, materials moved from regions of high energy to regions of lower energy. Concentration gradients, pressure gradients, osmotic gradients, and voltage gradients are all examples of *energy gradients* (more precisely, free energy gradients—see plate 9). Our discussion can be generalized: *energy gradients are forces that generate movements.*

CN: Use light blue for D.
1. Begin at the top of the page and complete each panel before proceeding to the next.
2. Flow, flux, diffusion, osmosis, and ionic currents are all examples of movement (A).
3. Energy gradients, pressure gradients, concentration gradients, osmotic gradients, osmotic pressure, and voltage gradients are all examples of forces that cause flow (B).

FLOW (FLUX) ∝ FREE ENERGY GRADIENT.

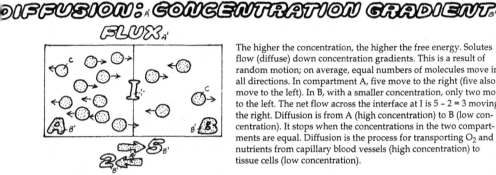

HIGH ENERGY. SOLUTE. FLOW. GRADIENT. LOW ENERGY STATE.

Substances flow down free energy gradients (i.e., from regions of high free energy toward regions with low free energy). Free energy gradients are the forces that propel the flow. The steeper the free energy gradient (i.e., the greater the difference in energies), the faster the flow (flux).

DIFFUSION: CONCENTRATION GRADIENT.

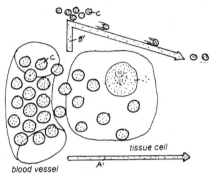

FLUX.

The higher the concentration, the higher the free energy. Solutes flow (diffuse) down concentration gradients. This is a result of random motion; on average, equal numbers of molecules move in all directions. In compartment A, five move to the right (five also move to the left). In B, with a smaller concentration, only two move to the left. The net flow across the interface at I is 5 – 2 = 3 moving to the right. Diffusion is from A (high concentration) to B (low concentration). It stops when the concentrations in the two compartments are equal. Diffusion is the process for transporting O_2 and nutrients from capillary blood vessels (high concentration) to tissue cells (low concentration).

blood vessel   tissue cell

BULK FLOW: PRESSURE GRADIENT.

FLUID. MERCURY (BACK PRESSURE).

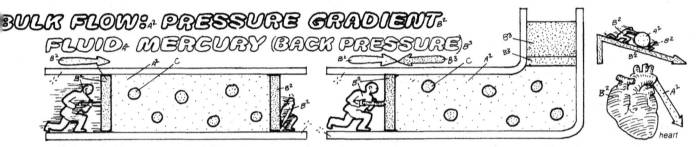

heart

Pressure (mechanical "push") also contributes to free energy. Fluid flows down pressure gradients, from regions where the pressure is high to where it is low. On the right, the pressure is balanced by a column of mercury (heavy fluid), which pushes in the opposite direction. The height of the mercury required to stop the flow is a measure of the original pressure. In the body, contractions of the heart supply the pressure gradients to propel blood.

OSMOSIS: OSMOTIC GRADIENT.

WATER (SOLVENT). MEMBRANE.

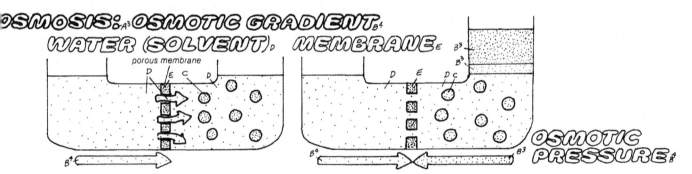

porous membrane   OSMOTIC PRESSURE.

The presence of solutes lowers the free energy of water. When a membrane separating two solutions allows water to pass through but restrains the solute, water flows down its free energy gradient toward the solute by a process called osmosis. This osmotic flow can be prevented by applying a pressure (column of mercury) to push in the opposite direction. This pressure measures the osmotic pressure. Note that by this definition, osmotic water flow goes from regions of LOW to HIGH osmotic pressure. Osmotic flow is responsible for swelling and shrinking of tissue.

IONIC CURRENT: VOLTAGE GRADIENT.

POSITIVE IONS. NEGATIVE IONS. ELECTRIC FORCE.

EQUILIBRIUM.

voltage gradient (voltage)   VOLTAGE.

nerve cell   BIOELECTRICITY. muscle cell

Solutes that carry an electrical charge (ions) also move under the influence of electrical forces that arise because ions of like sign repel and ions of unlike sign attract. When positive and negative ions are separated, they tend to move back together. Gradients associated with this attraction (or repulsion) are easy to measure. They are called voltage gradients. Positive ions move down voltage gradients; negative ions move in the opposite direction (up voltage gradients).

To deal with movements through membranes, we require a "common denominator" that allows us to compare magnitudes of forces and predict motions. Free energy provides that concept. Free energy is the amount of energy that can be "set free" to do work. When substances move from regions where their free energy is high to regions where it is low, down the free energy gradient, we call the movement passive because it can occur without any "aid" or work done by an external agency. The substance simply loses some of its energy to the environment. However, substances cannot move in the opposite direction (from low to high free energy) without obtaining energy (work) from the environment. When substances move uphill, from low to high free energy, we call the process active. One of the major problems of membrane physiology is to identify the source of energy supplied by the environment and to describe in detail how it is utilized.

Favorable free energy gradients by themselves are not sufficient to ensure transport. It doesn't matter how large a gradient is if the membrane does not allow the substance to pass through. In addition to a favorable gradient, there must also be a pathway. The common pathways described in this plate have not been fully identified; our understanding is incomplete, and descriptions of mechanisms are oversimplified.

PROTEIN CHANNELS & CARRIERS PROVIDE TRANSPORT PATHS
Some solutes, particulary steroid hormones, fat-soluble vitamins, oxygen, and carbon dioxide, are *lipid soluble*. They simply dissolve in the *lipid bilayer* portions of the membrane and diffuse to the other side (1). Many other important solutes, including ions, glucose, and amino acids, are more polar; they are soluble in water, but not in lipids. These substances move through special pathways provided by *proteins* that span the membrane. Small solutes like Na^+ pass through *channels* (2). Larger ones like glucose enter the cell by *facilitated diffusion* (3). They bind to a protein carrier that "rocks" back and forth or moves in some other way, exposing the binding site first to one side, then to the other side of the membrane. The solute hops on or off the site, depending on the concentration. If there is a higher concentration outside the cell, then the binding site will have a greater chance of picking up a solute on the outside, and more solutes will move in than out. This will continue until the concentrations on both sides are equal. At this point, movement in one direction is just balanced by movement in the opposite direction; net movement ceases. It is called *passive transport* because any solute movement is always down its *concentration gradient*. Similar facilitated diffusion systems exist for many different substances.

PRIMARY ACTIVE TRANSPORT: AGAINST GRADIENTS
Proteins also provide pathways for solute movements against concentration gradients ("uphill"). *Primary active transport* (4) is probably similar to facilitated diffusion. The transported molecule binds to a site on a protein that can "rock" or otherwise expose the binding site first to one side then to the other side of the membrane. Now, in contrast to the passive facilitated diffusion described above, suppose the binding site's

properties change and depend on which side of the membrane the site faces. In the extreme case, assume the solute can bind on only one side of the membrane—say, on the surface facing the inside of the cell—then transport is in only one direction, from inside to out, but not the reverse. If the concentration is less inside than out, our protein will transport against a gradient; it will be an active transport system. Energy is needed for the transport in order to change the binding site properties during each cycle. This energy is generally derived from the splitting of ATP.

SECONDARY ACTIVE TRANSPORT: PASSIVE-ACTIVE COUPLING
Under some conditions, solutes can also move against gradients by co- and counter-transport. Both utilize the passive transport of one solute to transport a different solute. Our example of co-transport (5) is similar to facilitated transport, but now the protein carrier has binding sites for two different solutes, Na^+ (represented by circles) and glucose (triangles). The carrier will not "rock" if only one of the sites is occupied. In order to "rock," both sites have to be empty or both sites occupied (both a Na^+ and a glucose have to be bound). Outside the cell, Na^+ is much more concentrated than glucose, but inside the cell, the concentration of Na^+ is very low because it is continually pumped out by an active transport process operating elsewhere in the membrane. Both Na^+ and glucose will move into the cell, but few molecules will come back out because the low concentration of intracellular Na^+ makes it difficult for glucose to find a Na^+ partner to ride the co-transport system in the reverse direction. By this mechanism, glucose can be pulled into the cell even against its concentration gradient. The energy for transporting glucose uphill against its concentration gradient comes from the energy dissipated by Na^+ as it moves down its concentration gradient. The concentration gradient for Na^+ is maintained by a primary active transport pump, which is driven by energy released by the splitting of ATP, so that ATP is indirectly involved in this co-transport example. Similar co-transport systems exist for other solutes such as amino acids.

Counter-transport (6) is similar to co-transport, but now the two solutes move in opposite directions. In our example, there are binding sites for two different solutes, say Na^+ (circles) and Ca^{++} (triangles). Again the carrier will not "rock" if only one of the sites is occupied. In order to "rock," both sites have to be occupied (both Na^+ and Ca^{++} have to be bound). Because the Na^+ concentration is much higher than Ca^{++}, it tends to dominate and keeps the counter-transporter moving in a direction that allows Na^+ to flow down its gradient (into the cell). It follows that Ca^{++} will flow out of the cell, even though the Ca^{++} concentration is higher outside than in. Once again the energy dissipated by Na^+ moving down its gradient is coupled to the uphill transport of another solute. Any process that uses the energy of a solute moving downhill to pump another solute uphill is called *secondary active transport*.

For simplicity, we have neglected the influence of *electrical forces* on the ions. The combination of electrical and concentration gradients is covered in plate 11.

CN: Use a dark color for F.
1. Begin with example number one, lipid solubility, and note that the arrows representing flow (F) receive a different color. Color the summary of this process in the central cell diagram. The various transport mechanisms in these summaries are given a circular shape for diagrammatic purposes.

2. Color the other five methods of transport. The flow arrows at (F), which go against the gradient (uphill), are drawn with bolder outline. Note that in 5 and 6 the wider gradient arrow (B') is the dominant gradient. A gradient of outside to inside the cell is used in most of these examples. If it were reversed, the flow would be in the opposite direction.

Solutes move passively (down concentration gradients) through cell membranes, provided there is a pathway. Some solutes such as steroids are lipid soluble (1). They dissolve in the membrane and diffuse to the other side. Most solutes are not lipid soluble; they move through special pathways provided by proteins that span the membrane (2). Small solutes like Na^+ pass through channels. Larger ones like glucose enter the cell by facilitated diffusion (3). They bind to a protein carrier that "rocks" back and forth exposing the binding site first to one side, then to the other side of the membrane. The solute hops on or off the site depending on the solute concentration. Solutes move through the membrane until the concentrations on both sides are equal. At this point, movement in one direction is just balanced by movement in the opposite direction; net movement ceases.

PASSIVE TRANSPORT (DOWNHILL):

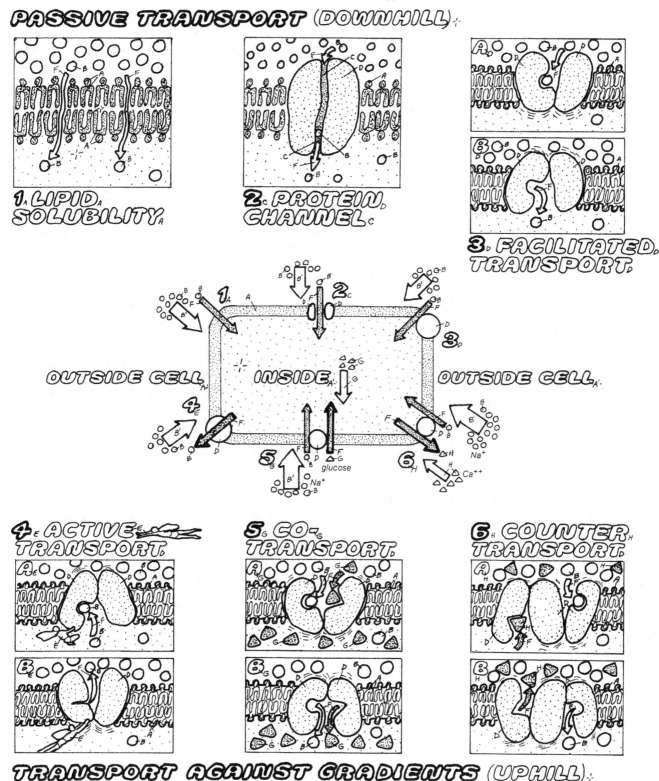

1. LIPID SOLUBILITY.

2. PROTEIN CHANNEL.

3. FACILITATED TRANSPORT.

OUTSIDE CELL · INSIDE · OUTSIDE CELL·

glucose

Na^+

Ca^{++}

4. ACTIVE TRANSPORT.

5. CO-TRANSPORT.

6. COUNTER TRANSPORT.

TRANSPORT AGAINST GRADIENTS (UPHILL):

Proteins also provide pathways for solute movements against concentration gradients (uphill). Primary active transport is probably similar to facilitated diffusion except that the binding site properties change and depend on which side of the membrane the site faces. In the extreme case, where the solute can bind on only one side of the membrane, transport is in only one direction. Energy required to change the site properties comes from splitting ATP (4). Solutes can also move uphill by co- and counter- transport. Both use the passive transport of one solute to transport a different solute. In co-transport, one solute (triangle) rides piggyback on the carrier of another that is moving down its gradient; the triangles may move uphill (5). Counter-transport is similar but now the two solutes move in opposite directions (6).

The Na^+-K^+ pump refers to an active transport system that continually pumps Na^+ out of and K^+ into cells. Generally, three Na^+ are pumped out for every two K^+ pumped in. This pump is found in the plasma membranes of all body cells, and it is one of the major energy-consuming processes of the body. The pump may account for more than a third of the resting energy consumption of the entire body. What functions does the pump perform to warrant this huge investment of energy?

FUNCTIONS OF THE PUMP

Osmotic stability — Proteins and many other smaller intracellular substances exert an osmotic pressure that is balanced by extracellular solutes, of which Na^+ and Cl^- are the most abundant. But both Na^+ and Cl^- leak into cells. If nothing intervened, this leak would create a continuous osmotic gradient, drawing water into the cell, which would swell and burst. In plant cells, this is prevented by a stiff wall. Animal cells are more flexible and mobile; they have no wall. Instead animal cells have a Na^+-K^+ pump that pumps Na^+ out, keeping internal solute concentrations low enough to prevent cells from swelling and bursting. (Some Cl^- follows to maintain electrical neutrality.) A steady low level of internal Na^+ is maintained because it is pumped out as fast as it leaks in. If the Na^+-K^+ pump is poisoned, animal cells swell and eventually burst.

Bioelectricity — The K^+ gradient maintained by the pump is a major determinant of the voltage gradient (negative inside) across the membrane. In many cells, this is primarily because K^+ leaks out of cells (carrying positive charges out) much faster than other ions leak in either direction. The pump also contributes to this voltage because it pumps three Na+ out for each two K^+ in, delivering one net positive charge to the outside each time the pump cycles.

Secondary active transport — (plate 9) The Na^+ gradient generated by the pump is used to drive transport of other solutes. Transport of glucose and amino acids by cells of the intestine and kidney are good examples of co-transport. In these cells, the solute (glucose or amino acid) enters or leaves the cell only when accompanied by Na^+. Outside the cell (high Na^+), the solute easily finds a Na^+ partner; inside it is difficult because Na^+ has been pumped out. The result: more solute enters than leaves even when solute is higher inside than out. Counter-transport is exemplified by Na^+ and Ca^{++} exchange in the heart. In this case, the energy of Na^+ moving down its gradient is coupled to Ca^{++} moving out.

Metabolism — The pump creates an intracellular environment that is rich in K^+ and poor in Na^+. These conditions are optimal for the operation of a variety of cellular processes including those involved in protein synthesis and in the activation of some enzymes.

PROPERTIES OF THE PUMP

The Na^+-K^+ pump transports Na^+ out of and K^+ into the cell. Na^+ is more than ten times more concentrated in the plasma than inside the cell. The reverse is true for K^+. As a result, Na^+ leaks in and K^+ leaks out. Nevertheless, a steady level of these two ions is achieved because as fast as they leak in (or out), they are pumped back out (or in); they are in a *steady state*. Under normal conditions, both ions are pumped against a concentration gradient, and the energy for this active transport is obtained from degradation of ATP. This latter fact is easily demonstrated in isolated cells whose metabolic machinery has been artificially inactivated or removed. These cells fail to actively pump Na^+ or K^+ unless ATP is introduced into the cytoplasm; other substrates will not work. Detailed studies show that for each ATP split, three Na^+ are pumped out and two K^+ are pumped in:

$$3Na^+_{in} + 2K^+_{out} + ATP_{in} \rightarrow 3Na^+_{out} + 2K^+_{in} + ADP_{in} + P_{in}.$$

Given this scheme, we can think of the pump not only as an ion-pumping machine, but also as an ATP-splitting machine. That is, it behaves like an enzyme that splits ATP; for this reason, we call it an ATPase. Further, the reaction written above will not proceed unless both Na^+ and K^+ are present; for this reason, we call it a Na^+-K^+ ATPase.

The pump appears to operate as follows. Three Na^+ ions arising from inside the cell bind to sites on the inner surface of the pump, which show a strong preference for Na^+ over K^+. This triggers the breakdown of ATP to ADP, and in the process, a high-energy phosphate is transferred to the pump protein (i.e., the pump is phosphorylated). Next the pump changes shape (conformation); the bound Na^+ now face outside, and the binding sites are altered. They no longer favor Na^+ over K^+, but just the reverse is true. Accordingly, they release the three Na^+ and bind two K^+ instead. Bound K^+ favors dephosphorylation; the proteins revert to their original shape with the binding sites back on the inside releasing K^+ because they have been transformed back to the Na^+-preferring state. The cycle repeats itself up to 100 times per second under optimal pumping conditions.

OTHER ATP-POWERED ION PUMPS

There are a number of other ion pumps that operate as an ATPase with characteristics similar to the Na^+-K^+ pump. One is a Ca^{++} *ATPase*, present in all cell membranes as well as some endoplasmic reticulum; it functions to maintain intracellular Ca^{++} at extremely low levels. Others are an H^+ *ATPase* that maintains an acid interior in lysosomes and endocytotic vesicles, and a H^+-K^+ *ATPase* that regulates acidity in the stomach and kidney.

CN: Use red for C and dark colors for F and G.
1. Begin with the upper four panels, completing each one before going on to the rest in clockwise order. Then color the summary diagram between the panels.
2. Color three functions of the pump, beginning with osmotic stability.

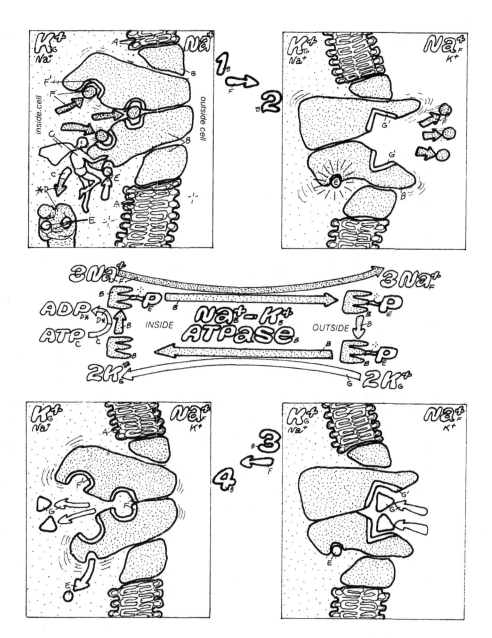

CELL MEMBRANE BILAYER_A SODIUM PUMP_B

(Labels:)
CELL MEMBRANE BILAYER$_A$
SODIUM PUMP$_B$
ATP$_C$/ADP$_{D*}$
PHOSPHATE$_E$
SODIUM (Na$^+$)$_F$
BINDING SITE$_{F'}$
POTASSIUM (K$^+$)$_G$
BINDING SITE$_{G'}$

The Na$^+$-K$^+$ pump transports Na$^+$ out of and K$^+$ into the cell. Both are transported against a concentration gradient: the energy is derived from degradation of ATP. Beginning in the upper left panel we see one version of the pump. It consists of two pairs of protein subunits. Three Na$^+$ ions arising from inside the cell bind to sites on the inner surface of the pump. This triggers the breakdown of ATP to ADP; in the process, a high-energy phosphate is transferred to the pump protein (i.e., the pump is phosphorylated). Next the pump changes shape (conformation): the bound Na$^+$ now face outside, and the binding sites are altered. They release the three Na$^+$ and bind two K$^+$ instead. Bound K$^+$ favors dephosphorylation: the proteins revert to their original shape, with the binding sites back on the inside releasing K$^+$ because they have been transformed back to the Na$^+$-preferring state. The pump behaves like an enzyme that splits ATP, but only in the presence of Na$^+$ and K$^+$, hence the name Na$^+$-K$^+$ ATPase.

The net result: $3Na^+_{in} + 2K^+_{out} + ATP_{in} \rightarrow 3Na^+_{out} + 2K^+_{in} + ADP_{in} + P_{in}$

FUNCTIONS OF THE SODIUM PUMP.

OSMOTIC STABILITY*-

Osmotic pressure exerted by solutes trapped in the cell (like proteins) draws water in. The Na$^+$-K$^+$ pump compensates by pumping Na$^+$ out so that internal solute concentrations are kept low enough to prevent cells from swelling and bursting. A steady low level of internal Na$^+$ is maintained because it is pumped out as fast as it leaks in.

GRADIENT FOR CO-TRANSPORT*.

The Na$^+$ gradient generated by the pump is used to drive transport of other solutes (e.g., in intestinal cells, glucose enters or leaves the cell only when accompanied by Na$^+$). Outside the cell (high Na$^+$), it easily finds a Na$^+$ partner; inside it is difficult because Na$^+$ has been pumped out. The result: more glucose enters than leaves when glucose is higher inside than out.

GLUCOSE CARRIER*

BIOELECTRICITY*-

POSITIVE CHARGE$_J$
K$^+$ CHANNEL$_{G2}$
NEGATIVE CHARGE$_K$

The K$^+$ gradient generated by the pump creates a voltage gradient (negative inside) across the membrane because K$^+$ leaks out of cells (carrying positive charges out) much faster than Na$^+$ leaks in. The pump also contributes to this voltage because it pumps three Na$^+$ out for each two K$^+$ in (see next plate).

Plates 8 and 9 focused on concentration gradients as the driving force for transport through membranes. When charged particles (ions) are involved, *electrical driving forces* become equally effective. These forces appear as voltage differences (voltage gradients) across cell membranes. The forces arise from separation of *positive* and *negative charge* by the membrane. The inside surface of a cell membrane generally has slightly more negative charge in its immediate vicinity, and the outside surface has slightly more positive charge. This creates an electrical force (voltage difference) that may be as large as 0.1 volt (100 mv), attracting positive charge inward and negative charge outward. This voltage difference is called the *membrane potential*, and it is always expressed as the voltage inside the cell minus the voltage outside. Because the inside is generally negative (and the outside positive), the normal membrane potential will be negative. It ranges from –10 mv in red blood cells to about –90 mv in heart and skeletal muscle. The magnitude of these forces can be appreciated if we realize that when K^+ is ten times more concentrated on one side of the membrane, its diffusion to the less concentrated side can be completely stopped by opposing it with a membrane potential of only 60 mv.

Membrane potentials are measured by making an electrical contact between each side of the membrane and an electrode (properly conditioned metallic wire). When these electrodes are connected through a metering device, electrons will flow through the wire from the negative to the positive side. In principle, the magnitude of the flow detected by the meter is proportional to the voltage difference. In practice, the measurement may not be so direct, as special precautions are taken to ensure that the drainage of charge through the electrodes is so small that it does not disturb the original voltage.

EQUILIBRIUM POTENTIALS REFLECT CONCENTRATION GRADIENTS

How does the charge separation responsible for membrane potentials arise? For simplicity, consider the impermeable membrane shown in panel A of the middle diagram. KCl is more concentrated on the left than on the right. Because both sides are electrically neutral ($[K^+]$ = $[Cl^-]$ on each side), the meter shows no voltage difference between the two sides. In panel B, *K^+ channels* are patched into the membrane, allowing K^+ but not Cl^- to go through. K^+ begins to diffuse to the right, building up positive charge on the right and abandoning negative charge on the left; a *voltage gradient* is created across the membrane, which tends to move K^+ in the opposite direction (from left to right). Each time a K^+ moves, the charge separation becomes larger so that the voltage opposing diffusion becomes larger. Finally (actually after a very short time), the voltage is just able to balance the concentration gradient, and net K^+ movement ceases (panel C). At this point, the system is in equilibrium; the voltage gradient required to stop diffusion of the K^+ ion is called the K^+ *equilibrium potential*. The larger the concentration gradient, the larger the equilibrium potential. The development of a non-zero equilibrium potential depended on the fact that Cl^- is assumed impermeable. If both

K^+ and Cl^- were permeable, then KCl would simply diffuse across until the concentrations were equal on both sides and there would be no membrane potential at equilibrium.

NET CHARGE IS ALWAYS CLOSE TO MEMBRANE

Notice in panel C that the excess ions charging up the membrane hover close to the membrane; the excess positive and negative ions attract each other. These excess ions are confined to a very thin layer adjacent to the membrane, and their numbers are very small compared to the number of ions present in the remaining bulk solution. Nevertheless, they produce significant electrical forces. *Within the bulk of the solution (away from the membrane), the numbers of positive and negative charges are equal.*

Instead of using K^+ channels, we could use Cl^- channels. Our analysis would be similar, only now the diffusing ion is negatively charged; it leaves a positive ion behind on the left, and the right-hand side becomes negatively charged. Equilibrium ensues when the membrane potential has the same magnitude as before, but is oriented in the opposite direction (negative on the right). In this case, the Cl^- equilibrium potential is equal and opposite to the K^+ equilibrium potential. When dealing with more complex mixtures of ions, their concentration gradients are more independent and each ion would have its own equilibrium potential.

CELL MEMBRANE POTENTIALS: STEADY STATE POTENTIALS

Similar effects occur in cell membranes, but now we deal with more ions, and the system does not settle down to equilibrium. The Na^+-K^+ pump maintains a K^+ gradient—high K^+ on the inside, low on the outside. (It also maintains an oppositely directed Na^+ gradient, but this is not as important because there are many more operative K^+ channels than Na^+ channels.) K^+ diffuses through K^+ channels to establish a membrane potential with the inside negative. The magnitude of the membrane potential is not quite equal to the K^+ equilibrium potential for two reasons: (1) other ions besides K^+ (e.g., Na^+ and Cl^-) can also permeate the membrane, and (2) the Na^+-K^+ pump may also make a direct contribution by pumping charge. Recall that each time the pump cycles, three Na^+ move out, but only two K^+ move in; this results in a net movement of one positive charge out. Thus, the pump not only plays an indirect role by setting up the original concentration gradient of K^+ so that it can diffuse out through K^+ channels, but it also pumps positive charge out. This latter (direct) contribution varies with circumstances. It is often small.

The concentrations of Na^+ and K^+ are fairly constant, but they are *not in equilibrium*. They settle down to steady state values because as fast as they leak in (or out), they get pumped back out (or in). If the pump is poisoned, the concentrations of Na^+ and K^+ will change as they drift toward equilibrium, and the membrane potential diminishes. Final equilibrium is then determined primarily by negatively charged ions inside the cell (e.g., protein anions) that cannot permeate the membrane. These are represented by *A-* in the large diagram.

CN: Use very light colors for A and B.
1. Begin with the upper panel.
2. Color the stages in the development of a membrane

potential. Include gradient symbols under each stag
3. Color the large panel in the lower right corner illustrating membrane potentials in cells.

ELECTRICAL FORCES
POSITIVE CHARGE_A
NEGATIVE CHARGE_B

Some atoms or molecules carry a net electrical charge, [+] or (–). They are called ions. [+] charged ions repel each other; (–) ions also repel each other, but [+] ions attract (–) ones and vice versa. Voltages are high around [+] ions and low around (–) ions. [+] ions flow from high to low voltages. (–) ions flow in the opposite direction.

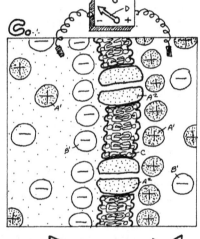

FLOW
FLOW_A'
LOW VOLTAGE_B
HIGH VOLTAGE_A

MEMBRANE POTENTIAL
MEMBRANE BILAYER_C
POTASSIUM (K⁺) ION_A'
K⁺ CHANNEL PROTEIN_A²
CONCENTRATION GRADIENT_A³
CHLORIDE (Cl⁻) ION_B'
VOLTAGE GRADIENT_D

A. An impermeable membrane separates two solutions of K^+, Cl^-. The left side is more concentrated than the right. Nothing happens because the ions cannot permeate the membrane.   **B.** Now K^+ channels are introduced; K^+ can get through, but Cl^- cannot. K^+ begins to diffuse to the right, building up [+] charge on the right and abandoning (–) charge on the left. A voltage gradient is created across the membrane, which tends to move K^+ in the opposite direction (from left to right).   **C.** As more K^+ diffuses, the voltage gradient grows larger until it is just able to balance the concentration gradient. At this point, net K^+ movement ceases; the system is in equilibrium. The voltage gradient required to stop the diffusion of any ions is called its equilibrium potential.

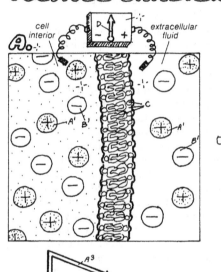

cell interior

extracellular fluid

A.

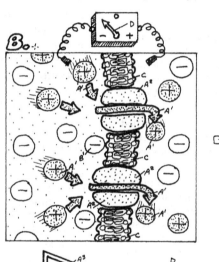

B.

C.

EQUILIBRIUM POTENTIAL

THE LIVING CELL
SODIUM (Na⁺) ION_E
Na⁺ CHANNEL_E
SODIUM-POTASSIUM ATPASE PUMP, ATP_G
ANION (A⁻)_H

WHEN PUMP STOPS

Similar events occur across cell membranes. The Na^+-K^+ pump maintains a *steady state* K^+ gradient — high K^+ on the inside, low outside. Minute amounts of K^+, the most permeable ion, diffuse through K^+ channels to establish a membrane potential with the inside negative (outside positive). The magnitude of the membrane potential is not quite equal to the K^+ equilibrium potential because other ions (e.g., Na^+ and Cl^-) also can permeate the membrane, and because the Na^+-K^+ pump can also transport net charge. When the pump stops, the ions equilibrate and the membrane potential decreases to values determined by A^-.

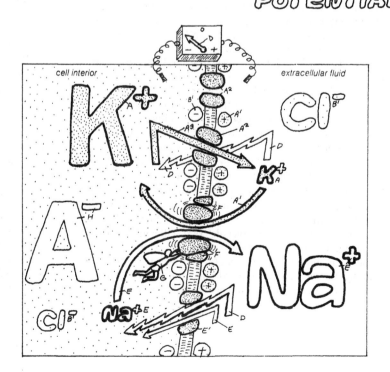

cell interior

extracellular fluid

K⁺
Cl⁻
A⁻
K⁺
Cl⁻
Na⁺
Na⁺

Cells of our body do not exist in isolation. They communicate by liberating signal molecules (e.g., hormones and neurotransmitters) to control vital processes including metabolism, movement, secretion, and growth. Although these actions are complex, a detailed study is essential because the control points are strategic targets for pathological toxins as well as for therapeutic drug interventions.

Signal molecules initiate their actions by binding to proteins called receptors. Although some important signal molecules (e.g., steroids, thyroid hormone) are lipid soluble and diffuse into the cell interior, most signal molecules are not. They are impermeable and restricted to reacting with receptors contained in the cell membrane. How can a reaction at the cell surface control actions within the cell interior? And how are small numbers of signal molecules amplified into a large response involving huge numbers of molecules?

TURNING THE SIGNAL ON

Resolution of these problems is illustrated in the first figure, in which a receptor spans the membrane. We follow the important case where the receptor is linked to a membrane-bound protein, the G protein, which in turn is linked to the production of an internal second "messenger," cyclic AMP. We recognize nine typical steps:

1. First messenger — The signal molecule (frequently called the first messenger) binds to the external surface of a receptor protein. This long protein folds back on itself, making seven passes through the membrane.

2. Receptor conformation — The binding causes a conformational change in the receptor (i.e., subtle changes in the 3-dimensional shape of the protein). Effects of the change are propagated to distant regions of the protein, where they result in activating part of the receptor exposed to the cytoplasm. Thus, the external signal is transmitted to the interior surface of the membrane.

3. G protein — The activated receptor can bind to and activate another membrane protein called G protein. This G protein has a binding site for GTP/GDP, which are analogs of ATP and ADP (see plate). At rest the GDP is bound by the G protein, but when the receptor and G protein complex, the bound GDP is exchanged for GTP. This allows the G protein to dissociate from the receptor and split into two activated parts. One part, called alpha, consists of a single subunit that contains the bound GTP. The other, composed of two subunits, is called beta-gamma. Both parts are free to move along the membrane surface and either the alpha or beta-gamma fragments will activate other membrane proteins called effectors. This leaves the activated receptor free to activate additional G proteins. One signal molecule can result in the activation of several G proteins. The effect is beginning to multiply; this is the first stage of amplification.

4. Effector — In this case the effector is an enzyme, adenylate cyclase. Each enzyme activated by a G protein, which will catalyze the formation of many intracellular cAMP (cyclic adenosine monophosphate) molecules that will continue to carry the signal and act as second messengers. At the same time, the complementary enzyme, cAMP-phosphodiesterase, continually converts cAMP into ordinary (noncyclic) AMP. The net amount of cAMP depends on the balance between the rates of formation and removal. cAMP production is a further stage of amplification; one effector catalyzes the formation of many cAMPs.

5. Second messenger — Second messengers (e.g., cAMP) carry the signal into the cytoplasm, where they may diffuse to any site and activate their targets. The signal is no longer confined to the cell membrane.

6. Protein kinase — cAMP acts by binding to and activating an enzyme called protein kinase A, which catalyzes the phosphorylation (addition of phosphate groups) of other enzymes or target proteins. The addition of the electrically charged phosphate group is sufficient to change the target's activity by changing its conformation — i.e., by "bending" the protein into or out of its active shape. Thus the phosphorylation can act as a "molecular switch" turning the protein either on or off. Other enzymes called phosphatases have the opposite effect; they dephosphorylate, turning the protein off (or on).

7. Final Target Protein — The final target is a functional protein within the cell that causes a specific response (e.g., secretion, division).

TURNING THE SIGNAL OFF

Like most body constituents, each member of a signal sequence is under a constant state of renewal; there are devices for removing it as well as producing it. In the case of target proteins activated by phosphorylation, the protein kinase, responsible for catalyzing activation, is generally accompanied by a complementary phosphatase enzyme, which removes the phosphate and turns off the activation. The status of each activated member of a cascade at any particular time reflects the balance between rates of production and rates of removal. Removal of the first messenger will upset this balance. If the first step is shut down, the production rate for the second step is shut down, which in turn shuts down the production of the third step, and so on.

8. G protein acts as a GTPase — G protein has a novel mechanism for self-deactivation. The dissociated alpha portion of the protein serves as its own phosphatase and catalyzes the hydrolysis of its bound GTP to GDP within seconds. This allows the alpha unit to reunite with the beta-gamma unit to form the original deactivated complex.

9. Removal of signal — If the delivery of signal molecules falls, their local concentration decays and they dissociate from the receptor.

EXAMPLES OF CAMP ACTION

The G protein/cAMP system can produce diverse effects depending on the tissue. In skeletal muscle and liver, the system is used by epinephrine (adrenalin) to initiate the breakdown of glycogen into glucose; in the heart it increases the rate and force of contraction; in fat cells it promotes the breakdown of fat for energy.

The importance of shutting signals off is illustrated by the disease cholera. Cholera toxin, produced by intestinal bacteria, catalyzes a chemical alteration of the alpha subunit, maintaining it in a permanently activated state. This particular subunit acts by stimulating a continuous secretion of Cl^-, Na^+, and water into the lumen of the intestine, causing a severe watery diarrhea (up to 20 liters per day). The resulting dehydration can be fatal if untreated.

CN: Use light colors for C and F.
1. Follow steps 1–9, coloring each step number as well.
2. Color all the chemical structures in the lower left panel.

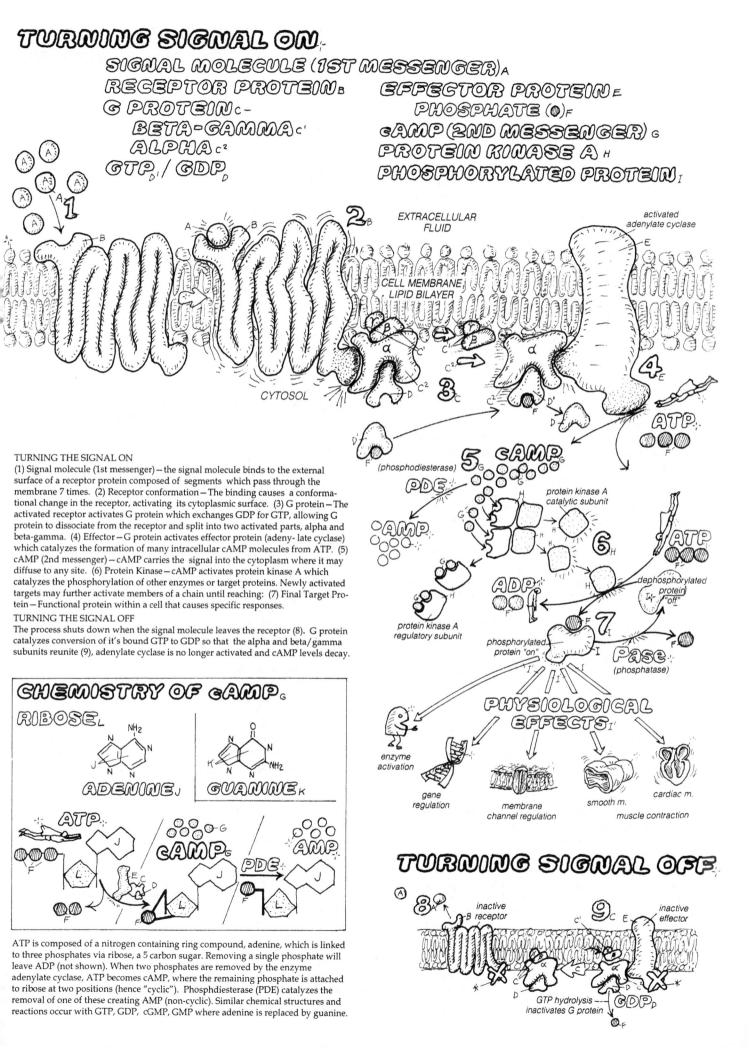

TURNING SIGNAL ON

SIGNAL MOLECULE (1ST MESSENGER) A
RECEPTOR PROTEIN B
G PROTEIN C -
BETA-GAMMA C¹
ALPHA C²
GTP D / GDP D'

EFFECTOR PROTEIN E
PHOSPHATE (℗) F
cAMP (2ND MESSENGER) G
PROTEIN KINASE A H
PHOSPHORYLATED PROTEIN I

EXTRACELLULAR FLUID

activated adenylate cyclase

CELL MEMBRANE LIPID BILAYER

CYTOSOL

(phosphodiesterase)

cAMP

PDE

AMP

ATP

protein kinase A catalytic subunit

ADP

protein kinase A regulatory subunit

dephosphorylated protein "off"

phosphorylated protein "on"

Pase (phosphatase)

PHYSIOLOGICAL EFFECTS

enzyme activation

gene regulation

membrane channel regulation

smooth m.

muscle contraction

cardiac m.

TURNING THE SIGNAL ON

(1) Signal molecule (1st messenger) – the signal molecule binds to the external surface of a receptor protein composed of segments which pass through the membrane 7 times. (2) Receptor conformation – The binding causes a conformational change in the receptor, activating its cytoplasmic surface. (3) G protein – The activated receptor activates G protein which exchanges GDP for GTP, allowing G protein to dissociate from the receptor and split into two activated parts, alpha and beta-gamma. (4) Effector – G protein activates effector protein (adeny- late cyclase) which catalyzes the formation of many intracellular cAMP molecules from ATP. (5) cAMP (2nd messenger) – cAMP carries the signal into the cytoplasm where it may diffuse to any site. (6) Protein Kinase – cAMP activates protein kinase A which catalyzes the phosphorylation of other enzymes or target proteins. Newly activated targets may further activate members of a chain until reaching: (7) Final Target Protein – Functional protein within a cell that causes specific responses.

TURNING THE SIGNAL OFF

The process shuts down when the signal molecule leaves the receptor (8). G protein catalyzes conversion of it's bound GTP to GDP so that the alpha and beta/gamma subunits reunite (9), adenylate cyclase is no longer activated and cAMP levels decay.

CHEMISTRY OF cAMP G

RIBOSE L

ADENINE J

GUANINE K

ATP

cAMP G

PDE

AMP

ATP is composed of a nitrogen containing ring compound, adenine, which is linked to three phosphates via ribose, a 5 carbon sugar. Removing a single phosphate will leave ADP (not shown). When two phosphates are removed by the enzyme adenylate cyclase, ATP becomes cAMP, where the remaining phosphate is attached to ribose at two positions (hence "cyclic"). Phosphdiesterase (PDE) catalyzes the removal of one of these creating AMP (non-cyclic). Similar chemical structures and reactions occur with GTP, GDP, cGMP, GMP where adenine is replaced by guanine.

TURNING SIGNAL OFF

inactive receptor

inactive effector

GTP hydrolysis inactivates G protein

GDP

CALCIUM IS AN IMPORTANT INTRACELLULAR SIGNAL

Calcium ions act to trigger responses involving secretion, contraction, and enzyme activation. To work, Ca^{++} must be presented to the site of action to turn the response on and removed to turn it off. Unlike the control of signals like cAMP, these processes do not take place through simple chemical production and destruction (Ca^{++} is a chemical element). Rather, they are accomplished by tapping into storage reservoirs for Ca^{++} to turn the process on and pumping Ca^{++} back into the reservoirs or out of the cell to turn it off. To be effective, cells need to maintain a very low resting intracellular Ca^{++} concentration, on the order of 0.1 μM or less. In this way only small amounts of Ca^{++} will need to move in order to make large differences (say, to increase the local concentration by a factor of 10 to 1 μM). The plate shows two prominent mechanisms used by the cell to maintain this low background of intracellular Ca^{++} concentration. There is an ATP-driven Ca^{++} pump that pumps Ca^{++} against gradients out of the cell into the extracellular spaces and out of the cytosol into the endoplasmic reticulum (sarcoplasmic reticulum in muscle). The second mechanism utilizes an Na-Ca exchanger (counter-transporter; see plate 9) in the plasma membrane, which transports 3 Na^+ into the cell for each Ca^{++} out. Again, Ca^{++} is pumped out of the cell against its gradient, but this time the energy for pumping comes from Na running down its gradient. The $3Na^+/Ca^{++}$ ratio provides a transfer charge of 3+ in and 2+ out = net of one positive charge inward for each cycle. This charge transfer, promoted by the negative internal membrane potential, contributes to the transport proficiency.

TURNING THE SIGNAL ON

In some cells—e.g., nerve endings—Ca^{++} enters the cytosol through Ca^{++} channels that are opened (activated) by changes in the membrane potential. These are taken up in context in later plates. In many other cells—e.g., smooth muscle—liberation of Ca^{++} into the cytosol from internal storage reservoirs is mediated by the G protein system described in this plate.

IP₃, DAG, AND Ca⁺⁺

Follow the numerical sequence in the top portion of the plate, which illustrates the formation and action of three related second messengers—IP_3, DAG, and Ca^{++}. A signal molecule activates a receptor protein (1) that activates a G protein (2). The receptor, like all those that are G protein linked, passes through the membrane 7 times. Instead of targeting adenylate cyclase, this G protein activates the enzyme phospholipase C (3). This novel enzyme attacks a special inositol phospholipid component (phosphatidylinositol bisphosphate) of the cell membrane itself, splitting its polar head group, called IP_3 (inositol triphosphate), from its non-polar tail, DAG (diacylglycerol), which remains behind in the membrane (1a). Splitting this special phospholipid does not threaten the structural integrity of the membrane because there are only minute quantities present.

DAG diffuses laterally on the inner leaflet of the membrane (4) where it participates in the activation of a phosphorylating enzyme, protein kinase C (9). The IP_3 is water soluble and free to diffuse throughout the cytoplasm. Its targets are Ca^{++} channels that are imbedded in the membrane of the endoplasmic

reticulum (5). IP_3 opens these channels and allows Ca^{++} ions, which have been sequestered at high concentration within the endoplasmic reticulum by a Ca^{++} pump (6), to escape into the cytosol. (5) Ca^{++} can then bind to, and activate, several types of Ca^{++} binding proteins (e.g., calmodulin) (7), which are involved in further actions (8). In addition, Ca^{++} binds to protein kinase C (C for Ca^{++} binding) and together with DAG activates this enzyme (9).

Protein kinase C then acts in ways similar to those of protein kinase A by catalyzing the phosphorylation and activation of other proteins (10).

PHOSPHORYLATION -DEPHOSPHORYLATION: A MOLECULAR SWITCH

This plate, as well as the last, illustrates a theme that recurs repeatedly throughout cell physiology: Proteins (enzymes, channels, motor molecules, etc.) are frequently controlled by the insertion or removal of phosphate (*phosphorylation* or *dephosphrylation*) into or from the molecule. Introducing the highly charged phosphate changes the shape of the protein. Depending on the protein, it is "bent" into or out of it's active state -- it is turned "on" or "off." Similarly removing the phosphate can "bend" it into or out of an active state. In this way, the phosphate acts as a molecular switch. The source of the phosphate is usually, but not always, ATP. An enzyme that catalyzes phophorylation is called a *kinase*. One that catalyzes dephophorylation is called a *phosphatase*.

G PROTEINS FORM A LARGE FAMILY

The term "G proteins" refers to a large family of regulatory molecules that bind GTP. Currently, over 20 different alpha subunits, 5 different beta subunits, and 12 different gamma subunits have been identified. Theoretically, these can combine to form over 1000 different alpha-beta-gamma members of the G protein family. Thus, it is not surprising to find G proteins involved in a large array of diverse cellular actions. The specific effect that follows G protein activation depends on which G protein is activated and on which effectors are available (e.g., which cell).

Some G proteins activate, others inhibit—Although our examples show G proteins activating cellular processes, cases where G proteins inhibit are also important. For example, in adipose (fat) cells, the hormones epinephrine and ACTH stimulate the enzyme adenylate cyclase via a G protein, while both prostaglandin and adenosine utilize a G protein to inhibit this enzyme.

Some G proteins act directly on membrane channels—Some G proteins that regulate ion channels do not rely on second messengers. Instead, they act directly, and more quickly, on the channel itself. For example, the plate shows how the neurotransmitter acetyl choline liberated by stimulation of the vagus nerve acts on the heart. Binding of acetyl choline by receptors on heart cell membranes activates a G protein that dissociates into the usual alpha and beta-gamma complex. In this case, the beta-gamma complexes play the active role by migrating to K^+ channels and opening them without any intervention of second messengers. This results in slowing of the heart rate (see plate 35).

CN: Use the same colors for A–D as were used on the preceding plate. Use a bright color for M.
1. Follow the numbered steps above. Only the title "Physiological Effects" is colored, not the drawings.
2. Color the five steps below, Note that a different signal molecule, acetylcholine (O) begins the process. 3. Note the diagram on phosphorylation (low right) showing how the addition of a phosphate (M turns a protein off—in step 10 phosphorylation ha the opposite effect.

G PROTEIN & 2ND MESSENGERS IP₃ DAG Ca⁺⁺

$$\text{G PROTEIN \& 2ND MESSENGERS } IP_3 \text{ DAG } Ca^{++}$$

SIGNAL MOLECULE (1ST MESS.)$_A$
MEMBRANE RECEPTOR PROTEIN$_B$
G PROTEIN$_C$
   GDP$_D$ **GTP**$_{D'}$
   PHOSPHOLIPASE C$_E$
   INISITOL PHOSPHOLIPID$_F$
IP₃ (2ND MESSENGER)$_G$
   ENDOPLASMIC RETICULUM$_H$
   Ca⁺⁺ CHANNEL$_I$ **PUMP**$_{I'}$
Ca⁺⁺ (2ND MESSENGER)$_{I^2}$
   CALMODULIN$_J$ (Ca⁺⁺ BINDING PROTEIN)
DAG (2ND MESSENGER)$_K$
   PROTEIN KINASE C$_L$

A signal molecule activates a receptor protein (1) which activates a G protein that activates the enzyme phospholipase C (2). This enzyme splits an inositol phospholipid component (phosphatidylinositol bisphosphate) (3) of the cell membrane into its polar head group: IP₃ (inositol triphosphate), and its non-polar tail, DAG (diacylglycerol), which remains behind in the membrane . DAG diffuses laterally on the inner leaflet of the membrane where it participates (4) in the activation of a phosphorylating enzyme, protein kinase C. The IP₃ diffuses to Ca⁺⁺ channels that are imbedded in the membrane of the endoplasmic reticulum (5). IP₃ opens these channels and allows Ca⁺⁺ ions, which have been sequestered at high concentration within the endoplasmic reticulum by a Ca⁺⁺ pump (6), to escape into the cytosol. Ca⁺⁺ can then bind to, and activate, several types of Ca⁺⁺ binding proteins (7) which are involved in further actions (8). In addition, Ca⁺⁺ binds to protein kinase C (9), and together with DAG activates this enzyme Protein kinase C then acts in similar ways to protein kinase A (see preceding plate) by catalyzing the phosphorylation and activation of other proteins involved in a diverse array of physiological functions (10).

Cells maintain a very low intracellular Ca⁺⁺ concentration primarily by two mechanisms: a. An ATP driven Ca⁺⁺ pump which pumps Ca⁺⁺ against gradients out of the cell into the extracellular spaces and out of the cytosol into the endoplasmic reticulum. b. A Na⁺- Ca⁺⁺ exchanger (counter-transporter) in the plasma membrane transports 3 Na⁺ into the cell for each Ca⁺⁺ out. Again Ca⁺⁺ is pumped out of the cell against its gradient, but this time the energy for pumping comes from Na⁺ running down its gradient (high outside, low inside).

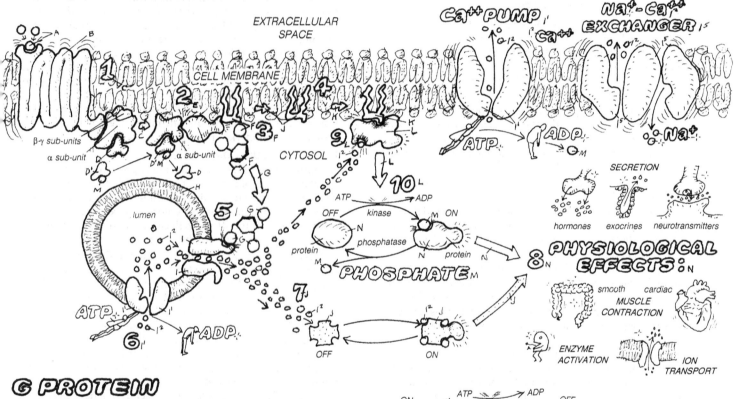

G PROTEIN ACTING DIRECTLY

Some G proteins that regulate ion channels do not rely on second messengers. This example shows that binding of neurotransmitter acetyl choline by receptors on heart cell membranes (11) activates a G protein (12) which dissociates into an alpha and a βγ complex (13). The complex migrates to K⁺ channels and opens them (14) without any intervention of second messengers. *This results in slowing of the heart rate.* When acetyl choline is removed, the G protein is deactivated *by its own GTPase activity* which splits a phosphate off the bound GTP leaving a bound GDP (15). The channel closes and the heart speeds up.

ACETYLCHOLINE$_O$
K⁺$_P$ **K⁺ CHANNEL**$_{P'}$

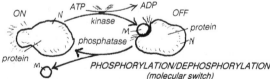

PHOSPHORYLATION/DEPHOSPHORYLATION
(molecular switch)

Proteins (enzymes, channels, motor molecules, etc.) are frequently controlled by the introduction or removal of phosphate (*phosphorylation* or *dephosphrylation*) into or from the molecule. Introducing the highly charged phosphate changes the shape of the protein and turns some proteins "on" (10) and others "off" (above diagram); phosphate acts as a molecular switch. The source of the phosphate is usually, but not always, ATP. An enzyme that catalyzes phosphorylation is called a *kinase*. One that catalyzes dephophorylation is called a *phosphatase*.

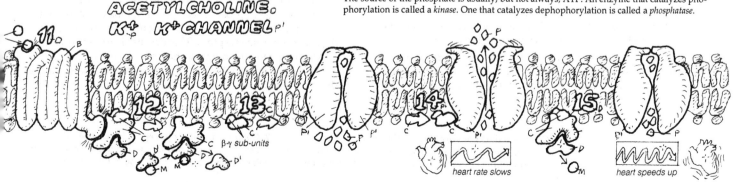

There are many receptors that are *not* linked to G-proteins. These include intracellular receptors for lipid-soluble hormones (steroids and thyroxine), ion channels, and receptors that are enzymes (catalytic receptors). Intracellular receptors are covered in plate 108, and ion channels in plates 13 and 16. This plate concerns enzyme receptors and receptor regulation.

ENZYME RECEPTORS: TYROSINE KINASES

Catalytic receptors are enzymes that are embedded in the membrane and activated by extracellular signal molecules when they bind to the receptor site. The largest class of enzyme receptors includes those whose cytoplasmic site acts as a protein kinase. They mediate cell differentiation, growth, and movements. Response to growth factors is generally slow, on the order of several hours.

Signal molecules for these receptors include:

Epidermal growth factor (EGF) — acts on a variety of cells, especially epithelial cells, stimulating them to enter the cell cycle and divide.

Platelet-derived growth factor (PDGF) — stimulates connective tissue cells to divide.

Nerve growth factor (NGF) — helps axon growth and survival of sympathetic plus some central nervous system neurons.

Fibroblast growth factor 2 (FGF-2) — stimulates cell division for many cell types, including fibroblasts, endothelial cells, and primitive muscle cells.

Insulin — stimulates glucose transport, metabolism, and growth (see plates 123, 124, 132).

Protein kinases act by catalyzing the phosphorylation of amino acid components of their target proteins. The protein kinases of in the last two plates — protein kinase A (activated by cAMP) and protein kinase C (activated by Ca^{++}) — phosphorylate the amino acids serine and threonine. In contrast, the major class of enzyme receptors phosphorylates the amino acid tyrosine. These enzymes are known as *tyrosine kinases*.

Phosphorylated tyrosines serve as initial sites for activation of further phophorylation cascades — Like other protein kinase molecules (e.g., protein kinase A), these receptors have a regulator site and a catalytic site. Unlike other protein kinases, the regulator and catalytic site are joined by a single hydrophobic alpha helix strand that passes through the membrane separating the two sites; the regulator site is exterior, while the catalytic site is inside the cell. Signal molecules are activators of the enzyme, but conformational changes rarely take place within a single chain. When signal molecules bind to receptor sites, the receptors pair up, joining together to form a dimer and making conformational changes feasible. Each receptor of the pair phosphorylates tyrosine components of its partner. These phosphorylation sites then become docking ports for adapter molecules that provide binding and activation sites for other intracellular signaling proteins. They may, for example, activate phospholipase C and unleash second messengers IP_3 and Ca^{++}). Alternatively, a cascade may be directed toward controlling genes in the nucleus.

From plasma membrane to nucleus: the *ras protein*-activated pathway — The path from tyrosine kinase to nucleus, where protein synthesis, cell proliferation, and differentiation are controlled, involves a membrane bound protein, the *ras* protein. Although smaller, ras resembles the alpha sub-unit of G-protein. At rest it binds GDP. Following interaction with an adaptor protein it becomes activated by exchanging the GDP for GTP. Further, like the alpha sub-unit, ras is a GTPase and will deactivate itself when it degrades its bound GTP to GDP. The ras protein activates a cascade of serine-threonine kinases that eventually phosphorylate an enzyme, called mitogen-activated protein kinase (MAPK). MAPK enters the nucleus, where it activates transcription factors; these are proteins that initiate copying of DNA into its complementary RNA sequence. Depending on which genes are activated, the final result can be cell growth, proliferation, or differentiation.

Hyperactive ras proteins are implicated in cancer — Given that tyrosine kinases are a major control of cell proliferation and differentiation, it is not surprising to find that abnormalities along the path from receptor to nucleus are associated with cancer. In fact, 30% of human cancers are associated with mutations in the genes that code the ras protein. If ras protein is activated but cannot deactivate itself by losing the ability to degrade its bound GTP to GDP — then the cell behaves as though it is under unrelenting stimulation by the corresponding growth factor. The result? Uncontrollable cell division and growth, the hallmarks of cancer.

RECEPTOR REGULATION

Receptors in general — those that are G-protein linked as well as catalytic — are not static. Both their numbers and nature change in reaction to conditions. Following continuous stimulation by a signal molecule, receptors may become desensitized. The size of the stimulus (local concentration of signal molecule) remains constant, but the response diminishes.

Occupation of binding site by signal molecule increases receptor removal by endocytosis — In some instances — e.g., EGF, insulin, and PDGF — the number of receptors actually diminishes. At "rest," the number of receptors represents a steady-state balance between the insertion in the cell membrane of "new" receptors arriving via exocytotic vesicles and the removal of "old" receptors via endocytotic vesicles. Occupation of receptors by signal molecules increases the rate at which they are removed, so that following the application of the signal molecule receptors spend less time on the plasma membrane. The resting balance is upset until a new steady state is reached, with fewer receptors on the membrane, a phenomenon known as *down regulation*. The system is desensitized because there are fewer receptors available to signal molecules. Once internalized, the receptor may be degraded by lysosomes and may need replacement by newly synthesized receptors; if not degraded, it will be recycled back to the membrane. When the signal molecule is removed, a new steady state will show an increased number of membrane receptors.

Receptors are often subject to feedback inhibition — In other cases, the number of receptors remains constant, but they don't work as effectively. For example, with cAMP-linked adrenergic receptors. This desensitization is due to a cAMP activated enzyme that phosphorylates the receptor. Once phophorylated, the receptor binds a "blocking" protein that prevents further interaction of receptor and G-protein.

CN: Use a dark color for F and light colors for P and Q.
1. Components of tyrosine kinase receive 3 colors (C–E) in the upper illustration, but only one color (B) in the lower.
2. Color the numbers (except 1) for each step.
3. Do not color cell membrane, but in the lower illustration color the vesicle (P) and lysosome (Q) membranes.

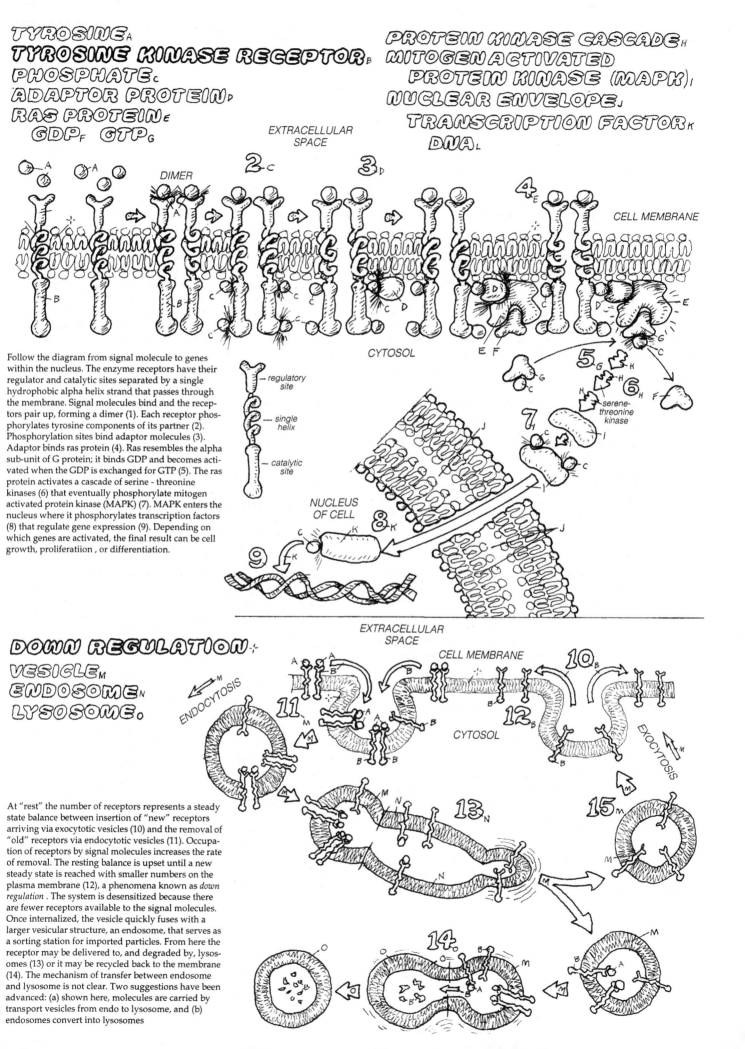

TYROSINE A
TYROSINE KINASE RECEPTOR B
PHOSPHATE C
ADAPTOR PROTEIN D
RAS PROTEIN E
GDP F GTP G

PROTEIN KINASE CASCADE H
MITOGEN ACTIVATED
PROTEIN KINASE (MAPK) I
NUCLEAR ENVELOPE J
TRANSCRIPTION FACTOR K
DNA L

EXTRACELLULAR
SPACE

DIMER

2 C

3 D

4 E

CELL MEMBRANE

CYTOSOL

E F

5 G

6 H

serene-
threonine
kinase

7

regulatory
site

single
helix

catalytic
site

NUCLEUS
OF CELL

8 K

9

Follow the diagram from signal molecule to genes within the nucleus. The enzyme receptors have their regulator and catalytic sites separated by a single hydrophobic alpha helix strand that passes through the membrane. Signal molecules bind and the receptors pair up, forming a dimer (1). Each receptor phosphorylates tyrosine components of its partner (2). Phosphorylation sites bind adaptor molecules (3). Adaptor binds ras protein (4). Ras resembles the alpha sub-unit of G protein; it binds GDP and becomes activated when the GDP is exchanged for GTP (5). The ras protein activates a cascade of serine - threonine kinases (6) that eventually phosphorylate mitogen activated protein kinase (MAPK) (7). MAPK enters the nucleus where it phosphorylates transcription factors (8) that regulate gene expression (9). Depending on which genes are activated, the final result can be cell growth, proliferatiion , or differentiation.

DOWN REGULATION
VESICLE M
ENDOSOME N
LYSOSOME O

EXTRACELLULAR
SPACE

CELL MEMBRANE

10 B

ENDOCYTOSIS

11

CYTOSOL

12 B

EXOCYTOSIS

13 N

15 M

14

At "rest" the number of receptors represents a steady state balance between insertion of "new" receptors arriving via exocytotic vesicles (10) and the removal of "old" receptors via endocytotic vesicles (11). Occupation of receptors by signal molecules increases the rate of removal. The resting balance is upset until a new steady state is reached with smaller numbers on the plasma membrane (12), a phenomena known as *down regulation* . The system is desensitized because there are fewer receptors available to the signal molecules. Once internalized, the vesicle quickly fuses with a larger vesicular structure, an endosome, that serves as a sorting station for imported particles. From here the receptor may be delivered to, and degraded by, lysosomes (13) or it may be recycled back to the membrane (14). The mechanism of transfer between endosome and lysosome is not clear. Two suggestions have been advanced: (a) shown here, molecules are carried by transport vesicles from endo to lysosome, and (b) endosomes convert into lysosomes

Cells that transmit "messages" or impulses throughout the nervous system are called neurons. Although the size and shape of these cells show large variations, a typical neuron consists of three characteristic parts. (1) *Dendrites* specialize in receiving stimuli from other neurons, from sensory epithelial cells, or simply from their environment. These are narrow extensions of the cell and are often short and branched. (2) The *cell body* also can receive impulses. It contains the nucleus, mitochondria, and other standard cell equipment. (3) The *axon,* a single, cylindrical extension of the cell, specializes in conducting impulses over large distances to other nerve, muscle, or gland cells. The final portion of the axon is generally branched, and each branch ends in an *axon terminal,* a swollen bulbous structure that is important in transmitting information to the next cell. *Nerves* — the whitish cables, easily seen in gross dissection, that connect the brain or spinal cord to various parts of the body — are made up of numerous axons held together by connective tissue.

NERVE IMPULSES TRAVEL AT SPEEDS FROM 0.5 TO 120 M/SEC

The existence of the "messages" or "signals" that are most often called *nerve impulses* is easily demonstrated. If the axon to a limb is cut, the limb becomes paralyzed. The muscles in the limb remain healthy and will move if they are stimulated directly with weak electrical shocks, but otherwise the muscles never get the message to move. After some time, the paralysis subsides and the return of normal activity coincides with the regeneration of the cut axons, so that the original connections are reestablished.

Using a single axon together with its associated muscle, we can study properties of the "messages." If we stimulate the axon with an electrical shock, the subsequent movement of the muscle will reveal whether it got the "message." This primitive strategy has been used to study how fast messages travel. Suppose we stimulate an axon at a specific point, A, and measure the time it takes before the muscle contracts. Next we stimulate at point B, 5 cm closer to the muscle. This time the muscle responds about .001 sec (1 msec) sooner because the message does not have as far to travel, and the difference in the two times (.001 sec) reflects the time taken for the message to travel 5 cm from A to B. If it takes .001 sec to travel 5 cm, then the velocity of the message must be $5/0.001 = 5000$ cm/sec $= 50$ m/sec. Other nerves have impulse velocities between 0.5 and 120 m/sec — i.e., about 1 to 268 mph.

NERVE IMPULSES ARE ACTION POTENTIALS

What are these "messages"? When an axon is stimulated, many changes occur, but electrical changes have been the easiest to measure and interpret. At rest, the axon behaves like any cell; its interior bears a negative electrical charge compared to the outside. Electrodes in contact with the two sides of the membrane register a membrane potential of about –70 mv (with the inside negative). When the axon is stimulated, this membrane potential reverses itself, moving momentarily to positive inside and then quickly back to the resting negative state. This sudden change in membrane potential that accompanies activity is called an *action potential.* It occurs first at the point of stimulation and moments later at each position along the axon; the farther from the stimulation point, the longer the delay before the action potential appears. In other words, the action potential begins at the stimulus and travels along the axon toward the muscle. The speed of travel is identical to the speed of the "message" (measured as described above). In fact, comparing the properties of "messages" with those of action potentials leads us to conclude that they are one and the same; "messages" are action potentials. At the height of the action potential, the outside of the membrane is negatively charged; this negative region moves along the axon. Viewed from outside the axon, the action potential appears to be a *wave of electrical negativity* traveling down the axon.

ACTION POTENTIALS ARE ALL-OR-NONE

In order to produce an action potential, the intensity of the stimulus has to exceed a critical value called the *threshold.* Beyond that level, all action potentials are the same size regardless of the size of the stimulus. The response is all-or-none. The magnitude of the signal sent along an axon cannot be contained in a single impulse; rather, it is contained in the *frequency* of impulses. The more impulses per second, the larger the signal.

ACTION POTENTIALS ARISE FROM NA & K MOVEMENT

How do action potentials arise? At rest, the membrane potential is about –70 mv. Because of this polarity (negative inside, positive outside), the membrane is said to be *polarized.* Recall (plate 11) that the intracellular fluid (e.g., in axons) has high K^+ and low Na^+ concentrations, while extracellular fluid is rich in Na^+ and poor in K^+. At rest, most operative channels allow passage of K^+ but not of Na^+. K^+ diffuses down its concentration gradient, delivering a positive charge on the outside surface while leaving negatively charged "partners" behind on the inner surface. The membrane becomes polarized, with the inside negative.

A stimulus causes a brief increase in the number of open Na^+ channels. If the stimulus is weak, only a few channels open, and the membrane potential is hardly perturbed. However, if the stimulus is stronger than *threshold,* then the number of open Na^+ channels becomes substantial. Na^+ ions, poised at high concentration outside the axon, leave their negatively charged "partners" behind on the outside and rush in fast enough to overwhelm the K^+ moving out. The inside surface of the cell membrane is inundated with positive charge so that the polarity is reversed; now the inside is positive and the outside is negative. A moment later the Na^+ channels close and extra K^+ channels open. K^+ moves out, making the membrane potential even more negative than it was at rest, driving it very close to the K^+ equilibrium potential. Finally (after several milliseconds), the extra K^+ channels close, and the membrane returns to its resting condition.

CN: Use the same colors as on the previous page for K^+ (G) and Na^+ (L).
1 Begin with elements of the nerve cell.
2. Color each membrane state, including the chart to its right, before going on the next.

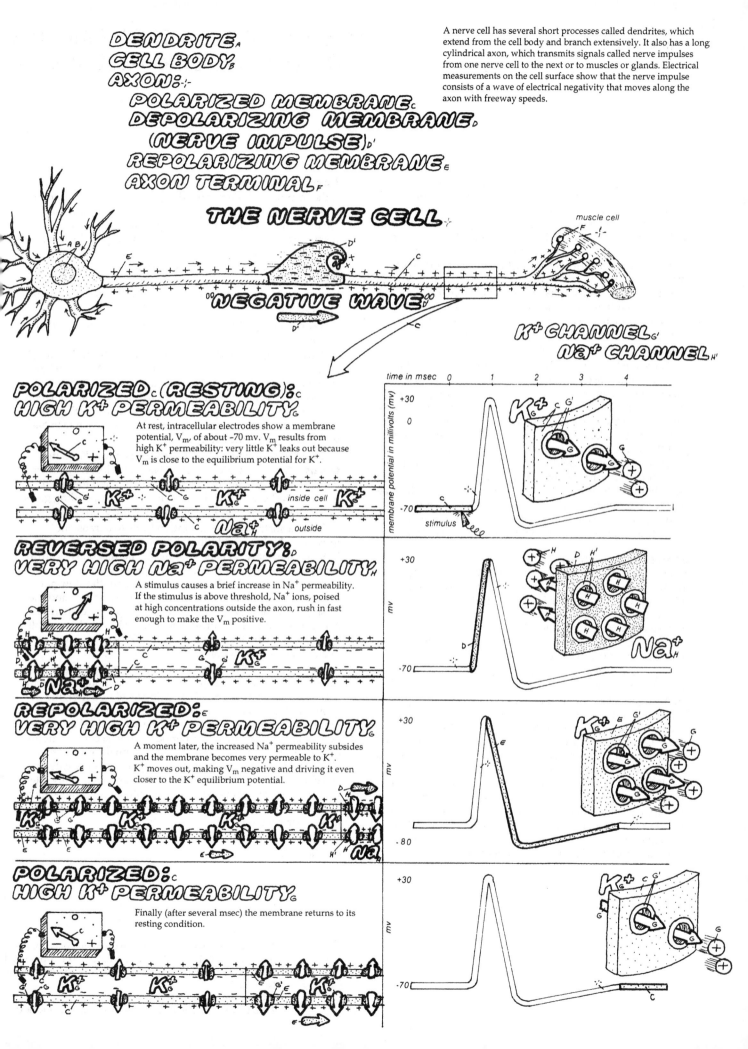

DENDRITE (A)
CELL BODY (B)
AXON: (C)
　POLARIZED MEMBRANE (C)
　DEPOLARIZING MEMBRANE (D)
　(NERVE IMPULSE) (D')
　REPOLARIZING MEMBRANE (E)
　AXON TERMINAL (F)

THE NERVE CELL

muscle cell

A nerve cell has several short processes called dendrites, which extend from the cell body and branch extensively. It also has a long cylindrical axon, which transmits signals called nerve impulses from one nerve cell to the next or to muscles or glands. Electrical measurements on the cell surface show that the nerve impulse consists of a wave of electrical negativity that moves along the axon with freeway speeds.

"NEGATIVE WAVE"

inside cell
outside

K^+ CHANNEL (G')
Na^+ CHANNEL (H')

POLARIZED (RESTING): HIGH K^+ PERMEABILITY

At rest, intracellular electrodes show a membrane potential, V_m, of about –70 mv. V_m results from high K^+ permeability: very little K^+ leaks out because V_m is close to the equilibrium potential for K^+.

time in msec

membrane potential in millivolts (mv)

+30
0
-70
stimulus

REVERSED POLARITY: VERY HIGH Na^+ PERMEABILITY

A stimulus causes a brief increase in Na^+ permeability. If the stimulus is above threshold, Na^+ ions, poised at high concentrations outside the axon, rush in fast enough to make the V_m positive.

+30
mv
-70

REPOLARIZED: VERY HIGH K^+ PERMEABILITY

A moment later, the increased Na^+ permeability subsides and the membrane becomes very permeable to K^+. K^+ moves out, making V_m negative and driving it even closer to the K^+ equilibrium potential.

+30
mv
-80

POLARIZED: HIGH K^+ PERMEABILITY

Finally (after several msec) the membrane returns to its resting condition.

+30
mv
-70

Modern interpretations of excitation in nerve and muscle are based on the sequential opening and closing of *ion channels* in the membrane. There are separate channels for different ions. Somehow each channel "recognizes" the appropriate ion (e.g., K^+) and allows it to pass while restraining others (e.g., Na^+). The channel mechanism responsible for this selectivity is called a *selectivity filter*. The status (i.e., open or closed) of many channels depends on the membrane potential. When the membrane potential changes, electrically charged portions of the protein that forms the channel may move slightly. These movements create small changes in shape of the channel protein, which act like "gates" that can open or close in response to membrane potential (voltage).

There are two types of K^+ channels. One type is voltage activated, but most of these are closed during rest, when the membrane potential is about –70 mv. The other type is not voltage activated; it is always open and provides the pathway for the small but continuous K^+ leakage that creates the resting potential. A resting membrane potential of –70 mv implies that the inside is negative while the outside is positive. Because of this electrical distinction between the membrane's inner and outer surfaces, the membrane is said to be *polarized*. By definition, the membrane is *depolarized* whenever the magnitude of this membrane potential becomes smaller than the resting potential (i.e., close to zero); conversely, when the magnitude is increased, the membrane *is hyperpolarized*.

STIMULI DEPOLARIZE THE MEMBRANE

When nerves are excited with electrical shocks, the impulse always arises at the negatively charged electrode (the cathode). By attracting positive ions and repelling negative ones, the electrode reduces the membrane potential so that the nerve membrane becomes depolarized (see plate). This simple observation can be generalized: *A stimulus will be effective only if it depolarizes the membrane.*

Depolarizing the membrane works because the permeability of nerve membranes to ions is very sensitive to the voltage gradient (membrane potential). The crucial relationship between membrane potential and ion flow has been studied in detail by an ingenious electrical method called a *voltage clamp*. Using this method, we can set the membrane potential to any desired value and keep it at that value for a prolonged period. At the same time, we can estimate the amounts of Na^+ and K^+ that flow through the membrane in response to the imposed membrane potential. These results are then interpreted in terms of opening and closing of Na^+ and K^+ channels, and they form the basis of our current understanding. In particular, they allow us to ask what happens when the membrane is stimulated (i.e., when it is depolarized).

EFFECTS OF DEPOLARIZATION

If the membrane is depolarized (e.g., the membrane potential is changed from the resting value of –70 mv to a. new value of -50 mv and *maintained* at this new potential), then the response of the ion channels can be arbitrarily divided into two phases: (1) an early response (< 1 msec) when the Na^+ channels open; and (2) a late response (> 1 msec) when the Na^+ channels close and the K^+ channels open. During this late period, the Na^+ channels appear to be inactivated; they will not respond to further depolarization.

We can interpret these changes in terms of our hypothetical gates as follows:

Early Response: Fast Na^+ Gate Opens—The Na^+ channel contains two gates, a slow one and a fast one. At rest (polarized membrane), in most channels, the slow gate is open but the fast gate is closed, so that most channels are closed. Upon depolarization, the fast gate opens quickly; now both gates are open and many channels become freely permeable to Na^+ which rushes into the axon.

Late Response: Slow Na^+ Gate Closes, Slow K^+ Gate Opens—A moment later the slow Na^+ gate closes. The membrane is no longer highly permeable to Na^+, and the rapid inflow of Na^+ ceases. In addition, the slowly responding gates in K^+ channels open and K^+ flows out of the axon.

Thus, a sustained depolarization leads to a transient increase in Na^+ permeability followed by a sustained increase in K^+ permeability. The increase in Na^+ permeability is attributed to the presence of two gates that give opposite responses to the depolarization. The fast gate opens, but the slow gate closes. The time between the opening of the fast gate and the closing of the slow gate corresponds to the period of increased Na^+ permeability. In contrast, a K^+ channel has only one voltage-activated gate that opens (slowly). Once open, it will stay open as long as the depolarization is sustained.

SLOW GATES ARE SLOW TO OPEN AND SLOW TO CLOSE

Immediately following depolarization, even though the membrane potential has returned to resting level (–70 mv), the axon is not fully recovered because the slow gates require a millisecond or two to respond to the newly established resting potential. If, during this brief period, a rapid second stimulus (depolarization) is delivered, the Na^+ channels will fail to open. The fast gates respond and open, but the slow gates are still closed as a result of the original depolarization. Only after a recovery period of one or two msec will the slow gates open and allow a second stimulus to trigger the transient increase in Na^+ permeability.

Application of these results to action potentials is detailed on plate 17.

CN: Use the same colors as on the preceding page for Na^+ (D) and K^+ (H). Use dark colors for E, F, and G. Although th positive charges have a separate color, they could either be Na^+ or K^+.
1. Color the upper diagram.
2. Complete each stage below before going on to the next.

STIMULUS: DEPOLARIZES MEMBRANE.

NEGATIVE ELECTRODE.ᴀ
MEMBRANE.ʙ
POSITIVE CHARGE.c
NEGATIVE CHARGE.ᴀ'

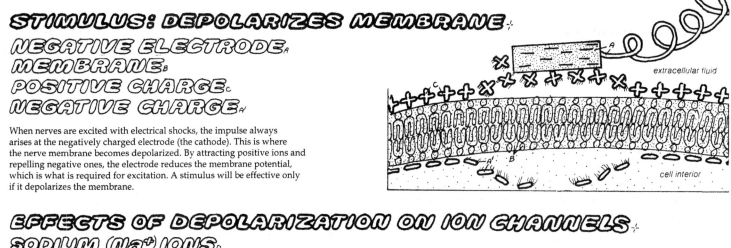

extracellular fluid

cell interior

When nerves are excited with electrical shocks, the impulse always arises at the negatively charged electrode (the cathode). This is where the nerve membrane becomes depolarized. By attracting positive ions and repelling negative ones, the electrode reduces the membrane potential, which is what is required for excitation. A stimulus will be effective only if it depolarizes the membrane.

EFFECTS OF DEPOLARIZATION ON ION CHANNELS.

SODIUM (Na⁺) IONS.ᴅ
CHANNEL.ᴅ'
SELECTIVITY FILTER.ᴇ
FAST GATE.ꜰ
SLOW GATE.ɢ
DEPOLARIZING Na⁺ FLUX.ᴅ²
POTASSIUM (K⁺) IONS.ʜ
CHANNEL.ʜ'
SELECTIVITY FILTER.ᴇ'
SLOW GATE.ɢ
POLARIZING K⁺ FLUX.ʜ²
LEAK CHANNEL.ʜ³

Nerve cell membranes contain separate channels for different ions. Each channel "recognizes" the appropriate ion (e.g., K⁺) and allows it to pass while restraining others (e.g., Na⁺ ions). The channel mechanism responsible is called a selectivity filter. In addition, many channels contain voltage-activated "gates" that open or close the channel, depending on the membrane polarization. The response of an excitable membrane to a sustained depolarization (stimulus) can be arbitrarily broken into two phases: (1) an early response (< 1 msec) when the Na⁺ channels open, and (2) a late response (> 1 msec) when the Na⁺ channels close and the K⁺ channels open. During this period, the Na⁺ channels will not respond to further depolarization.

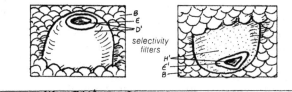

selectivity filters

NORMAL *
(RESTING POTENTIAL).*

There are two types of K⁺ channels. One is always open; it is the pathway for the small K⁺ leakage that creates the resting potential. The other is voltage activated; it is mostly closed when the membrane is highly polarized. Voltage-activated Na⁺ channels are also closed in the highly polarized state.

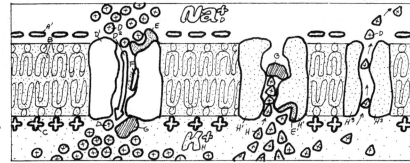

extracellular fluid

cell interior

DEPOLARIZATION.ᴅ²
EARLY (<1msec).ᴅ²

The Na⁺ channel contains two gates, a slow one and a fast one. At rest (polarized membrane), the slow gate is open, but the fast gate is closed, so that the channel is closed. Upon depolarization, the fast gate opens quickly, making the membrane permeable to Na⁺.

SUSTAINED
DEPOLARIZATION.ʜ²
LATE (>1msec).ʜ'

A moment later the slow Na⁺ gate closes, and the membrane is no longer permeable to Na⁺. In addition, a slow K⁺ gate opens, making the membrane more permeable to K⁺ than it was at rest.

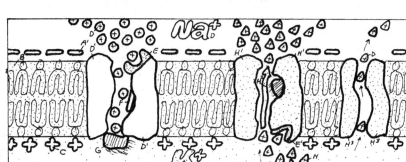

In this plate, we define and interpret three of the most important characteristics of excitation: a threshold stimulus, an all-or-none response, and a refractory period.

THRESHOLD: MINIMAL STIMULUS STRENGTH FOR EXCITATION
If a nerve axon is stimulated with weak electrical shocks, nothing seems to happen. When the stimulus is repeated many times with each stimulus a little stronger, eventually a point will be reached where an action potential appears. The strength of the stimulus just barely able to excite is called the threshold. Stimuli below threshold do not work; stimuli above threshold produce action potentials.

ALL-OR-NONE ACTION POTENTIALS
A stimulus above threshold excites the nerve, but the size of the response is independent of the stimulus strength. All action potentials are the same, no matter how large the stimulus; the response is all-or-none. This behavior is similar to a fuse; once lit, the size of the spark that travels along is independent of the size of the flame that initiated it.

CHANGES IN [Na$^+$] AND [K$^+$] ARE NEGLIGIBLE
To interpret these properties, recall that the inside of the axon has high K$^+$, the outside high Na$^+$, and that the membrane potential, abbreviated by V_m, is a measure of the electrical force on a positive charge. Also recall from plate 11 that the amount of charge movement necessary to make substantial changes in V_m (membrane potential) is very small. During the short time of a single action potential, the amounts of Na$^+$ and K$^+$ that move into or out of the axon are very small; they have significant effects on V_m, but the change in concentration of Na$^+$ or K$^+$ is so small that it is undetectable by chemical tests.

At rest, the axon is permeable mostly to K$^+$, but not much K$^+$ leaks out because the opposing membrane potential, V_m, is close to the K$^+$ equilibrium potential (i.e., the concentration gradient of K$^+$ is almost balanced by V_m pushing in the opposite direction).

EFFECTS OF DEPOLARIZATION ARE TIME DEPENDENT
Now the nerve is stimulated. Depolarization (stimulation) has two effects:
1. early on, voltage-activated Na$^+$ channels open (fast gates).
2. Later, Na$^+$ channels close and K$^+$ channels open (slow gates, delayed for about 1 msec).

With a weak, subthreshold stimulus (panels 1 and 2), not enough Na$^+$ flows in to overcome the outflow of K$^+$, and the axon repolarizes.

DURING EARLY EXCITATION Na$^+$ INFLOW IS REGENERATIVE
With a stronger, suprathreshold stimulus (panels 3 and 5), more Na$^+$ channels open so that Na$^+$ inflow exceeds K$^+$ outflow; the net flow of charge is now positive inward, and the axon is depolarized even further. But this opens even more Na$^+$ channels, which causes more depolarization. A vicious cycle ensues; the membrane potential takes off in the positive direction with an explosive velocity as the interior of the axon becomes more and more positive. But this rapid upward movement of the membrane potential does not persist. Soon (panel 5) V_m becomes positive and large enough to oppose Na$^+$ entry

despite the open channels (i.e., V_m approaches the Na$^+$ equilibrium potential, where the concentration gradient moving Na$^+$ inward is just balanced by V_m pushing Na$^+$ out). At the same time, the delayed effects (slow gates) begin to appear (panel 6). Na$^+$ channels close and voltage-activated K$^+$ channels open, K$^+$ outflow exceeds Na$^+$ inflow, and the net flow of charge is now positive outward. V_m plummets toward its resting value, overshoots momentarily, and comes very close to the K$^+$ equilibrium potential because the voltage-activated K$^+$ channels are still open, making the membrane even more K$^+$ permeable than it was at rest. Finally (panel 7), the repolarized membrane closes the voltage-activated K$^+$ channels and V_m returns to its resting value.

From this description, we see that the threshold is determined by the stimulus strength that causes inward Na$^+$ flow to just barely exceed outward K$^+$ flow. From that point onward, the stimulus plays no further role because the seeds of the positive feedback (vicious cycle) reside in the axon itself. The all-or-none response arises naturally out of this positive feedback; once the response is triggered, the positive feedback drives the membrane potential to its maximum value (given by the Na$^+$ equilibrium potential). The size of the action potential is determined by the concentration gradients of Na$^+$ and K$^+$ because the concentration gradient of K$^+$ limits the resting potential (K$^+$ equilibrium potential), and the concentration gradient of Na$^+$ limits the height of the action potential (Na$^+$ equilibrium potential). Just as a stick of dynamite contains its own explosive energy, the axon membrane is "loaded" with "explosive" energy in the form of ion gradients.

REFRACTORY PERIOD: SLOW GATES (Na$^+$ CLOSED, K$^+$ OPENED)
For a millisecond or two following excitation, the axon is no longer excitable. This recovery phase, called the refractory period, is divided into two parts. The earlier phase is the absolute refractory period, where the threshold appears to be infinite, and no stimulus will suffice. In the later phase, the relative refractory period, the threshold returns to normal. The basis for the refractory period is found in the "delayed effects." After the first millisecond of excitation, the slow Na$^+$ gates close and remain closed for a brief time despite the fact that V_m is near resting value. These gates are slow to respond to the initial depolarization, and they are equally slow in responding to the repolarized membrane. In addition, the voltage-activated K$^+$ gates are still open. With slow Na$^+$ gates closed and K$^+$ gates open, it is difficult if not impossible for Na$^+$ inflow to exceed K$^+$ outflow (i.e., to reach threshold).

Na$^+$-K$^+$ PUMP IS SWAMPED DURING ACTION POTENTIAL
How do the activities of the Na$^+$-K$^+$ pump influence the action potential? They don't, at least not directly. Contributions by the pump to V_m are overpowered by the more massive movements of the ions through the channels. The pump does not cycle often enough to make a difference during activity. However, action potentials are very brief, and the axon is at rest most of the time. During rest, there is ample time for the slow cycling of the pump to restore the small amounts of Na$^+$ and K$^+$ that have leaked through channels activated during the action potential.

CN: Use the same colors as on the preceding page for cell membrane (B), K$^+$ (C), Na$^+$ (D), positive charge (E), and negative charge (F).
1. Color each of the panels completely before going on.
2. Color the Na$^+$ (vicious) cycle.
3. Color the graph in the lower left corner; color the dar numerals gray, and note their relationship to the panels on the right.

STIMULUS (DEPOLARIZATION)ᴬ
CELL MEMBRANEʙ
POTASSIUM (K⁺)ᶜ
CONCENTRATION GRADIENTᶜ
LEAKAGE CHANNELᶜ²
VOLTAGE GATED CHANNELᶜ³
SODIUM (Na⁺)ᴰ
CONCENTRATION GRADIENTᴰ
VOLTAGE GATED CHANNELᴰ²
POSITIVE CHARGEᴱ
ELECTRICAL GRADIENTᴱ
NEGATIVE CHARGEꜰ

At rest the axon is permeable mostly to K^+, but not much K^+ leaks out because the opposing electrical gradient V_m is close to the K^+ equilibrium potential. Depolarization (stimulation) has two effects. A. Early on (fast), voltage-activated Na^+ channels open. B. Later (after 1 msec — slow gates) Na^+ channels close and K^+ channels open. **With a weak stimulus** (1–2), not enough Na^+ flows in to overcome the outflow of K^+ caused by the stimulus-induced reduction of V_m. As a result, the axon tends to repolarize. **With a stronger stimulus** (3–5), more Na^+ channels open so that Na^+ inflow exceeds K^+ outflow; the net flow of charge is now positive inward, and the axon is depolarized even further. But this opens even more Na^+ channels, which causes more depolarization. A vicious cycle ensues; the interior of the axon becomes more and more positive. The process begins to slow (5) as V_m becomes large enough to significantly oppose Na^+ entry. Now (6) the Na^+ channels close and voltage-activated K^+ channels open. V_m plummets toward its resting value, overshoots momentarily, and comes very close to the K^+ equilibrium potential before (7) the repolarized membrane closes the voltage-activated K^+ channels and V_m returns to its resting value.

Identify portions of the action potential (below) with relevant panels on the right. The action potential is all-or-none; its size is limited by the K^+ equilibrium potential (bottom) and the Na^+ equilibrium potential (top). Note the refractory period that follows the action potential where the threshold becomes infinite and slowly returns to normal. This arises partly because the slow K^+ gates are still open, but more importantly because the slow Na^+ gates are still closed, which inactivates the Na^+ channels, making them non-responsive until the slow gates open.

more Na⁺ in

VICIOUS CYCLE

more Na⁺ channels open

more depolarization

THE ACTION POTENTIALᴬ
RESTING POTENTIALᴳ
THRESHOLD POTENTIALᴴ
Na⁺ EQUILIBRIUM POTENTIALᴰ³
K⁺ EQUILIBRIUM POTENTIALᶜ⁴

WEAK STIMULUSᴬ
subthreshold depolarization — K⁺ leakage increases causing

repolarization — return to resting potential

STRONG STIMULUSᴬ²

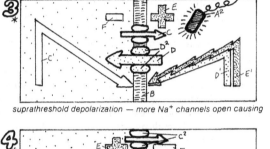

suprathreshold depolarization — more Na⁺ channels open causing

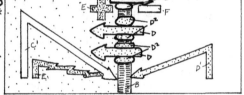

more depolarization — even more Na⁺ channels open

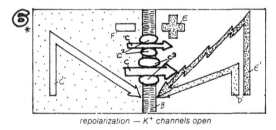

depolarization ends at Na⁺ equilibrium potential.

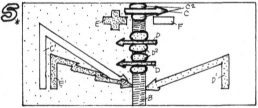

repolarization — K⁺ channels open

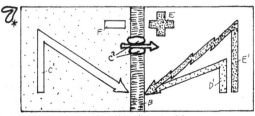

return to resting potential

Plate 17 described how depolarization at a particular location on an axon membrane leads to the formation of an action potential *at that location*. But once an axon is excited, the action potential moves! It is propagated along the entire length of the axon. The first diagram in the plate shows an axon with an impulse located at B that is traveling from left to right. The excited region at B is at the height of the action potential, where the membrane polarity is reversed. This discrepancy in charge between excited and unexcited regions of the axon will cause the charge to move along the axon. For simplicity, we will describe only the movement of the positive charge. (If the same arguments are applied to negative charge movements, the same conclusion will be reached.)

EXCITED REGIONS DEPOLARIZE ADJACENT RESTING REGIONS
On the external surface of the axon, the positive charge will be attracted to the negative charge of the excited region so that the adjacent regions, A and C, on either side will lose some of their positive charge. Further, on the inner surface, the positive charge at B will be attracted to the negative charge at adjacent regions. These actions take positive charge away from the external surface at A and C and add positive charge to the internal surface. The net result is to depolarize the membrane at both A and C. But depolarization stimulates! The stimulus is effective at C but not at A because A is still recovering from the impulse that just passed it (i.e., A is in a refractory period). Nerve impulse conduction is analogous to a spark traveling along a fuse. The heat of the spark ignites the powder in the region ahead of it. The spark does not travel backward because the trailing edge has been burned out. In a nerve, the impulse excites the region ahead of it, releasing some of the energy contained in ion gradients; it does not travel backward because of the refractory period. If you light a fuse in the middle, the spark will travel in both directions. Similarly, if you excite an axon in the middle, the refractory argument no longer applies, and the impulse will travel in both directions (away from the source of excitation).

LARGER AXONS CONDUCT FASTER
Axons vary in diameter as well as length. The larger the diameter, the faster it will conduct nerve impulses. This follows because the speed of conduction depends primarily on how far downstream the electrical effects of the excitatory impulse reach. The farther they reach, the quicker distant regions become excited. These electrical effects are propagated by charge movement (i.e., electrical current) inside the axon as well as out, and the narrower the axon, the more resistant it becomes to these movements. As a result, the electrical disturbance created by an action potential in a narrow axon is confined to regions close by; velocity of conduction is small.

Rapid reflexes require fast impulse conduction along nerves. Invertebrates acquire rapid responses by using very large nerve axons. However, their behavior is uncomplicated, and they do not require very many of these nerves. But vertebrates have complex behavior and require many more axons. If these were all large, they would be cumbersome and create

a packaging problem (see below). The problem is solved by keeping the axon diameters small and by using another means, *myelin sheaths*, to achieve rapid conduction velocities.

MYELIN SHEATHS SPEED UP CONDUCTION IN NARROW AXONS
Most axons are encased in a white, fatty, myelin sheath that is broken at intervals called *nodes of Ranvier*. These nodes are about 1 to 2 mm apart, and they are the only place that the bare axon membrane is exposed to the external solution. Myelin sheaths are formed by satellite *Schwann cells* that are wrapped spirally around the axons; the sheath is made of layers of ion-impermeable Schwann cell membranes. (In the central nervous system, myelin sheaths are formed by *oligodendrogliocytes* rather than Schwann cells.)

Impulse conduction in myelinated and non-myelinated axons differs because the myelin sheaths alter the charge distribution along the axon. At the nodal regions, positive and negative charges are separated by the thin plasma membrane; they are close enough to partially cancel each other. Between the nodes, the myelin sheath, which may contain as many as 300 tightly packed membranes, imposes a much larger distance between intra- and extracellular charge, and the partial cancellation is considerably reduced. As a result, for a given membrane potential of, say, –70 mv, there will be much less charge piled up at the internodal regions (along the sheath between the nodes) than at the nodes. Similarly, it will require much less charge removal to depolarize the internodes. Thus, when a node is excited, it quickly depolarizes the adjacent internodal region and reaches much farther downstream to the next node for more charge. The neighboring node becomes depolarized, and the impulse jumps from node to node. The internodal region does not become excited because the depolarization has to be "shared" by the many membranes that are stacked in series and also because there are few if any Na^+ channels in the internodal regions. These factors result in faster conduction because the impulse now jumps from node to node and does not have to wait for each (internodal) section of the membrane to become excited. Vertebrate nerves often contain some small, slow-conducting, non-myelinated nerves mixed with faster myelinated ones.

MYELINATED NERVES CONSERVE SPACE AND ENERGY
Myelin sheaths are important. When they are destroyed in crippling diseases like multiple sclerosis and Guillain-Barré syndrome, the resulting delay or blockage of impulse conduction accounts for the large array of symptoms. In normal axons, myelin sheaths conserve space. This is illustrated by calculations showing that if fast-conducting mammalian axons had to perform the same job without their sheaths, they would have to be 38 times larger—e.g., without myelin, a 1 mm nerve would have a diameter of 38 mm = 1½ inches! In addition, myelin sheaths help conserve energy. By restricting transmembrane ion movements to the nodes of Ranvier, the myelin sheath minimizes the dissipation of the Na^+ and K^+ gradients each time the nerve fires. Consequently, less energy is required to restore these gradients.

CN: Use the same colors as on the preceding page for Na^+ (B), positive charge (E), and negative charge (F). Use dark colors for C and D, light colors for G and H.
1. Color the upper panel, beginning on the right-hand side with the resting potential and working toward the left.

2. Color the upper panel cross-section of a myelinated axon and then the longitudal view below it. Although they have different colors, the myelin sheath (H) is simply made of flattened membranes of the Schwann cell (G) (i.e., the sheath is part of the Schwann cell).

NON MYELINATED AXON

AXON MEMBRANE A
Na^+ CHANNEL B, IONS B'
FAST GATE C
SLOW GATE D
POSITIVE CHARGE E
NEGATIVE CHARGE F
NERVE IMPULSE F'

Once an axon is excited, an impulse is propagated along its entire length. Area B below shows a patch of axon that is excited and is at the height of the action potential where the membrane polarity is reversed. Follow the flow of [+] charge from, and to, the adjacent membrane area, C. Because [+] charge is attracted to [−] charge, it is removed from the external surface and added to the internal surface at C. Both effects depolarize the membrane at C. But depolarization stimulates! An advancing impulse (B) stimulates the membrane ahead of it (C), but not the region behind (A) because that region is in a refractory period.

RESTING C → (cannot be excited) REFRACTORY D → (excited) IMPULSE F → (can be excited) RESTING C'

extracellular fluid

Na^+

cell interior

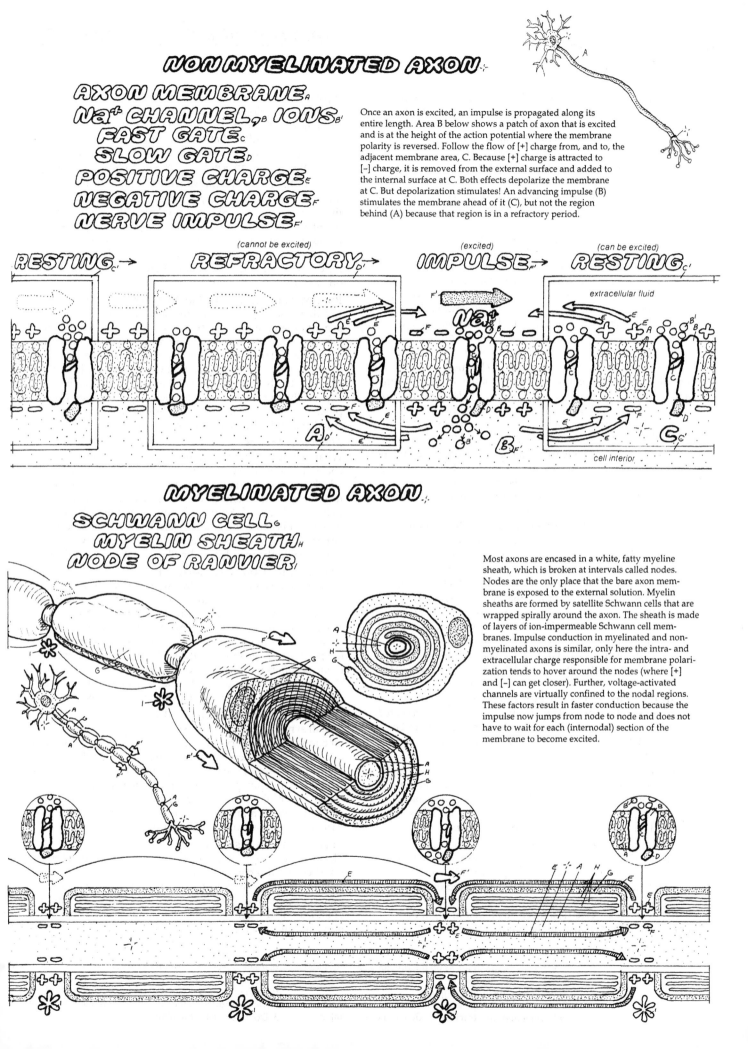

MYELINATED AXON

SCHWANN CELL G
MYELIN SHEATH H
NODE OF RANVIER I

Most axons are encased in a white, fatty myeline sheath, which is broken at intervals called nodes. Nodes are the only place that the bare axon membrane is exposed to the external solution. Myelin sheaths are formed by satellite Schwann cells that are wrapped spirally around the axon. The sheath is made of layers of ion-impermeable Schwann cell membranes. Impulse conduction in myelinated and non-myelinated axons is similar, only here the intra- and extracellular charge responsible for membrane polarization tends to hover around the nodes (where [+] and [−] can get closer). Further, voltage-activated channels are virtually confined to the nodal regions. These factors result in faster conduction because the impulse now jumps from node to node and does not have to wait for each (internodal) section of the membrane to become excited.

Nerve impulses are transmitted both along axons and from cell to cell. Transmission from nerve cell to nerve cell, from nerve to muscle, or from nerve to gland is called synaptic transmission, and the sites of this transmission are called *synapses*. A typical synapse consists of a terminal branch of an incoming nerve axon—the *presynaptic* cell—in close contact with the target *postsynaptic* cell (nerve, muscle, or gland). The distance between these two cells at a synapse is only about 20 nm (1 nm = 1 millionth of a mm), and the space between them is called the *synaptic cleft*.

CHEMICAL TRANSMISSION SEQUENCE
In some cases, transmission is electrical; the arriving impulse forces ion flow through gap junctions that connect the two cells, and the resulting electrical disturbance depolarizes and stimulates the postsynaptic cell. More often, transmission is quite different, operating by release of a chemical called a *neurotransmitter*. This arrangement allows transmission in only one direction: from pre- to postsynaptic cell. The sequence of events in chemical transmission is as follows.

1. The action potential opens Ca^{++} channels—The impulse arrives at the terminal branch of the incoming axon and depolarizes the *presynaptic membrane*. This depolarization opens Ca^{++} *channels* in the presynaptic membrane; Ca^{++} flows down its gradient from outside the cell, where its concentration is high, to inside, where it is very low.

2. Ca^{++} promotes release of neurotransmitter into synaptic cleft—The raised intracellular Ca^{++} promotes the fusion of *synaptic vesicles* with the presynaptic membrane. This process, called *exocytosis*, releases neurotransmitters that had been stored within the *vesicles* into the synaptic cleft.

3. Neurotransmitter activates postsynaptic receptors—The neurotransmitter molecules diffuse across the synaptic cleft and bind to proteins called *receptors* on the postsynaptic membranes. The time required for this diffusion over this very short distance is negligible (less than a microsecond).

4a. Some receptors open ion channels directly—The transmitter-receptor complex promotes the opening of specific *postsynaptic ion channels*. In this case the channels are part of the receptor itself. Here, the time between excitation of the nerve terminal and the opening of the channels (*synaptic delay*) is minimal—about 0.5 msec. This delay is primarily due to the time required for release of neurotransmitter.

4b. Other receptors activate second messengers/ enzymes—In this case, when opening of channels is mediated by second messengers (plate 13), the synaptic delay can be considerably longer—up to 1 sec, or 2000 times longer than a direct action on the channel.

5a. Some ion channels excite (EPSP)—Ions flow through the open channels and if *excitatory* channels are opened, the postsynaptic membrane is *depolarized*. The resulting membrane potential generated across the postsynaptic membrane is called an *EPSP (excitatory postsynaptic potential)*.

This depolarization (EPSP) stimulates voltage-activated channels adjacent to the synaptic region. If enough of these channels are activated, the postsynaptic cell membrane becomes excited, and the impulse is propagated out from the synaptic region over the surface of the postsynaptic cell membrane by the same electrical mechanism that brought the impulse into the synapse on the presynaptic axon. EPSPs

typically last longer than action potentials; when several impulses are delivered to the axon terminal in rapid succession, they may summate, giving a combined EPSP that is higher than the EPSP that results from an individual impulse. Similarly, EPSPs produced by action potentials arriving simultaneously at adjacent synapses also may summate.

5b. Other channels inhibit (IPSP)—If the open channels are *inhibitory*, the postsynaptic membrane *hyperpolarizes*. Now the membrane potential generated across the postsynaptic membrane is called an *IPSP (inhibitory postsynaptic potential)* because the hyperpolarization spreads to some extent to the adjacent voltage-activated channels, making it more difficult for them to respond to a stimulus (depolarization) from any other source (i.e., they are inhibited).

POSTSYNAPTIC CHANNELS ARE NOT VOLTAGE GATED
In either case (EPSP or IPSP), the postsynaptic channels in the synaptic cleft are different from the ordinary excitation channels that populate the other portions of nerve and muscle cell membranes. The postsynaptic channels are *not* activated by depolarization; instead, activation will occur only if a specific chemical binds to their associated receptor. Once activated chemically, they *produce* the electrical depolarization (hyperpolarization) required to excite (inhibit) ordinary voltage-activated channels that lie in adjacent areas.

EPSP: Na^+ MOVES IN
What distinguishes an "excitatory" from an "inhibitory" channel on the postsynaptic membrane? It all depends on which ions pass freely through the channel. In a typical excitatory synapse, the chemically activated channels are permeable to both Na^+ and K^+. More Na^+ moves into the cell than K^+ moves out because the gradient (electrical + concentration) is larger for Na^+. As a result, net positive charge moves in, and the postsynaptic membrane depolarizes. We have an EPSP.

IPSP: K^+ MOVES OUT, Cl^- MOVES IN
In an inhibitory synapse, the transmitter reacts with the postsynaptic membrane and opens chemically activated channels that are permeable to K^+ and Cl^-, but not to Na^+. K^+ moves out of the cell; Cl^- movements may contribute, but they usually are more limited because its gradient is smaller. Thus, net positive charge moves out, and the postsynaptic membrane becomes more polarized (hyperpolarized). We have an IPSP, making it more difficult for any excitatory impulse to depolarize the membrane. The postsynaptic cell is inhibited.

DIFFERENT SYNAPSES, DIFFERENT TRANSMITTERS
Not all synapses are alike. Those that occur at neuromuscular junctions between nerve and skeletal muscle use *acetylcholine* as a neurotransmitter; they are always excitatory. Those occuring in visceral organs (i.e., autonomic synapses—plates 20, 29) use norepinephrine or acetylcholine and are either excitatory or inhibitory. The synapses that occur between neuron and neuron in the central nervous system are the most varied; they use a multitude of neurotransmitters (plate 87).

TURNING THE TRANSMITTER OFF
The action of a neurotransmitier does not persist for a long time because it is continuously removed from the synaptic cleft either by enzymatic attack or by re-uptake by nerve terminals. A persistent response of the postsynaptic cell can be obtained only by delivery of an equally persistent barrage of nerve impulses to the synapse.

CN: Use the same colors as on the previous page for J and K, and a dark color for C.
1. Begin with the upper diagram.
2 Color the EPSP sequence on the left. This represents an enlargement (with structures added to the action taking place in the central illustration above). Complete stage 1 before going on to 2. Color everything in all four examples even though only stage 1 has been completely labeled. (Labels have been added to the other three only where necessary).
3. Do the same for the IPSP sequence.

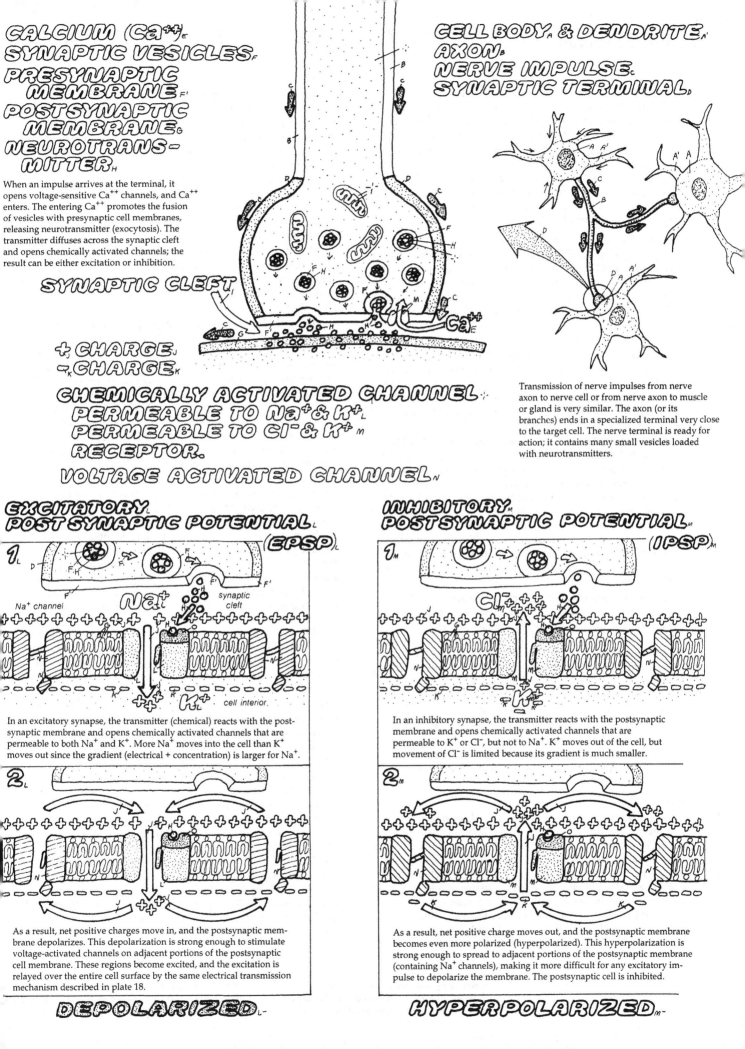

CALCIUM (Ca++) E
SYNAPTIC VESICLES F
PRESYNAPTIC MEMBRANE F'
POSTSYNAPTIC MEMBRANE G
NEUROTRANSMITTER H

When an impulse arrives at the terminal, it opens voltage-sensitive Ca^{++} channels, and Ca^{++} enters. The entering Ca^{++} promotes the fusion of vesicles with presynaptic cell membranes, releasing neurotransmitter (exocytosis). The transmitter diffuses across the synaptic cleft and opens chemically activated channels; the result can be either excitation or inhibition.

SYNAPTIC CLEFT I

+ CHARGE J
– CHARGE K

CHEMICALLY ACTIVATED CHANNEL
PERMEABLE TO Na+ & K+ L
PERMEABLE TO Cl– & K+ M
RECEPTOR O

VOLTAGE ACTIVATED CHANNEL N

CELL BODY & DENDRITE A'
AXON B
NERVE IMPULSE C
SYNAPTIC TERMINAL D

Transmission of nerve impulses from nerve axon to nerve cell or from nerve axon to muscle or gland is very similar. The axon (or its branches) ends in a specialized terminal very close to the target cell. The nerve terminal is ready for action; it contains many small vesicles loaded with neurotransmitters.

EXCITATORY POST SYNAPTIC POTENTIAL L
(EPSP) L

1 L

Na+ channel

Na+

synaptic cleft

K+ L

cell interior

In an excitatory synapse, the transmitter (chemical) reacts with the postsynaptic membrane and opens chemically activated channels that are permeable to both Na^+ and K^+. More Na^+ moves into the cell than K^+ moves out since the gradient (electrical + concentration) is larger for Na^+.

2 L

As a result, net positive charges move in, and the postsynaptic membrane depolarizes. This depolarization is strong enough to stimulate voltage-activated channels on adjacent portions of the postsynaptic cell membrane. These regions become excited, and the excitation is relayed over the entire cell surface by the same electrical transmission mechanism described in plate 18.

DEPOLARIZED L

INHIBITORY POST SYNAPTIC POTENTIAL M
(IPSP) M

1 M

Cl–

K+

In an inhibitory synapse, the transmitter reacts with the postsynaptic membrane and opens chemically activated channels that are permeable to K^+ or Cl^-, but not to Na^+. K^+ moves out of the cell, but movement of Cl^- is limited because its gradient is much smaller.

2 M

As a result, net positive charge moves out, and the postsynaptic membrane becomes even more polarized (hyperpolarized). This hyperpolarization is strong enough to spread to adjacent portions of the postsynaptic membrane (containing Na^+ channels), making it more difficult for any excitatory impulse to depolarize the membrane. The postsynaptic cell is inhibited.

HYPERPOLARIZED M

In Plate 19 we saw that synaptic transmission followed the Ca^{++}-stimulated release of neurotransmitters from vesicles into the synaptic cleft. Release of transmitter does not take place uniformly over the terminal surface; rather, there are preferred areas called *active zones* where vesicles, Ca channels, and receptors appear to congregate. In this plate we take up these processes within the nerve terminal in more detail. Although we use the most widely studied synapse, the neuro-muscular junction, as a convenient example, the general scheme applies to most fast-acting synapses.

SETTING UP THE VESICLES FOR ACTION

Synthesizing acetylcholine (ACh) in the cytosol — The *acetylcholine (ACh)* transmitter is formed in the cytosol by a combination of *acetate* and *choline*. The acetate is formed from ordinary metabolism and first reacts with co-enzyme A (CoA) to form an activated *acetyl-CoA* (plates 5, 6), which then readily reacts with choline to form acetylcholine in the presence of a specific enzyme.

Concentrating ACh in vesicles — The ACh is then concentrated within vesicles by a counter-transport carrier that couples the uphill transport of transmitter into the vesicle with the downhill transport of H^+ out. This requires the significant H^+ gradient from vesicle to cytosol that is provided by an ATP-driven H^+ pump.

Distributing the loaded vesicles: synapsin and SNAREs — Most loaded vesicles are held in reserve; they are anchored near the active zone by a protein, synapsin I, that tethers them to actin filaments. Other loaded vesicles are "docked" at active zones of the terminal membrane. They are believed to dock at these specific locations because of the interactions of a family of membrane proteins called SNAREs: v-SNARES on the *vesicle* membrane form complexes with complementary t-SNARES on their *target* membranes — in this case the target is in the active zone. This interaction ensures that vesicles dock at the correct location.

Cytosolic proteins are required for fusion — A vesicle delivers its contents to the synaptic cleft when it fuses with the plasma membrane. This will happen only when the lipid bilayers come within about 1.5 nm of each other — much closer than the docking distance. This is facilitated by a set of cytosolic fusion proteins that are positioned after the initial docking. How they function is not clear. The vesicle, containing numerous ACh molecules, is now set and awaits a large Ca^{++} signal to fuse and deliver its contents to the cleft.

Our description of docking and membrane fusion is fairly general. It applies to most exocytosis processes — e.g., secretion of proteins. However, unlike synaptic transmission, in some of these processes the fusion follows docking without any regulation by Ca^{++}.

RAPID FIRING

Organization of the active zone expedites rapid response — The pre-synaptic terminal can transmit impulses via ACh release in a steady state at high frequencies. To keep up with these frequencies, all of the pre- and post-synaptic events must take place very rapidly. This is reflected by the existence of docked vesicles. Each is capable of releasing 5000 ACh molecules in less than one msec, a feat that could not be matched by carriers or channels where the transmitters would have to diffuse through the cytosol. Rapid response

also requires the organization of the active zone. Docked vesicles lie in very close proximity to Ca^{++} channels and to the ACh receptors, which lie sequestered in the postsynaptic membrane, directly across the synaptic cleft. When the action potential arrives at the nerve terminal and activates Ca^{++} channels, the adjacent docking proteins are momentarily exposed to a torrent of concentrated Ca^{++}. The high concentration dissipates rapidly as the Ca^{++} diffuses through the cytosol and is taken up by other proteins. However, this exposure is sufficient to directly trigger membrane fusion and release of ACh.

Slightly later, the increased cytosolic Ca^{++} has another role. It initiates a sequence (via calmodulin-activated protein kinase (plate 13) that phosphorylates synapsin I, causing it to release tethered reserve vesicles so that they can move into docking positions.

Acetylcholinesterase turns the response off — Diffusing across the cleft, released ACh reaches receptor sites in about 2 μsec. When two molecules of ACh bind, the receptor channel opens and remains open for about 1.5 msec. After ACh dissociates from the receptor it can bind again, but more generally it is attacked by a fast enzyme, acetylcholinesterase (AChase), which is confined to the cleft. The ACh is split into two inactive components, choline and acetate. AChase also attacks some of the ACh before it even gets to the receptor for the first time, but the initial gush of ACh is so large that it overwhelms the enzyme and most of it escapes.

RECLAMATION OF VESICLES AND TRANSMITTER

At any given time, there are not enough vesicles or transmitter in the terminal to keep up with a rapid firing rate of substantial duration, and there is no time to synthesize more. The problem is solved by recycling the components. Most of the choline that has been split from ACh is taken back up into the terminal via a cotransporter that couples Na^+ running downhill into the cell with choline running uphill in the same direction. Discharged vesicles first become part of the plasma membrane, then are reclaimed by endocytosis. The nascent vesicle membrane is coated by clathrin, a protein; the clathrin-coating molecules associate with each other and tend to form a geodesic dome–type structure that drives the invagination of the membrane. Another protein, dynamin, is responsible for constricting the neck. Once the complete vesicle has been internalized, it loses its clathrin coat and often fuses with a larger vesicle, the endosome. Vesicles budding off from the endosome can begin a new cycle of loading with ACh.

TOXICOLOGY OF THE NEUROMUSCULAR SYNAPSE

With all of the processes taking place at rapid rates in the neuromuscular synapse, it is not surprising to find it vulnerable to drugs and paralytic toxic agents. Some of these are indicated in the plates.

Botulinum toxin degrades docking proteins so that ACh is not released; black widow spider venom releases all of the ACh. Nerve gases (Tabun, Sarin, Soman) as well as pesticides (Parathion, Malathion) inhibit AChase, keeping the cleft flooded with transmitter, while curare (a South American Indian arrow poison) and bungarotoxin (from Cobra venom) block the ACh receptor, making ACh release totally ineffective. Tetrodotoxin (from puffer fish) and saxitoxin (from "red tide" dinoflagellates) block Na^+ channels so that the action potential never reaches the terminal.

CN: Use light colors for A, B and C.
1. Begin with the motor neuron in the upper right and then follow the numbered steps below.

2. Color the membrane fusion diagram which details the events occuring in step 4 of the main illustration.
3. Color the toxicology events affecting the synapse.

NEUROMUSCULAR SYNAPSE

AXON TERMINAL A
PRESYNAPTIC MEMBRANE A'
POSTSYNAPTIC MEMBRANE B
ACETYLCHOLINE (ACh) C
SYNAPTIC VESICLE D
ACTIN FILAMENT E
SYNAPSIN F

ACTIVE ZONE:
  DOCKING PROTEIN (v-SNARE) G
  DOCKING PROTEIN (t-SNARE) H
  CALCIUM CHANNEL', Ca++ I'

NERVE IMPULSE J
ACH RECEPTOR C'
ACETYLCHOLINESTERASE K
CLATHRIN L   (AChase) K
ENDOSOME M

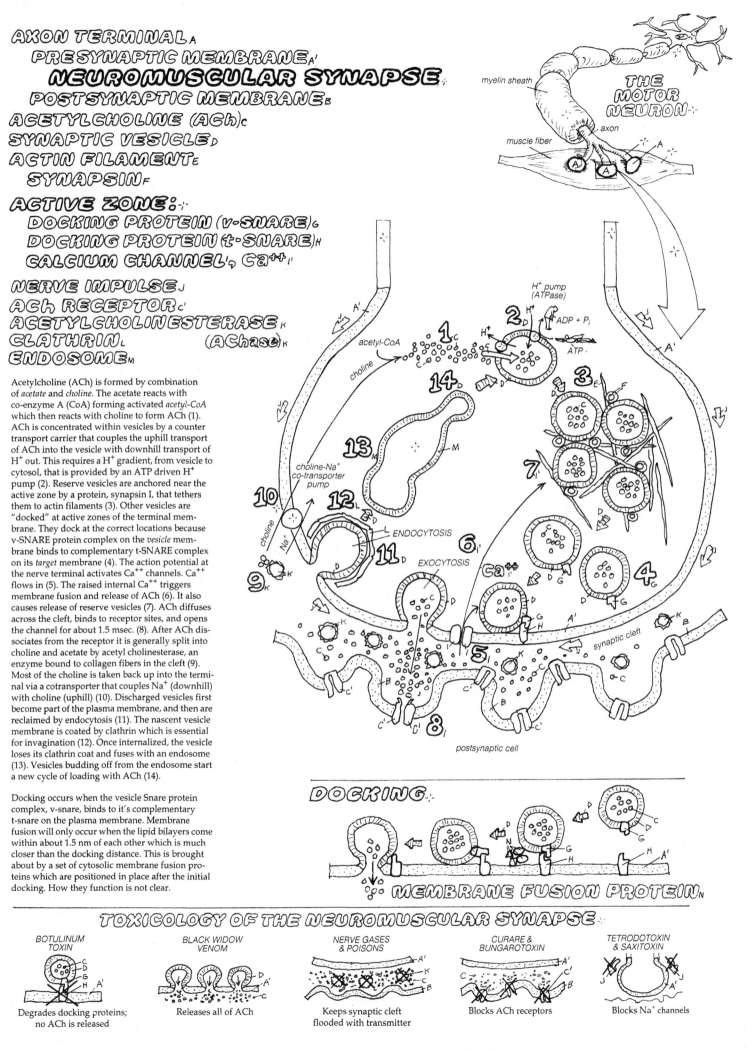

Acetylcholine (ACh) is formed by combination of *acetate* and *choline*. The acetate reacts with co-enzyme A (CoA) forming activated *acetyl-CoA* which then reacts with choline to form ACh (1). ACh is concentrated within vesicles by a counter transport carrier that couples the uphill transport of ACh into the vesicle with downhill transport of H$^+$ out. This requires a H$^+$ gradient, from vesicle to cytosol, that is provided by an ATP driven H$^+$ pump (2). Reserve vesicles are anchored near the active zone by a protein, synapsin I, that tethers them to actin filaments (3). Other vesicles are "docked" at active zones of the terminal membrane. They dock at the correct locations because v-SNARE protein complex on the *vesicle* membrane binds to complementary t-SNARE complex on its *target* membrane (4). The action potential at the nerve terminal activates Ca^{++} channels. Ca^{++} flows in (5). The raised internal Ca^{++} triggers membrane fusion and release of ACh (6). It also causes release of reserve vesicles (7). ACh diffuses across the cleft, binds to receptor sites, and opens the channel for about 1.5 msec. (8). After ACh dissociates from the receptor it is generally split into choline and acetate by acetyl cholinesterase, an enzyme bound to collagen fibers in the cleft (9). Most of the choline is taken back up into the terminal via a cotransporter that couples Na$^+$ (downhill) with choline (uphill) (10). Discharged vesicles first become part of the plasma membrane, and then are reclaimed by endocytosis (11). The nascent vesicle membrane is coated by clathrin which is essential for invagination (12). Once internalized, the vesicle loses its clathrin coat and fuses with an endosome (13). Vesicles budding off from the endosome start a new cycle of loading with ACh (14).

Docking occurs when the vesicle Snare protein complex, v-snare, binds to it's complementary t-snare on the plasma membrane. Membrane fusion will only occur when the lipid bilayers come within about 1.5 nm of each other which is much closer than the docking distance. This is brought about by a set of cytosolic membrane fusion proteins which are positioned in place after the initial docking. How they function is not clear.

DOCKING:
MEMBRANE FUSION PROTEIN N

TOXICOLOGY OF THE NEUROMUSCULAR SYNAPSE

BOTULINUM TOXIN
Degrades docking proteins; no ACh is released

BLACK WIDOW VENOM
Releases all of ACh

NERVE GASES & POISONS
Keeps synaptic cleft flooded with transmitter

CURARE & BUNGAROTOXIN
Blocks ACh receptors

TETRODOTOXIN & SAXITOXIN
Blocks Na$^+$ channels

Our description of synaptic transmission shows the axon terminal teeming with activities associated with the secretion of neurotransmitters. Although neurotransmitters can be synthesized within the terminal, associated proteins and organelles required for their synthesis cannot. These include metabolic enzymes, membrane carriers, pumps and channels, proteins required for exocytosis, mitochondria, and synaptic vesicles. Some of these items recycle but sooner or later they all degrade and need to be replaced. The axon and the terminal do not contain ribosomes, and the only source of protein synthesis or organelle formation is in the cell body, which may lie as far as 1 meter (spinal cord to toe) from the terminal. This dependence on the cell body is apparent when an axon is severed: that portion of the axon beyond the cut (furthest from the cell body) degenerates, while the cell body and the attached axon segment survive.

The time required for diffusion of even small proteins over the 1 meter distance from spinal cord to toe is prohibitive; it would take on the order of 150 years. Movements of organelles by random Brownian motion would take even longer. Instead, the cell uses a more efficient mechanism for these critical movements; it provides interactions of cell products with *molecular motor molecules* that travel along microtubular "tracks." This system of intracellular transport is common to most cells; we illustrate it within the axon, where its necessity is readily apparent and where it has been studied in detail.

AXONAL TRANSPORT HAS FAST AND SLOW COMPONENTS
Axonal transport mechanisms are capable of moving material (cargo) in the *anterograde* direction, away from the cell body, or in the *retrograde* direction, toward the cell body. In addition, the cargo may move at substantially different speeds. Fast axonal transport mechanisms move organelles and vesicles at velocities ranging from 250–400 mm/day anterograde and 100 to 200 mm/day retrograde. In addition, there is *intermediate* transport, clocked at 50 mm/day, and *slow* transport, which carries cytoskeletal proteins at 2–4 mm/day. Utilizing the fast component, we see that our long, 1 meter trip from spinal cord to toe could be accomplished in 2.5 to 4 days (not as fast as a nerve impulse, but not 150 years either!).

MICROTUBULES ARE BUILT OF TUBULIN SUBUNITS
Microtubules are made of subunit proteins, called *tubulin*, which are strung together end on end to form linear strands named *protofilaments*. A complete microtubule is the cylinder formed by 13 parallel protofilaments (illustrated in the plate). Each tubulin subunit is in turn composed of two smaller subunits called α- and β-tubulin, which confer a polarity to the tubulin — an α end and β end. This axial polarity is preserved in the assembled microtubule with the α- and β-tubulins alternating along the length of the chain so that one end has only α- tubulin exposed, while the opposite end has the β exposed. Microtubules are dynamic structures; they may grow in length by adding tubulin or retract by removing it. Both addition and removal usually take place at one end, called the (+) end, thought to correspond to where the β-tubulin is exposed. The other end is the (–) end.

The ability to assemble and disassemble is an important factor allowing microtubules to play crucial roles in organizing the cell and in formation of the mitotic spindle in cell division. Inhibition of the microtubular assembly-disassembly process in the mitotic spindle, for example, is the primary focus of anti-cancer drugs such as *colchicine* and *taxol*, which prevent cell division. On the other hand, some microtubules are naturally stabilized by binding to an associated "capping" protein that blocks off the (+) end. Stable microtubules are principal components, for example, of cilia and flagella, where, together with a molecular motor, they produce the bending movements characteristic of these organelles.

In axons, many microtubules grow outward from an organizing center located in the cell body near the basal part of the axon. Axon microtubules are oriented in the same direction; the (–) ends located closest to the cell body, the (+) end closest to the terminal. These microtubules provide the tracks for transport of proteins, vesicles, and organelles from the cell body to distal parts of the axon and to the axon terminal.

MOTOR MOLECULES "WALK" ON MICROTUBULES
Molecular motors are a class of protein ATPases that can bind to microtubules or to actin filaments and move steadily in a single direction along the tubule (or filament). Motors utilized in the axonal transport belong to two families — kinesins, which always move toward the positive end of the microtubule and *dyneins*, which walk in the opposite direction. Typically these molecules have two globular heads and a tail region. The head regions form cyclic attachments to the tubule as the molecule undergoes cyclic changes in conformation (shape) that propel the complex forward. At the same time, the tail region, which is attached to a particular protein, vesicle, or organelle, drags its cargo along. These movements require the input of energy, which is supplied by repetitive cycles of ATP splitting in the motive head region.

KINESIN MOVES TO (+), DYNEIN TO (–) END
In axons, kinesin always marches in 8 nm steps (the dimension of one tubulin subunit) away from the cell body (anterograde) in the direction of the (+) end. Apparently the binding site for kinesin is asymmetric and oriented so that the kinesin can bind only when it "faces" the (+) direction. Some molecules may carry cell products such as vesicles containing Na^+ channels to be inserted along the way in nodes of Ranvier, while others may carry mitochondria and synaptic vesicles destined for the terminal.

Dynein moves in the opposite direction, toward the cell body; it carries nerve growth factors and other extracellular products that are picked up by endocytosis . In addition, it carries old membrane components from the terminal to the cell body for recycling. Unfortunately, it also carries pathological factors such as tetanus toxin and polio virus.

MYOSINS "WALK" ON ACTIN FILAMENTS
A third family of molecular motors, the *myosins*, move along actin filaments in a similar manner to kinesin and dynein moving on microtubules. Actin is densely distributed within the axon terminal. When the vesicle reaches the beginning of the terminal, it may continue its journey bound to a myosin motor walking on actin filaments. The motion produced by actin and myosin is taken up in the next section on muscle.

CN: Use bright colors for G and H.
1. Begin with the parts of the long neuron and color the enlargement of a section of the axon displaying three microtubules (D). Color the entire enlarged section of the microtubule, and one of its 13 protofilaments with alternating colors (E and F).
2. Color the transportation of vesicles and related illustrations at the enlarged lower end of the axon.

AXON TRANSPORT

CELL BODY A
AXON B
TERMINAL C

Movements of proteins and vesicles from their site of synthesis in the cell body to the axon terminals which may be as far as 1 meter distant, is accomplished by *molecular motor molecules* that carry these products along microtubular "tracks."

MICROTUBULE D
PROTOFILAMENT I
α TUBULIN E
β TUBULIN F

Microtubules are made of subunit proteins, called *tubulin*, which are strung together end on end to form linear strands named *protofilaments*. The complete microtubule is the cylinder formed by 13 parallel protofilaments. Each tubulin subunit is composed of two smaller subunits called α- and β-tubulin, which confer a polarity to the tubulin—an α end and β end. This axial polarity is preserved in the assembled microtubule, with the α- and β-tubulins alternating along the length of the chain so that one end has only α- tubulin exposed while the opposite end has β exposed. Microtubules may change their length by adding or removing tubulin, generally at the (+) end.

MOTOR MOLECULE
KINESIN G
DYNEIN H

Molecular motors utilized in axonal transport belong to two families—*kinesins* which move away from the cell body toward the positive end of the microtubule, and *dyneins*, which walk in the opposite direction. These molecules have two globular heads and a tail region. The head regions form cyclic attachments to the tubule as the molecule undergoes cyclic changes in conformation (shape) that propel the complex forward. At the same time, the tail region, which is attached to a particular protein, vesicle, or organelle, drags its cargo along. These movements require the input of energy, which is supplied by repetitive cycles of ATP splitting in the motive head region.

MYOSIN I
ACTIN FILAMENT J

When the vesicle reaches the beginning of the terminal, it may continue its journey bound to a myosin motor walking on actin filaments. This is feasible if both myosin and kinesin are bound to the same vesicle. On the microtubules the kinesin carries the vesicle together with the attached myosin, but when the vesicle gets to the actin the myosin does the walking as it carries the vesicle together with the bound kinesin. A similar cooperative transport in the retrograde direction on oppositely oriented actin filaments would occur when the vesicle binds both myosin and dynein.

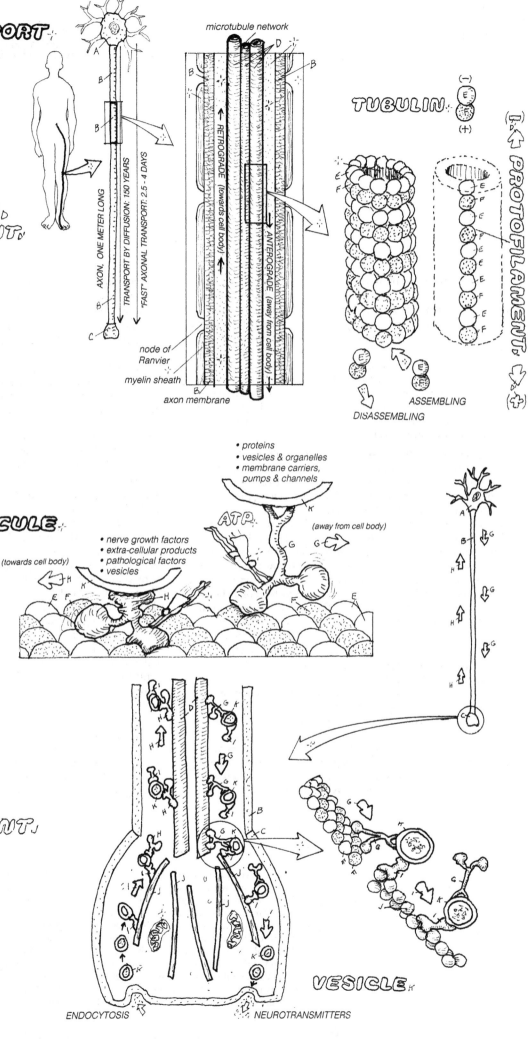

microtubule network

AXON, ONE METER LONG

TRANSPORT BY DIFFUSION: 150 YEARS

"FAST" AXONAL TRANSPORT: 2.5 - 4 DAYS

RETROGRADE (towards cell body)

ANTEROGRADE (away from cell body)

node of Ranvier

myelin sheath

axon membrane

TUBULIN

(−)

(+)

PROTOFILAMENT

ASSEMBLING

DISASSEMBLING

• proteins
• vesicles & organelles
• membrane carriers, pumps & channels

ATP

(away from cell body)

• nerve growth factors
• extra-cellular products
• pathological factors
• vesicles

(towards cell body)

VESICLE K

ENDOCYTOSIS

NEUROTRANSMITTERS

Beating of the heart, blinking an eye, breathing fresh air —
these obvious signs of life are all brought about by muscular
contraction. How do muscles shorten? Something "inside"
must move, but what? Years ago, many physiologists believed
that muscles contract because the proteins of which they are
made actually shorten, either by folding or by changes in the
pitch or diameter of helical molecules. In the 1950s, they were
startled to discover that this is not the case at all. True, the
contractile machinery is made of protein, but contraction does
not occur by protein folding. Rather than changing their
dimensions, the proteins simply slide past each other and
change their relative positions.

CONTRACTION: A BAND SHORTENS, I BAND DOES NOT
An important clue came from early studies of the striped
pattern of living skeletal muscle that could be seen under the
light microscope. The stripes are localized in long fibrous
cylinders called myofibrils that run the length of the muscle
cell. The muscle cell contracts because the myofibrils contract;
they contain the contractile machinery. Each myofibril is
punctuated with alternating light and dark bands called A
and I bands. These bands are "lined up" so that an A band on
one myofibril is closest to an A band on its neighbor. When
you look at the whole cell, you see stripes instead of a check-
erboard. When a muscle contracts, the I band shortens, but the
A band does not change size. The mystery of contraction
seemed to reside in the I band. Soon after the electron microscope became available, however, a new picture emerged.

ACTIN AND MYOSIN FILAMENTS ARE THE
CONTRACTILE MACHINE
Examination with an electron microscope reveals that each
myofibril contains many fibers, called filaments, which run
parallel to the myofibril axis. Thicker filaments are confined
to the A band; the other, thinner, ones seem to arise in the
middle of the I band, at the Z line (a structure that runs perpendicular to the myofibril through the I band, connecting
neighboring myofibrils). The thin filaments run the course of
the I band and part way into the A band, where they

overlap (interdigitate) with the thick filaments. The next
step is to identify the filaments and to determine their role in
contraction.

The chemical identity of the filaments can be determined
by using concentrated salt solutions that selectively extract
muscle proteins. When the protein called actin is extracted,
the thin filaments disappear, and when the protein called
myosin is extracted, the thick filaments disappear. Moreover,
when the cell membrane is destroyed and substances other
than these two proteins are leached out, the thick and thin filaments remain intact, and the muscle can still contract (if it is
provided with ATP as an energy source). These results imply
that the thick and thin filaments are the contractile machinery
and that the thick filaments are made of myosin, the thin
ones of actin.

FILAMENTS SLIDE DURING CONTRACTION
A thick band consists of a lighter middle region (the H zone)
with denser regions on each side. The denser edges are where
thick myosin and thin actin filaments overlap; the middle (H
zone) contains only myosin. The I bands contain only actin.
Whenever a muscle or myofibril changes length, either by
contracting or stretching, neither myosin nor actin filaments
change length, yet they are the contractile machine! It follows
that they must slide past each other, increasing their area of
overlap during contraction and decreasing it during stretching. During contraction, the I band decreases as more and
more of the actin filaments are "buried" in the region of overlap with myosin. The A band cannot change because it represents the length of the myosin filaments, which are invariant.
However, if this picture is correct, you might expect the H
zone to decrease upon contraction and lengthen on stretching.
And it does!

Because the motive force for contraction is provided by
actin and myosin filaments sliding together, there must be
some "connecting" elements that allow them to interact.
These are the cross bridges, made of globular myosin heads
(see figure) taken up in the next plate.

CN: Use dark colors for G and H.
1. Begin at the top with skeletal muscle (A) and
work your way down the right side of the page to
its molecular components. Note that the end surface of each cylindrical example receives the
color of its components. In the case of the myofibril (D), only the title and the myofibrils making
up the end of the cell (C) receive the color D. The
length of the myofibril receives the colors of the
various bands of contractile elements.
2. Color the diagrams of the contractile elements

on the left side of the page. Note that the first diagram attempts to show how the two kinds of filaments actually make up the bands you previously
colored. Note that the thin filaments (E) actually
penetrate the A band. This wasn't shown in the
drawing of the myofibril on right. Note, too, that the
lower two diagrams represent a vertical enlargement (in order to show cross-bridge activity) of the
upper diagram, but not a horizontal enlargement.
The Z lines (H) still coincide with the upper
diagram.

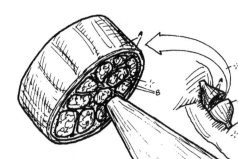

SKELETAL MUSCLE. 40% BODY WEIGHT.

CONTRACTILE ELEMENTS

A BAND
   THICK FILAMENT
   CROSS BRIDGE
I BAND
   THIN FILAMENT
H ZONE
Z LINE
SARCOMERE

Myofibrils are composed of repeating dark A and light I bands, which are responsible for the striations (stripes). Electron microscopy shows finer detail; as illustrated in the lower two diagrams, each fibril is composed of thick and thin filaments. Thick filaments run the length of the A band; thin filaments run through the I band and peripheral portions, but not the central H zone, of the A band. Thin filaments are anchored in the center of the I band by the Z line. That portion of the myofibril (2.5μ long) between the two Z lines is called a sarcomere. Thick and thin filaments interact through cross bridges which are bud-like extensions of thick filaments. The cross bridges are given a separate color for identification purposes.

When living muscle contracts, the I band shortens and the H zone shortens, but the length of the A band does not change. Thus, neither thick nor thin filaments change length; they simply slide past each other, increasing the area of overlap.

BUNDLE OF FIBERS.

Whole muscles are made of bundles of cylindrical striated cells called fibers.

CELL (MUSCLE FIBER).

Cells (muscle fibers) range from 5 to 100μ in diameter but may be several thousand times longer, as they extend from one bone to another.

MYOFIBRIL.

Hundreds of banded cylindrical myofibrils run the length of each cell; they are the contractile elements of the cell.

ACTIN FILAMENT (THIN).

Thin filaments are highly ordered assemblies of protein molecules called actin.

ACTIN MOLECULE.

Actin molecules are pear-shaped (approx. 4nm in diameter). In thin filaments they are joined together like two strings of beads intertwined at regular intervals. (Note: Thin filaments also contain other proteins in addition to actin).

MYOSIN FILAMENT (THICK).

Thick filaments are highly ordered assemblies of protein molecules called myosin.

MYOSIN MOLECULE.

Myosin molecules have long (160nm) rod-shaped tails with globular heads. The heads form cross bridges between thick and thin filaments.

In relaxed muscle, the myosin *cross bridges* are detached from *actin filaments*. During contraction, they attach and provide the contractile force. How does this come about? *Thick filaments* are ordered assemblies of *myosin* molecules; each molecule contains a long rod-shaped tail, a shorter rod-shaped neck, and two globular heads, which form the cross bridges. Only one head is shown in the drawings. The head attaches to the actin filament, forming a cross bridge between actin and myosin filaments. The head then undergoes a conformational change (changes its shape), which propels the actin a distance of about 10 nm. Following this movement, the head detaches and then repeats the cycle farther upstream. Each myosin filament contains about 300 heads, and each head can cycle about 5 times per second, moving the filaments at velocities up to 15μm per sec. This speed can move a muscle from its fully extended to fully contracted state well within 0.1 sec.

The cycles of individual head attachments are not synchronized as shown. They are out of phase, some attaching while others are detaching. Thus, at each moment, some of the heads are entering the motive "power stroke" while others leave. The movement is not jerky, and there is no tendency for the filaments to slip backward.

ATP SUPPLIES THE ENERGY FOR CONTRACTION

Gross muscle movements are brought about by a cyclic reaction of the cross bridges: *attachment* (to actin) → *tilting* (producing movement) → *release, attachment* (to the next site) → etc. By repeating the cycle many times the small movements add up to the smooth, coordinated, macroscopic motions we all enjoy. But cyclic reactions cannot occur without an energy source (if they could, we would be able to build perpetual motion machines). Further, muscle can do physical work (i.e., lift a weight), and work requires energy. The immediate source of this energy is *ATP*. When we incorporate ATP in our scheme, the details of each cycle become more complex as we are able to distinguish more steps. These are shown in the set of diagrams in the plate. Attachment of ATP to the myosin head groups allows the myosin heads to release the actin. Further, a *"high-energy" phosphate* is transferred from the ATP to the myosin, which becomes "energized," while the original ATP, having lost a phosphate, becomes *ADP*. The energized cross bridge is now ready for action. If the muscle is stimulated, the cross bridge will attach to the actin, tilt, and move the actin along (the *power stroke*). Following the power stroke, the myosin and actin remain attached until the beginning of the next cycle, when ATP once again binds, releases the attachment, and energizes the myosin cross bridge. Note that if ATP has been used up, the myosin heads will remain locked to the actin filaments, and no sliding can take place. The muscle will become rigid, resisting both contraction and stretching. This is the condition known as *rigor mortis*, which is common after death when ATP has degenerated. Also note that ATP splitting is not directly involved in the power stroke.

Its energy is used to "prime" the myosin head so that it can attach to the actin and repeat the cycle.

CA⁺⁺ IS REQUIRED FOR CONTRACTION

If ATP is present, why doesn't the muscle continue to contract until all the ATP is used up? The answer involves an additional substance, Ca^{++}, which is required for the attachment phase of the cycle. If sufficient Ca^{++} is present, attachment can occur; at lower levels, it cannot. The action of Ca^{++} as a trigger for contraction and its removal for relaxation are taken up in plate 24.

MYOSIN MOTORS

Muscle myosin is designated as myosin II; it is one member of a family of myosin molecular motors. Just as kinesins and dyneins "walk" on microtubules, myosins walk on actin filaments. During muscle contraction myosin II walks on actin, but the myosin filament does not move because the two ends of the filament are walking in opposite directions. Instead the actin filaments move. (Think about walking in a boat while you hold on to the dock; you don't move, the boat does!)

The secret of contraction appears to lie within the myosin heads. (Isolated myosin heads whose tails have been digested away are capable of walking on actin with unimpaired velocities.) Detailed studies of the molecular structure of myosin shows a prominent cleft in the head region that has been identified with the ATP binding site. Another cleft 3.5 nm (a large molecular distance) away is thought to be the actin binding site. These clefts may provide the malleable spaces needed to initiate conformational changes involved in binding and movement.

Actin and myosin II as well as the other two prominent myosins, myosin I and myosin V, are found in most cells One important difference between the myosins is that myosin I and V have shorter tails that contain binding sites for membranes. In many cells, these motors often transport vesicles on actin filaments (plate 21). On the other hand, the longer myosin tails are especially suited for interacting with other tails to form the filaments we see in muscle. Myosin II is prominent in driving cytokinesis, the final stage of cell division (plate 3), in addition to muscle contraction.

ACTIN FILAMENTS

Actin filaments are found in all cells. They are linear *polymers* – i.e., they are formed by linking many identical smaller units (G-actin) in a repetitive fashion. Like microtubules they grow primarily at one end called the (+) end, and myosin only will walk toward that end (which is located at the Z line in muscle). By interacting with other proteins actin filaments become instrumental in a number of diverse cell functions, such as the formation of the cell cortex, a meshwork of actin filaments just under the plasma membrane that gives the cell shape and mechanical strength. Rearrangements of actin within the cortex are responsible for the ability of some cells – like white blood cells – to crawl. (Also see plate 2, microvilli.)

CN: Use same colors for A, B, C and D that were used for those structures on the previous page.
1. Begin with upper diagrams that demonstrate how the contraction of myosin draws actin filaments inward.
2. Completely color each step of the contraction cycle before going on to the next.

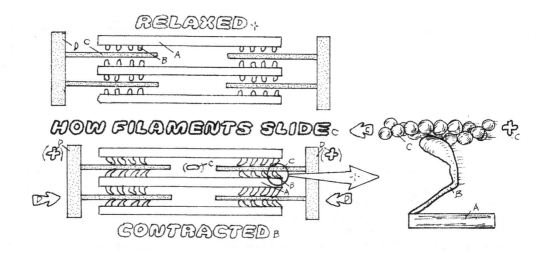

RELAXED +

D C
B A

HOW FILAMENTS SLIDE c

(+) (-) c (+)

CONTRACTED B

MYOSIN (THICK FILAMENT) A
MYOSIN (CROSS BRIDGE) B
ACTIN (THIN FILAMENT) C
Z LINE D
ATP E   ADP F
PHOSPHATE G

In the relaxed muscle the cross bridges are detached from actin filaments. During contraction they attach and provide contractile force. Thick filaments are made of myosin molecules; each molecule consists of a long rod shaped tail, a shorter rod shaped neck and 2 globular heads which form the cross bridge (only one is shown). During contraction the heads attach to actin, tilt, release and then attach to the next position as though they were walking on the filament. But, the actin filaments are anchored with their (+) ends in the Z line and myosin heads can only "walk" toward the (+) end. The myosin heads on the right "walk" toward the Z line on the right, while heads on the left walk toward the left Z line. As a result the thick myosin filaments do not move, but the actin filaments are pulled in.

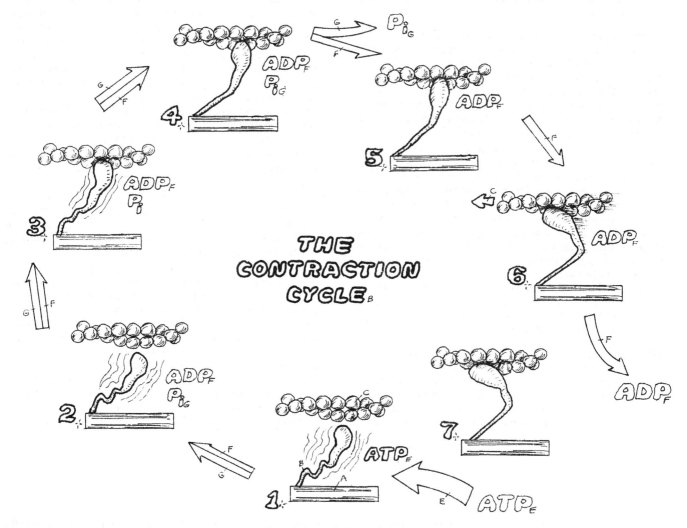

THE CONTRACTION CYCLE B

1–2 Relaxed: Following a contractile movement, the myosin binds an ATP allowing it to enter a relaxed state where it is detached from actin and has some degrees of freedom to wiggle about. The bound ATP is short lived because myosin, itself, is an ATPase (ATP splitting enzyme) Myosin splits the ATP and (2) the products ADP and P_i remain bound to the myosin. There is still freedom to wiggle. **3–4 Attachment:** Myosin makes contact with the actin. At first it is weak (3), but as the attachment gets stronger (4) the mobility of myosin diminishes.

and myosin increases. The binding becomes stronger and the myosin becomes more rigid, as stress is applied to the neck region. The power stroke is initiated. **5-6 Sliding Filament:** The myosin head tilts propelling the associated actin forward. **6-7 Rigor:** Following the sliding motion, ADP is released and the myosin is stuck to the actin - but only momentarily. **7- 1 Release:** ATP binds to the myosin head releasing it from actin and making the muscle pliable. If no ATP is available, myosin heads remain stuck to actin and the muscle becomes stiff. This is the rigidity of rigor mortis which follows death.

If, in the presence of ATP, the cross bridges can enter repeated cycles of attachment, propulsion (tilting), and release, how does this process stop? How do muscles relax? Two key discoveries provided important clues. One was the realization that the presence of minute quantities of *free* Ca^{++} ions was essential for contraction. This fact had escaped detection for many years because it was virtually impossible to remove small traces of Ca^{++} from laboratory chemicals or even from distilled water. Apparently, these traces were sufficient for the contractile process. After learning to control traces of Ca^{++}, we now know that raising the cytoplasmic Ca^{++} (inside the muscle cell) to concentrations as low as .0001 mM is sufficient to support contraction. (This is twenty thousand times more dilute than the free Ca^{++} level in the plasma!) When Ca^{++} is at this level or above, contraction ensues. When Ca^{++} is somewhat below this level, contraction cannot take place and the muscle relaxes.

TROPOMYOSIN COVERS MYOSIN BINDING SITES
How does Ca^{++} exert its influence? An important clue was the discovery that the thin filaments contain other proteins besides actin. In particular, they contain *tropomyosin* and *troponin*. These proteins can be removed from the actin in highly purified artificial systems. When this is done, the requirement for Ca^{++} disappears! The system contracts in the presence of ATP and in the absence of Ca^{++}.

In order for muscle to contract, the energized cross bridges must first attach to the actin filaments. During relaxation, this does not occur because the *sites for myosin attachment* on the actin filaments are covered by tropomyosin molecules; in this state, the sites are masked and not available for the cross bridges. Another protein, troponin, is bound to and serves as a "handle" on the tropomyosin. Troponin can bind Ca^{++} and change shape. When Ca^{++} is bound, the troponin moves the tropomyosin out of the way. The sites are now exposed, attachment of cross bridges can occur, and contraction ensues. When Ca^{++} is absent, the tropomyosin reverts back to its original position and blocks attachment; relaxation follows. But what controls Ca^{++}? How does its concentration rise to trigger contraction and fall to allow relaxation?

SR STORES Ca^{++}, RELEASES IT FOR CONTRACTION
Although the free Ca^{++} concentration in relaxed muscle is extremely low in the cytoplasm, other vesicular structures within the cell may contain an abundance. This is particularly true of the *sarcoplasmic reticulum (SR)*, a compartment containing Ca^{++} ions that are separated from the cytoplasm by the

membranes forming the compartment walls. Each myofibril is surrounded by a sheath of sarcoplasmic reticulum, which resembles a net stocking stretching from one Z line to the next. It is the movement of Ca^{++} from the SR interior to the cytoplasm and back that controls contraction and relaxation.

T TUBULES CARRY ACTION POTENTIALS TO INTERIOR
When nerve impulses activate muscles, the excitation is transmitted through the *motor endplate*, and a muscle *action potential* quickly spreads over the surface of the muscle cell. Contraction of all myofibrils, including those in the cell interior, follows within milliseconds. This all-or-none response is possible because a system of tiny tubes, the *T tubules* (transverse tubules), extends from the surface membrane deep into the interior of the muscle and encircles the perimeter of each myofibril at the level of the Z line in some muscles (frog skeletal muscle, mammalian heart) or at the level of the junction of A and I bands in mammalian skeletal muscle. The lumens of T tubules are continuous with extracellular spaces, and the membranes that form the walls conduct the surface action potential deep into the cell to each sarcomere, where the tubules are close to the SR. A voltage-sensitive tubule membrane protein (*dihydropyridine* receptor) changes conformation when it is depolarized by the advancing action potential. This protein lies in contact with Ca^{++} channels of the SR and when it takes on its depolarized shape, it forces the SR channels open, releasing Ca^{++} into the cytoplasm.

Upon entering the cytoplasm, Ca^{++} reacts with troponin, tropomyosin moves (exposing actin binding sites for myosin), and contraction occurs. Following the excitatory wave, Ca^{++} is pumped back into the SR by an ATP-driven active transport system for Ca^{++}. This lowers cytoplasmic Ca^{++}, and when it falls low enough, binding to troponin is no longer supported. At this point, tropomyosin returns to mask the actin binding site for myosin, and relaxation occurs.

Ca^{++} TRIGGERS MANY DIFFERENT PROCESSES
The role of Ca^{++} ions in muscle contraction is only one example of the ubiquitous role of intracellular Ca^{++} as regulator of cellular processes. In addition to skeletal, cardiac, and smooth muscle contraction, these include ciliary activity, amoeboid movement, exocytosis, synaptic transmission, enzyme activation, and cell cleavage. In the above example, the Ca^{++} level was increased by releasing it from intracellular stores. Having Ca^{++} stored near its site of action allows the very rapid response characteristic of skeletal muscle. Sometimes, the Ca^{++} level is raised by simple opening of Ca^{++} channels in the cell membrane, allowing Ca^{++} to flow in from the outside.

CN: Use same colors as on previous page for actin (C) and myosin (D).
1. Begin with the muscle cell in the Upper right corner. Note that the titles for the axon (G) and axon terminal (H) are listed below. The myofibrils within the cell are left uncolored here and below.
2. Complete the enlarged strand of actin filament and the stages of cross-bridge activation.

3. Begin the lower three drawings with the anatomical illustration in the lower left corner. Then color the upper enlargement of the neuromuscular junction and the release and withdrawal of Ca^{++}. Note that in the lower enlargement, the depiction of the release of Ca^{++} into the cytoplasm (steps 4A, 4B, 4C) is more accurate than the generalized view (4) in the illustration above.

FREE CALCIUM TRIGGERS CONTRACTION.

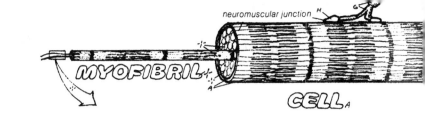

neuromuscular junction

MYOFIBRIL

CELL

ACTIN. MYOSIN. CROSS BRIDGE. TROPOMYOSIN. TROPONIN.

Tropomyosin is a long, two-stranded helical protein aligned almost parallel to the axis of the thin actin filaments. Troponin is a protein complex of three globular subunits, located at regular intervals (spaced approximately seven actin molecules apart) along the thin filament. One of the subunits attaches to tropomyosin, another to actin, and the third subunit can bind Ca++ ions

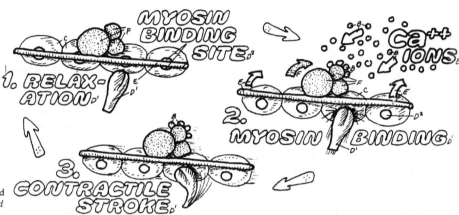

MYOSIN BINDING SITE

Ca++ IONS

1. RELAXATION

2. MYOSIN BINDING

3. CONTRACTILE STROKE

1. Relaxation: *Myosin cross bridges* cannot attach to thin filament because the site of attachment is blocked by *tropomyosin.* **2. Myosin binding:** *Ca++* ions appear on the scene. Four Ca++ bind to each troponin and the complex moves the tropomyosin away from the binding sites. Myosin can now bind to *actin.* **3. Contractile stroke:** Once *energized myosin* binds to actin, the head tilts and propels the thin filament.

SARCOPLASMIC RETICULUM & Ca++ STORAGE.

AXON. AXON TERMINAL. ACTION POTENTIAL. ACETYLCHOLINE. MOTOR END PLATE. CELL MEMBRANE. T TUBULE. VOLTAGE-SENSITIVE PROTEIN.

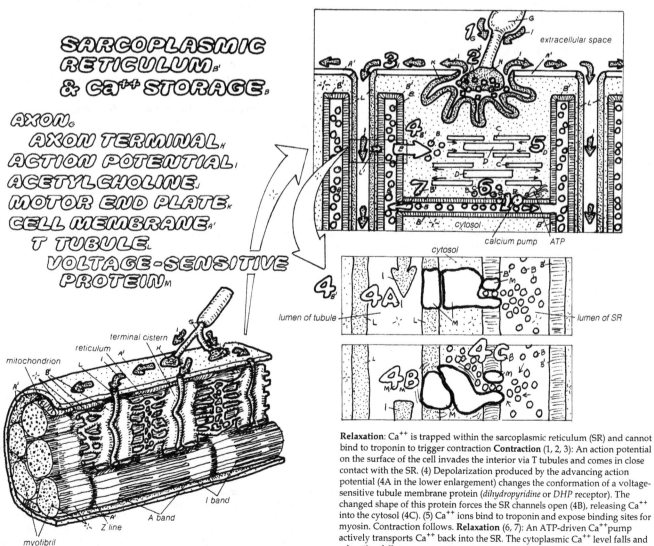

extracellular space

calcium pump   ATP

cytosol

cytosol

lumen of tubule

lumen of SR

terminal cistern

reticulum

mitochondrion

myofibril

Z line

A band

I band

Relaxation: Ca++ is trapped within the sarcoplasmic reticulum (SR) and cannot bind to troponin to trigger contraction **Contraction (1, 2, 3):** An action potential on the surface of the cell invades the interior via T tubules and comes in close contact with the SR. (4) Depolarization produced by the advancing action potential (4A in the lower enlargement) changes the conformation of a voltage-sensitive tubule membrane protein (*dihydropyridine* or *DHP* receptor). The changed shape of this protein forces the SR channels open (4B), releasing Ca++ into the cytosol (4C). (5) Ca++ ions bind to troponin and expose binding sites for myosin. Contraction follows. **Relaxation (6, 7):** An ATP-driven Ca++ pump actively transports Ca++ back into the SR. The cytoplasmic Ca++ level falls and relaxation follows.

Although the contraction of each muscle cell is all-or-none, it is obvious that body movements are not. Sometimes they are forceful, other times slight. This is easily accounted for by realizing that body movements are brought about by whole muscles (groups of muscle cells), not by single cells acting alone. Increasing the force of movement may simply be a matter of recruiting more and more cells into cooperative action. However, there are also more subtle means for changing the performance of individual cells.

NUMBER OF CROSS-BRIDGE CONTACTS (STRENGTH) DEPENDS ON MUSCLE LENGTH

The strength or, more precisely, the force a muscle is capable of exerting depends on its *length*. For each muscle cell, there is an optimum length or range of lengths where the contractile force is strongest. This is easily explained by the sliding-filament theory. The strength of contraction depends on the number of cross bridges that can make contact with actin filaments. When the muscle is too long, few cross bridges can make contact, and contraction is weak. When the muscle is too short, cross-bridge contact can be made, but the filaments begin to get in each other's way and jam up. Again, contraction is weak. Maximal force develops only at a small range of lengths where recruitment of operable cross bridges is maximal and where filaments do not interfere. For the human biceps muscle, this optimum length is attained when the forearm and upper arm are at right angles. When the arm is extended so that the angle between forearm and upper arm is 180°, the biceps is stretched, and contraction is weaker. This explains a common experience of weight lifters: when performing a "curl," it is most difficult to raise the weight from the bottom position with the arm extended. Progress is much easier once the weight has been lifted so that the fore- and upper arm are at right angles.

CONTRACTING MUSCLE STRETCES THE SERIES ELASTICITY

When picking up a light weight, the muscles shorten and move the skeleton. We call this *isotonic contraction*. What happens if you attempt to pick up a weight that is too heavy? The muscle tenses but does not shorten. This is called an *isometric contraction*, a contraction with no change in length! How is this contradiction in terms resolved? Actually, when a muscle undergoes an isometric contraction, the *contractile machinery* really does shorten slightly; the actin and myosin filaments slide past each other. But other passive parts of the cell attached to the contractile machinery—the tendon and connective tissue—are stretched, so there is no net movement. Those parts of the muscle stretched by the contractile machine are called the *series elasticity (SE)*. The exact identity of the SE is a bit vague, but it is known to include the tendons, connective tissue, and elasticity of the cross-bridge neck regions.

The SE stretches a little even when muscle undergoes isotonic contractions. At the beginning of a contraction, the SE is slack, and as the contractile machine shortens, this slack is taken up until the SE can support the load that is to be moved. From this point on, the muscle shortens.

CONTRACTILE STRENGTH IS INCREASED BY SUMMATION AND RECRUITMENT

Changing the length of a muscle is not the only way to alter the strength of contraction. If a rapid succession of stimulating impulses is delivered to a muscle, the cumulative effect will show a stronger contraction than the contraction resulting from a single impulse; the contractions summate. The contraction of a whole muscle can also be increased simply by stimulating more and more muscle cells, a process called recruitment. Summation and recruitment are described in plate 26.

MUSCLE TENSION-LENGTH

Contractile force (strength) depends on the number of cross bridges that can be recruited for "power strokes." This depends on the length of muscle because cross bridges must contact actin to be effective. When muscle is stretched, contact is poor and contraction is weak (4). When muscle is too short, filiaments jam up, interfering with each other's movements (1). Maximal strength of contraction occurs when all cross bridges can contact actin filaments and where there is still room for sliding without interference by actins running into each other (2-3).

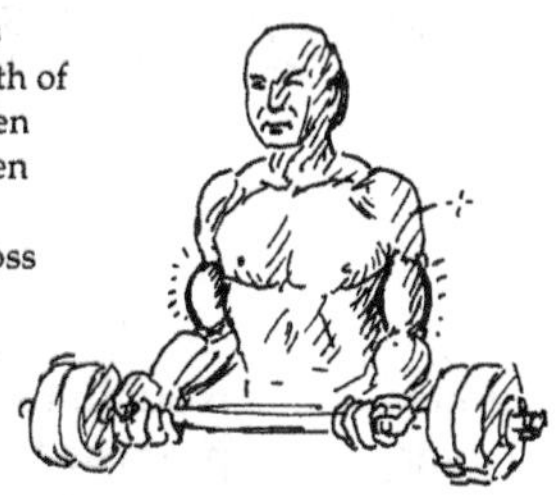

ACTIN_A
MYOSIN_B
CROSS BRIDGE_B'
Z LINE_C
MUSCLE TENSION_D
MUSCLE LENGTH_E

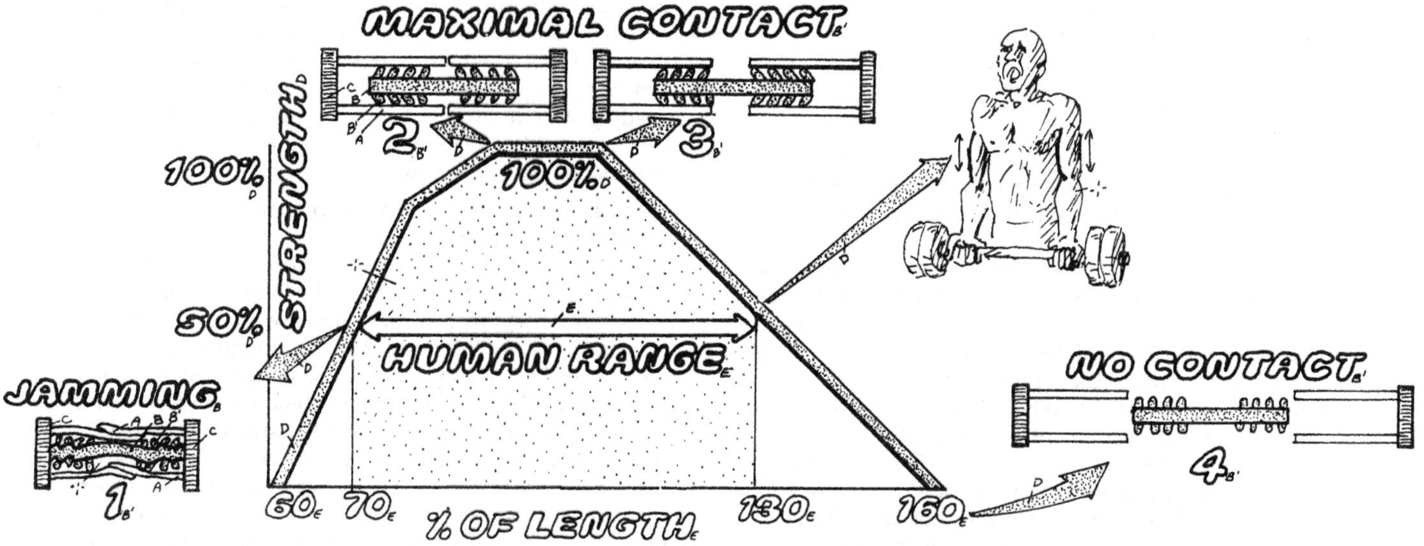

MAXIMAL CONTACT_B

STRENGTH_D

100%_D

50%_D

100%_E

HUMAN RANGE_E

JAMMING_B

NO CONTACT_B

1_B'

2_B'   3_B'

4_B'

60 70_E   % OF LENGTH_E   130_E   160_E

ISOTONIC VS. ISOMETRIC CONTRACTION

tendon

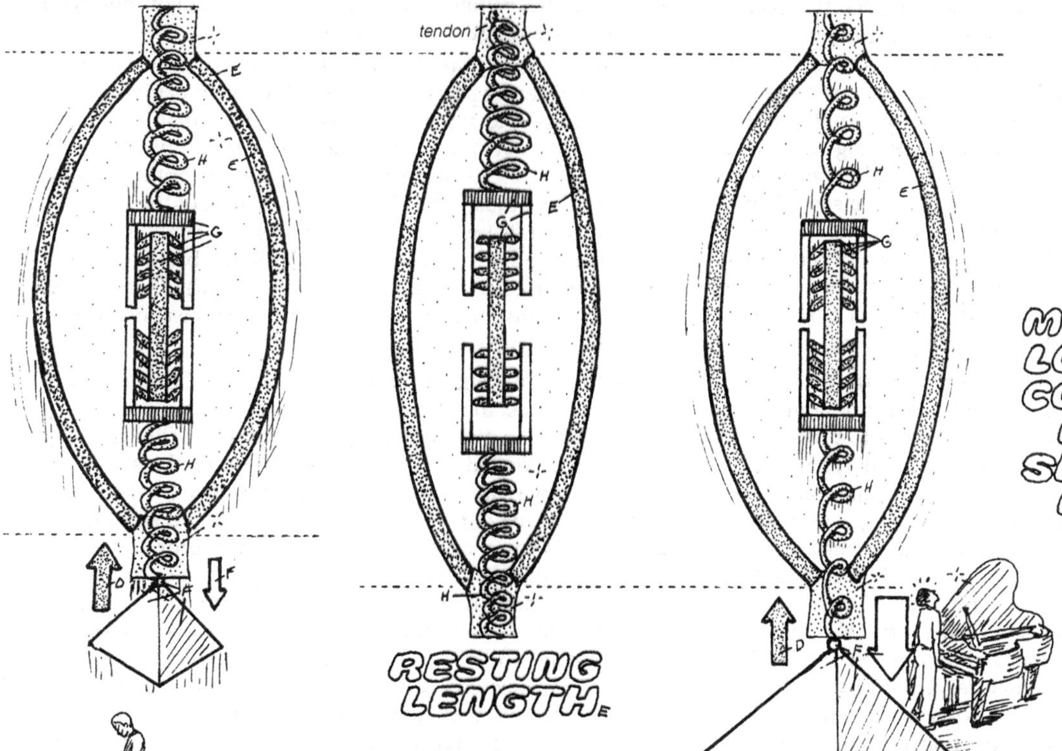

MUSCLE LENGTH_E
LOAD_F
CONTRACTILE ELEMENT_G
SERIES ELASTICITY_H

RESTING LENGTH_E

During contraction the contractile machinery stretches the series elasticity (SE). Sufficient slack is removed until SE can support the weight. Muscle then shortens in an isotonic contraction.

If the weight is too large, the contractile machinery is incapable of stretching the SE (removing slack) sufficiently so that SE can support weight. Contractile machinery simply stretches SE as much as it can, but no net movement occurs. This is an isometric contraction; increase in length of SE = decrease in length of contractile machinery. (The amount of movement of contractile elements shown in the figure is exaggerated for purposes of illustration.)

**TENSION_D > LOAD_F
(MOVEMENT)
ISOTONIC CONTRACTION**
(constant tension)

**TENSION_D < LOAD_F
(NO MOVEMENT)
ISOMETRIC CONTRACTION**
(constant length)

Different tasks call for different types of motion. Sometimes our movements are rapid and vigorous, other times they may be slow, and still other times they may be fine and precise. In this plate, we explore mechanisms in muscles that are utilized to vary the strength and pattern of contractions.

SIMPLE TWITCH: LATENT, CONTRACTION, AND RELAXATION PERIODS

A muscle can be stimulated by an electrical shock applied directly to a muscle cell or by an action potential arriving at the neuromuscular junction. When a single threshold stimulus is delivered to a muscle cell by either of these routes, the muscle responds with a *twitch* that has three phases. (1) The *latent period* consists of the brief delay of 2 or 3 msec between the delivery of the stimulus and the first moment when some contraction can be observed. During this time, Ca^{++} is released, which activates the contractile machinery, and this stretches the series elasticity. When changes in muscle length are measured in an isotonic contraction, no change will be observed until the developing tension matches and just begins to exceed the load (weight). In contrast, when changes in tension are measured for isometric contractions, the change will be observable as soon as the series elasticity is stretched. It follows that the isotonic latent period will be longer than the isometric one, and the duration of the isotonic latent period will increase with increasing load. (2) During the *contraction period* in an isotonic contraction, once the tension matches the load, the continued contraction of the contractile machinery causes a net shortening or contraction of the entire muscle. In an isometric contraction, the contraction phase begins as soon as tension begins to rise. In both cases, the recorded contraction phase lasts anywhere from 5 to 50 msec, depending on the muscle. In the isotonic case, the speed of shortening decreases when the load increases. (3) The *relaxation phase* sets in when the Ca^{++} level subsides as Ca^{++} is pumped back into the sarcoplasmic reticulum (SR). Ca^{++} leaves the troponin so that attachment sites for cross bridges on actin are covered by tropomyosin, and the actin and myosin cannot interact. The filaments slide passively back to their original position.

TWITCHES CAN SUMMATE FOR SUSTAINED CONTRACTION

A single twitch does not express the full potential of a muscular contraction. The twitch is brief and begins to subside before the maximum force or shortening has had a chance to develop. If several twitches are excited in rapid succession, they summate and give a combined contraction greater than a single one. In an isometric contraction, this summation occurs because, in a single twitch, contractile activity is terminated by pumping Ca^{++} back into the SR before the contractile machinery has had time to fully stretch the series elasticity. If another stimulus follows on the heels of the first, it too will initiate a twitch, but this latter twitch reaps the benefit of the first. It finds the SE already partially extended, and its efforts are added to this. In an isotonic contraction, a single twitch has enough time to develop the tension (otherwise the muscle

wouldn't shorten), but the extent of shortening has been compromised by the brief twitch duration. Again, a second twitch arising before the muscle has had time to fully relax will reap benefits from the first. It will find the muscle partially shortened and will summate. In either case (isotonic or isometric), when the frequency of stimuli is sufficiently rapid, each succeeding stimulus arrives before the twitch from the preceding one has even begun to relax. The result is a smooth, sustained contraction called a tetanus.

MOTOR UNIT RECRUITMENT INCREASES CONTRACTILE STRENGTH

The strength of contraction of a single muscle cell can be altered by changing its length or the frequency of stimulation (frequency of nerve impulses). Because a whole muscle is a collection of cells, it follows that the strength of contraction also can be increased simply by engaging the simultaneous contraction of more cells, a phenomenon known as *recruitment*. Each motor nerve axon that transmits impulses to a muscle branches several times before making synaptic connections with muscle cells. Branches from one axon innervate many muscle cells. Each muscle cell in a mammal receives branches from only a single axon. Thus (in the body), the muscles innervated by a single axon will all contract if and only if that axon transmits an impulse. A single motor neuron and the muscle fibers connected to it act as a unit; it is called a *motor unit.* Recruitment of motor units is the major means for varying the strength of contraction.

The number of muscle cells (fibers) in each motor unit varies in different parts of the body. Motor units involved in finely graded and skilled motions (e.g., those that move the fingers or the eyes) contain a small number of muscle cells (as low as ten). This gives the nervous system the option of making very tiny adjustments in the performance of these muscles. Those units involved in more gross contractions (e.g., those that control the postural muscles in the back) have many muscle cells (perhaps over two hundred) for each nerve axon. In this instance, the nervous system makes powerful adjustments with only a few nerves.

ASYNCHRONOUS FIRING PROMOTES SMOOTH CONTRACTIONS

Muscles fatigue; their activity cannot be sustained indefinitely. Yet some muscles (e.g., postural muscles) are called upon for prolonged, sustained, smooth contractions. In the laboratory, we can demonstrate smooth contractions (shown on the right side of the chart) by delivering rapid stimuli and producing a complete tetanus. In the body, motor units are rarely stimulated at those frequencies. Contractions of a motor unit often involve a train of twitches that are not completely fused; the muscle cells have moments to relax, so the motion of any individual motor unit appears jerky. Contraction of the whole muscle is smooth because many motor units are involved, and they do not fire in synchrony. When one unit is beginning to relax, another will start contracting. Individual motions summate, and the net motion is smooth.

CN: Use dark colors for A, O, and P.
1. Begin with the examples of a twitch on the left, noting that this measurement of muscle shortening receives the colors of its three phases, whereas in the myogram to the right the twitch receives its single color, A (drawn on the revolving drum).
2. Color the instrument measuring the twitch

material in the upper right corner.
3. Color the chart distinguishing between various contractions.
4. Color the recruitment panel beginning with the motor unit. Color only the motor unit drawn in bold outline (O, P, and G). The unit left uncolored represents inactivity.

SUMMATION OF CONTRACTIONS

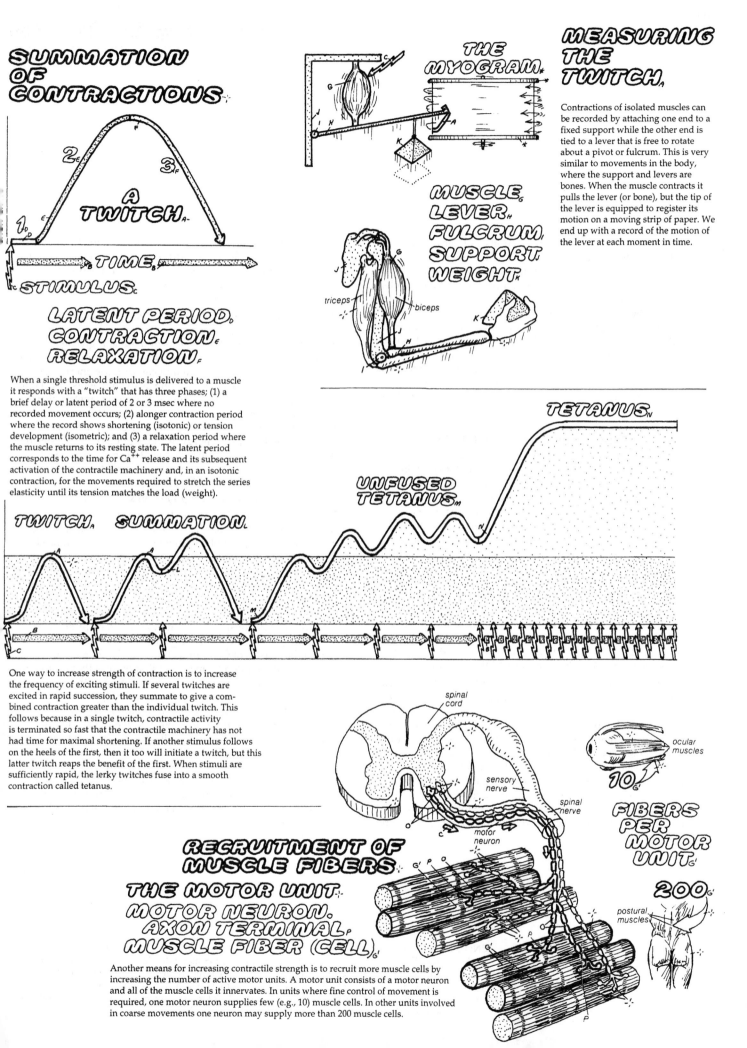

A TWITCH

2 3

1

TIME

STIMULUS

LATENT PERIOD
CONTRACTION
RELAXATION

When a single threshold stimulus is delivered to a muscle it responds with a "twitch" that has three phases; (1) a brief delay or latent period of 2 or 3 msec where no recorded movement occurs; (2) a longer contraction period where the record shows shortening (isotonic) or tension development (isometric); and (3) a relaxation period where the muscle returns to its resting state. The latent period corresponds to the time for Ca^{++} release and its subsequent activation of the contractile machinery and, in an isotonic contraction, for the movements required to stretch the series elasticity until its tension matches the load (weight).

THE MYOGRAM

MUSCLE
LEVER
FULCRUM
SUPPORT
WEIGHT

triceps   biceps

MEASURING THE TWITCH

Contractions of isolated muscles can be recorded by attaching one end to a fixed support while the other end is tied to a lever that is free to rotate about a pivot or fulcrum. This is very similar to movements in the body, where the support and levers are bones. When the muscle contracts it pulls the lever (or bone), but the tip of the lever is equipped to register its motion on a moving strip of paper. We end up with a record of the motion of the lever at each moment in time.

TWITCH   **SUMMATION**   **UNFUSED TETANUS**   **TETANUS**

One way to increase strength of contraction is to increase the frequency of exciting stimuli. If several twitches are excited in rapid succession, they summate to give a combined contraction greater than the individual twitch. This follows because in a single twitch, contractile activity is terminated so fast that the contractile machinery has not had time for maximal shortening. If another stimulus follows on the heels of the first, then it too will initiate a twitch, but this latter twitch reaps the benefit of the first. When stimuli are sufficiently rapid, the jerky twitches fuse into a smooth contraction called tetanus.

spinal cord

sensory nerve

spinal nerve

motor neuron

RECRUITMENT OF MUSCLE FIBERS
THE MOTOR UNIT
MOTOR NEURON
AXON TERMINAL
MUSCLE FIBER (CELL)

Another means for increasing contractile strength is to recruit more muscle cells by increasing the number of active motor units. A motor unit consists of a motor neuron and all of the muscle cells it innervates. In units where fine control of movement is required, one motor neuron supplies few (e.g., 10) muscle cells. In other units involved in coarse movements one neuron may supply more than 200 muscle cells.

ocular muscles

10

FIBERS PER MOTOR UNIT

200

postural muscles

All cells use *ATP* to fuel their reactions and perform work (plate 5). The concentration of ATP within most cells is generally around 5 mM; it is kept at this steady-state level because new ATP is synthesized as fast as it is utilized. Muscle cells present a special case because they are called upon for both sudden bursts and long, sustained periods of intense activity. During endurance exercise, a muscle may utilize a hundred to a thousand times as much ATP as it does during rest. Somehow the supply has to adjust and meet these enormous demands. ATP (as shown in the upper panel) is supplied via three separate sources: *creatine phosphate* (2), *the glycolysis-lactic acid system* (4), and *aerobic metabolism, or oxidative phosphorylation* (3).

THE HIGH-ENERGY PHOSPHATE SYSTEM (20-25 SEC)
The amount of ATP present in muscle cells at any given moment is small. By itself, it is barely enough to sustain 5–6 seconds of intense activity, say a 50-m dash. But as ATP is utilized, it is quickly replenished by the small reserve of energy stored as creatine phosphate. Creatine phosphate very rapidly donates its high-energy phosphate to *ADP* the moment ADP forms, converting it back to ATP. This extra source of ATP is easily mobilized and is very effective as long as it lasts. Unfortunately, it is limited because the store of creatine phosphate is small, only about four to five times the original store of ATP. Normally, the supply of creatine phosphate is replenished by oxidative metabolism via the ATP produced by the citric acid cycle (plate 6). But during sustained, intense exercise, there is not enough time for this to occur. Thus, after some 20–25 seconds of intense activity, we are back in the same place—no ATP. We require additional sources.

THE GLYCOLYSIS–LACTIC ACID SYSTEM (NO O_2 REQUIRED)
ATP can be supplied in a hurry through the *anaerobic breakdown of glucose* (or stored *glycogen*). Each time a glucose is chopped up by this anaerobic path, 2 ATP are formed. The advantage is that it produces the ATP without O_2, and it produces it fast. Though half as fast as the creatine phosphate system, it is two to three times faster than aerobic metabolism. It is limited, however, because on this path the hydrogens stripped off glucose that are normally bound for O_2 to form water are taken up instead by pyruvate to form lactic acid. For each new ATP, a lactic acid also is formed. Energy production

via this pathway is limited by this accumulation of lactic acid, which produces fatigue. In addition, anaerobic glycolysis produces very small amounts of ATP—two per glucose consumed—compared to oxidative phosphorylation, which yields 36 ATP per glucose.

AEROBIC METABOLISM—OXIDATIVE PHOSPHORYLATION (SLOWER, BUT MORE BOUNTIFUL)
This system utilizes *fats* as well as glucose and glycogen, in contrast to creatine phosphate or glycolysis. *Aerobic metabolism* is fairly slow, but it is efficient and can provide energy for almost unlimited durations, as long as the nutrients last. Typically, it takes about 0.5 to 2 minutes for aerobic metabolism to adjust to the increased demands of exercise. Thus, anaerobic processes are required not only for brief peak physical exertion, but also to supply energy at the beginning of long-term muscular activity before aerobic metabolism becomes fully mobilized. Once this has occurred, an exhausted runner may experience a "second wind."

DIFFERENT MUSCLE TYPES FOR DIFFERENT TASKS
Not all skeletal muscle cells are the same. The three types—*red/slow, red/fast, and white/fast*—differ in their capacity to generate ATP, their speed of contraction, and their resistance to fatigue. These and related properties are illustrated in the plate. In general, whole skeletal muscles in humans contain all three types, but in different proportions. Postural muscles of the back, for example, are continually active and have a high proportion of red/slow fibers. These fibers are specialized for aerobic metabolism. They contain the red respiratory pigment *myoglobin*, which stores O_2 and facilitates the diffusion within the muscle of O_2 to mitochondria. Further, the fibers are small, surrounded by many capillaries, and they contract slowly so the blood supply of O_2 can keep up with demand. Red/fast fibers are intermediate between red/slow and white/fast. White/fast fibers are abundant in muscles that have rapid, intense bursts of activity. Myoglobin is absent, mitochondria are sparse, and capillaries are less profuse. Glycolysis is well developed so that ATP is produced rapidly, but the muscle fatigues quickly when the limited glycogen stores are depleted. Muscles of the arms, which may be called upon to produce strong contractions over short periods of time (e.g., weight lifting), have a relatively large proportion of white/fast fibers.

CN: Use purple for A, red for J, and a dark or bright color for E.
1. Begin by first coloring the boundaries of the capillary (A) and muscle fiber cell (B). Then color each title in the column on the right, beginning with number one, and coloring each process in the cell that the title refers to. Note that the group of sports figures along the right margin is not to be colored. The upper figures are representative of an exercise in which ATP is formed aerobically.
2. Color the characteristics of the three skeletal muscle fibers.

SOURCES OF ENERGY IN MUSCLE FIBER

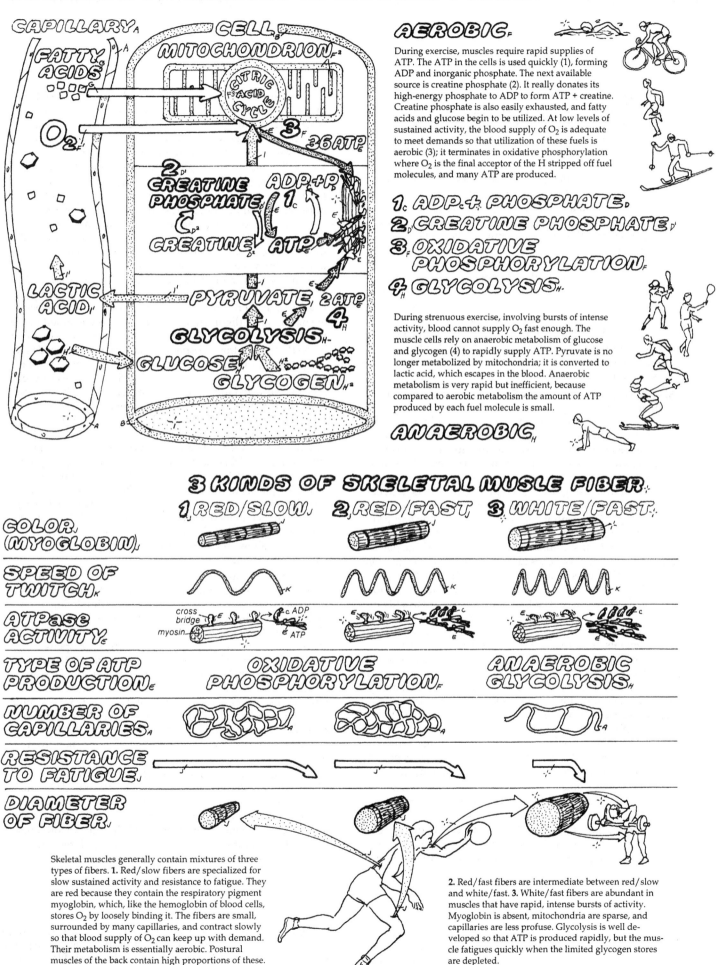

CAPILLARY

CELL

MITOCHONDRION

CITRIC ACID CYCLE

FATTY ACIDS

O_2

36 ATP

CREATINE PHOSPHATE

ADP+P

CREATINE   ATP

LACTIC ACID

PYRUVATE 2 ATP

GLYCOLYSIS

GLUCOSE

GLYCOGEN

AEROBIC

During exercise, muscles require rapid supplies of ATP. The ATP in the cells is used quickly (1), forming ADP and inorganic phosphate. The next available source is creatine phosphate (2). It really donates its high-energy phosphate to ADP to form ATP + creatine. Creatine phosphate is also easily exhausted, and fatty acids and glucose begin to be utilized. At low levels of sustained activity, the blood supply of O_2 is adequate to meet demands so that utilization of these fuels is aerobic (3); it terminates in oxidative phosphorylation where O_2 is the final acceptor of the H stripped off fuel molecules, and many ATP are produced.

1. ADP + PHOSPHATE
2. CREATINE PHOSPHATE
3. OXIDATIVE PHOSPHORYLATION
4. GLYCOLYSIS

During strenuous exercise, involving bursts of intense activity, blood cannot supply O_2 fast enough. The muscle cells rely on anaerobic metabolism of glucose and glycogen (4) to rapidly supply ATP. Pyruvate is no longer metabolized by mitochondria; it is converted to lactic acid, which escapes in the blood. Anaerobic metabolism is very rapid but inefficient, because compared to aerobic metabolism the amount of ATP produced by each fuel molecule is small.

ANAEROBIC

3 KINDS OF SKELETAL MUSCLE FIBER

	1. RED/SLOW	2. RED/FAST	3. WHITE/FAST
COLOR (MYOGLOBIN)			
SPEED OF TWITCH			
ATPase ACTIVITY	cross bridge, myosin		
TYPE OF ATP PRODUCTION	OXIDATIVE PHOSPHORYLATION		ANAEROBIC GLYCOLYSIS
NUMBER OF CAPILLARIES			
RESISTANCE TO FATIGUE			
DIAMETER OF FIBER			

Skeletal muscles generally contain mixtures of three types of fibers. **1.** Red/slow fibers are specialized for slow sustained activity and resistance to fatigue. They are red because they contain the respiratory pigment myoglobin, which, like the hemoglobin of blood cells, stores O_2 by loosely binding it. The fibers are small, surrounded by many capillaries, and contract slowly so that blood supply of O_2 can keep up with demand. Their metabolism is essentially aerobic. Postural muscles of the back contain high proportions of these.

2. Red/fast fibers are intermediate between red/slow and white/fast. **3.** White/fast fibers are abundant in muscles that have rapid, intense bursts of activity. Myoglobin is absent, mitochondria are sparse, and capillaries are less profuse. Glycolysis is well developed so that ATP is produced rapidly, but the muscle fatigues quickly when the limited glycogen stores are depleted.

Smooth muscles are responsible for movements of the viscera and blood vessels. Unlike skeletal muscles, they are involuntary and are adapted for long, sustained contraction. Although these contractions are slower, they can generate forces of the same magnitude as skeletal muscle without fatigue and with little energy consumption. The structure of the two types also differs. Spindle-shaped smooth muscles are smaller (about 50–400 μm long and 2–10 μm thick) contain a single nucleus and a poorly developed *sarcoplasmic reticulum,* and have no obvious motor endplate. Autonomic nerve axons innervating smooth muscle have numerous swollen *varicosities* containing *neurotransmitters.* Although smooth muscle utilizes myosin "walking" on actin to contract, the filaments are not held in register, so they do not show cross-striations. There are no Z lines; instead, actin filaments appear to be anchored to small *dense bodies* that are scattered throughout the cytoplasm. This lack of rigid organization probably accounts for the ability of smooth muscle to be stretched four to five times its length and still contract; the bladder, for example, stretches enormously while filling, but still develops substantial tension on emptying.

SINGLE-UNIT MUSCLES SHOW SPONTANEOUS SUSTAINED CONTRACTIONS

Smooth muscles are divided into *single-* and *multi-unit* classes. *Single-unit* muscles act together in groups because they are interconnected by *gap junctions* that are capable of transmitting excitation from one cell to the next at speeds of about 5–10 cm/second. These muscles often show spontaneous activity, with slowly rising resting potentials (*pacemakers)* that culminate in *action potentials* that are entirely independent of the nerve supply. The contractions that result are slow and prolonged; a single twitch elicited by an action potential can last several seconds. If stimuli occur at a frequency of one per second, the individual twitches fuse into a sustained tetanic contraction. This differs from skeletal muscle only in the remarkably low frequency of stimuli that produce continuous tension. However, as a result, smooth muscles are usually in a state of partial contraction or tension, which is called *tone* or *tonus.* The nerve supply does not initiate this activity, it simply augments or inhibits it. When acetylcholine (transmitter for parasympathetic nerves) is applied to smooth muscle in the large intestine, the pacemaker cells are depolarized to near threshold levels so that the frequency of action potentials increases, and individual twitches fuse and summate. The greater the frequency, the stronger the net contraction. If norepinephrine (transmitter for sympathetic nerves) is applied, pacemaker cells hyperpolarize, lowering the frequency of action potentials and the tonus (tension generated).

The smooth muscle response to stretch is not always predictable. Sometimes it shows plastic behavior; when it is stretched, it releases tension. In other cases, the stretch acts as a stimulus for contraction; when the muscles are stretched, they produce more tension. In these instances, stretch appears to depolarize the pacemaker cells, and they respond by discharging action potentials at an increased rate. This response

is implicated in autoregulation of blood vessels (plate 62) and in automatic emptying of a filled urinary bladder in absence of neural regulation (e.g., after spinal cord injuries). Single-unit smooth muscle often occurs in large sheets and is found in the walls of hollow visceral organs like the intestine, uterus, and bladder as well as small blood vessels and the ureters.

INTRACELLULAR CA^{++} CONTROLS CONTRACTION

Intracellular Ca^{++} is the trigger for both skeletal and smooth muscle contraction. However, smooth muscle does not contain *troponin* and the primary mechanism of Ca^{++} action appears to be different in smooth muscle. Instead of troponin, the rising Ca^{++} (step 5 in bottom diagram) reacts with an intracellular protein called *calmodulin,* forming a complex that in turn activates an inactive form of an enzyme, *myosin light chain kinase (MLCK).* The active MLCK catalyzes the *phosphorylation* (transfer of a phosphate group to an organic molecule) of specific small (light) chains of amino acids contained in the myosin head groups. In this process, the phosphate is donated by ATP. Without this phosphorylation, the myosin heads are incapable of forming cross bridges with actin. This phosphorylation activates the myosin ATPase, which enables further ATP to be utilized in the formation of active cross bridges so that contraction occurs. As long as the level of intracellular Ca^{++} concentration is maintained, myosin remains phosphorylated, and tension is sustained. When Ca^{++} is reduced (e.g., by a membrane Ca^{++} ATPase, which pumps Ca^{++} out of the cytoplasm), MLCK is inactivated and myosin is dephosphorylated. The expected muscle relaxation follows, but often not nearly as quickly as anticipated. Somehow smooth muscle myosin can hang on to its attachment and sustain the tension (as though it contained a *latch*) even when its ATPase activity has been depressed or shut down. This poorly understood mechanism allows the muscle to maintain tension with minimal energy expenditure.

In addition to spontaneous action potentials, contraction of smooth muscle can be initiated or modified by nerve excitation, stretch, hormones, or direct electrical stimulation. In each case, the stimulation results in an increase in intracellular Ca^{++}, either from extracellular sources via Ca^{++} channels or from the sarcoplasmic reticulum. Depolarization opens voltage-activated Ca^{++} channels and stretch opens stretch-activated Ca^{++} channels in the plasma membrane, allowing an influx of extracellular Ca^{++}. Acetylcholine enhances smooth muscle tension by increasing intracellular Ca^{++} via the second messenger phospholipase-IP3 system (plate 13). Norepinephrine inhibits tension, via the second messenger cAMP, by inactivating MLCK and by enhancing Ca^{++} extrusion from the cytoplasm.

MULTI-UNIT ACTIVITY DEPENDS ON NERVE SUPPLY

Multi-unit smooth muscle is more like skeletal muscle because it shows no inherent activity but depends on its nerve supply. However, this nerve supply is more diffuse, extending over a larger area of the muscle membrane. Multi-unit smooth muscles are found in the large airways to the lungs, large arteries, seminal ducts, irises, and some sphincters.

CN: Use dark colors of D, J, and L.
1. Begin with the upper panel; note that varicosity (C) is left uncolored.
2. Color the comparison between single-unit and multi-unit smooth muscle.

3. Color the lower diagram of how Ca^{++} triggers contraction, following the numbered sequence. Note that the size of the myosin filaments (F) has been reduced in comparison to the MLCK (O).

THE SMOOTH MUSCLE CELL.

ACTIN.
MYOSIN.
DENSE BODY.
SARCOPLASMIC RETICULUM.

Smooth muscles are responsible for movements of the viscera. Unlike skeletal muscle, they are involuntary and are adapted for long sustained contraction without fatigue and with little energy consumption. They have a poorly developed sarcoplasmic reticulum and have no obvious motor endplate. Autonomic nerve axons innervating smooth muscle have numerous bulb-like varicosities containing neurotransmitters. Although smooth muscle contains actin and myosin, these filaments are not held in register so they do not show cross-striations. There are no Z lines; instead, actin filaments appear to be anchored to small dense bodies that are scattered throughout the cell. This lack of rigid organization probably accounts for the ability of smooth muscle to be stretched 4 to 5 times its length and still contract.

AUTONOMIC NERVE AXON.
VARICOSITY.
NEUROTRANSMITTER.

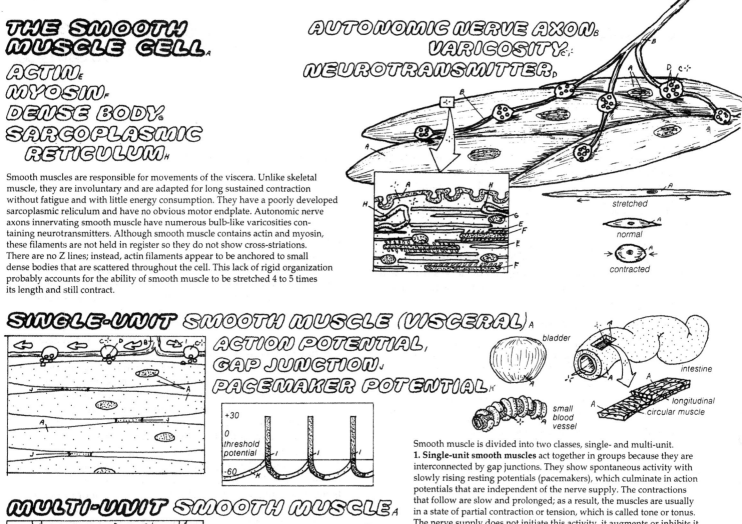

stretched

normal

contracted

SINGLE-UNIT SMOOTH MUSCLE (VISCERAL).

ACTION POTENTIAL.
GAP JUNCTION.
PACEMAKER POTENTIAL.

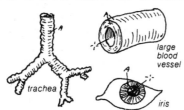

+30

0

threshold potential

-60

bladder

intestine

small blood vessel

longitudinal circular muscle

MULTI-UNIT SMOOTH MUSCLE.

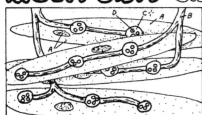

large blood vessel

trachea

iris

Smooth muscle is divided into two classes, single- and multi-unit.
1. Single-unit smooth muscles act together in groups because they are interconnected by gap junctions. They show spontaneous activity with slowly rising resting potentials (pacemakers), which culminate in action potentials that are independent of the nerve supply. The contractions that follow are slow and prolonged; as a result, the muscles are usually in a state of partial contraction or tension, which is called tone or tonus. The nerve supply does not initiate this activity, it augments or inhibits it. The response of smooth muscle to stretch is not always predictable. Sometimes it shows plastic behavior; when it is stretched it releases tension. In other cases the stretch acts as a stimulus for contraction. Single-unit muscle often occurs in large sheets and is found in the walls of hollow visceral organs like intestine, uterus, and bladder. It is also found in small blood vessels and ureters. **2. Multi-unit smooth muscle** is more like skeletal muscle because it shows no inherent activity, but depends on its nerve supply. However, this nerve supply is more diffuse, extending over a larger area of the muscle membrane.

Ca⁺⁺ & MUSCLE CONTRACTION.

MEMBRANE RECEPTOR.
CALCIUM CHANNEL.
CALMODULIN.
MYOSIN LIGHT CHAIN KINASE (MLCK).

In addition to spontaneous action potentials, contraction of smooth muscle can be initiated or modified by nerve excitation, stretch, hormones, or electrical stimulation (1). In each case, stimulation results in increased intracellular Ca^{++}, either from extracellular sources via Ca^{++} channels (2) or from the sarcoplasmic reticulum (3). The rising Ca^{++} (4) reacts with an intracellular protein, calmodulin (5), forming a complex (6) that in turn activates an inactive form of an enzyme, myosin light chain kinase (MLCK) (7). The active MLCK catalyzes the transfer of phosphates to specific small (light) chains of amino acids contained in the myosin head groups. In this process the phosphate is donated by the reaction $ATP \rightarrow ADP + Pi$. Somehow this phosphorylation allows actin to activate myosin to further utilize ATP in the formation of active cross bridges, and contraction occurs. As long as the intracellular Ca^{++} concentration remains above threshold, myosin remains phosphorylated and tension is maintained. When Ca^{++} is reduced, MLCK is inactivated, myosin is dephosphorylated, and the muscle relaxes.

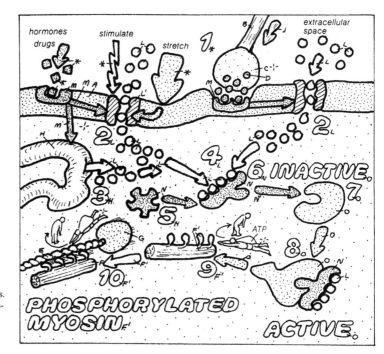

hormones drugs

stimulate

stretch

extracellular space

INACTIVE.

ATP

PHOSPHORYLATED MYOSIN.

ACTIVE.

The *autonomic nervous system (ANS)*, together with the endocrine (hormone) system, controls the body's internal organs. It innervates smooth muscle, cardiac muscle, and glands, controlling the circulation of blood, the activity of the gastrointestinal tract, body temperature, and a number of other functions. Most of this control is not conscious.

The ANS is divided into two parts, the *sympathetic* and the *parasympathetic nervous systems*, whose actions are mostly antagonistic. Many organs are supplied by nerves from each division, but some are not. The following table summarizes some of these actions.

AUTONOMIC EFFECTS OF SELECTED ORGANS

Organ	Effect of sympathetic stimulation	Effect of parasympathetic stimulation
Heart:		
Muscle	Increases rate Increases force of contraction	Slows rate Decreases force of contraction (primarily atrium)
Coronaries	Dilates (β); constricted (α)	Dilates
Systemic arterioles:		
Abdominal	Constricts	None
Muscle	Constricts (α) Dilates (β)	None
Skin	Constricted	None
Lungs:		
Bronchi	Dilates	Constricts
Blood vessels	Constricts (slightly)	Dilate?
Adrenal medullary secretion	Increases	None
Liver	Released glucose	Small glycogen synthesis
Sweat glands	Copious sweating	None
Glands:		
Nasal, lacrimal, salivary, gastric	Vasoconstriction and some secretion	Copious secretion
Gut:		
Lumen	Decreases peristalsis and tone	Increases peristalsis and tone
Sphincter	Increases tone	Relaxes
Gallbladder and bile ducts	Relaxes	Contracts
Kidney	Decreases urine and renin secretion	None
Bladder:		
Detrusor	Relaxes (slight)	Excites
Trigone	Excites	Relaxes
Penis	Ejaculates	Erection
Basal metabolism	Increases	None
Eye:		
Pupil	Dilates	Constricts
Ciliary muscle	Relaxes slightly	Constricts

Modified from A.C. Guyton and J.E. Hall, *Textbook of Medical Physiology*, 9th Edition, 1996, W.B. Saunders.

THE SYMPATHETIC NERVOUS SYSTEM IS ACTIVATED IN EMERGENCIES

If we examine the effects of sympathetic stimulation, a useful pattern emerges. In many instances, sympathetic stimulation appears to prepare the animal for emergencies—for running or fighting. For example, air passages to the lungs (bronchi) dilate, making rapid breathing easier; the heart beats faster and stronger, and the liver releases glucose into the bloodstream. In addition, although it is not evident from the table, the constriction of blood vessels is most prominent in the intestinal tract and least in skeletal and heart muscle; so blood is shifted to the heart and skeletal muscle; where it is most needed. Viewed from this perspective, the often antagonistic parasympathetic nervous system appears to serve a vegetative function. However, the generalization that the sympathetic nervous system prepares the animal for emergencies has several important exceptions; for example, the sympathetic control of blood vessels to the skin is primarily responsive to changes in body temperature. Nevertheless, the generalization serves as a useful aid for remembering the diverse functions of the two divisions of the ANS.

ANS SIGNALS ARE RELAYED IN GANGLIA

The diagrams at the bottom of the plate illustrate that autonomic nerves differ from those going to skeletal muscle. Instead of going directly to their target, the autonomic nerves first make synaptic connections with other neurons, which then relay the impulses to the organs. These synaptic relay stations are called ganglia. Nerves conveying impulses into the ganglia are called *preganglionic fibers*; those that relay impulses to the organs are called *postganglionic fibers*. Both divisions of the ANS use the same neurotransmitter—acetylcholine—to transmit impulses over synaptic connections, from pre- to postganglionic fibers, within the ANS ganglia. However, the two divisions liberate different chemical transmitters at their postganglionic terminals making connection with the organs. Parasympathetic postganglionic transmission, again, uses acetylcholine; sympathetic postganglionic transmission employs norepinephrine (plate 20).

THE ADRENAL MEDULLA SECRETES NEUROTRANSMITTER INTO THE BLOOD

The adrenal medulla gland resembles a sympathetic ganglion. It is stimulated by the acetylcholine liberated by sympathetic preganglionic nerves, which make direct synaptic connections with the gland. However, no postganglionic fibers arise from the ganglion-like gland. Instead, the activated cells secrete mixtures of norepinephrine and the closely related epinephrine directly into the bloodstream. More details about the adrenal medulla are given in plate 125; the brain centers controlling the ANS are covered in plates 107 and 108.

CN: Use a dark color for D.
1. Begin in the upper panel and work your way down the central portion, coloring the plus or minus symbols representing effects of both sympathetic and parasympathetic nerves on the glands and muscles of the body. Color the tiny circles that represent parasympathetic ganglia (B¹) in these various organs. The bottom diagram (under the genitals) shows a general rule for the location of the ganglia of each system (titles are in the lower left corner). Note that the upper three ganglia of the parasympathetic system are an exception; they lie outside the effector organ.
2. The lower panel expands the upper diagram by introducing preganglionic and postganglionic neurotransmitters. Note that the titles for the various effector cells are colored, but the cells themselves are not.

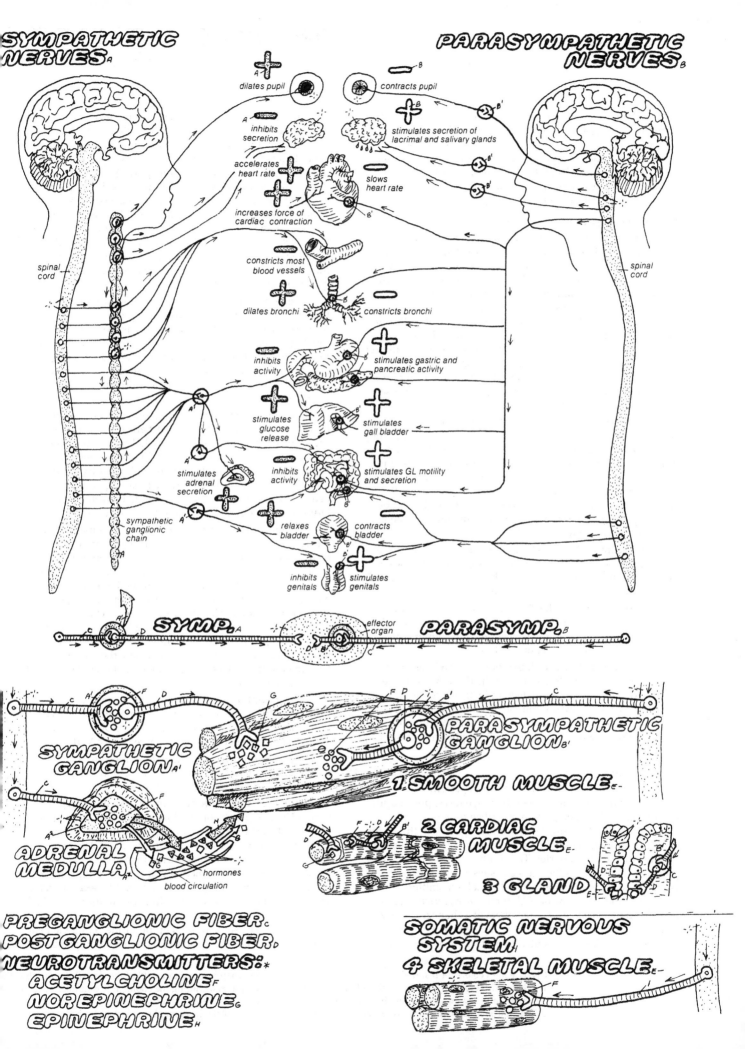

In this plate, we continue our discussion of the autonomic nervous system (ANS), paying attention to the diversity of receptors. The number of identified receptors is growing rapidly. Since each receptor type initiates its own characteristic response, the details of their actions are germane to interpretations of physiological events and crucial for the development of highly selective drugs.

CHOLINERGIC SYNAPSES

Fast nicotinic receptors are found in ANS preganglionic synapses as well as the neuromuscular synapse — Presynaptic events in *cholinergic* (acetylcholine-liberating) synapses have been discussed in detail in plate 20. Responses to drugs show that not all cholinergic receptors are identical. They are classified in two groups — *nicotinic* and *muscarinic*. Nicotinic receptors respond to nicotine as though it were acetylcholine but are insensitive to muscarine, a poison found in some mushrooms and in rotten fish. Nicotinic receptors are all excitatory, and their response is rapid, coming to completion within milliseconds. They are found in neuromuscular junctions and in the preganglionic synapses of the ANS. These receptors are blocked by curare.

Muscarinic receptors form a family of subtypes — Muscarinic receptors respond to muscarine, but not to nicotine. They are G-protein-coupled receptors that can be excitatory or inhibitory, and their response is often prolonged, lasting for seconds. They are found in cardiac muscle, smooth muscle, and exocrine glands as well as on neurons. Unlike nicotinic receptors, they are insensitive to curare, but are blocked by the drug atropine. There are three major members of the muscarinic receptor family.

1. **M_1 receptors ("neural")** are found on CNS neurons and sympathetic postganlionic neurons as well as on gastric cells that secrete stomach acid. They act by formation of IP_3, DAG, and increased intracellular Ca^{++} (plate 13).

2. **M_2 receptors ("cardiac")** are found on cardiac and smooth muscle. They function by opening K^+ channels (plate 13) and by inhibition of adenyl cyclase, with consequent reduction of cAMP production (plate 12).

3. **M_3 receptors ("glandular/smooth muscle")** produce mostly excitatory effects. Acting via the IP_3, DAG, Ca^{++} second messengers, they stimulate glandular secretions and contraction of visceral smooth muscle. In addition, some M_3 receptors mediate the relaxation of vascular smooth muscle (vasodilation) via nitric acid release in neighboring endothelial cells (plate 43).

ADRENERGIC SYNAPSES

Presynaptic events follow a common pattern — In *adrenergic* (norepinephrine-liberating) synapses, presynaptic events follow the same pattern described for cholinergic axons (plate 20). Transmitter synthesis begins with the amino acid *tyrosine*, which is taken up by the nerve terminal via an Na^+-tyrosine co-transporter. The synthesis uses a common pathway followed in other nerve tissues and the adrenal gland for the formation of two additional transmitters, dopamine and epinephrine. (All three — *norepinephrine, epinephrine,* and *dopamine* — are members of a chemical family called *catecholamines*.) Storage of norepinephrine in intracellular synaptic vesicles is essential to protect it from the action of

an intracellular degrading enzyme, *monoamine oxidase* (MAO), attached to the outside surface of mitochondria. Once it is secreted into the synaptic cleft, norepinephrine continues to act until it is taken back up into the presynaptic axon terminal (by a Na^+-norepinephrine co-transporter) or diffuses away. About 70% of the liberated norepinephrine is recaptured intact by the uptake mechanism and returned to synaptic vesicles, where it can be reused. Although there is an extracellular degrading enzyme, *COMT (catechol-O-methyl transferase)* that inactivates norepinephrine, it is not concentrated in the synaptic region and appears to operate chiefly on catecholamines that have escaped or have been secreted (by the adrenal gland) into the circulation.

Adrenergic receptors form a family — Adrenergic receptors are also classified into two major groups — *alpha* and *beta* receptors. The α receptors are more sensitive to norepinephrine than to epinephrine, while the reverse is true for β receptors. Both α and β receptors have distinct subtypes — α_1, α_2, β_1, and β_2. Each subtype initiates a second-messenger sequence that begins with the activation of a G-protein. Examples of their actions are given below.

Effects of α_1 **receptor** activation often reflect the "flight or fight" characteristics of the sympathetic nervous system. They cause constriction of blood vessels, helping to maintain blood pressure, and they inhibit motility in the gut by contracting sphincter muscles but relaxing non-sphincter tissue. In addition, they help mobilize energy by breaking down liver glycogen to glucose. These receptors operate by formation of IP_3, DAG, and increased intracellular Ca^{++} (plate 13).

α_2 **receptors** are found in presynaptic terminals of adrenergic nerves, where they are an essential element in an apparent feedback control of neurotransmitter secretion. When adrenergic transmitter is released it diffuses in all directions. Some molecules reach their target receptor on the post-synaptic membrane, but others activate the α_2 receptors on the presynaptic membrane. Here they inhibit Ca^{++} channels (decreasing Ca^{++} influx) and reducing further release of transmitter (norepinephrine) (see plate). These receptors are also found on vascular smooth muscle, where they induce contraction. In general, they reduce production of cAMP by inhibiting adenylate cyclase.

The β_1 **receptors** are well known for their effects in the heart, where they cause an increased rate and force of contraction. They also induce smooth-muscle relaxation in the gut. Their actions are mediated by an increase in cAMP.

The β_2 **receptors** act via cAMP to induce bronchodilation, vasodilation, relaxation of visceral smooth muscle, and conversion of glycogen to glucose in the liver.

A single tissue may contain more than one receptor type — The two graphs on the lower right-hand side of the plate illustrate how understanding of receptor types can resolve observations that seem contradictory. The first graph shows that injection of epinephrine increases blood pressure, because the epinephrine, acting primarily through alpha receptors, constricts small blood vessels. The second graph shows what happens when the alpha receptors are blocked with a drug. Now the action of epinephrine on beta receptors is unmasked. The same dose of epinephrine acting via beta receptors dilates blood vessels and lowers blood pressure.

CN: Use light colors for B and H.
1. Begin with Tyrosine (A) entering the adrenergic axon in the upper left drawing and follow the numbered sequence.
2. Note the enlargement of the **a** receptor (F1), from the

presynaptic membrane (B) of the main drawing; the oth receptors, including the cholinergic examples are all in th the postsynaptic membrane (H).
3. Complete the epinephrine graphs in the lower right.

In adrenergic (norepinephrine liberating) synapses transmitter synthesis begins with the amino acid tyrosine, which is taken up by the nerve terminal via a Na⁺-tyrosine co-transporter (1). The synthetic pathway for norepinephrine produces two intermediates, dopa and dopamine (also a transmitter) (2), (The neurotransmitters dopamine and norepinephrine, as well as the adrenal hormone epinephrine, are members of a chemical family called *catecholamines*.) Dopamine is transported into vesicles where it is converted to norepinephrine (3). Storage of norepinephrine in intracellular vesicles protects it from the action of an intracellular degrading enzyme MAO (4) attached to the outside surface of mitochondria. An action potential activates Ca^{++} channels (5), and the entering Ca^{++} induces exocytosis of synaptic vesicles releasing norepinephrine (6). Once in the synaptic cleft norepinephrine continues to act until either it is taken back up into the axon terminal via a Na⁺ co-transporter (7), or it diffuses away to be inactivated by an extracellular enzyme COMT (8). Norepinephrine taken back up into the cell enters the synaptic vesicles by the same vesicular transporter that imports dopamine.

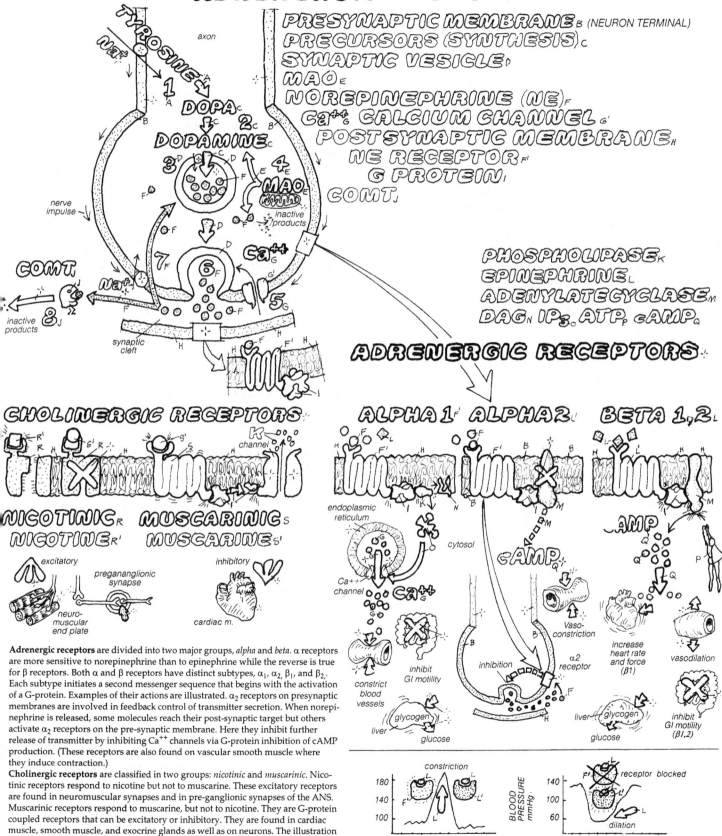

ADRENERGIC NEURON

PRESYNAPTIC MEMBRANE B (NEURON TERMINAL)
PRECURSORS (SYNTHESIS) C
SYNAPTIC VESICLE D
MAO E
NOREPINEPHRINE (NE) F
Ca⁺⁺ CALCIUM CHANNEL G
POSTSYNAPTIC MEMBRANE H
NE RECEPTOR F'
G PROTEIN I
COMT J

PHOSPHOLIPASE K
EPINEPHRINE L
ADENYLATE CYCLASE M
DAG N IP₃ O ATP P cAMP Q

ADRENERGIC RECEPTORS

CHOLINERGIC RECEPTORS

NICOTINIC R   MUSCARINIC S
NICOTINE R'   MUSCARINE S'

excitatory
preganglionic synapse
neuromuscular end plate
inhibitory
cardiac m.

ALPHA 1 F'   ALPHA 2 I   BETA 1,2 L

endoplasmic reticulum
cytosol
Ca⁺⁺ channel
Ca⁺⁺
constrict blood vessels
liver glycogen
glucose
inhibit GI motility
α2 receptor
inhibition
cAMP
Vaso-constriction
liver glycogen
glucose
increase heart rate and force (β1)
inhibit GI motility (β1,2)
vasodilation

Adrenergic receptors are divided into two major groups, *alpha* and *beta*. α receptors are more sensitive to norepinephrine than to epinephrine while the reverse is true for β receptors. Both α and β receptors have distinct subtypes, α_1, α_2, β_1, and β_2. Each subtype initiates a second messenger sequence that begins with the activation of a G-protein. Examples of their actions are illustrated. α_2 receptors on presynaptic membranes are involved in feedback control of transmitter secretion. When norepinephrine is released, some molecules reach their post-synaptic target but others activate α_2 receptors on the pre-synaptic membrane. Here they inhibit further release of transmitter by inhibiting Ca^{++} channels via G-protein inhibition of cAMP production. (These receptors are also found on vascular smooth muscle where they induce contraction.)

Cholinergic receptors are classified in two groups: *nicotinic* and *muscarinic*. Nicotinic receptors respond to nicotine but not to muscarine. These excitatory receptors are found in neuromuscular synapses and in pre-ganglionic synapses of the ANS. Muscarinic receptors respond to muscarine, but not to nicotine. They are G-protein coupled receptors that can be excitatory or inhibitory. They are found in cardiac muscle, smooth muscle, and exocrine glands as well as on neurons. The illustration shows an M_2 receptor found on cardiac cells that opens K⁺ channels slowing the heart rate. Other muscarinic receptors found on neurons, smooth muscle, and exocrine glands operate via $IP_3/Ca^{++}/DAG$ or by inhibition of cAMP production.

constriction
BLOOD PRESSURE mmHg.
180 140 100
140 100 60
receptor blocked
dilation
minutes   minutes

EFFECT OF EPINEPHRINE L

At first glance, the anatomy of the circulation is a mess! Its basic functional simplicity is obscured by the fact that the heart appears to be a single anatomical organ, when it is actually composed of two separate, functionally distinct pumps. The right heart pumps blood to the lungs, where blood takes up oxygen and gives up carbon dioxide; the left heart pumps blood to tissues, where just the reverse happens. Further, although the lungs appear to be two organs (right and left), both lungs do exactly the same thing; they are really just one organ. We can use these ideas to untangle the circulation.

Begin by separating the heart into its two functional units by an imaginary slice through the thick septum (wall) that divides the heart into right and left pumps. Pull these pumps apart, and place all vessels entering or leaving each heart in neat parallel arrays. This can be accomplished without really compromising the functional pathway taken by the blood. Finally, represent the lungs as a single organ, and you will arrive at the simple circle shown in the illustration in the upper right corner. Here, all *pulmonary arteries* (i.e., arteries that leave the right heart and go to the lungs) are collected into a single functional path, as are all *pulmonary veins* (veins that leave the lungs and enter the left heart). Similarly, all *systemic arteries* (i.e., those that leave the left heart bound for all non-lung tissues of the body) are collected, as are all *systemic veins* (veins leaving the non-lung tissues and emptying into the right heart). As illustrated in the functional diagram, the northern hemicircle (between right and left heart) supplying the lungs is called the *pulmonary circulation*. The southern hemicircle (between left and right heart) supplies the rest of the body tissues; it is called the *systemic circulation*. The eastern hemicircle contains oxygen-rich blood. In the western hemicircle, the blood is oxygen poor.

PROPERTIES OF FLOW AND VELOCITY

Follow the diagrams on the opposite page that show important properties of blood flow through this circular path.

1. Steady-state blood flow is the same through any total cross-section of the circulation—During each minute of a steady state, the amount of blood flowing out of the right heart equals the amount flowing into the left heart. If this were not true—for example, if the output of the right heart were greater than the input to the left heart—then blood

would be piling up continuously in the lungs. Conversely, if the right heart output were less than the input to the left heart, then blood would drain from the lungs. Momentarily, some fluid shifts can occur, but on average, over a substantial period of time, our conclusion is correct. This argument can be applied to any section of the diagram. When the average adult is at rest, this flow amounts to about 5000 mL per min.

2. Blood flow = blood velocity x cross-sectional area— Blood *velocity* represents the speed of a blood "particle" in the stream—i.e., how far the particle moves in one minute. *Blood flow* represents how many particles pass (more precisely, the volume of these "particles" passing) a given cross-section in one minute; it is measured in mL/min. The diagram illustrates the relationship between these two quantities.

3. Total cross-sectional area of the vascular tree is greatest in the capillaries—By total cross-sectional area, we mean the sum of the cross-sections of all branches of a similar type— e.g., major arteries, minor arteries, capillaries, major veins. Beginning in the aorta and progressing toward the tissue, the total cross-section of the vascular tree gets larger and larger until it becomes maximal in the capillaries. Although each branch is smaller than its parent, the number of branches increases so rapidly that it more than compensates for the reduction in size of any individual branch. Progressing from capillaries to venules to veins and back to the heart, the reverse occurs.

4. Blood velocity is slowest in the capillaries—This follows because the blood flow is constant throughout the vascular tree, and the total cross-sectional area is largest in the capillaries. The equation *total blood flow = blood velocity x total cross-sectional area* shows that as cross-sectional area increases, blood velocity decreases, so that the product of the two does not change. (The same argument applies to a river, where a widening of the river bed is accompanied by a slowing of the stream.) It follows that the velocity will be smallest where the area is largest (in the capillaries). This is important because capillaries are very short (approximately 0.1 cm), and if the blood didn't slow down, there wouldn't be enough time for exchange (e.g., of O_2) between blood and tissues to occur. For example, blood normally spends about 1 sec. in a capillary; if it traveled at the same speed as it does in the aorta, this time would be reduced to only 0.001 sec., a 1000-fold reduction.

CN: Use blue for B, purple for C, and red for D.
1. Begin with the anatomical description of blood flow. Follow the numbered sequence starting with 1 in the right atrium (marked by an asterisk).
2. Color the functional description of blood flow, starting again with the right heart.
3. Color the diagrams describing the relationship between flow, area, and velocity.
4. Color the chart in the lower right corner demonstrating these physical principles as they occur in the body.

MYOCARDIUM A.
DEOXYGENATED BLOOD B:
SYSTEMIC VEINS B'
PULMONARY ARTERIES B²
CAPILLARIES C
OXYGENATED BLOOD D
SYSTEMIC ARTERIES D'
PULMONARY VEINS D²

FUNCTIONAL FLOW

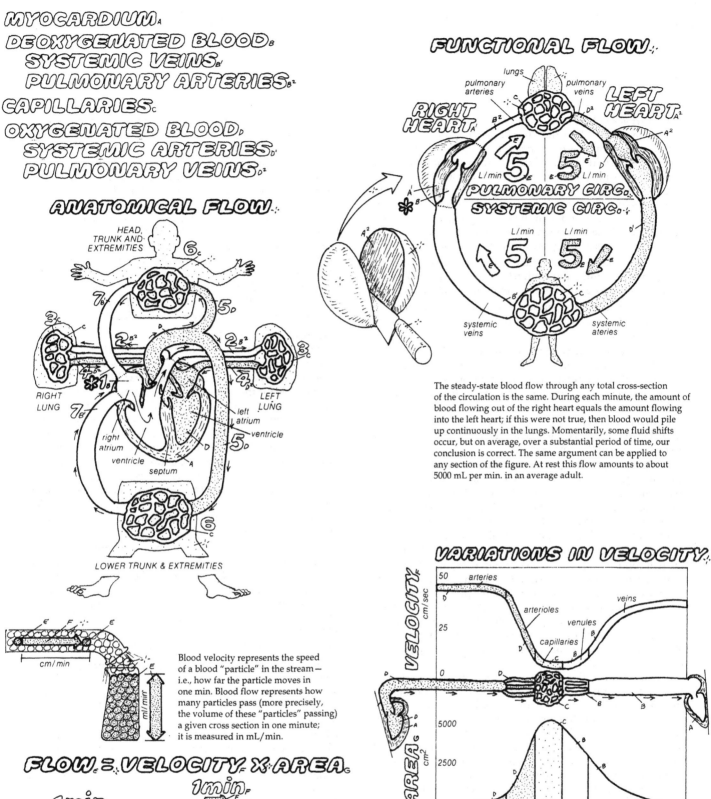

The steady-state blood flow through any total cross-section of the circulation is the same. During each minute, the amount of blood flowing out of the right heart equals the amount flowing into the left heart; if this were not true, then blood would pile up continuously in the lungs. Momentarily, some fluid shifts occur, but on average, over a substantial period of time, our conclusion is correct. The same argument can be applied to any section of the figure. At rest this flow amounts to about 5000 mL per min. in an average adult.

ANATOMICAL FLOW

HEAD, TRUNK AND EXTREMITIES

RIGHT LUNG

LEFT LUNG

left atrium

ventricle

right atrium

ventricle

septum

LOWER TRUNK & EXTREMITIES

Blood velocity represents the speed of a blood "particle" in the stream— i.e., how far the particle moves in one min. Blood flow represents how many particles pass (more precisely, the volume of these "particles" passing) a given cross section in one minute; it is measured in mL/min.

VARIATIONS IN VELOCITY

VELOCITY cm/sec

arteries

arterioles

capillaries

venules

veins

AREA cm²

Beginning in the aorta and progressing toward the tissues, the total cross section of the vascular tree gets larger and larger until it becomes maximal in the capillaries. Although each branch is smaller than its parent, the number of branches increases so rapidly that it more than compensates for the reduction in size of any individual branch. Progressing from capillaries to venules to veins and back to the heart, the reverse occurs. Because area is largest in the capillaries, and blood flow is the same in all sections, it follows that blood velocity must be slowest in the capillaries. This is important because capillaries are very short (approx. 0. 1 cm), and if the blood didn't slow down, there wouldn't be enough time for exchange (e.g., of O_2) between blood and tissues to occur. For example, blood normally spends about 1 sec. in a capillary. If it traveled at the same speed as it does in the aorta, this time would be reduced to only 0.001 sec, a 1000-fold reduction.

FLOW = VELOCITY X AREA G

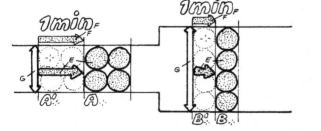

1 min F

Blood flow = velocity x cross-sectional area. You can verify this in the special case shown above. The total cross section of the tubing at B is twice that at A, so that a given length of tubing near B holds twice as many cubic elements as a similar length near A. With constant flow, the number of cubic fluid elements passing through the plane at A in one min. must equal the number passing through plane at B in one min. (= 4 in our example). For this to occur, the elements at A must have passed the point A^1 one minute ago, while those at B passed the point B^1. During the same minute, those at A have traveled twice the distance as those at B; their velocity is twice as great.

The heart is a hollow organ with walls made from *cardiac muscle*. When excited, these muscles shorten, thicken, and squeeze on the hollow cavities of the heart, forcing blood to flow in directions permitted by the heart valves. Cardiac and skeletal muscles are similar in many ways: both are striated, containing actin and myosin filaments (see plate 22), which interdigitate and slide closer together during contraction; both can be electrically excited; and both show action potentials that propagate along the surface membrane, carrying excitation to all parts of the muscle. But there also are significant differences.

1. The duration of the cardiac action potential is very long, lasting throughout contraction — Action potentials in cardiac muscle are 100 times longer than in skeletal muscle.

2. A long refractory period — associated with the prolonged action potential also lasts throughout the contraction. This implies that:

3. Cardiac muscle contractions are always brief twitches — In skeletal muscle, contractions resulting from rapid repetitive stimulation can summate or "fuse" to provide smooth, sustained contractions. This cannot happen in cardiac muscle because the long refractory period "cancels" any stimulus that occurs before the heart has a chance to relax. Relaxation between beats is essential for the heart to fill with blood to be pumped at the next beat.

4. Cardiac muscles are interconnected by gap junctions (nexus) — These are channels that allow action potentials to pass from one cell to the next and ensure that the entire heart participates in each contraction. The heartbeat is *all or none*. In contrast, skeletal muscle cells are electrically isolated; one cell may contract while its neighbor remains quiescent.

5. Cardiac muscle excites itself — Normally, skeletal muscle will contract only if it receives a nerve impulse. Nerves that carry impulses to the heart influence the rate and strength of contraction, but they do not initiate the primitive heartbeat. When these nerves are destroyed, the heart continues to beat without any external prompt. In contrast, when nerves to skeletal muscle are destroyed, the muscle is paralyzed.

THE PROLONGED ACTION POTENTIAL IS MAINTAINED BY CA^{++} MOVING IN

The shape of the action potential varies in different parts of the heart. The top figure shows an intracellular recording from a *Purkinje fiber*. These cardiac muscle fibers are particularly adept at conducting impulses. They also can excite themselves; when a Purkinje fiber is isolated, it continues to beat at its own rhythm. Notice that the resting potential is not level; it slowly rises to a threshold and initiates an action potential. The initial spike (very rapid rise in potential) is similar to those observed in nerve and skeletal muscle. In each case, the rise is due to the opening of Na^+ channels, which allow positively charged Na^+ ions to rush into the cell from outside, where they are highly concentrated. In all three cases, the opening of the Na^+ channels is caused by membrane depolarization so that a *positive feedback* (depolarization → opening of Na^+ channels → depolarization) is activated. In nerve and skeletal muscle, this is followed by an inactivation of the Na^+ channels together with an opening of the K^+

channels, which repolarizes the membrane very quickly. Cardiac muscle is different; its Na^+ channels inactivate, but the opening of its K^+ channels is delayed. Meanwhile, the membrane potential is held in a suspended plateau by small amounts of Ca^{++} flowing through Ca^{++} channels that have opened in response to the depolarization. The small amounts of Ca^{++} that enter just balance the small amounts of K^+ that are leaking out. Finally, after 0.2 to 0.3 sec., K^+ channels open, the Ca^{++} channels close, and the membrane is rapidly depolarized. The potential falls to a minimum and then begins to slowly rise toward threshold as the cycle repeats. This type of "resting" membrane potential, spontaneously rising toward excitation, is called a *pacemaker potential*. Action potentials recorded from other areas of the ventricle are similar, but their resting potential remains level. When isolated, they do not beat.

THE SA NODE IS THE PACEMAKER

Action potentials recorded from the *SA* or *AV nodes* are different. Instead of Na^+ channels, Ca^{++} channels are activated by membrane depolarization, and the inward flow of Ca^{++} is responsible for the rising phase of the action potential. Further, the rise in pacemaker potential is rapid and reaches threshold quickly; when SA node cells are isolated, they beat at fast rates — faster than those from the AV node, and these beat faster than Purkinje fibers. In the intact heart, cells of the SA node set the rhythm for the entire heart; the SA node is the *pacemaker*. These rapidly beating cells become excited first and transmit their excitation to all others. Although many cells are capable of beating at their own (slower) rate, they never do because they are driven at a faster rate by impulses originating in the SA node.

The ionic flow during the pacemaker potential is small and poorly understood. It is believed to result from three different channels: (1) special Na^+ channels (called "funny" channels) that open as the potential returns to its "resting" value; (2) Ca^{++} channels that open with depolarization; and (3) the very slowly closing K^+ channels responsible for ending the plateau of the previous beat. Purkinje fibers have a similar pacemaker, but the Ca^{++} channels are not as significant.

PURKINJE FIBERS PROVIDE RAPID CONDUCTION IN VENTRICLES

The pacemaker can initiate a coordinated beat only if there is rapid impulse conduction to all parts of the heart. This is important because the atria and ventricles are separated by a band of connective tissue that does not conduct impulses. The required pathway is provided by the AV node and the *Purkinje system*, shown in the bottom figure. The AV node supplies the only normal conductive bridge between atria and ventricles. It takes only about 0.04 sec. for the impulse to travel from its origin in the SA node to the beginning of the AV node, but by the time the impulse finally leaves the AV node to emerge in the bundle, there is an additional delay of about 0.11 sec. This AV delay provides time for the atria to complete their beat before the ventricles begin. Once past the AV node, the impulse is rapidly conveyed via the Purkinje network to all parts of the ventricle, ensuring that all parts beat in unison to impart maximal thrust to the blood.

CN: Use dark colors for J and K structures and very light colors for H and I.
1. Begin in the upper right corner. Color the membrane diagram at the top as you color each phase of the action potential. Color the smaller chart.

2. Color the two graphs representing excitation from other sites in the heart.
3. Color the large heart drawing, following the title sequence in the lower left corner. Color the diagram to the right.

PACEMAKER POTENTIAL $_A$

ACTION POTENTIALS: $_-$
HIGH Na⁺ PERMEABILITY. $_B$
HIGH Ca⁺⁺ PERMEABILITY. $_C$
HIGH K⁺ PERMEABILITY. $_D$
REFRACTORY PERIOD $_E$
CONTRACTILE RESPONSE $_F$
ELECTRODE $_G$

An action potential in a Purkinje heart cell begins with a resting potential that drifts upward until it reaches threshold, when there is a rapid rise in Na⁺ permeability and Na⁺ entry into the cell. The high Na⁺ permeability is quickly inactivated and a prolonged plateau period follows where the stagnant potential is produced by a slow Ca⁺⁺ entry and slow K⁺ exit that almost balance. Finally K⁺ permeability increases, K⁺ exit dominates, and the potential rapidly returns to rest levels. The action potential lasts throughout contraction, providing a long refractory period that prevents tetanic contractions and ensures relaxation for filling of the heart.

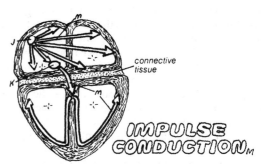

OUTSIDE CELL

membrane

INSIDE CELL

CARDIAC MUSCLE FIBER ACTION POTENTIAL (PURKINJE)

+20
0
-90
millivolts

SLOW

depolarization
plateau
repolarization

milliseconds 0   100   200   300

VENTRICLE $_I$

Action potentials from ordinary ventricle muscle cells are similar to those from Purkinje cells, with the exception that their resting potentials do not show any slow depolarizing drifts. These cells do not excite themselves.

NODE $_J$ $_K$

In the SA or AV nodes, Na⁺ channels no longer play a role. Instead, Ca⁺⁺ channels are activated and inward flow of Ca⁺⁺ is responsible for the rising phase of the action potential. Further, the rise in diastolic potential is well developed. Cells of the SA node are the heart's pacemaker.

SKELETAL MUSCLE FIBER ACTION POTENTIAL

VERY FAST

0
-70
millivolts

depolarization
repolarization

0 1 2 3 4 5 6 7 8 milliseconds

The duration of the action potential is very brief in skeletal muscle. In cardiac muscle it is some 100 times more prolonged.

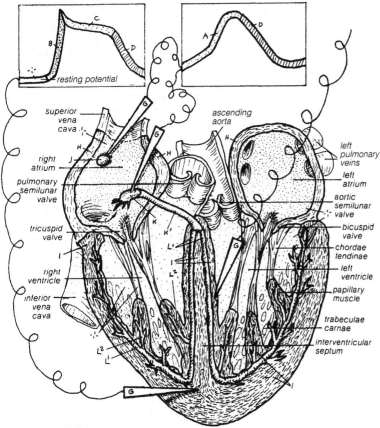

resting potential

superior vena cava
right atrium
pulmonary semilunar valve
tricuspid valve
right ventricle
inferior vena cava

ascending aorta
left pulmonary veins
left atrium
aortic semilunar valve
bicuspid valve
chordae tendinae
left ventricle
papillary muscle
trabeculae carnae
interventricular septum

IMPULSE CONDUCTION $_M$

Although the atria and ventricles are almost entirely separated by a band of connective tissue that does not conduct impulses, a conductive path is provided by the AV node and the Purkinje system. Purkinje fibers leading from the AV node into the ventricles conduct impulses very fast so that all parts of the heart beat in unison to impart maximal thrust to the blood.

connective tissue

ARTIFICIAL PACEMAKER $_{J'}$

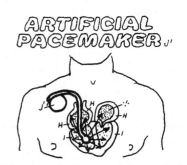

When the conduction system fails, an artificial pacemaker can be used. The ventricle is stimulated directly via leads that convey electrical impulses from a battery implanted under the skin.

ATRIUM WALL $_H$
VENTRICLE WALL $_I$
SINOATRIAL NODE, (SA NODE) (PACEMAKER) $_J$
ATRIOVENTRICULAR NODE (AV NODE) $_K$

PURKINJE SYSTEM: $_L$ $_-$
RIGHT & LEFT BUNDLE BRANCHES $_{L'}$
PURKINJE FIBERS $_{L'}$

The accurate recording of cardiac membrane potentials described in the previous plate requires the insertion of a microelectrode into the cytoplasm of a single cell. However, there are other, non-invasive methods for assessing the electrical activity of the heart. Although these methods are less precise, they have the advantage of providing global information about how activities of various portions of the heart are integrated into a coherent beat.

ELECTRICAL ACTIVITY OF HEART CAN BE RECORDED BY SURFACE ELECTRODES
To understand these measurements, first consider the simple case illustrated in the top figure, where two electrodes are placed near the heart's surface. The cells on the left are active and their extracellular surfaces are negatively charged while surfaces of resting cells on the right are positive. This difference is picked up by surface electrodes. The *electrode* on the left, close to the negative charge, is more negative than the electrode on the right (close to the positive charge). The meter detects only the *difference* between the electrodes. When the circumstances are reversed, with cells on the left at rest while cells on the right are active, then the right-hand electrode is close to the negative charge, and it will be negative with respect to the one on the left.

How close do the electrodes have to be to make the measurement? Fortunately, the heart is large, and body fluids contain ions that conduct electricity so that electrodes can be placed at some distance from the heart, anywhere on the body surface, as long as they are in good electrical contact with body fluids.

The middle figure shows a typical ECG. Although it is not obvious, this recording represents the sum of all action potentials of all cardiac muscle cells during one beat. Remember that the recording is made some distance from the heart, that various heart cells are oriented in different directions, and that they are excited at different times and recover at others. As "seen" by an electrode on the body surface, the electrical signal from one cell may easily augment or detract from the signal of another. No wonder the composite ECG bears no obvious resemblance to the action potential of a single cell. Nevertheless, years of careful observations and correlations have established a basis for interpreting ECGs. The landmarks on a typical record are designated by the letters P, QRS, and T.

P wave — The P wave signals the beginning of the heartbeat. It corresponds to the spread of excitation over both atria.

P–R interval — The time from the beginning of the P wave to the beginning of the R wave measures the time for impulse conduction from atria to ventricles. Although the heart appears to be "electrically silent" during this time, a wave of electrical depolarization is propagated; the time includes passage of the impulse to the AV node, the delay imposed by the AV node, and passage through the AV bundle, the bundle branches, and the Purkinje network. Disturbances of AV conduction induced by inflammation, poor circulation, drugs, or nervous mechanisms are often revealed by an abnormal prolongation of the P–R interval.

QRS complex — This corresponds to the invasion of the ventricular musculature by excitatory impulses. It is higher than the P because the ventricular mass is much larger than the atria. The duration of the QRS complex is shorter than the P wave because impulse conduction through the ventricles (partly via the Purkinje network) is very rapid.

S–T segment — During the interval between S and T, the ECG registers zero. All of the ventricular muscle is in the same depolarized state (recall the long plateau of the action potential of ventricular fibers), and no differences are recorded.

T wave — The T wave results from ventricular repolarization as different parts of the ventricle repolarize at different times.

These are only the rudiments of information buried in an ECG. By examining these records, a cardiologist learns about the anatomical orientation of the heart, disturbances of heart rate and impulse conduction, the extent and location of damaged tissue, and the effect of disturbances in plasma electrolytes.

HEART BLOCK: DISSOCIATION OF ATRIAL AND VENTRICULAR EXCITATION
Heart block and *fibrillation* are pathological conditions that are easy to detect in ECG recordings. In *heart block,* impulse propagation through the AV node is impeded. In first degree block, the impulse is merely slowed so that there is an abnormally long P–R interval. In one form of second-degree block, the AV node fails to pass some impulses. Only one out of two or one out of three impulses passes, and the ECG contains two or three P waves for every QRS. In more severe cases (third degree or complete block), the AV node fails completely, no impulses get through, and the atria and ventricles are electrically isolated. Ventricular pacemakers then take over, and the atria and ventricles beat independently of one another. The ECG in this case shows no correlation between the appearance of P waves and QRS complexes.

FIBRILLATION: ANARCHY IN THE VENTRICLE
In *ventricular fibrillation,* individual portions of the heart beat independently, without coordination. The heart is reduced to a quivering mass with no obvious excitation period and no obvious resting period. Blood is no longer pumped. Ventricular fibrillation appears to result from rapid and chaotic pacemaker activities that develop in different locations, together with long, circuitous conduction pathways. Once started, an impulse may continue to go around in a circular path and never die out. Fibrillation confined to the atria can cause serious disturbances in ventricular rhythm, but can be tolerated because, at rest, the atrial contribution to filling of the ventricle is small. In contrast, ventricular fibrillation is always fatal unless it can be immediately arrested.

ANTI-ARRHYTHMIC DRUGS
Drugs commonly used to help control pathological excitation or conduction in the runaway heart act at different strategic points in the excitatory cycle. Lidocaine, for example, is a Na^+ channel blocker, propranolol is an adrenergic β receptor blocker, and diltiazem is a Ca^{++} channel blocker. Other drugs (e.g., amiodarone) prolong refractory periods, increasing the likelihood that a circulating impulse will be extinguished as it enters a region that is still refractory. Some Na^+ channel blockers are very interesting, because their binding sites within the channel are much more exposed when the gates are open. This implies that they will be more effective on rapidly firing channels (i.e., the ones that are creating the problem).

CN: 1. Begin with the upper third of the page. 2. Color the heart and ECG. As you color each title, complete corresponding structures in the heart diagrams and in the ECG. 3. Color the lower material on heart block and fibrillation.

MEASURING ELECTRICAL ACTIVITY
CELL ACTIVITY.

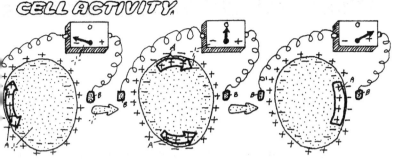

Two electrodes are placed on the surface of the heart. The meter detects only the **difference** between the two electrodes (left side–right side). If cells on the left are active while those on the right are at rest, the electrode on the left will be negative with respect to the electrode on the right. When cells on the left rest while cells on the right are active, then the right-hand electrode will be negative with respect to the one on the left; or in other words, the left electrode will be positive with respect to the right.

TAKING THE ECG
ELECTRODE.

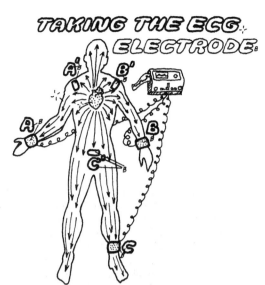

In the ECG, electrodes are placed on the arms and left leg. Body fluids conduct electrical signals from the surface of the heart to the electrodes. Measurements are taken as the difference between two of the three electrodes, the legs and arms serving as simple extensions of the electrodes. Measurements from the ankle (C) approximate electrical variations that would be measured with an electrode placed in the groin (C^1). Similarly for A and A^1, and B and B^1

THE ECG.

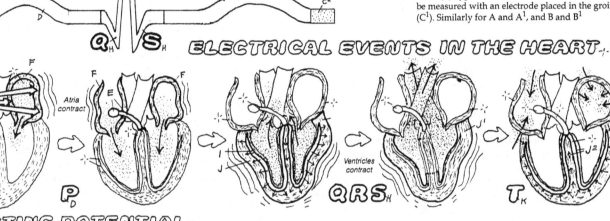

ELECTRICAL EVENTS IN THE HEART.

In a typical ECG, the P wave corresponds to atrial depolarization, the QRS complex occurs as the impulse invades the ventricle, the flat interval between S and T signifies complete ventricular depolarization, and the T wave corresponds to repolarization of the ventricle.

RESTING POTENTIAL c*
P WAVE b
SA NODE EXCITATION e
ATRIAL DEPOLARIZATION f
AV NODE, AV BUNDLE EXCITATION g
QRS COMPLEX h
PURKINJE EXCITATION i
VENTRICLE DEPOLARIZATION i
TOTAL DEPOLARIZATION i
T WAVE k
VENTRICLE REPOLARIZATION j2

HEART BLOCK *
ATRIAL BEAT e
VENTRICULAR BEAT g

In second degree heart block, the A-V node fails to pass some impulses so that only one out of two or one out of three impulses passes. In these cases the ECG contains two or three P waves for every QRS complex.

FIBRILLATION k

In fibrillation, individual portions of the heart appear to beat independently and without any coordination. The heart is reduced to a quivering mass of tissue. This fatal condition is reflected in a chaotic ECG.

Just as in excitation, the contractile properties of the heart are very similar to those of skeletal muscle (see plate 24). Both are striated. Like skeletal muscle contraction, heart muscle contraction is based on *actin* and *myosin* filaments that interdigitate and slide closer together in the presence of *free Ca*$^{++}$ in the *cytosol*. In both cases, the sliding is mediated by myosin cross bridges, which reach out to contact special sites on the actin filaments. Both skeletal and cardiac muscles contain *T tubules*, which conduct impulses perpendicular to the cell surface toward its interior; both contain a well-developed tubular network, the *sarcoplasmic reticulum* (SR), which releases Ca^{++} to trigger contraction and sequesters it for relaxation; and both contain the regulatory proteins *troponin* and *tropomyosin*, which keep the actin and myosin cross bridges apart in the absence of free Ca^{++}.

CARDIAC MUSCLE CAN INCREASE STRENGTH OF CONTRACTION BY INCREASING FREE CA^{++}

There are also important differences. When skeletal muscle is excited, enough Ca^{++} is released to react with all of the troponin so that all reactive sites on actin become available and all *cross bridges* become activated. Normally in contracting cardiac muscle, this is not the case; troponin is not fully covered by the Ca^{++} released on excitation. This is important because it implies that anything that increases Ca^{++} availability inside the cell will increase the number of cross bridges that can form; just as increasing the number of persons pulling on a rope in a "tug of war" will increase the tension or pull on the rope, this increase in the number of active cross bridges will increase the strength of cardiac contraction. Whatever controls internal Ca^{++} will control cardiac performance.

CA^{++} INDUCES ITS OWN RELEASE

How is free Ca^{++} controlled? Free Ca^{++} levels in the cytosol are about 20,000 times lower than external free Ca^{++}. Most Ca^{++} inside the cell either is bound to proteins or is sequestered inside mitochondria or the SR. Ca^{++} is poised at higher concentrations both outside the cell and in the SR waiting to enter the cytosol, where it will have easy access to the troponin and the contractile filaments. During activity, action potentials travel over the surface membrane and invade the T tubules, where the excitatory waves come in close proximity to the SR. Here, voltage-sensitive Ca^{++} channels in the tubules open and external Ca^{++} flows into the small spaces between the tubules and SR; this Ca^{++} induces the release of large amounts of SR Ca^{++} through channels in the SR membrane. *Ca^{++}-induced Ca^{++} release is essential!*

If Ca^{++} induces its own release, then we might expect a positive feedback: the more Ca^{++} release, the more stim-

ulation for further release; the process would stop only when *all* of the Ca^{++} was released. This does not happen. The current explanation is that Ca^{++} interacts with its channel at more than one site; one fast-acting site is responsible for opening the channel while a slower site closes it.

FREE CA^{++} IS PUMPED OUT OF cytosol BY A CA^{++} PUMP AND A NA^{+}–CA^{++} EXCHANGER

During relaxation, the level of internal free Ca^{++} is reduced primarily because it is pumped back into the SR by an *ATP-driven Ca^{++} pump*. If there is more Ca^{++} in the cell, more Ca^{++} will be loaded back into the SR, and more Ca^{++} will be released at the next beat to cause a more forceful contraction. The Ca^{++} that continually leaks into the cytosol (e.g., via the action potential) is removed by three routes.

1. An ATP-driven Ca^{++} pump in the SR membrane pumps it into the SR.

2. An ATP-driven Ca^{++} pump in the plasma membrane pumps it out of the cell.

3. A *Na^{+}–Ca^{++} exchanger* in the plasma membrane pumps it out of the cell.

DIGITALIS INDIRECTLY INCREASES CA^{++} BY INCREASING INTERNAL NA^{+}

Although a complete description of free Ca^{++} balance within the heart cell is essential for understanding cardiac performance, the details still elude us. The story of the common cardiac drug *digitalis* provides an example of how known details have been used to interpret clinical experience. This drug has been successfully used for many years in cardiac patients to strengthen the force of cardiac contraction, yet experiments failed to reveal any effect of the drug on the contractile machinery. All that could be shown was that digitalis is a potent inhibitor of the *Na^{+}–K^{+} pump*. What does the Na^{+}–K^{+} pump have to do with cardiac contraction? Our current interpretation involves the Na^{+}–Ca^{++} exchanger. This exchanger works because Na^{+} is more concentrated outside the cell than inside and because Na^{+} movements into the cell along this route are tightly coupled to Ca^{++} movements out. The energy for moving Ca^{++} from low internal to high external concentrations (i.e., for pumping the Ca^{++} out) is provided by the energy loss accompanying the movement of Na^{+} from high external to low internal concentrations. When digitalis is administered, it inhibits the Na^{+}–K^{+} pump, so less Na^{+} is pumped out of the cell, its internal concentration rises, and the Na^{+}–Ca^{++} exchanger (which requires low internal Na^{+}) is inhibited. Internal Ca^{++} rises so that more is available to activate the cross bridges, resulting in more forceful contractions.

CN: Use red for M and dark colors for A and K.
1. Begin with the second illustration from the top, an anatomical description of cardiac muscle cells.
2. Color the schematic diagram of the same subject just below it. Color the cartoon on the left.
3. Go to bottom of the page and work your way up through the relaxation sequence. The two rectangles at the top of this panel illustrate the operation of the Na^{+}–K^{+} pump and the Na^{+}–Ca^{++} exchanger. Do the left one first, then the other, which suggests how digitalis enhances cardiac muscle contraction by preventing Ca^{++} removal.

ACTION POTENTIAL A
PLASMA MEMBRANE B
T TUBULE C
SARCOPLASMIC RETIC. D
Ca++ IN CYTOSOL ∧ E'
TROPONIN F TROPOMYOSIN G
ACTIN H Z LINE I
MYOSIN J CROSS BRIDGE ∧ J'

MUSCLE CONTRACTION ÷

CARDIAC MUSCLE ÷

intercalated disk

intercalated disk

mitochondrion

myofibril

During activity, action potentials travel over the cell membrane and invade T tubules, where the excitatory waves come close to the SR, (1) and external Ca++ enters the cytosol through voltage-activated channels. This amount of external Ca++ by itself is not sufficient to initiate contraction. However, it does stimulate the release of more Ca++ from internal stores in the SR, (2). Generally the Ca++ released with each excitation is insufficient for maximal contraction, making the release of SR CA++ an important determinant of cardiac performance.

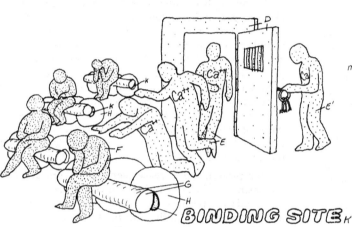

outside cell

in cell

myofibril

myocardium

BINDING SITE K

Somehow Ca++ triggers its own release. The Ca++ level rises and, by interacting with the troponin-tropomyosin complex, uncovers the sites on actin that form attachments for myosin cross bridges.

Ca++ PUMP E² (ATP) M
Na+ – K+ PUMP L (ATP) M
Na+ – Ca++ EXCHANGER E³
Ca++ IN CYTOPLASM ∨ E'
CROSS BRIDGE ACTIVITY ∨

MUSCLE RELAXATION ÷

NORMAL STATE E³

DIGITALIS N

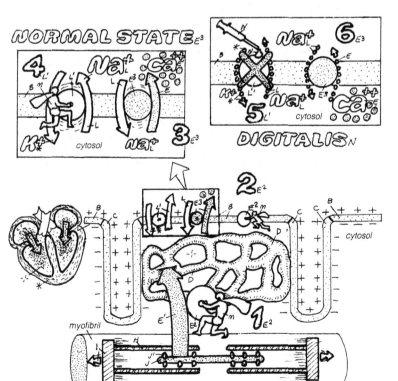

After the action potential, the muscle relaxes as the level of internal free Ca++ is reduced via three routes: (1) most prominently, it is pumped back into the SR by an ATP-driven Ca++ pump; (2) it is pumped out of the cell by an ATP-driven Ca++ pump; and (3) it is pumped out of the cell by the a Na+–Ca++ exchanger, where energy required for the uphill movement of Ca++ out of the cell is supplied by the coupled downhill flow of Na+ into the cell. The Na+ gradient promoting this downhill flow is created by the Na+ –K+ pump (4).

The drug digitalis promotes the strength of cardiac contraction. It appears to work by inhibiting the Na+–K+ pump (5) and lowering intercellular Na+ so that the downhill gradient for Na+ entry is depressed. Thus, the Na+–Ca++ exchanger (6) is less effective and Ca++ accumulates inside the cell, where it is pumped into the SR, making more Ca++ available for release upon excitation. More Ca++ release → more cross-bridge activation → stronger contraction.

myofibril

cytosol

cytosol

During activity, the heart beats faster and stronger; during rest, it slows down. These alterations occur largely through the action of *sympathetic* and *parasympathetic nerves* on the heart, and to a lesser extent by *catecholamine* (epinephrine) secretions from the *adrenal medulla* gland. The physiological effects of these two nerves are simple: the sympathetic nerves liberate norepinephrine (noradrenaline), which stimulates the heart, increasing its rate and force of contraction; the parasympathetic nerves (vagus nerves) liberate acetylcholine, which inhibits the heart, slowing its rate. Both sets of nerves innervate the *SA* and *AV nodes,* and both affect the heart rate by their influence on the primitive pacemaker activity of the SA node. However, unlike the sympathetic nerves, parasympathetic nerves have no substantial effect on the strength of ventricular contraction; stimulating the vagus nerve at *maximal* rates yields only a 15 to 25% decrease in the strength of ventricular contractions

How can these nerves (or more precisely these neurotransmitters) control the heart rate? Recall (plate 32) that heart excitation occurs in the SA node as a result of Ca^{++} and Na^+ ions moving down their concentration gradient into the cell and depolarizing the cell (i.e., making the inside less negative) until the *membrane potential* reaches *threshold*. Any K^+ leakage out of the cell during this time does just the opposite; it tends to repolarize the cell, driving the membrane potential away from threshold. The key to our problem lies in the balance between the opposing actions of K^+ moving out and Na^+ and Ca^{++} moving into the cell to produce the pacemaker potential.

PARASYMPATHETIC NERVES

Acetylcholine slows the heart rate by increasing K^+ permeability of the SA nodal cells — Acetylcholine liberated by the vagus nerve acts on muscarinic receptors of cells in the SA node activating a G protein. The G protein's βγ subunit acts directly, without the intervention of a second messenger, to open K^+ channels, increasing nodal *permeability to K^+* (plate 13). Consequently, the resting potential of the SA node becomes more negative, pushing it further away from the threshold potential. The outward-moving K^+ slows the normal rate of depolarization (the pacemaker potential) caused

by inward-moving Ca^{++} and Na^+. This lengthens the time required to reach threshold and slows the heart. In addition to its effect on the rate of firing of the pacemaker (SA node), the outward-moving K^+ impedes excitability in other cells, and this tends to slow conduction of the impulse through the atrium and AV node.

SYMPATHETIC NERVES

Norepinephrine increases the heart rate by increasing slope of pacemaker potential — Both norepinephrine and catecholamines secreted by the adrenal medulla will react with β receptors in the heart and increase the level of the second messenger, cAMP (plate 12). cAMP activates enzymes that phosphorylate Ca^{++} channels, which is necessary for the channels to function. In any case, the effect of norepinephrine is to increase Ca^{++} permeability by recruiting more channels. In addition it appears that norepinephrine will increase slow pacemaker Na^+ flux. In the SA node, both of these factors will increase the rate of rise of the pacemaker potential, shortening the time to reach threshold and increasing the heart rate.

Norepinephrine increases ventricular contraction by increasing Ca^{++} permeability — In ventricular muscle, the increased Ca^{++} permeability induced by norepinephrine leads to an increased Ca^{++} influx during the plateau of each action potential. The larger amounts of entering Ca^{++} result in a larger induced Ca^{++} release from the SR, making each contraction stronger.

Norepinephrine shortens contraction duration by increasing rate of Ca^{++} uptake by SR — Finally, norepinephrine also increases the rate of *re-uptake of Ca^{++}* by the *sarcoplasmic reticulum*; this speeds up the relaxation process and consequently shortens the duration of contraction. With a fast heart rate, it is important to curtail the contraction period to allow sufficient time for the heart to fill between beats.

Again, this action of norepinephrine is mediated by the increased levels of cAMP. (In this case cAMP activates a phosphokinase, which phosphorylates a membrane protein of the SR — *phospholamban*. Phospholamban, in its phosphorylated form, activates the SR Ca^{++} pump.)

CN: Use red for J and dark colors for A, B, H, and I.
1. Begin with the upper right diagram.
2. Color the three stages of autonomic nerve activity.
3. Finish with the two phases of Ca^{++} involvement in the contraction process.

PARASYMPATHETIC NERVES. A
SYMPATHETIC NERVES. B
ADRENAL MEDULLA. C
CATECHOLAMINES. D
VENTRICULAR MUSCLE. E
SA NODE. F AV NODE. G
MEMBRANE POTENTIAL F'
THRESHOLD POTENTIAL. H

HEART RATE

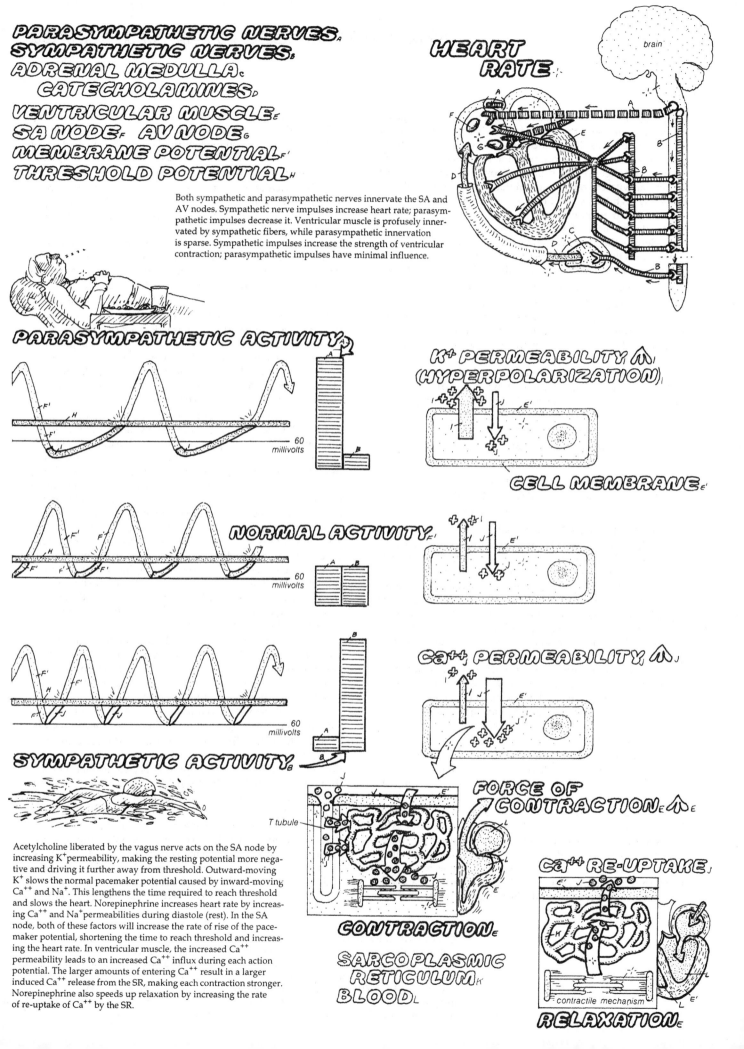

Both sympathetic and parasympathetic nerves innervate the SA and AV nodes. Sympathetic nerve impulses increase heart rate; parasympathetic impulses decrease it. Ventricular muscle is profusely innervated by sympathetic fibers, while parasympathetic innervation is sparse. Sympathetic impulses increase the strength of ventricular contraction; parasympathetic impulses have minimal influence.

PARASYMPATHETIC ACTIVITY.

60 millivolts

K⁺ PERMEABILITY (HYPERPOLARIZATION).

CELL MEMBRANE E'

NORMAL ACTIVITY.

60 millivolts

Ca⁺⁺ PERMEABILITY. J

SYMPATHETIC ACTIVITY. B

60 millivolts

T tubule

FORCE OF CONTRACTION E

Ca⁺⁺ RE-UPTAKE. J

Acetylcholine liberated by the vagus nerve acts on the SA node by increasing K⁺ permeability, making the resting potential more negative and driving it further away from threshold. Outward-moving K⁺ slows the normal pacemaker potential caused by inward-moving Ca⁺⁺ and Na⁺. This lengthens the time required to reach threshold and slows the heart. Norepinephrine increases heart rate by increasing Ca⁺⁺ and Na⁺ permeabilities during diastole (rest). In the SA node, both of these factors will increase the rate of rise of the pacemaker potential, shortening the time to reach threshold and increasing the heart rate. In ventricular muscle, the increased Ca⁺⁺ permeability leads to an increased Ca⁺⁺ influx during each action potential. The larger amounts of entering Ca⁺⁺ result in a larger induced Ca⁺⁺ release from the SR, making each contraction stronger. Norepinephrine also speeds up relaxation by increasing the rate of re-uptake of Ca⁺⁺ by the SR.

CONTRACTION E

SARCOPLASMIC RETICULUM K
BLOOD L

contractile mechanism

RELAXATION E

The pumping action of the heart is reflected in the changes in volume and pressure that occur in each heart chamber and in the great arteries as the heart completes a single cycle. This plate shows changes that occur on the left (systemic) side of the heart. Changes on the right (pulmonary) side of the heart are similar with the one exception that pressures are only about one-eighth as large. Five curves are shown. The top three are obtained by inserting pressure-measuring instruments into the *aorta, left atrium,* and *left ventricle.* The next curve depicts the volume of the left ventricle, and the last curve shows the *ECG.* Our object is to appreciate the interrelationships of these curves and how they relate to blood flow at each moment during the cardiac cycle.

ONE-WAY VALVES CONTROL DIRECTION OF FLOW

To interpret these curves, we note that each of the heart valves normally "points" in the direction of flow, working to prevent backflow. Whenever the downstream pressure builds up and exceeds upstream pressure (a condition for backflow), the pressure difference forces the valve closed. Similarly, when upstream pressure is greater than downstream, fluid flows forward and the valves are forced open. In the heart, pressure conditions are as follows:

VALVES	STATE	CONDITION
AV	open	P (atrium) > P (ventricle)
AV	closed	P (atrium) < P (ventricle)
aortic	open	P (ventricle) > P (aorta)
aortic	closed	P (ventricle) < P (aorta)
		(see exception below)

THE CYCLE IS DIVIDED INTO 5 PERIODS

1. Atrial contraction — Atrial contraction is signaled by the P wave of the ECG. As atrial pressure rises, blood is thrust into the ventricles through the open *AV valves.* These valves are open (as they have been throughout the diastole) because pressure in the atrium is higher than pressure in the quiescent ventricle. Blood enters the ventricle but cannot leave because the *aortic valves* are closed ($P_{aortic} > P_{ventricular}$). Note that the resulting volume increase on the ventricular volume curve appears as a small "bump." The atrium serves as a "booster" pump, but its contribution to ventricular filling is small; most of the ventricular filling occurred earlier, when both atrium and ventricle were at rest. When the heart rate goes up, as in exercise, there is less time between beats for filling, and the atrial contribution becomes more significant. Atrial contraction is followed by:

2. Isovolumetric ventricular contraction — Now the impulse invades the ventricles (QRS in the ECG), and, after a short delay, they begin to contract. This is the beginning of systole. Ventricular pressure builds up steeply and quickly exceeds atrial pressure. The AV valves snap shut, producing the first heart sound—"LUPP." Following closure of the AV valves, ventricular pressure continues to rise steeply until it exceeds aortic pressure. Pressure rises rapidly because both sets of heart valves are closed; the heart continues to contract, but there is no place for the blood to go to relieve the ascending pressure. (Contraction of the heart during this period is similar to an *isometric* contraction in skeletal muscle.) During this period, the ventricular volume cannot change—note the flat horizontal trace on the ventricular volume curve. The constant ventricular volume is the reason for naming this period "isovolumetric ventricular contraction."

3. Ventricular ejection — As soon as the ventricular pressure exceeds aortic pressure, the aortic valves are thrust open, and blood is ejected into the aorta. Pressure in the aorta begins to rise because blood is entering from the ventricles faster than it can leave through the smaller arteries. Prior to this time, pressure in the aorta had been falling because the aortic valves were closed; blood continued to leave the aorta though smaller arteries, but none could enter from the ventricle.

Blood leaving the ventricles is reflected in the ventricular volume curve, which drops precipitously as soon as ejection begins. Soon afterward, the contractile force of the ventricle wanes; the ventricular pressure ascent slows and begins to reverse while the initial rapid change in ventricular volume begins to level off. As the ventricles begin to repolarize (T wave of ECG) and relax, the ventricular pressure curve crosses the aortic curve and goes below it.

Shortly thereafter, the aortic valve snaps shut, producing a sharp "DUP" sound (the second heart sound) and bringing the ventricular ejection period, as well as the period of systole, to an end. (Systole = isovolumetric ventricular contraction period + ventricular ejection period.) It also produces a bump notch on the aortic pressure curve. The aortic valve closure is not simultaneous with the crossover of the ventricular and aortic pressure curves because the blood flowing through the valves has an appreciable momentum (mass × velocity) in the direction of forward flow. Applying a force (pressure difference) in the opposite direction requires a small amount of time to stop or reverse the motion. (Imagine trying to stop a rolling automobile with a hand push in the opposite direction.) Notice that not all of the blood contained within the ventricle is ejected with each beat. The residual blood is almost equal to the amount ejected.

4. Isovolumetric ventricular relaxation — Now, as in isovolumetric contraction, both valves are closed, and blood cannot enter or leave the ventricles. This time, however, the ventricular muscles relax; it is the beginning of diastole. Pressure falls precipitously, but ventricular volume does not change. Soon the ventricular pressure falls below atrial pressure, the AV valves open, and isovolumetric relaxation ends.

5. Ventricular filling — In this period, atrial pressure is higher than ventricular pressure because blood continues to flow into the atrium from the pulmonary veins. Blood flows through the open AV valve from atrium to ventricle. Ventricle filling continues throughout diastole, not just when the atrium contracts. The ventricular volume curve during diastole shows that early ventricular filling is most prominent and that contraction of the atrium contributes only a minor portion to the ventricular contents. Toward the end of this period, atrial contraction ensues, and this period, as well as diastole, ends with closure of the AV valves. (Diastole = isovolumetric ventricular relaxation period + ventricular filling period.)

CN: Use red for F and dark colors for B and D.
1. Begin in the upper left corner by coloring titles 1–5. Color all structures in each left heart illustrated across the top. Note that the pulmonary arteries have been left out of the last three figures. Include the two heart sounds and the sound bars surrounding the relevant valve.

2. Color the pressure section and refer to the relevant phase above.
3. Do the same for the blood volume of the left ventricle.
4. Color the ECG results gray.
5. Color the bottom figures representing the time intervals.

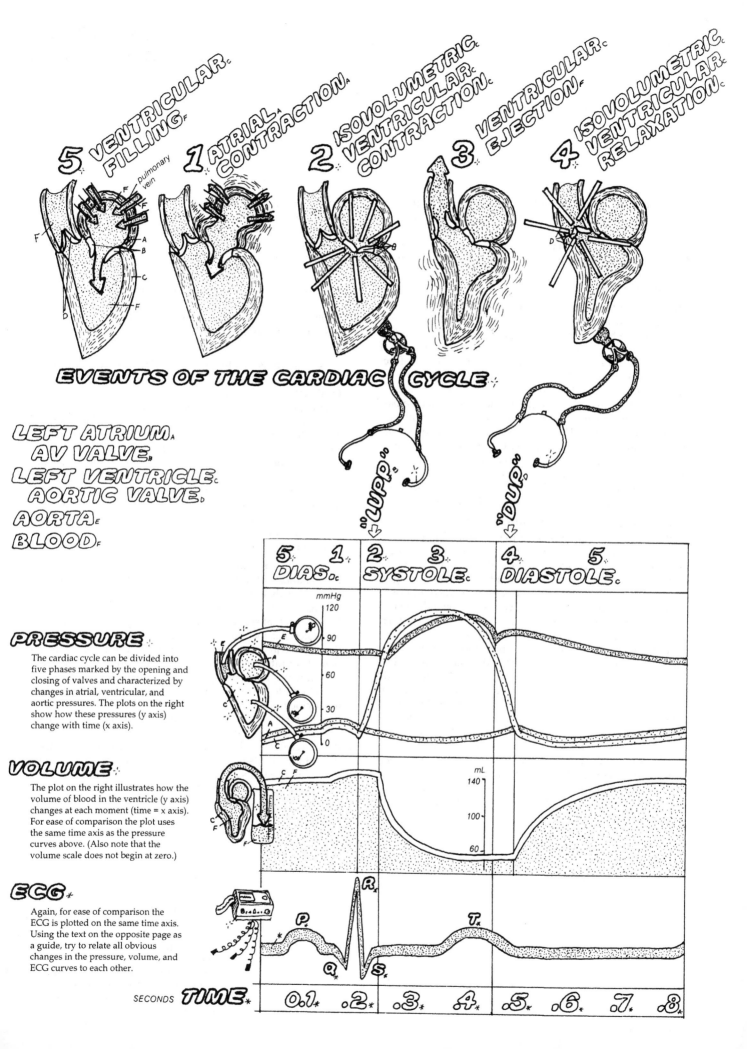

5 VENTRICULAR FILLING

1 ATRIAL CONTRACTION

2 ISOVOLUMETRIC VENTRICULAR CONTRACTION

3 VENTRICULAR EJECTION

4 ISOVOLUMETRIC VENTRICULAR RELAXATION

pulmonary vein

EVENTS OF THE CARDIAC CYCLE

LEFT ATRIUM
AV VALVE
LEFT VENTRICLE
AORTIC VALVE
AORTA
BLOOD

"LUPP"

"DUP"

5	1	2	3	4	5
DIAS		SYSTOLE		DIASTOLE	

PRESSURE

The cardiac cycle can be divided into five phases marked by the opening and closing of valves and characterized by changes in atrial, ventricular, and aortic pressures. The plots on the right show how these pressures (y axis) change with time (x axis).

mmHg
120
90
60
30
0

VOLUME

The plot on the right illustrates how the volume of blood in the ventricle (y axis) changes at each moment (time = x axis). For ease of comparison the plot uses the same time axis as the pressure curves above. (Also note that the volume scale does not begin at zero.)

mL
140
100
60

ECG

Again, for ease of comparison the ECG is plotted on the same time axis. Using the text on the opposite page as a guide, try to relate all obvious changes in the pressure, volume, and ECG curves to each other.

P Q R S T

SECONDS TIME

0.1 .2 .3 .4 .5 .6 .7 .8

Blood flows from the arteries through the capillaries to veins because the pressure is higher in the arteries than in the veins. Pressure is the force exerted on each square centimeter; it is a measure of "push." Blood flows from arteries to veins because the blood in the arteries "pushes" harder than the blood in the veins. It is the *difference in pressure* that constitutes the driving force for movement.

MEASURING PRESSURE WITH A COLUMN OF FLUID
How do we measure this force or push? Think about a fluid particle at the bottom of a tube where there is no motion (see diagram). It is subjected to the force exerted by the weight of the column of fluid on top of it. If, for some reason, this force was less than the pressure on the particle, it would move upward. The fact that there is no motion means that the weight of the column is just equal to the pressure on the particle. We use the height of a fluid column as a convenient way to measure pressure. (The weight of the column depends on what the fluid is. *Mercury* is denser than water, so a given weight will require a much smaller column of mercury as compared to water. For this reason, it is simply more convenient to use mercury as a reference. To convert a millimeter of mercury to a millimeter of water, multiply by 13.6.)

FLOW = PRESSURE DIFFERENCE/RESISTANCE
Pressure drop is proportional to resistance — In the illustration with the stopcock closed, the fluid rises to the same level in all standpipes; the pressure is the same at each point in the horizontal tube — no pressure difference, no driving force, no motion. When the stopcock is open, the fluid flows out of the tube and the different levels in each standpipe indicate different pressures at each point along the horizontal. The pressure falls uniformly from left to right along the horizontal. In the figure below, a partially open stopcock is placed closer to the left, where it obstructs but does not stop the flow. Now fluid piles up behind the stopcock until the pressure difference across the stopcock becomes large enough to maintain the flow despite the obstruction. The pressure still falls from left to right, but the fall is no longer uniform. The greatest decrease in pressure occurs across the obstructing stopcock. When we examine the pressures in the circulation, we find that the greatest fall in pressure occurs across those terminal arteries, the *arterioles*, that enter the capillaries. In blood circulation, the arterioles offer the greatest *resistance* to flow.

The idea of frictional *resistance* can be made quantitative.

For any given vessel or system of vessels, we simply divide the pressure difference between any two points by the flow; the quotient is defined as resistance between those points. (Think of pressure difference as "cost" and flow as "payoff.") It follows that:

$$flow = pressure\ difference \div resistance.$$

Resistance is very sensitive to the radius of a tube — much more so than to its length. For a given difference in pressure, doubling the radius of a tube will increase the flow (decrease the resistance) sixteen times! The flow is proportional to the radius raised to the fourth power.

Resistance is due to viscous friction as layers of blood slide past each other — Resistance to flow occurs because of frictional forces that oppose the motion of two layers of fluid sliding past each other. The more viscous the fluid, the larger the frictional forces. In a blood vessel, the blood layer immediately adjacent to the vessel wall is held back as it adheres to the stationary wall. This layer retards the next layer, which retards the next and so on (see figure). This results in telescoping layers of fluid. Fluid in the center of the vessel moves the fastest, while fluid at the wall does not move at all. The overall resistance to flow arises from the frictional interactions of these layers as they slide past each other. The drop in pressure that occurs during flow through a vessel reflects the energy lost to these frictional interactions.

Arterioles are the "bottlenecks" of the circulation — Examination of the pressure in different parts of the circulation shows that the greatest drop occurs across the arterioles. Because the flow is the same through all sections of the circulatory tree, that section with the greatest pressure drop, the arterioles, has the most resistance to flow. (This follows from the above equation.) In other words, the arterioles are the bottleneck, or rate-limiting step, in the circulation. Bottlenecks are strategic places for regulation, and the arterioles appear to be a chief site for regulation of both blood pressure and flow to specific tissues. This is accomplished by smooth muscles that are wrapped circularly around the walls of the arterioles. These muscles are controlled by nerves and hormones. When the muscles relax, the radius increases; when they contract, the radius decreases. Remember that the resistance of a tube is very sensitive to its radius. By controlling the radii of arterioles, the body exercises tight control over the flows and pressures in its own circulation.

CN: Use red for C.
1. Begin at the top and work from left to right.
2. Color the next series of diagrams, noting the long bars (B') that reflect the overall level of pressure.
3. In the large chart below, the pressure line (B') represents mean blood pressure. Note the change (for emphasis) in color of blood at the arteriole level.

MERCURY$_A$ FORCE$_F$; PRESSURE$_{B'}$ FLUID$_C$; BLOOD$_{C'}$ RESISTANCE$_D$.

Pressure is a measure of "push" or force per unit area. Pressures can be measured with a "U" tube half filled with mercury (Hg). The harder the man pushes on the mercury, the higher the mercury rises. Its height (mm Hg) measures his push (pressure). Similarly, blood pressure can be measured by connecting an artery to a "U" tube; the "push" arises in the heart.

MEASUREMENT OF PRESSURE$_{B'}$

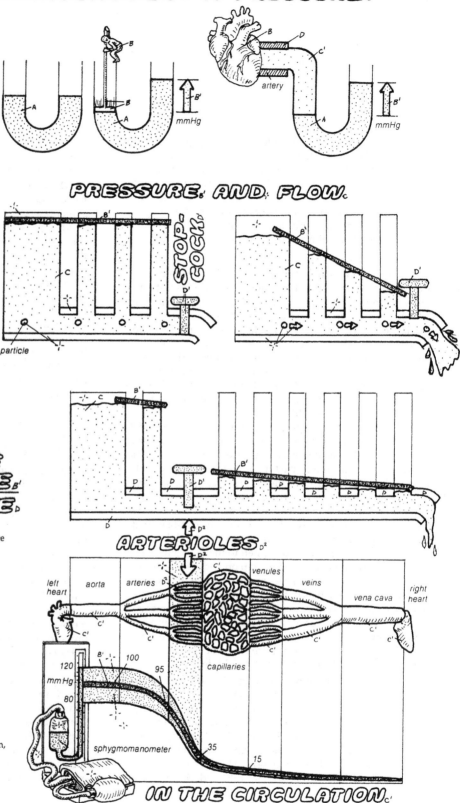

When fluid in a horizontal pipe is not moving, the pressure is the same at each point along the horizontal. We verify this by placing standpipes along the path; the level of fluid rises to the same height in each standpipe (now we are using the fluid itself instead of Hg to measure pressure). When we open the stopcock, fluid flows and the pressure falls uniformly along the horizonal. The difference in pressure from point to point pushes the fluid along. When a half open stopcock is placed in the middle of the pipe it resists the flow. Fluid piles up behind the stopcock until the pressure difference across the stopcock becomes large enough to maintain the flow despite the resistance. Pressure still falls from left to right, but the fall is no longer uniform. The greatest fall in pressure occurs across the obstructing stopcock. To maintain flow, fluid must be continually forced through a pipe to overcome frictional resistances (depicted in bottom panel) that oppose the motion.

PRESSURE$_B$ AND FLOW$_C$

FLOW$_C$ = $\dfrac{\text{PRESSURE DIFFERENCE}_{B'}}{\text{RESISTANCE}_D}$

Resistance to flow is increased by increasing viscosity, increasing pipe length, and decreasing pipe radius. Resistance is most sensitive by far to changes in radius. Halving the radius decreases flow 16 times!

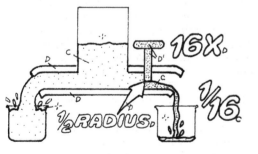

In blood circulation, flow is constant in each total cross-section, and the largest fall in pressure occurs in the arterioles. This means that arterioles offer the greatest resistance to flow; they are "the bottleneck." Arterioles act like stopcocks. Contraction of smooth muscles in the arteriole walls changes the vessel radius and alters resistance.

ARTERIOLES$_{D^2}$

IN THE CIRCULATION$_{C'}$

Resistance to flow occurs because of frictional forces that oppose the motion of two layers of fluid sliding past each other. In a blood vessel, the blood layer immediately adjacent to the vessel wall is held back as it tends to adhere to the stationary wall. This layer retards the next layer, which retards the next and so on, resulting in telescoping layers of fluid. Fluid in the center of the pipe moves the fastest, fluid at the wall does not move at all. The overall resistance to flow arises from the frictional interactions of these layers as they slide past each other. The drop in pressure that occurs during flow through a vessel reflects the energy lost to these frictional interactions.

LAYERS OF FLUID DURING FLOW$_C$

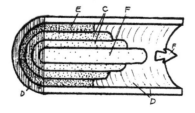

STATIONARY$_E$
MOVING$_C$
FASTEST$_F$

The heart pumps blood intermittently; during *systole*, some 70 mL of blood is thrust into the aorta, but during *diastole*, no blood leaves the heart. Despite this choppy, discontinuous flow of blood through the root of the aorta, blood flows out of the arteries into the capillaries in a smooth and continuous motion because the aorta and other arteries are not rigid pipes. Instead, they have *elastic walls* that can passively *expand* or *recoil*, much like a simple rubber band.

ARTERIAL PRESSURE

Blood vessel elasticity buffers cyclic changes in blood pressure — During systole, blood enters the arteries faster than it leaves through the capillary beds. Packing more fluid into the arteries tends to increase arterial pressure, which forces the elastic arterial walls to expand the same as a balloon does when more air is forced into it. The excess fluid is taken up by the expanding arteries, and this relieves some of the pressure increase that would have occurred if the walls were more rigid and could not expand as much. In contrast, during diastole, blood still leaves the arteries for the capillaries, but none enters from the heart. At this time, blood stored in the expanded arteries leaves, propelled in part by the recoil of the arterial walls; it is this stored blood that prevents the pressure from falling as low as it would if the arteries were rigid. The elastic arterial walls minimize fluctuations in pressure that would otherwise occur — i.e., they buffer changes in pressure. In a system with rigid walls, the pressure would rise to very high values during systole and fall to near zero during diastole. Similarly, the blood would spurt into the capillary bed with each systole and virtually come to a standstill during diastole. Imagine what would happen to the flow out of a faucet if you intermittently turned the tap on and off. In a healthy arterial system, the arterial pressure fluctuates with each beat, but not nearly as much as in the rigid system. The maintenance of a reasonable pressure level throughout the entire cycle has two advantages: first, it sustains a smooth and continuous flow into the capillaries; and second, it relieves the heart of work that would be required to eject blood against the enormous systolic pressure that would develop.

"Normal Blood Pressure": Systolic/Diastolic = 120/80 mm Hg — Arterial pressure does pulsate. With each heartbeat, the arterial pressure in a resting, normal young adult generally varies between 80 and 120 mm Hg. The minimum pressure occurs just at the end of diastole; it is called *diastolic pressure* (80 mm Hg in our example). The maximum occurs midway into systole; it is called *systolic pressure* (120 mm Hg in our example). The difference between systolic and diastolic is called the *pulse pressure* (120 – 80 = 40 mm Hg). From the discussion above, it follows that a person with more rigid arteries (e.g., an older person) will have a higher pulse pressure.

Mean pressure is a common estimate of the driving force — Rather than dealing with fluctuating pressures, it is sometimes useful to have a single measure that represents the average pressure or driving force within the arterial tree. (Plate 37 shows that this number is approximately the same throughout the arterial tree — until the arterioles are approached.) Simply taking the average between systolic and diastolic pressures ([120 + 80]/2 = 100 mm Hg) is not the best estimate, because inspection of the pressure contour shows that arterial pressure spends more time around diastolic pressure than around systolic pressure. This is taken into account by the mean pressure, which is represented by the horizontal line in the upper left-hand figure. The position of this line can be determined because it splits the area under the pressure contour into two equal parts, one area lying above the horizontal line drawn at the diastolic pressure level and below the mean pressure line, the other area lying above the mean pressure and contained by the upper parts of the pressure curve. The mean pressure can be approximated by the formula: *mean pressure = diastolic pressure + (1/3) pulse pressure.*

MEASURING ARTERIAL PRESSURE

To measure human arterial blood pressure (lower figure), an inflatable rubber bag constrained by a cloth cuff is wrapped around an arm. The bag is inflated and compresses the blood vessels in the arm. It is assumed that the pressure in the bag is transmitted to the arm so that the bag pressure equals the actual pressure within the tissue of the arm. It follows that when the pressure in the bag just exceeds the pressure in the artery, the compression will be sufficient to collapse the artery. The procedure is to inflate the bag above the arterial pressure so that blood flow stops. Air is then released from the bag so that bag pressure falls very slowly. At a certain bag pressure, flow resumes, but only for the short time that arterial pressure is at its maximum. During this time, sounds are produced that can be easily heard with the aid of a stethoscope placed near the artery. The pressure at which the sounds first occur is a measure of systolic pressure. As more and more pressure is released, a point is reached where the sounds become very muffled. This pressure is the diastolic pressure. The sounds arise from the turbulent blood flow through the narrowed (partially collapsed) artery under the cuff, just as sounds arise from the turbulent flow in a river when it passes through a narrow bed.

CN: Use red for A, dark colors for B, C, and D.
1. Begin with the comparison of rigid versus elastic tubes in the upper right corner.
2. Color the arterial pressure contour on the left.
3. Color the measurement of pressure, working from left to right. Complete each section before going on to the next.

ARTERIAL PRESSURE

BLOOD (A)
TUBE, ARTERY, RECOIL (B, B¹, B²)
SYSTOLE (C)
DIASTOLE (D)
PULSE PRESSURE (E)
MEAN PRESSURE (F)

The heart pumps blood intermittently; during systole, blood is thrust into the aorta, but during diastole no blood leaves the heart. In vessels with rigid walls the pressure would rise to very high values during systole and fall to near zero during diastole. Blood would spurt into the capillaries with each systole and almost stop during diastole. But arteries are not rigid pipes; they have elastic walls. During systole, part of the stroke volume is stored by the expanding arteries; during diastole, blood stored in the expanded arteries leaves for the capillaries, propelled in part by the recoil of arterial walls. The elastic walls buffer changes in pressure and flow caused by the intermittent heartbeat.

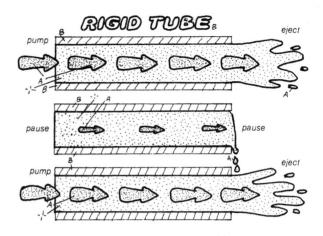

RIGID TUBE (B)

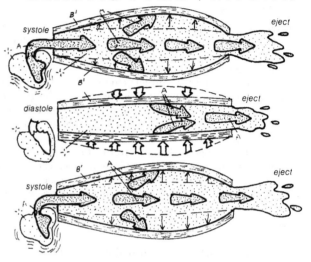

ELASTIC TUBE (ARTERY) (B¹)

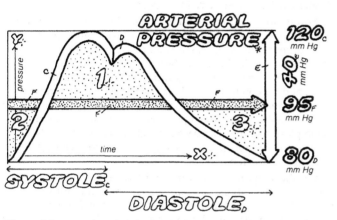

ARTERIAL PRESSURE

120 mm Hg (C)
40 mm Hg (E)
95 mm Hg (F)
80 mm Hg (D)

SYSTOLE (C)
DIASTOLE (D)

The arterial pressure (y axis) is shown through time (x axis). With each heartbeat, the arterial pressure in a young adult varies between 80 (diastolic) and 120 (systolic) mm Hg. Pulse pressure = systolic – diastolic pressure = 120 – 80 = 40 mm Hg. The mean pressure represented by the horizontal line is determined by splitting the area under the pressure curve into two equal parts (area 1 = 2 – 3 in the illustration).

MEASUREMENT OF PRESSURE

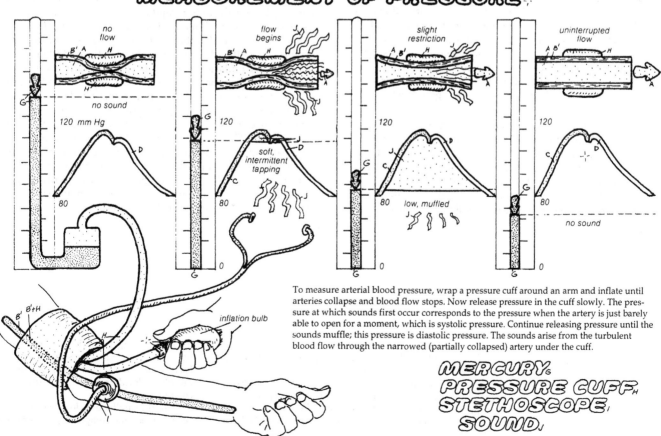

To measure arterial blood pressure, wrap a pressure cuff around an arm and inflate until arteries collapse and blood flow stops. Now release pressure in the cuff slowly. The pressure at which sounds first occur corresponds to the pressure when the artery is just barely able to open for a moment, which is systolic pressure. Continue releasing pressure until the sounds muffle; this pressure is diastolic pressure. The sounds arise from the turbulent blood flow through the narrowed (partially collapsed) artery under the cuff.

MERCURY (G)
PRESSURE CUFF (H)
STETHOSCOPE (I)
SOUND (J)

CAPILLARY STRUCTURE & SOLUTE DIFFUSION

The amount of blood that fills the *capillary bed* at any moment is only about 5% of the total blood volume. Nevertheless, this is where the "business" of the circulation is transacted. It is the place where exchange of O_2 and nutrients for CO_2 and wastes occurs.

CAPILLARY CIRCULATION

Capillary walls are thin and porous — Exchange takes place in the capillaries because their walls are composed of only one layer of very thin, porous *endothelial cells*, which permit solutes smaller than proteins to rapidly diffuse between capillary blood and interstitial tissue spaces. (The granular *basement membrane* that surrounds each capillary does not present any special barrier to diffusion.) Before entering the capillaries, blood must pass through *arterioles*, the resistance vessels. They range in diameter from 5 to 100 μm and are surrounded by thick, smooth muscular walls that can contract to constrict the arteriole and regulate blood flow delivery to the capillary bed. Blood leaving the capillaries enters the *venules*, which serve as collecting vessels. Their walls are thinner than arterioles, but thicker and much more impermeable than capillaries.

Smooth muscles control distribution of blood to capillaries — In some tissues, blood goes directly from arteriole to capillary; in others, the blood is delivered to *metarterioles*, which then give rise to capillaries. Metarterioles can serve as supply vessels to the capillaries, or they can bypass the capillaries and convey blood directly into venules. Capillaries that arise as side branches of arterioles or metarterioles have muscle cells around their origin that act as gates or *precapillary sphincters*, the last control on local blood flow before it enters the capillary bed. Sometimes a second type of bypass vessel called an *AV shunt* occurs. These are direct connections between arterioles and venules that do not give rise to capillaries.

Not all capillary beds are open at any one time. Smooth muscles regulating the microcirculation are controlled by both nerves and local metabolites (chemicals involved in metabolism). Arteriolar smooth muscle has a rich supply of nerves and is less sensitive to metabolites; metarterioles and precapillary sphincters have a poor nerve supply and are largely governed by local metabolites. The combined action of these muscular controls produces an intermittent flow through any particular capillary bed. First one bed opens, then closes while another opens.

CAPILLARY-TISSUE SOLUTE EXCHANGE

Solutes *diffuse* from blood to tissue — Most solutes diffuse freely through the capillary walls. The concentration of O_2 and nutrients physically dissolved in the blood plasma is higher than that in the tissues because the solutes are *consumed* in the tissue; the concentration gradient promotes nutrient diffusion from blood plasma to tissue. In contrast, CO_2 and waste products are constantly *produced* in the tissues; their concentration gradient promotes diffusion from tissue to blood plasma. If they can pass through the capillary walls, there is no need for special transport systems to exchange materials between blood and tissue.

Capillary permeability varies with the tissue — How do solutes get through the capillary walls? The respiratory gases O_2 and CO_2 are lipid soluble so that permeation is no problem; they permeate all cell membranes (including the endothelial cells that make up the capillary walls) with ease. In addition, capillary walls behave as though they contain large pores to permit anything smaller than a protein to pass through. In many tissues, such as skeletal, cardiac, and smooth muscle, the *junction between endothelial cells* is loose enough to allow passage of most molecules, but not proteins. This situation is different in the brain. Here the junctions are very tight and restraining; capillaries in the brain are impermeable to many small molecules as well as protein. This barrier to exchange, called the *blood-brain barrier*, is circumvented by special facilitated transport systems in the endothelial cells of the brain capillaries that transport such required nutrients as glucose and amino acids. In contrast, capillary endothelial cells in the intestines, kidneys, and endocrine glands are riddled with large "windows" called *fenestrations*, which provide large surface areas for permeation. These fenestrations are not simple holes. They are covered by a thin, very porous, very permeable membrane that permitts passage of relatively large molecules. Finally, capillaries in the liver can be extremely porous. The endothelial cells do not provide a continuous covering, leaving large gaps between cells that are easily traversed by large molecules, including proteins.

CN: Use red for A and a closely related color for red blood cells (G). Use purple for D and blue for F. Use a dark color for C.
1. Begin with the large illustration of a capillary bed. Note that in the right half, the capillary network (D) is left uncolored because precapillary sphincters (C) are tightened. Next color the inset.

2. Color the diagram of the capillary. Note that titles for the various arrows shown entering ar leaving are listed at the bottom and only the heavily outlined portion of the arrows should colored.
3. Color the capillaries below, starting with "continuous" and completing it before going o

ARTERIOLE A
METARTERIOLE B
PRECAPILLARY SPHINCTER C
CAPILLARY D
AV SHUNT E
VENULE F

In many tissues, blood is delivered to metarterioles, which serve as supply vessels to capillaries. Capillaries that arise as side branches of either arterioles or metarterioles have muscle cells around their origin, which act as gates (precapillary sphincters). If the gates are shut, blood in the metarterioles is conveyed directly to the venules. Sometimes a second type of bypass vessel called an AV shunt occurs. These are direct connections between arterioles and venules that do not give rise to capillaries. Note the occurrence of smooth muscle in the different vessels.

CAPILLARY BED

from artery

to vein

RED BLOOD CELL G

CAPILLARY

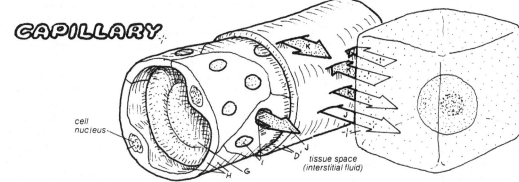

cell nucleus

tissue space (interstitial fluid)

ENDOTHELIAL CELL H
BASEMENT MEMBRANE D'
FENESTRATION I

Most solutes diffuse freely through the capillary walls. The concentration of O_2 and nutrients dissolved in the blood plasma is higher than that in the tissues because solutes are consumed in the tissue; the concentration gradient promotes nutrient diffusion from blood plasma to tissue. In contrast, CO_2 and waste products are constantly produced in the tissues; their concentration gradient promotes diffusion from the tissue to blood plasma.

TYPES OF CAPILLARIES

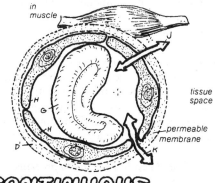

in muscle

tissue space

permeable membrane

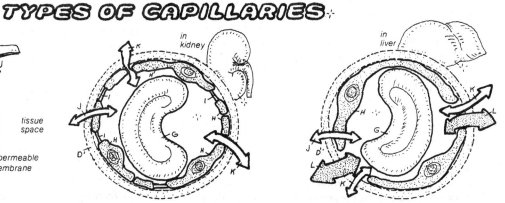

in kidney

in liver

CONTINUOUS H

In many tissues, such as skeletal, cardiac, and smooth muscle, the junction between endothelial cells is loose enough to allow passage of most molecules, but not proteins.

FENESTRATED I

Capillary endothelial cells in the intestines, kidney, and endocrine glands are riddled with large "windows" called fenestrations, which provide large surface areas for permeation. These fenestrations are not simple holes. They are covered by a thin, porous, very permeable membrane.

DISCONTINUOUS H

Capillaries in the bone marrow, liver, and spleen can be extremely porous. The endothelial cells do not provide a continuous covering, leaving large gaps between cells that are easily traversed by large molecules, including proteins.

SOLUTES
LIPID SOLUBLE MOLECULE (O_2, CO_2) J
WATER SOLUBLE MOLECULE (GLUCOSE, IONS, H_2O) K
MACROMOLECULE (PLASMA PROTEIN) L

Because capillary walls are porous, you might expect fluid to filter through the walls from the lumen of the capillary, where pressure is higher, to tissue spaces, where pressure is lower. Tissues would fill with fluid, swelling and stretching surrounding structures until pressures in tissue spaces built up to a level where they balanced the blood pressure, preventing further flow. Normally this does not happen because an additional important force, the *osmotic pressure* exerted by *plasma proteins*, counteracts the *blood pressure*.

STEADY-STATE FLUID BALANCE

Plasma proteins produce an osmotic gradient — The figure at the top of the plate reviews osmotic pressure, with emphasis on plasma proteins. Strictly speaking, all dissolved solutes contribute to the osmotic pressure of a solution. However, the solutions on the two sides of the capillary wall (i.e., blood plasma in capillaries and intercellular fluids in tissue spaces) are practically identical in composition, with the major exception that plasma contains large quantities of protein and intercellular spaces have very little. This follows because the capillary pores are very permeable to small molecules and hinder only giant molecules like proteins from passing. The small molecules do not contribute much to water movements because (1) they are almost evenly distributed on both sides of the capillary and (2) in porous membranes, extremely permeable molecules hardly influence osmotic water flow even when they are not uniformly distributed. This simplifies our problem. For osmotic flow across capillaries, we have to consider only the proteins. The osmotic pressure exerted by the plasma proteins is sometimes called *oncotic pressure* or *colloid osmotic pressure*. The other important force, *capillary blood pressure*, is often called *hydrostatic pressure*.

Osmotic gradient pulls fluid in, hydrostaic gradient pushes fluid out of capillaries — Fluid flow through the capillay wall depends on balance between the hydrostatic (blood pressure) pressure gradient, ΔP_{hydro}, forcing fluid out of the capullary, and the osmotic (oncotic) pressure gradient, ΔP_{osm}, pulling fluid in. (The symbol Δ represents "difference" or "gradient.") As shown by the top plate, the oncotic pressure of the plasma is approximately 25 mm Hg. Most of this pressure is due to albumin, the smallest, most abundant protein in plasma. Because of its small size, a tiny amount of albumin leaks out into tissue spaces, so that the oncotic pressure of tissue spaces may be approximately 2 mm Hg. The net osmotic gradient drawing fluid into the capillary is 25 – 2 = 23 mm Hg. How does this compare with the hydrostatic pressure gradient pushing fluid out?

Balancing the two gradients — The hydrostatic pressure gradient, ΔP_{hydro}, is equal to the difference between the blood pressure within the capillary, P_{cap}, and the hydrostatic pressure in tissue spaces, P_{tiss}. Both of these vary from tissue to tissue, and P_{cap} even varies at different locations within the same capillary. A typical situation where P_{tiss} = 2 mm Hg is illustrated in the plate. At the arterial end of the capillary, $\Delta P_{hydro} = P_{cap} - P_{tiss} = 35 - 2 = 33$ mm Hg. Therefore, a net force of 33 – 23 = 10 mm Hg pushes fluid out into the tissues at the arterial end of the capillary. Now examine the venous end, where blood pressure is lower, about 15 mm Hg. (It has been reduced as a result of friction encountered in pushing blood through the narrow capillaries.) The properties of the tissue space as well as the composition of the plasma remain practically the same. Now we have only ΔP_{hydro} = 15 – 2 = 13 mm Hg pushing fluid out while the same oncotic pressure gradient of 23 mm Hg pulls fluid in. In other words, at the venous end of the capillary, we have 23 – 13 = 10 mm Hg pulling fluid in! In this capillary, fluid flows out at the arterial end but flows right back in at the venous end; as a result, blood leaves the capillary containing the same amount of fluid as when it entered. The tissue has neither gained nor lost fluid from this capillary.

Not every capillary is as precisely balanced as the one in our example. The figures used for the pressures were rough averages, and there can be considerable variation from vessel to vessel. In one capillary, filtration may dominate while in another nearby capillary, reabsorption prevails. On the whole, in most organs, they nearly balance, and whatever small imbalance exists (usually filtration is slightly larger than reabsorption) is taken care of by the lymphatic vessels (plate 41). These vessels drain tissue spaces of excess fluid together with plasma proteins that leak out of capillaries and return them to the circulation via the large veins. The final result is no net fluid exchange.

Fluid balance varies with the tissue — Capillary beds in the lungs are a notable exception. Blood pressure in the pulmonary circulation, including lung capillaries, is low. This favors reabsorption of fluid, which is advantageous because it keeps the lungs drained of congesting fluids, which could hamper respiration. In the kidneys, one set of capillaries (glomerular capillaries) has an unusually high blood pressure; they filter large quantities of fluid along their entire length. Another set (peritubular capillaries) has an unusually low blood pressure together with elevated oncotic pressure; they reabsorb fluid.

EDEMA

Fluid balance can be upset by a number of factors, resulting in the accumulation of large amounts of fluid in tissue spaces, a condition called *edema*. Edema poses a threat because it compromises the circulation by increasing distances that substances have to diffuse in going to and from capillaries. This is particularly true in the lungs, and pulmonary edema is serious.

Causes of edema are revealed by a review of factors involved in fluid transfer.

1. *Increase in capillary blood pressure.* This could result from dilation of arterioles or from venous congestion extending back to the capillaries, as may occur in heart failure.

2. *Decreased plasma oncotic pressure.* This can result from starvation (dietary protein deficiency), disturbed protein synthesis, as in liver disease, or loss of plasma protein caused by kidney disease.

3. *Increased permeability of capillary walls to proteins.* This reduces the oncotic pressure gradient across the capillary and decreases reabsorption. Increased capillary permeability occurs in response to allergies, inflammation, and burns.

4. *Disturbance of lymph drainage.* May be caused by obstruction of lymph vessels and occasionally follows operations.

CN: Use very light colors for A and C and dark colors for B and D.
I. Begin with the upper diagram. Color the background areas of blood plasma (A) and interstitial fluid (C) first, and then color the proteins (D^1) and pressure numbers (B & D) afterwards, with darker colors. Follow this procedure for the other diagrams. Don't color the capillary membrane.
2. Color the large diagram beginning with the three examples above the main drawing.

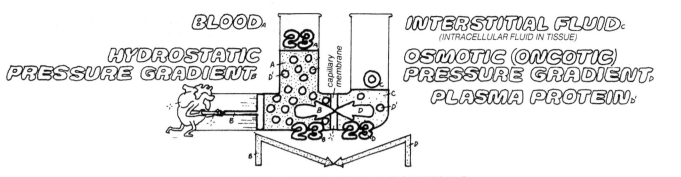

PRESSURE GRADIENTS

Blood plasma and interstitial fluid are almost identical in composition except for the presence of considerable amounts of protein in the plasma and very little in the interstitial fluid. With plasma on one side and interstitial fluid on the other side of a membrane made up of a capillary wall, we find fluid flowing from interstitial fluid to plasma as a result of inequality of plasma proteins in the two fluids. To prevent flow, pressure equal to 23 mm Hg must be applied to the plasma. This is the osmotic gradient caused by unequal amounts of proteins (the oncotic gradient). In the body, this pressure is applied by the heart; it is the capillary blood pressure.

FILTRATION AND REABSORPTION IN THE CAPILLARY

Capillary blood pressure is not the same in all parts of a capillary; it is higher at the arterial end and falls continuously through the capillary until reaching the lowest value at the venous end. Keeping track of fluid balance in an average capillary shows that at the arterial end, hydrostatic pressure gradient = blood pressure – tissue pressure = 35 – 2 = 33 mm Hg pushing fluid out (filtration). Oncotic pressure gradient = plasma oncotic pressure – tissue oncotic pressure = 25 – 2 = 23 mm Hg pulling fluid in (reabsorption). The net result at the arterial end is 33 – 23 = 10 mm Hg pushing fluid out. At the venous end only the capillary pressure has changed. Now it is 13 mm Hg. Repeating the analysis shows a net force of 10 mm Hg pulling fluid in at the venous end.

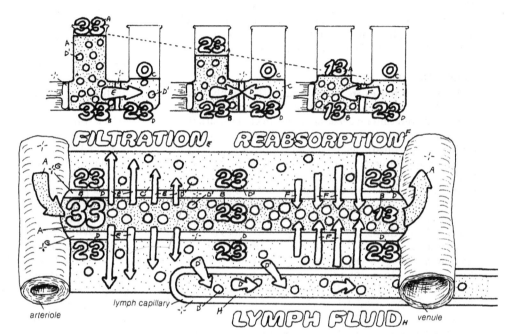

EDEMA DUE TO CAPILLARY PRESSURE

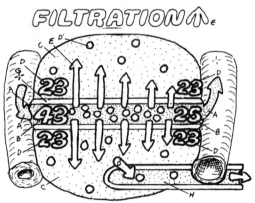

Edema (swollen tissue) often results from increased capillary blood pressure. Forces favoring filtration are increased while oncotic pressures may not change very much, so that filtration prevails along the entire length of the capillary. This could result from dilation of the arterioles or from venous congestion extending back to the capillaries, as may occur in heart failure.

EDEMA DUE TO HIGH TISSUE PROTEINS

Edema can also result from disturbances of lymph drainage. Lymphatics serve as an overflow system, returning excess tissue fluid and proteins to the circulation. When lymph drainage is poor, proteins accumulate in tissue spaces and compromise the oncotic pressure gradient while the hydrostatic pressure gradient remains unchanged. These disturbances follow obstruction of lymph vessels, which can result from operations or from parasitic invasion (and obstruction) of the lymph vessels of the limbs, a condition known as elephantiasis.

Most tissues contain enormous numbers of tiny lymph vessels called *lymph capillaries*. One end of each lymph capillary is blind (closed off); the open ends coalesce into larger vessels called collecting ducts, which in turn merge into still larger vessels (called *lymph ducts*), and so on, until the largest ducts drain into the circulation via connections with large veins (e.g., the subclavians at the junction with the jugular veins). This system functions to return plasma proteins that have leaked into tissue spaces as well as excess tissue fluid back to the circulation. Reclamation of these proteins is essential; without it we would die within a day or two.

LYMPHATIC CIRCULATION

Lymph capillaries are very porous – The lymph ducts resemble veins: both have smooth muscle embedded in their walls, and both contain *one-way valves* directed toward the heart. Although both lymphatic and ordinary circulatory capillaries are constructed of similar endothelial cells, there are important differences between them. Lymph capillaries have no basement membranes, and the junctions between their endothelial cells are often open, with no tight intercellular connections. This makes them very permeable to proteins as well as smaller molecules and water. When the tissue spaces fill with fluids, the increased pressure does not compress and close the lymph capillaries because they are held open by *anchoring filaments* attached at one end to the endothelial cells and to surrounding connective tissue at the other end. The edges of the endothelial cells overlap slightly so that they form "flap valves," which allow fluid to enter the lymph capillary but not to leave it.

One-way valves direct lymph flow toward the great veins – Lymph flow is propelled through the periodic contractions of the smooth muscle embedded in the walls of the ducts. These contractions, which "milk" the lymph along, occur on the average some two to ten times per minute. One-way flow (out of the tissues and toward the veins) is ensured by the numerous valves that occur every few millimeters. In addition to these contractions, lymph flow is aided by the same factors that promote venous return (plate 42). These include contractions of nearby skeletal muscle compressing lymph vessels (the skeletal muscle pump), periodic changes in intrathoracic pressure associated with breathing (respiratory pump), and possibly the pulsations of nearby

arteries. Finally, the high velocity of flow in the veins where the lymph ducts terminate produces a suction that draws lymph toward it.

Lymphatic system allows plasma proteins access to tissue spaces – During each twenty-four hours the heart pumps 8400 L of blood. Of this, 20 L filter out of the capillaries into the tissues; and of this 20 L, some 16 to 18 L are returned to the circulation via capillary reabsorption, leaving 2 to 4 L to be returned by the lymphatic system. Compared to the blood, lymph flows very slowly. The lymphatic system also returns plasma protein that has leaked into the tissues from capillaries. This leakage is slow on a time scale of minutes, but when viewed over an entire day, the amount of protein returned by the lymphatics is equal to 25 to 50% of all the plasma proteins of the body. Seen from this perspective, the lymphatics seem to be a simple overflow system for correcting fluid imbalance and for recovering protein that capillaries were unable to restrain. However, this perspective is limited; the lymphatic circulation fulfills a real function and is not simply a means of compensating for apparent capillary inefficiency. Passage of plasma proteins into tissue spaces is an important means of transport for antibodies and for many hormones that are bound to plasma albumin. In addition, the lymphatic circulation provides a path for transport of long-chain fatty acids and cholesterol that have been absorbed from the intestine and for the entrance of lymphocytes (a type of white blood cell) into the circulation.

LYMPH NODES FILTER PATHOGENS

As lymph travels from the tissues toward the veins, it flows through one or more enlarged structures called *lymph nodes*. The lymph nodes contain phagocytic cells, which engulf and destroy foreign matter that is brought to them in the circulating lymph. The nodes also sequester lymphocytes, which can transform into antibody-producing plasma cells. Lymph nodes are powerful defense stations, guarding against foreign materials and invading bacteria. When there is a local infection, the regional lymph nodes frequently become inflamed as a result of the accumulation of toxins or bacteria carried to the nodes by the lymph. The efficiency of this system has been demonstrated by animal experiments in which bacteria have been injected into a lymph duct leading to a node; lymph collected from the duct leading away from the node is virtually free of bacteria.

CN: Use red for A, purple for B, blue for C, and dark colors for E, G, and I.
1. Begin with upper simplified diagram for blood and lymph circulation, including the two enlargements of the fluid exchange process.
2. Color the larger diagram of blood circulation, below, starting with the left heart and ending with

the right heart, then color the lymphatic portion beginning with the lymph capillary (D) at the bottom.
3. On the person on the right, color the three areas of superficial lymph nodes (H) and the cutaway diagram of the deeper lymph structures. It isn't necessary to color the superficial vessels, drawn in very light lines.

LYMPHATIC SYSTEM

LYMPH CAPILLARY_D
ANCHORING FILAMENT_E
LYMPH FLUID_F
LYMPH VESSEL_F'
VALVE_G
LYMPH NODE_H
FLUID EXCHANGE_I

During each 24 hours the heart pumps 8400 liters of blood. Of the 20 liters that filter out of the capillaries into the tissues, some 16 to 18 liters are returned to the circulation via capillary reabsorption, leaving 2 to 4 liters to be returned by the lymphatic system.

When the tissue spaces fill with fluids, the lymph capillaries are held open by anchoring filaments attached at one end to the endothelial cells and to surrounding connective tissue at the other end. Tissue fluid containing small amounts of plasma protein that has escaped from the blood capillaries freely permeates the gaps between these endothelial cells of the lymph capillaries.

The network of lymph vessels returns fluid to the circulation via connections with large veins. Lymph flow is propelled largely through the periodic contractions of the smooth muscle embedded in the walls of the ducts. One-way flow (out of the tissues and toward the veins) is ensured by the numerous valves that occur every few mm. As lymph travels through the lymphatics, it flows through lymph nodes, which serve as powerful defense stations guarding against foreign materials and invading bacteria.

LYMPH NODE_H

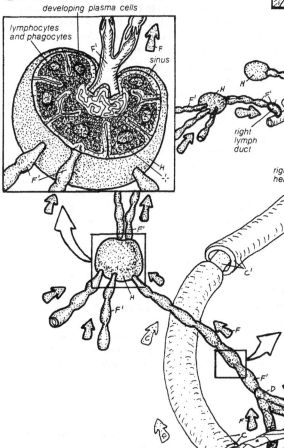

developing plasma cells

lymphocytes and phagocytes

sinus

right lymph duct

right heart

left heart

thoracic duct

BLOOD CIRCULATION

LEFT HEART_A
ARTERIES_A'
ARTERIOLES_A²
CAPILLARIES_B
VENULES_C
VEINS_C'
RIGHT HEART_C²

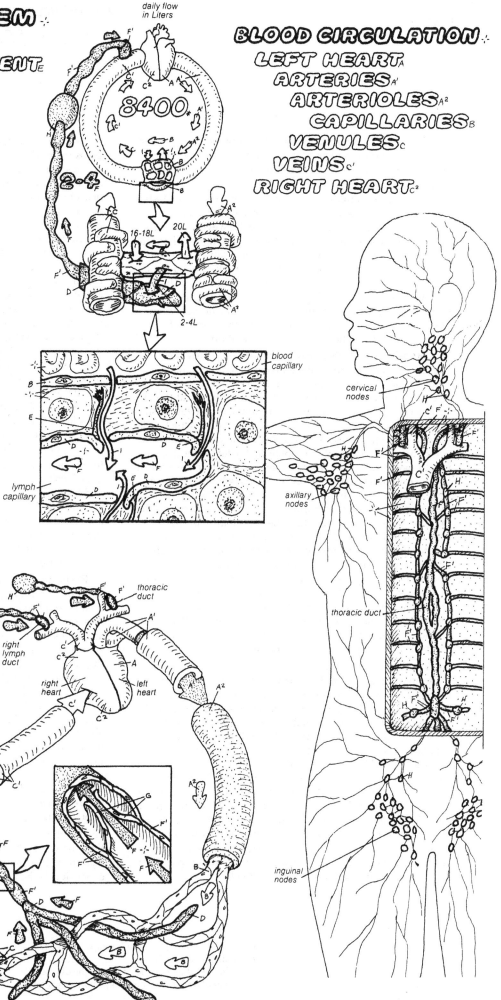

daily flow in Liters

8400*

2-4

16-18L   20L

2-4L

blood capillary

lymph capillary

cervical nodes

axillary nodes

thoracic duct

inguinal nodes

Veins have two main functions: (1) they provide a *low-pressure storage system* for blood and (2) they serve as *low-resistance conduits* to return the blood to the heart.

LOW-PRESSURE STORAGE

The walls of the veins are thin; they contain very little elastic tissue and may appear difficult to stretch or contract to accommodate changes in blood storage. However, veins are normally easy to distend because they are partially collapsed. On the other hand, the volume of the veins can be reduced through the contraction of smooth muscle cells embedded in their walls. At rest, veins contain about two-thirds of the blood in the body, although this can vary. In response to hemorrhage or exercise, for example, sympathetic nerves stimulate venous smooth muscles, constricting the veins and shunting blood to other parts of the circulation.

Compliance reflects storage capacity — To study the capacity of veins to store blood, take isolated segments of veins, tie off all possible exits, and inflate them with fluid as you would inflate a bicycle tire with air. (See figure at top of plate.) The question is, "How much fluid (air) can you push into the vein (tire) before the pressure in the vein (which will oppose your effort) rises 1 mm Hg?" This amount is called the compliance. The larger the compliance, the larger the capacity for storage. You can see from the plate that veins have a much larger compliance than arteries; at normal venous pressures, they are only partially filled and can easily distend. When they are filled far beyond normal, the slack is taken up and the pressure begins to rise very fast; the compliance falls. The easy distensibility (high compliance) of normal veins serves the storage function very well; however, it can create problems. For example, when we rise from the supine to the upright position, blood tends to pool in the veins (especially in the legs and feet); it is essentially withdrawn from the arterial side of the circulation. Without any compensatory response, such as activation of sympathetic nerves, the result would be disastrous. The nature of these challenges and the compensatory responses are covered in plate 46.

AIDS TO VENOUS RETURN

A small pressure gradient drives blood toward heart — Blood flows from regions where its mechanical energy is high to regions where it is low. When we are in a recumbent position, most of this energy is in the form of pressure. As blood passes through the narrow arterioles and capillaries, the pressure falls substantially. In many venules, blood pressure is around 15 mm Hg. In the atria, the average pressure is close to 0 mm Hg. It follows that there is a small but definite *pressure gradient* available to force blood back to the heart. The fact that this small gradient (approximately 15 mm Hg) is sufficient to drive large volumes of blood demonstrates the low resistance of the venous pathway. Even veins that appear to be collapsed have a low resistance because the apparent "creases" in the vessel are never really flat; they always leave some space that can be easily traversed by circulating blood.

Muscle contraction pushes venous blood — In addition to pressure gradients, there are other mechanisms that aid venous return of blood to the heart. These include "pumping actions" of noncardiac muscles as well as movements of the heart itself, and they depend on the *valves* in the veins, which point in the direction of the heart. This orientation ensures a forward flow toward the heart: blood flowing forward forces the valves open; backflow snaps them shut. The third figure on the plate shows this action in a vein lodged between two *skeletal muscles*. When the muscles are relaxed, blood flows forward because of the pressure gradient described above, and the vein fills with blood. When the muscles contract, they squeeze on the vein, forcing blood in all directions. Blood flowing backward closes the bottom valve, but forward-flowing blood keeps the upper valve open so that blood spurts in the forward direction. When the muscle relaxes, there is no longer any external force pushing on the venous walls: the presure gradient from below (farthest from the heart) forces blood flow in the forward direction, opening the lower valve and reestablishing our initial condition. Thus, each time the muscle contracts and relaxes, a spurt of venous blood is sent toward the heart. This action is called the *muscle pump*.

A good illustration of the importance of the muscle pump in exercise is provided if a runner remains motionless just after finishing a strenuous race. His cardiac output is still high and his capillaries and small blood vessels are still dilated in response to the exercise. Without the muscle pump the veins are quickly drained, venous return to the heart decreases, and the cardiac output may falter sufficiently to compromise the blood supply to the brain. Fainting can be avoided if the runner continues mild exercise for a few minutes.

Respiration draws blood into the thoracic veins — An additional pumping action, *the respiratory pump*, occurs during breathing. Each breath is preceded by an enlargement of the *chest cavity* (the *thorax*). The enlarging thorax acts as a bellows and "sucks" air into the lungs (see plate 49). The same expansion also "sucks" blood into the thoracic veins. The thoracic cage not only expands its lateral dimensions (a result of skeletal muscles pulling the rib cage upward), it also expands its vertical dimensions as a result of the dome-shaped *diaphragm* contracting and pushing downward on the *abdomen*. Pushing on the abdominal contents then squeezes the veins in the abdomen. Thus, each time a breath is drawn, the expanding thorax sucks and the compressed abdomen squeezes blood toward the heart. During expiration, the reverse occurs, and although there are no valves in the great veins of the thorax or abdomen, backflow is checked by valves in the large veins of the extremities.

Expansion of the atria draws blood into the heart — Finally, the motion of the heart itself aids venous return. Each time the ventricles beat, the upper portions of the ventricles near the valves move downward toward the apex, resulting in expansion of the atria, which draws blood into the heart.

CN: Use blue for A, red for C, and purple for D. Use a dark color for H.
1. Begin with the top panel, coloring the diagrams of the circulatory system with a person at rest and then during exercise. Color the graph to the right, comparing the relationship between pressure and volume.
2. Color the remaining panels.

VENOUS STORAGE

VEINS~A~
HEART~B~
ARTERIES~C~
CAPILLARIES~D~
SYMPATHETIC NERVE~E~

Veins provide a low-pressure storage system for blood. At rest the veins contain about 63% of the blood in the body, but this can vary in response to hemmorhage or exercise; for example, sympathetic nerves become active and stimulate the smooth muscles, which constrict the veins. As a result, blood is shunted to arteries and other parts of the circulation.

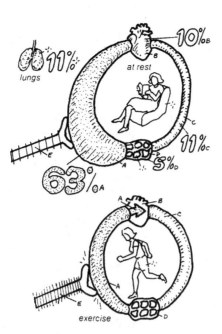

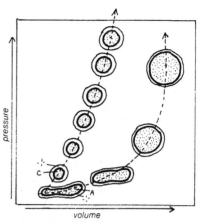

Pressure in arteries and veins (y axis) plotted against volume (x axis). Veins have a much larger storage capacity than arteries; at normal venous pressures, they are only partially filled and can easily distend. Filling them far beyond normal takes up the slack and pressure rises fast.

VENOUS RETURN

PRESSURE GRADIENT~A~

ARTERIAL PRESSURE~C'~
VENOUS PRESSURE~A'~

There is a small but definite pressure gradient available to force blood back to the heart. The fact that this small gradient (about 15 mm Hg) is sufficient to drive large volumes of blood to the heart demonstrates the low resistance to the venous pathway.

MUSCULAR PUMP~G~

VALVE~F~
SKELETAL MUSCLE~G~

When the muscles are relaxed, the venous pressure gradient drives the blood forward toward the heart, filling the vein. When the muscles contract they squeeze the thin-walled vein, squishing blood in all directions. Blood flowing backward closes the bottom valve, but the forward-flowing blood keeps the upper valve open. Blood spurts forward toward the heart. A pulsating artery next to the vein may exert the same "milking" action as the muscle.

RESPIRATORY PUMP~H~

DIAPHRAGM~H~
CHEST CAVITY & AIR PASSAGES~I~
ABDOMINAL CAVITY~J~

With each breath, the thorax acts as a bellows. During inspiration it "sucks" air into the lungs and blood into the thoracic veins. In addition, the diaphragm pushes on the abdominal contents, squeezing the abdominal veins and forcing blood into the thoracic veins. During expiration the reverse occurs. Backflow is checked by valves in veins of the extremities.

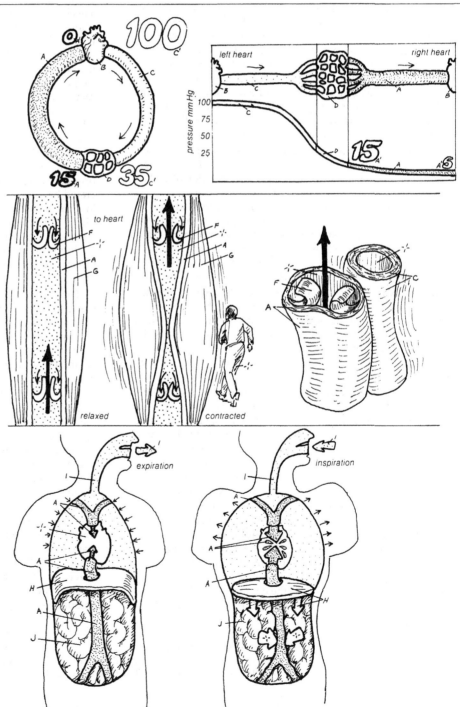

Control of the diameter of small blood vessels is critical for at least two primary reasons; (1) blood vessel diameter controls the blood flow, and thereby the nourishment of regions that lie downstream; and (2) the diameter of small blood vessels controls the flow resistance of the vascular bed. This is one of the major determinants of blood pressure.

LOCAL REGULATION OF TISSUE PERFUSION

When a tissue becomes active, its blood vessels dilate. Local blood flow increases, bringing the active tissue more nourishment and carrying away waste products. This response is clearly beneficial; more blood is supplied to active tissue, less to quiescent tissue. It is a local response to local changes in local conditions.

Vasodilating metabolites are released by ischemic (poorly perfused) tissue — Local regulation of vessel diameter is illustrated by a simple experiment. A tourniquet is applied to an arm, blocking blood flow for a few minutes, and then released. Following release, local blood vessels dilate and blood flow to the impoverished arm is temporarily much higher than normal, as if to compensate for the period when flow was obstructed. If extracellular fluids collected from the starved tissues during the blockage are injected into the opposite (normal) arm, vessels dilate and blood flow increases, as though it too were recovering from the tourniquet. The impoverished tissue releases into its surrounding fluids substances that diffuse to local blood vessels causing dilation.

The search for dilating substances shows that any of the following characteristics of the fluid can cause *dilation of arterioles* and *precapillary sphincters:* high concentration of *acids*, CO_2, *potassium*, and *adenosine* and low concentration of O_2. Common to all these conditions is that they are produced by cells when their blood supply doesn't support their activity. This provides a very simple *negative feedback* scheme to match local blood flow to cellular activity: *increased tissue activity (inadequate blood supply) → accumulation of metabolites (acids, CO_2, low O_2, etc.) → dilation of blood vessels (increased blood flow).*

NITRIC OXIDE: ENDOTHELIUM-DERIVED RELAXING FACTOR

Nitric oxide (NO), a smog pollutant, recently has been implicated in the regulation of bloodflow. It is formed during lightning storms from atmospheric nitrogen and oxygen. It is also formed in the human body (from the amino acid argenine and molecular oxygen) where it plays important roles in the regulation of cardiovascular, nervous, and immune systems.

Nitric oxide production in endothelial cells is stimulated by Ca^{++} — In the cardiovascular system nitric oxide, a potent *vasodilator*, is produced by endothelial cells that form the inner lining of blood vessels. Its production is stimulated by intracellular Ca^{++} (via calmodulin), which activates the enzyme (NO synthase) responsible for NO synthesis. Anything that can raise intracellular Ca^{++} will stimulate NO production. These factors include a number of vasodilators, such as acetylcholine (a neurotransmitter) and bradykinin (a circulating hormone), as well as the shear stress of flowing blood.

NO vasodilates by relaxing neighboring smooth muscle — NO is a paracrine (plate 113). It exerts its vasodilating action by easily permeating cell membranes and diffusing from its origin in endothelial cells to neighboring smooth muscle cells. There it activates the enzyme guanyl cyclase to catalyze the formation of cGMP from GTP (plate 12). cGMP then initiates the final steps for smooth-muscle relaxation (dephosphorylation of MLCK; plate 28).

NO plays several cardiovascular roles — Continuous production of NO has been implicated in the maintenance of normal blood pressure. Further, it appears to be responsible for up to 50% of the increased blood flow following ischemia. Apart from its vasodilator action, NO is a potent inhibitor of white blood cell adhesion to vascular walls and of blood platelet aggregation.

NEURAL REGULATION OF BLOOD FLOW

Sympathetic nerve impulses constrict blood vessels — Smooth muscle in blood vessel walls is controlled by *sympathetic nerves* in addition to local chemical control. These nerves are generally active, constantly barraging the blood vessels with impulses liberating norepinephrine, which, in turn, binds to an alpha receptor in the vascular smooth muscle. This liberates free Ca^{++} into the cytosol via the G protein–phospholipase C-IP3–released Ca^{++} sequence (plate 13). The free Ca^{++} causes muscular contraction, resulting in vessel constriction. When the frequency of sympathetic impulses increases, blood vessel constriction is more intense; when the frequency decreases, the vascular smooth muscle is more relaxed and blood vessels dilate. This description has covered the primary mechanism for neural control of blood vessels. In addition, there appear to be a number of minor pathways that operate on a different basis. Parasympathetic fibers supply blood vessels of the head and viscera, but not skeletal muscle or skin. These fibers release acetylcholine, stimulating NO production and causing vasodilation. Acetylcholine is also liberated by a small number of vasodilator fibers that are carried in the sympathetic nerve trunks going to skeletal muscle. Their significance is not apparent; they are activated by excitement and apprehension, and it has been suggested that they are involved in the vasodilation that occurs with the anticipation of exercise. They have been found in cats and dogs and are probably present in humans.

The density of sympathetic innervation varies widely from tissue to tissue. Arterioles and veins of the viscera and skin have a rich supply of nerves and show intense vasoconstriction upon sympathetic stimulation. In contrast, blood vessels in the brain and coronary circulation are nonresponsive to sympathetic stimulation. Their circulation is rarely compromised by vasoconstriction; neither the brain nor the heart can sustain O_2 deprivation for significant amounts of time.

Local and systemic requirements are not always compatible — While chemical control matches blood flow to metabolic activity, sympathetic vasoconstrictors play a major role in the control of vascular resistance (and therefore blood pressure). Situations in which these two mechanisms oppose each other (e.g., blood pressure falling while an organ has inadequate blood supply) can easily arise. Through reflexes discussed in plate 45, the sympathetic nerves are activated in response to low blood pressure, causing vasoconstriction. At the same time, the deprived organ starts producing vasodilator substances. Although the net result depends on the particular organ, vasodilator responses most often predominate. There is evidence that vasodilator substances act not only on blood vessels, but directly on sympathetic nerve endings, inhibiting amounts of norepinephrine released by sympathetic impulses.

CN: Use red for A and purple for G.
1. Begin with the upper left panel, note that the capillary (G) is left uncolored because its sphincter (H) is closed.

The actions of metabolites (J) open them up on the right.
2. Color the actions of Nitrous Oxide in the arteriole (A).
3. Color the actions of neural control below.

LOCAL (CHEMICAL) CONTROL

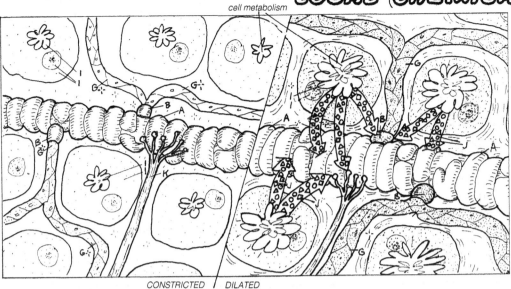

cell metabolism

CONSTRICTED | DILATED

When a tissue becomes active, it releases a number of metabolic products that dilate local arterioles and precapillary sphincters. Local blood flow increases, bringing the active tissue more nourishment and carrying away waste products. High concentration of acids, CO_2, K^+, and adenosine and low concentration of O_2 are produced by cells when their blood supply is inadequate to support their activity. These substances all dilate blood vessels, providing a negative feedback to match local blood flow to cellular activity.

ARTERIOLE A
ENDOTHELIAL CELL B
CALCIUM C
NITRIC OXIDE D
SMOOTH MUSCLE CELL E
cGMP F
CAPILLARY G
PRECAPILLARY SPHINCTER H
TISSUE CELL I
METABOLITES J
SYMPATHETIC NERVE K
GANGLION K'
FREQUENCY OF IMPULSE K²
NOREPINEPHRINE L
VESSEL RECEPTOR L'

Nitric oxide (NO), a vasodilator, is produced by endothelial cells in response to a rise in intracellular Ca^{++} which activates the enzyme responsible for NO synthesis. NO is a lipid soluble paracrine. It permeates membranes with ease and diffuses to neighboring smooth muscle cells where it promotes the formation of cGMP from GTP. cGMP then initiates steps for smooth muscle relaxation. Stimuli which raise Ca^{++} and evoke this response include acetylcholine (a neurotransmitter) and bradykinin (a circulating hormone), as well as the mechanical shear stress of flowing blood. The cardiac drug nitroglycerine is effective for treating angina because it is broken down to NO and dilates the coronary arteries.

ACTIONS OF NO D

CONSTRICTED

CONTRACTED

RELAXED

SYSTEMIC (NEURAL) CONTROL

CONSTRICTED

LOCAL OVERRIDE J

DILATED

In addition to acting directly on blood vessels, metabolic vasodilator substances also act directly on sympathetic nerve endings to inhibit the amount of norepinepherine released by sympathetic impulses.

Smooth muscles, in blood vessel walls are also controlled by sympathetic nerves. These nerves constantly barrage blood vessels with impulses that liberate norepinephrine, causing the vascular smooth muscle to contract and constrict the vessels. When the frequency of sympathetic impulses increases, blood vessel constriction is more intense; when the frequency is decreased, the muscle is more relaxed and blood vessels dilate.

ACTIVATED A
CONSTRICTED

NORMAL A

INHIBITED A
DILATED

The cardiovascular system is intricate and complex, yet its function is simple: it moves blood. The most important index of cardiovascular performance is: "How much blood is moved to the tissues during each minute?" This quantity is called the *cardiac output*. Cardiac output equals the amount of blood expelled from one ventricle during a single beat (*stroke volume*) times the number of beats per minute (*cardiac output = stroke volume × heart rate*). In a steady state, the cardiac output of the left heart equals that of the right heart (flows in the systemic and pulmonary circulations are equal).

In an average-sized person at rest, the cardiac output is about 5 L per minute. However, this figure fluctuates; it rises with activity, reaching as high as 25–30 L per minute during heavy exercise, and even higher in athletes. Cardiac output can change by alterations in either stroke volume or heart rate. During exercise, for example, the stroke volume may show a moderate increase while the heart rate rises about by three times. These changes in stroke volume and heart rate are brought about by intracardiac mechanisms (response of the contractile machinery to *stretch*) and to extracardiac mechanisms (action of *sympathetic* and *parasympathetic* nerves).

INTRACARDIAC MECHANISMS

The strength of contraction of a heart muscle fiber depends on its length — The tension developed by cardiac muscle, like skeletal muscle, depends on length. Under normal resting conditions, the length of an average heart muscle fiber may be only about 20% of its optimum length for maximal force. Stretching the fiber beyond its norm reveals a reserve of additional power for forceful contractions. This response to stretch, called the *Frank-Starling mechanism*, has important implications. If more blood is returned to the heart, the walls of the ventricles are stretched, and the Frank-Starling mechanism ensures that the heart can develop the extra strength required to empty itself. If arterial pressure suddenly rises, the stroke volume will decrease because the ventricle will not have sufficient force to overcome the increased arterial pressure. The extra blood that remains in the heart (the *residual volume*) just following the beat will increase, and this increased blood will help stretch the walls prior to the next beat. Consequently, the force of the next beat will increase, helping the heart meet the increased load imposed by increased arterial pressure. This will increase the stroke volume back toward normal.

The Frank-Starling mechanism is particularly important in adjusting the output of the right and left hearts. If, for example, your right heart output was just 1 mL/min. greater than your left, then after about 15 min. some 1000 mL of fluid would accumulate in the pulmonary circulation. The increased pressure would force fluid out of the capillaries into the lungs, and you would drown!

EXTRACARDIAC MECHANISMS

Parasympathetic impulses slow the heart rate — The action of the autonomic cardiac nerves has been considered in plate 35. The *parasympathetic* nerves to the heart are carried in the vagus nerve. The vagus nerve is generally active, discharging a continuous barrage of impulses at the SA and AV nodes and slowing the basic heart rate. When the parasympathetic nerve supply to the heart is interrupted, the heart speeds up. Increasing the frequency of parasympathetic

impulses slows the heart; decreasing the frequency speeds it.

Sympathetic impulses increase the heart rate — The *sympathetic* nerves to the heart are also continually active, but the their effect on rate is the opposite of that of the parasympathetic nerves. Sympathetic impulses increase the heart rate, and when this nerve supply is interrupted, the heart slows. Increasing the frequency of sympathetic impulses speeds the heart; decreasing the frequency slows it. Generally, the activities of these two opposing sets of nerves are coordinated; when the sympathetic nerves are excited, the parasympathetic are inhibited, and vice versa.

Sympathetic impulses increase the stroke volume — In addition, sympathetic impulses increase stroke volume by increasing the force of contraction of the ventricular muscle. Thus, there are two independent mechanisms for changing stroke volume: (1) changing the initial length of the cardiac fibers (i.e., changing the end diastolic volume) and (2) increasing the barrage of sympathetic impulses to the ventricular musculature and/or, similarly, by releasing catecholamine hormones from the *adrenal medulla*.

THE FICK PRINCIPLE — MEASUREMENT OF CARDIAC OUTPUT

Blood flow through any organ can be measured by a simple application of the conservation of matter known as the *Fick principle*. We apply the principle to oxygen consumption using steady-state blood flow through the lung as our example. Let F denote blood flow in L/min, $[O_2]$ denote oxygen concentration, and $\{O_2/min\}$ denote oxygen consumption (i.e., O_2 delivered to the blood/min). Then with each minute:

$\{O_2/min\}$ = O_2 consumed = O_2 taken up by blood
O_2 taken up by blood = O_2 carried away by veins
$\qquad\qquad\qquad\qquad$ – O_2 carried in by arteries

O_2 carried away by veins = F × $[O_2]_{ven.}$
O_2 carried in by arteries = F × $[O_2]_{art.}$

Putting these steps together:
$\{O_2/min\}$ = F × $[O_2]_{ven.}$ – F × $[O_2]_{art.}$
Solving for F :
F = $\{O_2/min\}$ / ($[O_2]_{ven.}$ – $[O_2]_{art.}$)

To measure cardiac output, simply measure the blood flow through the lungs. (This is the flow out of the right heart, which equals the flow into and out of the left heart.) In this case, the $\{O_2/min\}$ — the oxygen consumed — is obtained by measuring the difference between the O_2 content in the inspired air and the O_2 content in the expired air. The measurement also requires analysis of $[O_2]$ in blood samples from the pulmonary artery and vein. A sample representing pulmonary venous blood can be obtained from any systemic *artery*. Pulmonary veins and systemic arteries have the same O_2 content because blood has no opportunity to exchange O_2 until it reaches the systemic capillaries via the left heart and systemic arteries. Taking a sample of pulmonary arterial blood is much more difficult. It requires passing a catheter (a narrow, flexible, hollow tube) into a vein and threading it into the right heart, a nontrivial routine! The right heart and pulmonary artery have the same $[O_2]$. Typical figures for these quantities are: $\{O_2/min\}$ = 250 ml/min; $[O_2]_{ven.}$ = 150 ml/L; and $[O_2]_{art.}$ = 200 ml/L , so that F = 250/(200–150) = 5L/min.

CN: Use red for B and blue for G.
1. Begin with the upper panel.
2. Color the heart rate panel.
3. Color the stroke volume panel.
4. Color the elements of cardiac content measurement below. Note the letter labels.

CARDIAC OUTPUT

$$HR_B \times SV_C = CO_A$$

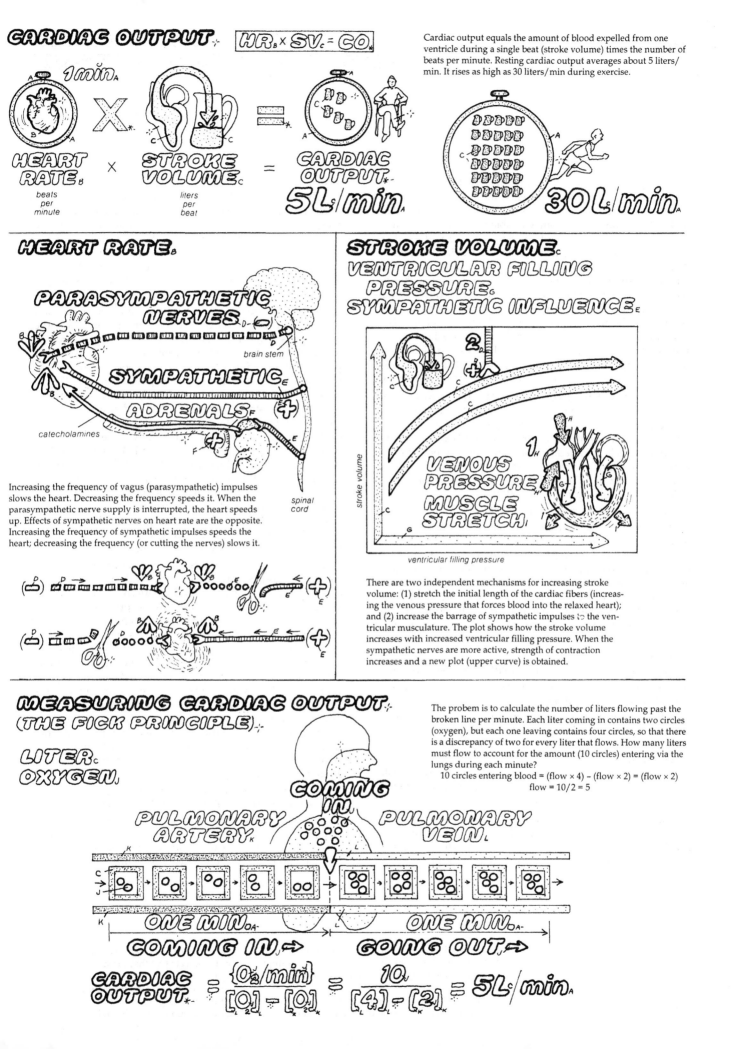

Cardiac output equals the amount of blood expelled from one ventricle during a single beat (stroke volume) times the number of beats per minute. Resting cardiac output averages about 5 liters/min. It rises as high as 30 liters/min during exercise.

1 min

HEART RATE × **STROKE VOLUME** = **CARDIAC OUTPUT 5 L/min**

beats per minute · liters per beat

30 L/min

HEART RATE

PARASYMPATHETIC NERVES

brain stem

SYMPATHETIC

ADRENALS (+)

catecholamines

spinal cord

Increasing the frequency of vagus (parasympathetic) impulses slows the heart. Decreasing the frequency speeds it. When the parasympathetic nerve supply is interrupted, the heart speeds up. Effects of sympathetic nerves on heart rate are the opposite. Increasing the frequency of sympathetic impulses speeds the heart; decreasing the frequency (or cutting the nerves) slows it.

STROKE VOLUME

VENTRICULAR FILLING PRESSURE. SYMPATHETIC INFLUENCE.

stroke volume

VENOUS PRESSURE MUSCLE STRETCH

ventricular filling pressure

There are two independent mechanisms for increasing stroke volume: (1) stretch the initial length of the cardiac fibers (increasing the venous pressure that forces blood into the relaxed heart); and (2) increase the barrage of sympathetic impulses to the ventricular musculature. The plot shows how the stroke volume increases with increased ventricular filling pressure. When the sympathetic nerves are more active, strength of contraction increases and a new plot (upper curve) is obtained.

MEASURING CARDIAC OUTPUT
(THE FICK PRINCIPLE)

LITER. OXYGEN.

The problem is to calculate the number of liters flowing past the broken line per minute. Each liter coming in contains two circles (oxygen), but each one leaving contains four circles, so that there is a discrepancy of two for every liter that flows. How many liters must flow to account for the amount (10 circles) entering via the lungs during each minute?

10 circles entering blood = (flow × 4) – (flow × 2) = (flow × 2)

flow = 10/2 = 5

COMING IN

PULMONARY ARTERY. **PULMONARY VEIN.**

ONE MIN. **ONE MIN.**

COMING IN → **GOING OUT →**

$$\text{CARDIAC OUTPUT} = \frac{\{O_2/min\}}{[O]_2 - [O]_2} = \frac{10}{[4]-[2]} = 5 \, L/min$$

A tissue gets its nutrients only from its blood supply. During activity, it consumes more nutrients, and it will be able to sustain the increased activity only if it receives more blood. Plate 43 illustrated how each tissue was able to regulate its own perfusion (i.e., the amount of blood flowing through it) to satisfy its metabolic requirements. When metabolism increases, its products, potent dilators in the microcirculation, accumulate. The result is an opening of local capillary beds so that the tissue receives more blood. The opposite occurs during quiescence. This scheme will work only if there is reasonably high pressure in the arteries; opening or widening blood vessels hardly helps if there is no pressure head to propel the blood. Further, the blood supply to a particular tissue can increase only by compromising the blood supply to other tissues or by increasing the cardiac output or both. There are no alternatives.

MEAN ARTERIAL PRESSURE = CARDIAC OUTPUT × RESISTANCE
How does the heart "know" to speed up or to increase its stroke volume? How does smooth muscle in the arterioles of quiescent tissue "know" to contract and constrict vessels so that blood can be shunted to more active areas where it is needed? Somehow the nervous system must be involved, but how does it "know"? The missing piece to this puzzle is provided by the arterial pressure. Recall that

mean pressure difference = flow × resistance.

Apply this to the entire circulation (top illustration in plate). The flow is simply the cardiac output. If we neglect the small pressure in the blood just prior to its entering the ventricles, we can equate the pressure difference (arteries – right atrium) with pressure in the arteries. In this case the resistance will refer to the total resistance of the entire systemic circulatory tree (from the beginning of the aorta to the right atrium). Finally, recall that the primary bottleneck or resistance in the vascular tree is in the arterioles, so that the approximate formula becomes

mean arterial pressure = cardiac output × arteriolar resistance.

Although only an approximation, this expression is fundamental to our interpretations.

When a tissue becomes active, metabolic products accumulate and dilate the local microcirculation, which reduces the resistance. Looking at our formula, we should expect the reduced resistance to produce a decrease in arterial pressure. This does not happen (or at least the drop in pressure is small) because the body has a number of mechanisms to maintain a relatively constant blood pressure.

BARORECEPTORS MAINTAIN ARTERIAL PRESSURE
The primary control over sudden changes in blood pressure involves reflexes that originate in special areas (called baroreceptors) in the walls of the aortic arch (blood supply to

systemic circulation) and the internal carotid arteries (blood supply to the brain). Receptors in these areas are sensitive to stretch. At normal pressure, the walls are stretched and the receptors are active, sending impulses via sensory nerves to centers in the brain that are responsible for coordinating information and regulating the cardiovascular system. These cardiovascular centers, located in the medulla, control the autonomic nerve supply to the heart and blood vessels.

When arterial pressure drops, arterial walls are subjected to less stretch, and the sensory nerves coming from the carotid sinus (sinus nerve) and from the aortic arch (depressor nerve) become less active and send fewer impulses. Upon receiving fewer impulses from the baroreceptors (signaling the fall in pressure), the cardiovascular centers respond by exciting sympathetic and inhibiting parasympathetic nerves. This results in (1) an increased heart rate, (2) an increased strength of contraction (stroke volume) so that cardiac output increases, (3) a general increased constriction of arterioles (but not in the brain or heart or in impoverished areas where effects of local vasodilators prevail), and (4) an increased constriction of veins. All these factors contribute to a compensatory raising of the blood pressure back toward normal. Factors 1 and 2 act to raise the cardiac output (flow), factor 3 raises the resistance, and factor 4 raises venous return to the heart as it redistributes blood, shifting it from the venous reservoir to the arterial side of the circulation. When the pressure rises, just the reverse occurs (see plate).

The cardiovascular centers receive detailed information about the high-pressure (arterial) side of the circulation. Careful study of the nerve impulse patterns on the sinus and depressor nerves shows that the baroreceptors respond not only to the actual pressure in the carotid sinus and aortic arch, but also to the rate of change of that pressure. It appears that the signals (patterns of nerve impulses) sent to the cardiovascular centers contain information about the mean pressure, the steepness of rise of the pulse curve, and the heart rate. In addition, cardiovascular centers receive information regarding the low-pressure side of the circulation. Baroreceptors similar to those of the arteries are also found in the atria and pulmonary arteries, but their significance in rapid regulation of blood pressure is not clear. (They appear to be more involved in slower, long-term regulation of blood volume and pressure—see plate 47).

The cardiovascular centers are also influenced by higher brain centers. The hypothalamus sends impulses that are connected with vascular responses to temperature regulation, defense, and rage. Examples of influence from the cerebral cortex appear in the vascular responses of fainting at the sight of blood and blushing from embarrassment.

CN: Use red for A and dark colors for H and I.
1. Begin with the upper panel, coloring each example completely before going on to the next line of illustrations.
2. Color the large illustration beginning with (1) in the aortic arch above the heart. Note that the

sympathetic nerves (I) are represented by dotted lines to indicate that they are not active. Although the parasympathetic nerves (H) are active, they are represented by a broken line to emphasize that they exert an inhibiting effect on the heart.
3. Color the summary diagram below.

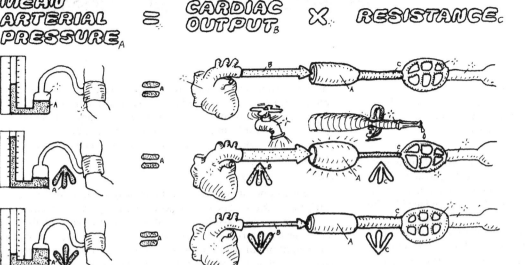

MEAN ARTERIAL PRESSURE $_A$ = CARDIAC OUTPUT $_B$ × RESISTANCE $_C$

In the circulation, total flow is equated to cardiac output, the driving force for this flow is equated to arterial pressure, and resistance is attributed to the arterioles. The arterial pressure is given by the product of cardiac output × arteriolar resistance. It is easy to see how an increase of either cardiac output or arteriolar resistance will cause an increased arterial pressure. Imagine pinching a garden hose. Turning more water on causes the water to pile up behind the pinch and raises the force or pressure. Pinching harder causes a similar pileup. Decreasing either cardiac output or resistance decreases pressure.

HIGHER BRAIN CENTERS $_D$ HYPOTHALAMIC THERMO-REGULATION CENTER $_E$

When arterial pressure rises (1), arterial walls are stretched and the sensory nerves coming from the carotid sinus (sinus nerve) and from the aortic arch (depressor nerve) become more active and send more impulses (2). Upon receiving more impulses from the baroreceptors (signaling the rise in pressure), the cardiovascular centers (3) respond by exciting parasympathetic (4) and inhibiting sympathetic (5) nerves. This results in (6) a decreased cardiac output (via decreased heart rate and stroke volume), (7) a general decreased constriction of arterioles, and (8) a decreased constriction of veins. All of these factors contribute to a compensatory lowering of the blood pressure back toward normal.

(diagram labels:) cerebral cortex; external carotid artery; internal carotid artery; left common carotid; right common carotid; aortic arch; venodilation; arteriolar dilation

1 BLOOD PRESSURE RISE $_A$
2 BARORECEPTORS: $_F$
AORTIC ARCH $_{F'}$
DEPRESSOR NERVE $_{F^2}$
CAROTID SINUS $_{F^3}$
SINUS NERVE $_{F^4}$
3 MEDULLA CARDIO-VASCULAR CENTER $_G$
4 PARASYMPATHETIC NERVE (VAGUS) $_H$
5 SYMPATHETIC NERVE $_I$
6 CARDIAC OUTPUT $_B$
7 PERIPHERAL RESISTANCE $_C$
8 VENOUS RESERVOIR $_J$
9 BLOOD PRESSURE DROP $_A$

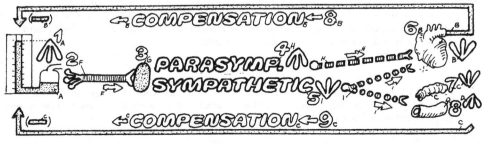

COMPENSATION ← 8 $_B$

PARASYMP. SYMPATHETIC

COMPENSATION ← 9 $_C$

This summary of the baroreceptor reflex (described above) emphasizes its negative feedback character. Begin on the left and follow the two loops that feed back to compensate for the disturbed (in this case increased) pressure.

The response to a sudden loss of blood and the response to sudden redistribution of blood that occurs upon suddenly rising from a recumbent position provide two good examples of fast regulation in the cardiovascular system.

HEMORRHAGE

Significant loss of blood tends to lower blood pressure and elicits baroreceptor reflexes — The "leak" causing loss of blood could occur in the arteries or in the veins. When blood is withdrawn from the arteries faster than it is replaced by the heart, the mean arterial pressure falls. When blood is suddenly withdrawn from the veins faster than it is replaced by capillary flow, the mean venous pressure falls. A drop in venous pressure results in decreased venous return, and this decreases cardiac output, which in turn decreases mean arterial pressure. In both types of leak, the mean arterial pressure falls unless some regulatory compensation occurs. This fall in arterial pressure is rapidly offset by the mechanisms described in plate 45. The following familiar sequence is set into motion: *decreased arterial pressure → reduced stretch of baroreceptors in aortic arch and carotid sinus → reduced frequency of nerve impulses traveling on sensory nerves to the cardiovascular centers in the medulla → inhibition of parasympathetic and activation of sympathetic nerves.*

Parasympathetic nerve impulses slow the heart; inhibiting them will speed it up. Activating the sympathetic nerves also speeds the heart and makes it beat more forcefully. In addition, the sympathetic nerves cause intense constriction of the arterioles (raising resistance) and veins (reducing the volume of the vascular tree). All these responses occur within moments after blood is lost, and all tend to raise the arterial pressure back toward normal.

Fluid shifts from tissue spaces to blood — Constricting veins reduces the total vascular volume so that less blood is required to fill the system. Further, venous constriction shifts blood into the arteries and raises arterial pressure. Constriction of the arterioles increases peripheral resistance to raise blood pressure. It also diminishes flow into the capillary beds so that blood pressure in the capillaries falls. Fluid balance across the capillary walls (plate 40) is upset, and fluid filters from tissues into capillaries. After several minutes, the total amount of fluid transferred becomes significant; it helps replace blood lost during the hemorrhage. Of course, this tissue fluid is not blood; it doesn't contain plasma proteins or blood cells, and it dilutes the plasma proteins and blood cells.

Huge blood loss, prolonged intense vasoconstriction: The system fails — Constriction of the blood vessels is most intense in organs like the skin, kidneys, and liver, but hardly occurs at all in the brain, heart, or lungs. Nourishment is maintained in organs whose moment-to-moment performance is essential. Should the intense vasoconstriction persist or more blood be lost, dire consequences called *circulatory shock* may result. When the oxygen supply to any organ is inadequate, metabolic acids, which accumulate and impair organ function, are produced. Tissue damage can occur, vasodilator substances are released, and capillary walls may become leaky, allowing protein to seep into tissue spaces. Vasodilator substances expand the vascular tree, pooling blood in tissues and veins and thus reducing venous return,

cardiac output, and arterial pressure. Loss of plasma protein into tissue spaces again upsets fluid balance across the capillary walls, but now in the direction from capillary to tissues. Thus, fluid is *lost* from the vascular tree, and the blood becomes more viscous, and eventually may even stop as a result of intravascular coagulation.

CHANGING POSTURE

Venous pooling of blood and edema tend to occur in the upright position — The simple act of changing from a recumbent to an upright position presents some of the same challenges as a hemorrhage! When we are in a standing position, the weight of our blood becomes important; blood in a capillary in a foot, for example, may have to support the weight of the column of fluid contained within the veins reaching all the way from the foot to the heart. The pressure on a fluid particle within that capillary will rise to the same level it would experience at the bottom of a water tank filled to the same height (several feet). It is important to realize that this does not directly influence flow within the closed circulatory system. The increase in pressure on the particle, tending to push it upward where the pressure is less, is just counterbalanced by the weight of the column of fluid. Thus, the forces operative in propelling the blood in a recumbent subject are not disturbed. However, the increased pressures due to gravity are significant because they redistribute fluids in two ways. (1) Veins are more extensible than arteries. As shown in the figure, the increased pressure expands the venous system, and blood pools in the systemic veins. (As much as 600 mL may pool in the lower extremities upon quiet standing.) (2) The high hydrostatic pressures in the capillaries force fluid out of the capillaries into the tissue spaces.

Because of venous pooling, the sudden change in position from recumbent to upright resembles hemorrhage; the subject bleeds into his own vascular system. The same compensatory responses (activation of sympathetics, inhibition of the parasympathetics) occurs. In this situation, however, in contrast to hemorrhage, filtration of fluid from capillaries to tissues occurs. Venous pooling and edema can be counteracted by moving about (plate 42). Contracting muscles compress veins and lymph vessels to help empty them and temporarily relieve local venous pressures. Valves close, preventing backflow and supporting the weight of blood above them until the vein refills with blood from the capillaries. This provides temporary relief from the high hydrostatic capillary pressure and begins to alleviate the edema.

Orthostatic (postural) hypotension — Some people find it difficult to stand up after lying down for a long time. They experience dizziness, impaired vision, and buzzing in the ears, all signs of the inadequate cerebral circulation that arises from the drop in blood pressure following a sudden change to the upright position. In more severe cases, fainting may occur (a fortunate response in this case, because it restores the recumbent position and relieves the stress). Similar reactions may occur even in healthy persons, especially when blood vessels in the skin or muscles are dilated because of heat or exercise. In those cases, the regulatory responses may fail because the intense demands of heat regulation and metabolism have priority.

CN: Use red for A and dark colors for G and H.
1. Begin with the upper right diagram and the systemic response to hemorrhage, directly beneath it.

Then color the diagrams to the left.
2. Color the four examples below, completing them left to right.

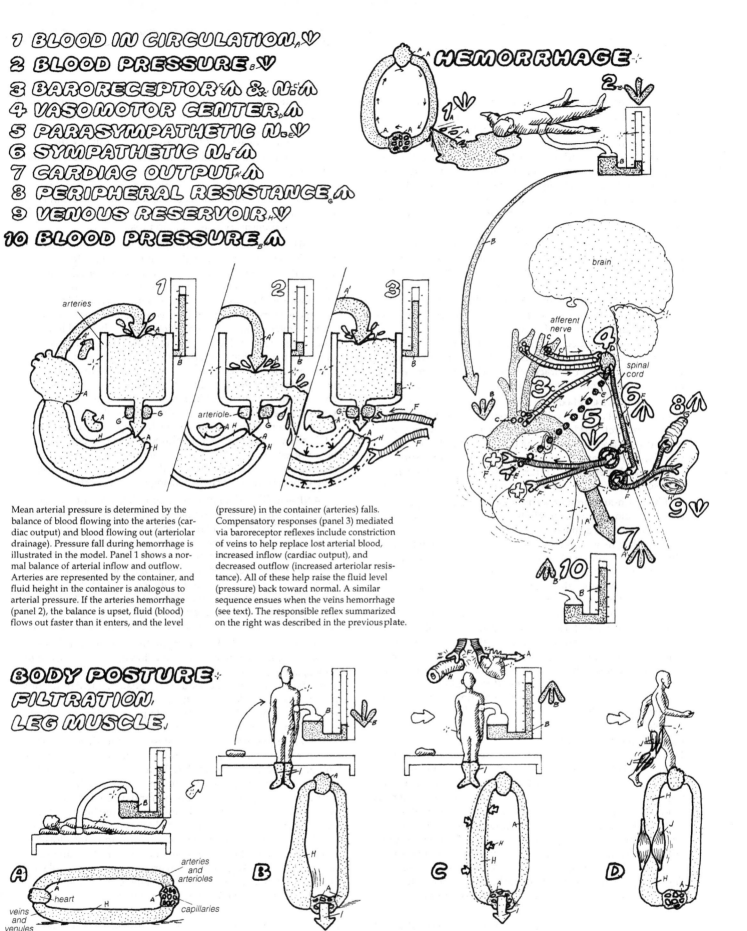

1 BLOOD IN CIRCULATION $_A$ ⩔
2 BLOOD PRESSURE $_B$ ⩔
3 BARORECEPTOR ⩓ & N. $_C$ ⩓
4 VASOMOTOR CENTER $_D$ ⩓
5 PARASYMPATHETIC N. $_E$ ⩔
6 SYMPATHETIC N. $_F$ ⩓
7 CARDIAC OUTPUT $_A$ ⩓
8 PERIPHERAL RESISTANCE $_G$ ⩓
9 VENOUS RESERVOIR $_H$ ⩔
10 BLOOD PRESSURE $_B$ ⩓

HEMORRHAGE

arteries

brain

afferent nerve

spinal cord

arteriole

Mean arterial pressure is determined by the balance of blood flowing into the arteries (cardiac output) and blood flowing out (arteriolar drainage). Pressure fall during hemorrhage is illustrated in the model. Panel 1 shows a normal balance of arterial inflow and outflow. Arteries are represented by the container, and fluid height in the container is analogous to arterial pressure. If the arteries hemorrhage (panel 2), the balance is upset, fluid (blood) flows out faster than it enters, and the level

(pressure) in the container (arteries) falls. Compensatory responses (panel 3) mediated via baroreceptor reflexes include constriction of veins to help replace lost arterial blood, increased inflow (cardiac output), and decreased outflow (increased arteriolar resistance). All of these help raise the fluid level (pressure) back toward normal. A similar sequence ensues when the veins hemorrhage (see text). The responsible reflex summarized on the right was described in the previous plate.

BODY POSTURE
FILTRATION
LEG MUSCLE

A
heart
arteries and arterioles
capillaries
veins and venules

B

C

D

A. In a recumbent position the circulation of the blood is hardly influenced by gravitational forces. B. In a standing position the weight of the blood becomes important. It increases pressures below the level of the heart, particularly in the lower parts of the body. The increased pressures redistribute fluid in two ways: (1) they expand the venous system and blood pools in systemic veins; (2) they force fluids out of capillaries into tissue spaces. Because of venous pooling, the sudden change in position from recumbent to upright resembles

hemorrhage—the subject bleeds into his own vascular system.
C. The same compensatory response (activation of sympathetics, inhibition of the parasympathetics) occurs. In this situation, however, in contrast to hemorrhage, we now have filtration of fluid from capillaries caused by the high hydrostatic pressure. The extremities become edematous. D. Venous pooling and edema are counteracted by moving about. Contracting muscles compress veins and lymph vessels to help empty them and temporarily relieve local venous pressures.

Regulation of blood pressure by the pressure receptors described in plate 45 is very effective and rapid. It does not persist, however, and if the alteration in blood pressure persists for hours the pressure receptor mechanism "adapts" to the new conditions and becomes less responsive. Fortunately, there are a number of mechanisms, classified as rapid, intermediate, and long-term regulators, that help stabilize blood pressure. Rapid regulators work within seconds; *arterial pressure receptor reflexes* are the most important example of these.

INTERMEDIATE-TERM REGULATION

These regulators begin in a matter of minutes following a sudden change in pressure, but they may not be fully effective until hours later. We include three mechanisms under this classification: (1) transcapillary volume shifts, (2) vascular stress relaxation, and (3) the renin-angiotensin mechanism.

Transcapillary volume shifts (plate 40) occur when capillary blood pressure rises. If capillary pressure is high, fluid leaves the vascular tree, which tends to lower the blood pressure. When capillary pressure is low, the reverse happens. With this mechanism, extracellular fluid in tissue spaces forms a reserve pool of fluid available to the vascular system.

Vascular stress relaxation refers to a peculiar property of blood vessels that is well developed in veins. When these vessels are stretched by increased pressure, they very slowly expand with pressure becoming correspondingly less. Conversely, as intravascular volume decreases, the opposite occurs. The net effect is to return pressures toward normal after some 10-60 min. following a change in vascular volume.

The renin-angiotensin system (plate 70) is activated whenever blood flow through the kidneys is reduced, as would occur with a sharp drop in arterial blood pressure. The response begins with secretion of the hormone *renin* by the kidney. Renin splits a plasma protein called *angiotensinogen* (produced in the liver), producing a small peptide called *angiotensin I*. A converting enzyme present in plasma changes angiotensin I into a smaller peptide called *angiotensin II*, which gives rise to an even smaller peptide, *angiotensin III*. Angiotensins II and III cause intense constriction of arterioles, raising vascular resistance. To a lesser extent, they also constrict veins, reducing vascular volume. Both increased vascular resistance and reduced volume raise arterial pressure. The renin-angiotensin system becomes effective after about 20 min., and its effects can persist for a long time. Further, angiotensins II and III also stimulate *thirst* as well as the secretion of *aldosterone* (see below) by the *adrenal cortex*.

LONG-TERM REGULATION

This occurs in the *kidney*, which regulates the *volume of body fluids*. This volume represents the balance between fluid intake and excretion. When arterial pressure rises, the kidney responds by excreting more urine, reducing the volume of body fluids (including both plasma and interstitial volume). Diminished plasma volume decreases venous return to the heart, reducing the cardiac output so that elevated blood pressure is brought back toward normal. A decrease in blood pressure elicits the opposite response; urine excretion is decreased. These long-term responses to disturbances are mediated by two hormones—aldosterone and ADH (vasopressin).

Aldosterone (plates 69, 70) is secreted by the adrenal cortex in response to angiotensins II and III. It acts on kidney tubules to retain sodium that would have been excreted in the urine. In these sections of the kidney, water follows the sodium, maintaining osmotic equilibrium. The result is water retention—i.e., an increase in body (and blood) fluid volume. A drop in arterial pressure → renin secretion → angiotensins II and III production → aldosterone secretion → sodium retention by the kidney → water retention → increased blood volume → compensatory rise in blood pressure.

ADH (anti-diuretic hormone) is produced in the *hypothalamus* (plates 66, 69). It travels through nerve fibers to storage sites in the pituitary gland, where it can be released into the circulation. This hormone acts on the kidney (independently of aldosterone) to promote water retention. When blood volume is markedly higher, the resulting increased venous return stretches the atria. *Stretch receptors* embedded in the *atrial walls* are stimulated, sending to the hypothalamus impulses that inhibit the formation and secretion of ADH. With less ADH present, there is more urine excreted (less water retention). Body fluid volume decreases, and this helps compensate for the initial increase in blood volume. Conversely, a decrease in stretch of the atrial walls will withdraw any inhibitory effects of these stretch receptors and promote ADH release. Because blood volume is closely related to blood pressure, regulating blood volume often regulates blood pressure. ADH is also called "vasopressin" because, in high concentrations, it causes strong vasoconstriction.

ANP—ANP (atrial natriuretic peptide), a recently discovered hormone, involved in volume regulation, is secreted by the atria of the heart. When extracellular fluid volume is expanded, the plasma concentration of this hormone increases and causes an increase in Na^+ excretion by the kidney. It also inhibits the secretion of renin and ADH, and it desensitizes the adrenal cortex to stimuli that increase aldosterone secretion. All of these promote water excretion, helping to compensate for the original disturbance (increased extracellular fluid). The magnitude of these effects remains to be clarified.

Hypertension: a failure to regulate at normal levels.—Hypertension refers to chronic elevation of blood pressure in which, at rest, either systolic pressure > 140 or diastolic pressure > 90 mm Hg. Blood pressure is regulated, but at unhealthy high levels. The deleterious effects of hypertension are related to the extra load that the heart has to push against in order to eject blood, and to the damage to and within blood vessels, which are subjected to the high pressures and shear stresses. Compensating for the extra work load, the heart walls stretch and grow thicker (hypertrophy). With this extra burden on coronary circulation a heart attack may be precipitated when it cannot keep up with the demand for oxygen. Damage within blood vessels can be dramatic when the high pulsating pressures rupture atherosclerotic plaques (commonly found in arteries) exposing their contents and precipitating clots and blockage of blood flow. Blockage of the coronaries can cause heart attacks; blockage of cerebral arteries can cause strokes. It is estimated that 20% of the people in the U.S. have hypertension. In most cases (90%), the cause is unknown.

CN: Use red for A, purple for E, and blue for G.
1. Begin with the short-term regulation column and work your way down. For consistency, all the examples on this page (with the exception of the final one) show the response to a decrease in blood pressure. The bottom right illustration (natriuretic hormone) shows the response to an increase in pressure.
2. Color the intermediate regulators. At the bottom of this column the adrenal cortex is shown secreting aldosterone (K). Color that arrow as it leads to an action in the third column. Color the title, aldosterone (K), but then go to the top of the column and work downwards.
3. In long-term regulators (response 1), atrial receptors respond to a drop in blood pressure by shutting off afferent nerves that inhibit the neurosecretory cells of the hypothalamus (i.e., the cells are released from inhibition).

SHORT-TERM REGULATORS

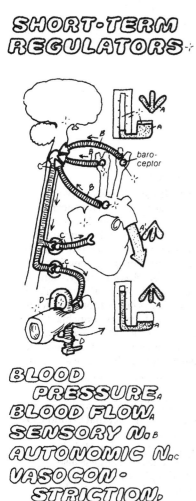

BLOOD PRESSURE, BLOOD FLOW, SENSORY N., AUTONOMIC N., VASOCONSTRICTION.

Short-term changes in blood pressure are buffered by arterial pressoreceptor (baroreceptor) reflexes. These work within a few seconds but they do not persist and are ineffective in regulating chronic changes.

Intermediate regulators begin in minutes but are not fully effective until hours later. These include **(1)** transcapillary volume shifts in response to changes in capillary blood pressure; **(2)** vascular stress relaxation, the slow constriction (expansion) of blood vessels in response to decreased (increased) stretch; and **(3)** secretion of renin in response to decreased blood flow to the kidney. Renin produces angiotensins from plasma proteins; they cause intense vasoconstriction, raising vascular resistance and blood pressure.

Long-term regulation is accomplished by the kidney, which regulates body fluid volume by tipping the balance between fluid intake and excretion. When arterial pressure rises, the kidney excretes more urine, reducing body fluid volume, decreasing venous return and cardiac output so that the elevated blood pressure is reduced. A decrease in blood pressure elicits the opposite response. These responses are mediated by **(1)** ADH (secreted by the posterior pituitary in response to a decrease in nerve impulses arising from stretch receptors in arterial walls), which reduces water excretion by the kidney, and **(2)** aldosterone (secreted by the adrenal cortex in response to angiotensins), which promotes Na^+ and water retention by the kidney. **(3)** In addition, natriuretic hormone, a newly discovered factor secreted by the atrium in response to increased volume, also plays a role. This hormone promotes Na^+ (and water) excretion and inhibits renin, ADH, and aldosterone secretion. If pressure and volume decrease, the hormone secretion slows and its water-excreting actions diminish.

INTERMEDIATE REGULATORS

1 TRANS-CAPILLARY SHIFT

CAPILLARY, BODY FLUID

2 VASCULAR STRESS RELAXATION

VENOUS SYSTEM

3 RENIN: ANGIOTENSIN SYSTEM

ANGIOTENSINOGEN

ANGIOTENSIN I

ANGIOTENSIN II, III

LONG-TERM REGULATORS

EFFECT OF BODY FLUID VOLUME

THIRST

HYPOTHALAMUS

ATRIAL RECEPTOR

ADH

ALDOSTERONE

Na^+ H_2O

ATRIAL ENDOCRINE CELLS

ANP (NATRIURETIC HORMONE)

RENIN, ADH, ALDOSTERONE

excretion

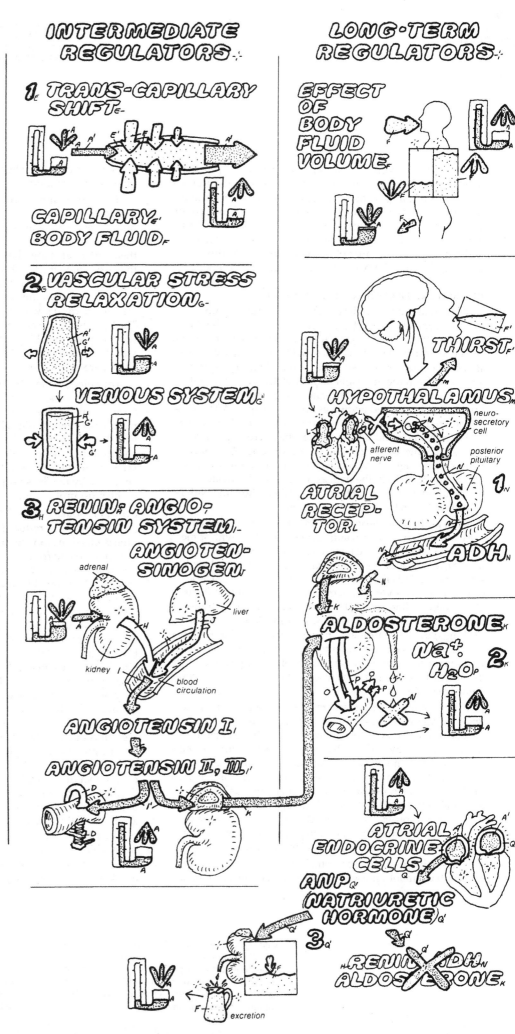

We live at the bottom of a vast sea of air composed primarily of oxygen and nitrogen. By living in air rather than water, we enjoy surroundings 50 times richer in oxygen. By breathing, we give our body fluids access to this reservoir as they continually exchange both oxygen (O_2) and carbon dioxide (CO_2) with air. There are no long-term storage sites for oxygen within the body; they are not necessary, as long as this exchange between body fluids and air remains unimpeded.

RESPIRATORY TRACT BRANCHES MANY TIMES

Alveoli are the final destination of a tortuous path — Efficient contact of body fluids with air is mediated by the *respiratory tract*, which begins in the *nasal* and *oral cavities* and ends in a vast system of microscopic blind-ended sacs called *alveoli* in the deepest recesses of the lungs. During inspiration, air travels from the atmosphere through the nasal (or oral) passages, through the *pharynx*, and into the *trachea*. During this time, it is warmed and takes up water vapor. After passing down the trachea, it flows through the *bronchi, bronchioles,* and *alveolar ducts* and finally reaches the microscopic alveoli, where exchange of oxygen and carbon dioxide takes place. Following a single O_2 molecule along this tortuous route, we find about 23 forks in the path as the airways bifurcate into finer and finer branches. During expiration, the same path is traversed, but in the opposite direction.

Cilia and mucus serve as filters — The widest tubes (trachea and bronchi) contain stiff cartilage together with some smooth muscle. They are lined by a layer of epithelial cells (not shown on opposite page) that often have minute hair-like structures, called cilia, projecting from their surface. These cells also secrete over their surface a *mucus* sheet that floats above a thin saline layer. The mucus is continuously transported like an escalator in an upward direction (away from the lungs) by the coordinated movement of the cilia. This process serves as an efficient filter for dust particles that strike the walls as turbulent air flows in and out of the air passage. Once the upward-traveling mucus reaches the pharynx we unconsciously swallow it.

Failure of this filtering system occurs when cilia are paralyzed by cigarette smoke. The system also fails in the fatal genetic disease *cystic fibrosis*, in which there is a congenital loss of cAMP-activated Cl⁻ channels in airway epithelial cells. Without this channel, secretion of the thin saline layer between cilia and mucus is grossly impaired. Mucus becomes thicker and cilia, now trapped in the sticky mucus, are no longer able to function. This leads to repeated pulmonary infections and progressive destruction of the lungs. Cystic fibrosis is one of the most common genetic disorders in Caucasians, occurring in about 1 out of 2000 births.

The smaller branches (bronchioles) of the respiratory tract also contain smooth muscle, but no cartilage, cilia, or mucus

glands. Particles deposited in the bronchioles and alveoli are removed by wandering alveolar macrophages.

Total alveolar surface area is huge — The extensive branching pattern of the air passages results in an enormous number of alveoli — approximately 300 million. The diameter of each alveolar sphere is only about 0.3 mm, but adding all their surface area together gives a total of about 85 sq. m (close to the size of half a tennis court!) available for gas exchange with blood. Yet this enormous surface is contained within a maximum total volume of only 5 to 6 L, which fits very nicely into the thorax. However, this device is not without problems. The tiny alveoli are at the dead end of narrow bronchial tubes in a complex branching tubular network. Left to itself, air would stagnate within them. This does not occur because the alveoli are intermittently flushed with fresh air as we breathe.

The enlarged view of a single alveolus in the plate shows the actual interface between body fluids and air where *gas exchange* takes place. Alveolar walls, like blood capillaries, are made of extremely thin cells. Despite the fact that O_2 and CO_2 have to traverse two cell layers in passing between alveolus and capillary, the total distance is very short, and diffusion is correspondingly rapid. Efficient gas exchange is also enhanced by the dense supply of capillaries in the lungs, one of the most profuse networks of blood vessels in the body.

LOW PRESSURES IN THE PULMONARY CIRCULATION

Low pressures reduce the work of the right heart and provide protection from edema — The pulmonary circulation that transports blood from the right heart to this alveolar exchange interface also has peculiarities that appear well adapted to its function. Most notably, the pressures in the pulmonary circulation are small; the mean pressure in the pulmonary artery is about 15 mm Hg, only about one-seventh the 100 mm Hg mean pressure in the aorta. Thus, the forces driving blood through the pulmonary circulation are relatively small, and because the blood flows through the pulmonary and systemic circulations are equal, it follows that the resistance of the pulmonary circulation must also be small. Keeping the pulmonary pressures and resistance low so that flow can be maintained reduces the work required of the right heart. In addition, the low pressure in the pulmonary capillary pushing fluids out into the alveolar spaces is overbalanced by the oncotic pressure of the plasma proteins (plate 40) drawing fluids in. The net force favors reabsorption of fluid from the alveolus so that the normal lung has no tendency to fill with fluid. Further, blood vessels in the lungs have an atypical response to low concentrations of O_2 dissolved in blood plasma. Unlike arterioles of the systemic circulation, which dilate, lung arterioles constrict in response to low local plasma O_2 concentrations. This has the advantage of shunting blood away from areas of the lung that are poorly ventilated and cannot serve as adequate sources of O_2.

CN: Use red for A, blue for B, purple for N, and a very light color for L.
1. Begin with the large illustration. The edge of the right lung is colored gray, and the left lung is colored completely gray. Only the bronchi (I) are colored in the right lung.

2. Color the enlargement on the lower left showing circulation to and from the alveoli.
3. Color the enlargement (lower right) of the gas exchange between an alveolus and a lung capillary.
4. Color the schematic diagram (to the left of the large figure) of external and internal respiration.

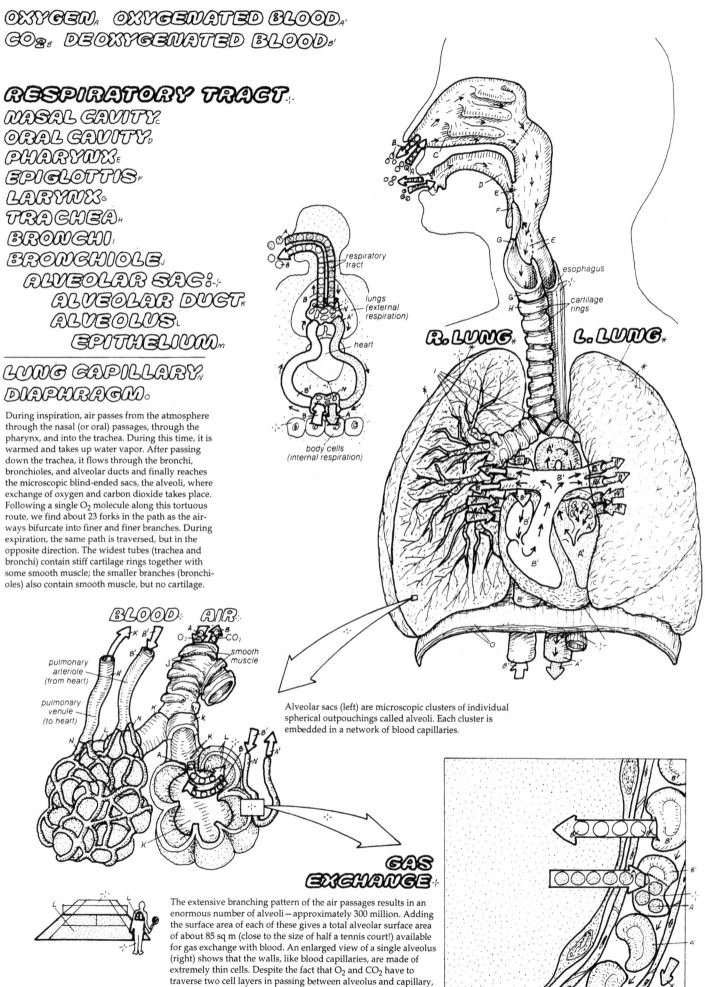

OXYGEN. OXYGENATED BLOOD.'
CO₂. DEOXYGENATED BLOOD.'

RESPIRATORY TRACT.
NASAL CAVITY.
ORAL CAVITY.
PHARYNX.
EPIGLOTTIS.
LARYNX.
TRACHEA.
BRONCHI.
BRONCHIOLE.
ALVEOLAR SAC:
ALVEOLAR DUCT.
ALVEOLUS.
EPITHELIUM.

LUNG CAPILLARY.
DIAPHRAGM.

During inspiration, air passes from the atmosphere through the nasal (or oral) passages, through the pharynx, and into the trachea. During this time, it is warmed and takes up water vapor. After passing down the trachea, it flows through the bronchi, bronchioles, and alveolar ducts and finally reaches the microscopic blind-ended sacs, the alveoli, where exchange of oxygen and carbon dioxide takes place. Following a single O_2 molecule along this tortuous route, we find about 23 forks in the path as the airways bifurcate into finer and finer branches. During expiration, the same path is traversed, but in the opposite direction. The widest tubes (trachea and bronchi) contain stiff cartilage rings together with some smooth muscle; the smaller branches (bronchioles) also contain smooth muscle, but no cartilage.

respiratory tract

lungs (external respiration)

heart

body cells (internal respiration)

esophagus

cartilage rings

R. LUNG

L. LUNG

BLOOD   AIR

O_2 CO_2

smooth muscle

pulmonary arteriole (from heart)

pulmonary venule (to heart)

Alveolar sacs (left) are microscopic clusters of individual spherical outpouchings called alveoli. Each cluster is embedded in a network of blood capillaries.

GAS EXCHANGE

The extensive branching pattern of the air passages results in an enormous number of alveoli — approximately 300 million. Adding the surface area of each of these gives a total alveolar surface area of about 85 sq m (close to the size of half a tennis court!) available for gas exchange with blood. An enlarged view of a single alveolus (right) shows that the walls, like blood capillaries, are made of extremely thin cells. Despite the fact that O_2 and CO_2 have to traverse two cell layers in passing between alveolus and capillary, the total distance is very short; diffusion is rapid.

Efficient gas exchange will occur only if the alveoli are regularly flushed with fresh air; this happens with each breathing cycle as air is pumped into and out of the lungs. The lungs themselves are passive structures. They are in contact with the chest wall through a thin layer of fluid, the pleural fluid, which allows them to glide easily on the chest wall. The lungs resist being pulled off the wall in the same way that two pieces of wet plate glass slide on each other but are not easily separated. The pressure within this pleural fluid, the intrapleural pressure, is subatmospheric (between 3 and 6 mm Hg less than atmospheric) at the end of respiration, when the system is at rest.

DIAPHRAGM AND CHEST BREATHING

During inspiration, the thoracic cage enlarges, and the lungs expand so that air is drawn in via the air passages. This enlargement is produced primarily by contraction of the dome-shaped *diaphragm*, which flattens and increases the vertical length of the cage. In normal breathing, the diaphragm moves about 1 cm, but during forced breathing, this excursion may reach 10 cm. Contraction of *external intercostal muscles* also contributes to the expansion of the chest by pulling the sagging rib cage upward into a more horizontal position, increasing its width.

During quiet breathing, expiration is passive. The inspiratory muscles relax; the diaphragm assumes its resting curved shape and pushes upward on the thoracic cage while the relaxing external intercostals allow the rib cage to sag downward under its own weight. The lungs and chest wall are elastic, so they return to their former position, driving air out of the lungs. During forced expiration, new sets of muscle become active. Muscles of the abdominal wall contract, pushing the diaphragm upward while the contraction of the *internal intercostal muscles*, whose orientation is opposite to that of the external intercostals, pulls the rib cage downward. These actions accelerate the expulsion of air.

PRESSURE CHANGES DURING BREATHING

If we follow the pressure changes in the intrapleural and alveolar spaces during a single quiet breathing cycle, we arrive at the results shown at the bottom of the plate. At rest, the pressure of the thin layer of pleural fluid is about –3 mm Hg relative to atmospheric pressure (i.e., it is 3 mm Hg below atmospheric); pressure in the lungs is atmospheric (0 mm Hg). This negative intrapleural pressure (–3 mm Hg) reflects the elastic recoil properties of the lungs. As the lungs attempt to pull away from the chest wall, there is no air to fill the potential gap, and the slightest move away from the wall creates a negative pressure ("pulls a vacuum"). The "suction," reflected by the –3 mm Hg, pulls the lungs toward the chest wall, and it

just balances the elastic recoil pulling them away. (If air is introduced into this space, raising the intrapleural pressure to atmospheric—e.g., by opening the chest wall—the lungs pull inward and collapse.) Alveolar pressure is atmospheric at this time, reflecting the fact that there is no pressure gradient between the atmosphere and the lungs so that there is no air flowing in or out. Now inspiration begins. The thoracic cage expands, and the falling intrapleural pressure pulls the lungs along with it. As a result, intrapleural pressure falls to –5 mm Hg, and intrapulmonary pressure falls to –1 mm Hg. Air flows down the pressure gradient from the atmosphere (0 mm Hg) to the lungs (–1 mm Hg) until this gradient is finally dissipated at the height of inspiration, when 0.5 L of air has been added. Now expiration begins; the lungs become compressed, raising intrapulmonary pressure and forcing air out until the added 0.5 L is expelled. The system returns to its initial state.

These figures for normal quiet breathing are subject to great variation. For example, at the end of a deep inspiration, intrapleural pressure may be as low as –14 mm Hg, and during a particularly forceful expiration, it may reach as high as +50 mm Hg. Nevertheless, it is remarkable that a pressure gradient of only 1 mm Hg is sufficient to move the required 0.5 L of air in and out of the lungs during normal quiet breathing. It illustrates the lungs' easy distensibility (high compliance). In contrast, a toy balloon may require up to 200 mm Hg for the same increase in volume.

AIRWAY RESISTANCE

The alveolar spaces are in constant contact with the atmosphere via the airways (nose and mouth, trachea, bronchi, and bronchioles). The fact that the pressure in the alveoli is not equal to that in the atmosphere at various times in the respiratory cycle (i.e., at the beginning of inspiration and at the beginning of expiration) reflects the *resistance* to air flow offered by the airways. The major site of resistance lies in the medium-sized bronchi. (Although the smaller bronchioles have narrower tubes, they are much more numerous, and this factor more than compensates for their small size.)

Airway resistance changes during the normal respiratory cycle. During inspiration, both lungs *and* airways expand in response to the decreased intrapleural pressure; the widened airways offer less resistance. During expiration, the reverse occurs, and resistance increases. This explains why persons with constricted airways (e.g., *asthma*) have much more difficulty exhaling than inhaling. The resistance can also be altered by contractions of bronchial smooth muscle, which narrow the passages and increase resistance. These muscles are under the control of autonomic nerves: sympathetic stimulation (norepinephrine) dilates them, and parasympathetic stimulation (acetylcholine) constricts them.

CN: Use dark colors for C and D.
1. Begin with the diagram in the upper left corner.
2. Color the figure labeled "inspiration" on the right, and include the diagram showing intercostal muscles below it. Do the same for "expiration."
3. Color the lower diagrams, beginning on the left and completing each in sequence before moving on to the next.

RESPIRATORY STRUCTURES

STERNUM·F
SPINAL COLUMN·G
EXTERNAL INTERCOSTAL MUSCLES·H
INTERNAL INTERCOSTAL MUSCLES·I

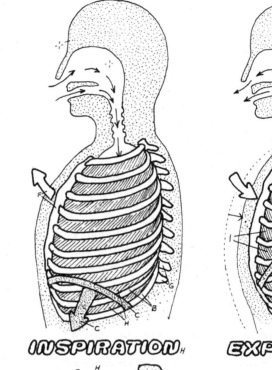

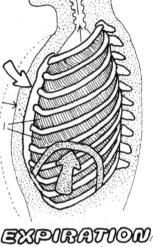

LUNGS·A

RIB CAGE·B
DIAPHRAGM·C
PLEURAL MEMBRANE·D
INTRAPLEURAL FLUID·E

During inspiration, the thoracic cage enlarges; air is drawn into the lungs via the air passages. This enlargement is produced by contraction of the dome-shaped diaphragm, which flattens and increases the vertical length of the cage, and by contraction of external intercostal muscles, which pull the sagging rib cage upward into a more horizontal position, increasing its width. During expiration, these muscles relax: the diaphragm assumes its resting curved shape and pushes upward on the thoracic cage while the relaxing external intercostals allow the rib cage to sag downward under its own weight. Air is driven out of the lungs. During forced expiration, a new set of muscles, the internal intercostals, becomes active. Their orientation is opposite to that of the external intercostals. When they contract, they pull the rib cage downward and accelerate the expulsion of air.

INSPIRATION·H

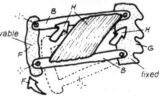

movable / fixed

EXPIRATION·I

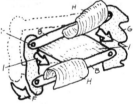

RESPIRATORY PRESSURES·

INTRAPLEURAL PRESSURE·E'
INTRAPULMONARY PRESSURE·A'
NORMAL ATMOSPHERIC PRESSURE·J
RESTING LUNG VOLUME·K (FUNCTIONAL RESIDUAL CAPACITY)

INSPIRATION·J   **EXPIRATION·A'**

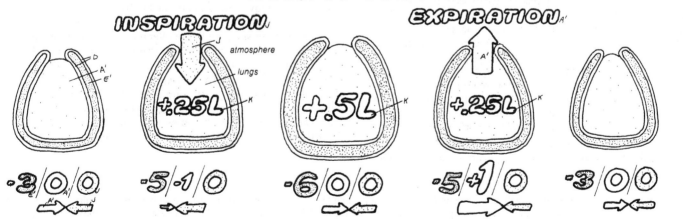

atmosphere / lungs

+.25L   +.5L   +.25L

-3/0/0   -5/-1/0   -6/0/0   -5/+1/0   -3/0/0

The lungs and chest wall are lined by pleural membranes. At rest, the pressure of the thin layer of pleural fluid that lies in between these membranes is about –3 mm Hg relative to atmospheric pressure (i.e., it is 3 mm Hg below atmospheric). Pressure in the lungs is atmospheric (0 mm Hg). This pressure difference makes the lungs adhere to the chest wall, keeping them inflated. During inspiration, the thoracic cage expands, intrapleural pressure falls (to –5 mm Hg), and so does intrapulmonary pressure (to –1 mm Hg). Air flows down the pressure gradient from the atmosphere (0 mm Hg) to the lungs (–1 mm Hg) until this gradient is finally dissipated at the height of expiration when 0.5 L of air has been added. Now expiration begins: the lungs become compressed, raising intra-pulmonary pressure and forcing air out until the added 0.5 L is expelled, and the system returns to its initial state.

SURFACTANT, SURFACE TENSION, AND LUNG COMPLIANCE

Although it is important that the lungs can be distended by small forces, it is equally important that they show elastic behavior and return to their original volume when distending forces are relaxed. Two components are responsible for this elastic behavior. First, *elastic tissue,* consisting of elastic and collagen fibers embedded in alveolar walls and around bronchi, resists stretching. Second, *surface tension,* which arises at any air-water interface, resists expansion of the surface.

SURFACE TENSION IS A FORMIDABLE FORCE IN THE LUNGS

Surfactants account for 2/3 of the lung's elasticity — The importance of these two components is illustrated in the top panel, which shows that it requires much less pressure (force) to inflate the lungs with water (more precisely, with physiological saline) than with air. When inflating with water, there is no air-water interface, therefore no surface tension; the only resisting force comes from the elastic tissue. When inflating with air, both forces are operative. By taking the difference of the two measurements, we can estimate that forces arising from surface tension account for two-thirds of the lung's elastic behavior; the remaining one-third arises from elastic tissue.

Surface tension arises from intermolecular attractions — How does surface tension arise? As shown in the second panel, water molecules attract one another. If they did not, the molecules would fly apart, and water would not be a liquid; it would be a gas. Those water molecules in the bulk of the fluid have neighbors in every direction, and they are pulled in every direction. Molecules on the surface have neighbors only in the interior of the fluid. Accordingly, they are continually pulled off the surface toward the interior. In other words, the water molecules tend to avoid the surface, and as a result, the surface behaves like a thin sheet of rubber that resists expansion. This property is called surface tension; it is a force that acts tangential to the surface and resists expansion of it.

Natural surfactants reduce surface tension — Surface tension can be reduced by introducing solute molecules called surface active agents or *surfactants.* Unlike water, surfactants are attracted to the surface; they displace water molecules there and allow the surface to expand. Phospholipids are common surfactants; they have a polar, hydrophilic head that is attracted to the water and a hydrophobic tail that is squeezed out of the water phase (plate 7). Unless they form micelles or bilayers, the only place that can accommodate both the hydrophobic and hydrophilic properties of the molecule is the air-water interface, with the heads immersed in the water and the tails in the air, allowing for easy expansion of the surface. The surface tension is determined by the relative proportions of water and surfactants that occupy the surface. Surfactants, especially phospholipids, are secreted by some of the cells lining the alveoli. These secretions are important because they reduce surface tension in the alveolar air-water interface, decreasing resistance to stretch and the work of breathing.

For the same tension, the smaller the sphere the larger the required pressure — Further complications arise from the relation between the surface tension and the internal pressure required to keep an alveolus inflated. In a spherical structure like an alveolus or a soap bubble, surface tension acts to collapse the bubble, and the pressure required to keep it inflated depends on both the surface tension and the *size* of the bubble. The smaller the bubble, the larger the pressure — remember how difficult it is to begin blowing up a balloon, but once it attains a reasonable size, the task is much easier. This follows because the curvature of the bubble modifies the surface force so that part of it pulls inward toward the center of the sphere. The smaller the sphere (the greater the curvature), the larger the force pulling inward. This inward component operates to compress the bubble and requires an oppositely directed pressure. If you imagine a small patch on the surface (see plate), you will notice that the larger the bubble, the less curved the patch will be and the less inward pull there will be from surface forces. As the bubble gets very large, the patch becomes practically flat, and there is no inward-directed component. The mathematical relation between sphere size (radius R), tension T, and pressure P is $P = 2T/R$.

Without surfactants, alveoli would collapse — The lungs can be regarded as a collection of 300 million minute bubbles connected to each other. If there were no surfactant, the surface tension in each bubble would be the same, and the system would be unstable because, as shown in the bottom panel (top figure), the smaller bubbles would have a larger pressure and would blow up the larger ones and collapse in the process. When surfactant is present (lower figure), this does not occur, because the smaller bubbles have a higher proportion of surfactant on their surfaces and, therefore, smaller surface tensions than larger ones. As bubbles become smaller, their surface areas decrease, largely by losing surface water molecules (not surfactant) to the interior. Thus, the proportion of surfactant to water in the surface increases so that the decrease in alveolar size is accompanied by a decrease in surface tension. By this mechanism, the surface tension of the smaller alveoli is lowered so that the pressure need not rise to keep it inflated. In our example with no surfactant, the surface tension T is 20 (arbitrary units) in both bubbles. The pressure of each bubble is given by $P = 2T/R$; so the large bubble (R = 2) has P = 20, and the smaller bubble (R = 1) has P = 40. Air will move from the small bubble to the large one. Further, the more air that moves, the smaller the bubble gets and the greater the imbalance. With surfactant, both bubbles have a larger surface tension, but the smaller one has less (T = 5) than the larger (T = 10). Now the two pressures balance at 10 each, and the system is stable.

RESPIRATORY DISTRESS SYNDROME

The importance of lung surfactant is apparent in infants born with deficient secretion of it, giving rise to *respiratory distress syndrome.* In these cases, the lungs are "stiff," areas are collapsed, and breathing requires extraordinary effort.

CN: Use light blue for E and dark colors for A and H.
1. Begin with the upper panel and the chart on the left. Then color the two diagrams of the lungs being filled with water and air. Note the enlargement of a lung alveolus (G) in which a band of surface tension (A) separates the water-lined alveolus and the air (F).
2. Color the next panel. Note that the band of surface tension in the upper left beaker represents the alignment of water molecules (E) along the surface of the enlarged area. Also note that only the upper band of molecules and a single molecule in the center are colored. In the example to the right, the band of tension is weakened by the presence of surfactant molecules (H) that displace water along the surface of the enlarged portion.
3. Color the chart in the next panel, noting the great amount of pressure required to fill a small balloon. Color the two examples of how surface tension is affected by bubble size.
4. Color the effects of surfactant below.

LUNG'S ELASTIC BEHAVIOR

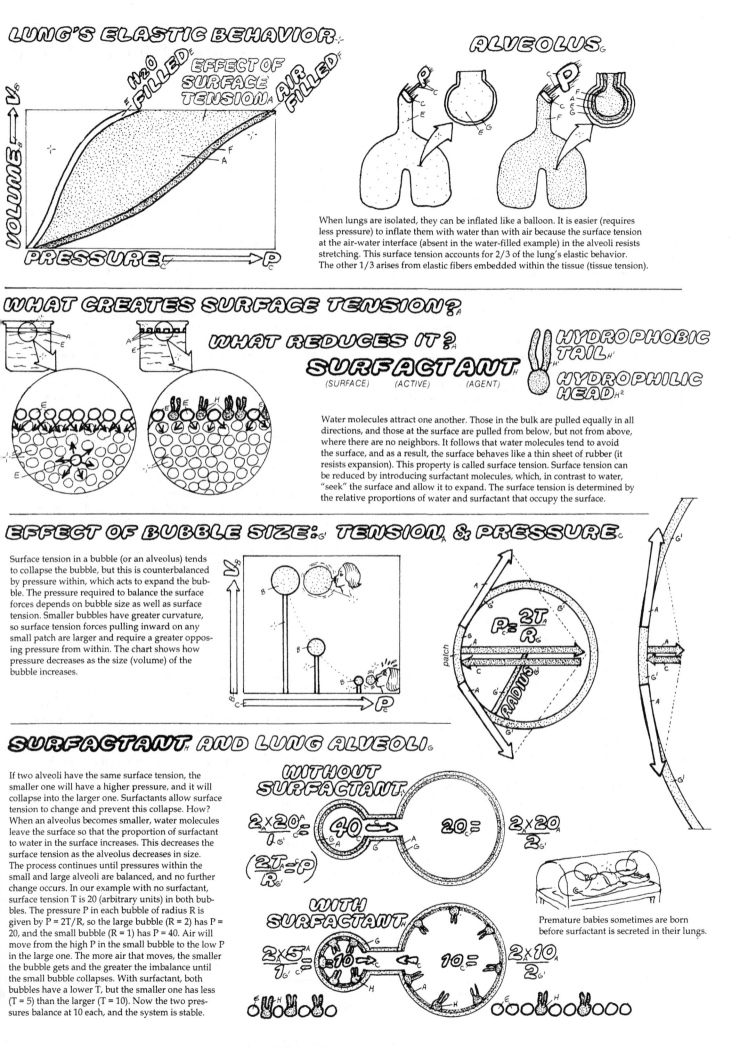

H₂O FILLED   **EFFECT OF SURFACE TENSION**   **AIR FILLED**

VOLUME → V

PRESSURE → P

ALVEOLUS

When lungs are isolated, they can be inflated like a balloon. It is easier (requires less pressure) to inflate them with water than with air because the surface tension at the air-water interface (absent in the water-filled example) in the alveoli resists stretching. This surface tension accounts for 2/3 of the lung's elastic behavior. The other 1/3 arises from elastic fibers embedded within the tissue (tissue tension).

WHAT CREATES SURFACE TENSION?

WHAT REDUCES IT?
SURFACTANT
(SURFACE)   (ACTIVE)   (AGENT)

HYDROPHOBIC TAIL

HYDROPHILIC HEAD

Water molecules attract one another. Those in the bulk are pulled equally in all directions, and those at the surface are pulled from below, but not from above, where there are no neighbors. It follows that water molecules tend to avoid the surface, and as a result, the surface behaves like a thin sheet of rubber (it resists expansion). This property is called surface tension. Surface tension can be reduced by introducing surfactant molecules, which, in contrast to water, "seek" the surface and allow it to expand. The surface tension is determined by the relative proportions of water and surfactant that occupy the surface.

EFFECT OF BUBBLE SIZE: TENSION & PRESSURE

Surface tension in a bubble (or an alveolus) tends to collapse the bubble, but this is counterbalanced by pressure within, which acts to expand the bubble. The pressure required to balance the surface forces depends on bubble size as well as surface tension. Smaller bubbles have greater curvature, so surface tension forces pulling inward on any small patch are larger and require a greater opposing pressure from within. The chart shows how pressure decreases as the size (volume) of the bubble increases.

$$P = \frac{2T}{R}$$

patch   RADIUS

SURFACTANT AND LUNG ALVEOLI

If two alveoli have the same surface tension, the smaller one will have a higher pressure, and it will collapse into the larger one. Surfactants allow surface tension to change and prevent this collapse. How? When an alveolus becomes smaller, water molecules leave the surface so that the proportion of surfactant to water in the surface increases. This decreases the surface tension as the alveolus decreases in size. The process continues until pressures within the small and large alveoli are balanced, and no further change occurs. In our example with no surfactant, surface tension T is 20 (arbitrary units) in both bubbles. The pressure P in each bubble of radius R is given by P = 2T/R, so the large bubble (R = 2) has P = 20, and the small bubble (R = 1) has P = 40. Air will move from the high P in the small bubble to the low P in the large one. The more air that moves, the smaller the bubble gets and the greater the imbalance until the small bubble collapses. With surfactant, both bubbles have a lower T, but the smaller one has less (T = 5) than the larger (T = 10). Now the two pressures balance at 10 each, and the system is stable.

WITHOUT SURFACTANT

$$\frac{2 \times 20}{1} = 40 \rightarrow 20 = \frac{2 \times 20}{2}$$

$$\left(\frac{2T}{R} = P\right)$$

WITH SURFACTANT

$$\frac{2 \times 5}{1} = 10 \rightarrow 10 = \frac{2 \times 10}{2}$$

Premature babies sometimes are born before surfactant is secreted in their lungs.

If the function of breathing is to flush the alveoli with fresh air, it is natural to ask how much air is moved. How efficient is the ventilation of the alveoli? What common disturbances result from this scheme?

PARTITIONING AIR VOLUMES IN THE LUNGS

The volume of air that moves in (or out) of the lungs per minute is called the *pulmonary ventilation* or sometimes the *minute volume*. It is the product of the amount taken in with each breath (tidal volume) and the number of breaths per minute. During normal quiet breathing, this is about 6 L/min. (a tidal volume of 0.5 L per breath × 12 breaths per min.), but both the depth of each breath and the rate of breathing can vary greatly, depending on the body's needs.

At rest, the tidal volume is a small fraction of the total lung capacity, and even the deepest expiration cannot expel all the air; some always remains in the alveoli and in the air passages. To evaluate these relations in both health and disease, we divide the changes in air volume within the lungs at different stages of breathing into the following categories:

1. Tidal volume — the amount of air that moves in and out with each normal breath.

2. Inspiratory reserve volume — the maximal additional volume of air that can be inspired at the end of a normal inspiration.

3. Expiratory reserve volume — the maximal additional quantity of air that can be expired at the end of a normal expiration.

4. Vital capacity — the greatest volume of air that can be moved in a single breath. The largest portion that can be expired after maximal inspiration, it is the sum of 1, 2, and 3.

5. Residual volume — the amount of air that remains within the lungs after maximal expiration.

6. Functional residual capacity — the "resting volume." The volume of the system just before a normal inspiration, it is the sum of 3 and 5.

7. Total lung capacity — the lung volume at its maximum (i.e., after a maximal inspiration). It is the sum of 4 and 5.

VOLUME CHANGES IN VENTILATION DEFECTS

Measuring these quantities is relatively easy (see plate) and often provides diagnostic clues for respiratory tract disturbances that interfere with ventilation. These can be divided into two types:

1. Restrictive disturbances — those cases where the lungs' ability to expand is compromised (reduced *compliance*). This occurs, for example, in pulmonary fibrosis or in fusion of the pleurae. Restrictive disturbances are often indicated by an abnormally low *vital capacity*.

2. Obstructive disturbances — caused by constriction of the airway (increased *resistance* to airflow). These contractions

often result from mucus accumulation, swollen mucus membranes, and bronchial muscle spasms, as occur in bronchial asthma or in spastic bronchitis. Because these disturbances are due to changes in resistance, identifying them requires measuring flow rather than volume (i.e., a rate rather than an equilibrium property). This can be accomplished by measuring the volume expelled from the lungs by forced expiration *in 1 sec.* This quantity, called the FEV_1 *(forced expiratory volume)*, is abnormally low in obstructive disease.

THE DEAD SPACE DOES NOT EXCHANGE O_2 OR CO_2

In addition to lung volumes, the space occupied by the conducting airways — the trachea, the bronchi, and the bronchioles, or the *anatomical dead space* — also requires attention. The 150 mL of air contained within this "dead" space moves in and out with each breath. But unlike alveolar air, it is not in close contact with the capillaries, so it has no opportunity to exchange O_2 or CO_2 with blood. Each time a tidal volume of 500 mL of air is exhaled, 500 mL leaves the alveoli, but only 500 − 150 = 350 mL reaches the atmosphere. The trailing 150 mL is still contained within the airways, in the anatomical dead space. When a fresh breath is inhaled, 500 mL of air enters the alveoli, but the first 150 mL that enters is not atmospheric. It is the "old" alveolar air from the last exhalation that never reached the atmosphere and was trapped within the dead space. Thus, with each inspiration, only 350 mL of fresh air enters the alveoli; the last 150 mL of the fresh inspired air never makes it because it is held up in the dead space and will be expelled at the next expiration.

It follows that only 350/500 = 70% of the normal tidal volume is used to ventilate the alveoli. Instead of using *pulmonary ventilation = tidal volume × breaths per min.* as a physiological index of effective lung ventilation, we more accurately use *alveolar ventilation = (tidal volume − anatomical/ dead space) × breaths per min.* The following example illustrates why. Consider two subjects with the same pulmonary ventilation: subject A has a small tidal volume (say 250 mL) but a fast breathing rate of 24 per min.; subject B, with a tidal volume of 500 mL and a rate of 12 per min., breathes twice as deep but half as often. In both cases, the pulmonary ventilation is 6000 mL/min. (250 × 24 and 500 × 12). But B has an alveolar ventilation of (500 − 150) × 12 = 4200 mL/min. A has only (250 − 150) × 24 = 2400 mL/min. Clearly, B is better off; most of A's effort goes into moving air back and forth in the dead air space. This result holds in general: given the same pulmonary ventilation, alveolar ventilation will increase by deeper breaths (although they will be less frequent). In extreme cases (e.g., as during circulatory shock), breathing becomes so shallow and so rapid that hardly any ventilation takes place, and the subject is in acute danger. Dogs, however, can use this rapid shallow breathing (i.e. panting) in a controlled way to lose heat by evaporation from the airways without *over-ventilating*.

CN: Use a dark color for I.
1. Begin with the upper drawing, coloring all the cubes; each one represents 500 mL of air.
2. Color the chart, including the three vertical bars.

3. Color the spirometer on the right.
4. Color the anatomical dead space. Note that th drawings on the right are schematics of the mo accurate anatomical drawing on the left.

NORMAL, QUIET BREATHING:
TIDAL VOLUME (500 mL)ₐ

DEEPEST INSPIRATION:
INSPIRATORY RESERVE VOLUME (2500-3500 mL)ᵦ

DEEPEST EXPIRATION:
EXPIRATORY RESERVE VOLUME (1000 mL)c

REMAINING AIR:
RESIDUAL VOLUME (1000 mL)ᴅ

AIR VOLUMES DURING RESPIRATION

The volume of air (500 mL) that moves in (or out) of the lungs with each inspiration (or expiration) during quiet breathing is called the tidal volume. During strenuous breathing, the amount of air moving with each breath increases. The maximum amount of additional air that can be inspired above the tidal volume is called the inspiratory reserve volume; the maximal volume of additional air that can be expired is called the expiratory reserve volume. The maximal amount of air that can be moved with each breath, the vital capacity, equals the sum of the inspiratory reserve, tidal, and expiratory reserve volumes. However, the lungs never empty completely. The volume of remaining air following a maximal expiration is called the residual volume. Finally, the total lung capacity equals the sum of all these volumes.

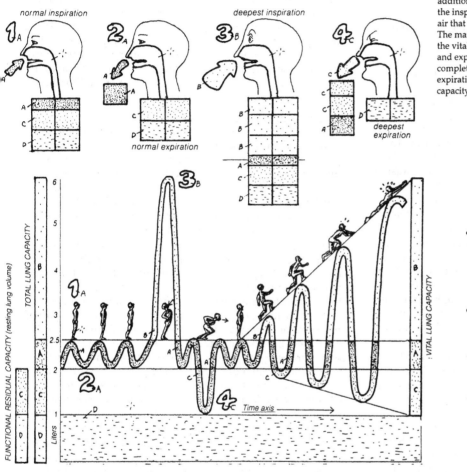

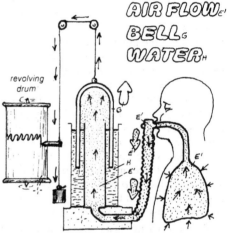

AIR FLOW ᴇ'
BELL ɢ
WATER ʜ

SPIROMETER ✱

Respiratory volumes are measured with a spirometer, consisting of an inverted container (bell) floating on water. Using a connecting hose, the subject expires (inspires) into (from) the bell as if it were a partially inflated balloon. The bell moves up (or down) with each breath, and its movements, which are proportional to changes in volume, are recorded on a rotating drum.

ALVEOLAR AIR AFTER RESPIRATION ᴀ'
ANATOMIC DEAD SPACE ₁   150 mL ₁
FRESH AIR ⱼ   350 mL ⱼ
TIDAL VOLUME ₐ   500 mL ₐ

During inspiration, some stale air reaches the alveoli. Close to 1/3 of the tidal volume is nonfunctional and is required simply to fill the air passages of the head, neck, bronchi, etc. The total volume of these passageways (about 150 mL) is called the anatomical dead space. Each time 500 mL of air is drawn into the lungs, the first 150 mL comes from the dead space, with the following 350 ml arising from fresh atmospheric air. If your tidal volume were only 150 mL, you would never get any fresh air! You would simply exchange the 150 mL back and forth between dead space and alveoli. Similarly, if you use a snorkel tube with a 350 mL volume, then you will increase your dead space to 500 mL! In this case, the normal tidal volume of 500 mL will be useless. Dogs lose heat by fluid evaporating from their dead space during panting. By restricting the amount of air moved, they bring fresh dry air to the dead space without allowing it to reach the alveoli. Thus, their rapid breathing movements do not interfere with normal respiration; they do not over-ventilate.

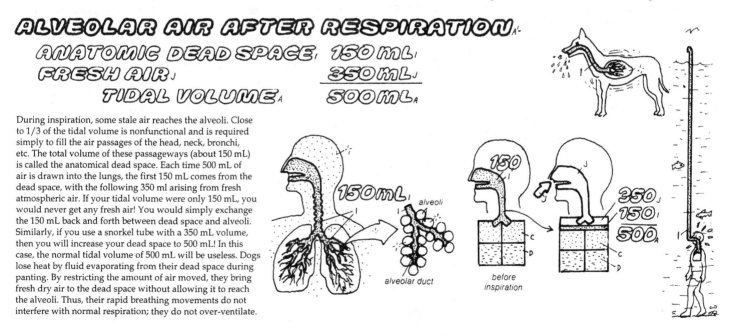

Diffusion of O_2 and CO_2 in the lung alveoli is complicated because these molecules move across an air-water interface. To describe these movements, we need a vocabulary equally applicable to both the liquid and air (gaseous) phases. We begin with a review of the properties of a gas.

MEASURING GAS CONCENTRATIONS

Ideal gas molecules act independently of each other — In a gas, *pressure* (force/unit area) results from gas molecules colliding with the walls of the container. It is determined by the frequency and force of the collisions. Each gas molecule is oblivious to the presence of any other; it strikes the container walls just as frequently as if it were all alone. Increasing the temperature of a gas raises pressure, because the higher the temperature the greater the velocity of the molecules, causing more frequent collisions and greater force to be imparted. Decreasing the volume occupied by the gas also increases pressure, because the gas molecules are confined to a smaller space and collide with the walls more frequently.

Measuring gas pressure — The pressure of air (or any gas) is measured by bringing it in contact with a pool of mercury (Hg) connected to a closed-ended tube containing no air (or gas). Force exerted by air pressure is not opposed by the vacant tube; therefore, the Hg rises until its weight just counterbalances the air pressure. The height of this column (mm Hg) is a measure of the pressure of the air (gas). Atmospheric air has a pressure of 760 mm Hg at sea level.

Partial pressure is a measure of gas concentration — In a mixture of gases, each component acts independently of the others, and each molecule makes the same contribution to the pressure. Air (a mixture of gases) consists of approximately 20% O_2 and 80% N_2. If we remove the N_2, we measure a pressure of 20% of 760 = 152 mm Hg. Similarly, retaining the N_2 but removing the O_2 yields a pressure of 80% of 760 = 608 mm Hg. In the mixture of the two, O_2 contributes 152 mm and N_2 contributes 608 mm Hg pressure. These are the *partial pressures* of O_2 and N_2, respectively. They are abbreviated as PO_2 and PN_2. Knowing the partial pressure of a gas is useful because at constant temperature (which is always the case in the alveoli) the partial pressure is a measure of the concentration of the gas and indicates the driving force available to dissolve the gas in a liquid.

Now suppose we bring the air in contact with a gas-free liquid, say water. The higher the partial pressure of O_2 (PO_2) in the gas, the more often O_2 will strike the surface of the water, and the more often some of the O_2 molecules will enter and dissolve in the liquid. But the dissolved O_2 molecules will also strike the surface from below, and some of these will tend to escape into the gas phase. As the concentration of O_2 builds up in the liquid, more and more will tend to escape until we reach an *equilibrium,* where the number leaving exactly balances the number entering the liquid. The O_2 concentration in the liquid is *directly proportional* to the partial pressure of the O_2 that it is equilibrated with, and we often use partial pressure as a measure of the concentration of O_2 in the liquid. If the PO_2 in the air were 152 mm Hg, then after equilibration, the PO_2 in solution would also be 152 mm Hg.

PARTIAL PRESSURE GRADIENTS

What has been described for O_2 applies equally to all gases, particularly CO_2. When the partial pressures between any two points are not equal, the two points are not in equilibrium; given the opportunity, gas will diffuse from one to the other. If the partial pressure in a gaseous phase (e.g., alveolus) is greater than in the water (e.g., plasma), gas will move into the water; if it is less, gas will move out of the water. Gas molecules move down partial pressure gradients.

With each inspiration, air moves by bulk flow into the alveoli, as described in plate 49. From the alveoli, O_2 diffuses down its partial gradient into the blood while CO_2 diffuses in the opposite direction. Alveolar air loses O_2 and gains CO_2, together with some water vapor that evaporated from the walls of the moist respiratory passages. As a result, the partial pressures of these gases in the alveoli differ from those in the atmosphere, as shown in the illustration. The circulation (bulk flow) carries O_2 contained in the blood to systemic capillaries, where once again it diffuses down its partial pressure gradient, this time into the tissues. Again CO_2 diffuses in the opposite direction, this time into the blood, which will carry it by bulk flow via the venous system and the pulmonary artery to the lungs.

Three important variables determine the speed of gas diffusion in the body: (1) the gradient in partial pressure, (2) the surface area available for diffusion, and (3) the magnitude of the diffusion distance. Although the bottom diagram shows that O_2 always moves down its partial pressure gradient from the atmosphere to mitochondria, movement over long distances (between atmosphere and alveoli and between lungs and systemic tissue) is driven by the pumping action of respiratory and cardiac muscles. In these cases, transport occurs by bulk flow. Diffusion is the effective transport mechanism only over the short distances between alveolus and blood and between blood and tissue. The situation is similar for CO_2. Gas transport can be compromised if diffusion distances are lengthened, as in *pulmonary edema,* and if the surface area available for diffusion is reduced, as in *emphysema.*

CN: Use red for B and a dark color for I.
1. Begin with the upper panel, top line first. The titles for B and C are O_2 and N, in the equation on the right.
2. Do the middle panel next.
3. Color the lower panel, following the numbered sequence.
4. Color the numbers in the diagram of PO_2 at various stages in its journey to its site of utilization the mitochondria. Do the same for PCO_2.

WHAT IS PRESSURE? GAS MOLECULE.

In a gas, pressure (force/unit area) results from gas molecules colliding with the walls of the container. It is determined by the frequency of collisions and the force imparted by each collision.

WHAT CHANGES PRESSURE?

Increasing temperature raises pressure because it increases the velocity of molecules, which increases frequency of collision and the force imparted. Compression also raises pressure because gas molecules collide with the walls more frequently.

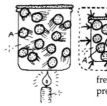

PARTIAL PRESSURE (P) OF GAS

MERCURY.

760 mm Hg AIR AT SEA LEVEL

152 mm Hg AIR LESS NITROGEN

Force (pressure) exerted by air is not opposed by the vacant (air-free) tube. Therefore the Hg rises until its weight just balances the air pressure. The height at this column (mm Hg) is the pressure of the air. Atmospheric air has a pressure of 760 mm Hg at sea level.

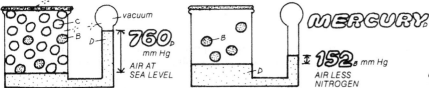

$$PO_2 = 1/5 \times 760 = 152$$
$$PN_2 = 4/5 \times 760 = 608$$
$$\overline{\text{TOTAL PRESSURE} = 760}$$

The total gas pressure reflects the sum of the collisions exerted by all gas molecules. If air has 20% O_2, then O_2 exerts $.20 \times 760 = 1/5 \times 760 = 152$ mm Hg. O_2 has a partial pressure (PO_2) of 152 mm Hg. Partial pressure is proportional to the concentration of the gas.

SOLUBILITY OF GAS: P IN AIR VS. P IN GAS.

When O_2-free water is first brought into contact with air, O_2 enters (dissolves in) the water until equilibrium is reached, when the rate of O_2 leaving the water just equals the rate of O_2 entering. We measure concentrations of gas in solution in terms of partial pressure. If the final PO_2 in the air were 152 mm Hg, then PO_2 in solution would also be 152 mm Hg. By definition, the partial pressure of a gas in solution equals the partial pressure of the same gas in the gaseous phase that would be required if the water and gaseous phases were in equilibrium.

ALVEOLUS. BLOOD CAPILLARY.

PO_2

CO_2

If the partial pressure in the gaseous phase (e.g., alveolus) is greater than in the water (e.g., plasma), gas will move into the water. If it is less, gas will move out of the water. Gas molecules move down partial-pressure gradients.

FACTORS AFFECTING TRANSPORT OF GAS IN LUNG AND TISSUES

1. P GRADIENTS

152 mmHg PO_2
32 mmHg PCO_2

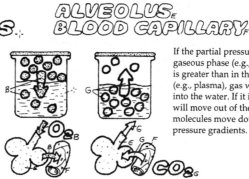

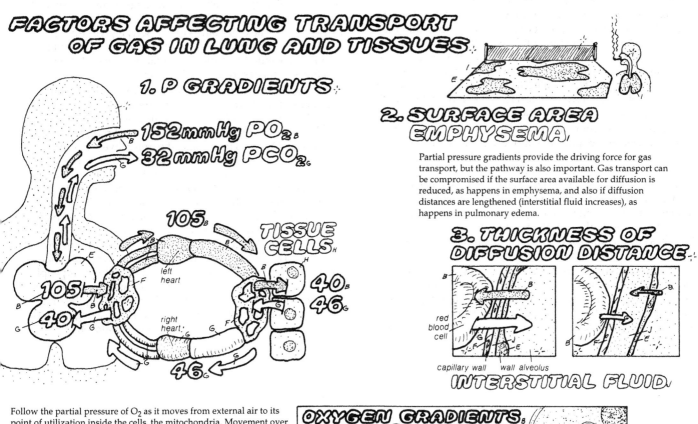

105 TISSUE CELLS

left heart

105
40

right heart

40
46

46

2. SURFACE AREA EMPHYSEMA.

Partial pressure gradients provide the driving force for gas transport, but the pathway is also important. Gas transport can be compromised if the surface area available for diffusion is reduced, as happens in emphysema, and also if diffusion distances are lengthened (interstitial fluid increases), as happens in pulmonary edema.

3. THICKNESS OF DIFFUSION DISTANCE.

red blood cell

capillary wall   wall alveolus

INTERSTITIAL FLUID.

Follow the partial pressure of O_2 as it moves from external air to its point of utilization inside the cells, the mitochondria. Movement over the long distances between atmosphere and alveoli and between lungs and tissue occurs by bulk flow. Diffusion is the effective transport mechanism over the short distances between alveolus and blood and between blood and tissue. The situation is similar for CO_2, but recall that PCO_2 is highest in the cells, where it is produced, and lowest in the atmosphere, so it moves in the opposite direction.

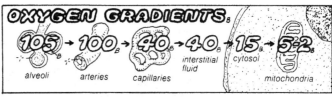

OXYGEN GRADIENTS.

105 → 100 → 40 → 40 → 15 → 5-2

alveoli   arteries   capillaries   interstitial fluid   cytosol   mitochondria

Like any solute, O_2 can simply dissolve in the watery fluids of the blood, but the amount that can dissolve is very small. At the PO_2 (partial pressure of O_2) = 100 mm Hg that exists in arterial blood and with a normal cardiac output, the amount dissolved could supply only about 6% of the body's requirements at rest. During activity, it would fall even shorter. Clearly, there has to be, and is, another way. Most O_2 carried by the blood is combined with *hemoglobin* (Hb), an iron-containing protein within the red blood cell. Hb can carry nearly 70 times the O_2 held in simple solution.

CO_2 IS CARRIED AS HCO^{3-} AND AS CARBAMINOHEMOGLOBIN

Although CO_2 is more soluble than O_2, it too is carried primarily in different combined forms in the plasma and red cells. Most CO_2 reacts with water to form carbonic acid (H_2CO_3), which dissociates into H^+ and bicarbonate (HCO_3^-) according to the reaction

$$H_2O + CO_2 \rightarrow H_2CO_3 \rightarrow H^+ + HCO_3^-.$$

Another fraction of CO_2 combines with some of the amino groups on polypeptide portions of Hb to form *carbamino-hemoglobin*.

O_2 BINDS COOPERATIVELY TO HEMOGLOBIN

Oxygen binds to the ferrous iron in heme — Hb's ability to bind O_2 depends on the presence of a *heme* group within the molecule. Heme, a nonpolypeptide, consists of an organic part and an *iron* atom; it gives Hb (and red cells) characteristic red color. The iron can be in one of two states: the ferrous state (charge = +2) or the ferric state (charge = +3). O_2 binds only to Hb with iron in the ferrous state. Hb in the ferric state, called *methemoglobin*, is a darker color and cannot bind O_2.

Hb consists of four subunits — The heme group is embedded in a large polypeptide chain, and together (heme + polypeptide chain) they are called a subunit. The entire Hb molecule, which has a molecular weight of 64,450, consists of four of these subunits. The size of an O_2 molecule makes up only about 0.1% of the size of one of these subunits. It is natural to wonder whether the large structure has any significance and whether the combination of subunits into groups of four has any advantage. Would an iron molecule or a heme by itself suffice? How about a subunit by itself?

Water exclusion protects the ferrous iron — When isolated heme is dissolved in water, it binds O_2, but only momentarily because it is rapidly converted from the ferrous (+2) to the ferric (+3) state. But this does not happen to the heme in Hb or even in a subunit because the heme is embedded in a crevice that has a distinctive nonpolar character so that water is excluded. Apparently, the polypeptide protects the heme from water and helps keep it in the reduced ferrous (+2) state. Even here some conversion takes place at a slow rate, but the red cell contains an enzyme that can keep pace and convert the methemoglobin back to Hb.

The ideal oxygen-storage molecule binds tightly in the lungs and loosely in the tissues — Given that iron in a subunit will be reasonably stable in the ferrous state and bind O_2, why bother to string four of them together? The answer appears to be "too much of a good thing"; a subunit binds O_2 too well. This can be demonstrated by studying *myoglobin*, a very close relative of Hb containing heme and a similar polypeptide chain, but consisting of only one subunit. Myoglobin takes up O_2 well at a very low PO_2, much lower than the PO_2 of venous blood. But this also means that the myoglobin won't give it up until the PO_2 is correspondingly low. Myoglobin functions well as an O_2 storage compound in muscle, where it releases its O_2 only when the PO_2 drops very low during strenuous exercise, but it would not suffice as an O_2 carrier in the blood. We could imagine other single subunits that have lower affinities for O_2, but they would present a new problem: they would not pick up enough O_2 in the lungs. Thus, one type of molecule binds too tightly; it works well in the lungs but not in the tissues. The other binds too loosely; it gives up O_2 readily in the tissues but can't pick up enough in the lungs. Ideally, we would like a molecule that switches between the two types as it goes from lungs to tissue.

Hb approximates the ideal by stringing four subunits together so that the heme sites can interact — Hb exists in more than one state. When none of the ironbinding sites are occupied, Hb is in a T ("tense") state and not receptive to O_2. But once an O_2 binds to one site, the iron moves slightly and so do parts of the polypeptide chain attached to it. This loosens the structure, making it easier for the next O_2 to attach to one of the remaining empty sites. The sequence repeats, making it still easier for the next O_2, etc., until (in the lungs) all four sites are occupied by O_2, and the Hb is in an R ("relaxed") state. Conversely (in the tissues), as one O_2 frees itself from the Hb, the Hb changes slightly, making it easier for the next to unload. This behavior is called *cooperative*. A simple analogy in the plate shows a boat (Hb) with room for four people (O_2). They start off in the water, but as one gets on the boat, he helps the next, etc. Similarly, when the boat docks, the first one off the boat helps the next. The physiological significance of this cooperative behavior is discussed in plate 54.

CN: Use red for B and blue for C (both for venous blood and CO_2 transport).
1. Begin with the upper panel. Note that the symbol for oxyhemoglobin is a further simplification of the symbol used at the bottom of the page, which is a simplification of the hemoglobin model (the large illustration below).

2. Color the lower panel, beginning with the large illustration. Note that the lower left alpha chain (E) shows the polypeptide chain of which it and the other chains are composed. Color the three examples below. Note that the boat on the far right is given the oxyhemoglobin red because it is holding the four O_2 molecules.

O_2 SATURATION IN BLOOD CIRCULATION

Like any solute, O_2 can simply dissolve in blood plasma, but the amount that can dissolve is very small and cannot supply the body's needs. Most O_2 carried by the blood is not in simple solution; rather, it is combined with hemoglobin (Hb), an iron-containing protein within red blood cells. Hb is represented by the squares in the beaker.

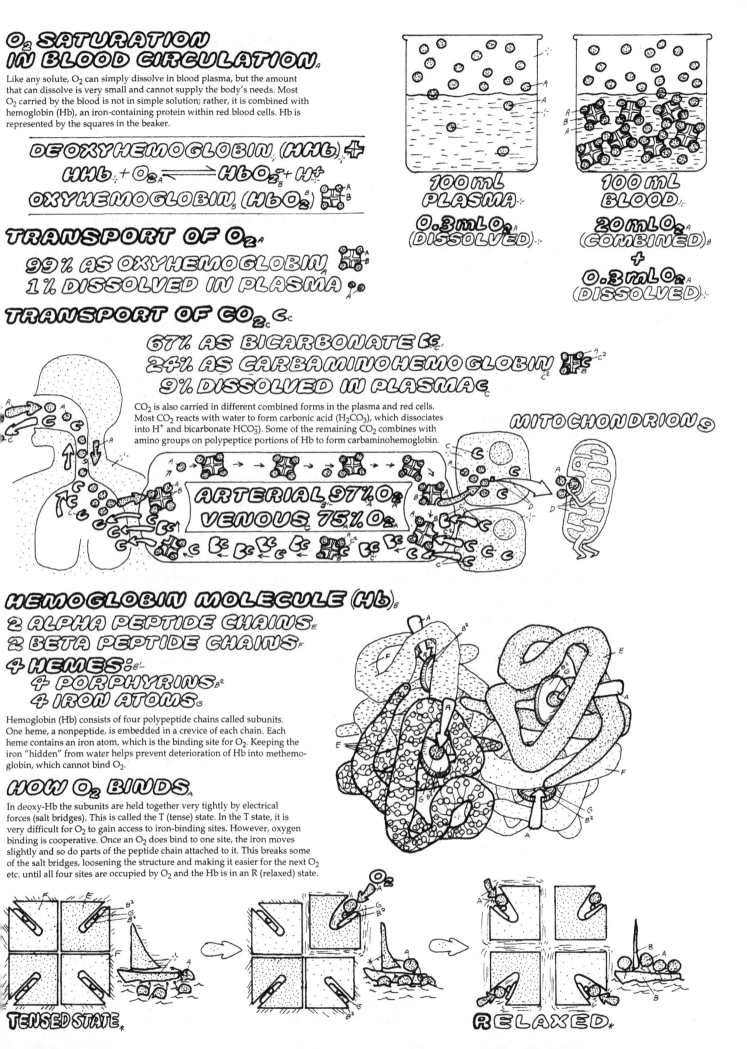

100 mL PLASMA
0.3 mL O_2 (DISSOLVED)

100 mL BLOOD
20 mL O_2 (COMBINED)
+
0.3 mL O_2 (DISSOLVED)

DEOXYHEMOGLOBIN (HHb) +

$$HHb + O_2 \rightleftharpoons HbO_2^- + H^+$$

OXYHEMOGLOBIN (HbO$_2$)

TRANSPORT OF O_2

99% AS OXYHEMOGLOBIN
1% DISSOLVED IN PLASMA

TRANSPORT OF CO_2

67% AS BICARBONATE
24% AS CARBAMINOHEMOGLOBIN
9% DISSOLVED IN PLASMA

CO_2 is also carried in different combined forms in the plasma and red cells. Most CO_2 reacts with water to form carbonic acid (H_2CO_3), which dissociates into H^+ and bicarbonate HCO_3^-. Some of the remaining CO_2 combines with amino groups on polypeptice portions of Hb to form carbaminohemoglobin.

MITOCHONDRION

ARTERIAL 97% O_2
VENOUS 75% O_2

HEMOGLOBIN MOLECULE (Hb)

2 ALPHA PEPTIDE CHAINS
2 BETA PEPTIDE CHAINS
4 HEMES:
4 PORPHYRINS
4 IRON ATOMS

Hemoglobin (Hb) consists of four polypeptide chains called subunits. One heme, a nonpeptide, is embedded in a crevice of each chain. Each heme contains an iron atom, which is the binding site for O_2. Keeping the iron "hidden" from water helps prevent deterioration of Hb into methemoglobin, which cannot bind O_2.

HOW O_2 BINDS

In deoxy-Hb the subunits are held together very tightly by electrical forces (salt bridges). This is called the T (tense) state. In the T state, it is very difficult for O_2 to gain access to iron-binding sites. However, oxygen binding is cooperative. Once an O_2 does bind to one site, the iron moves slightly and so do parts of the peptide chain attached to it. This breaks some of the salt bridges, loosening the structure and making it easier for the next O_2 etc. until all four sites are occupied by O_2 and the Hb is in an R (relaxed) state.

TENSED STATE

RELAXED

When hemoglobin (Hb) is exposed to O_2, the O_2 molecules continually collide with it. If there is an empty binding site on the Hb, a colliding O_2 may bind to it. But bound O_2s are continually shaking loose from their sites. Equilibrium is reached when the number being bound just equals the number shaking loose. In Hb, this equilibrium is reached very fast, and its position is determined largely by the PO_2. The higher the PO_2 (the more concentrated the O_2), the more frequent the collision with Hb and the more frequently an O_2 will bind. As the O_2 concentration increases, more and more binding sites are filled, until finally every site is filled, with each Hb molecule containing four bound O_2 molecules. At this point, we say the Hb is 100% *saturated;* when only half are occupied, the Hb is 50% saturated.

Hb HAS AN S-SHAPED O_2 UPTAKE CURVE
The large illustration in the plate shows how Hb takes up O_2 at the partial pressures that exist in the lungs and in the tissues. In the lungs, PO_2 = 105 mm Hg; the curve shows that Hb is 97% saturated. The illustration also shows that Hb will unload O_2 in the tissue capillaries where PO_2 averages about 40 mm Hg and may fall even lower to 20 mm Hg in active muscle capillaries. The vertical arrows show the difference between the percentage of Hb saturation of blood just after leaving the lungs and the percentage of Hb saturation in the tissues. This difference is the O_2 delivered to tissues.

The S shape reflects cooperative interaction of four subunits — Hb "works" because its saturation curve is S shaped; it unloads most of its O_2 in a very narrow range of PO_2 — between 20 and 40 mm Hg. This behavior is due to the fact that Hb is made of four interacting subunits that "cooperate" in binding O_2. The first portion of the curve, at very low PO_2, is flat because Hb is in the tense state and not receptive to O_2. As more O_2 molecules are introduced, the likelihood of one of them binding goes up. Once it binds, it influences the other vacant binding sites on the same Hb molecule, increasing the probability of binding a second O_2, which will increase the chances for a third, etc. Thus, the binding (saturation) curve rises very steeply — and fortunately in just the right region!

Contrast this behavior with that of *myoglobin,* the O_2 storage protein in muscle cells. It is similar to Hb, but it contains only one subunit; one molecule binds only one O_2, and there is no possibility of a T state or of *cooperative binding.* Its binding curve is not S shaped, and rather than giving up its O_2 at the PO_2 found in the venous blood, it takes it up. But this fits its function; myoglobin stores O_2 and will give it up in the tissues only when the PO_2 falls very low.

CO_2, H^+, AND 2,3 DIPHOSPHOGLYCERATE (DPG)

CO_2, H^+, and 2,3 DPG "tense" Hb and release O_2 — The PO_2 is not the only variable that influences the binding of O_2 to Hb. The last diagram in the plate shows several percentage-of-saturation curves for Hb under different conditions. In one of them, the concentration of CO_2 has increased and the O_2 saturation curve for Hb has shifted to the right (i.e., it lies below the "normal" curve). In this case, a higher PO_2 is required to achieve the same percentage of saturation; this means the Hb has a lower affinity for O_2. If the Hb were just sitting there, exposed to a constant PO_2, and CO_2 suddenly increased, shifting the curve to the right, then the Hb would release some of its O_2. This actually happens as blood passes through a capillary, and CO_2 diffuses into the blood from the tissues. In addition to CO_2, two other important substances shift the curve to the right. These are H^+ and a phosphorus-containing metabolite, 2,3 DPG. These bind at separate locations on the Hb molecule, but they all act in similar ways by strengthening linkages between Hb subunits, which promotes the tense state with low O_2 affinity. Tissues commonly produce CO_2 and H^+, helping to drive O_2 off the Hb and making it more available to tissue cells.

Fetal Hb's strong affinity for O_2 results from its low uptake of 2,3 DPG — When the curve is shifted to the left, above the "normal" curve, the Hb has more affinity for O_2; it takes some up. This will occur whenever the 2,3 DPG level falls. When all the 2,3 DPG is removed, Hb enters a more relaxed state where its affinity for O_2 increases to such an extent that it begins to resemble myoglobin. The *Hb in fetal red cells* is different from adult Hb; in particular, fetal Hb does not bind 2,3 DPG as readily as adult Hb. In other words, it is less sensitive to 2,3 DPG. As a result, the O_2 saturation curve for fetal Hb lies above the curve for maternal Hb, showing that fetal Hb has a greater affinity for O_2. This is an advantage for the fetus because when fetal Hb comes in proximity to maternal Hb (in the placenta), it draws O_2 from the maternal blood.

2,3 DPG plays a regulatory role in O_2 transport — The role of 2,3 DPG has attracted a good deal of attention because it is not simply an essential "ingredient" whose presence is required for normal Hb function. Its level can vary considerably, and it is involved in regulating O_2 transport in both health and disease. Its level rises when O_2 uptake in the lungs is reduced, and this helps the Hb unload a larger portion of the O_2 that it does carry, when it gets to the tissues. This rise in 2,3 DPG occurs, for example, during the first day's adaptation to high altitude (plate 57) and during obstructive lung diseases.

CN: Use the same color for O_2 (B) as used on previous plates. Note that the Hb is shown by two different symbols, a dump truck and a four-unit structure.
1. Begin with the graph in the upper right. First color the percentage and PO_2 coordinates.

Then color the curve and the corresponding O_2 concentrations below on the horizontal axis.
2. Color the myoglobin example (F).
3. Color the factors influencing the curve. Note that the dump truck receives a different color in two of the examples.

HEMOGLOBIN / OXYGEN DISSOCIATION CURVE

The more concentrated the O_2 (i.e., the higher the PO_2), the more it will fill up empty sites on Hb. When all possible sites are occupied by O_2, we say the Hb is 100% saturated; when only half are occupied, Hb is 50% saturated. The illustration shows that Hb will take up O_2 in the lungs; here PO_2 = 105 mm Hg, and the curve shows that Hb is 97% saturated. The figure also shows that Hb will unload (dump) O_2 in the tissues, where PO_2 averages about 40 mm Hg but may fall as low as 20 mm Hg in active muscles. Hb "works" because its saturation curve is S shaped; it unloads most of its O_2 in a very narrow range of PO_2 (between 20 and 40 mm Hg). This behavior reflects the co-operative nature of O_2 binding to Hb. The first portion of the curve, at very low PO_2, is flat because Hb is in the tense state and not receptive to O_2. As more O_2 are introduced, the likelihood of one of them binding goes up. Once it binds, it increases the probability of a second one, which increases the chances for a third, etc. Thus, the binding (saturation) curve rises very steeply—and fortunately in just the right region!

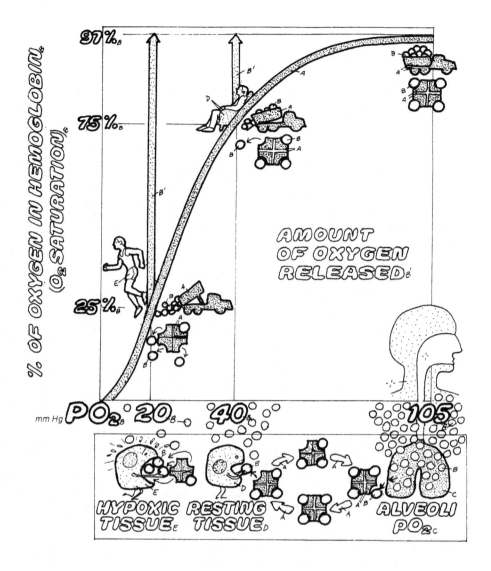

MYOGLOBIN

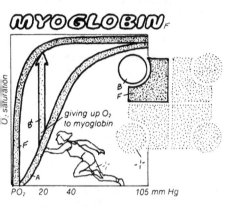

giving up O_2 to myoglobin

Contrast this with myoglobin, the O_2 storage protein in muscle cells. It is similar to Hb but contains only one subunit. One molecule binds only one O_2, and there is no possibility of a T state or of cooperative binding. Its binding curve is not S shaped, and rather than giving up its O_2, it takes it up. But this fits its function; it stores O_2 and will give it up in the tissues only when the PO_2 falls very low.

FACTORS AFFECTING THE CURVE

NORMAL CURVE: MATERNAL Hb

SHIFT TO THE RIGHT: pH$^+$, CO_2, & DPG

When an O_2 saturation curve for Hb is shifted to the right (i.e., when it lies below the "normal" curve), it has a lower affinity for O_2 and it gives some up. Three important substances shift the curve to the right—H$^+$, CO_2, and a phosphorus-containing metabolite, 2,3 DPG. They all act in similar ways by creating bridges between the Hb subunits, promoting the tense state with low O_2 affinity. Tissues commonly produce CO_2 and H$^+$, helping to drive O_2 off the Hb and making it available to tissue cells.

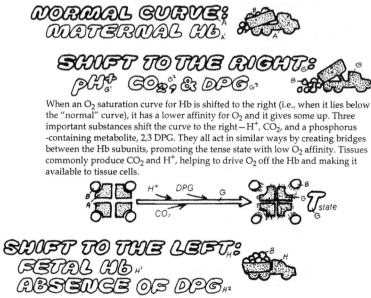

SHIFT TO THE LEFT: FETAL Hb ABSENCE OF DPG

When the curve is shifted to the left, the Hb has more affinity for O_2; it takes some up. The O_2 saturation curve for fetal Hb lies above the curve for normal maternal Hb; the fetal Hb has a greater affinity for O_2. This is an advantage for the fetus because when fetal Hb comes in proximity to maternal Hb (in the placenta), it will draw O_2 from the maternal blood. Fetal blood has this property because it is less reactive to affinity-lowering 2,3 DPG than normal adult Hb.

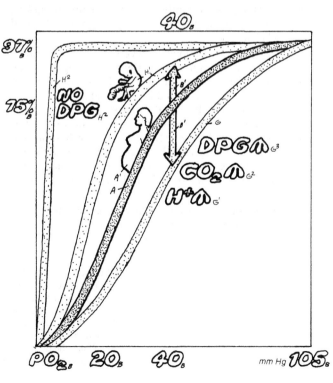

In plate 54, we saw how the subunit structure of Hb introduces into the molecule new properties that are not shared by the simpler, single-unit analog, myoglobin. In particular, increasing the concentrations of CO_2 and H^+ drives O_2 off the Hb molecule. The converse also holds: increasing the concentration of O_2 drives off both CO_2 and H^+. At first, this unusual sensitivity of Hb to its environment may seem undesirable in a molecule whose function is to stabilize the PO_2 in body fluids. However, the function of Hb goes beyond this; it not only transports O_2, it also transports both CO_2 and H^+. Further, Hb reacts with these three substances in a remarkable way so that just the "right" thing happens at the "right" time.

CO_2 IS CARRIED AS BICARBONATE

Like O_2 transport, CO_2 transport is passive. PCO_2 is high in the tissues because it is produced there. It is low in the lung alveoli because it is swept out with each breath, and therefore it is also low in the arterial blood entering tissue capillaries. CO_2 moves down its partial pressure gradient from tissue to capillary blood to lung alveoli (plate 53). Although blood holds a small amount of CO_2 (about 9%) in simple solution and another fraction (about 27%) in combination with Hb, the major portion (64%) reacts with water, forming *bicarbonate* (HCO_3^-) and *hydrogen ions* (H^+).

$$CO_2 + H_2O \rightleftharpoons H_2CO_3 \rightleftharpoons HCO_3 + H^+$$

HEMOGLOBIN BUFFERS THE H^+

Because PCO_2 is high in the tissues, this reaction proceeds to the right, and CO_2 is carried as bicarbonate. However, there is a major problem: this reaction leads to the accumulation of H^+ ions. Not only are H^+ ions acid, but their accumulation will slow down and block the reaction of CO_2 with water, which severely limits the amounts of CO_2 that can be carried. The dilemma is resolved by substances in the blood that "soak up" or *buffer* excess H^+ ions. Hb is one of the most important of these buffers; its reaction with H^+ is represented as follows:

$$H^+ + HbO_2 \rightleftharpoons HHb + O_2$$

where the HbO_2^- represents Hb with O_2 attached (*oxyhemoglobin*) and the (–) sign signifies one of the many charges carried by the Hb molecule. Similarly, HHb represents Hb with an extra H^+ attached.

Notice that both reactions are reversible (i.e., they can proceed from left to right or from right to left depending on the concentrations of reactants and products). At *equilibrium*, the reaction proceeds in both directions, but at equal rates so that no noticeable change takes place. However, when concentrations of substances on the right are decreased, the reaction gets "pulled" from left to right. Increasing concentrations on the left will "push" the reaction from left to right. Conversely, decreasing the concentrations of substances on the left, or increasing them on the right, moves the reaction from right to left.

Hydration of CO_2 is catalyzed by carbonic anhydrase

Ordinarily the reaction of CO_2 and water is sluggish, taking many seconds in water or plasma. However, it is accomplished within milliseconds inside the red blood cells where the reaction is catalyzed by a powerful enzyme, *carbonic anhydrase*. The lipid-soluble CO_2 passes through the membrane bilayer with ease and enters the red cell. The newly formed HCO^-_3 leaves by a special counter-transporter that very rapidly exchanges Cl^- for HCO_3^-. The net effects of this very rapid enzyme action and transport are to ensure that the reaction is completed within the short span of time (approximately 1 sec) that the blood spends in a capillary, and to allow both red cells and plasma to share the carriage of bicarbonate. Keeping the carbonic anhydrase within the red cell (rather than the plasma) has the advantage that the red cell environment protects it from oxidative damage.

H^+ IS JUGGLED BETWEEN HB AND HCO_3^-

Tissues: H^+ combines with Hb and aids release of O_2
In the tissues, the reactions involving Hb and bicarbonate are coupled because H^+ ions are a common participant in both.

$$CO_2 + H_2O \rightarrow H_2CO_3 \rightarrow HCO_3^- + H^+$$
$$H^+ + HbO_2^- \rightarrow HHb + O_2$$

The first reaction proceeds in the indicated direction because (1) CO_2 is produced in tissues so its concentration is high, and (2) as soon as excess H^+ begins to accumulate, it is consumed by the second reaction. The second reaction proceeds in the indicated direction because (1) a steady supply of H^+ is liberated by the first reaction, (2) a steady supply of HbO_2^- at high concentration is coming from the lungs, (3) HHb is continually swept away in the venous blood, and (4) O_2 is consumed by the tissues, so its concentration is low. Note that as soon as H^+ is produced, it is picked up by the Hb, so free H^+ does not accumulate to dangerous levels. In the process, the tissues receive an extra dividend: more O_2 is driven off the Hb than would be without the H^+ binding.

Lungs : H^+ combines with HCO_3^- and aids release of CO_2
In the lungs, these same reactions occur, but now in reverse:

$$O_2 + HHb \rightarrow HbO_2^- + H^+$$
$$H^+ + HCO_3^- \rightarrow H_2CO_3 \rightarrow H_2O + CO_2$$

The first reaction proceeds in the direction of the arrow because (1) PO_2 is high in the lungs, (2) there is a steady supply of HHb at high concentration coming from the tissues (via systemic venous blood), and (3) as soon as excess H^+ accumulates, it is consumed by the second reaction. The second reaction proceeds as shown because (1) there is a steady supply of H^+ liberated by the first reaction, (2) there is a steady supply of HCO_3^- at high concentration coming from the tissues, and (3) breathing keeps CO_2 at a low level.

Thus, H^+ ions, which at first appeared to be a problem, actually play a very useful role: in the tissues they drive O_2 off of Hb, and in the lungs they help drive CO_2 off of HCO_3^-. They never accumulate in the free state because they pass back and forth like a "hot potato" between Hb and HCO_3^-.

CN: Use the same colors as on the previous page for O_2 (I). Use red for C, blue for D, and light blue for F. Use a dark color for H.
1. Begin by coloring the tissue cell and the titles at the top of the page and the lung alveolus and titles at the bottom. Then color the red blood cell section. Color the two horizontal bands (where gas exchanges occur) gray, and color the vertical bands, arterial (C) and venous (D) circulation.
2. Start with number 1 at the top (under "CO_2 Produced") and follow the numbered sequence. Continue down the right side, coloring all symbols. Then color all the processes of gas exchange in the lungs, beginning with number 5 in the lower right corner.
3. Color the overview diagram within the rectangle.

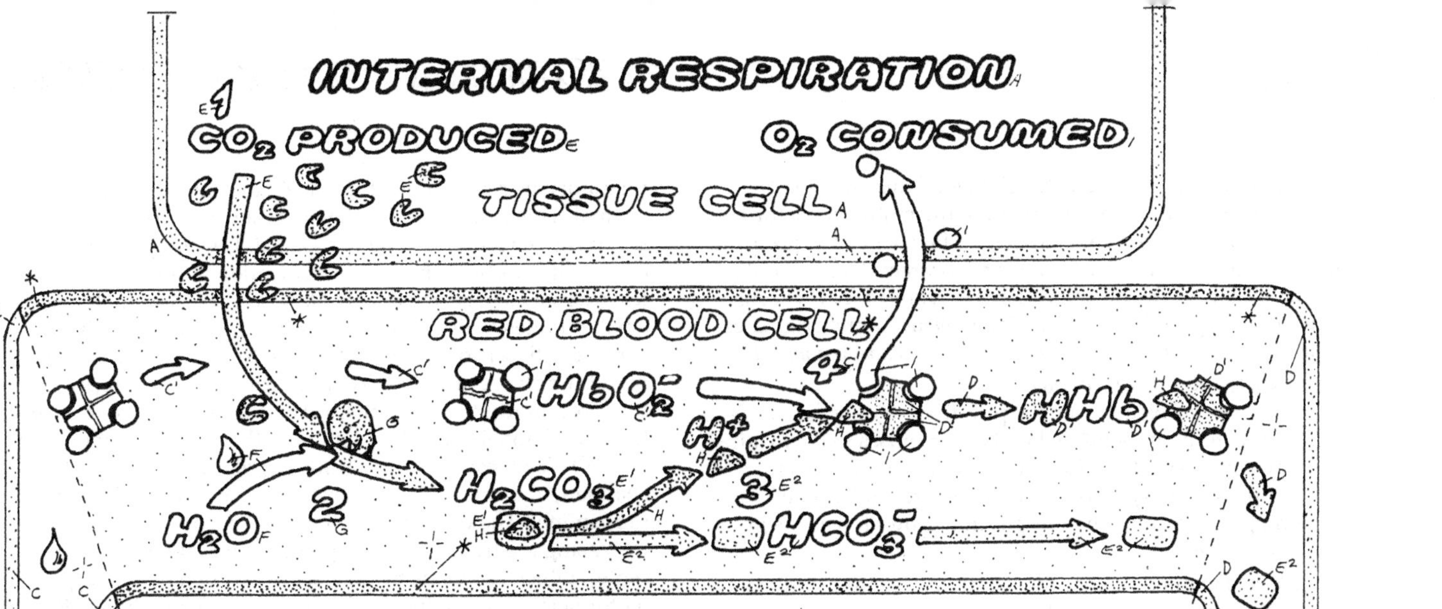

INTERNAL RESPIRATION

CO_2 PRODUCED O_2 CONSUMED

TISSUE CELL

RED BLOOD CELL

HbO_2^- HHb

H^+

H_2CO_3 HCO_3^-

H_2O

In tissues, CO_2 production promotes the reaction $CO_2 + H_2O \rightarrow H_2CO_3^- + H^+$.
The consumption of O_2 promotes the reaction $H^+ + HbO_2^- \rightarrow HHb + O_2$.

This is shown above as (1) CO_2 diffuses from tissue cells where it is produced into
the plasma and then into red blood cells; (2) in the red cells, combination of the CO_2
with water to form H_2CO_3 is accelerated by the enzyme carbonic anhydrase; (3) the
H_2CO_3 rapidly dissociates into HCO_3^- (bicarbonate) and H^+ ions (acid); (4) the H^+ are
not left free, as a large portion of them combine with oxyhemoglobin. This provides
two advantages: blood does not become intolerably acid, and the combination of H^+
with oxyhemoglobin helps unload the O_2 in the tissues.

CARBON DIOXIDE
WATER
CARBONIC ANHYDRASE
CARBONIC ACID
BICARBONATE
HYDROGEN ION
OXYHEMOGLOBIN
OXYGEN
DEOXYHEMOGLOBIN

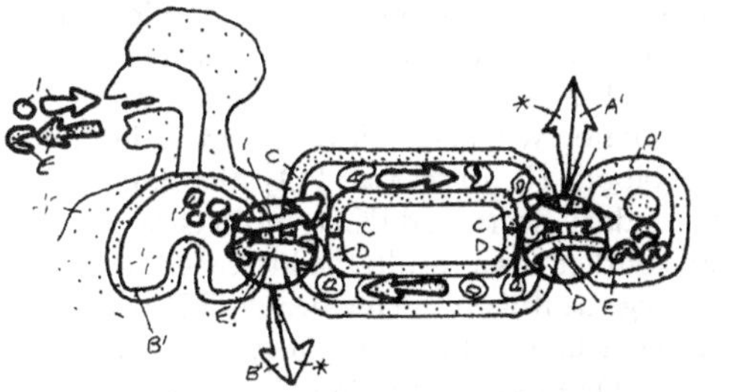

ARTERIAL CIRCULATION

VENOUS CIRCULATION

In the alveoli, high O_2 and low CO_2 are continually maintained through the act of breathing, so the reactions described above are
reversed. Here:
$$O_2 + HHb \rightarrow HbO_2^- + H^+$$
and $H^+ + HCO_3^- \rightarrow H_2CO_3 \rightarrow CO_2 + H_2O$.

This is shown as (5) O_2 diffuses from the alveoli into the plasma and then into red cells (6)
O_2 combines with HHb to form HbO_2^-,
releasing H^+; (7) H^+ combines with HCO_3^-,
forming H_2CO_3 and then (8) H_2O and CO_2.
Again the liberated H^+ does not accumulate;
it reacts with HCO_3^- and helps drive off CO_2,
which is expelled from the alveoli (9) with
each breath.

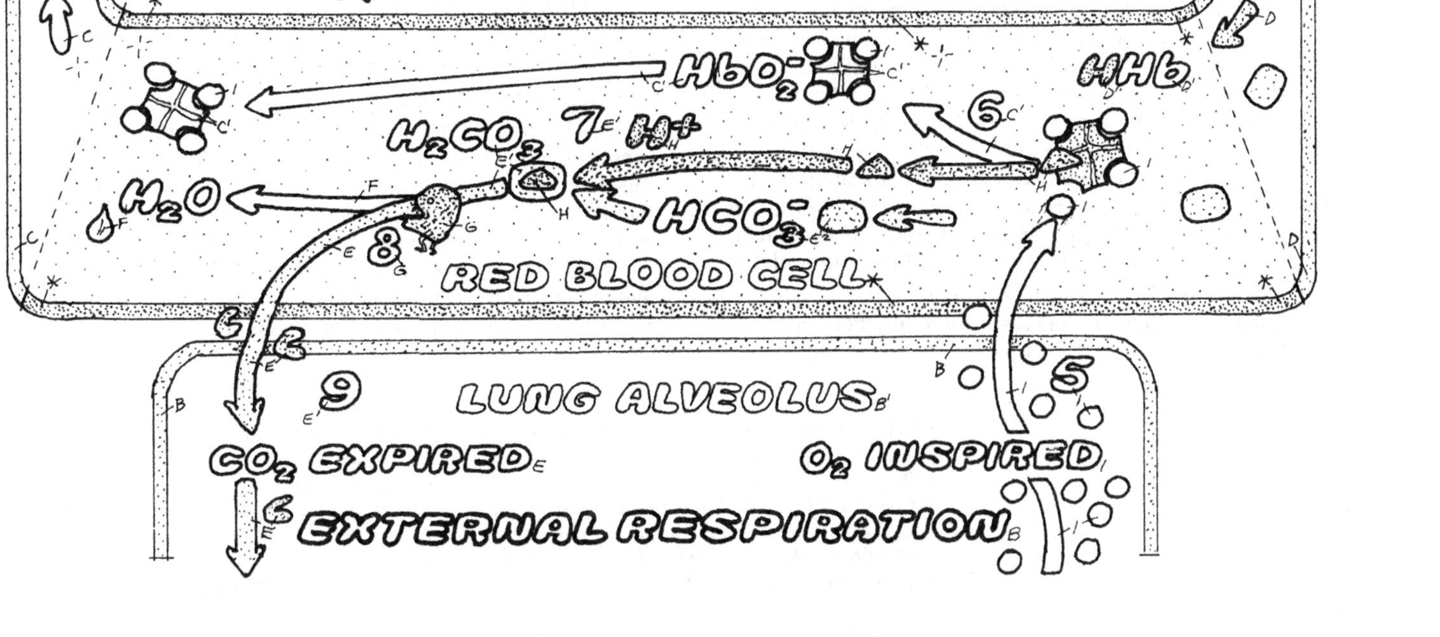

HbO_2^- HHb

H_2CO_3 H^+

H_2O HCO_3^-

RED BLOOD CELL

LUNG ALVEOLUS

CO_2 EXPIRED O_2 INSPIRED

EXTERNAL RESPIRATION

How does breathing originate? Unlike cardiac or smooth muscle, skeletal muscles that provide the motive force for respiration have no pacemaker activity. They depend entirely on the nervous system for a stimulus to contract. Two separate neural systems control respiration: (1) *voluntary control* originates in the *cerebral cortex*, and (2) *automatic control* originates in lower brain *respiratory centers*, in the *pons* and the *medulla*.

AUTOMATIC NEURAL CONTROL
Medullary neurons generate the primitive rhythm for involuntary breathing by sending bursts of impulses to the inspiratory muscles about 12 to 15 times/min. Other neurons that send impulses to expiratory muscles are generally quiet, becoming active only when respiration grows forced, as in heavy exercise.

Several factors influence the respiratory centers. Stretch receptors respond to overinflation of the lungs and send impulses to the medullary centers, inhibiting inspiration and protecting the lungs from mechanical damage. Other reflexes originate in receptors (proprioceptors) that are located in muscles, tendons, and joints and are sensitive to movement. They send to the respiratory centers stimulating impulses that help increase ventilation during exercise. Since respiration is primarily a means for maintaining the PO_2, pH, and PCO_2 levels of body fluids its success depends largely on reflexes initiated by low PO_2, low pH, and high PCO_2 in the plasma.

CHEMICAL CONTROL (PCO_2, PO_2, AND pH)

PCO_2 is the most important — Breathing is regulated by reflexes that respond to blood chemicals. Of these, PCO_2 is the most important. Whenever plasma PCO_2 rises (as it does during increased metabolism), it is met by a compensatory increase in ventilation, which returns the PCO_2 toward normal. Conversely, when PCO_2 falls, ventilation slows, allowing CO_2 to accumulate until PCO_2 approximates the normal level. This regulation is very sensitive and precise; an increase of arterial PCO_2 by only 1 mm Hg will stimulate an increase in ventilation of about 3 L/min. In common daily activities of rest and exercise, arterial PCO_2 does not appear to vary by more than 3 mm Hg.

Central chemoreceptors respond to PCO_2 via H^+ — The response to PCO_2 is mediated by special areas called *central chemoreceptors* located on the ventral surface of the medulla. These are anatomically distinct from the respiratory centers and are bathed in cerebrospinal fluid, which is separated from blood by the *blood-brain barrier* (i.e., blood capillary membranes that are highly permeable to CO_2, O_2, and water, but only slowly permeable to most other substances). Local application of H^+ ions to these areas rapidly stimulates ventilation. The connection with CO_2 arises because CO_2 easily diffuses through the barrier into the cerebrospinal fluid, where it is converted into HCO_3^- and H^+. A rise (fall) in CO_2 is followed by a rise (fall) in H^+ ion concentration in the cerebrospinal fluid. The CO_2 level in the blood regulates respiration by its effect on the H^+ ion concentration in cerebrospinal fluid. (The effect of arterial CO_2 is much stronger than that of arterial H^+ concentration, presumably because CO_2 diffuses through the blood-brain barrier more easily than H^+.)

Peripheral chemoreceptors respond to low PO_2 and to low pH — When arterial PO_2 drops to very low levels, com-

pensatory increases in ventilation act to return PO_2 toward normal. This response is mediated by a reflex that begins in O_2-sensitive receptors called *peripheral chemoreceptors* located close to the aortic arch and the bifurcation of the carotid arteries. Known as the *aortic* and *carotid bodies*, these receptors are small nodules of tissue containing neuro-epithelial cells in contact with sensory nerve terminals, together with a profuse blood supply. A drop in PO_2 in the arterial blood supplying these receptors stimulates them. This occurs because the cells have K^+ channels in their membranes that close in response to low PO_2. The resultant membrane depolarization opens Ca^{++} channels, and this initiates secretion of a neurotransmitter (dopamine). The dopamine stimulates sensory nerves, increasing the frequency of impulses sent to the respiratory center, which responds by increasing its discharge along motor nerves, which increase ventilation. Normally, the PO_2 in alveolar blood can be reduced considerably before this reflex becomes activated so that it does not appear to play a significant role in the day-to-day management of ventilation. In cases where arterial PO_2 is markedly reduced ($PO_2 < 60$ mm Hg) — for example, at high altitudes, in lung disease, or in hypoventilation — this reflex becomes significant.

Increasing the H^+ ion concentration in the plasma also stimulates ventilation. In practice, it is difficult to separate the effects of H^+ ions from PCO_2 because the reaction of H^+ with HCO_3^- produces CO_2. However, experiments where the PCO_2 is artificially maintained at a constant level while the H^+ ion concentration is changed leave no doubt that H^+ ions by themselves stimulate ventilation. This response is believed to be mediated by the peripheral chemoreceptors.

Responses to CO_2, O_2, and H^+ can conflict — Under normal circumstances, we rarely encounter a situation where only one of the three chemicals (CO_2, O_2, and H^+) that drive respiration changes. Each time ventilation changes, we can anticipate changes in all three. Because the response to CO_2 is so strong, its regulation most often dominates and sometimes obscures other responses. For example, if the PO_2 of inspired air is suddenly depressed, there will be increased ventilation due to the peripheral chemoreceptor reflex, but this increased ventilation will also "blow off" CO_2, depressing the PCO_2 in the blood. The decreased PO_2 stimulates respiration, but the secondary decreased PCO_2 inhibits respiration; the two stimuli conflict. As a result, the increased respiration is not nearly as large as it would have been if PCO_2 were held constant. In some instances, the respiratory gases interact in synergistic ways. Depressing PO_2 and elevating PCO_2 both stimulate respiration, but somehow the response to the two stimuli is greater than the sum of the responses to each alone.

PO_2 and PCO_2 hardly change during exercise — We might anticipate that the large increase in ventilation during exercise is brought about by lower arterial PO_2 and elevated PCO_2, but this does not seem to be the case. Careful measurements show that PO_2 and PCO_2 remain nearly constant during exercise and can hardly provoke the immense increases in ventilation. Somehow, during exercise, ventilation keeps pace with metabolism so that CO_2 is eliminated as fast as it is produced, and arterial O_2 is supplied as fast as it is consumed. The detailed mechanism for this response is not known.

CN: Use red for D (blood plasma found to the left of the second panel).
1. Begin with the upper panel.
2. Color the CO_2 control of ventilation, beginning with the blood vessel in the upper left corner and continuing clockwise. Note that the curve superimposed near the bottom of the ribs is the

diaphragm — a breathing muscle, as are the intercostals.
3. Color the bottom panel, starting at the upper left and continuing clockwise but excluding the illustration on the far right. Note that the numbers 1 and 2 (but not their titles) are colored gray.
4. Color the summary diagram on the far right.

VENTILATION

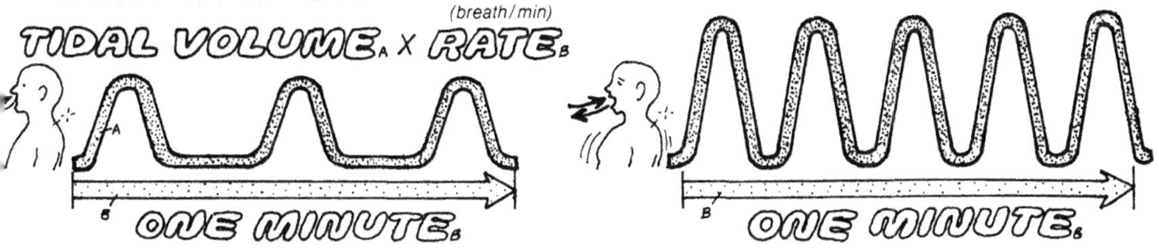

TIDAL VOLUME$_A$ × RATE$_B$ (breath/min)

ONE MINUTE$_B$

ONE MINUTE$_B$

The amount of air moved in and out of the lungs during each minute is called pulmonary ventilation, or sometimes minute respiratory volume. It is increased by increasing the amount taken in with each breath, by increasing the number of breaths per minute, or both.

CONTROL OF VENTILATION BY CO_2 & $H^+_{C'}$

METABOLISM*

IN PLASMA$_D$

$\uparrow\uparrow PCO_{2C}$

compensation

PCO_2

return toward normal

CEREBRO-SPINAL FLUID$_E$

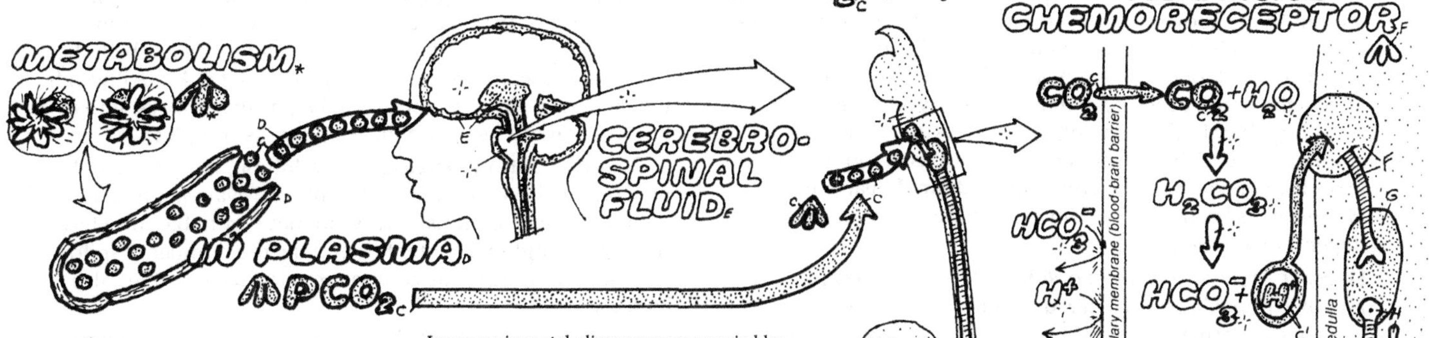

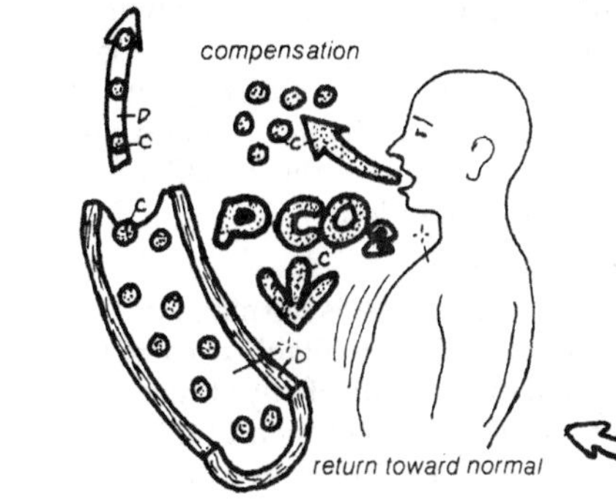

Increases in metabolism are accompanied by increases in ventilation. Plasma PCO_2 is responsible. Whenever plasma PCO_2 rises (as it does during increased metabolism), it is met by a compensatory increase in ventilation, which returns the PCO_2 toward normal. The highly permeable CO_2 diffuses easily into the cerebrospinal fluid and is converted into HCO_3^- and H^+. Special areas on the ventral surface of the medulla respond to increases in H^+ and stimulate the respiratory center to increase the rate and depth of breathing.

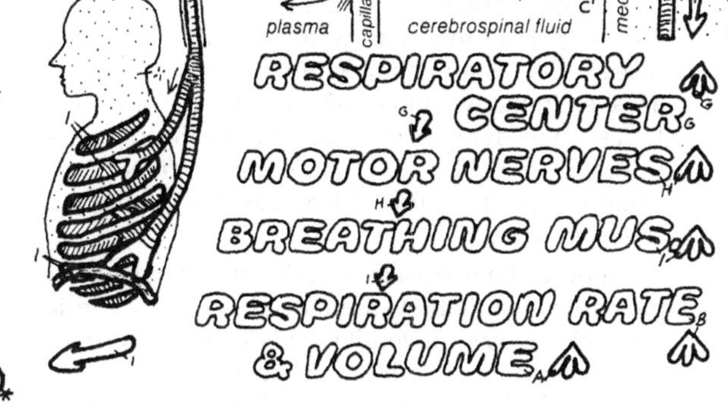

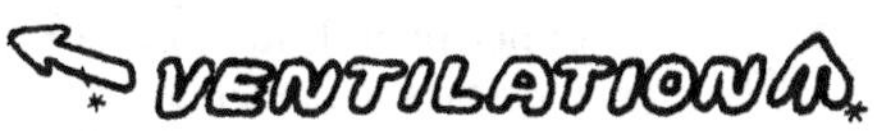

VENTILATION$\uparrow$*

CENTRAL CHEMORECEPTOR

$$CO_{2C} \rightarrow CO_2 + H_2O$$

$$H_2CO_3$$

$$HCO_3^-$$

HCO_3^- H^+

plasma   cerebrospinal fluid   medulla

capillary membrane (blood-brain barrier)

RESPIRATORY CENTER$_G$

MOTOR NERVES$_H$

BREATHING MUS.$_A$

RESPIRATION RATE$_B$ & VOLUME$_A$ $\uparrow$

CONTROL OF VENTILATION BY O_{2J}

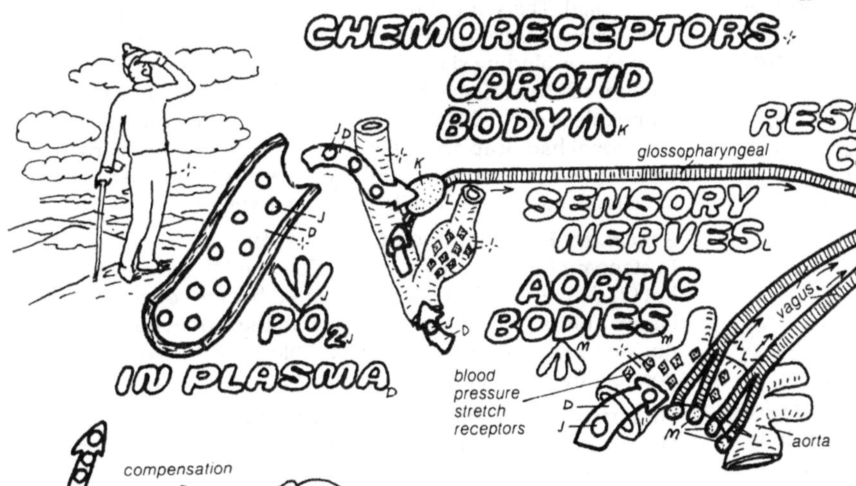

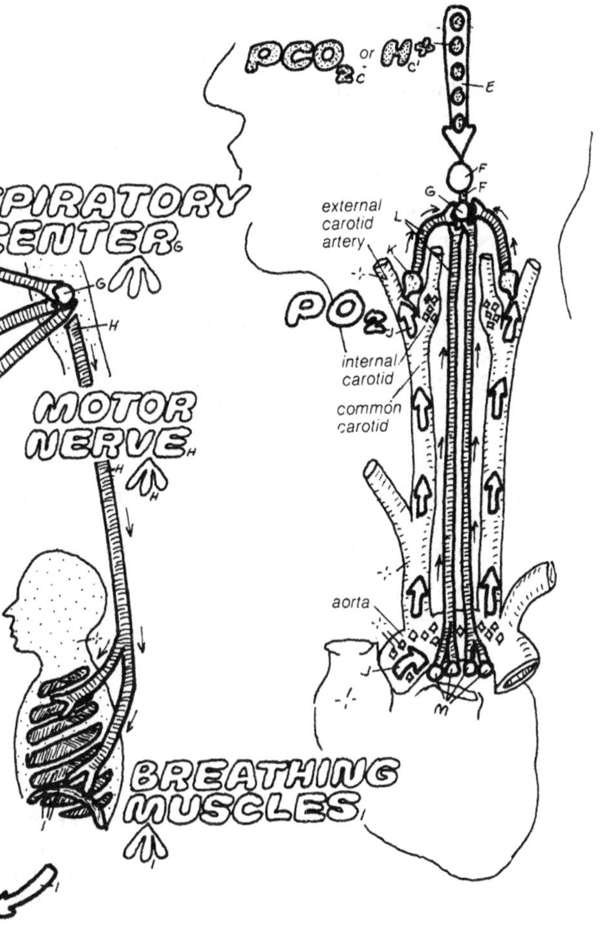

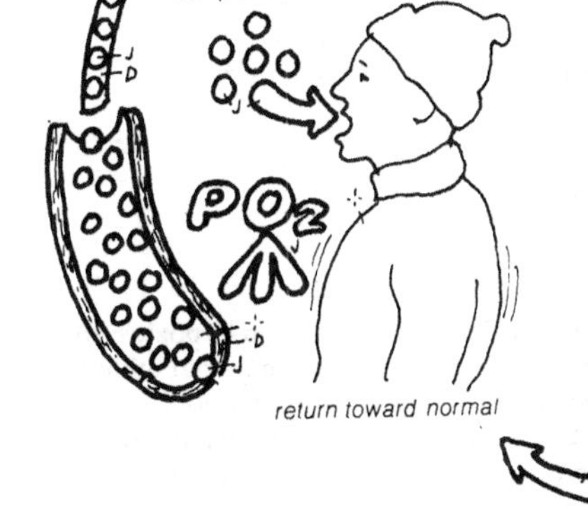

CHEMORECEPTORS

CAROTID BODY$_K$

AORTIC BODIES

SENSORY NERVES

glossopharyngeal

vagus

blood pressure stretch receptors

aorta

IN PLASMA$_D$

PO_{2J}

compensation

PO_2

return toward normal

RESPIRATORY CENTER$_G$

MOTOR NERVE$_H$

Whenever arterial PO_2 drops very low (as happens at very high altitudes), compensatory increases in ventilation act to return PO_2 toward normal. This response is mediated by a reflex that begins in O_2-sensitive receptors located close to the aortic arch (aortic bodies) and the bifurcation of the carotid arteries (carotid bodies). A drop in PO_2 stimulates these receptors; an increased frequency of impulses is sent to the respiratory center, which responds by increasing its discharge along those motor nerves that increase ventilation. The diagram to the right illustrates the combined influence of plasma PCO_2 and PO_2.

PCO_{2C} or $H^+_{C'}$

external carotid artery

internal carotid

common carotid

PO_2

aorta

BREATHING MUSCLES

VENTILATION$\uparrow$*

Hypoxia means there is an O_2 deficiency in the tissues. In most cases of severe hypoxia, the brain is the first organ to be affected. If, for example, cabin pressure is suddenly lost in an aircraft flying above 50,000 ft., the inspired PO_2 will fall to less than 20 mm Hg, consciousness will be lost in about 20 sec., and death will follow 4–5 min. later. Less severe hypoxia also affects the brain, producing an inebriated type of behavior, including impaired judgment, drowsiness, disorientation, and headache. Other, non-mental symptoms of hypoxia may include anorexia, nausea, vomiting, and rapid heart rate. Hypoxia has been classified into four different types, depending on the cause.

HYPOXIC HYPOXIA: REDUCED PO_2 IN ARTERIAL BLOOD
Hypoxic hypoxia refers to a reduced PO_2 in arterial blood. It occurs in normal people at high altitudes, where the O_2 content per volume of air is low, and it is also found in lung diseases like pneumonia. Symptoms of "mountain sickness" are seen in many people 8–24 hr. after they arrive at high altitudes. These symptoms, which include headache, irritability, insomnia, breathlessness, nausea, and fatigue, gradually disappear in the course of 4–8 days through a process called *acclimatization*.

Ventilation increases — Acclimatization begins with an increase in ventilation stimulated by the low arterial PO_2. At first, this increase in ventilation is small because it drives off CO_2, so the normal stimulating action of PCO_2 on ventilation is diminished. However, ventilation steadily increases over the next four days as the central chemoreceptor response to low PCO_2 gradually subsides. The basis for this gradual reduction of sensitivity to the lowered PCO_2 is not clear.

Kidneys excrete more bicarbonate — Reducing the normal stimulus for breathing is not the only problem created by low PCO_2. Reducing CO_2 shifts the following reaction to the left:

$$CO_2 + H_2O \rightleftharpoons H_2_2O_3 \rightleftharpoons HCO_3 + H^+$$

H^+ ions are used up in the reaction, causing body fluids to become alkaline (plate 63). Fortunately, this problem is also handled within the next few days, this time by the kidneys as they excrete more HCO_3^-. Loss of HCO_3^- compensates because it shifts the reaction back to the right, toward its original position (plate 64).

2,3 DPG increases — Acclimatization also involves the enhanced production of 2,3 DPG in red cells. Recall (plate 54) that 2,3 DPG lowers the O_2 affinity of hemoglobin (shifting the saturation curve to the right) so that it releases more O_2 to the tissues. This shift occurs within a day. However, in severe hypoxia, its usefulness is limited because the lowered affinity also makes it harder for Hb to pick up the O_2 in the lungs.

Red cell production increases — An increase in red blood cell concentration of the circulating blood also begins during the first few days of acclimatization. This raises the concentration of Hb in the blood, thus increasing the blood's capacity to carry O_2 even though the PO_2 is low. The stimulus for the enhanced production and release of red cells by the bone marrow is provided by a hormone, *erythropoietin*, which the kidneys secrete in response to hypoxia (plate 143). Although the increased red cell production begins in 2–3 days, it may take several weeks before this response is complete.

Vascularization increases — In addition, long-term acclimatization also involves a *growth of new capillaries*, thus reducing the distance O_2 must diffuse to move from blood to tissue cell. The myoglobin content of muscle, the number of mitochondria, and the tissue content of oxidative enzymes also increase.

In summary, acclimatization promotes the O_2 supply to tissues in 3 ways: (1) greater delivery of O_2 to the blood via increases in ventilation, (2) enhanced O_2-carrying capacity of the blood due to increases in red cell production, and (3) easier delivery of O_2 to the tissues by means of the 2,3 DPG response and the increased vascularization.

ANEMIC HYPOXIA: Hb DEFICIENCY
Anemic hypoxia occurs when arterial PO_2 is normal, but there is a deficiency in the amount of Hb available to carry O_2. Because arterial PO_2 is normal, there is little if any stimulation of peripheral chemoreceptors. However, compensatory increases in 2,3 DPG are often sufficient to remove hypoxia distress during rest. Difficulties arise during exercise because the ability to enlarge O_2 delivery to active tissues has been reduced. Anemias arise from a variety of causes: some are nutritional (e.g., *iron deficiency*), others are genetic (e.g., *sickle cell anemia*). The symptoms of anemic hypoxia also appear in *carbon monoxide poisoning* because carbon monoxide competes with O_2 for binding sites on the Hb molecule, reducing the amount of Hb available to carry O_2. (Hb's affinity for carbon monoxide is about 200 times larger than its affinity for O_2.) An additional handicap arises because in the presence of carbon monoxide, any "surviving" HbO_2 binds its O_2 more tenaciously, making it less available to the tissues.

STAGNANT HYPOXIA: POOR CIRCULATION
In stagnant (or ischemic) hypoxia, PO_2 and Hb are normal, but O_2 delivery to the tissue is impaired because of poor circulation. This is particularly a problem in the kidneys and heart during shock and may become a problem for the liver and possibly the brain in congestive heart failure.

HISTOTOXIC HYPOXIA: POOR UTILIZATION
Histotoxic hypoxia arises when the tissue cells are poisoned and cannot utilize the O_2, even though the O_2 delivery rate is adequate. Cyanide poisoning, which inhibits oxidative enzymes, is the most common source of this syndrome.

CN: Use the same color as on previous page for O_2 (D).
1. Begin in the upper panel by coloring all four elements (A–D); starting on the right with O_2 loading onto the boat representing Hb.

2. Color the vertical panel on hypoxic hypoxia on the left. Then color the material within the rectangle dealing with acclimation.
3. Color the remaining cartoon panels; only the significant element is to be colored.

NORMAL OXYGEN TRANSPORT,

TISSUE CELLS, HEMOGLOBIN, CARDIAC OUTPUT, OXYGEN,

Oxygen is loaded from the lungs (dock) onto hemoglobin molecules (boats), which are moved by the circulation (stream) to the tissues (dock), where it is unloaded.

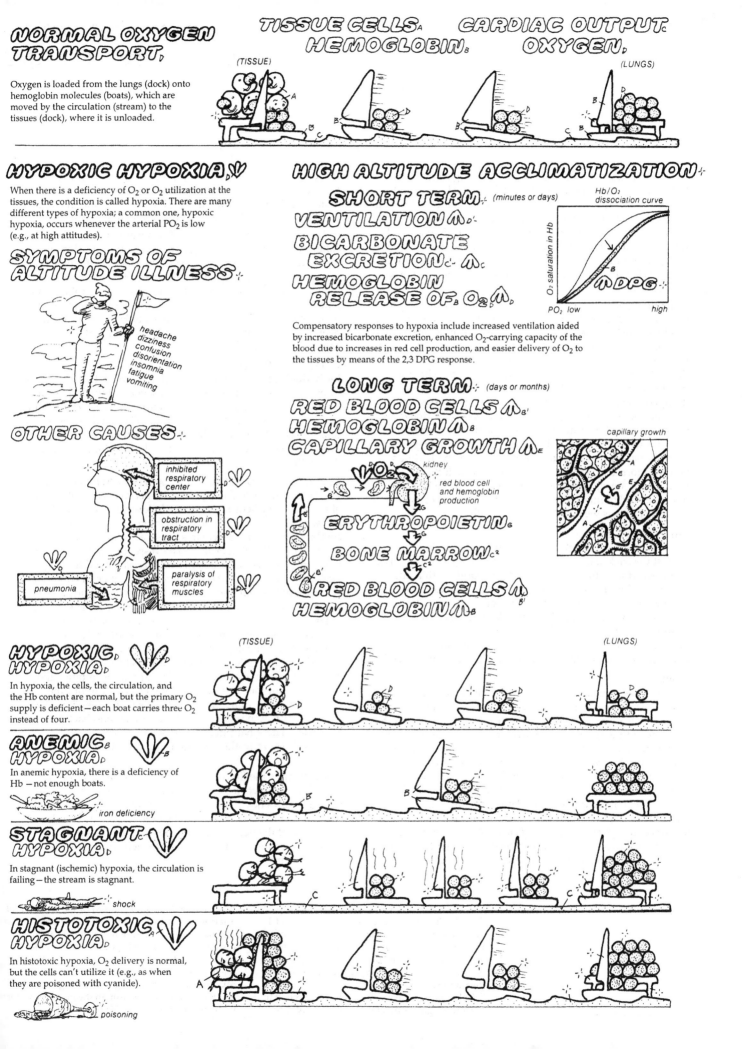

(TISSUE) ... (LUNGS)

HYPOXIC HYPOXIA,

When there is a deficiency of O_2 or O_2 utilization at the tissues, the condition is called hypoxia. There are many different types of hypoxia; a common one, hypoxic hypoxia, occurs whenever the arterial PO_2 is low (e.g., at high altitudes).

SYMPTOMS OF ALTITUDE ILLNESS,

headache
dizziness
confusion
disorientation
insomnia
fatigue
vomiting

OTHER CAUSES,

inhibited respiratory center

obstruction in respiratory tract

pneumonia

paralysis of respiratory muscles

HIGH ALTITUDE ACCLIMATIZATION,

SHORT TERM, (minutes or days)

VENTILATION

BICARBONATE EXCRETION

HEMOGLOBIN RELEASE OF, O_2

Hb/O_2 dissociation curve

O_2 saturation in Hb

PO_2 low — high

DPG

Compensatory responses to hypoxia include increased ventilation aided by increased bicarbonate excretion, enhanced O_2-carrying capacity of the blood due to increases in red cell production, and easier delivery of O_2 to the tissues by means of the 2,3 DPG response.

LONG TERM, (days or months)

RED BLOOD CELLS

HEMOGLOBIN

CAPILLARY GROWTH

O_2 kidney

red blood cell and hemoglobin production

ERYTHROPOIETIN

BONE MARROW

RED BLOOD CELLS

HEMOGLOBIN

capillary growth

HYPOXIC HYPOXIA,

In hypoxia, the cells, the circulation, and the Hb content are normal, but the primary O_2 supply is deficient—each boat carries three O_2 instead of four.

ANEMIC HYPOXIA,

In anemic hypoxia, there is a deficiency of Hb —not enough boats.

iron deficiency

STAGNANT HYPOXIA,

In stagnant (ischemic) hypoxia, the circulation is failing—the stream is stagnant.

shock

HISTOTOXIC HYPOXIA,

In histotoxic hypoxia, O_2 delivery is normal, but the cells can't utilize it (e.g., as when they are poisoned with cyanide).

poisoning

Kidneys produce urine. Under normal resting conditions, the kidneys, which comprise less than 0.5% of the body weight, receive 25% of the cardiac output! Each minute some 1300 ml of blood enters the kidneys through the *renal arteries*, and 1298–1299 ml leaves via *renal veins*, with the difference, 1–2 ml, leaving as urine via the *ureter*. Why all this fuss (claiming one quarter of the body's blood supply) for a measly 2 ml of urine? What does urine contain and why is its formation so important?

At first glance, the composition of the urine is not impressive: water, salt, small amounts of acid, and a variety of waste products, such as urea. What is impressive is how urine composition and volume *change* to compensate for any fluctuation in volume or composition of body fluids. The composition of the body fluids is apparently determined not by what the mouth takes in but by what the kidneys keep. While the design of the gastrointestinal tract appears to maximize absorption, its role in regulation is minimal. The kidneys are the guardians of the internal environment; they rework the body fluids fifteen times a day. When the body is dehydrated, the volume of water excreted decreases; when body fluids become more acid, kidneys excrete more acid; if the K$^+$ content of body fluids rises, the kidneys excrete more K$^+$. "We have the kind of internal environment we have because we have the kidneys we have" — Homer Smith.

THE WHOLE KIDNEY
The kidneys are about the size of a clenched fist. They lie against the back abdominal wall, just above the waistline. The outer covering of the kidney, called the *capsule*, is thin but tough and fibrous. When it is cut open, two regions appear: an outer zone (the *cortex*) and an inner region (the *medulla*). A microscopic view reveals the unit of kidney function, the *nephron*. Each kidney has about 1 million nephrons, which are tubular structures about 45 to 65 mm long and about .05 mm wide. Their walls are made of a single layer of epithelial cells.

NEPHRONS
A funnel-like structure about 0.2 mm in diameter, called *Bowman's capsule*, comprises the top end of the nephron. These capsules are always found in the cortex. Fluid flows through the lumen of the tubule from Bowman's capsule into the next section, the *proximal tubule*, which has a "curly" or convoluted portion and then a straight segment that dips into the medulla. This section, about 15 mm long, is called the proximal tubule because it is near the origin of the nephron (Bowman's capsule). Fluid then flows into a long, thin tube that plummets straight toward the depths of the medulla. This is the descending limb of the *loop of Henle*. At its lowest point, the loop makes a hairpin turn and begins to ascend out of the medulla back toward the cortex, becoming considerably

thicker toward the latter portions of its ascent. In the cortex, the ascending limb of the loop becomes continuous with the distal tubule. Finally, the distal tubule empties into the *collecting duct*, a tube that gathers fluid from several nephrons.

There are two major classes of nephrons. The majority, called cortical nephrons, originate in the outer portions of the cortex and are characterized by short loops of Henle that reach only the outer regions of the medulla. The remaining nephrons, which comprise only about 15% of the total, originate closer to the medulla and are known as juxtamedullary nephrons. These have very long loops of Henle that reach deep into the medulla; they are important for water conservation in the body.

COLLECTING DUCTS
Individual collecting ducts coalesce into larger tubular structures, and this pattern repeats until several of the larger tubes empty into a still larger funnel structure, the *renal pelvis*. Fluid in the renal pelvis is identical to urine. The renal pelvis is continuous with the ureter, which leaves each kidney to convey urine to the bladder, where it is stored until elimination via the urethra.

BLOOD SUPPLY TO NEPHRONS: 2 CAPILLARY BEDS IN SERIES
The blood supply to the nephrons is special because it consists of two capillary beds in series. Each Bowman's capsule has its own capillary bed, called a *glomerulus*. (Sometimes the combined structure, Bowman's capsule + glomerulus, is referred to as the glomerulus.) The vessel bringing blood to the glomerulus is called the *afferent arteriole*. Blood leaving the glomerulus does not enter a venule; rather, it enters another arteriole, the *efferent arteriole*, which serves as a conduit to the second capillary bed, called *peritubular capillaries*. The peritubular capillaries are so interconnected that it is difficult to tell which capillary came from which efferent arteriole; the tubules of any one nephron probably receive blood from several efferent arterioles. Efferent arterioles from juxtamedullary nephrons also form peritubular capillaries in much the same way, but they also send off branches—straight tubes that follow descending limbs of Henle's loops deep into the medulla, turn at the bend of the loop, and ascend back toward the cortex. These hairpin loops of blood vessels are called *vasa recta*; their design is important for water conservation.

By the time the fluid in the nephron has passed through the collecting ducts to reach the pelvis, it has become urine. Plate 59 shows how fluid simply filters out of the glomerular capillaries into Bowman's capsule. From here, it flows along the lumen of the nephron and is modified by the epithelial cells of the tubules and the collecting ducts until it finally becomes urine.

CN: Use red for A structures, blue for B, purple for R, and yellow for H. Use dark colors for J and T.
1. Begin with the cut-away drawing of the kidney in the upper right. Color the section of a kidney showing the location of two types of nephrons. Note that these have been greatly enlarged for diagrammatic purposes.
2. Color the enlarged view of a kidney section in the lower right corner. Begin with the entry of

blood (A^1) at the bottom and color the arteries and arterioles. Note that the afferent and efferent arterioles have been given different colors (P & Q) to distinguish them from the other vessels. Color the blood circulation before coloring the structures of the nephron—K–O.
3. Color the lower left diagram of the glomerulus and the flow of filtrate through the nephron. Color in the Bowman's capsule (K) first.

RENAL ARTERY A
RENAL VEIN B
KIDNEY c
CAPSULE c
CORTEX d
MEDULLA e
RENAL PELVIS f
URETER g
URINE h
BLADDER i
NEPHRON j
BOWMAN'S CAPSULE k
PROXIMAL TUBULE l
LOOP OF HENLE m
DISTAL TUBULE n
COLLECTING DUCT o

ARTERY A'
ARTERIOLE A²
AFFERENT ARTERIOLE p
EFFERENT ARTERIOLE q
PERITUBULAR CAPILLARY R
VASA RECTA R'

The kidney fabricates urine from the blood that passes through it. Each minute 1300 ml of blood enters the kidneys through renal arteries, 1298–1299 leaves via renal veins, and the difference 1–2, leaves as urine.

1300 mL/min

1299 mL/min (both kidneys)

inferior vena cava

aorta

JUXTA-MEDULLARY NEPHRON

CORTICAL NEPHRON

1 mL/min

Each kidney has about 1 million tubular nephrons, functional units that produce urine from a filtrate of blood. Each nephron contains a filtering part — Bowman's capsule — followed by a long tubular part consisting of the proximal tubule, loop of Henle, distal tubule, and collecting duct. Fluid in Bowman's capsule is protein-free plasma. Fluid at the end of the collecting duct is urine.

Blood supply to the nephrons (seen on the right) consists of two capillary beds in series. The afferent arteriole conveys blood to the glomerulus (seen below) lying in Bowman's capsule. Blood then flows through the efferent arteriole and empties into the peritubular capillaries to supply the proximal and distal tubules in the cortex. The medulla is supplied by branches of the efferent arteriole from juxtamedullary nephrons. These branches, the vasa recta, plunge into the medulla, and following the loop of Henle, make hairpin turns and return to the cortex.

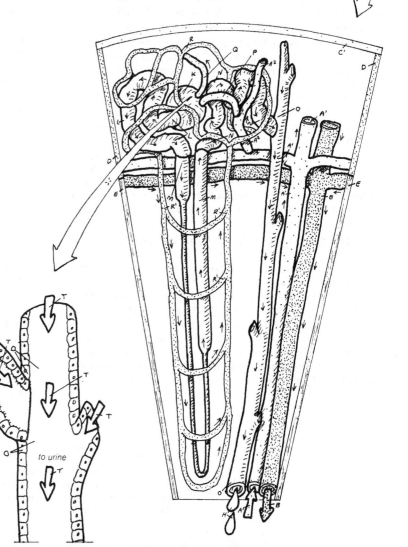

GLOMERULUS s
FILTRATE t
BOWMAN'S CAPSULE k

Fluid filters through the glomerulus into Bowman's capsule of the nephron. The filtrate then continues through the tubules on its way to the collecting duct. During this process, nutrients and most of the fluid are withdrawn; the composition of the remaining fluid is further modified until it becomes urine at the end of the collecting duct.

to urine

By the time fluid in the nephron passes through the collecting ducts to reach the pelvis, it has become urine. What is this fluid in the nephron and how did it get there? The unusual pattern of blood circulation to the kidney provides a clue. Rarely do we find one capillary bed (the *glomerulus*) leading into an arteriole (*efferent arteriole*), which in turn leads into another capillary bed (*peritubular capillaries*). The pressure in a typical capillary located elsewhere in the body begins around 35 mm Hg and falls some 20 mm until it reaches 15 mm Hg at the venous end of the vessel. If glomerular capillaries were typical, pressure of the blood delivered to the efferent arteriole would be only 15 mm Hg, hardly enough to drive blood through the next vessel, the narrow efferent arteriole. Thus, the pressures in these capillary beds must be atypical.

BLOOD PRESSURE: HIGH IN GLOMERULUS, LOW IN PERITUBULAR CAPILLARIES

Pertinent data for normal kidney blood pressures in humans are unavailable, and it is difficult to obtain accurate values even in animals, where measurements are compromised by anesthesia, surgical trauma, and blood loss. The best figures for monkeys, dogs, and rats indicate that glomerular blood pressure is high, around 50 mm Hg. The pressure drop in passing through these capillaries is small, only a few mm Hg; apparently, glomerular capillaries have a low resistance. But like that of any arteriole, the resistance of the efferent arteriole is considerable; by the time blood reaches the peritubular capillaries, the pressure has fallen to about 15 mm Hg.

Glomerular filtration: ultra filtrate of blood plasma enters the nephron — What physiological significance can we attach to these aberrant pressures? Glomerular pressures are abnormally high; peritubular pressures are abnormally low. Fluid transfer across capillary walls is determined by the balance of capillary *blood pressure* and *oncotic pressure*, so the high glomerular pressure suggests that a net fluid *filtration* occurs at these capillaries and that fluid within Bowman's capsule is simply a *filtrate* of blood plasma (i.e., fluid that would be obtained from blood if it were strained through a porous filter, in this case the porous walls of the glomerular capillary). This has been verified experimentally. Fluid at the beginning of the nephron does not arise out of any active transport process; proteins and cells are simply separated from the plasma by a passive filtration process.

High pressure and high permeability ensure glomerular filtration — This filtration is the first step in modifying a portion of blood plasma that will eventually be excreted as urine. As the fluid flows along the nephron past the cells making up the tubular walls, substances may be withdrawn from the fluid and returned to the blood via the peritubular capillaries; this process is called *reabsorption*. Alternatively, some substances may be removed from the blood and added to the tubular fluids in a process called *secretion*. Glomerular filtration followed by tubular reabsorption and secretion are the fundamental processes by which the kidney regulates the internal environment.

The funnel-like structure of *Bowman's capsule* allows it to collect the filtrate and convey it to the lumen of the proximal tubule. Fluid that filters from the blood into the lumen of the nephron must pass through three potential barriers: (1) the thin cell layer making up the capillary wall (the capillary endothelium), (2) the *basement membrane* associated with the capillary, and (3) the epithelial cell layer making up Bowman's capsule. The capillary endothelium is riddled with *fenestrations*, which are easily penetrated by most molecules but not by cells. The basement membrane and the outer surfaces of both cell layers are embedded with glycoproteins that contain a strong negative charge. The last barrier, those epithelial cells of Bowman's capsule that are in direct contact with the capillaries, have a peculiar structure; they are called *podocytes*. Podocytes send out foot processes that interdigitate with foot processes of other cells. Vacant spaces or *slits* remain between these foot processes. The filtration pathway through the capillary fenestrations, across the basement membrane, and through the slit passages does not hinder the passage of small molecules like salts, glucose, and amino acids. As the size of the molecules increases, the pathway begins to offer some resistance, depending on the molecule's size, shape, and electrical charge. Electrically neutral molecules, the size of the plasma protein albumin, can permeate this barrier to a limited extent. However, albumin, which is negatively charged, is restrained by the negative charge on the basement membrane and cell surfaces.

The glomerular capillaries not only have a higher pressure, they are also more permeable than many capillaries. Both factors promote filtration. Glomerular capillaries filter twenty times more fluid than ordinary capillaries. Fully one-fifth of the fluid entering the capillary is delivered to the nephron via the filtration path. This loss of fluid from the blood concentrates the remaining proteins. By the time blood enters the efferent arteriole and peritubular capillaries, the oncotic pressure (osmotic pressure exerted by the plasma proteins) has risen from a normal value of 25 to 30 mm Hg.

LOW PRESSURE FAVORS PERITUBULAR REABSORPTION

Just as high pressure prepares the glomerulus for filtration, low pressure in the peritubular capillary promotes fluid reabsorption over its entire length. The major force for reabsorption from interstitial spaces arises from the osmotic pressure of the plasma proteins, and we have seen that this is unusually high in peritubular capillaries. Because the opposing filtration pressure (blood pressure in the capillary) is low, the peritubular capillary is well adapted to reabsorb fluid from its environment. However, these arguments apply only to reabsorption between capillary and interstitial fluids. Fluid reabsorption between nephron and interstitial space is governed by a more complex set of osmotic forces determined by the concentrations of small solutes, especially NaCl. Compared to capillaries, nephron walls are much less permeable to small solutes, so the contribution of these small solutes to effective osmotic pressure gradients becomes large.

CN: Use red for A, light blue for G, and dark colors for H, J, and L.

1. Begin with the diagram of filtration, reabsorption, and secretion by coloring the horizontal blood vessels, blood, and plasma proteins before coloring the upper diagrams.

2. Color the elements of the renal corpuscle, beginning with the small exterior diagram, then the functional diagram on the far left, and then the interior view. Notice that the slit pore (P), represented by arrows, refer to a narrow space between the cells of the podocytes (P).

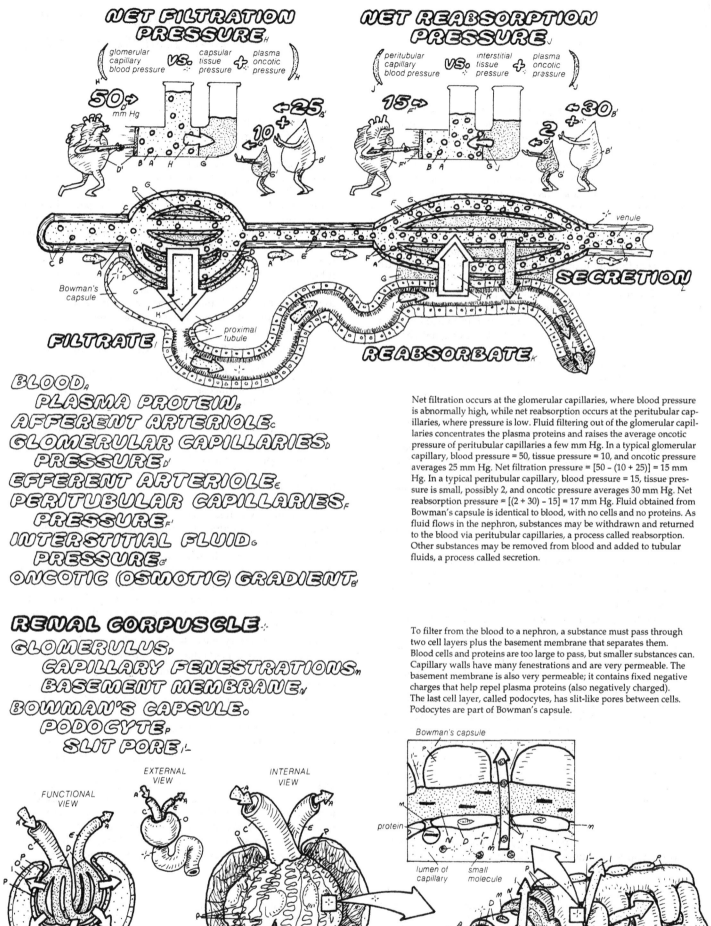

NET FILTRATION PRESSURE

glomerular capillary blood pressure vs. capsular tissue pressure + plasma oncotic pressure

50 mm Hg   10 + 25

NET REABSORPTION PRESSURE

peritubular capillary blood pressure vs. interstitial tissue pressure + plasma oncotic pressure

15   2 + 30

venule

SECRETION

Bowman's capsule

FILTRATE

proximal tubule

REABSORBATE

BLOOD
PLASMA PROTEIN
AFFERENT ARTERIOLE
GLOMERULAR CAPILLARIES
PRESSURE
EFFERENT ARTERIOLE
PERITUBULAR CAPILLARIES
PRESSURE
INTERSTITIAL FLUID
PRESSURE
ONCOTIC (OSMOTIC) GRADIENT

Net filtration occurs at the glomerular capillaries, where blood pressure is abnormally high, while net reabsorption occurs at the peritubular capillaries, where pressure is low. Fluid filtering out of the glomerular capillaries concentrates the plasma proteins and raises the average oncotic pressure of peritubular capillaries a few mm Hg. In a typical glomerular capillary, blood pressure = 50, tissue pressure = 10, and oncotic pressure averages 25 mm Hg. Net filtration pressure = [50 – (10 + 25)] = 15 mm Hg. In a typical peritubular capillary, blood pressure = 15, tissue pressure is small, possibly 2, and oncotic pressure averages 30 mm Hg. Net reabsorption pressure = [(2 + 30) – 15] = 17 mm Hg. Fluid obtained from Bowman's capsule is identical to blood, with no cells and no proteins. As fluid flows in the nephron, substances may be withdrawn and returned to the blood via peritubular capillaries, a process called reabsorption. Other substances may be removed from blood and added to tubular fluids, a process called secretion.

RENAL CORPUSCLE
GLOMERULUS
CAPILLARY FENESTRATIONS
BASEMENT MEMBRANE
BOWMAN'S CAPSULE
PODOCYTE
SLIT PORE

To filter from the blood to a nephron, a substance must pass through two cell layers plus the basement membrane that separates them. Blood cells and proteins are too large to pass, but smaller substances can. Capillary walls have many fenestrations and are very permeable. The basement membrane is also very permeable; it contains fixed negative charges that help repel plasma proteins (also negatively charged). The last cell layer, called podocytes, has slit-like pores between cells. Podocytes are part of Bowman's capsule.

FUNCTIONAL VIEW

EXTERNAL VIEW

INTERNAL VIEW

Bowman's capsule

protein

lumen of capillary

small molecule

Approximately 120 ml of protein-free plasma filters into the nephrons each minute. If this fluid were excreted as urine, it would take only 25 minutes (3000 ml plasma/120) to exhaust the entire plasma volume. This fluid would carry with it everything dissolved in the plasma (glucose, amino acids, minerals, vitamins, etc.). The fact that you are reading this page is living proof that this does not happen. The tubules recapture (reabsorb) most of the fluid, practically all the nutrients, and some minerals before the filtrate reaches the ends of the collecting ducts.

The nephron is primarily a regulatory organ. Faced with a torrent of fluid at its origin, its first job is to reduce the volume of filtrate to manageable levels and to reclaim essential nutrients. The responsibility for this task falls primarily on the proximal tubule. By the time the filtrate reaches the end of the proximal tubule, two-thirds of the water and virtually all of the nutrients have been reabsorbed. Of the original 120 ml of fluid that entered through the filter, only 40 ml passes on to the loop and distal tubule where more subtle regulatory processes take place.

TUBULAR CELLS ARE ASYMMETRIC
This massive transport requires asymmetric tubular cells. Note in the bottom diagram that the cell membrane facing the lumen is covered with fingerlike projections called *microvilli*. They resemble bristles in a brush; hence the membrane is called the brush border. The membrane surrounding the remaining three-quarters of the cell has no microvilli; it is called the *baso-lateral* membrane (plate 2). These two membranes have different properties—they contain different proteins, enzymes, and transport systems. The two membranes are separated by *tight junctions* that prevent migration of any proteins from one membrane to the next. The baso-lateral membrane resembles membranes of most cells—e.g., it contains many Na^+-K^+ *pumps* and *facilitated diffusion* systems for glucose and amino acids (plates 9, 10). The brush border does not contain these transporters, but it contains others.

ACTIVE Na+ TRANSPORT DRIVES WATER REABSORPTION
The prime mover for most proximal tubular transport is the active transport of Na^+ (via the Na^+-K^+ pump), which keeps intracellular Na^+ concentration more than ten times lower than extracellular. Because Na^+ concentration is higher in the lumen than in tubular cells, it moves down its concentration gradient via several different co- and counter-transporters into the cell (see below). But it cannot be pumped back out into the lumen because the brush border has no Na^+ pump; it can be pumped out of the cell into the interstitial spaces only by the baso-lateral membrane. The result is a stream of Na^+ moving from lumen to cell, only to be pumped out into the interstitial space, where it can diffuse into the peritubular capillary. In other words, Na^+ is reabsorbed. But Na^+ carries a positive charge; it attracts negatively charged ions that, one way or another, move along with it. Because Cl^- is the most abundant, easily transported negative ion, we end up reabsorbing large quantities of Na^+ and Cl^-.

Both Na^+ and Cl^- are important determinants of the effective osmotic gradients across the tubular cell. Each time a Na^+ and Cl^- are transported from lumen to interstitial space, the lumen loses two osmotically active particles, while the interstitial space gains two. This creates an osmotic gradient favoring reabsorption of water. For each Na^+ and Cl^- moved, about 370 water molecules follow to maintain osmotic equilibrium. Once the water arrives in the interstitial space, the high oncotic pressure (and low blood pressure) in the peritubular capillaries is sufficient to absorb the water back into the blood. The loss of water from the tubular lumen concentrates the remaining solutes, and those that are freely permeable to tubular membranes will diffuse down the resulting concentration gradient from lumen to interstitial space. So in addition to reabsorption of Na^+, the asymmetric Na^+ transport is also responsible for reabsorption of Cl^-, copious amounts of water, and some fraction of other diffusible solutes.

SECONDARY ACTIVE TRANSPORT OF GLUCOSE, AMINO ACIDS, LACTATE, AND PHOSPHATE
In addition to Cl^- and water, Na^+ transport is also coupled to the reabsorption of glucose, amino acids, lactate, and phosphate. The brush border contains a different system to *cotransport* Na^+ with each of these substances (plate 9). Since these systems operate similarly, we shall describe only Na^+ and glucose. This system transports Na^+ and glucose together, but will not operate with either alone. The system is symmetric; it does not require ATP, and it is capable of transporting the pair into or out of the cell. In practice, the co-transport system always transports the pair into the cell because of the Na^+-K^+ pump, which keeps intracellular Na^+ scarce and makes it difficult for glucose to find an Na^+ partner to ride the co-transport system in the reverse direction. By keeping intracellular Na^+ low, the cell creates a one-way system for glucose transport. As a result, glucose accumulates inside the cell even above its concentration in the lumen or plasma; it is as though glucose has been actively transported into the cell. And it has, in a way, only now the energy has come from the Na^+ gradient and only indirectly from the splitting of ATP. It is an example of secondary active transport.

Once glucose is inside the cell in higher concentrations, it moves out through the baso-lateral membrane toward the blood via a facilitated transport system that does not require Na^+. The transport of amino acids, lactate, and phosphate are similar.

OTHER FUNCTIONS OF THE PROXIMAL TUBULE
Proximal tubules also play a role in acid-base balance (plate 64) and in regulating calcium, magnesium, and phosphorus. In addition, they have active transport systems for *secretion* of organic acids and bases from blood to lumen. This system is important clinically because many drugs afect it. The secretory transporter is often located on the baso-lateral membrane, so the secreted material is accumulated in the cell. The brush border passively transports these substances; they move from the cell, where they are concentrated, to the lumen.

CN: Use the same colors as on the previous plate for proximal tubule (A), filtration (B), secretion (E), and capillary (C).
1. Begin with the upper diagram, coloring the filtrate (B^1) entering the proximal tubule (A) on the far left.
2. In the lower diagram, first color the filtrate arrow, the tubular cells (A^2) with their brush borders (F), and the capillary wall on the right. Then follow the order of the titles and color the various transport mechanisms. Note that the substances being transported receive the color of that particular mode of transport, so that it is possible for Na^+ to receive four different colors. Start with the ATP-driven sodium pump at the asterisk in the lower right corner of the center cell.

PROXIMAL TUBULE

The proximal tubule invariably reabsorbs 2/3 of the water and virtually all of the nutrients in the filtrate. It also secretes organic acids and bases into the lumen. Of the original 120 ml of fluid that enters through the filter, only 40 mL is passed on to the loop and distal tubule, where more subtle regulatory processes take place.

PERITUBULAR CAPILLARY

BLOOD

SECRETION

fatty acids, prostaglandins, uric acid, bile salts

cyclic AMP, acetyl choline, epinepherine, histamine, dopamine

saccharin, aspirin, penicillin, morphine, cimetidine

others

loop of Henle

FILTRATION

Bowman's capsule

PROTEIN-FREE PLASMA:

water
salts and minerals
glucose, amino acids
vitamins, hormones
urea & other small solutes

FILTRATE

VOLUME IN
120 mL / min

VOLUME OUT
40 mL / min

REABSORPTION

water (2/3)
glucose, amino acids, vitamins
some salts and minerals
some urea and other permeable
solutes

80 mL / min

TUBULAR CELL

BASOLATERAL MEMBRANE
BRUSH BORDER (MICROVILLI)
ACTIVE TRANSPORT
CO-TRANSPORT
DIFFUSION
FACILITATED DIFFUSION
COUNTER TRANSPORT
OSMOSIS

Most reabsorptive processes are coupled to movements of Na^+. Na^+ is pumped out of the tubular cell by an Na^+-K^+ pump located in the basolateral membrane, but not in the brush border. This provides a one-way movement from lumen to blood. Cl^- follows Na^+ because of electrical attraction. Movement of Na^+ and Cl^- creates an osmotic gradient that drags water with it. By linking with downhill Na^+ movement into the cell, other solutes are pumped uphill. Glucose and amino acids are co-transported with Na^+ into the cell. H^+ is countertransported out of the cell (in exchange for Na^+).

tight junction

channel

CAP.

tubular lumen

FILTRATE

GL

GL

GL

red blood cell

Na^+
GL

Na^+

Na^+

also amino acids, lactate, and phosphate

Na^+

Na^+

H^+

H^+

K^+

Na^+

ATP

interstitial fluid

large protein

K^+

capillary lumen

NaCl

H_2O

Na^+

Cl^-

H_2O

H_2O

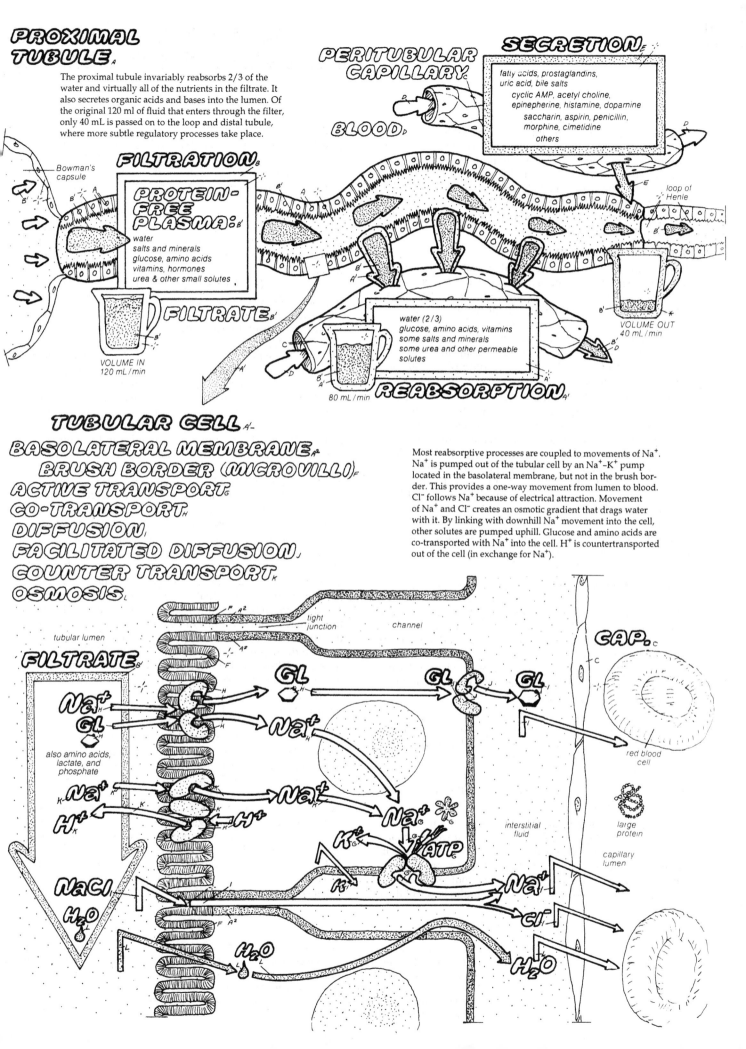

In modern times it has been possible to micro-dissect the kidney in anesthetized animals, collect fluid at different positions in individual nephrons, and tease out portions of nephrons to study them in detail. However, techniques for the study of quantitative aspects of the whole kidney have been available for many years. The latter techniques are particularly valuable because they are non-invasive and can be readily applied to unanesthetized humans.

BOOKKEEPING IN THE KIDNEY

The principle used to measure filtration, reabsorption, or secretion is simple: what goes in must come out. It is an application of the conservation of matter. Suppose you knew that 100 mg of a sugar were filtering into the nephrons each minute, but only 60 mg were appearing in the urine. Unless the nephrons were destroying the sugar, 40 mg (100 – 60) were reabsorbed. Alternatively, if 120 mg appeared in the urine, you would conclude that 20 mg (100 – 120 = –20) had been secreted during that minute.

Amount excreted per minute = $U_s \times V$ — How do we estimate how much goes through the filter and how much comes out in the urine during each minute? The latter is easy. Collect the urine over a period of time — say an hour. Analyze it to find out how much sugar there is in each milliliter (i.e., determine the concentration of the sugar in the urine), then multiply this figure by the total number of milliliters of urine collected. This gives the amount excreted per hour. To find the amount excreted per minute, divide by 60. Letting E = the amount of a solute excreted per minute, U_s = the concentration of the solute in the urine, and V = the volume of urine that is excreted per minute, we have

$$(1) \quad E = U_s \times V.$$

Amount filtered per minute = $P_s \times GFR$ — Estimating the amount of solute going through the filter each minute (called *filtered load*) is trickier. A related quantity, the number of milliliters of fluid flowing through the filter each minute, is called the glomerular filtration rate, abbreviated as GFR. If we knew the GFR, the problem would be easier. Let P_s = the concentration of any solute, such as sugar, in the blood plasma. Then the amount of sugar coming through the filter each minute (i.e., the filtered load), F, will be given by

$$(2) \quad F = P_s \times GFR.$$

For our final bookkeeping on tubular activities (reabsorption or secretion), we let RS_s denote the amount that is reabsorbed (or secreted) during each minute. Then

$$(3) \quad RS_s = F - E = [P \times GFR] - [U_s \times V].$$

If RS_s is positive, it represents reabsorption. If it is negative, it represents secretion.

Inulin clearance measures the GFR — Using equation (3), we can calculate RS_s, provided we can measure all the quantities on the right-hand side. Three of these — P_s, U_s, and V — are routine. The fourth, GFR, is not. Turning the problem inside out, if we knew RS_s for any substance, we could solve for GFR. Fortunately, these substances exist; inulin is one of them.

Inulin is a nontoxic polysaccharide that is small enough to pass freely through the filter but too large to pass through solute channels in cell membranes or through the tight junctions between tubular epithelial cells. Further, inulin is not lipid soluble, so it won't permeate the lipid bilayer portion of the cell membrane. Finally, inulin is not produced or metabolized in the body; there are no special transport systems that will carry it. In particular, the tubules neither secrete nor reabsorb inulin; $RS_{inulin} = 0$. Using this fact, we rewrite equation (3) for the special case of inulin: $0 = [P_{in} \times GFR] - [U_{in} \times V]$. Solving for GFR:

$$(4) \quad GFR = [U_{in} \times V] / P_{in}.$$

In practice, GFR is measured by injecting inulin, collecting and analyzing blood and urine samples at intervals, and using this last expression for calculation. For historical reasons, the ratio $[U_s \times V] / P_s$ for any substance s is called the *clearance* of S. The GFR is equal to the *inulin clearance*. Notice that GFR is simply the amount of fluid flowing through the filter per minute. It really has nothing to do with inulin. Inulin is merely an artificial substance that we use to trace the filtrate so we can measure its volume. To study a more interesting solute — call it S — we follow the same routine, only now we analyze the blood and urine for both inulin and S. Inulin data are used to calculate GFR from equation (3) as before, and this result, together with the blood and urine data for S, is used in equation (3) to calculate RS_s. These procedures have been used both clinically to test renal function and experimentally to study renal mechanisms.

Renal tubular cells "renew" the plasma every 25 min — Through the use of inulin clearance, an estimate of a normal value for GFR = 120 ml/min (both kidneys) has been obtained. This means that each day $120 \times 60 \times 24 = 172,800$ ml of fluid passes through the glomerular filter into the lumens of the nephrons, a space that is essentially outside the body. That is an enormous amount of fluid, a volume approximately three times the total volume of all the body fluids. It means that the entire plasma volume (approximately 3000 ml) passes through the nephrons every 25 min (3000 /120)! That is, by selective reabsorption and secretion, the renal tubular cells renew the plasma every 25 min.

Glucose reabsorption is limited — A typical example of the use of clearance techniques is provided by the glucose excretion experiment illustrated in the lower diagram. Various amounts of glucose were administered to systematically change plasma glucose concentrations from normal to very high. At normal levels (70–110 mg/100 ml) and below, no glucose is excreted; the entire filtered load is reabsorbed. As plasma concentration is increased, so is the filtered load. Eventually, we reach a threshold plasma concentration where almost all reabsorption sites are working to maximal capacity, and some glucose escapes reabsorption, spilling over into the urine. The maximal capacity to reabsorb glucose is called the *TM (tubular max)*. The diagram shows how reabsorption for glucose (RS_{GL}) changes with plasma concentration. It is obtained by subtracting E from F at each concentration.

CN: Use the same colors as on the previous page for filtration (C), reabsorption (E), and excretion (D). Use light blue for A and a dark color for B. Note that the two shapes at the top are Bowman's capsule (filtration) and a drop of urine (excretion).
1. Begin with the formula at the top.

2. Color the three central panels, starting on the left. Note that in the first two panels the number of boxes filtering is kept artificially small for purposes of simplicity.
3. Color the procedure for the measurement of glucose reabsorption into the blood stream.

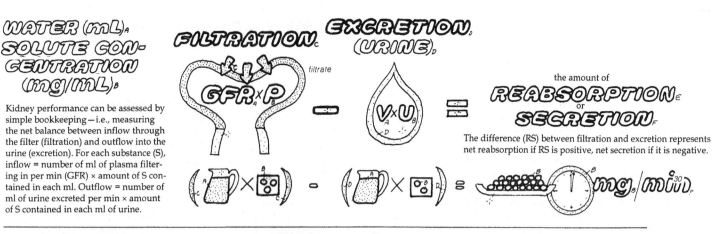

WATER (mL)ₐ

SOLUTE CON-CENTRATION (mg/mL)ʙ

Kidney performance can be assessed by simple bookkeeping—i.e., measuring the net balance between inflow through the filter (filtration) and outflow into the urine (excretion). For each substance (S), inflow = number of ml of plasma filtering in per min (GFR) × amount of S contained in each ml. Outflow = number of ml of urine excreted per min × amount of S contained in each ml of urine.

FILTRATIONᴄ EXCRETION (URINE)ᴅ

REABSORPTIONᴇ or SECRETIONꜰ

The difference (RS) between filtration and excretion represents net reabsorption if RS is positive, net secretion if it is negative.

FILTRATION & REABSORPTION

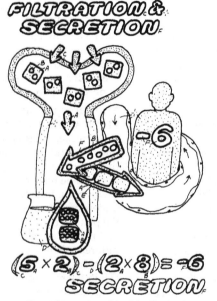

$$(5 \times 2) - (2 \times 3) = 4$$

REABSORPTION

Count the number of solute particles (circles) filtering in. Each box of water contains two, and there are five boxes per min (GFP), so that filtered load = 5 × 2 = 10 per min. Similarly, there are 2 × 3 = 6 particles leaving (excreted) per min. The difference, four particles, is reabsorbed (RS = 10 – 6 = 4 particles per min).

FILTRATION & SECRETION

$$(5 \times 2) - (2 \times 8) = -6$$

SECRETION

Count the number of particles filtering in: 5 × 2 = 10 per min. Similarly there are 2 × 8 = 16 particles leaving per min. The balance (difference) is negative (RS = – 6). Six particles are secreted per min.

FILTRATION ALONE

INULIN

$$(120 \times 2) - (1 \times 240) =$$

REABSORPTION = 0 SECRETION = 0

Inulin is a special substance; it is not reabsorbed and not secreted (RSₐᵢₙ = 0), and filtered load = excretion. We use this fact to measure GFR. Inulin is injected into plasma, and samples of plasma and urine are taken for analysis. If 240 inulin particles are excreted per min and each ml of plasma contains only two, then we require 120 ml of plasma per min to deliver the 240 particles being excreted. GFR = 120 ml per min. The algebra is shown below.

$$(GFR \times P) - (V \times U) = 0$$

$$GFR = \frac{V \times U}{P}$$

$$GFR = C_{in}$$

(CLEARANCE OF INULIN)

MEASURING GLUCOSE REABSORPTION (RS_gl)

To study how glucose reabsorption depends on plasma glucose:

1 INJECT: GLUCOSE + INULIN

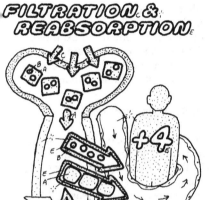

3 $GFR = \dfrac{V \cdot U_{IN}}{P_{IN}}$

2 TAKE SAMPLES & MEASURE

plasma   urine

P_{GL} U_{GL} V
P_{IN} U_{IN}

Calculate the GFR from measured values of P_{in}, U_{in}, and V. Use this value of GFR together with measured values of P_{gl}, U_{gl}, and V to calculate RS_{gl}.

4 $(GFR \times P_{gl}) - (V \times U_{GL}) = RS_{GL}$

A plot of results of these measurements at different plasma glucose concentrations (P_{gl}) is shown here. RS_{gl} is obtained by subtracting the amount excreted from the amount filtered at each concentration. At normal levels (A) and below, no glucose is excreted; the entire F is reabsorbed. Increasing P_{gl} increases F and RS_{gl}. Eventually a P_{gl} is reached where almost all reabsorption sites are working to maximal capacity (B) and some glucose spills over into the urine (C). Increasing F beyond this level cannot increase RS_{gl}.

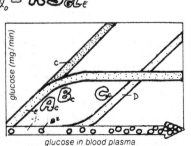

glucose (mg/min)

glucose in blood plasma

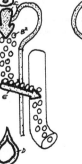
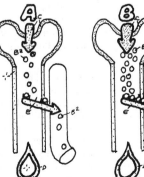
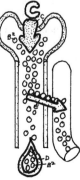

Control of the *glomerular filtration rate (GFR)* is crucial to kidney performance. Abnormally fast filtration will swamp the tubules, allowing filtrate to speed by the cells before they have time to modify the fluid. Abnormally slow rates will compromise the kidneys' ability to process adequate amounts of fluid during each minute. Nevertheless, blood pressure and blood flow to the kidney do change, often in response to stresses not directly related to kidney function (e.g., a sudden drop in arterial pressure; see plate 45). How can changes in GFR be modulated under these circumstances?

SYMPATHETIC NERVE IMPULSES CONSTRICT RENAL ARTERIOLES

Panel A shows that renal blood flow is reduced by, *sympathetic nerve* impulses, which constrict arterioles, but the effect of these impulses on GFR depends on which arterioles are most constricted. Constricting the *afferent arteriole* reduces renal blood flow, causing downstream (glomerular) pressure to decrease, thereby decreasing GFR. Constricting the *efferent arteriole* also reduces renal blood flow, but now the glomerulus is upstream. Its pressure rises and GFR increases. Because the afferent arterioles contain more smooth muscle, their constriction is generally the more forceful. However, the simultaneous constriction of efferent arterioles can be expected to diminish changes in GFR that might otherwise occur.

RENAL BLOOD FLOW AND GFR ARE INSENSITIVE TO ARTERIAL PRESSURE

Myogenic mechanism — Panel B illustrates an important property of renal blood vessels: both renal blood flow and GFR are very insensitive to changes in systemic arterial blood pressure in the range of 80 to 180 mm Hg. (Compare the flat part of the curves with the dotted diagonal line that would be expected if the blood vessels were simple passive structures.) Shared by most vascular beds, this behavior is most pronounced in the kidneys. Due to properties of the smooth musculature of the vessel walls, this behavior persists when all nerve supplies are cut but disappears when the smooth muscle is paralyzed with drugs. Apparently, the blood vessel smooth muscles are sensitive to pressure. Stretching the muscle opens special stretch-sensitive ion channels that depolarize the cells and the muscle contracts. When pressure rises, flow would ordinarily increase, but contraction of the smooth muscle in the arteriolar wall, reduces the radius of the vessel and increases its resistance. As a result, flow does not increase as much, and energy (pressure) is lost flowing through the high resistance. Thus, capillary pressure and the ensuing GFR do not increase as much.

MATCHING GFR TO REABSORPTION CAPACITY OF INDIVIDUAL NEPHRONS

The kidneys' capacity to regulate body fluids is especially sensitive to the rate at which fluid is delivered to the *distal tubule*. This is where regulation of salt, water, and acidity occurs. If the flow is too fast, the distal tubule cells will be overwhelmed; if it is too slow, there is danger of overcompensation. The lower diagram on the left shows a *feedback* mechanism that adjusts the GFR *in each single nephron* to maintain a constant load delivery to the distal tubule.

The beginning of the distal tubule of each nephron is located next to its corresponding glomerulus and makes contact with the afferent arteriole in a specialized structure called the *juxtaglomerular (JG) apparatus.* As flow increases, solute delivery (probably NaCl) to the JG apparatus increases and in some unknown way stimulates constriction of the afferent arteriole so that GFR in the same nephron decreases. Conversely, as flow decreases, GFR increases. In this way, the GFR is matched to the reabsorption capacity of the proximal tubule. The mechanism, called *tubuloglomerular feedback*, is particularly interesting because, unlike the two mechanisms described above, it is a discrete, local regulation; each nephron has its own independent control system. If, for example, the glomerulus of a particular nephron becomes damaged and leaky so that the filtration rate in that nephron increases, the feedback will constrict the afferent arteriole of that nephron and no others.

MATCHING PROXIMAL FLUID REABSORPTION TO GFR: PHYSICAL MECHANISMS

Increased GFR is accompanied by increased peritubular oncotic pressure — There are two simple physical mechanisms that operate to match proximal fluid reabsorption to the GFR. If, for some reason, GFR increases, so does *proximal tubular reabsorption.* If GFR goes down, reabsorption decreases. To understand the first mechanism, illustrated on the bottom of the plate, recall that fluid reabsorption is determined by net Na^+ reabsorption. But net Na^+ reabsorption is given by the difference between active pumping of Na^+ (lumen to interstitial space) and the back-leak of Na^+ through tight junctions (TJ) in the reverse direction. If GFR decreases, compensatory reductions in proximal fluid reabsorption occur because of the following. With a small GFR, less fluid is removed from glomerular capillaries, so the plasma proteins become less concentrated as they flow through the glomerulus. This means that the oncotic pressure delivered to the peritubular capillaries is lowered, reducing the forces favoring fluid reabsorption by these capillaries from the interstitial fluid. The buildup of fluid in the interstitial space will increase the tissue pressure, which may force the seal between cells (the tight junction) to leak so that both water and Na^+ leak back into the tubular lumen. The steps are reversed when GFR increases, resulting in a compensatory increase in reabsorption.

Distant portions of the proximal tubule provide a reserve for reabsorption under increased loads — The second mechanism that helps match changes in tubular reabsorption to changes in GFR depends on the coupling of fluid reabsorption to solute reabsorption, particularly to Na^+, which is co-transported with glucose, amino acids, and other solutes. With normal GFR, these co-transported nutrients are completely reabsorbed before they reach the end of the proximal tubule. With higher GFR, more solute is filtered and the more distant reaches of the tubule begin to be utilized. More solute is reabsorbed so more fluid is also reabsorbed. Those distant portions of the proximal tubule not used to transport glucose or amino acids during normal GFR supply a reserve for reabsorption under increased loads.

CN: Use red for the D letters including RBF (renal blood flow — D^3). Use dark colors for A and B.
1. After noting the question posed at the top of the page, color the titles 1 (above) and 2 (below) titles 1 (above) and 2 (below), which are the two major categories covered. Begin coloring in the upper left corner, which defines the area of the nephron that deals with the subject matter of this page. Note the area enclosed in the rectangle in the upper part of the illustration. It represents the cells of the juxtaglomerular apparatus (I), which is composed of part of the distal tubule (G) and the afferent arteriole (C).
2. Color the systemic sympathetic (J) control of the GFR.
3. Color example B, noting the sharp rise in both RBF and GFR when the arterial blood pressure exceeds 180 mm Hg.
4. Color the response of distal tubule feedback to a rise in the GFR. Begin with the GFR.
5. Color the bottom panel, completing the left example first.

HOW IS FLOW TO THE DISTAL TUBULE REGULATED?

1 REGULATE GLOMERULAR FILTRATION RATE (GFR)

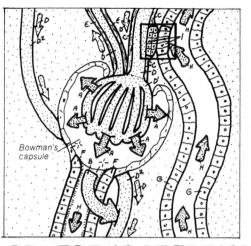

Bowman's capsule

AFFERENT ARTERIOLE c
BLOOD d **GLOMERULUS** d'
EFFERENT ARTERIOLE e
PERITUBULAR CAPILLARY d²
PROXIMAL TUBULE f
DISTAL TUBULE g
SOLUTES h
JUXTAGLOMERULAR
APPARATUS (JG) i

A SYSTEMIC SYMPATHETIC CONTROLS

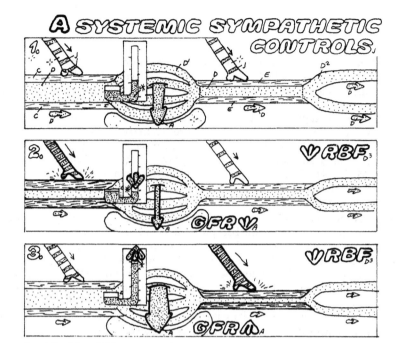

Renal blood flow (RBF) is controlled by sympathetic nerve impulses that constrict arterioles, but their effect on GFR depends on which arterioles are constricted the most. Constricting the afferent arteriole reduces RBF, causing downstream (glomerular) pressure to decrease, thereby decreasing GFR (2). Constricting the efferent arteriole also reduces RBF, but now the glomerulus is upstream. Its pressure rises and GFR increases (3).

C TUBULOGLOMERULAR FEEDBACK g

GFR in each single nephron is adjusted to maintain a constant delivery to the distal tubule by negative feedback. As flow increases, solute delivery to the JG apparatus increases and stimulates constriction of the afferent arteriole, so that GFR in the same nephron decreases.

FEEDBACK LOOP *

cells of the juxtaglomerular apparatus

B LOCAL RESPONSE TO ARTERIAL PRESSURE CHANGE

RBF and GFR are insensitive to changes in systemic arterial pressure (compare with dotted line that would be expected if blood vessels were simple passive structures). Blood vessels regulate flow by constricting when pressure within goes up. This provides more resistance and prevents the elevated pressure from causing a proportional increase in flow.

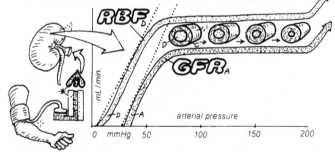

2 REGULATE PROXIMAL TUBULE REABSORPTION B

Fluid reabsorption is determined by net Na^+ reabsorption, which is the balance between active pumping of Na^+ (lumen to interstitial fluid, ISF) and back-leak of Na^+ through tight junction (TJ) in the reverse direction. If GFR decreases, compensatory decreases in proximal fluid reabsorption occur because less filtration → lower concentration of plasma proteins (oncotic pressure) delivered to peritubular capillaries → less capillary reabsorption of fluid from ISF → increased pressure in ISF → increased back-leak of Na^+ through TJ → less net Na^+ reabsorption → less fluid reabsorption. Conversely, an increase in GFR induces a compensatory increase in reabsorption.

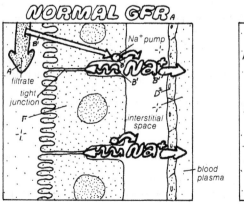

NORMAL GFR a

Na^+ pump

filtrate

tight junction

interstitial space

blood plasma

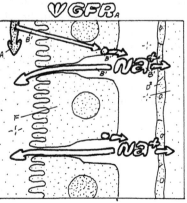

∨ GFR a

Acid-base balance refers to the complex array of mechanisms employed to regulate the concentration of *hydrogen ions* in the body fluids, even though H^+ ions are present only in trace amounts. For each H^+ ion, there are more than a million Na^+ ions in blood plasma, yet H^+ is important because it is so reactive. It is simply a positive charge (a proton) that can easily attach to a variety of molecules, especially proteins, changing their charge and how they interact. Pure water contains 0.0001 mM H^+ (pH = 7.0). Any aqueous solution that contains more H^+ is *acidic*; if it contains less, it is called *alkaline*. Blood contains 0.00004 mM H^+ (pH = 7.4); it is slightly alkaline.

Although free H^+ ions are scarce, there is a huge amount of potential H^+ lurking in the background, bound to other substances. To understand acid-base balance, we must take into account substances that are sources of H^+ and those that may absorb H^+, as well as following the concentration of free H^+ ions. Sources are called acids; an acid is a substance that gives up H^+. A strong acid gives up most of its H^+, whereas a weak one gives up only part. A base is a substance that takes up H^+.

BUFFERS BIND/RELEASE H^+ AND RESIST CHANGES IN PH

A *buffer* is a pair of substances that resist changes in acidity of a solution. It works by storing (binding) the H^+. When H^+ is added to a solution containing buffer, it is "soaked up" by empty storage sites on some of the buffer molecules. When it is removed, it is replaced by H^+ that had been stored on other buffer molecules. In order to work, a buffer must have some molecules with storage sites that are occupied by H^+ while other molecules have empty (storage) sites. Those buffer molecules with occupied sites are acids (they can give up H^+); those that are unoccupied are bases (they can take up H^+).

LUNGS & KIDNEY KEY TO THE BICARBONATE BUFFER SYSTEM

The pair *bicarbonate / carbonic acid* forms an important buffer system in the body:

$$H^+ + HCO_3^- \rightleftharpoons H_2CO_3 \rightleftharpoons H_2O + CO_2$$

H_2CO_3 (carbonic acid) is the acid member of the pair because it can release H^+. Bicarbonate, HCO_3^-, is the base member because it can bind H^+. In water, this step takes about a minute, but in the kidney and red blood cells, it is catalyzed by the enzyme *carbonic anhydrase* and is completed within a fraction of a second. The reaction is so rapid that we often identify CO_2 with H_2CO_3. This system is especially important because two of its components are rigorously controlled by the body: the *lungs control CO_2* and the *kidneys control HCO_3^-*. Although there are other significant buffers in the body (notably phosphate and proteins), this simple chemical reaction is crucial because it links the lungs and kidneys, allowing them to maintain a viable H^+ concentration in the body fluids.

Respiration compensates for metabolic acids by blowing off CO_2 — Each day an average person on a mixed diet produces 60 mM of H^+ in the form of sulfuric, phosphoric, and organic acids. These are called *metabolic acids* because they do not arise from CO_2, and the disturbances in H^+ they create must eventually be corrected by the kidneys. When metabolic H^+ is produced in any organ, most of it is picked up by HCO_3^-

in the blood to form CO_2. The increased CO_2 and the increased H^+ stimulate respiration, which helps eliminate the increased CO_2. In this case, the bicarbonate reaction shown above moves to the right, downhill, because one of the reactants, H^+, is continually produced while one of the products, CO_2, is continually removed.

Kidneys compensate for metabolic acids by excreting H^+ and reabsorbing HCO_3^- — The respiratory regulation of H^+ described above will work only if the bicarbonate that is used can be replenished. This task is accomplished by the kidneys, where the bicarbonate reaction takes place in the reverse direction. This reversal in direction occurs because the kidneys remove the H^+ and excrete it in urine. In the process, the newly formed HCO_3^- is reabsorbed. Thus, the kidney manufactures new HCO_3^- without retaining the attendant H^+ (see plate 64). The result is that for every H^+ produced by metabolism, one H^+ is excreted in the urine and one new bicarbonate is reabsorbed, replacing the bicarbonate that was used up in the respiratory response.

Respiratory acid-base disturbances are reflected by changes in CO_2 — An H^+ concentration below 0.00002 mM (pH = 7.7) or above 0.0001 mM (pH = 7.0) is incompatible with life. If plasma becomes more acid than normal, the condition is called *acidosis*; when it is less acid, the condition is *alkalosis*. In either case, it is useful to recognize whether the disturbance arises from respiratory or other (metabolic) causes. The best clues come from studies of the buffer pair HCO_3^-/ H_2CO_3. An increase of H_2CO_3 (or CO_2) will tend to increase H^+; an increase in HCO_3^- will "soak up" free H^+ and reduce its concentration.

Respiratory acid-base disturbances are reflected by changes in plasma CO_2 or the equivalent H_2CO_3. If these are depressed, as in rapid breathing, there is a diminution of suppliers of H^+ and the H^+ concentration goes down; this condition is respiratory alkalosis. Compensation by the kidneys requires excretion of HCO_3^- to rid the plasma of a disproportionate amount of substances that soak up the scarce H^+. Conversely, in pneumonia or polio, there is a failure to eliminate CO_2 (and H_2CO_3) and plasma acidity rises; this condition is respiratory acidosis. Renal compensation elevates the plasma HCO_3^- to a level commensurate with the elevated H_2CO_3.

Metabolic acid-base disturbances are reflected by changes in HCO_3^- — Non-respiratory acid-base disturbances are called metabolic disturbances. When plasma HCO_3^- decreases and plasma H^+ increases, the condition is called metabolic acidosis. The increased H^+ signifies acidosis; the decreased HCO_3^- implicates its non-respiratory origin. These occur, for example, in renal failure and in diabetes. Respiratory compensation occurs because the increased H^+ stimulates breathing, which reduces CO_2 and H_2CO_3. Finally, in metabolic alkalosis, which sometimes occurs during vomiting of HCl from the stomach, there is increased HCO_3^- with decreased H^+. Respiratory compensation consists of CO_2 (and H_2CO_3) retention.

CN: Use a dark color for A. Notice that three different carbon substances receive the B color (B^1, B^2, B^3).
1. Color the two examples of variation of blood pH level.
2. Color the bicarbonate reaction at the top of the lower panel.

3. Starting at the asterisk in the liver, follow hydrogen and color the downhill reaction to the release of CO_2 into the kidneys; color the downhill process to the release of acid in the urine and the passage of bicarbonate to the plasma.

STRONG ACID
HYDROGEN ION H⁺ H^+

BUFFER PAIR
BICARBONATE HCO₃⁻ HCO_3^-
CARBONIC ACID H₂CO₃ H_2CO_3

WHAT IS A BUFFER?

A buffer is a pair of substances that resists changes in acidity of a solution.

When acid (H^+) is added to the blood, the pH decreases. Then increased acidity (decreased pH) is minimized by buffers, which bind some of the added H^+.

holding one's breath

When acid is taken away, blood becomes more alkaline (pH increases). This change is minimized by buffers, which release H^+ and replace some of the acid that was lost.

hyperventilation

NORMAL PLASMA pH LEVEL 7.4

PLASMA pH LEVEL
BUFFER ACTION

ACID pH 7.4–7.1

BASE pH 7.4–7.7

$$H^+ + HCO_3^- \rightleftharpoons H_2CO_3 \rightleftharpoons H_2O + CO_2$$

The pair bicarbonate/carbonic acid forms an important buffer system. H_2CO_3 (carbonic acid) is the acid member of the pair because it can release H^+. HCO_3^- is the base member of the pair because it can accept H^+. This system is especially important because two of its components are rigorously controlled by the body: the lungs control CO_2 and the kidneys control HCO_3^-.

RESPIRATORY REGULATION (MINUTES)
LUNGS
RESPIRATORY CENTER
NERVE

Each day an average person on a mixed diet produces 60 mm of H^+ in the form of sulfuric, phosphoric, and organic acids. These are called metabolic acids because they do not arise from CO_2. The figure shows metabolic H^+ produced in the liver. Most of it is picked up by HCO_3^- in the blood and forms CO_2. The increased CO_2 and the increased H^+ stimulate respiration, which helps eliminate the increased CO_2. The bicarbonate reaction moves to the right (downhill) because H^+ is continually produced while CO_2 is continually removed.

RENAL REGULATION (HOURS)
KIDNEYS

Respiratory regulation shown above will work only if the bicarbonate that is used can be replenished. This task is accomplished by the kidneys, where the bicarbonate reaction takes place in the reverse direction. This reversal in direction can occur because the kidneys continually remove the newly formed H^+ and excrete it in the urine. In the process, the newly formed HCO_3^- is reabsorbed. Thus, the kidney is able to manufacture HCO_3^- without retaining the attendant H^+.

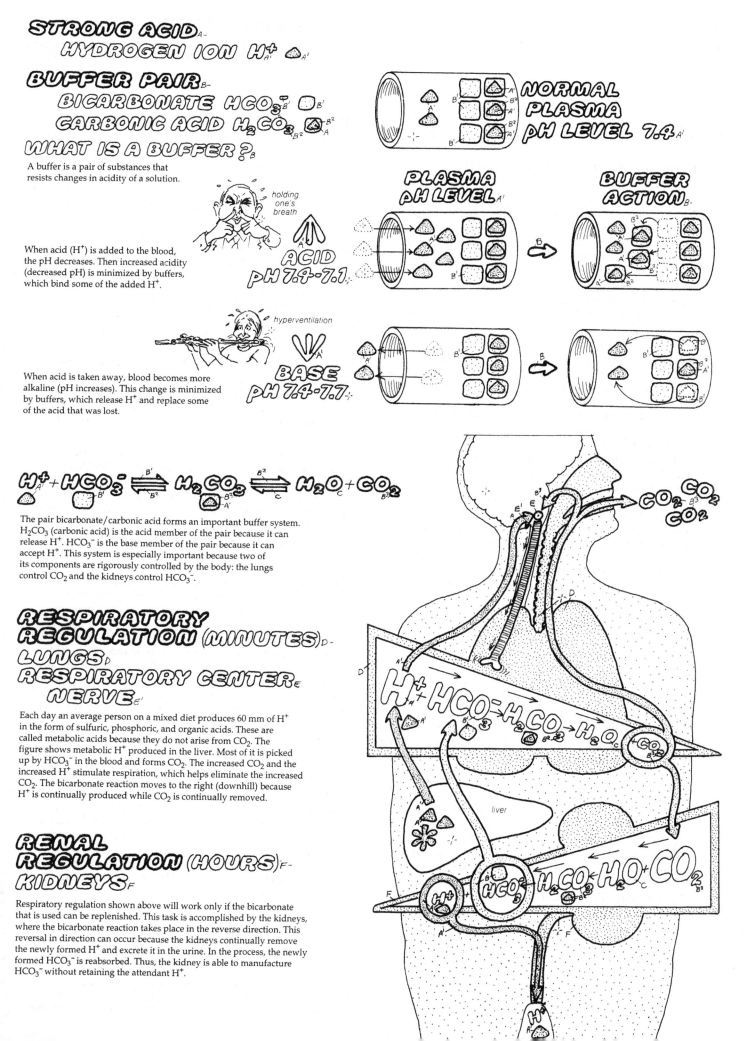

Although several systems contribute to the buffering of H^+ fluctuations induced by metabolic acid production, the ultimate responsibility for restoring the body fluids to their original state falls on the kidney.

PROXIMAL TUBULE REABSORBS BICARBONATE

Renal regulation of plasma acidity begins in the *proximal tubule* (top diagram), where some 80–90% of the filtered HCO_3^- is reabsorbed, through the following processes.

1. CO_2 and water combine to form carbonic acid inside the cell. The reaction is catalyzed by the enzyme *carbonic anhydrase*. Carbonic acid dissociates into H^+ and HCO_3^-, two products with separate fates.

2. HCO_3^- is reabsorbed by moving down its concentration gradient through an Na^+ co-transport system that is confined to the basolateral membrane (the transporter moves 3 HCO_3^- for each Na^+).

3. H^+ moves in the opposite direction and crosses the brush border (luminal) membrane via an Na^+– H^+ exchanger (counter-transporter) located in the luminal, but not the basolateral, membrane. Intracellular H^+ is exchanged for luminal Na^+ and is *counter-transported* into the lumen of the nephron. This counter-transporter works in the indicated direction because of the high concentration gradient for Na^+, tending to drive it into the cell.

4. The secreted H^+ then combines with HCO_3^- that has filtered into the nephron, forming H_2CO_3. The external membrane surfaces of the proximal tubule contain carbonic anhydrase, so the H_2CO_3 is quickly converted to water and CO_2. The CO_2 simply diffuses down its gradient into the cell, where it can enter the cycle at step 1. Note that the H^+ that gets secreted in the above cycle never finds its way into the urine. The principal role of the proximal tubule in acid-base balance is to reclaim the HCO_3^- that comes through the filter.

To focus on HCO_3^- reabsorption, it is helpful to use the plate (after coloring) to trace the fate of the carbon dioxide in its various guises, starting with its entrance in the filtrate in the form of HCO_3^-. The HCO_3^- is converted to H_2CO_3 and then to CO_2 in the lumen. Moving into the cell, this or other CO_2 is then transformed back into H_2CO_3 and finally into HCO_3^-, which is reabsorbed. The diagram shows that the process is driven by the steady flow (secretion) of H^+, which in turn is driven by the steady flow (reabsorption) of Na^+. Energy for the entire process comes from the ATP energy expended by the Na^+– K^+ pump, which is responsible for the Na^+ gradient.

Ammonium synthesis produces bicarbonate — In addition to the above, in times of high plasma H^+ (acidosis), the proximal tubule synthesizes ammonium (NH_4^+) from the amino acid glutamine. For each glutamine consumed, two NH_4^+ and two HCO_3^- are formed. The two HCO_3^- are added to the pool of bicarbonate reabsorbed in the proximal tubule and will be available as buffer in the plasma. The ammonium is in equilibrium with its buffer partner NH_3 (i.e., $NH_4^+ \rightleftharpoons NH_3 + H^+$). Both are secreted into the lumen, only to be reabsorbed into the interstitial spaces of the medulla by the ascending limb of the loop of Henle. From here the lipid soluble, freely permeable, NH_3 diffuses into the cells, and finally into the lumen of

the collecting duct where it can soak up H^+ (see below).

DISTAL NEPHRON EXCRETES H^+ TO PRODUCE NEW HCO_3^-

Similar processes are at work inside specialized *intercalated A cells* located in the *distal nephron* (distal tubule and collecting duct). However, filtrate reaching these cells no longer contains much HCO_3^- because most of it has been reabsorbed in the proximal tubule. Further, H^+ is still secreted, but now instead of being driven by Na^+ exchange, it is directly coupled to ATP splitting via an H^+-ATPase. The presence of an H^+ pump in this location allows H^+ to be excreted without using up HCO_3^-. However, the ability of kidney cells to secrete H^+ is limited; if the free H^+ concentration in the lumen gets too high (pH <4.5), H^+ secretion stops. Luminal buffers are needed, in place of HCO_3^-, to "soak up" the secreted H^+ ions. These buffers take the form of filtered *phosphate buffers*, which are present in the filtrate, and of *ammonia* manufactured in the proximal tubule. Phosphate buffers bind free H^+ as:

$$H^+ + HPO_4^{2-} \rightleftharpoons H_2PO_4^-$$

Ammonia, like CO_2, is soluble in lipids and passes through cell membranes with ease; it diffuses into the lumen, where it binds H^+ to become NH_4^+. Ammonium has no transporters in this region of the nephron; once it is formed, it is trapped in the lumen and will be excreted along with the attached H^+.

$$H^+ + NH_3 \rightleftharpoons NH_4^+$$

The H^+ excreted into the urine bound to ammonia and to phosphate compensates for metabolic acids. Note that for each H^+ excreted, an HCO_3^- that wasn't present in the filtrate is reabsorbed. It is called "new" HCO_3^- to distinguish it from the reabsorbed HCO_3^- that simply replicates the HCO_3^- that came through the filter.

In acidosis, plasma H^+ rises, and the kidneys compensate by excreting acid in the urine. (In respiratory acidosis, there is increased CO_2, which promotes H^+ formation and secretion by the kidney cells. In metabolic acidosis, the filtered load of HCO_3^- is reduced so that it is less effective in trapping secreted H^+ in the lumen. Once secreted, H^+ has more of a chance to escape into the urine.) All the HCO_3^- is reabsorbed, and most of the HPO_4^{2-} is converted to $H_2PO_4^-$. Chronic acidosis stimulates the kidneys to synthesize more NH_3. This provides more buffering capacity in the filtrate so that more secreted H^+ can be carried (in the form of NH_4^+) without substantially decreasing the pH of the filtrate. Thus, the H^+ gradient from cell to lumen does not increase to prohibitive levels despite the increased H^+ secretion.

In alkalosis, both plasma and intracellular H^+ fall. In kidney cells, less intracellular H^+ is available for secretion. As a result, HCO_3^- reabsorption does not go to completion and some HCO_3^- escapes into the urine. Further, intercalated B cells may become active. These cells are similar to A cells with the exception that the positions of the H^+ and the HCO_3^- transporters are reversed. The H^+ transporters are located in the baso-lateral membrane, whereas the HCO_3^- transporters are in the brush border. Consequently, these cells excrete HCO_3^- and reabsorb H^+. The final result is that urine becomes alkaline so that blood leaving the kidney is more acidic than blood entering.

CN: Use red for B and the same colors that were used on the previous page for the same structures labeled here: D, F, C, C^1, C^2.

1. Begin with the proximal tubule by first coloring the cell membrane (A^2 and A^3) and capillary (B). Next color the role of Na^+ (G) in this process by following it down the tubule lumen as part of the filtrate, through the membrane via a counter transport mechanism (G^1) and out of

the cell via the sodium-potassium pump (G^3). Then follow the numbered sequence starting at the top of the ce[ll].
2. Do the same with the distal nephron by coloring the Na^+ material first. Note that because of space limitation[s] the step preceding the formation of carbonic acid (C^1), the combination of carbon dioxide and water plus enzym[e] action, has been deleted in the bottom cell. Where a pro[cess] is repeated the number next to it is also repeated.

PROXIMAL TUBULE A

CAPILLARY B
CELL MEMBRANE A'
BRUSH BORDER A²
BASOLATERAL A³
CARBON DIOXIDE C
WATER D
CARBONIC ANHYDRASE E
CARBONIC ACID C'
BICARBONATE C²
HYDROGEN ION F
Na⁺/H⁺ COUNTER-TRANSPORT G'
Na⁺/K⁺ PUMP G³

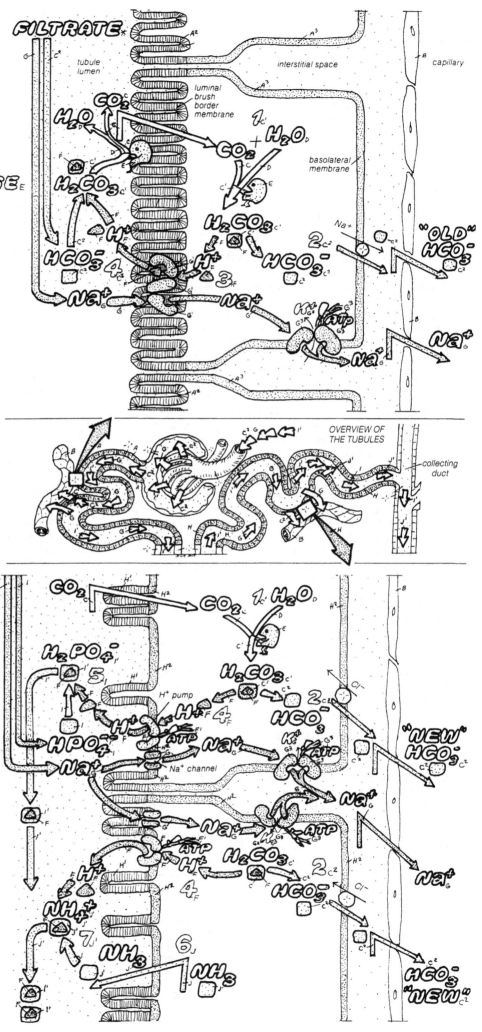

Regulation of plasma acidity begins in the proximal tubule where most of the filtered HCO_3^- is reabsorbed. The process involves (1) CO_2 and water combined to form carbonic acid inside the cell. The reaction is catalyzed by an enzyme, carbonic anhydrase. Carbonic acid dissociates into H^+ and HCO_3, two products with separate fates. (2) HCO_3^- moves down its gradient through an Na^+ – HCO_3^- co-transporter in the basolateral membrane; it is reabsorbed. (3) H^+ is exchanged for Na^+ and is counter-transported into the lumen of the nephron. (4) This H^+ then combines with the HCO_3^- that has filtered into the nephron. Water and CO_2 are formed and the CO_2 simply diffuses down its gradient into the cell where it can enter the cycle at step 1. The reaction proceeds in the indicated direction because H^+ is continually transported out of the cell into the lumen. Energy for this H^+ transport comes from the Na^+ gradient.

DISTAL NEPHRON H

BRUSH BORDER H'
BASOLATERAL H²
BUFFERS: I
MONOHYDROGEN PHOSPHATE I'
DIHYDROGEN PHOSPHATE I²
AMMONIA J
AMMONIUM ION J'

Similar processes are at work in the distal nephron, but the filtrate no longer contains much HCO_3^-. Further, H^+ is still secreted, but instead of being driven by Na^+ exchange, it is directly coupled to ATP splitting (4). If the free H^+ concentration in the lumen gets too high H^+ secretion stops. Luminal H^+ is prevented from rising too high by buffers (e.g. phosphate) that are present in the filtrate and bind free H^+ (5). During acidosis, when these buffers are exhausted, the kidneys adapt by manufacturing NH_3 The NH_3 follows a circuitous path (see text) but finally diffuses into the collecting duct lumen (6) where it binds H^+ to become ammonium (7). Ammonium is "trapped" in the lumen and is excreted along with the attached H^+. The H^+ excreted compensates for production of metabolic acids. For each H^+ excreted a HCO_3^- is reabsorbed that wasn't present in the filtrate, hence the description "new" HCO_3^-.

Potassium is the most abundant solute inside cells. Its high concentration is required for optimal growth and DNA and protein synthesis; it is an important factor in the performance of many enzyme systems and plays a role in maintaining cell volume, pH, and membrane potentials. Because most of the body's K^+ lies within cells, with only about 2.5% in the extracellular fluid, a small K^+ shift between intra- and extracellular fluids could cause a huge change in extracellular K^+. If, for example, only 5% of the body K^+ moved into the extracellular fluids, the extracellular K^+ would triple—a lethal result.

BODY CELLS BUFFER ACUTE CHANGES IN PLASMA K^+

Alterations of extracellular fluid or plasma K^+ are important because cell *excitability* (membrane potential) is sensitive to extracellular K^+. Increasing extracellular K^+ depolarizes membranes and raises excitability; in the heart, fibrillation may occur. Decreasing K^+ hyperpolarizes and lowers excitability. Skeletal and smooth muscle disturbances may include flaccid paralysis, abdominal distention, and diarrhea. Shifts of K^+ in and out of cells can easily occur in acid-base disorders or disturbances of hormone balance, and in response to drugs. Further, on a normal diet, the amounts of K^+ absorbed from the intestine into the plasma each day exceeds the total K^+ content of the entire extracellular fluid! This potentially disastrous rise in plasma K^+ does not occur because the increased plasma K^+ stimulates increased *insulin* secretion from the pancreas, *aldosterone* secretion from the adrenal cortex, and *epinephrine* release from the adrenal medulla. All three of these hormones increase K^+ uptake by liver, bone, skeletal muscle, and red blood cells. (They act by directly or indirectly stimulating the $Na^+ - K^+$ ATPase.) Although the kidneys act more slowly, over the course of hours, they provide the ultimate means for removal or retention of K^+ from the body and thereby maintain the long-term plasma level within the narrow range of 3.5 to 5 mM/l.

THE PROXIMAL TUBULES REABSORB MOST K^+

Renal handling of K^+ begins by reclamation of bulk amounts from the filtrate. Most of the filtered K^+ is reabsorbed in the proximal tubule and in the loop of Henle; by the time the filtrate reaches the distal tubule, only 5 to 10% of the filtered load remains. From here on, depending on conditions, it may be reabsorbed further, but most often it is *secreted*. Most regulatory changes in excretion are due to variations in secretion in these latter portions of the nephron.

THE DISTAL NEPHRONS REGULATE K^+ BY SECRETION

Distal tubule and collecting duct cells accumulate high concentrations of intracellular K^+ via the $Na^+ - K^+$ *pump*, which is located primarily in the basolateral membrane. The *electrical gradient* (membrane potential) across the basolateral membrane is sufficiently high (70 mv) to oppose the K^+ *concentration gradient* and prevent leakage from cell to interstitial space, but the membrane potential across the apical membrane (facing the lumen) is smaller and can't stop leakage. The result is a simple pathway for K^+ secretion; K^+ is pumped into the cell from the blood side and leaks out the lumen side. How can

this mechanism explain variations in K^+ excretion?

1. **Rise in intracellular K^+ is accompanied by increased K^+ excretion.** According to the previous discussion, K^+ excretion will increase whenever distal tubular (or collecting duct) cell K^+ increases because the concentration gradient driving K^+ leakage into the lumen will increase. But the K^+ content of these cells often reflects the K^+ content of body cells in general. This provides a mechanism for regulating general body intracellular K^+; changes that increase internal K^+ will increase leakage and secretion.

2. **Aldosterone secretion provides feedback regulation of K^+.** Intracellular K^+ (and consequently K^+ secretion) has a tendency to rise and fall with plasma K^+, providing some regulation of plasma K^+. However, plasma K^+ is guarded by another potent feedback mechanism. A rise in plasma K^+ stimulates the *adrenal cortex* to secrete *aldosterone*, which promotes secretion and excretion of K^+ (and reabsorption of Na^+). Unlike other feedback paths that involve aldosterone secretion, K^+ stimulates the adrenal cortex directly and does not utilize the renin-angiotensin system as an intermediary.

3. **Excretion of Na^+ and K^+ are often correlated**—Tubular cell leakage of K^+ also helps explain the frequent positive correlation between excretion of Na^+ and K^+. As more Na^+ is delivered to the distal tubule, the excess Na^+ causes an increase in both Na^+ reabsorption and Na^+ excretion. K^+ leaks faster because more positive charge (in the form of Na^+) is available to exchange with K^+ across the luminal membrane; this allows more K^+ to escape down its concentration gradient without building up a membrane potential.

4. **Increased fluid flow increases K^+ excretion**—When fluid flow in the distal tubule increases, there is generally an increase in K^+ excretion. This can be explained by the more efficient "washing away" of the secreted K^+ by the faster-moving tubular stream, which reduces the K^+ concentration in the luminal fluid adjacent to the tubular cells and promotes leakage from cells to lumen.

5. **$K^+ - H^+$ exchange is apparent during acidosis and alkalosis**—K^+ excretion commonly increases during acute alkalosis and decreases during acute acidosis. This is consistent with the fact that alkalosis is often associated with K^+ entry into cells and acidosis with its departure. It appears as though K^+ and H^+ are exchanged across the cell membranes. For example, in acidosis, H^+ enters the cell and reacts with negatively charged proteins, reducing the charge on the protein. K^+, the most abundant intracellular cation (positively charged ion), suddenly finds itself in excess; it is in an environment with too few negative charges to support all the K^+. Since K^+ is the most permeable cation, some of it moves out. Renal cells are no exception. In acidosis, distal tubular and collecting duct cells lose K^+ to the plasma; the intracellular K^+ decreases, as does the leakage and secretion into the tubular lumen. During alkalosis, the reverse occurs.

In addition to the above, a recently identified $H^+ - K^+$ ATPase that exchanges H^+ and K^+ across intercalated cell membranes may play a role.

CN: Use red for blood vessels (E), including both arterioles and capillaries.

1. Begin with the upper panel. Note the large area within the cell on the left marked by the letter K^+, symbolically representing larger amounts of K^+ present.

2. Color the illustration (top of bottom panel) of filtrate flow through the nephron, beginning in the upper left corner of the rectangle. Note the separation of K^+ (A)

from the filtrate (F) as it is first reabsorbed into the proximal tubule (uncolored).

3. Color the enlargement of the secretion process. Note the two symbols representing K^+ concentration gradients in the lower tubule cell and the way in which they compare to the two electrical potentials (C).

4. Color the summary below, beginning with the circle representing elevated K^+ in the blood.

ROLE OF K⁺ IN NERVE & MUSCLE EXCITABILITY.

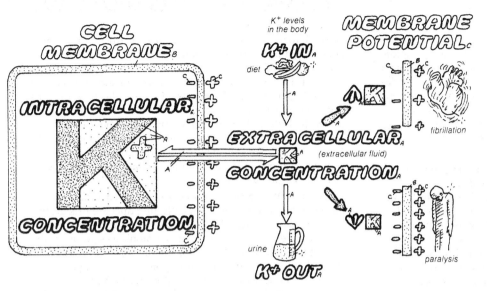

CELL MEMBRANE

INTRACELLULAR K CONCENTRATION.

K⁺ levels in the body

K⁺ IN.
diet

EXTRACELLULAR (extracellular fluid) **CONCENTRATION.**

urine

K⁺ OUT.

MEMBRANE POTENTIAL.

fibrillation

paralysis

Most body K⁺ (about 90%) lies within cells, with only about 2.5% in the extracellular fluid (the rest is in bone). A small decrease — say 5% — in cellular K⁺ could cause a huge change in extracellular K⁺ (in our example it could triple, going from 2.5 to 7.5%). These large changes do not occur, however, because the kidney regulates plasma K⁺.

Alterations of plasma K⁺ are important because cell excitability (membrane potential) is sensitive to extracellular K⁺. Increasing extracellular K⁺ depolarizes membranes and raises excitability in the heart, and fibrillation may occur. Decreasing K⁺ hyperpolarizes membranes and lowers excitability. Skeletal and smooth muscle disturbances may include flaccid paralysis, abdominal distension, and diarrhea.

K⁺ LEVELS REGULATED BY SECRETION. IN DISTAL NEPHRON.

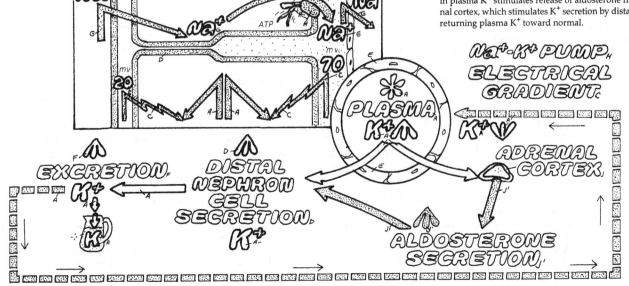

peritubular capillary. efferent & afferent arterioles. capillary

proximal tubule. distal tubule. collecting duct

DISTAL NEPHRON = distal tubule & collecting duct

lumen

extracellular fluid

K⁺ **Na⁺** ATP **Na⁺** **K⁺** **Na⁺**

70 mv.

20 mv.

PLASMA K⁺

BLOOD VESSEL.
FILTRATE.
K⁺ REABSORPTION.
K⁺ SECRETION.

K⁺ is reabsorbed in the proximal tubule and in the loop of Henle; by the time it reaches the distal tubule only 5–10% of the filtered load remains. From there on, depending on conditions, it may be reabsorbed further, but most often it is secreted. Most regulatory changes in excretion are due to variations in secretion in these latter portions of the nephron.

The distal nephron (distal tubule and collecting duct cells) accumulates high concentrations of intracellular K⁺ via the Na⁺ – K⁺ pump, which is located primarily in the basolateral membrane. The electrical potential across the basolateral membrane is sufficiently high (70 mv) to oppose the K⁺ concentration gradient and prevent leakage from cell to interstitial space, but the potential across the apical membrane (facing the lumen) is smaller and cannot prevent leakage. The result is a simple pathway for K⁺ secretion; K⁺ is pumped into the cell from the blood side and leaks out the lumen side. Cell K⁺ is regulated because changes that increase (decrease) internal K⁺ will increase (decrease) leakage and secretion. Extracellular K⁺ is regulated by a negative feedback path. An increase in plasma K⁺ stimulates release of aldosterone from the adrenal cortex, which stimulates K⁺ secretion by distal tubule cells, returning plasma K⁺ toward normal.

Na⁺-K⁺ PUMP.
ELECTRICAL GRADIENT.

EXCRETION. K⁺

DISTAL NEPHRON CELL SECRETION. K⁺

ADRENAL CORTEX.

ALDOSTERONE SECRETION.

Animals living in fresh water are continuously challenged with water balance problems. Their plasma has a high solute concentration (osmolarity) and tends to draw water by osmosis from its surroundings. They cope with a continuous inundation of water by excreting large volumes of it. Animals living on land—including humans—have the opposite problem. Their environment is arid, and they face the threat of drying up. To conserve water, birds and mammals excrete very small volumes of concentrated urine, but how?

LOOP OF HENLE CREATES A HYPERTONIC INTERSTITIAL SPACE

Only birds and mammals excrete urine that is *hypertonic* (more concentrated than their plasma). Only birds and mammals have long loops of Henle. Further, those species with more highly developed loops are capable of excreting more concentrated urine. These observations led earlier investigators to suggest that the formation of hypertonic urine takes place in the loops of Henle. This idea was shattered by the first micropuncture studies of the distal tubule, which contains the fluid just after it has passed through the loop of Henle. This fluid is always *hypotonic* or at most isotonic, but never hypertonic, as required by the hypothesis. Apparently, hypertonic urine must be formed in the collecting duct. The loops of Henle are involved in a more subtle way. By actively pumping NaCl without allowing water to follow, the loops of Henle create a unique *hypertonic interstitial fluid* in the deep portions of the *medulla*. Collecting ducts pass through this fluid on the way to the ureter and take advantage of their hypertonic surroundings by allowing water to be withdrawn by osmosis from the lumen of the duct to the interstitial space.

Ascending limb of loop of Henle actively reabsorbs Na⁺, leaves water behind to deliver hypotonic fluid to the distal tubule—The loops of Henle of juxtamedullary nephrons plunge into the depths of the medulla. These descending limbs are fairly permeable to Na⁺ and water and do not appear to have any special properties. Once around the bend in the loop, however, the tubules become *water impermeable*, a property that extends well into the distal tubule. Further, the ascending limb actively transports NaCl from the lumen into the interstitial fluid. The major portion of this transport takes place in the thick (upper) portions of the ascending limb, which has cells richly endowed with mitochondria—i.e., ATP producers. (Na⁺ crosses the luminal membrane via a cotransport system that moves Cl⁻ and K⁺ as well as Na⁺ into the cell. Once inside, Na⁺ is pumped out by the Na⁺ – K⁺ pump in the basolateral membrane. Cl⁻ follows through a passive channel, preserving electroneutrality; the result is transport of NaCl.)

Final adjustment of urine volume and concentration occurs in the collecting ducts —Although the ascending limbs transport NaCl, their membranes prevent the usual concomitant transport of water, so fluid delivered to the distal tubule is hypotonic regardless of the final composition of the urine. This transport of NaCl (out of the water-impermeable ascending limb) without water creates a unique hypertonic interstitial fluid in the medulla. Collecting ducts from *all* nephrons pass through these fluids on their way to the ureter. If the hormone *ADH (antidiuretic hormone, vasopressin)* is present, the latter portions of the distal tubule and the entire collecting duct become water permeable. As fluid flows

through these sections of the *distal nephron* (distal tubule and collecting ducts), water equilibrates with the surrounding interstitial fluids. Therefore, as fluid descends via the collecting ducts into the medulla, it becomes more and more concentrated (hypertonic) until urine leaving the collecting duct has the same hypertonic osmolarity as the interstitial fluid in the lower medulla. In fact, the osmolarity of the medulla sets the limit to which urine can be concentrated. ADH also makes the last portions of the collecting duct permeable to urea, which becomes trapped in the interstitial fluid and makes a substantial contribution to its osmolarity.

When ADH is absent, as occurs in the disease *diabetes insipidus*, the distal tubule and collecting duct are practically impermeable to water. The hypotonic fluid delivered to the distal tubule becomes even more hypotonic as salts are reabsorbed (without water being able to follow). Fluid reaching the end of the collecting duct is hypotonic, resulting in a large volume of dilute urine. In the disease, lack of ADH causes the kidneys to produce as much as 30 liters of urine per day. One would have to drink more than 120 glasses of water per day simply to keep from drying up.

ADH CONSERVES THE OSMOTIC PRESSURE OF BODY FLUIDS

In times of water deprivation, the kidneys conserve water; with the aid of high ADH levels they excrete a low-volume, concentrated (hypertonic) urine. With water intoxication, ADH levels are minimal and the kidneys release the excess water by excreting a high-volume, dilute (hypotonic) urine.

ADH is produced in neural cells of the hypothalamus and is carried via axonal transport to axon terminals in the posterior pituitary, where it is stored. It is secreted from the terminals at the "right time" because neural osmoreceptors, also in the hypothalamus, are very sensitive to the osmotic pressure in the plasma. They respond to small increases in osmotic pressure by increasing their frequency of stimulating impulses sent to the ADH-producing cells. These impulses are relayed down to the axon terminals in the posterior pituitary where they activate channels, allowing Ca⁺⁺ to enter, and stimulate secretion (by exocytosis—plate 20) of the hormone. Thus, a rise in osmotic pressure of the extracellular fluids (reflected in the blood plasma) stimulates cells of the hypothalamus to increase ADH production and to release ADH from the posterior pituitary. ADH travels via the bloodstream to the kidney, where it acts (via cAMP/protein kinase) to insert water channels into the membranes of the distal tubule and collecting duct. This promotes water retention and relieves the rise in osmotic pressure. Hypothalamic osmoreceptors also send excitatory signals to thirst centers in the hypothalamus that initiate sensations of thirst. Drinking more water dilutes the concentration of body fluids. Conversely, when plasma osmotic pressure decreases, secretion of ADH and sensations of thirst diminish.

Regulation of body-fluid osmotic pressure is important. When it is low, water is drawn into cells that swell—posing a danger to brain cells, which are confined by the rigid walls of the cranium. Symptoms may include nausea, malaise, confusion, lethargy, seizures, and coma. When osmotic pressure is high, cells shrink; neurological symptoms include lethargy, weakness, seizures, and coma. Sometimes it can even be fatal.

CN: Use light blue for D, a dark color for G, yellow for H, red for I, and purple for J.
1. Begin by coloring the rectangular borders (A and B) of the two large diagrams representing a kidney nephron. They demonstrate which part of the nephron lies within the cortex (A) and which part is in the medulla (B).
2. Color the state of low osmolarity on the left by beginning with the cartoon figure above and then coloring the borders

of the nephron itself (D¹ and E). Color all circles and arrows.
3. Do the same for the diagram on the right, noting the ADH influence (G¹) on the collecting duct and a portion of the distal tubule (making the membranes water permeable).
4. Starting with step one, color the lower right illustration, which shows how a rise in osmolarity results in ADH secretion and water reabsorption. Do the summary diagram to the left. The numbers refer to the illustration on the right.

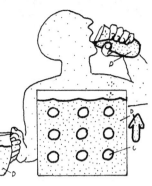

In times of water deprivation (right diagram), the kidneys conserve water; they excrete a low-volume, concentrated (hypertonic) urine. With water intoxication (left diagram), they release the excess water by excreting a high-volume, dilute (hypotonic) urine. To accomplish this task the loops of Henle of juxtamedullary nephrons plunge into the depths of the medulla. The ascending limbs actively reabsorb NaCl, but prevent the usual concomitant reabsorption of water (due to impermeable membranes) so that fluid delivered to the distal tubule is hypotonic regardless of the final composition of the urine. Further, this transport of NaCl (without water following) creates a unique hypertonic interstitial fluid in the medulla. Collecting ducts pass through these fluids on their way to the ureter. If the hormone ADH is present (right diagram), the latter portions of the distal tubule and the entire collecting duct become water permeable. Water equilibrates with interstitial fluid in these sections of the nephron; fluid leaving the distal tubule is isotonic, while fluid leaving the collecting duct has the same hypertonic osmolarity as the interstitial fluid in the lower medulla. When ADH is absent (left diagram), the distal tubule and collecting duct are practically impermeable to water. The hypotonic fluid delivered to the distal tubule becomes even more hypotonic as salts are reabsorbed (without water being able to follow). Urine reaching the end of the collecting duct is hypotonic. ADH also makes the last portions of the collecting duct permeable to urea, which becomes trapped in the interstitial fluid and makes a substantial contribution to its osmolarity.

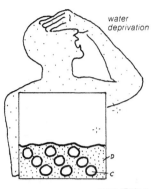

water deprivation

LOW OSMOLARITY. ∨ ADH.

HIGH OSMOLARITY. ∧ ADH.

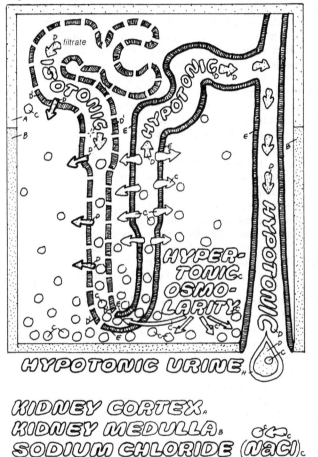

HYPOTONIC URINE

HYPERTONIC URINE

KIDNEY CORTEX.

KIDNEY MEDULLA.

SODIUM CHLORIDE (NaCl).

WATER.

WATER PERMEABLE MEMBRANE.

IMPERMEABLE MEMBRANE.

UREA.

ANTIDIURETIC HORMONE (ADH).

INFLUENCE ON MEMBRANE.

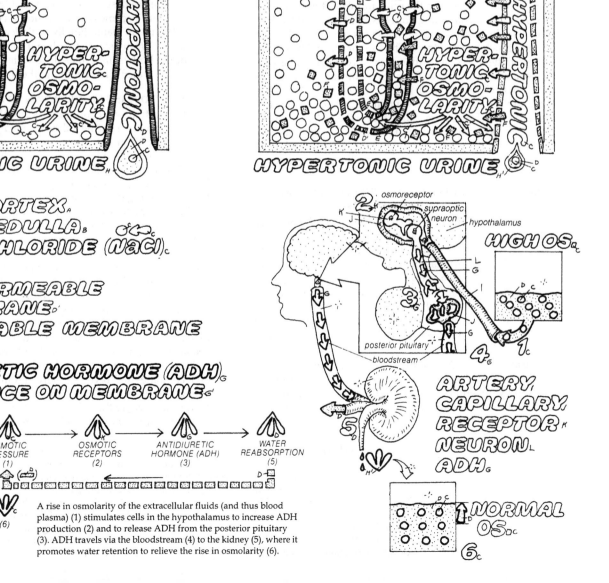

osmoreceptor
supraoptic neuron
hypothalamus
HIGH OS.
posterior pituitary
bloodstream

ARTERY.
CAPILLARY.
RECEPTOR.
NEURON.
ADH.

NORMAL OS.

OSMOTIC PRESSURE (1) → OSMOTIC RECEPTORS (2) → ANTIDIURETIC HORMONE (ADH) (3) → WATER REABSORPTION (5)

(6)

A rise in osmolarity of the extracellular fluids (and thus blood plasma) (1) stimulates cells in the hypothalamus to increase ADH production (2) and to release ADH from the posterior pituitary (3). ADH travels via the bloodstream (4) to the kidney (5), where it promotes water retention to relieve the rise in osmolarity (6).

The kidney regulates the internal environment by judicious excretion of ions and water. It also excretes waste products, the most notable being *urea.* Urea is produced in the liver and contains the nitrogen derived from amino acids or proteins. When these compounds are broken down by metabolism, they yield ammonia. Free ammonia is very soluble in water and very toxic. Fortunately, the liver quickly converts it to the relatively harmless urea. Metabolism of protein produces about 30 g of urea per day, which is excreted in the urine. Because ions and urea are water soluble, their excretion necessarily draws water with them. Excretion of water in the urine is obligatory, and it behooves the kidney to conserve water whenever it is in short supply by excreting a concentrated urine. What do we mean by "concentrated" urine?

TOTAL OSMOTIC CONCENTRATION IS MEASURED IN MILLIOSMOLES

Ordinarily, we express the concentration of a solute like urea by the number of moles (1 mole = 6×10^{23} molecules) of urea contained in each liter of solution. This is the *molar concentration* of urea. When this number is small, we reduce the unit by 1000 and call it a millimole (mM; 1000 millimoles = 1 mole). In every solution, each specific solute has its own molar (or millimolar) concentration. When we are dealing with osmotic water movements, all molecules and ions make an almost equal contribution to osmotic pressures. A 100 mM solution of urea exerts the same osmotic force as a 100 mM glucose solution because they contain the same number of molecules per liter. A solution containing both (100 mM urea + 100 mM glucose) contains twice as many molecules per liter and exerts twice as much osmotic force. The sum of the molar concentration of all the molecules and ions in a given solution is called the *osmolar* concentration (Osm). Sometimes we use *milliOsmolar* (mOsm) instead (1000 mOsm = 1 Osm). The "total solute concentration" of a solution containing 100 mM urea + 100 mM glucose is 200 mOsm. (Note that 100 mM NaCl would be 200 mOsm because it contains 100 mM Na^+ + 100 mM Cl^-.) The total concentration of blood plasma is consistently about 300 mOsm: urine is commonly around 950 but can range from 50 to 1400 mOsm.

CONCENTRATION OF INTERSTITIAL FLUID IN THE LOWER MEDULLA CAN BE AS HIGH AS 1400 mOSM

Na^+ pumps in the loop of Henle create gradients of only 200 mOsm — Excretion of concentrated urine requires an *interstitial fluid space* in the *medulla* 4 to 4.5 times more concentrated (1200 to 1400 mOsm) than blood plasma. To create this space, the kidneys rely on Na^+ pumps in the *ascending loop of Henle* that create only 200 mOsm gradients across the tubular cells. Because proximal tubule fluid delivered to the loop is isotonic (300 mOsm), the most concentrated interstitial space possible should be about 500 mOsm. How does the kidney manage to get 1400 mOsm?

The ability of the Na^+ pump to create a 200 mOsm gradient is called the *"single effect."* The single effect is multiplied several-fold by imbedding the pumps in the ascending limb of the two streams moving in opposite directions (*counter-current*) through the loop of Henle.

The descending limbs deliver concentration boosts to pumps in the ascending limbs — The ascending limb is impermeable to water; NaCl is pumped out into the interstitial fluid (*ISF*), but water cannot follow. The NaCl that has been pumped creates a small gradient, 200 mOsm, so the ISF becomes slightly hypertonic. The descending limb is permeable to both NaCl and water; NaCl diffuses down its concentration gradient from the ISF into the descending limb, while water is drawn out of the descending limb into the hypertonic environment. This loss of water and gain of solute makes the contents of the descending limb hypertonic, like the ISF. But the slightly concentrated fluid in the descending limb moves! It flows toward the ascending limb, where pumps create the same 200 mOsm gradient — only this time they begin with a higher concentration and create a correspondingly higher concentration in the ISF. The cycle repeats, with elevated concentrations delivered to the descending limb, which in turn delivers these elevated concentrations to pumps in the ascending limb; *the single effect is multiplied.* The concentration of solutes in the ISF builds up until a steady state is reached, whereby amounts delivered to the ISF are just balanced by the amounts taken away by the blood supply.

The diagram on the right illustrates the scheme in a steady state. Note that the proximal tubule continues to deliver isotonic fluid (300 mOsm) to the loop, but as it descends, the fluid becomes more concentrated as NaCl enters and water leaves. The greatest concentration is at the tip. Upon ascending, the fluid becomes less concentrated as NaCl is pumped out without any water. Finally, fluid leaves the loop less concentrated (100 mOsm) than when it came in. Because the ascending limb is impermeable to water, relatively more NaCl than water is left behind in the medullary ISF.

UREA'S IMPORTANT CONTRIBUTION TO MEDULLARY ISF

In addition to its effect on water permeability, ADH also increases the urea permeability of the lower medullary collecting duct. With ADH present, urea makes a substantial contribution to the ISF solute concentration in the medulla. Urea becomes trapped in the lower medullary ISF as it flows in a circle along the following path (lower left illustration): lower collecting duct → lower medullary ISF → thin ascending limb → thick ascending limb → distal tubule → collecting duct → lower collecting duct → ... This circulation and trapping occurs because the upper portions of the collecting duct are impermeable to urea, and as water is reabsorbed, the remaining urea becomes concentrated. When it reaches the lower portions, the collecting duct becomes permeable (ADH) and urea diffuses to the ISF. From there, some of it diffuses into the lower thin ascending limb, which is urea permeable. The thick ascending limb and distal tubule are not permeable to urea. As water is withdrawn from these portions, the urea becomes even more concentrated, only to be delivered to the collecting duct, where the cycle begins anew. In this way, the urea circulates and becomes more and more concentrated in all sections of its route (including the ISF) until it reaches a steady state where the delivery of "new" urea is just balanced by the amounts of urea the blood circulation carries away.

CN: Use light blue for E, darker colors for B, C, and D.
1. Begin by coloring the entire title in the upper left corner, noting that the NaCl pump (B) receives a different color. Color the anatomical drawing of the loop of Henle.
2. Color the counter-current multiplier by starting in the upper left corner with the NaCl solutes (A) entering the descending limb. Work your way down the limb, coloring the numbers and solutes in the interstitial fluid and the diffusion/gradient symbol (A) moving into the

descending limb. The broken line suggests that the membranes of the cells of descending limbs are permeable to water (E) (arrows entering ISF). Then work your way up the ascending limb.
3. Color the way in which urea becomes trapped in the ISF. The drawing of the ascending limb of Henle, distal tubule, and collecting duct is highly simplified. Note that here the water-permeable membranes (broken lines) are not colored.

PROBLEM:

HOW TO CREATE A 300-1200 mOsm CONCENTRATION GRADIENT IN THE MEDULLA WITH ONLY A 200 mOsm NaCl PUMP?

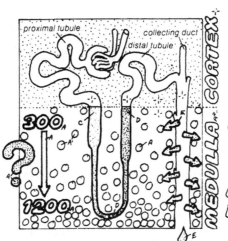

CORTEX

MEDULLA

300

1200

LOOP OF HENLE:
DESCENDING LIMB.
ASCENDING LIMB.
NaCl SOLUTE.
WATER.

By excreting concentrated urine, the kidney conserves body water. This requires an interstitial fluid (ISF) space in the medulla some 4 to 4.5 times more concentrated (1200 to 1400 mOsm) than blood plasma (300 mOsm). To create this space, the kidneys rely on Na^+ pumps that create only 200 mOsm gradients across the cells. Since proximal tubule fluid delivered to the loop is isotonic (300 mOsm), it would appear that the most concentrated ISF is 500 mOsm. How does the kidney manage to get 1200 mOsm?

SOLUTION:

THE COUNTER-CURRENT MULTIPLIER.

The ability of the Na^+ pump to create a 200 mOsm gradient is multiplied several-fold by embedding the pumps in the ascending limb of the two streams moving in opposite directions (counter-current) through the loop of Henle. The ascending limb is impermeable to water. NaCl is pumped out into the ISF, creating a small gradient (200 mOsm) and making the ISF hypertonic. The descending limb is permeable to both NaCl and water. They equilibrate passively so that the contents of the descending limb match the ISF. But the slightly concentrated fluid in the descending limb moves! It flows toward the ascending limb, where the pumps are located, giving the pumps an opportunity to create the same 200 mOsm gradient—only this time they begin with a higher concentration and can create a correspondingly higher concentration in the ISF. The cycle repeats, with elevated concentrations delivered to the descending limb, which in turn delivers these elevated concentrations to pumps in the ascending limb, the single effect is multiplied. The diagram on the right illustrates the scheme in a steady state. Note that the loop continues to receive isotonic fluid (300 mOsm), but as fluid descends it becomes more concentrated as NaCl enters and water leaves. The greatest concentration is at the bend. Upon ascending, the fluid becomes less concentrated because NaCl is pumped out without water. Finally, fluid leaves the loop less concentrated (100 mOsm) than when it came in. Relatively more NaCl than water is left behind in the medullary ISF because the ascending limb is impermeable to water. (Note: the diagram has been simplified. Pumps are actually located exclusively in the thick portion of the ascending limb).

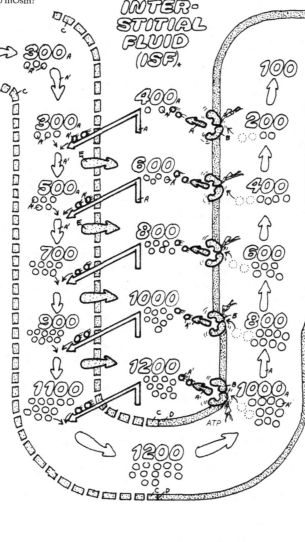

INTER-STITIAL FLUID (ISF)

300
400
300
600
500
800
700
1000
900
1200
1100
1200

100
200
400
600
800
1000

ATP

The desert rat does not need to drink water because its counter-current multiplier can establish a hypertonic medullary ISF that is 20 times more concentrated than blood plasma. Its urine is so concentrated that it can maintain body fluids with water obtained from carbohydrate breakdown.

ISF CONCENTRATION VARIATIONS.

Man is able to establish a medullary ISF concentration that is only four times as concentrated as blood plasma. Therefore, the highest urine concentration is also four times that of than blood plasma.

(300)
1 L
4
(1200)

DAILY FLUID INTAKE.

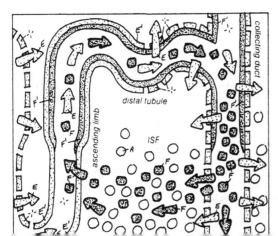

collecting duct

distal tubule

ascending limb

ISF

UREA TRAPPING.
H_2O BARRIER.
UREA BARRIER.

In the presence of ADH, urea is trapped in the ISF and contributes to its solute concentration. Follow the route taken by urea as it circulates from the collecting duct through the ISF to the ascending limb distal tubule, then back again to the collecting duct. Note places that are impermeable to urea and where water is reabsorbed, concentrating the urea.

Like any tissue, the renal medulla must be supplied with blood, and if solutes in the *medullary ISF* are highly concentrated, we might expect them to be washed away as blood within the capillary beds equilibrates with the ISF. A capillary exchange vessel entering an impermeable venule at the tip of the medulla would carry away fluid as concentrated as 1200–1400 mOsm! This does not happen, because of the peculiar shape of the exchange vessels, the *vasa recta*. These are long, highly permeable vessels that exchange materials with their surroundings along their entire length just as though they were capillaries. They enter from the cortex, descend into the medulla, form loops, and return to the cortex. The important point is that they leave the medulla at the level of the cortex. Few, if any, exchange vessels enter an impermeable collecting vein in the depths of the medulla, so few if any collecting veins contain highly concentrated (1200 mOsm) fluid.

THE UPPER VASA RECTA TRAPS WATER;
SOLUTE IS TRAPPED BELOW

Follow the exchange of solute and water in the diagram on the far right as the vasa recta travels from the cortex, makes a hairpin turn in the depths of the medulla, and returns to the cortex before entering an impermeable collecting vein. Solute (*NaCl* and/or *urea*) flows passively down the concentration gradient, from regions of high to regions of low concentration (i.e., from higher to lower numbers in the diagram). Water flows passively in the opposite direction, from regions where the solute is less concentrated to regions of higher concentration (i.e., from lower to higher numbers in the diagram). Note that water always moves from descending to ascending limb (left to right in the diagram), and solute always moves from ascending to descending limb (right to left in the diagram).

Fluid entering the descending vasa recta is isotonic; it comes from the general circulation. Fluid leaving the ascending vasa recta is slightly hypertonic; it has been in contact with the hypertonic fluids in the medullary ISF. Hence, water flows across from ascending to descending limb, and solutes flow in the reverse direction. Similar arguments apply to the two limbs at each level of the medulla: water takes a shortcut, flowing from descending to ascending limb so that not much of it reaches the depth where it could dilute the hypertonic ISF. Solutes take a similar shortcut in the reverse direction, flowing from ascending to descending limb so that not much is allowed to escape with the fluid entering the veins. Although both water and solute flow in the indicated directions at every level, some of the solute flow arrows have been

omitted from the top of the vasa recta to emphasize entering water that never reaches bottom. Similarly, some of the water flow arrows have been omitted from the bottom to emphasize the solutes trapped in the depths and do not escape.

Fluid leaving the medulla is slightly hypertonic—Note that fluid leaving the medulla at the top of the ascending vasa recta is slightly more hypertonic than fluid entering at the top of the descending vasa recta (350 mOsm compared to 300 mOsm). The counter-current exchange system is not 100% efficient. The vasa recta carries away more solute from the medulla than it brings in. It also carries away water that has been reabsorbed from the collecting duct, but because the nephron continually transfers more solute than water into the medullary ISF, the system will reach a steady state only when the blood supply carries this excess solute away as fast as it forms. Thus, fluid leaving the medulla in the ascending vasa recta must be hypertonic.

At first, the conclusion of the above paragraph (blood leaving the medulla is hypertonic) seems to challenge the assertion that the counter-current multiplier and exchanger act to conserve water. The apparent contradiction is resolved by the fact that the medulla receives only a tiny fraction of the total blood supply to the kidney and that considerable water reabsorption occurs in the distal tubule (see plate 67), which more than compensates for the small hypertonic blood flow that leaves the medulla.

COUNTER-CURRENT MULTIPLIER AND EXCHANGER
WORK TOGETHER

The bottom diagram on the plate shows how activities of the nephron (on the right) and its blood supply (on the left) are integrated to provide the hypertonic ISF required for water conservation. The nephron (more specifically, the loop of Henle) acts as a *counter-current multiplier;* it creates the hypertonic environment. This is an active process requiring metabolic energy that becomes apparent through the active transport of NaCl. The blood supply to the medulla (the vasa recta) acts as a *counter-current exchanger;* it maintains the stability of the hypertonic environment by minimizing the likelihood of excess solutes being washed away by the circulation. This is a purely passive process where much of the entering water is shunted across the top of the exchanger and never reaches the depths, while solutes are shunted across the bottom and are trapped as they simply recirculate from ascending vasa recta to ISF to descending vasa recta and around the loop again to ascending vasa recta.

CN: Use the same colors as were used on the previous page for water (B) and the NaCl concentration solutes (C). Use purple for A, red for D, and blue for F.
1. Begin with the problem in the upper left corner; note that a different color is given to each line of this title. Color the long, straight blood vessel (capillary) that illustrates the problem.
2. Color the title, "solution," and the counter-current exchanger illustration on the far right. Note that the numbers reflecting the osmolarity within the vasa recta and the numbers within the ISF are not to be

colored. Begin with the entry of arterial blood (D) in the upper left corner and first color the diffusion of water across to the ascending limb. Then color the diffusion of NaCl (C) in the opposite direction and the buildup of the concentration gradient.
3. Color the anatomical illustration on the left side, noting that only the vasa recta (A) and the loop of Henle (F) are colored.
4. Color the lower diagram, which summarizes the mechanisms discussed on this and the previous pages.

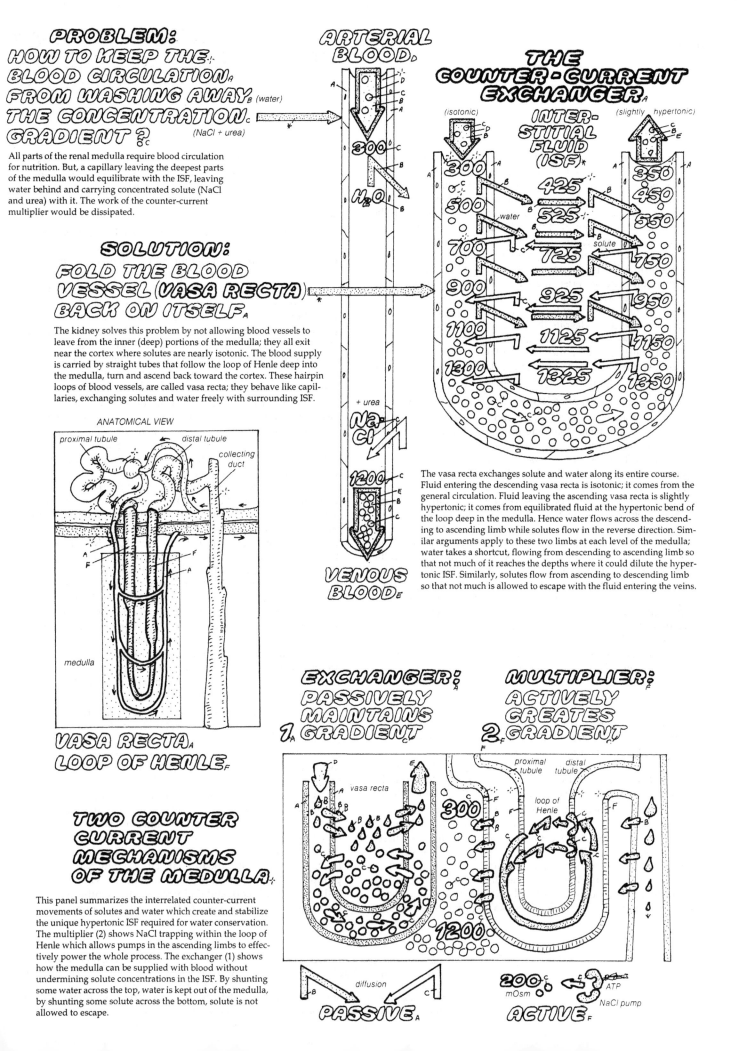

PROBLEM: HOW TO KEEP THE BLOOD CIRCULATION FROM WASHING AWAY (water) THE CONCENTRATION GRADIENT? (NaCl + urea)

All parts of the renal medulla require blood circulation for nutrition. But, a capillary leaving the deepest parts of the medulla would equilibrate with the ISF, leaving water behind and carrying concentrated solute (NaCl and urea) with it. The work of the counter-current multiplier would be dissipated.

SOLUTION: FOLD THE BLOOD VESSEL (VASA RECTA) BACK ON ITSELF

The kidney solves this problem by not allowing blood vessels to leave from the inner (deep) portions of the medulla; they all exit near the cortex where solutes are nearly isotonic. The blood supply is carried by straight tubes that follow the loop of Henle deep into the medulla, turn and ascend back toward the cortex. These hairpin loops of blood vessels, are called vasa recta; they behave like capillaries, exchanging solutes and water freely with surrounding ISF.

ANATOMICAL VIEW

proximal tubule   distal tubule

collecting duct

medulla

VASA RECTA LOOP OF HENLE

TWO COUNTER CURRENT MECHANISMS OF THE MEDULLA

This panel summarizes the interrelated counter-current movements of solutes and water which create and stabilize the unique hypertonic ISF required for water conservation. The multiplier (2) shows NaCl trapping within the loop of Henle which allows pumps in the ascending limbs to effectively power the whole process. The exchanger (1) shows how the medulla can be supplied with blood without undermining solute concentrations in the ISF. By shunting some water across the top, water is kept out of the medulla, by shunting some solute across the bottom, solute is not allowed to escape.

ARTERIAL BLOOD

300

H₂O

+ urea

Na Cl

1200

VENOUS BLOOD

THE COUNTER-CURRENT EXCHANGER

INTERSTITIAL FLUID (ISF)

(isotonic)   (slightly hypertonic)

descending	ISF	ascending
300		350
	425	450
500	525	550
700	725	750
900	925	950
1100	1125	1150
1300	1325	1350

water   solute

The vasa recta exchanges solute and water along its entire course. Fluid entering the descending vasa recta is isotonic; it comes from the general circulation. Fluid leaving the ascending vasa recta is slightly hypertonic; it comes from equilibrated fluid at the hypertonic bend of the loop deep in the medulla. Hence water flows across the descending to ascending limb while solutes flow in the reverse direction. Similar arguments apply to these two limbs at each level of the medulla; water takes a shortcut, flowing from descending to ascending limb so that not much of it reaches the depths where it could dilute the hypertonic ISF. Similarly, solutes flow from ascending to descending limb so that not much is allowed to escape with the fluid entering the veins.

EXCHANGER: PASSIVELY MAINTAINS 1. GRADIENT

MULTIPLIER: ACTIVELY CREATES 2. GRADIENT

vasa recta

300

proximal tubule   distal tubule

loop of Henle

1200

diffusion

PASSIVE

200 mOsm   ATP   NaCl pump

ACTIVE

One of the major functions of the kidney is to regulate the *volume of extracellular fluid*. This is important because plasma volume is largely determined by extracellular volume; plasma and other extracellular spaces continually exchange fluid across capillary walls. When plasma volume and extracellular volume fall, the amount of fluid filling the vascular tree can become inadequate, and despite short-term compensations (increase in heart rate and increase in vascular resistance), the long-term effect is likely to be a decrease in blood pressure. On the other hand, a rise in extracellular volume may fill the vascular tree with too much fluid—it becomes tense, and in the long run pressure will increase. Normally, these events do not occur because, despite the huge variations in daily water and salt intake, the extracellular fluid and plasma volumes remain fairly constant; they are regulated by the kidney, so that responsibility for long-term regulation of blood pressure also resides with the kidney (see plate 47).

EXTRACELLULAR VOLUME REFLECTS THE MASS OF NaCL

ADH matches extracellular volume to NaCl mass—The most important factor that determines extracellular volume is the total *amount* (not concentration) of NaCl in the extracellular spaces. This follows because the NaCl concentration is closely regulated by mechanisms illustrated in the plate. In sum, increasing NaCl causes water retention by the kidney, which dilutes the NaCl but raises extracellular fluid volume. Conversely, decreasing NaCl is accompanied by extra water excretion and decreased extracellular volume. These responses take place because (1) *NaCl is the most abundant solute in the extracellular fluids*, so it largely determines extracellular *osmotic* pressure (concentration of solutes), and (2) the *hormone ADH* closely regulates osmotic pressure. The *"quick osmotic response"* of the ADH system to an increase in salt is illustrated in the plate, where the response has been artificially broken into two steps for purposes of illustration. In stage B, NaCl is suddenly introduced so that there is an exaggerated increase in total amount of NaCl without change of fluid volume. The result is increased NaCl concentration and increased osmotic pressure. In stage C, the hypothalamic osmo-receptors respond (plate 66), releasing ADH, which promotes water reabsorption until the NaCl concentration is practically back to normal. The excess NaCl has not been removed, but the extracellular volume has been increased. In practice, these events take place continuously. Compensation by the ADH system is relatively rapid and precise, so the mass of NaCl and fluid volume generally appear to rise and fall together, with only small changes in NaCl concentrations.

Action of aldosterone regulates volume—The action of ADH explains the linkage between NaCl and extracellular volume, but it does not account for volume regulation. These are accounted for by the "slow volume response" illustrated in the plate. The increased fluid volume initiates a series of steps (described in plate 70) that results in the inhibition of aldosterone release by the adrenal cortex. Without aldosterone, reabsorption of NaCl by the distal nephron is reduced;

more NaCl spills over into the urine, carrying water along with it. The increased ADH secretion that caused the original water retention is no longer operative because the solute concentration has been corrected; the original stimulus for ADH secretion has been removed.

ADH plays a dual role—Release of ADH to correct disturbances in body fluid osmolarity is rapid and very sensitive; osmo-receptors respond significantly to changes as small as 1%. However, ADH-producing cells receive other neural inputs from pressoreceptors (plate 70) that become apparent when vascular volume changes begin to exceed 5 to 10%. From here, ADH is released to promote water retention independently of the plasma osmolarity. In other words, ADH can switch from being an osmolarity regulator to a volume regulator. In addition to a direct effect on increasing water reabsorption, ADH also stimulates Na^+ reabsorption in the collecting duct, increasing the driving force for fluid retention.

GFR and proximal reabsorption may change during severe volume expansion—During more severe volume expansion, still other factors become important as the autoregulation of GFR and tubuloglomerular feedback become compromised. Volume regulation will no longer be confined to the distal nephron as other sections begin to make adjustments. Some of these are launched by the decrease in sympathetic nerve activity that generally results from increased vascular pressure. We may find an increased glomerular pressure arising from dilation of the afferent arteriole as sympathetic impulses are withdrawn. As a result, GFR will increase. We may also find a decrease in Na reabsorption in the proximal tubule; sympathetic impulses are known to stimulate Na^+ reabsorption there and withdrawal of that stimulus will have its effect. These factors increase the load of NaCl delivered to the distal nephron, which, lacking its normal aldosterone stimulus, becomes overwhelmed, allowing more NaCl to spill over into the urine. In addition, the atrial natriuretic hormone (ANP) makes a tangible contribution. Recall that ANP is secreted when the volume of the atria expands and that it acts to promote Na^+ excretion and to inhibit renin, ADH, and aldosterone secretion—all of which reduce volume.

Aldosterone initiates synthesis of new Na^+ transporters; ADH initiates recruitment of water channels—How do ADH and aldosterone exert their characteristic effects on the cells of the kidney? ADH acts by inserting water channels in the collecting ducts and in the distal tubule. Aldosterone promotes Na^+ reabsorption in the distal tubule and collecting ducts. The hormone is lipid soluble; it passes through the plasma membrane and reacts with an intracellular receptor protein, which acts on the nucleus and leads to the synthesis of new protein. The new protein may be involved in the supply of (1) new Na^+ channels on the luminal membrane, promoting Na^+ entry into the tubular cells, (2) new $Na^+ - K^+$ pumps on the basolateral membrane to pump Na^+ out of the cell toward the blood, and (3) increasing the synthesis of enzymes that provide more ATP to power the pumps.

CN: Use the same colors as were used on the previous page for water (A) and NaCl (D).
1. Color the upper panel first.
2. Color the quick osmotic response to an increase to the solute concentration of the plasma.

3. Color the slow volume response, which deals with the resulting increase in body fluid volume shown in figures C and C^1.
4. In the lower panel, color the actions of ADH (E) and aldosterone (F).

BODY WATER CONTENT

The total body water constitutes 60% of the body weight. Two-thirds of this water lies within cells (intracellular) and one-third lies outside (extracellular). Most cell membranes permit free exchange of water between intra- and extracellular spaces.

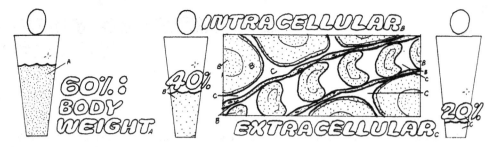

60%: BODY WEIGHT.
INTRACELLULAR.
40%
EXTRACELLULAR.
20%

EXPANSION & CONTROL OF EXTRACELLULAR VOLUME.

1 QUICK OSMOTIC RESPONSE: (Δ)ADH.

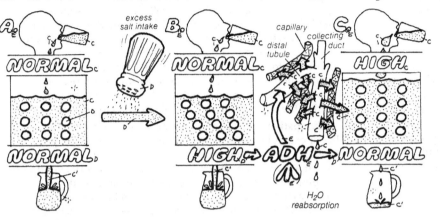

excess salt intake

capillary
collecting duct
distal tubule

NORMAL / NORMAL
NORMAL / HIGH → ADH →
HIGH / NORMAL

H₂O reabsorption

SOLUTE CONCENTRATION (NaCl)
URINE.
ADH.
ALDOSTERONE.

By responding quickly to alterations in the solute concentrations (osmolarity) of the plasma, ADH keeps the body fluids practically isotonic at all times. If the solute concentration goes up (more solute dissolved in the same water volume, as shown in the middle figure), ADH is secreted and less water is lost in urine, so that the isotonic condition is restored. But now the volume of body fluids has increased. This ADH mechanism ensures that a proportionate amount of water will be retained (or lost) whenever there is a gain (or loss) of solute (principally NaCl). Body fluid volume faithfully follows changes in total solute.

2 SLOW VOLUME RESPONSE: (∇)ALDOSTERONE.

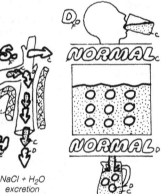

HIGH → ALDOS. → NORMAL
NORMAL / ADH / NORMAL

NaCl + H₂O excretion

The increased fluid volume initiates a series of steps that result in the inhibition of aldosterone release by the adrenal cortex. Without aldosterone, reabsorption of NaCl by the distal nephron is reduced and more NaCl spills over into the urine, carrying water along with it. The increased ADH secretion that caused the original water retention is no longer operative because the solute concentration has been corrected — the original stimulus for ADH secretion has been removed. (Increased fluid volume may also inhibit ADH secretion.)

ANTIDIURETIC HORMONE.
"WATER REABSORPTION"
ADH RECEPTOR.
ATP → CYCLIC AMP.

ADH acts by opening channels in the collecting ducts and in the distal tubule. The hormone reacts with a receptor on the basal membrane, activating adenyl cyclase, the enzyme that converts ATP to cyclic AMP. Cyclic AMP acts as a second messenger, initiating a sequence of steps that culminates in the insertion of water channels.

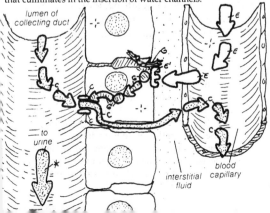

lumen of collecting duct
to urine
interstitial fluid
blood capillary

ALDOSTERONE.
"NaCl REABSORPTION"
RECEPTOR PROTEIN.
NUCLEUS.
SYNTHESIZED PROTEIN.

Aldosterone promotes Na⁺ reabsorption: it passes through the plasma membrane and reacts with an intracellular receptor protein that leads to synthesis of new protein. The new protein may be involved in the supply of (1) new Na⁺/K⁺ pumps on the basal membrane, (2) more ATP, and (3) new Na⁺ channels on the luminal membrane.

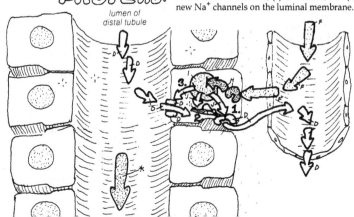

lumen of distal tubule

Plate 69 illustrated how the total amount of NaCl determines the extracellular volume. Attention is focused on Na^+ because regulatory mechanisms act through it and because changes in Cl^- are, to a large extent, secondary to Na^+ movements. Our example showed how the body fluids expand whenever the amount of Na^+ (or NaCl) increases, and how compensatory changes help return the volume toward normal. The theme is continued here as we examine in more detail how extracellular volume is regulated by the kidney through the hormonal control of Na^+ excretion. This time we have the reverse situation: the response to body fluid depletion.

Depletion of the extracellular volume is a common clinical event. It occurs in severe vomiting, in diarrhea, and in sweating response to intense heat (heat prostration). In each of these cases, considerable Na^+ is lost from the body, and compensatory processes are set in motion to restore the Na^+ and water loss. The plate emphasizes the *renin-angiotensin-aldosterone* response, one of the most important of these compensatory processes. This system is activated by several stimuli, all of which arise directly or indirectly from changes in extracellular volume (see below).

ACTIONS OF ANGIOTENSIN II, III: ALDOSTERONE, THIRST AND VASOCONSTRICTION

Renin begins the sequence by catalyzing the conversion of angiotensinogen to angiotensin I—When the extracellular volume is depleted, *renin* is released from specialized secretory cells in the wall of the afferent arteriole where it butts up against the distal tubule and forms a structure called the *juxtaglomerular apparatus* (see plate 62). The released renin is an enzyme that acts on the plasma protein *angiotensinogen* (produced by the *liver*) and splits off a small, ten-amino-acid fragment called *angiotensin I*.

Angiotensin II and III stimulate aldosterone secretion and thirst—Angiotensin I is converted into a smaller peptide (eight amino acids), *angiotensin II,* by action of an angiotensin converting enzyme (ACE) that is especially prominent in the lungs but also occurs elsewhere. Finally, angiotensin II is split into an even smaller peptide, *angiotensin III*. Angiotensins II and III are active products. In addition to vasoconstriction, they both stimulate secretion of aldosterone, and they both stimulate thirst. (ACE inhibitors have become a popular treatment for hypertension because of the vasoconstrictive actions of angiotensin II and III)

Aldosterone stimulates Na reabsorption—Aldosterone reaches the kidney via the circulation and promotes reabsorption of Na^+ by the distal tubule and the upper collecting ducts. Cl^- follows the Na^+, preserving electrical neutrality, and water follows, preserving osmotic equilibrium. The net result is the reabsorption of NaCl and water; angiotensins II and III also stimulate thirst. The volume of body water and the NaCl content rise toward normal. The relative proportions of NaCl and water gained are "finely tuned" by the *ADH* feedback mechanism, operating on water reabsorption to maintain a constant solute concentration in the body fluids.

RENIN SECRETION IS DRIVEN BY REDUCED VOLUME/PRESSURE AND BY THE JG APPARATUS

Low pressure volume receptors monitor extracellular volume—We have yet to account for the linkage between changes in extracellular volume and renin secretion. Several stimuli giving rise to renin secretion have been identified, but fine details of the steps leading from stimulus to final response have remained elusive and speculative. The first issue is to establish how the extracellular volume is monitored. This appears to be accomplished primarily by stretch receptors in the walls of the atria, near their junctions with the venae cavae and pulmonary veins. Although these receptors are similar to those found in the arterial pressoreceptors, they are considered *volume receptors* because unlike arteries, the atrial walls expand readily without building up pressure, making them more sensitive to volume than to pressure.

Sympathetic nerve impulses stimulate renin secretion—In our example, the depleted volume depresses the output from the atrial volume receptors, and if the volume change is large it will also depress arterial pressures. Both atrial volume receptors and arterial pressoreceptors normally send nerve impulses to the brain stem, where they inhibit sympathetic nerves. When pressures are lowered, pressoreceptors become less active, and sympathetic nerves to the kidney are released from their "braking" action. The kidney is thus showered with sympathetic impulses, which stimulate renin release.

Reduced pressure in the afferent arterioles stimulates renin secretion—A second important regulatory system for renin secretion is provided by the direct action of *pressure in the afferent arterioles* of the kidney itself. When this pressure rises, renin secretion is inhibited; when it falls (as in our example), secretion is enhanced. This arteriolar mechanism is independent of nerves. When they are cut, the response persists.

Decreased fluid delivery to the JG apparatus stimulates renin secretion—A third regulatory system is found in the *juxtaglomerular apparatus*. This composite structure consists of the secretory cells in the afferent arteriole and specialized cells of the distal tubule, called macula densa, which are in close contact with the secretory cells. A decrease in fluid delivery within the nephron to the macula densa results in stimulation of the secretory cells, and more renin is released into the circulation. The decrease in fluid delivery occurs when the glomerular filtration rate is lowered, which can happen in response to the lowered arterial pressure, especially if sympathetic nerve impulses constrict the afferent arterioles. (Note that the reduced glomerular filtration by itself will help compensate for fluid depletion because it reduces fluid excretion.) The mechanism secretes renin into the systemic circulation, where it catalyzes the formation of angiotensins II and III; these stimulate release of aldosterone, etc. The relation of this regulatory system to the mechanism described in plate 62, which utilizes the same juxtaglomerular apparatus for matching the glomerular filtration rate of each nephron to its tubular reabsorptive capacity, is not clear.

CN: Use the same colors as on the previous page for water (A), NaCl solute (B), ADH (H), and aldosterone (G). Use red for blood vessels (C).
1. Begin with the figure in the upper left showing extracellular volume depletion. Note the use of gray in coloring the symbols of increase and decrease in the chain of events leading to the release of renin (D) by the cells of the juxtaglomer- ulus (which receive the blood vessel color in the enlarged view in the center of page).
2. Color the role of renin (D) in hormonal regulation, in the material under the enlargement, going from the liver on the left to the adrenals.
3. Color the effects of aldosterone (G) in the lower right corner by following the numbered sequence that leads to the actions of ADH on the left.

DEPLETION AND CONTROL OF EXTRACELLULAR VOLUME.

WATER.
NaCl SOLUTE.
BLOOD VESSEL.
INCREASE ⇧
DECREASE ⇩

Depletion of the extracellular volume (as occurs in severe vomiting or diarrhea) sets in motion a number of processes that converge on the stimulation of renin release by the juxtaglomerular cells in the afferent arteriole. Atrial volume receptors and arterial pressoreceptors respond to the depleted volume by sending fewer nerve impulses to the brainstem, and this activates portions of the sympathetic nervous system. Sympathetic impulses arriving at the kidney stimulate renin release. Renin release is also stimulated by the reduced pressure and compromised flow in the renal artery.

HORMONES / SOURCE.
RENIN / KIDNEYS.
ANGIOTENSINOGEN / LIVER.
ANGIOTENSIN I, II, III.
ALDOSTERONE / ADRENALS.
ADH / HYPOTHALAMUS.

The released renin acts on the plasma protein angiotensinogen (produced by the liver) and splits off a small, 10-amino-acid fragment called angiotensin I. Angiotensin I is converted into a smaller peptide, angiotensin II, which in turn forms an even smaller peptide, angiotensin III. Angiotensin II and III are active products. In addition to vasoconstriction, they both stimulate secretion of aldosterone and they both stimulate thirst.

ACTIONS OF ALDOSTERONE AND ADH.

Aldosterone reaches the kidney via the circulation (1) and promotes reabsorption of Na^+ by the distal tubule and the upper collecting ducts (2). Cl^- follows the Na^+, preserving electrical neutrality, and water follows, preserving osmotic equilibrium. The net result is the reabsoption of NaCl and water (3). In addition, angiotensin II and III stimulate thirst (4). The volume of body water and the NaCl content rise toward normal (5). The relative proportions of NaCl and water gained are "finely tuned" by the ADH feedback mechanism (6), which operates on water reabsorption to maintain a constant solute concentration in the body fluids.

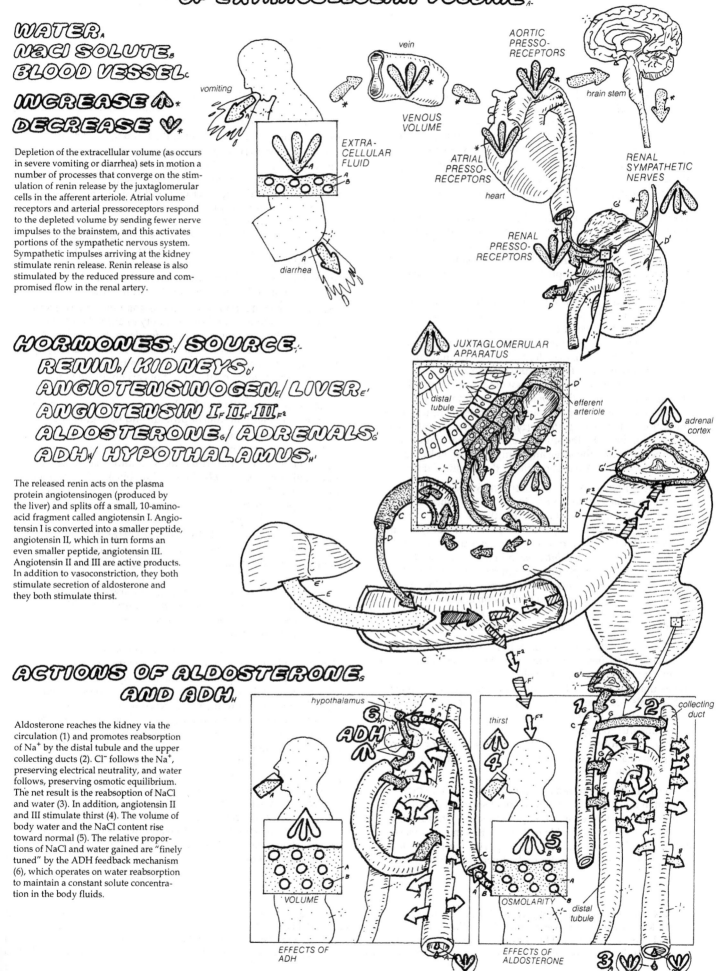

vomiting

diarrhea

EXTRA-CELLULAR FLUID

vein

VENOUS VOLUME

AORTIC PRESSO-RECEPTORS

brain stem

ATRIAL PRESSO-RECEPTORS

heart

RENAL PRESSO-RECEPTORS

RENAL SYMPATHETIC NERVES

JUXTAGLOMERULAR APPARATUS

distal tubule

efferent arteriole

adrenal cortex

hypothalamus

ADH

thirst

collecting duct

VOLUME

OSMOLARITY

distal tubule

EFFECTS OF ADH

EFFECTS OF ALDOSTERONE

The *digestive system* (also called *gastrointestinal system*, *digestive tract*, or *digestive tube*) is basically a long tube open at both ends, making the lumen of the digestive system an extension of the external environment. Food enters from the oral end (mouth) and is broken down mechanically and chemically by the aid of a variety of digestive structures; next the lining of the digestive tube absorbs the usable nutrients, and leftover materials leave from the anal end as waste products.

FOODSTUFF UNDERGOES MECHANICAL & CHEMICAL DIGESTION
Humans ingest food, usually in forms unsuitable for uptake and use by body cells. The digestive system transforms the ingested foodstuff to simpler nutrients, capable of uptake by body cells. This is accomplished by *mechanical* and *chemical* digestive processes that occur in the oral (mouth), gastric (stomach), and intestinal stations, in an orderly manner, resembling a food-processing plant. During mechanical digestion, solid food masses are torn apart and ground by the teeth and mixed with juices from the digestive glands (salivary, gastric, and intestinal), in order to dissolve the food particles and form a rich soup. This mixture is vigorously shaken during various gastrointestinal movements generated by the gut's muscular wall. *Chemical digestion* transforms the dissolved food particles into simple, absorbable nutrients by the action of various *digestive enzymes*, secreted mainly by the pancreas and also by the stomach and intestinal glands. These enzymes *hydrolyze* the large and complex food molecules into simpler forms that are absorbable by the intestinal lining.

DIGESTION BEGINS IN THE MOUTH AND STOMACH
The *salivary glands* secrete *saliva* to aid in mechanical digestion and dissolving of the food in the mouth. The *pharynx* and *esophagus* aid in *swallowing* and transport of the food into the *stomach*, which acts as a reservoir to receive a meal at once while delivering it to the *intestine* in intervals. In the stomach, food is subjected to vigorous movements that mix it with the *gastric juices* to form *chyme*. Gastric juices, containing *mucus*, *acid*, and *enzymes*, are secreted by the stomach glands. Some chemical digestion of proteins, but no absorption of any significance (except for alcohol), occurs in the stomach.

THE LIVER AND PANCREAS AID IN INTESTINAL DIGESTION
In the *small intestine*, the dissolved food particles in the gastric chyme are subjected to further shaking and mixing movements that mix them with the alkaline *intestinal juice*. Intestinal juice also contains the secretions of the large accessory digestive glands (the *pancreas* and *liver*). The pancreatic juice is alkaline due to high bicarbonate content and also rich in a variety of hydrolytic enzymes that are essential for chemical digestion of all food substances. The liver secretes the *bile*, which facilitates fat digestion.

SMALL INTESTINE ABSORBS NUTRIENTS INTO THE BLOOD
The small intestine is the main site of absorption of nutrients. This occurs across the inner lining of the small intestine. Upon absorption, all water-soluble material enters the *intestinal-hepatic* portal venous system (hepatic portal vein) and is taken to the liver for processing. From the liver the nutrients are transported by blood to the body cells, where they are taken up and consumed for energy and cellular metabolism. The absorbed fatty nutrients enter the lymph vessels, bypassing the liver, and enter the blood via the lymphatic circulation.

THE LARGE INTESTINE DEHYDRATES UNUSED CHYME
The last function of the digestive system, carried out by the *large intestine* (*colon*), is to remove and absorb the water from the remaining and unused chyme and treat the non-absorbable remnants of absorption (e.g., fiber). Dehydration produces solid fecal masses (feces) that, along with bacterial debris, are moved by *peristalsis* and *mass action* to the *rectum* and *anus* where the feces are excreted (*defecation*). The useful intestinal bacteria play a major role in colon function and fecal formation. Salts (sodium) and some vitamins of bacterial origin (e.g., vitamin K) are also absorbed in the colon.

ENZYMES TRANSFORM FOOD INTO ABSORBABLE NUTRIENTS
Humans consume foods from a variety of animal and plant sources. In the fresh form, all these foods contain different amounts of the main classes of nutrients: *proteins, carbohydrates*, and *fats*. For example, meats contain a lot of protein, some fat, and a very small amount of carbohydrate while breads, pasta and potato contain a lot of carbohydrates, some proteins and very little fat. Apples contains fiber, some carbohydrates, smaller amounts of protein, and negligible fat.

During chemical digestion, with the aid of a variety of *protease* enzymes, dietary proteins are broken down first into *oligopeptides*, which are further digested into smaller *peptides* and finally into *amino acids*, the building blocks of all peptides and proteins. Free amino acids are then in the form suitable for absorption by the intestinal mucosa and delivery to the liver and other body cells.

Dietary sources of carbohydrates are plant starches (polysaccharides) and *disaccharides* such as sucrose (table sugar) and lactose (milk sugar). Polysaccharides are broken down to *oligo-* and *disaccharides* with the help of *amylase enzymes*; more specific enzymes (e.g., sucrase and lactase) work on disaccharides to form *monosaccharides* (simple sugars) like glucose, fructose, and galactose, absorbable forms of carbohydrates.

Dietary fats are available mainly as *triglycerides* (*triacylglycerols*), which are broken down in the intestine by the action of lipases into their constituents—*glycerol* and *fatty acids*. Occasionally, *mono-* or *diglycerides* are also produced. The bile, an important digestive secretion of liver, plays an important role in fat chemical digestion. The simpler fats are then absorbed across the mucosa. Before entry into the blood, triglycerides are resynthesized and incorporated into lipoprotein particles called *chylomicrons*, which are then transported via the *lymphatic system* to the blood. Action of pancreatic nucleases (RNAase and DNAase) and related enzymes chemically digests the dietary nucleic acids to form nucleotides and then nucleosides and finally sugars, phosphoric acids, and pyrimidine and purine bases, which are absorbed. Dietary fiber is not absorbed; bacteria digest it in the large intestine.

CN: Use blue for L and light gray (or a single light color) for structures H–K. Notice the use of overlapping colors in the stomach region of the central illustration to suggest the presence of one organ in front of another.
1. Color the same structure in both the anatomic and the functional diagrams before going on to the next structure. Color the titles along the right edge of the page.
2. Color the inner edge of the doughnut, which demonstrates that the digestive tract (the mouth to the anus) is essentially outside the body

DIGESTIVE SYSTEM

DIGESTIVE TRACT

ORAL CAVITY A
PHARYNX B
ESOPHAGUS C
STOMACH D
SMALL INTESTINE E
LARGE INTESTINE F
RECTUM G

DIGESTIVE GLANDS

SALIVARY GLANDS H*
LIVER I*
PANCREAS J*

GALL BLADDER K*

The digestive system functions to ingest, digest, and absorb food substances into the bloodstream and to eliminate remaining wastes. The digestive structures in the mouth and stomach act primarily in the mechanical and chemical digestion of foods. The small intestine acts in chemical digestion of food substances and absorption of resultant nutrients. The large intestine (colon) absorbs remaining water and salts and excretes waste products of digestion (feces) through its exit end, the rectum and anus. To facilitate digestion, numerous exocrine glands secrete a variety of alkaline or acidic juices containing enzymes and mucus into the digestive lumen. The large and separately located pancreas, liver, and salivary glands constitute the accessory digestive glands, while the numerous small stomach and intestinal glands form an intrinsic part of the gut wall. During the absorption process, the breakdown products of proteins, carbohydrates, and nucleic acids, as well as water, minerals, and water-soluble vitamins, are transported across the intestinal mucosa into the hepatic portal circulation for delivery to liver and bloodstream. Fats and fat-soluble vitamins, however, are absorbed via the lacteals and lymph vessels for delivery to the blood via the lymphatic circulation.

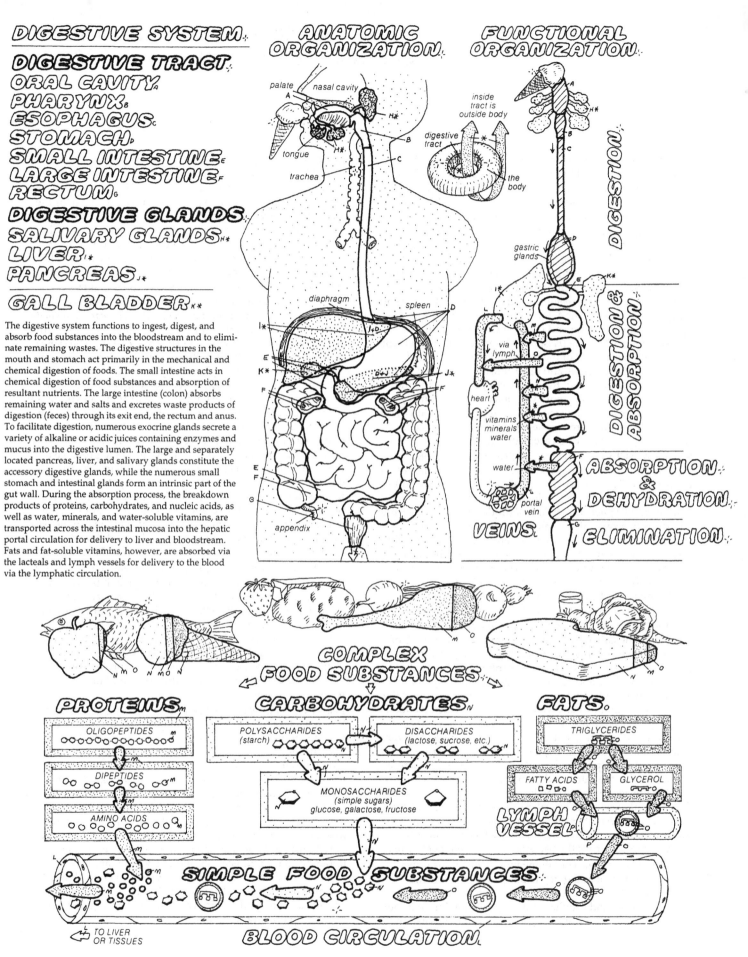

ANATOMIC ORGANIZATION.

palate · nasal cavity
tongue
trachea
diaphragm
spleen
appendix

FUNCTIONAL ORGANIZATION.

inside tract is outside body
digestive tract
the body
gastric glands
via lymph
heart
vitamins minerals water
water
portal vein

DIGESTION
DIGESTION & ABSORPTION
ABSORPTION & DEHYDRATION
VEINS.
ELIMINATION.

COMPLEX FOOD SUBSTANCES

PROTEINS M
OLIGOPEPTIDES M
DIPEPTIDES M
AMINO ACIDS M

CARBOHYDRATES N
POLYSACCHARIDES (starch)
DISACCHARIDES (lactose, sucrose, etc.)
MONOSACCHARIDES (simple sugars) glucose, galactose, fructose

FATS O
TRIGLYCERIDES
FATTY ACIDS · GLYCEROL

LYMPH VESSEL

SIMPLE FOOD SUBSTANCES

TO LIVER OR TISSUES

BLOOD CIRCULATION.

Dietary foods such as meats, fruits, dairy products, bread, and vegetables are rarely found in readily absorbable form. The complex dietary substances are proteins (meat, egg white, beans), carbohydrates (bread, rice, potato), and fats (milk fat, egg yolk, butter, oils). Digestive enzymes (not shown) secreted by the pancreas and other digestive glands hydrolyze complex dietary substances into simpler and smaller molecules that can be readily absorbed by the intestinal mucosa into the bloodstream. Proteins are digested into amino acids, complex carbohydrates (polysaccharides) into simple sugars (monosaccharides—e.g., glucose), fats (triglycerides = triacylglycerols) into fatty acids and glycerol, and nucleic acids (not shown) into purine and pyrimidine bases and ribose sugars. Dietary fibers facilitate digestion but are not absorbed.

The *mouth* is the first station in the digestive process. Here, solid food is subjected to a number of mechanical and chemical digestive processes such as chewing and mixing with saliva. As a result, the solid food pieces are converted into a *bolus*, a form that can be easily swallowed.

TEETH & CHEWING BEGIN MECHANICAL DIGESTION
Several structures in the mouth aid in food ingestion and its mechanical digestion: the *lips*, the *teeth*, the *tongue*, and the muscles of the *cheeks*. Adult humans have 32 teeth arranged in two sets attached to the upper and lower jaw bones. Human teeth are adapted to an *omnivorous* diet: the 8 front *incisors* are designed for cutting, the 4 *canines* or *cuspids* for tearing, the 8 *premolars* for crushing and the 12 *molars* for grinding of ingested food. Chewing (*mastication*) involves the movements of the jaws, the actions of the teeth, and the coordinated movement of the tongue and other muscles of the *oral cavity* (mouth). The activities of the masticatory muscles and the tongue are controlled by both voluntary and involuntary nervous control mechanisms. The mere placing of food in the mouth can activate some of the involuntary reflex mechanisms, the centers of which are in the *brain stem*.

SALIVA AIDS TASTE, CHEWING, & BOLUS FORMATION
The chewing actions would be extremely difficult without the aid of *saliva*, a viscous, sticky, and slippery fluid secreted by three pairs of *salivary glands*: the *parotid* in the cheeks secrete a watery (*serous*) juice, the *sublingual* (under the tongue) secrete a mucous (viscous, sticky) saliva, and the *submandibular* (under the lower jaw) secrete both serous and mucous fluid. The salivary glands are acinar exocrine glands. The acini are of either serous or mucous types. The *serous acini* secrete a watery saliva, and the *mucous acini* secrete a more viscous fluid containing the glycoprotein substance *mucin*, which gives the saliva its sticky, viscous, and slippery texture.

SALIVA CONTAINS WATER, IONS, MUCUS, & ENZYMES
Three pairs of salivary glands secrete an average of 1.5 L of saliva each day. Of this, 20% is secreted by the parotid, 70% by the submandibular, and 5% by the sublingual glands, the remaining 5% by minor buccal glands (not shown). The serous saliva, containing more than 99% water, dissolves the food particles and forms a wetter mold from which the food *bolus* is produced; it also keeps the mouth wet and aids in speech. The dissolving of food particles is also necessary for *taste sensation by the taste receptors* in the tongue's *taste buds*; taste receptors respond only to dissolved substances. Serous saliva contains the salivary digestive enzyme *ptyalin*, an *amylase* that breaks down the starches to maltose, a disaccharide; this may enhance sweet sensation and promote ingestion of carbohydrates. Another salivary enzyme is *lysozyme*, an antibacterial agent protecting the mouth and teeth against bacteria; lysozyme destroys the bacteria by lysing their cell walls.

The mucous saliva, containing mucin, functions principally as a lubricant and glue, helping the formation and movements of the food bolus in the mouth and its transport along the throat and esophagus during swallowing. Without saliva, chewing and swallowing are very difficult. The saliva contains *sodium*, *potassium*, and *calcium* as well as *bicarbonate* and *chloride*. The high levels of salivary calcium is thought to prevent calcium loss from the teeth. The bicarbonate acts as a buffer, helping to maintain a neutral pH of about 7.0 for the saliva under normal conditions and an alkaline pH of 8.0 during active secretion. Recently some hormones (e.g., steroids), antibodies (IgA), and drugs have been detected in salivary secretions in very minute amounts, opening the door to the use of saliva in laboratory diagnosis.

AUTONOMIC NERVES REGULATE SALIVARY SECRETION
The formation and secretion of saliva are under the control of the autonomic nervous system (plate 29). The *parasympathetic* nerves originating in the *salivary nuclei* of the brain stem stimulate both serous and mucous salivary secretion. The *sympathetic* nerves inhibit the secretion of serous saliva, mainly by vasoconstriction. This explains why the mouth becomes dry during fear and excitement (a sympathetic condition) and salivary juice flows profusely during relaxation or expectation of food and pleasure. During oral digestion, the presence of food, particularly dry or sour foods, in the mouth serves as a strong stimulus for salivary secretion. This reflex response is initiated by sensory nerves communicating the presence of the food stimuli in the mouth to the brain stem salivary centers. These in turn activate the parasympathetic nerves to the salivary glands, increasing their production of saliva. Similarly the sight of food and its odors acting through the olfactory (smell) sense and even thoughts of food can increase salivary flow. Salivary flow can be easily conditioned in humans and animals by learning, as shown by Pavlov in dogs.

After the food bolus is appropriately formed in the mouth, the movements of the tongue gradually push it backward. Presence of the bolus on the back of the tongue activates the swallowing (*deglutition*) reflexes, which are regulated by neural centers in the brain *medulla*. When the tongue moves back to force the food bolus into the throat's *pharynx*, the soft *palate* closes the nasal passage, and the *epiglottis* moves over the *glottis* to close the *larynx* and *trachea*. These protective reflexes prevent the bolus from entering the upper and lower respiratory passages.

PERISTALSIS PROPELS THE FOOD BOLUS IN THE ESOPHAGUS
When the food bolus arrives in the pharynx, other reflexes relax the upper esophageal sphincter, allowing the bolus to enter the esophagus, a tubular organ connecting the pharynx with the stomach. The muscular wall of the esophagus contains layers of *circular* and *longitudinal smooth muscles* whose coordinated contractions produce a special traveling wave of movements called *peristalsis*, which begins in the upper esophagus and moves toward the stomach. As a result, the food bolus is propelled from pharynx to stomach. Although gravity may aid bolus transport in the human esophagus under normal circumstances, it is not a necessary condition for esophageal transport. Food and water can be swallowed in a supine position as well, as seen in human babies, or against gravity, as in playful children and actors standing on their heads or in grazing ruminants.

CN: Use red for C and a dark color for Q.
1. Color the structure of the mouth including the three salivary glands and the diagrams of tooth function.
2. Color the chemical events in the mouth along the right side of the page.
3. Color the panel on swallowing.

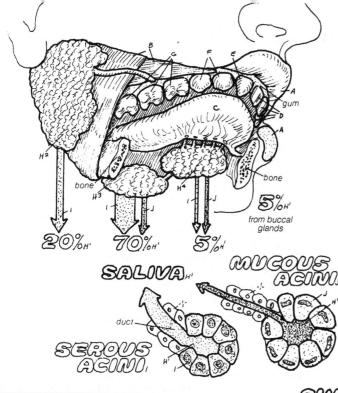

gum
bone
bone
H²
H³
H⁴
20%ₒₕ' 70%ₒₕ' 5%ₒₕ'
5%ₒₕ'
from buccal glands

SALIVA ₕ'

MUCOUS ACINI ⱼ

duct

SEROUS ACINI ₕ'

MECHANICAL EVENTS

LIPS ₐ
MUSCLES/CHEEKS ᵦ
TONGUE c
TEETH *
INCISORS 8 ᴅ
CANINES 4 ᴇ
PREMOLARS 8 ꜰ
MOLARS 12 ɢ

The chewing action (mastication) carried out by the mouth and its associated structures (lips, tongue, cheeks, teeth, jaws) mechanically breaks the food materials into smaller pieces and forms a bolus for swallowing.

CUT ᴅ   TEAR ᴇ   CRUSH ꜰ   GRIND ɢ

In the adult human, a total of 32 permanent teeth act to cut, tear, crush, and grind the food materials. Teeth are absent in the newborn. Deciduous (temporary) teeth (20 in number) form between 6 to 24 months of age as the infant begins to use solid foods. Permanent teeth appear from 6 to 21 years.

SWALLOWING & PERISTALSIS *

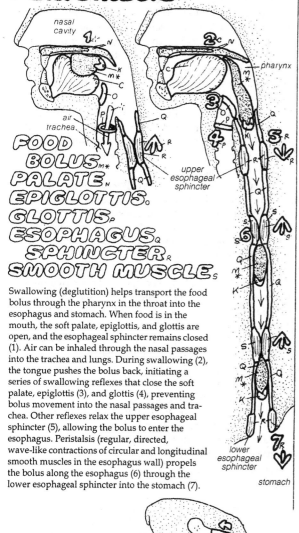

nasal cavity
1 2 c
pharynx
air trachea
3
4
5
upper esophageal sphincter
6

FOOD
BOLUS ₘ*
PALATE ₙ
EPIGLOTTIS ᴏ
GLOTTIS ᴘ
ESOPHAGUS ǫ
SPHINCTER ʀ
SMOOTH MUSCLE ꜱ

lower esophageal sphincter
7
stomach

Swallowing (deglutition) helps transport the food bolus through the pharynx in the throat into the esophagus and stomach. When food is in the mouth, the soft palate, epiglottis, and glottis are open, and the esophageal sphincter remains closed (1). Air can be inhaled through the nasal passages into the trachea and lungs. During swallowing (2), the tongue pushes the bolus back, initiating a series of swallowing reflexes that close the soft palate, epiglottis (3), and glottis (4), preventing bolus movement into the nasal passages and trachea. Other reflexes relax the upper esophageal sphincter (5), allowing the bolus to enter the esophagus. Peristalsis (regular, directed, wave-like contractions of circular and longitudinal smooth muscles in the esophagus wall) propels the bolus along the esophagus (6) through the lower esophageal sphincter into the stomach (7).

CHEMICAL EVENTS
SALIVARY GLANDS ⇒ SALIVA ₕ'
PAROTID ₕ²
SEROUS SALIVA ₁
SUBMANDIBULAR ₕ³
SEROUS S. ₁ MUCOUS S. ᴅⱼ
SUBLINGUAL ₕ⁴
MUCOUS S. ₇ⱼ SEROUS S. ₁

DAILY 1.5 LITERS ₕ'

Each day about 1.5 liters of saliva are secreted by the three pairs of salivary glands: parotid, sublingual, and submandibular (submaxillary). The serous saliva, containing more than 99% water, dissolves the food particles and forms a wetter mold from which the food bolus is produced; it also keeps the mouth wet and aids in speech. The dissolving of food particles is also necessary for taste sensation by the taste receptors in the tongue's taste buds; taste receptors respond only to dissolved substances. Serous saliva contains the salivary digestive enzyme ptyalin, an amylase that breaks down the starches to maltose, a dissacharide; this may enhance sweet sensation and promote ingestion of carbohydrates. Another salivary enzyme is lysozyme, an antibacterial agent that protects the mouth and teeth against bacteria; lysozyme destroys the bacteria by lysing their cell walls.

CONTENT & FUNCTION OF SALIVA
99%% WATER: DISSOLVES BOLUS FOR TASTE ₕ⁵

The water in saliva helps dissolve food particles and facilitates taste sensation. Dry foods and sour (acidic) juices induce copious salivary secretion.

MUCUS: LUBRICATION FOR BOLUS ⱼ

The glycoprotein mucin, secreted by the mucous acini, gives saliva its sticky and lubricating property. Without saliva, bolus formation is very difficult and swallowing is painful.

ENZYME: AMYLASE BEGINS STARCH DIGESTION ₖ

Salivary amylase begins the chemical breakdown of starches, forming oligo- and di-saccharides (maltose). Amylase action is important for the sensation of sweet.

LYSOZYME: ANTI-BACTERIAL ACTION ₗ

The antibacterial enzyme lysozyme and immunoglobulins IgA in the saliva help against bacterial infection in the mouth.

The *stomach* is a large muscular sac connected at its opening to the esophagus and at its end to the *duodenum* of the small intestine. Two *sphincters*, the *cardiac (lower esophageal)* and the *pyloric*, act as unidirectional flow valves permitting food to move into and out of the stomach, respectively. The stomach functions as a reservoir, receiving the ingested food in one portion and mixing it with gastric juice. This juice contains acid, which helps dissolve and disinfect the food, as well as protease enzymes (pepsin), which partially digest the proteins. Finally, the stomach's strong movements mix the food with gastric juice, producing a soupy *chyme* that is delivered, through the pyloric sphincter, to the small intestine, in regular intervals, for more enzymatic digestion and absorption.

THE STOMACH GLANDS PRODUCE GASTRIC JUICE
Numerous exocrine gastric *glands* or *pits*, located in the stomach wall, secrete a juice containing *acid*, *pepsin*, and *mucus* into the stomach lumen. Each gland contains three types of cells, which together produce the bulk of *gastric juice*. The cells near the gland's neck (*mucous cells*) and those lining the stomach's inner surface secrete the gastric mucus. In the gland's deeper zone, two other cell types, the *chief cells* and the *parietal cells*, secrete the proenzyme *pepsinogen* and *hydrochloric acid* (H^+Cl^-) respectively. The stomach glands also contain scattered endocrine cells that secrete the hormone gastrin and paracrine cells that release local hormones (e.g., prostaglandins, histamine) into the local tissue spaces.

Stomach acid has several functions. The acidic gastric juice acts as a superior solvent, dissolving foodstuff not soluble in water. Acid is a strong disinfectant, killing bacteria and other microorganisms in the food. Acid is also necessary to activate the gastric enzyme pepsin (see below). Finally, acid stimulates the duodenum to secrete hormones that release bile and pancreatic juice (plate 74) into the duodenum.

THE PARIETAL CELLS ACTIVELY SECRETE H^+ AND Cl^-
The parietal cells of the stomach glands can secrete an isotonic, essentially concentrated solution of *hydrochloric acid* (pH 1) into the stomach lumen. Acid secretion peaks within one to two hours after a meal. The parietal cells use active transport by H^+-K^+ *pumps* in their apical membrane to pump hydrogen ions (H^+), obtained from dissociation of intracellular water ($H_2O \rightarrow H^+ + OH^-$), into the gland lumen. The H^+-K^+ pump, an ATPase closely related to the Na^+-K^+ pump, exchanges K^+ (in) for H^+ (out). Since intracellular pH is neutral (pH 7), to achieve a pH of 1 in the extracellular space, the parietal cells must transport H^+ against a gradient of $1:10^7$ To accomplish this feat, the parietal cells are packed with mitochondria to supply the large amount of ATP used in acid pumping. Upon hormonal or nervous stimulation, intracellular vesicle membranes containing H^+-K^+ pumps fuse with the deep involutions (canaliculi) of the parietal cell apical membrane to begin secretion of acid, which drains from the canaliculi through the gland lumen into the stomach lumen.

The excess base left in the parietal cell cytoplasm is removed by a two-step process. The enzyme carbonic anhydrase in parietal cells promotes carbon dioxide hydration: ($CO_2 + H_2O \rightarrow [H_2CO_3] \rightarrow H^+ + HCO_3^-$). The CO_2 is readily available from oxidative metabolism throughout the body. The H^+ from this reaction neutralizes excess OH^- left behind by acid pumping. The weak base left over from this reaction—bicarbonate (HCO_3^-)—exits through chloride-bicarbonate exchangers at the serosal (blood side) border of the parietal cell to maintain neutral pH in the cytoplasm. Here, the exchange of Cl^- (in) for HCO_3^- (out) is a *secondary active transport* process, because it is indirectly driven by results of the *primary active acid pumping*. The Cl^- ions entering this way move through the cell and pass through the Cl^- channels into the canaliculi and finally enter into the gastric lumen to balance the charge of the pumped H^+ ions. These ions (H^+ and Cl^-) draw water with them, osmotically, into the lumen, producing liquid hydrochloric acid.

PEPSIN DIGESTS PROTEINS INTO SMALL PEPTIDES
Pepsin, a well-known protease, is the only digestive enzyme of any significance produced in the stomach. It cleaves food *proteins*, forming small *peptides*. This action probably is not crucial in overall protein digestion because the pancreatic protease, chymotrypsin, performs a similar function later in the small intestine. Pepsin may serve a regulatory function: the small peptides produced by pepsin action stimulate the sensory receptors in the gastric mucosa to initiate hormonal and nervous signals aimed to increase stomach motility and secretion (plates 74, 75). When secreted by the chief cells (*zymogen cells*), pepsin is in its inactive form, a larger protein called *pepsinogen*. Acid in the lumen promotes hydrolytic conversion of pepsinogen to pepsin. Pepsin, once formed, also attacks pepsinogen, producing more pepsin molecules, a process called *autocatalysis*.

The alkaline stomach *mucus* forms a thick protective coat covering the inner linings of the stomach in order to protect it from mechanical damage and also from the corrosive actions of the acid in the gastric juice. The breakdown of this coat is one of the causes of stomach *ulcers*.

GASTRIC MOVEMENTS CHURN FOOD TO PRODUCE CHYME
Shortly after food enters the stomach, when sufficient gastric juice has been produced, special weak contractions (*mixing waves*) begin in the stomach *fundus* and spread to the *pylorus*. These waves (occurring every 20 sec) help mix the food with the gastric juice. Later on, less frequent but much stronger peristaltic waves occur and force the *chyme* against the closed *pyloric sphincter*, resulting in chyme backflow. This movement vigorously mixes food with gastric juice, forming a soupy solution (chyme) that can now be processed by the intestinal enzymes. Gradually, the pyloric sphincter relaxes partially, allowing a small amount of soupy chyme to enter the small intestine's duodenum with each peristaltic wave.

The rate of stomach emptying depends on the (chemical) type of the food: carbohydrates empty rapidly, fats slowly, and protein-rich foods at an intermediate rate. Thus 30 min after ingestion of a purely carbohydrate meal, nearly 75% of the entire meal has emptied into the duodenum, while this amount would be 50% for a protein meal and 30% for a mainly fatty meal. This differential rate is regulated by hormones and nerves (plates 74, 75).

CN: Use dark colors for A, E, L, S, U.
1. Color the stomach in the upper right corner; notice the different secretory cells adjacent to the body and antrum portions, indicating their location. Color the stomach wall enlargement at the top.
2. Before coloring the gastric gland illustration in the center of the page, color the material on the four types of cells that surround the gland. Then color their location along the length of the gastric pit.
3. Color the gastric motility panel, coloring each figure completely before going on to the next. Next, color the four situations that determine whether or not the stomach will empty its contents into the duodenum.

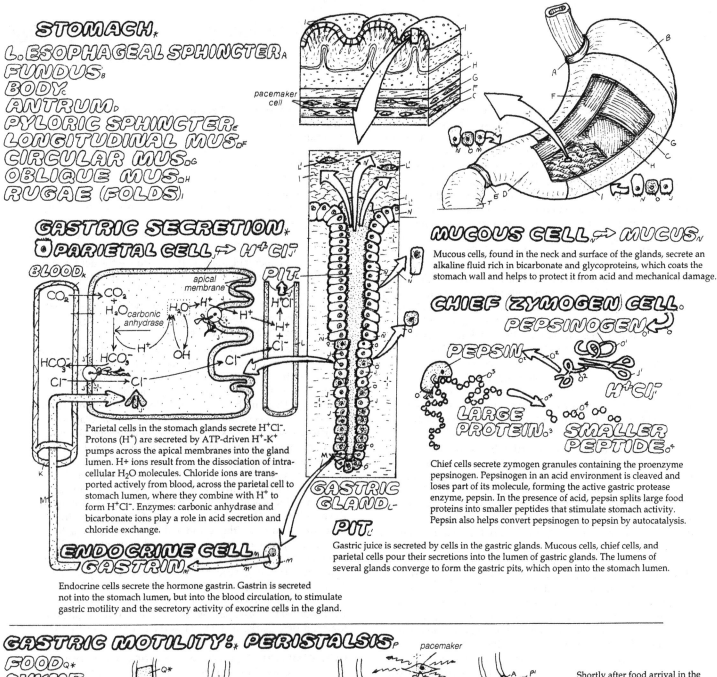

STOMACH.

L. ESOPHAGEAL SPHINCTER A
FUNDUS B
BODY C
ANTRUM D
PYLORIC SPHINCTER E
LONGITUDINAL MUS. F
CIRCULAR MUS. G
OBLIQUE MUS. H
RUGAE (FOLDS) I

pacemaker cell

GASTRIC SECRETION *

PARIETAL CELL ⇒ H^+Cl^- I

BLOOD K

apical membrane

PIT L'

CO_2 → CO_2
H_2O carbonic anhydrase
H_2O → H^+
H^+
H^+Cl^-

H^+ → OH
HCO_3^-

HCO_3^-
Cl^- → Cl^- → Cl^-
Cl^-

Parietal cells in the stomach glands secrete H^+Cl^-. Protons (H^+) are secreted by ATP-driven H^+-K^+ pumps across the apical membranes into the gland lumen. H+ ions result from the dissociation of intracellular H_2O molecules. Chloride ions are transported actively from blood, across the parietal cell to stomach lumen, where they combine with H^+ to form H^+Cl^-. Enzymes: carbonic anhydrase and bicarbonate ions play a role in acid secretion and chloride exchange.

ENDOCRINE CELL M
GASTRIN. m

Endocrine cells secrete the hormone gastrin. Gastrin is secreted not into the stomach lumen, but into the blood circulation, to stimulate gastric motility and the secretory activity of exocrine cells in the gland.

GASTRIC GLAND L
PIT L'

MUCOUS CELL N ⇒ MUCUS n

Mucous cells, found in the neck and surface of the glands, secrete an alkaline fluid rich in bicarbonate and glycoproteins, which coats the stomach wall and helps to protect it from acid and mechanical damage.

CHIEF (ZYMOGEN) CELL O
PEPSINOGEN. J'

PEPSIN o²
LARGE PROTEIN. o³
H^+Cl^-
SMALLER PEPTIDE. o⁴

Chief cells secrete zymogen granules containing the proenzyme pepsinogen. Pepsinogen in an acid environment is cleaved and loses part of its molecule, forming the active gastric protease enzyme, pepsin. In the presence of acid, pepsin splits large food proteins into smaller peptides that stimulate stomach activity. Pepsin also helps convert pepsinogen to pepsin by autocatalysis.

Gastric juice is secreted by cells in the gastric glands. Mucous cells, chief cells, and parietal cells pour their secretions into the lumen of gastric glands. The lumens of several glands converge to form the gastric pits, which open into the stomach lumen.

GASTRIC MOTILITY: * PERISTALSIS P

FOOD Q *
CHYME R
GASTRIC JUICES S
3 L/DAY s

pacemaker
contraction waves

50 ml
1,000 mL

STOMACH WALL P' DUODENUM T

Shortly after food arrival in the stomach, gastric juice is copiously secreted to mix with and digest the food. To enhance mixing and digestion, the stomach's muscular wall begins a series of regular mixing and peristaltic contractions. The peristaltic contractions are initiated by pacemaker cells in the muscular wall and travel from fundus to antrum, being strongest in the antrum. The hormone gastrin and the parasympathetic nerve (vagus) regulate the strength of these contractions.

GASTRIC EMPTYING RATE. * ⬆⬇ *

LIQUIDITY OF CHYME R
CHYME IN DUODENUM T
HIGH ACIDITY IN DUODENUM T
FATS IN DUODENUM T

Gastric peristalsis also functions to deliver, at regular intervals, the partially digested, soupy chyme into the duodenum of the small intestine (gastric emptying). Highly liquid chyme (with carbohydrate foods) increases the rate of emptying. Highly acid chyme (with protein foods) and fatty chyme decrease this rate. The reduction is to allow more time for digestion of the protein and fat in the stomach and the intestine.

Hormones play major roles in digestive control. The motility and secretory activities of the digestive system are under the control of both the autonomic nervous system and its enteric components, plus several gastrointestinal hormones. Here, we deal with control of digestion by the *gastrointestinal hormones*.

GASTRIN STIMULATES STOMACH SECRETION AND MOTILITY

Gastrin is a single-chain peptide hormone secreted by the *G-cells*, isolated flask-shaped endocrine cells in the lateral walls of *stomach glands* in the *antrum* region. Gastrin is secreted into the blood in response to stimulation by the bulk and composition (small peptides) of the ingested food. These peptides may directly stimulate the chemoreceptors of the G-cells, whose long necks protrude into the lumen. The food peptides also may act via specialized sensory cells (*chemo-* and *stretch-receptor cells*) sensitive to food peptides and bulk. Acting via intrinsic nerve connections, or through local hormones in the stomach mucosa, these receptor cells signal the G-cells to release gastrin into the blood. As the blood circulates, gastrin is returned to its target cells in the stomach fundus (main body) where it stimulates the stomach glands and smooth muscles to enhance gastric secretion and motility, respectively. The action of gastrin on the stomach is one reason that secretion and motility can continue in the absence of all external nerves to the stomach (stomach denervation).

The gastrin effect on acid secretion by the parietal cells is mainly indirect, through the release of histamine, which binds to histamine H_2 receptors on the parietal cells, stimulating them to secrete more acid. The histamine is released by enterochromaffin-like (ECL) cells in the gastric mucosa, which are activated by gastrin. The involvement of histamine in regulation of acid secretion is clinically very important, as shown by the widespread use of the "H_2 blockers," drugs that are administered to reduce excess acid secretion (e.g., cimetidine [Tagamet], ranitidine [Zantac], and famotidine [Pepcid]). Gastrin is also important clinically because excessive amounts of it are related to ulcer formation. The human fetal pancreas contains G-cells, but they are not normally active in adults. Occasionally, these cells develop tumors that secrete large quantities of gastrin, leading to excessive acid secretion in the stomach, which may result in gastric ulcers and bleeding.

DUODENAL MUCOSA SECRETES SEVERAL PEPTIDE HORMONES

The mucosal walls of the small intestine, particularly in the duodenum and jejunum, also produce several known and suspected gastrointestinal hormones. Of physiological significance are the peptide hormones *secretin* and *choleocystokinin* (CCK). These are secreted by isolated endocrine cells in the duodenum and jejunum. Other hormones are GIP (*glucose-dependent insulinotropic polypeptide*) and *motilin*. Secretin was the first hormone to be discovered in the history of endocrine research. In 1902, the English physiologists Bayliss and Starling noted that duodenal extracts injected into the blood of fasting dogs (in which all the nerves to the pancreas had been cut) augmented the secretion of pancreatic juice. The implication was that under normal conditions, the duodenum secretes into the blood a substance that, upon reaching the pancreas, stimulates its secretion of pancreatic juice (hence the name "secretin"). The term "hormone" was then adopted for such bloodborne humoral messengers. At the time of this discovery, all physiological regulations, including those of digestive activities, were thought to occur by the actions of nerves and the nervous system.

SECRETIN STIMULATES RELEASE OF PANCREATIC BICARBONATE

Secretin's target in the pancreas appears to be the cells lining the ducts of the *pancreatic acini* (aggregates of exocrine cells surrounding a cavity with a duct outlet), because secretin stimulates mainly the flow of a bicarbonate-rich juice, known to be produced by the pancreatic *duct cells*. The signal for secretion of the secretin hormone is the presence of acid in the duodenal lumen. This acidic chyme stimulates the sensory chemoreceptors in the duodenal mucosa, which in turn signal the *secretin cells* to release secretin. The highly alkaline, bicarbonate-rich pancreatic juice helps neutralize the acid in the duodenal chyme. This is important because the mucosa of the small intestine is much less protected against acid hazards compared to the stomach and because the enzymes of the intestine and pancreas work best in a neutral or slightly alkaline environment (plate 76).

While gastrin increases stomach motility and secretion, the duodenal hormone secretin opposes these effects by inhibiting stomach functions. This effect may protect the duodenum against excessive acid as well as regulate the rate of gastric emptying. Thus, high fat or acid content in the chyme stimulates release of secretin into the blood. Blood circulation returns the secretin back to its targets in the stomach, where the hormone exerts its inhibitory action. If the food is fatty, the reduced motility of the stomach results in slower chyme delivery to the duodenum, permitting increased time for digestion of what is already there. If the chyme is too acidic, secretin reduces acid secretion, diminishing acid damage to the duodenum. The gastric inhibitory effects of secretin were first attributed to another duodenal hormone, GIP (*gastric inhibitory peptide*), but it is now believed that only very high levels (pharmacological doses) of GIP can exert such effects and that normally gastric inhibition is achieved by secretin. To avoid confusion, the full name of the hormone GIP has now been changed (while keeping the same acronym) to *glucose-dependent insulinotropic peptide*, since GIP, in physiological doses, markedly stimulates the release of insulin in response to glucose in the small intestine.

CCK STIMULATES BILE AND EMPTIES THE GALLBLADDER

A third digestive hormone is choleocystokinin (CCK), a peptide hormone originating in the duodenal mucosa endocrine cells. The stimulus for CCK release into the blood is the arrival of chyme containing fat or acid from the stomach into the duodenum. CCK has two main targets: the *gallbladder* and the *pancreatic acinar cells*. Upon stimulation by CCK, the gallbladder contracts, releasing its stored *bile* into the duodenum. This effect is particularly striking after a fatty meal. The alkaline bile neutralizes the acid and *emulsifies* the fat in the chyme, facilitating its chemical digestion by the pancreatic enzyme lipase (plates 76, 77).

CCK also stimulates production and release of *pancreatic enzymes* from acinar cells of the pancreas. These enzymes are extremely important for the chemical digestion of various foodstuffs in the small intestine (plate 76). *Motilin* is yet another gastrointestinal hormone secreted by the duodenal mucosa to influence digestion. Motilin acts on the smooth muscles of the small intestinal wall to enhance intestinal contractions and movements, following stomach emptying and arrival of gastric chyme.

CN: Use red for E and dark colors for C, D, I, K, M, and Q.
1. Begin with the gastrin panel and follow the numbered sequence. Go on to the upper right panel, then the lower left and lower right.

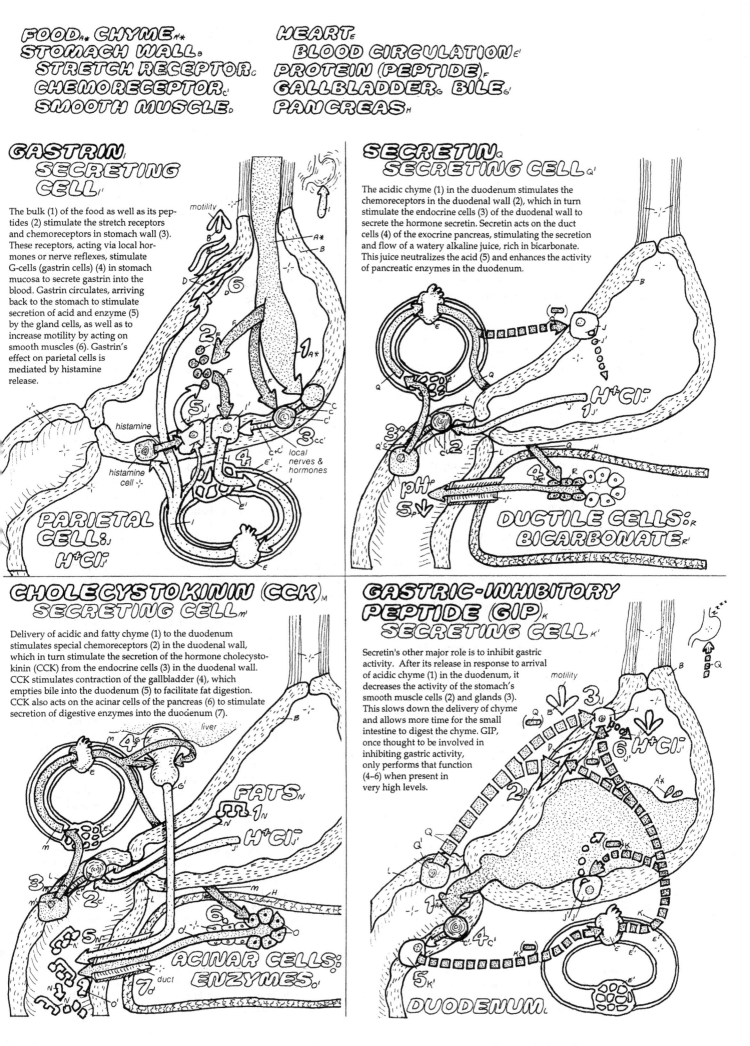

FOODₐ* **CHYME**ₐ'*
STOMACH WALLᵦ
STRETCH RECEPTOR_C
CHEMORECEPTOR_C'
SMOOTH MUSCLE_D

HEARTₑ
BLOOD CIRCULATIONₑ'
PROTEIN (PEPTIDE)_F
GALLBLADDER_G **BILE**_G'
PANCREAS_H

GASTRIN SECRETING CELL_I'

The bulk (1) of the food as well as its peptides (2) stimulate the stretch receptors and chemoreceptors in stomach wall (3). These receptors, acting via local hormones or nerve reflexes, stimulate G-cells (gastrin cells) (4) in stomach mucosa to secrete gastrin into the blood. Gastrin circulates, arriving back to the stomach to stimulate secretion of acid and enzyme (5) by the gland cells, as well as to increase motility by acting on smooth muscles (6). Gastrin's effect on parietal cells is mediated by histamine release.

motility

histamine

histamine cell

PARIETAL CELL_J
H⁺Cl⁻

local nerves & hormones

SECRETIN SECRETING CELL_Q'

The acidic chyme (1) in the duodenum stimulates the chemoreceptors in the duodenal wall (2), which in turn stimulate the endocrine cells (3) of the duodenal wall to secrete the hormone secretin. Secretin acts on the duct cells (4) of the exocrine pancreas, stimulating the secretion and flow of a watery alkaline juice, rich in bicarbonate. This juice neutralizes the acid (5) and enhances the activity of pancreatic enzymes in the duodenum.

H⁺Cl⁻

pH S

DUCTILE CELLS:_R
BICARBONATE_R'

CHOLECYSTOKININ (CCK)_M SECRETING CELL_M'

Delivery of acidic and fatty chyme (1) to the duodenum stimulates special chemoreceptors (2) in the duodenal wall, which in turn stimulate the secretion of the hormone cholecystokinin (CCK) from the endocrine cells (3) in the duodenal wall. CCK stimulates contraction of the gallbladder (4), which empties bile into the duodenum (5) to facilitate fat digestion. CCK also acts on the acinar cells of the pancreas (6) to stimulate secretion of digestive enzymes into the duodenum (7).

liver

FATS_N
H⁺Cl⁻

ACINAR CELLS:_O
ENZYMES_O'

duct

GASTRIC-INHIBITORY PEPTIDE (GIP)_K SECRETING CELL_K'

Secretin's other major role is to inhibit gastric activity. After its release in response to arrival of acidic chyme (1) in the duodenum, it decreases the activity of the stomach's smooth muscle cells (2) and glands (3). This slows down the delivery of chyme and allows more time for the small intestine to digest the chyme. GIP, once thought to be involved in inhibiting gastric activity, only performs that function (4–6) when present in very high levels.

motility

H⁺Cl⁻

DUODENUM

The digestive system is innervated profusely by the nerve fibers of both the *sympathetic* and *parasympathetic* divisions, but the regulatory role of the parasympathetic division, carried out primarily by the *vagus nerve*, seems to be paramount. In general, the parasympathetic system *increases* gastrointestinal activity (secretion and motility), but the sympathetic system has a net *inhibitory* effect. Our knowledge of control of digestive activities by the *autonomic nervous system* even precedes that of its hormonal control. Pavlov, the Russian physiologist and Nobel laureate, and his predecessors made many discoveries in this area.

The parasympathetic vagus nerve contains both motor and sensory fibers. The motor fibers enhance digestive activities by stimulating local neurons within the *gut wall*. These neurons in turn stimulate the gut wall's *smooth muscles* and *gland cells*. Acetylcholine is the neurotransmitter released by the motor fibers of the vagus nerve as well as by many of the target neurons of the vagus. The numerous afferent (sensory) fibers in the vagus nerve inform the brain digestive centers about the condition of the gut wall and its content. Although the sympathetic fibers directly influence the smooth muscle and secretory cells in certain instances, the sympathetic system's general inhibitory effects on digestion are mediated indirectly, by *constricting* the *blood vessels* in the digestive tract. The reduction in blood flow diminishes both secretory and contractile digestive activity.

THE GUT WALL HAS AN INTRINSIC NERVOUS SYSTEM

The numerous short axon and inter-neurons of the gut wall constitute its *intrinsic* or *enteric nervous system* (ENS). The ENS consists of two sets of *ganglia* or *plexi*: the superficial *submucosal plexus* (Meissner's plexus) mainly regulates the *digestive glands*, and the *myenteric plexus* (Auerbach's plexus), located deeper within the muscle layers, primarily regulates gut *motility*. The enteric plexi function, in part, as the peripheral ganglia of the parasympathetic system, within the gut (plate 29).

The plexi contain local sensory and motor neurons as well as interneurons. *Sensory neurons* are connected to the sensory *chemoreceptors*, which detect different substances in the gut lumen, as well as the *stretch receptors*, which respond to the tension in the gut wall, caused by the bulk of the food and chyme. The short effector *motor neurons* increase digestive gland activity or induce smooth muscle contraction. The myenteric and submucosal plexi in the same region communicate, through *interneurons*, with each other as well as with plexi farther in the gut.

The vast numbers of neurons and neuronal connections in the gut ENS carry out many digestive reflexes independently, as well as mediating brain influence on digestive functions. For example, the complex movements of peristalsis seen in the esophagus, stomach, and intestine are entirely initiated and regulated by the ENS. In addition to acetylcholine and norepinephrine, many other neurotransmitters including GABA (gamma-amino-butyric acid), serotonin, and NO (nitric oxide) as well as the peptide neurotransmitters like substance-P and VIP (vasoactive intestinal peptide) are released by the ENS neurons to control glandular secretion and muscle contraction in the gut wall.

NEURAL REGULATION OF DIGESTION OCCURS IN 3 PHASES

The regulation of digestive activity by the nervous system is traditionally divided into three consecutive phases: *cephalic*, *gastric*, and *intestinal*.

1. Cephalic Phase—The cephalic phase consists of the digestive responses that occur *before* the food is ingested and while it is still in the mouth. When one is hungry, odors or even thoughts of foods commonly evoke salivary secretion (mouth watering). Experiments have shown that this anticipatory response also involves the secretion of a small amount of gastric juice. When food is placed in the mouth, gastric juice production is substantially increased, as is salivary secretion. There is also a slight increase in the secretion of pancreatic juice. These gastric and pancreatic secretions during the cephalic phase prepare the gut lumen to receive food. For example, acid and pepsin in the stomach will help form peptides, which will stimulate further gastric juice production when food arrives in the stomach.

These anticipatory salivary, gastric, and intestinal responses are mediated by the higher brain centers, as implied by the term "cephalic" for this phase. Both the higher and lower brain centers play essential roles here. The main brain centers regulating digestive functions are in the *medulla oblongata*, where the afferent taste fibers also have their primary centers and where the cell bodies of the parasympathetic vagus and salivary nerves are located. The higher cortical and olfactory centers influence these medullary motor centers in order to exert their effects on digestive regulation during the cephalic phase. All the digestive responses in the cephalic phase are conducted by the parasympathetic outflow.

2. Gastric Phase—When food enters the stomach, the mechanical stretch receptors sense the increase in bulk, and the chemoreceptors detect the presence of peptides in the food. These sensors signal the information to two targets: (1) the effector neurons in the local enteric plexi and (2) the brain medullary digestive centers. The parasympathetic motor outflow from these centers increases the stomach's secretion and motility far above the level during the cephalic phase. In fact, most of the stomach's activity occurs during the gastric phase—for example, 80% of gastric juice is secreted in the gastric phase compared to 10% in the cephalic. Also during this *gastric* phase, the stomach hormone *gastrin* is released to further enhance gastric secretion and motility.

3. Intestinal Phase—The arrival of the chyme in the duodenum initiates the *intestinal* phase of autonomic nervous control, during which gastric secretion and motility are at first increased to promote further digestion and emptying. As the small intestine becomes filled with acidic and fatty chyme, inhibitory signals (mostly hormonal) decrease stomach activity to prolong emptying and allow time for intestinal digestion.

CN: Use dark colors for F & J.
1. Color the diagram of the sympathetic and parasympathetic nervous system in the upper right corner in order to familiarize yourself with their effect on the digestive process. Notice the presence of the

parasympathetic ganglia in the organs themselves. These have been deleted from the other diagrams for purposes of simplification.
2. Color the three phases of digestion (in the enclosed sections of the pag
3. Color the diagram of the intrinsic nervous system in the lower left corn

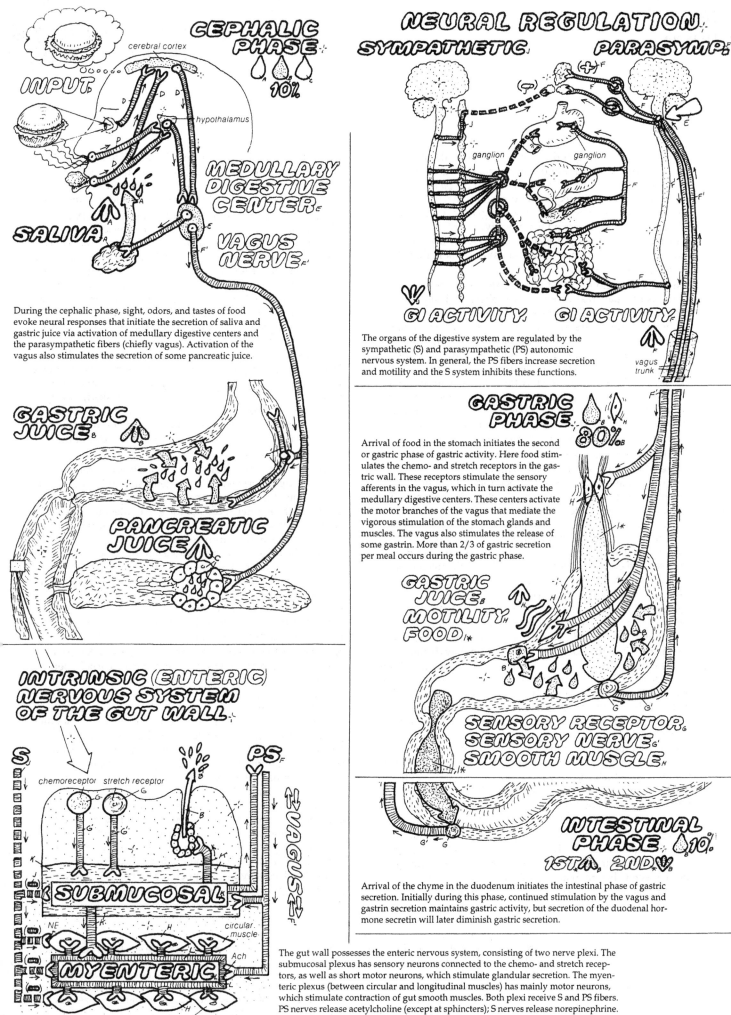

CEPHALIC PHASE
10%

INPUT

cerebral cortex

hypothalamus

MEDULLARY DIGESTIVE CENTER

VAGUS NERVE

SALIVA

During the cephalic phase, sight, odors, and tastes of food evoke neural responses that initiate the secretion of saliva and gastric juice via activation of medullary digestive centers and the parasympathetic fibers (chiefly vagus). Activation of the vagus also stimulates the secretion of some pancreatic juice.

GASTRIC JUICE

PANCREATIC JUICE

INTRINSIC (ENTERIC) NERVOUS SYSTEM OF THE GUT WALL

S   PS

chemoreceptor   stretch receptor

SUBMUCOSAL

NE

circular muscle

Ach

MYENTERIC

VAGUS

longitudinal muscle

NEURAL REGULATION
SYMPATHETIC   PARASYMP.

ganglion   ganglion

GI ACTIVITY.   GI ACTIVITY

vagus trunk

The organs of the digestive system are regulated by the sympathetic (S) and parasympathetic (PS) autonomic nervous system. In general, the PS fibers increase secretion and motility and the S system inhibits these functions.

GASTRIC PHASE
80%

Arrival of food in the stomach initiates the second or gastric phase of gastric activity. Here food stimulates the chemo- and stretch receptors in the gastric wall. These receptors stimulate the sensory afferents in the vagus, which in turn activate the medullary digestive centers. These centers activate the motor branches of the vagus that mediate the vigorous stimulation of the stomach glands and muscles. The vagus also stimulates the release of some gastrin. More than 2/3 of gastric secretion per meal occurs during the gastric phase.

GASTRIC JUICE.
MOTILITY.
FOOD

SENSORY RECEPTOR.
SENSORY NERVE.
SMOOTH MUSCLE

INTESTINAL PHASE 10%
1ST.   2ND.

Arrival of the chyme in the duodenum initiates the intestinal phase of gastric secretion. Initially during this phase, continued stimulation by the vagus and gastrin secretion maintains gastric activity, but secretion of the duodenal hormone secretin will later diminish gastric secretion.

The gut wall possesses the enteric nervous system, consisting of two nerve plexi. The submucosal plexus has sensory neurons connected to the chemo- and stretch receptors, as well as short motor neurons, which stimulate glandular secretion. The myenteric plexus (between circular and longitudinal muscles) has mainly motor neurons, which stimulate contraction of gut smooth muscles. Both plexi receive S and PS fibers. PS nerves release acetylcholine (except at sphincters); S nerves release norepinephrine.

The *pancreas*, a large gland located underneath the stomach, performs both endocrine and exocrine functions. Its endocrine function is carried out by the hormones of the *pancreatic islets* – *insulin* and *glucagon* – and their roles in regulating carbohydrate metabolism and blood sugar are discussed in plate 122. Here we focus on the digestive functions of the pancreas, namely, the production of pancreatic juice by the exocrine part of the gland, which constitutes more than 98% of the gland's mass.

EXOCRINE PANCREAS PRODUCES BICARBONATE & ENZYMES
The exocrine pancreas produces two physiologically important secretions, usually in response to the arrival of the chyme in the duodenum. First is a watery secretion rich in *sodium bicarbonate*. This effect is stimulated by the duodenal hormone secretin. The alkaline bicarbonate solution neutralizes the gastric acid arriving in the duodenum and provides a suitable alkaline environment for the function of pancreatic enzymes. The second secretion, produced by the pancreatic *acini*, consists of numerous *hydrolytic enzymes* for the chemical breakdown of most large molecules found in the diet. This process is stimulated by the duodenal hormone CCK.

The exocrine pancreas consists of numerous acini. Each of these acini consists of a single layer of epithelial cells surrounding a cavity into which the secretory cells pour their secretions. The acinar cells secrete the digestive enzymes. The acinar cavity opens into a duct through which the secretions of the acinar cells flow out. The ducts of the pancreatic acini are lined with specialized secretory cells, the *ductile* cells, which secrete a bicarbonate-rich solution. The smaller ducts coalesce and converge, finally connecting to the main *pancreatic duct*, which opens into the *duodenal lumen*.

BICARBONATE NEUTRALIZES THE STOMACH ACID
The presence of sodium bicarbonate ($Na^+HCO_3^-$) in the *pancreatic juice* gives this fluid an alkaline pH of 8, enabling it to neutralize the acidic chyme delivered from the stomach. Upon entry into the duodenum, the sodium bicarbonate reacts with the hydrochloric acid (H^+Cl^-), producing sodium chloride (Na^+Cl^-) and carbonic acid [H_2CO_3]. The latter acid is unstable and readily dissociates into carbon dioxide and water. This way the hydrogen ions are gradually and effectively removed from the chyme in the duodenum, resulting in reduced acidity and a neutral or even alkaline intestinal chyme. The reduced duodenal acidity has two important effects. (1) It reduces the noxious effects of acid on the duodenal mucosa, which has little protection against acid. (2) It makes the duodenal environment suitably alkaline for activation of the pancreatic and intestinal digestive enzymes.

Bicarbonate is secreted by the duct cells into the duct lumen through an active pumping mechanism. The duct cells contain high amounts of the enzyme *carbonic anhydrase*, which is thought to be involved in the active secretion of bicarbonate. To secrete bicarbonate, the duct cells may operate like the turned-around parietal cells of the stomach, which secrete

acid into the stomach lumen and bicarbonate into the blood (plate 73). Pancreatic duct cells carry out the same process but in the opposite direction, secreting bicarbonate ions (with lots of sodium ions) into the duct lumen and acid into the blood.

ENZYMES OF ACINAR CELLS HYDROLYZE THE FOODSTUFF
The acinar cells of the pancreas produce a viscous secretion rich in protein (enzymes), secreted in *zymogen granules*. The physiological stimulus for acinar cell secretion of pancreatic enzymes is the presence of *fat* and *protein* in the duodenum. These stimuli trigger secretion of the duodenal hormone cholecystokinin (CCK), which stimulates the acinar cells to secrete enzymes. The stimulation of the vagus nerve also increases enzyme production. Pancreatic enzymes comprise the *proteases, amylases, lipases*, and *nucleases*. The protease enzymes are initially secreted in their *inactive* forms (i.e., as larger proenzyme proteins). This is critical because the powerful pancreatic enzymes could easily digest the pancreas tissue in a short time if they were not inhibited (masked) during their transport from the acinar cavity to the intestinal lumen. In the disease *acute pancreatitis*, these enzymes are activated before reaching the intestine. As a result they digest the pancreas, causing death within days.

A key pancreatic proenzyme is *trypsinogen*, which is activated upon arrival in the duodenal lumen by the hydrolytic action of *enterokinase*, an enzyme present in the brush border of the epithelial cells of the small intestine. The activation produces *trypsin*, a well-known all-purpose *protease* that can attack and hydrolyze many kinds of dietary *proteins*. Among the targets of trypsin attack are the inactive proenzymes secreted by the pancreas—i.e., the proteases and lipases. Some of the pancreatic enzymes, such as *amylase*, do not require activation by trypsin. In some people, intestinal mucosa is deficient in enterokinase. As a result, trypsin is not formed, proteases are not activated, and dietary proteins remain undigested, causing protein deficiency and disease.

ACTIONS OF PROTEASES, LIPASES, AMYLASES, & NUCLEASES
Pancreatic proteases (trypsin, chymotrypsin, carboxypeptidase) attack peptide bonds located between different but specific amino acids. As a result, the pancreatic proteases convert all the dietary *proteins* into *dipeptides*. The final hydrolysis of dipeptides to free *amino acids* occurs by the action of other proteases (peptidases) secreted from the *intestinal mucosa*. Pancreatic amylase attacks the large dietary *polysaccharides* such as those found in the starches, forming smaller *oligo-* and *disaccharides* like dextrose and maltose (glucose-glucose). Further digestion of disaccharides into such monosaccharides as glucose, fructose, and galactose occurs by the action of the enzymes (e.g., maltase and lactase) secreted by the intestinal mucosa. *Lipases* of the pancreas attack *triglycerides (triacylglycerols)*, forming *glycerol* and *fatty acids* or *monoglycerides* and *fatty acids*. The type of conversion depends on the type of lipase. Nucleases hydrolyze nucleic acids (DNA, RNA) into purine and pyrimidine bases and ribose sugars.

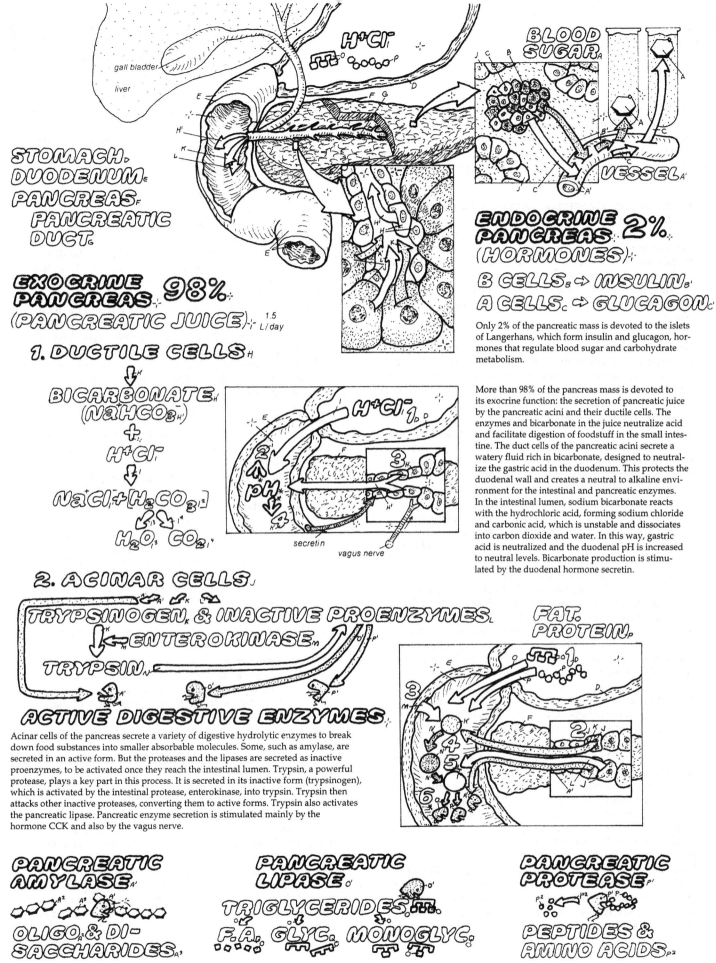

STOMACH. DUODENUM. PANCREAS. PANCREATIC DUCT.

gall bladder
liver

H⁺Cl⁻

BLOOD SUGAR.

EXOCRINE PANCREAS 98% (PANCREATIC JUICE). 1.5 L/day

1. DUCTILE CELLS.

⬇

BICARBONATE (NaHCO₃)
+
H⁺Cl⁻
⬇
NaCl + H₂CO₃
H₂O, CO₂

secretin

vagus nerve

H⁺Cl⁻

ENDOCRINE PANCREAS 2% (HORMONES):

B CELLS ⇨ INSULIN.
A CELLS ⇨ GLUCAGON.

Only 2% of the pancreatic mass is devoted to the islets of Langerhans, which form insulin and glucagon, hormones that regulate blood sugar and carbohydrate metabolism.

More than 98% of the pancreas mass is devoted to its exocrine function: the secretion of pancreatic juice by the pancreatic acini and their ductile cells. The enzymes and bicarbonate in the juice neutralize acid and facilitate digestion of foodstuff in the small intestine. The duct cells of the pancreatic acini secrete a watery fluid rich in bicarbonate, designed to neutralize the gastric acid in the duodenum. This protects the duodenal wall and creates a neutral to alkaline environment for the intestinal and pancreatic enzymes. In the intestinal lumen, sodium bicarbonate reacts with the hydrochloric acid, forming sodium chloride and carbonic acid, which is unstable and dissociates into carbon dioxide and water. In this way, gastric acid is neutralized and the duodenal pH is increased to neutral levels. Bicarbonate production is stimulated by the duodenal hormone secretin.

VESSEL.

2. ACINAR CELLS.

TRYPSINOGEN & INACTIVE PROENZYMES.
ENTEROKINASE
TRYPSIN
ACTIVE DIGESTIVE ENZYMES

FAT. PROTEIN.

Acinar cells of the pancreas secrete a variety of digestive hydrolytic enzymes to break down food substances into smaller absorbable molecules. Some, such as amylase, are secreted in an active form. But the proteases and the lipases are secreted as inactive proenzymes, to be activated once they reach the intestinal lumen. Trypsin, a powerful protease, plays a key part in this process. It is secreted in its inactive form (trypsinogen), which is activated by the intestinal protease, enterokinase, into trypsin. Trypsin then attacks other inactive proteases, converting them to active forms. Trypsin also activates the pancreatic lipase. Pancreatic enzyme secretion is stimulated mainly by the hormone CCK and also by the vagus nerve.

PANCREATIC AMYLASE

OLIGO & DI-SACCHARIDES

Pancreatic amylase attacks large polysaccharides (starches), converting them into smaller oligo- and disaccharides.

PANCREATIC LIPASE

TRIGLYCERIDES
F.A. GLYC. MONOGLYC.

Pancreatic lipase attacks triglycerides (triacylglycerols), forming mostly fatty acids and monoglycerides along with some glycerol.

PANCREATIC PROTEASE

PEPTIDES & AMINO ACIDS

Pancreatic proteases (trypsin, chymotrypsin, elastase, carboxypeptidase) attack various proteins, forming smaller peptides and amino acids. Each protease hydrolyzes peptide bonds between different amino acid residues of proteins.

The *liver*, a vital body organ and the largest gland in the body, has many functions including endocrine effects, control of metabolism, deactivation of hormones, and detoxification of drugs and toxins. Through its exocrine function—namely, the formation of *bile*—the liver also plays an important role in digestion, particularly of dietary *fats*.

LIVER LOBULES ARE SPECIALLY ARRANGED TO SECRETE BILE
The formation of bile (nearly 0.5 L per day) is the liver's major exocrine function. The liver has several *lobes*, each consisting of numerous *lobules*. Bile is secreted by the liver cells (hepatocytes) that form the units of the liver lobules. Each lobule acts as the liver's basic anatomical-functional unit and is a part of a hexagon in which the lobules are connected peripherally to the incoming blood and centrally to a vein that drains the blood.

The liver receives blood from two sources, the *hepatic artery* and the *hepatic portal vein*, bringing blood from the heart and the intestines, respectively. The liver thus has the unique ability to receive, sample, and process absorbed nutrients before they reach the general circulation. Blood flows out of the liver and into the heart via the *hepatic vein*.

HEPATOCYTES SECRETE BILE INTO THE BILE DUCTS
The *liver cells* (*hepatocytes*) are packed in walls (slabs) of cells, separated by blood *sinusoids,* a highly porous type of capillary. The incoming arterial and portal blood are mixed as they flow into these sinusoids. After the hepatocytes extract oxygen and nutrients from this pool, the blood flows into a centrally located branch of the hepatic vein. The hepatocytes form the bile and secrete it into small *canaliculi*, which coalesce to form first the smaller and then the larger *bile ducts*. In these ducts, bile flows in the opposite direction of blood, preventing their mixing.

The various bile ducts finally coalesce to form the larger *hepatic duct*, which emerges from the liver. The hepatic duct bifurcates to form the *cystic duct,* which leads to the *gallbladder*, and the *common bile duct*, which connects to the *duodenum*, together with the pancreatic duct. The *sphincter of Oddi* regulates the bile outflow from the common bile duct into the duodenum. When this sphincter is closed, the bile accumulates in the common bile duct, flowing back into the cystic duct and the gallbladder, where it is temporarily stored. After meals, in response to the release of duodenal hormone CCK, the gallbladder contracts, releasing bile into the duodenum to facilitate fat digestion (see below).

BILE CONSISTS MAINLY OF BILE SALTS & BILE PIGMENTS
Besides water (97%), bile contains two major organic constituents, *bile salts* and *bile pigments*. In addition, bile's inorganic salts (sodium chloride and sodium bicarbonate) provide for the bile's alkalinity. Bile salts (also called *bile acids*) such as *cholic acid* and *deoxycholic acid* are formed from *cholesterol* within the liver cells (hepatocytes). Liver cells are the major producers of cholesterol in the body, most of which is used to produce bile salts. Bile pigments are derived mainly from *bilirubin*, a compound produced following destruction of the red blood cells and catabolism of the heme in the hemoglobin. Bilirubin is taken up from the blood by the liver cells and conjugated to *glucuronic acid* to form *bilirubin glucuronide*, a water-soluble, golden yellow compound that is excreted in

the bile, giving it its characteristic yellow color. About 4 g of bile salts and 1.5 g of bile pigments are secreted every day in the bile. Most (90%) of the bile salts are recycled by reabsorption and delivery back to the liver via the entero-hepatic circulation. Most of the bile pigments are excreted with the feces; the remainder are reabsorbed and excreted in the kidney. The bile pigments give feces and urine their characteristic colors.

BILE AIDS IN FAT DIGESTION BY EMULSIFYING FATS
Bile salts such as cholate and deoxycholate play a major physiological role in digestion through facilitating fat digestion. The bile salts act as fat *solubilizing agents.* They have both a fat-soluble *hydrocarbon ring* and several *charged groups*, enabling them to mix with fat and water, respectively. Thus, the addition of bile salts to a fat and water mixture increases fat's solubility, similar to the action of a detergent. In the presence of bile salts, large fat droplets in the chyme become dispersed, forming smaller fat particles, a process called *emulsification.* In the emulsified form, fats can be much more readily and efficiently digested by the water-soluble enzyme *lipase* from the pancreas (plate 76). The products of lipase digestion (glycerides and fatty acids) form special fatty aggregates called *micelles* (plate 7), which can be readily absorbed by the intestinal mucosal cells. In the absence of bile, fat digestion diminishes markedly (by nearly one-half) even though the enzyme lipase is present.

GALLBLADDER STORES BILE FOR RELEASE AFTER MEALS.
The gallbladder is a storage sac for the bile, which is produced continuously by the liver but is delivered to the duodenum only after meals, particularly fatty meals. Before meals, the sphincter of Oddi, located at the opening of the common bile duct into the duodenum, is closed, causing the bile flow to back up, filling the gallbladder. While the bile is stored in the gallbladder, the bladder wall absorbs some of its water, concentrating the bile. The arrival of fatty food in the duodenum stimulates the release of the duodenal hormone CCK (plate 74), which acts on the gallbladder, causing it to contract. The bile is then released into the duodenum to emulsify fat. CCK also stimulates liver production of bile.

Two major problems and diseases are associated with abnormalities of the gallbladder function. One is *gallstones*, which are of two types, cholesterol (majority) and calcium bilirubinate. In certain individuals, excess amounts of the water-insoluble *cholesterol* (usually a minor bile constituent) in the bile precipitate, perhaps as a result of excessive removal of water, or supersaturation, forming gallstones. Gallstones in the bile duct may cause severe abdominal pain, requiring surgery. If the enlarged stones obstruct the common bile duct, bile flows back into the liver and eventually leaks into the blood, causing *jaundice,* a disorder characterized by a yellowish color of the skin and eyes due to deposition of bilirubin and related bile pigments in the capillaries and tissue spaces. Jaundice may also occur because of excessive hemolysis of the red cells, a condition that occurs in certain diseases and produces unusually large amounts of bilirubin. Liver damage caused by certain viral infections (hepatitis) also results in jaundice. Another form of jaundice seen in some newborns is due to immaturity of liver function and is often easily reversible.

CN: Use red for M, blue for L, yellow for F, and a light color for B.
1. Begin with the upper drawing of the various organs.
2. Color the enlargement of a liver section, beginning with blood from the digestive system entering from the portal vein (L).

Include the liver cell diagram, noting that some of the cholesterol forming the bile salts is made within the liver cell.
3. Color the illustrations depicting gallbladder function. Note that no. 8 is found in the material below.

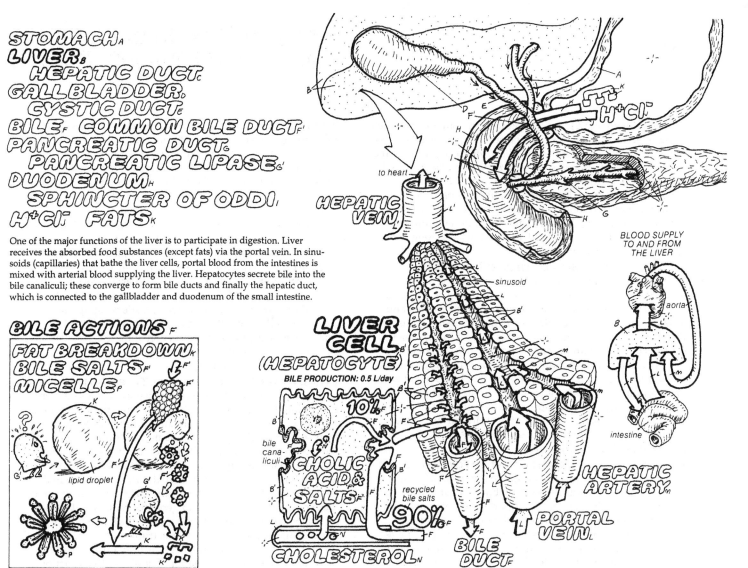

STOMACH_A
LIVER_B
HEPATIC DUCT_C
GALL BLADDER_D
CYSTIC DUCT_E
BILE_F COMMON BILE DUCT_F'
PANCREATIC DUCT_G
PANCREATIC LIPASE_G'
DUODENUM_H
SPHINCTER OF ODDI_J
H⁺Cl⁻_K FATS_K

One of the major functions of the liver is to participate in digestion. Liver receives the absorbed food substances (except fats) via the portal vein. In sinusoids (capillaries) that bathe the liver cells, portal blood from the intestines is mixed with arterial blood supplying the liver. Hepatocytes secrete bile into the bile canaliculi; these converge to form bile ducts and finally the hepatic duct, which is connected to the gallbladder and duodenum of the small intestine.

BILE ACTIONS_F

FAT BREAKDOWN_K
BILE SALTS_F'
MICELLE_P

lipid droplet

Bile salts emulsify large fat droplets, breaking them down into smaller particles on which pancreatic lipase can act more efficiently. Bile salts also combine with products of lipase digestion (monoglycerides and fatty acids), forming lipid micelles, which are readily absorbed by the intestinal mucosa.

LIVER CELL (HEPATOCYTE)

BILE PRODUCTION: 0.5 L/day

sinusoid

bile canaliculi

CHOLIC ACID & SALTS

recycled bile salts

CHOLESTEROL_N

BILE DUCT_F

PORTAL VEIN_L

HEPATIC ARTERY_M

10%

90%

BLOOD SUPPLY TO AND FROM THE LIVER

aorta

intestine

Bile contains bile salts and bile pigments. Bile salts (cholate and deoxycholate) are formed from cholesterol in the liver and secreted in the bile to aid in fat digestion. Most (90%) of bile salts are reabsorbed in the blood and recycled via the entero-hepatic circulation.

GALL BLADDER FUNCTION_D

Bile is continuously secreted from the liver (1). Before meals, the sphincter of Oddi is closed (2), directing the bile flow into the gallbladder for storage (3). Mucosal cells of the gallbladder actively reabsorb sodium, chloride, and water, concentrating the bile (4). After meals, the fat in the chyme (5) dilates the sphincter of Oddi and stimulates the release of the hormone CCK. CCK induces the contraction of the gallbladder (6), releasing bile into the duodenum (7). Increased cholesterol or decreased water in the bile favors formation of cholesterol-containing gallstones. Gallstones may obstruct bile flow, resulting in jaundice (8).

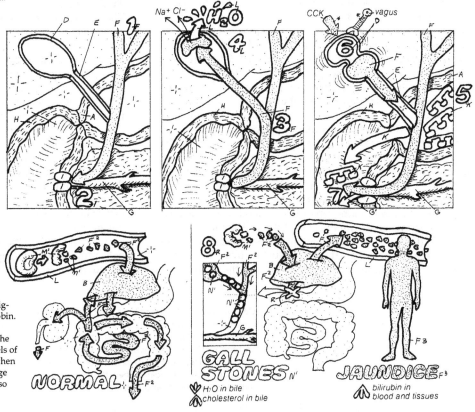

BILE PIGMENTS (BILIRUBIN)_F²
RED BLOOD CELL_M

The liver cells also secrete, for excretion, bilirubin glucuronide, which is derived from the yellowish pigment bilirubin, a metabolite of heme in the hemoglobin. Some of the bile pigments are excreted in the feces; others are reabsorbed in the blood for excretion by the kidneys. Jaundice, a disease caused by elevated levels of bilirubin in the blood, tissues, and skin, can occur when biliary flow is obstructed by gallstones. Liver damage or excessive hemolysis of the red blood cells may also cause jaundice.

NORMAL_F²

GALL STONES_N'

JAUNDICE_F³

⇑ H₂O in bile cholesterol in bile

⇑ bilirubin in blood and tissues

The *small intestine* is a long, convoluted tube, situated between the stomach and the large intestine, and is specialized for two major digestive processes: *chemical (enzymatic) digestion* of the foodstuff in the lumen and the *absorption* of nutrients into the bloodstream. The length of an animal's small intestine depends on its *dietary habits*. It is relatively short in meat eaters *(carnivores)* and long in grass eaters *(herbivores)*. In the omnivorous human, the small intestine is of medium length (about 3 m), although its length appears twice as long in the cadavers due to loss of muscle tone.

SMALL INTESTINE IS SEGMENTED INTO THREE PARTS
At its beginning, the small intestine is connected to the stomach where the *pyloric sphincter* controls *chyme* inflow into the *duodenum*, the first intestinal segment. The *jejunum* and *ileum* are the second and third segments. The ileum joins with the *cecum* of the large intestine by the *ileocecal valve*, which controls outflow from the small intestine. The different segments vary in their functions. The duodenum is highly secretory (mucus, enzymes, and hormones) and is the site of entry of the bile and pancreatic juice flow into the small intestine; the jejunum and ileum are specialized for nutrient absorption. Though absorption occurs across the entire surface of the jejunum and ileum, various substances are selectively absorbed in different segments (plate 79).

INTESTINAL WALL IS UNIQUELY ADAPTED FOR ABSORPTION
The structure of the intestinal wall broadly resembles that of the other parts of the digestive tract, but there are histological variations suitably adapted for its particular absorptive functions. The most superficial (facing the lumen) layer is the *mucosa*, which contains the epithelial *absorptive cells (enterocytes)*. Beneath the mucosa is the *submucosa*, containing the *intestinal glands* and small blood vessels. Two layers of smooth muscles, the *circular* and *longitudinal*, are found deep under the submucosa. These are responsible for intestinal motility. Groupings of neurons and nerves, the *submucosal and myenteric plexi*, are located within these layers (plate 75). A supportive layer of connective tissues, the *serosa*, forms the intestine's outer cover.

INTESTINAL VILLI ARE SPECIALIZED ABSORPTIVE UNITS.
The inner wall of the small intestine is extensively folded *(plicae circularis)*, tripling its surface area. Each fold in turn contains numerous microscopic structures called *villi* (fingers). The villi, totaling to 30 million (30 villi/mm^2), expand the absorptive surface area by another 10 fold. Each villus consists of a single layer of *surface epithelial cells* covering an inner core of very small *blood* and *lymph vessels, autonomic nerve fibers*, and *smooth muscle cells*. Most of the surface epithelial cells of the villi are *absorptive (enterocytes)*, but some are secretory. The absorptive cells are glued tightly together by *desmosomes* and *tight junctions* (plate 2) so that nutrients can pass only across, not between, the cells. The tight junctions also prevent entry of pathogens into the blood. The absorptive cells, occurring mostly on the hills of the villi, contain *microvilli* (brush border) on their luminal surface; these effectively increase each cell's absorptive surface by twenty times. Altogether, the plicae circularis, the villi, and the microvilli

increase the absorptive surface of the small intestine by 600 fold, providing a total area of 200 m^2, about the size of a tennis court, for nutrient absorption. The smooth muscle cells in the villi and the contractile protein *actin* in the microvilli enable the villi and microvilli to be motile, increasing absorption efficiency. The autonomic nerve fibers regulate these activities.

The secretory cells, another type of villus epithelial cells, are located deep in the intervilli valleys *(intestinal crypts)*, where they may form glands, secreting mucus or enzymes into the crypts. The entire population of surface epithelial cells continuously turns over, being replaced every few days. New cells are formed deep in the crypts and migrate up toward the tips of the villi, where they die by apoptosis (programmed cell death) and are shed into the intestinal lumen. The shedding and destruction of these cells (20 million cells/day in humans) provides one source for intestinal enzymes (e.g., enterokinase).

FATS & WATER-SOLUBLE NUTRIENTS TAKE DIFFERENT PATHS.
During absorption, nutrients are transported across the brush border into the absorptive cells and out into the villus core. Further movements of the different nutrients depends on their solubility. The water-soluble nutrients directly enter to the blood capillaries while the fat soluble ones move into the lacteals, small blind lymph vessels. The blood capillaries coalesce to form venules, which leave the villus to form small veins, leading finally into the *hepatic portal vein*, which carries these nutrients to the liver and general circulation. The lacteals coalesce to form the lymph vessels which join with the larger lymph vessels, to ascend the trunk and connect with the large veins in the chest, where the fatty nutrients are delivered to the blood circulation. The absorptive functions, motility and blood flow of the villus are regulated by the nerve fibers and smooth muscles of the villus as well as by release of local hormones.

INTESTINAL MOVEMENTS PROMOTE CHYME DIGESTION & TRANSPORT.
The small intestine shows two major movement patterns: *segmentation* and *peristalsis*. Segmentation movements are achieved by sustained contractions of the circular muscles. Superimposed on these sustained contractions, which tend to entrap the chyme within a small segment of the intestine, are other local *mixing* movements that shake the chyme, mix it with the intestinal juice, to facilitate enzymatic digestion and promote absorption of nutrients. The peristaltic movements, generated by the coordinated contractions of the circular and longitudinal muscles propel the intestinal chyme down the tract, toward the large intestine. It takes several hours for the chyme to traverse the entire length of the small intestine, from the duodenum to the end of the ileum. Intestinal movements are generated by the *myenteric nerve plexi* of the *enteric nervous system* in the intestinal wall (see plate 75) and do not require the autonomic nerves for their initiation. However, these nerves can regulate the intensity and frequency of the intestinal contractions. The peptide hormone motilin secreted from the duodenal endocrine cells also enhances intestinal motility.

CN: Use red for 1, purple for J, blue for K, and a very light color for G.
1. Begin with the diagram in the upper left corner and follow the enlargements down the page. Note that the internal structures of the three villi have been segregated (lacteal in the left, blood vessels in the center, nerves and muscles on the right) for

purposes of clarity.
2. Color the tiny diagram of a tennis court in the lower left corner, demonstrating the total absorptive area of the small intestine.
3. Color the motility panel. Note that only the portion of the intestinal wall which is contracted is to be colored.

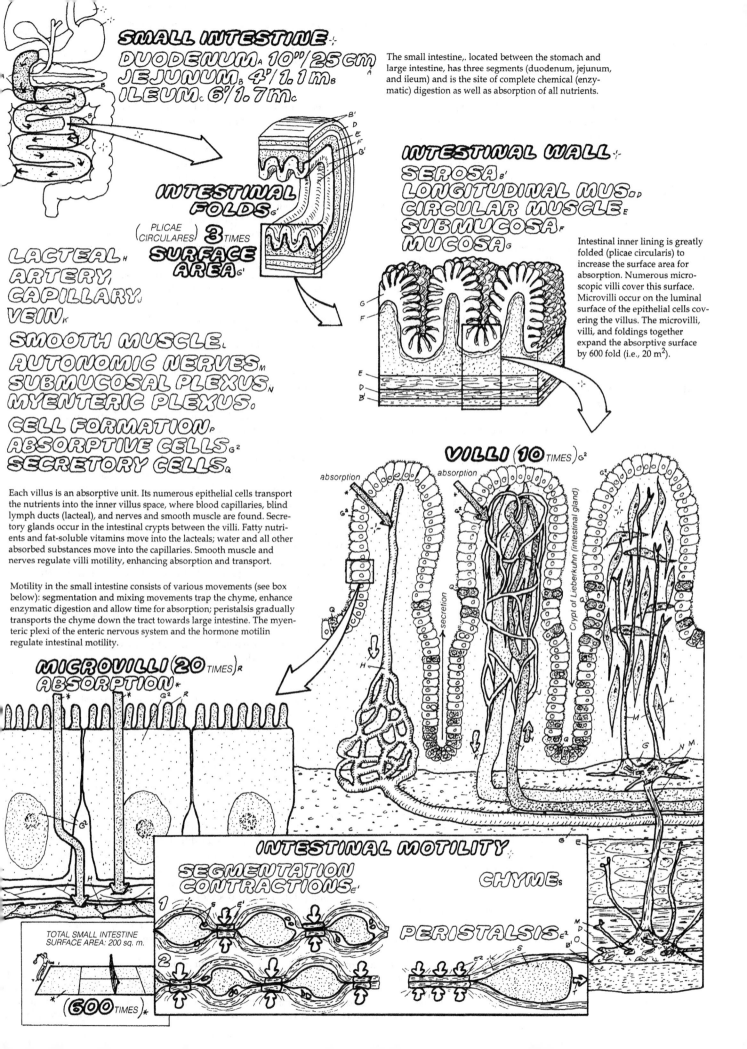

SMALL INTESTINE
DUODENUM. 10"/25CM (A)
JEJUNUM. 4'/1.1M (B)
ILEUM. 6'/1.7M (C)

The small intestine,. located between the stomach and large intestine, has three segments (duodenum, jejunum, and ileum) and is the site of complete chemical (enzymatic) digestion as well as absorption of all nutrients.

INTESTINAL FOLDS (G')
(PLICAE CIRCULARES) **3** TIMES
SURFACE AREA (G')

LACTEAL (H)
ARTERY (J)
CAPILLARY (J)
VEIN (K)

SMOOTH MUSCLE (L)
AUTONOMIC NERVES (M)
SUBMUCOSAL PLEXUS (N)
MYENTERIC PLEXUS (O)
CELL FORMATION (P)
ABSORPTIVE CELLS (G²)
SECRETORY CELLS (Q)

INTESTINAL WALL
SEROSA (B')
LONGITUDINAL MUS. (D)
CIRCULAR MUSCLE (E)
SUBMUCOSA (F)
MUCOSA (G)

Intestinal inner lining is greatly folded (plicae circularis) to increase the surface area for absorption. Numerous microscopic villi cover this surface. Microvilli occur on the luminal surface of the epithelial cells covering the villus. The microvilli, villi, and foldings together expand the absorptive surface by 600 fold (i.e., 20 m²).

Each villus is an absorptive unit. Its numerous epithelial cells transport the nutrients into the inner villus space, where blood capillaries, blind lymph ducts (lacteal), and nerves and smooth muscle are found. Secretory glands occur in the intestinal crypts between the villi. Fatty nutrients and fat-soluble vitamins move into the lacteals; water and all other absorbed substances move into the capillaries. Smooth muscle and nerves regulate villi motility, enhancing absorption and transport.

Motility in the small intestine consists of various movements (see box below): segmentation and mixing movements trap the chyme, enhance enzymatic digestion and allow time for absorption; peristalsis gradually transports the chyme down the tract towards large intestine. The myenteric plexi of the enteric nervous system and the hormone motilin regulate intestinal motility.

VILLI (10 TIMES) (G²)
absorption   absorption
Crypt of Lieberkuhn (intestinal gland)
secretion

MICROVILLI (20 TIMES) (R)
ABSORPTION *

TOTAL SMALL INTESTINE SURFACE AREA: 200 sq. m.
(600 TIMES)

INTESTINAL MOTILITY
SEGMENTATION CONTRACTIONS (E')
CHYME (S)
1
2
PERISTALSIS (E²)

After completion of enzymatic digestion in the intestinal lumen, the nutrients are *absorbed*, by a variety of transport mechanisms, across the *absorptive cells (enterocytes)* in the intestinal epithelium, into the bloodstream for delivery to the body cells. Intestinal absorption utilizes many of the common trans-membrane and transcellular transport mechanisms seen in other body cells as well as mechanisms that are unique to the intestinal absorptive cells. These mechanisms include both physical (e.g., diffusion) and physiological (e.g., active transport) processes (plates 8, 9). Upon passing across the intestinal mucosa, the water-soluble nutrients flow into the capillaries and venous blood of the villi, on to the hepatic portal vein, which delivers them to the liver and general circulation. The fatty and fat-soluble nutrients, however, flow into the *lacteals* and the lymphatic vessels before entering the bloodstream.

ABSORPTION INVOLVES PASSIVE & ACTIVE MECHANISMS
Some nutrients pass across the mucosa by simple physical diffusion or osmosis, mechanisms not requiring expenditure of cellular energy. For example, water is absorbed by *osmosis*, following the transport of salts and other osmotically active nutrients such as glucose and amino acids. *Potassium* passes primarily by diffusion while sodium is transported actively. Still major minerals like iron and calcium are transported with the aid of facilitated mechanisms involving transporter and binding proteins.

IRON & CALCIUM ABSORPTION INVOLVE BINDING PROTEINS
Iron is transported from the intestinal lumen, across the mucosa, and into the plasma by an iron-binding transporter protein called *transferrin or mobilferrin*. When dietary iron is abundant, iron is stored within the mucosal cells bound to *ferritin (apoferritin+iron)*, a ubiquitous *iron-binding* storage protein. When dietary iron is deficient, iron is released from ferritin and transported to the blood. *Calcium* is actively taken up by calcium transporters in the microvillar brush border, and moved across the cell by a *calcium-binding protein* to be released on the blood side. Synthesis of this transporter protein in the mucosal cells is stimulated by the hormone *calcitriol, a* derivative of *vitamin D* (plate 120). This is the mechanism by which vitamin D increases calcium absorption.

SODIUM IS TRANSPORTED ACTIVELY BY A TWO-STEP PROCESS
Sodium, the body's major extracellular electrolyte, is transported by a two step process, involving first a transporter protein and, second, an active, energy (ATP)-dependent mechanism. At the luminal side of the absorptive cells, sodium is taken up by specific sodium transporter proteins which move it across the brush border membrane and release it to the intracellular medium. This increases the concentration of intracellular sodium which in turn activates the Na-K-ATPase pumps, located on the *basolateral* borders of the absorptive cells. These pumps actively move the sodium uphill and out, into the intercellular space, from which it diffuses into the blood. This way sodium concentration inside the absorptive cells is kept low, allowing its continued inward movement from the intestinal lumen.

GLUCOSE & AMINO ACID ABSORPTION IS ACTIVE & NA-DEPENDENT
Enzymatic digestion of carbohydrates and proteins in the intestine produces glucose (and some fructose and galactose) as well amino acids respectively (Plate 76). Absorption of some of these essential nutrients such as glucose and certain amino acids occurs actively but mainly in conjunction with the active transport of sodium (cotransport, secondary active transport). Initially, glucose and amino acids are transported by a facilitated-diffusion mechanism, across the brush border membrane, bound to sodium-dependent transporters. This increases the concentrations of these nutrients inside the cell, permitting them to move across the basal membrane by facilitated diffusion (transporters), and flow into the blood by diffusion. Transport of glucose and amino acids is reduced in the absence of dietary sodium salt, presumably due to lack of activation of sodium dependent glucose/amino acid transporters. Similarly glucose/amino acid transport is reduced if the sodium pump is inhibited; in this condition the intracellular sodium rises, preventing its facilitated diffusion from the lumen; this prevents transport of glucose and amino acids.

FAT ABSORPTION INVOLVES HIGHLY COMPLEX EVENTS.
Dietary fat is mostly in the form of triglycerides (triacylglycerols) which are digested in the small intestine to form *monoglycerides, glycerol*, and *fatty acids*. Fatty acids are either short chained or long chained. *Short chained* fatty acids pass across the intestinal mucosa by diffusion and directly enter the blood capillaries, due to their fairly high water solubility. However, *long chained* fatty acids and other fatty nutrients, including *cholesterol*, undergo special processing during absorption. These fatty products diffuse across the brush border within micelles.

Once inside the mucosal cells, the *triglycerides (triacylglycerols)* are reformed (reesterified) from the fatty acids and monoglycerides, and packed, with cholesterol and other fatty substances (e.g., fat soluble vitamins), within certain *lipoprotein particles* called *chylomicrons*. The packaging process occurs within the Golgi apparatus (plate 1). Chylomicrons, like other lipoprotein particles, contain a coat of protein and a core of fat (plate 135), allowing the fat to float in the bloodstream without coalescing and clogging the blood vessels. Chylomicrons are extruded from the mucosal cells by exocytosis and moved into the lacteals and then to the larger lymph vessels, from which then pour into the veins in the upper trunk near the neck.

VITAMINS ABSORPTION FOLLOWS DIFFERENT PATHWAYS
Vitamins are divided into two categories, *water-soluble* and *fat-soluble*. The water soluble vitamins, such as the B family and C, pass across the mucosa by diffusion and/or bound to various specialized transporters. Fat-soluble vitamins such as vitamins A, D, and K are absorbed, together with the fatty nutrients, in the chylomicrons. *Vitamin B_{12}* (cyanocobalamine) has the largest molecular size among the vitamins, and its transport utilizes yet another mechanism involving a specific transporter protein called the *intrinsic factor*. This factor is a glycoprotein (proteins containing special polysaccharide component), secreted by the parietal cells of the pyloric glands in the stomach. In the chyme, the intrinsic factor binds with vitamin B_{12}, forming a complex which is taken up by the absorptive cells via endocytosis. Endocytotic vesicles release free vitamin B_{12} at the basal surface by exocytosis. In diseases of the stomach (e.g., *gastritis*) vitamin B_{12} absorption is reduced because of depletion of the intrinsic factor, causing *pernicious anemia* (plate 143). Endocytosis is also the mechanism for absorption of the maternal antibodies delivered to the neonates through the milk providing the suckling infant with passive immunity against some diseases.

CN: Use the same colors as were used on the previous page for lacteal (A), capillary (B), absorptive cell (C), and microvilli (D).
1. Color the four bold titles, beginning with lacteal (A), and the structures they refer to, before you color anything else. Then begin coloring the substances absorbed from the lumen of the intestine, starting with iron (Fe^{++}). Follow each substance as it enters the cell, moves through, and out again either into a capillary or lacteal.

TRANSPORT MECHANISMS -:-

active transport   ATP

co-transport   facilitated diffusion   diffusion

1   2   3   4

Nutrients, vitamins, minerals, and water are transported from the intestinal lumen to the blood and lymph by the absorptive cells of intestinal mucosa using a variety of cellular and membrane transport mechanisms discussed earlier in the book.

LACTEAL₄ CAPILLARY₈ ABSORPTIVE CELL꜀ MICROVILLI₀

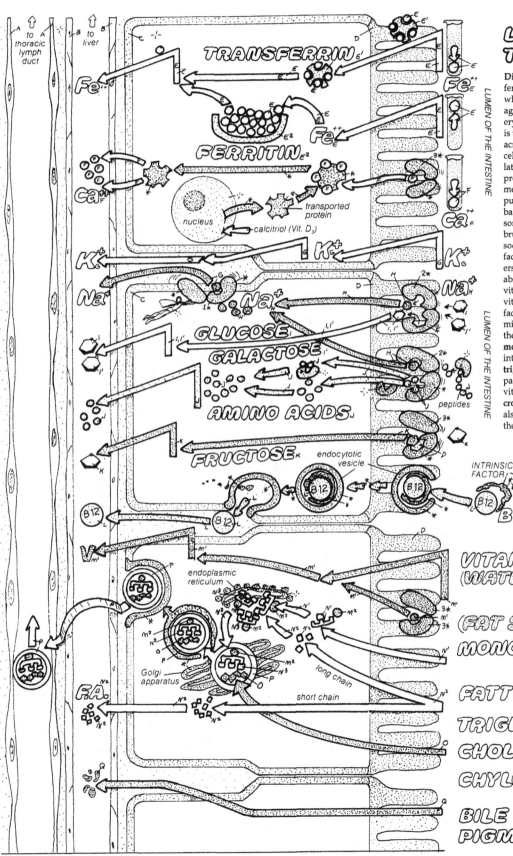

A to thoracic lymph duct

B to liver

LUMEN OF THE INTESTINE

TRANSFERRIN

FERRITIN

Fe

Ca

nucleus

transported protein

calcitriol (Vit. D₃)

K⁺

Na⁺

GLUCOSE

GALACTOSE

AMINO ACIDS

peptides

FRUCTOSE

endocytotic vesicle

B12

INTRINSIC FACTOR

stomach

B12

V

endoplasmic reticulum

Golgi apparatus

long chain

short chain

FA

LUMEN OF THE INTESTINE

Dietary iron is transported by transferrin (=mobilferrin), a transporter protein into the mucosal cell where it is stored bound to ferritin. Binding (storage) and release of iron from ferritin and its delivery to the blood depend on body needs. **Calcium** is transported actively by special transporters across the brush border and through mucosal cells. The hormone calcitriol (vitamin D₃) stimulates the synthesis of this calcium transporter protein. **Sodium** moves across the brush border membrane by sodium transporters and is actively pumped out of the absorptive cells across the basolateral membranes. **Glucose**, **galactose** and some **amino acids** are cotransported across the brush border by the same transporters that move sodium; **fructose** and other amino acids move by facilitated diffusion, utilizing different transporters. The **fat-soluble vitamins** (A, K, and D) are absorbed along with fats; the water-soluble vitamins diffuse across the mucosa. To transport vitamin B₁₂, the stomach secretes the intrinsic factor, a glycoprotein transporter that binds vitamin B₁₂, stimulates endocytosis and transport of the complex across the mucosa. **Glycerol, monoglycerides**, and **long chain fatty acids** move into mucosal cells, where they are reesterified into **triglycerides (triacylglycerols)** which are then packaged along with **cholesterol** and fat soluble vitamins into large lipoprotein particles (**chylomicrons**). These are then extruded out into the lacteals. **Short chain fatty acids** pass directly through the cell into the blood.

LUMEN OF THE INTESTINE

VITAMINS (WATER SOLUBLE)ₘ

(FAT SOLUBLE)ₘ

bile & lipase

MONOGLYCERIDES

FATTY ACIDS   LIPID

TRIGLYCERIDES

CHOLESTEROL

CHYLOMICRON

BILE SALTS & PIGMENTS

The *large intestine* (colon) is a tube 6 cm wide and 1.5 m long, extending between the small intestine and the *rectum* and processes the remaining undigested chyme into *feces* (stools), a relatively solid and bulky material that are excreted at intervals. To do so, the colon absorbs most of the chyme water. The colon's specific movements aid in processing of chyme and transport its solidifying content forward and finally into the rectum to enable fecal excretion in infrequent and prolonged intervals. The colon's exocrine glands secrete a viscous *mucus* that aids to mold the solid fecal matter and protect the colon wall from possible mechanical damage by the moving and churning solid contents.

The large intestine begins with the *cecum*, a short blind pouch formed by the ileum's protruded end into the colon. A unidirectional *ileocecal valve*, at the ileocecal border, permits discontinuous delivery of chyme to the large intestine; this provides the needed time for the colon to perform its functions and also prevents the colon bacteria from penetrating the normally aseptic small intestine. A *gastroileal reflex* controls the ileocecal valve. Increase in gut motility and gastrin release after meals relax the valve while colon *distention* closes it. Peristaltic waves are responsible for the ileocecal valve's periodic opening. The cecum and its vestigial attachment, the *vermiform appendix* contain *bacteria*.

The colon has four main segments, the *ascending, transverse, descending,* and *sigmoid*. The ascending and transverse colons are the sites of water and sodium *absorption* and *secretory activities*. Absorption of water dehydrates the colon content, facilitating the formation of solid fecal matter. The descending and sigmoid colons are the sites of *storage* of fecal matter. The sigmoid colon joins with the *rectum*, a muscular cavity performing short-term storage of feces and stimulation of *defecation* (fecal excretion, bowel movement). The *anus* is the end organ of the digestive tract, consisting of an internal smooth muscle sphincter and an external striated muscle sphincter, aiding in involuntary and voluntary control of defecation.

THE LARGE INTESTINE ABSORBS SODIUM & WATER
To form solid feces, the remaining chyme entering the colon, must be dehydrated. This is achieved by *absorption of water* across the large intestinal mucosa. About 1 L of water is absorbed daily in this process. Water absorption occurs in an *obligatory* manner by *osmosis* following the active sodium absorption of sodium. *Potassium*, however, is *secreted* in the large intestine, creating a major problem of potassium depletion during severe diarrhea (plate 81). Since the small intestine absorbs all of the organic dietary nutrients the chyme entering the colon lacks any glucose, amino acids or fatty acids. In fact the large intestine lacks the mechanisms for their absorption. Certain vitamins of bacterial origin and some drugs, however, can be absorbed (hence, rectal suppositories).

DIETARY FIBER & BACTERIA AID IN THE COLON FUNCTIONS
Diets rich in pectin and cellulose fibers (e.g., fruits and leafy vegetables) provide bulk and result in higher stool mass. The fiber in the chyme help retain water in the stool; this prevents formation of dry and hard stools which may damage the intestinal mucosa and cause constipation. The slow rate of movement in the colon permits the bacteria inhabiting the

large intestine to digest some of the chyme's mucus and fiber content, grow and proliferate. As the colon content nears the rectum, its proportion of digestible dietary fibers and mucus decrease while bacterial debris increase. Nearly a third of the stool's solid mass is of bacterial origin. The metabolism and turn-over of the colon bacteria provide a useful though minor source for some vitamins, such as the B family and K. An inportant source during dietary vitamin deficiency.

PERISTALSIS & MASS MOVEMENTS MOVE CHYME & FECES
Motility in the large intestine is slow, permitting long transit times of 1–3 days, compared to several hours in the small intestine. Three types of movements characterize the colon: segmentation, peristalsis, and mass movement. The *segmentation movements*, occurring infrequently (2/hr), entrap the colon contents within small segments or "haustra", characteristic pouches observed in the colon wall. These haustral segmentation turn and churn the colon content, exposing it to the epithelial cells for sodium and water absorption. The *peristalsis movements* occur slowly but in regular intervals, passing along as waves of contractions down the colon, gradually pushing the dehydrating feces toward the descending colon for storage. The descending colon exhibits another type of movement called *mass movement*. Here, and in the sigmoid colon, a strong peristaltic wave forces large "masses" of feces along the colon or into the rectum at once. Such contractions occur 2–3 daily, usually after meals.

THE ENS & PS NERVES CONTROL COLON MOTILITY
The myenteric plexi of the enteric nervous system (ENS) and the extrinsic parasympathetic nerves (PS) regulate colon motility and the defecation reflex; the vagus nerve controls the upper colon and sacral nerves the lower colon, rectum, and anus. The nervous control of segmentation and slow peristalsis is basically similar to that of the small intestine (i.e., under enteric plexi control but influenced in intensity by parasympathetic nerves). The mass movements are generated intrinsically by the plexi, but the brain and extrinsic nerves can regulate their intensity and frequency. Thus, factors like anxiety, and the presence of coffee and food in the mouth, can activate the colon and the urge for a bowel movement.

RECTAL FILLING INDUCES THE DEFECATION REFLEX
The occasional mass movements eventually force fecal matter into the rectum, distending this organ. Rectal distention triggers the defecation reflex which contracts the sigmoid colon and rectum to force the feces out through the anus (defecation); simultaneously the normally closed anal sphincters are relaxed to permit outflow. The involuntary defecation reflex occurs normally in infants, in whom voluntary neural control of defecation has not developed. In adults, rectal distention signals the brain's higher centers, creating the urge for a bowel movement. The external anal sphincter, a striated muscle, is voluntarily relaxed, permitting fecal outflow. The voluntary control over this sphincter develops by the end of infancy and early childhood. Other voluntary mechanisms such as pressure from the abdominal and respiratory muscles (diaphragm) also aid in defecation. In adults, the defecation reflex may be inhibited voluntarily, postponing the bowel movement.

CN: Use red for the blood capillary (T) and light blue for Q. Use dark colors for K, L, and O.
1. Begin with the upper left illustration.
2. Color the enlargement of the cecum (below

the upper illustration). Note that the circular muscle (O) refers to the three motility diagrams.
3. Color the absorption and secretion process.
4. Color the steps involved in defecation.

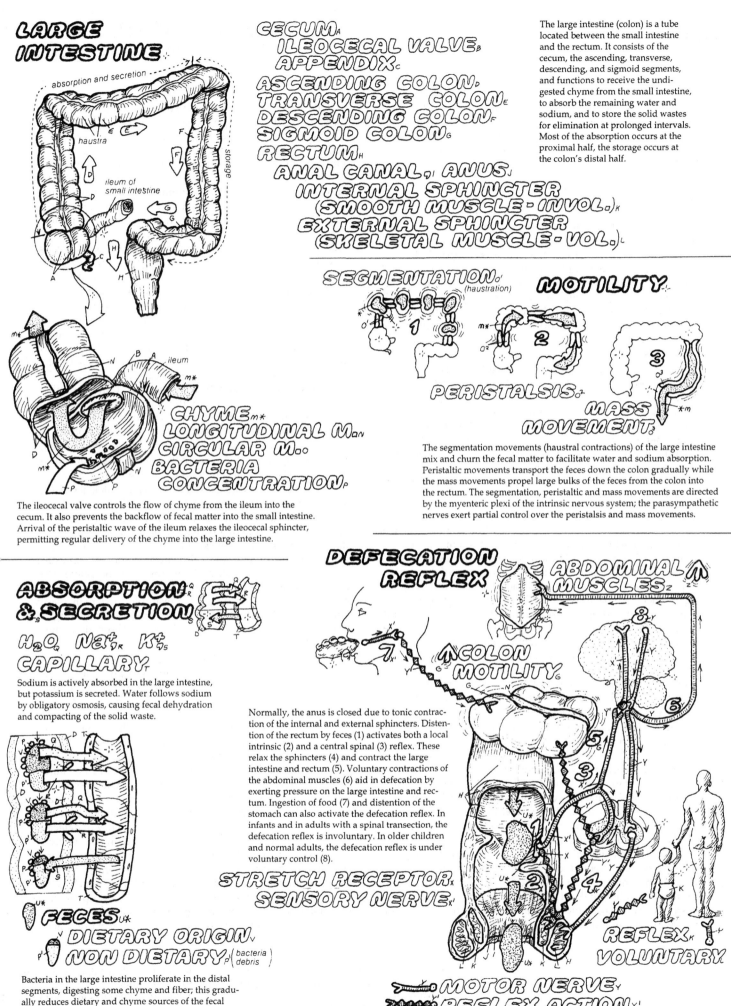

LARGE INTESTINE

absorption and secretion
haustra
ileum of small intestine
storage

CECUM A
ILEOCECAL VALVE B
APPENDIX C
ASCENDING COLON D
TRANSVERSE COLON E
DESCENDING COLON F
SIGMOID COLON G
RECTUM H
ANAL CANAL I, **ANUS** J
INTERNAL SPHINCTER
(SMOOTH MUSCLE - INVOL.) K
EXTERNAL SPHINCTER
(SKELETAL MUSCLE - VOL.) L

The large intestine (colon) is a tube located between the small intestine and the rectum. It consists of the cecum, the ascending, transverse, descending, and sigmoid segments, and functions to receive the undigested chyme from the small intestine, to absorb the remaining water and sodium, and to store the solid wastes for elimination at prolonged intervals. Most of the absorption occurs at the proximal half, the storage occurs at the colon's distal half.

CHYME m*
LONGITUDINAL M. N
CIRCULAR M. O
BACTERIA
CONCENTRATION P

The ileocecal valve controls the flow of chyme from the ileum into the cecum. It also prevents the backflow of fecal matter into the small intestine. Arrival of the peristaltic wave of the ileum relaxes the ileocecal sphincter, permitting regular delivery of the chyme into the large intestine.

SEGMENTATION O' (haustration)   **MOTILITY**

PERISTALSIS P'   **MASS MOVEMENT** Q'

The segmentation movements (haustral contractions) of the large intestine mix and churn the fecal matter to facilitate water and sodium absorption. Peristaltic movements transport the feces down the colon gradually while the mass movements propel large bulks of the feces from the colon into the rectum. The segmentation, peristaltic and mass movements are directed by the myenteric plexi of the intrinsic nervous system; the parasympathetic nerves exert partial control over the peristalsis and mass movements.

ABSORPTION & SECRETION

H_2O Q, Na^+ R, K^+ S
CAPILLARY T

Sodium is actively absorbed in the large intestine, but potassium is secreted. Water follows sodium by obligatory osmosis, causing fecal dehydration and compacting of the solid waste.

FECES U*
DIETARY ORIGIN V
NON DIETARY P' (bacteria debris)

Bacteria in the large intestine proliferate in the distal segments, digesting some chyme and fiber; this gradually reduces dietary and chyme sources of the fecal matter while increasing the bacterial portion (debris).

DEFECATION REFLEX   ABDOMINAL MUSCLES Z

COLON MOTILITY G

Normally, the anus is closed due to tonic contraction of the internal and external sphincters. Distention of the rectum by feces (1) activates both a local intrinsic (2) and a central spinal (3) reflex. These relax the sphincters (4) and contract the large intestine and rectum (5). Voluntary contractions of the abdominal muscles (6) aid in defecation by exerting pressure on the large intestine and rectum. Ingestion of food (7) and distention of the stomach can also activate the defecation reflex. In infants and in adults with a spinal transection, the defecation reflex is involuntary. In older children and normal adults, the defecation reflex is under voluntary control (8).

STRETCH RECEPTOR X
SENSORY NERVE X'

REFLEX K
VOLUNTARY

MOTOR NERVE Y
REFLEX ACTION Y'

Digestive disorders and diseases are among the most common problems in the body. Some of these, such as vomiting, are normal and useful responses to ingesting toxins or excess food; others, such as ulcers, may have complicated causes, including bacterial infection.

TOXINS & EXCESS FOOD BULK CAUSE VOMITING
Vomiting is a useful physiological defense response. The *vomiting reflex* rejects the undesirable food from the stomach by expelling it out through the esophagus and mouth. The vomiting reflex is initiated by activation of the *chemoreceptors and stretch receptors* in the stomach wall. Excessive food bulk expands the stomach wall, overstimulating the stretch receptors which initiate vomiting. Poisons and microbial toxins initiate vomiting by acting on the chemoreceptors of gastric mucosa. These sensory signals are communicated from the gastric mucosa via sensory fibers in the *vagus* to a *vomiting* center in the *brain medulla*. This center also responds to certain toxic substances in the blood.

Activation of the vomiting center results in a complex of reflex responses: the *glottis* closes to keep the vomit from entering the respiratory passages; the *lower esophageal sphincter* opens; massive contractions of the *abdominal* and *respiratory* muscles occur to exert external pressure on the stomach; the vagus nerve stimulates the stomach vigorously; and, finally, a strong wave of reverse *peristalsis* moves from the pylorus to the cardia. As a result, the stomach contents are expelled through the esophagus and mouth, eliminating the source of toxicity and discomfort.

BACTERIA ARE THE CHIEF CAUSE OF ULCERS
Ulcers are wounds occurring in the inner lining of the stomach and small intestine, particularly in the duodenum. In fact, only 10% of ulcers occur in the stomach (peptic ulcers), the majority occur in the duodenum, since this part of the gut is exposed to high acid from the stomach but has little resistance against acid. Stomach ulcers are, however, more dangerous as they involve bleeding. Ulcers are caused by the corrosive and noxious effects of acid on the gut wall. At first, these ulcers are superficial. If exposure to acid continues unchecked, the wound deepens, reaching the vascular layers deep in the wall, and bleeding occurs. The bleeding is worsened by digestion of food which increases acid secretion and stomach motility. This bleeding, which makes ulcers painful and dangerous, can be detected by the presence of fresh blood clots in the feces.

Several factors and conditions may contribute to ulcers. Recently infection of the stomach lining by the bacterium *Helicobacter pylori* is recognized as a major cause of peptic ulcers. This infection, now diagnosable by simple blood or breath tests, is found in about two thirds of ulcer cases in the United States. The bacteria disrupt the protective barrier of the stomach against acid. The treatment recommended in such cases is an acid inhibitor, such as histamine receptor (H_2) blocker or a proton pump inhibitor, together with antibiotics to eradicate the bacterial infection. Many of the remaining ulcer cases are attributed to high consumption of non-steroidal anti-inflammatory drugs (NSAID, e.g., aspirin and ibuprofen) or alcohol intake. These substances penetrate the gastric wall and erode the mucosal barrier against acid. Stress is now considered to be of minor importance.

Other causes of ulcers may be due the prolonged and excessive acid production. Most duodenal ulcers are due to excess acid which may result from increased vagus nerve activity or excess gastrin secretion. In the past, vagotomy, surgical cutting of the vagus nerve to the stomach, treated the ulcers. This procedure is not common now due to marked success of treatment by anti-histaminic agents which reduce excess acid. Gastrin-producing tumors of the pancreas also cause ulcers (Zollinger-Elllison syndrome). Psychological stress, once widely believed to causes peptic ulcer is now considered a minor factor. However excessive levels of corticosteroid hormones released in severe stresses or in Cushing's disease may weaken the gut wall's resistance to acid.

DIARRHEA IS CAUSED BY INCREASED INTESTINAL MOTILITY
Diarrhea is characterized by excessive and frequent discharge of watery feces. This condition is often due to increased intestinal motility, delivering large amounts of watery chyme to the large intestine. The colon's inability to absorb the excess water causes diarrhea. Different factors may be responsible for the increased motility. Certain fruits, such as prunes, contain substances that naturally increase intestinal motility. Diarrhea can also be caused by the effects of certain toxins on the intestinal mucosa. For example, *cholera toxin* causes the intestinal glands to secrete large quantities of electrolytes (sodium, chloride, bicarbonate) into the lumen. Water enters the lumen by osmosis. A cholera victim can lose about 10 L of water per day, a lethal condition if not treated.

Certain diarrheas are caused by enzyme deficiency in the small intestine. For example, most adult Asian, African and Native American people lack the intestinal enzyme *lactase*, and cannot therefore digest lactose, the sugar in milk and dairy products. The undigested lactose increases the lumen osmolarity, decreasing water absorption in the small intestine, resulting in increased rate and amount of chyme delivery to the colon which results in diarrhea. Diarrhea may also be of nervous (psychogenic) origin. For example, anxiety increases parasympathetic activity to the lower bowels, stimulating intestinal motility, which in turn decreases absorption time, leading to diarrhea.

CONSTIPATION IS OFTEN DUE TO LACK OF DIETARY FIBER
Reduced motility of the large intestine is responsible for *constipation*, a common digestive disorder. In this condition the storage time in the colon is increased, which in turn increases the amount of water absorbed from the feces. Dried feces have reduced bulk and therefore less likely to move forward and initiate bowel movements. Causes of constipation are not well understood. *Dietary habits* may be a major cause. Increased *fiber* content (leafy vegetables, fruits) in the diet helps retain water in the colon, increasing fecal bulk, which in turn stimulates colon motility and defecation. Learning to inhibit the defecation reflex during childhood may be another cause of constipation. Average frequency of defecation in adult human is about once or twice daily, but many have less frequent movements. Although mild and occasional constipation does not pose any problems, prolonged constipation is accompanied by abdominal discomfort, headaches, loss of appetite, and even depression. Hemorrhoids, outpouchings of rectal tissue and veins through the anus can also be caused by continued constipation and painful extrusion of dried stools.

CN: Use red for E and dark colors for A and F.
1. Begin with the upper panel, completing the diarrhea segment first.
2. Color the belching material, noting that the symbol for fermentation (J) is that of an undigested carbohydrate.
3. Color lactose intolerance and then vomiting
4. Color peptic ulcers.

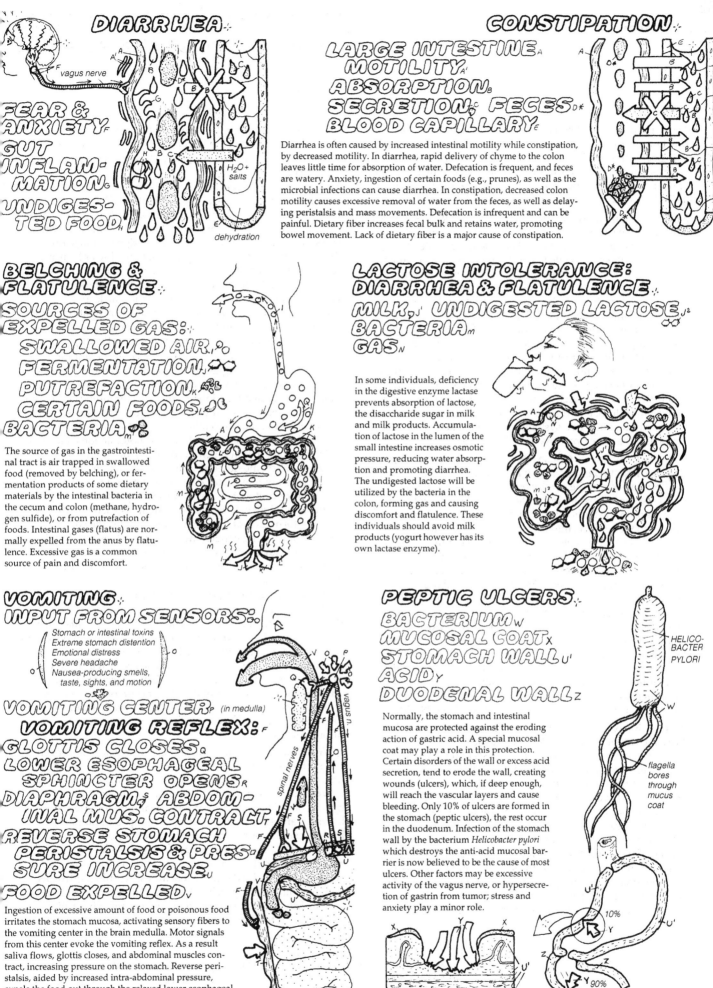

DIARRHEA

FEAR & ANXIETY_F
GUT INFLAM- MATION_G
UNDIGES- TED FOOD_H

vagus nerve
H_2O + salts
dehydration

CONSTIPATION

LARGE INTESTINE_A
MOTILITY_A'
ABSORPTION_B
SECRETION_C, FECES_D*
BLOOD CAPILLARY_E

Diarrhea is often caused by increased intestinal motility while constipation, by decreased motility. In diarrhea, rapid delivery of chyme to the colon leaves little time for absorption of water. Defecation is frequent, and feces are watery. Anxiety, ingestion of certain foods (e.g., prunes), as well as the microbial infections can cause diarrhea. In constipation, decreased colon motility causes excessive removal of water from the feces, as well as delaying peristalsis and mass movements. Defecation is infrequent and can be painful. Dietary fiber increases fecal bulk and retains water, promoting bowel movement. Lack of dietary fiber is a major cause of constipation.

BELCHING & FLATULENCE

SOURCES OF EXPELLED GAS:
SWALLOWED AIR_I'O
FERMENTATION_J
PUTREFACTION_K
CERTAIN FOODS_L
BACTERIA_m

The source of gas in the gastrointestinal tract is air trapped in swallowed food (removed by belching), or fermentation products of some dietary materials by the intestinal bacteria in the cecum and colon (methane, hydrogen sulfide), or from putrefaction of foods. Intestinal gases (flatus) are normally expelled from the anus by flatulence. Excessive gas is a common source of pain and discomfort.

LACTOSE INTOLERANCE: DIARRHEA & FLATULENCE

MILK_J' UNDIGESTED LACTOSE_J²
BACTERIA_m
GAS_N

In some individuals, deficiency in the digestive enzyme lactase prevents absorption of lactose, the disaccharide sugar in milk and milk products. Accumulation of lactose in the lumen of the small intestine increases osmotic pressure, reducing water absorption and promoting diarrhea. The undigested lactose will be utilized by the bacteria in the colon, forming gas and causing discomfort and flatulence. These individuals should avoid milk products (yogurt however has its own lactase enzyme).

VOMITING

INPUT FROM SENSORS:
Stomach or intestinal toxins
Extreme stomach distention
Emotional distress
Severe headache
Nausea-producing smells, taste, sights, and motion

VOMITING CENTER (in medulla)

VOMITING REFLEX:
GLOTTIS CLOSES_P
LOWER ESOPHAGEAL SPHINCTER OPENS_R
DIAPHRAGM_S, ABDOM- INAL MUS. CONTRACT_T
REVERSE STOMACH PERISTALSIS & PRES- SURE INCREASE_U
FOOD EXPELLED_V

vagus n.
spinal nerves

Ingestion of excessive amount of food or poisonous food irritates the stomach mucosa, activating sensory fibers to the vomiting center in the brain medulla. Motor signals from this center evoke the vomiting reflex. As a result saliva flows, glottis closes, and abdominal muscles contract, increasing pressure on the stomach. Reverse peristalsis, aided by increased intra-abdominal pressure, expels the food out through the relaxed lower esophageal sphincter, esophagus, pharynx, and the mouth.

PEPTIC ULCERS

BACTERIUM_W
MUCOSAL COAT_X
STOMACH WALL_U'
ACID_Y
DUODENAL WALL_Z

Normally, the stomach and intestinal mucosa are protected against the eroding action of gastric acid. A special mucosal coat may play a role in this protection. Certain disorders of the wall or excess acid secretion, tend to erode the wall, creating wounds (ulcers), which, if deep enough, will reach the vascular layers and cause bleeding. Only 10% of ulcers are formed in the stomach (peptic ulcers), the rest occur in the duodenum. Infection of the stomach wall by the bacterium *Helicobacter pylori* which destroys the anti-acid mucosal barrier is now believed to be the cause of most ulcers. Other factors may be excessive activity of the vagus nerve, or hypersecretion of gastrin from tumor; stress and anxiety play a minor role.

HELICO- BACTER PYLORI

flagella bores through mucus coat

10%
90%

FUNCTIONAL ORGANIZATION OF THE NERVOUS SYSTEM

The nervous system (NS) is responsible for sensory and motor functions, for instinctive and learned behaviors, and for regulating activities of the internal organs and systems. To appreciate its importance, consider the problems faced by the blind or deaf or the difficulties encountered by the motor-disabled individuals suffering from spinal injury or brain stroke.

NEURONS, GLIA, & SYNAPSES: PARTS OF THE NERVOUS SYSTEM

The various parts of the NS consist of numerous specialized and excitable *nerve cells* (neurons) and *synapses* (plates 19, 20) that connect the nerve cells to one another, to those in other centers, or to neurons in the periphery. The nerve cells and the neural centers operate on the basic principles of *excitation* and *inhibition*, determined by the type of neurons and the synapses that connect them (plate 87). Although the morphology (shapes) of the nerve cells may vary in different parts and may be important, it is the connections of the nerve cells that mainly determine their function. The variety of "glia" cells (astrocytes, oligodendrocytes, microglia) found in the neural tissue are not excitable like the neurons but perform critical supportive functions such as myelination, ionic regulation of extracellular fluid, and response to injury. The NS as a whole may be divided into two systems—the *central nervous system* (CNS) and the *peripheral nervous system* (PNS).

PERIPHERAL NERVOUS SYSTEM (PNS) CONSISTS OF NERVES, SENSORY RECEPTORS, & MOTOR EFFECTORS

The PNS consists of the *sensory receptors* or *sensory organs*, which are specialized to detect changes in the external environment or in the body interior and to communicate these changes to the CNS via the *afferent sensory nerves*. Another part of the PNS is the *motor effectors*. These consist of the *voluntary skeletal muscles*, responsible for body and limb movements, and the *smooth muscles* and *exocrine glands*, which effect changes in motility and secretions in the visceral organs. *Efferent motor nerves* extending from the CNS to these effector organs are also part of the PNS. On the basis of these different targets, the peripheral motor system has been divided into a *somatic* division, which regulates the voluntary skeletal muscles, and an *autonomic* division, which deals with the visceral effectors (glands and smooth muscle). Although the somatic and autonomic systems are distinct in terms of their motor output nerves and targets, they may share both peripheral sensors and certain central nervous centers (plates 29, 85).

CENTRAL NERVOUS SYSTEM (CNS) CONSISTS OF LOWER & HIGHER CENTERS IN BRAIN & SPINAL CORD

The central nervous system (CNS) consists of the *brain* and *spinal cord*, which process sensory information and integrate them with their inborn patterns of responses and with past experience to produce appropriate motor commands. The CNS operations are carried out by the *sensory, motor,* and *association (integrative)* centers in the brain and spinal cord. Different regions of the CNS are devoted wholly or partly to any one of these functions. Moreover, the neural centers are organized in a hierarchy; thus the sensory, motor, and association centers may be thought of as *lower* or *higher* centers. The lower centers are in direct contact with the peripheral neural structures via sensory and motor nerves. In order for the

higher centers to communicate with the peripheral effectors, they must go through the lower centers, and vice versa.

RESPONSES TO SOUND ILLUSTRATE CNS INTEGRATIVE FUNCTIONS

To understand the operations of different sensory, motor, associative, and integrative processes of the CNS, consider the human reaction to a loud or strange sound. The sound waves will be detected by the sound receptor cells in the ear, which transduces the sound waves into nerve signals and sends them, via the afferent auditory nerve, to the lower hearing center in the brain stem. At this point the signals are first processed and then sent to the lower motor centers in the brain stem and spinal cord to activate the startle and head-turning reflexes. At the same time, the activation of the autonomic centers results in increased heart and breathing rates, in preparation for eventual running or fleeing.

At the same time, the lower hearing centers communicate nerve signals to the higher hearing centers in the cortex, the highest hierarchical brain center, where other qualities of the sound are evaluated, and the results are communicated to the cortical association and integrative centers. Here the sound is examined in relation to other sensory stimuli (e.g., visual) converging simultaneously. If further motor actions, particularly voluntary actions such as running from the site of the loud noise, become necessary, appropriate commands are issued to the higher motor centers, which in turn signal the lower motor centers to activate the appropriate muscle groups. The signals from the brain stem also activate the reticular formation of the brain, which excites the cortex globally, increasing general awareness, alertness, and vigilance. The higher centers can also enhance the behavioral drives and autonomic responses necessary for carrying out these motor tasks.

CNS ORGANIZATION REFLECTS ITS EVOLUTION

Division of the CNS into higher and lower centers reflects its evolutionary development. The earliest nerve centers may have resembled the rudimentary operations of the spinal cord (i.e., direct contact between the lower sensory and motor components), enabling very fast spinal reflexes such as limb withdrawal in response to painful stimuli; these occur very rapidly without the involvement of the higher centers. The defensive reflexes ensure and optimize survival. The spinal cord structure and function in this respect remains fairly uniform throughout evolution.

With brain evolution, new higher sensory and motor centers such as the cerebral cortex emerged above the lower centers, allowing control over the lower centers as well as newer nervous abilities. Indeed, the cerebral cortex, the site of the highest and finest analysis of sensory and motor integration, learning, and skilled tasks, is well developed in humans but nonexistent or rudimentary in lower vertebrates. In the human cortex, the association and integrative areas have enlarged markedly, occupying most of the cortical area. This is the basis of such adaptive capacities as learning, introspection, planning, and speech in humans. The lower vertebrates, such as fish and amphibians, remain more reflexive and instinctual in their nervous functions and responses.

CN: Use a dark color for the A structures and a very light color for E.
1. Begin with the anatomical illustration at the top of the page. Because the peripheral nerves (B) are so numerous and so small in this drawing, color the many lines representing nerves.
2. Color the organizational diagram across the middle of the page.

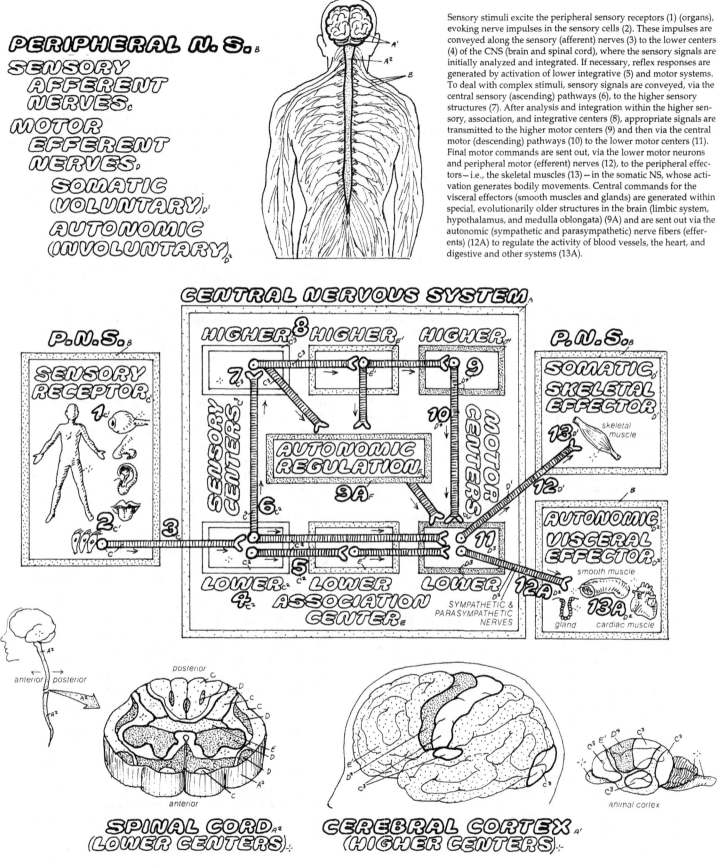

PERIPHERAL N.S.

SENSORY AFFERENT NERVES.

MOTOR EFFERENT NERVES.

SOMATIC (VOLUNTARY).

AUTONOMIC (INVOLUNTARY).

Sensory stimuli excite the peripheral sensory receptors (1) (organs), evoking nerve impulses in the sensory cells (2). These impulses are conveyed along the sensory (afferent) nerves (3) to the lower centers (4) of the CNS (brain and spinal cord), where the sensory signals are initially analyzed and integrated. If necessary, reflex responses are generated by activation of lower integrative (5) and motor systems. To deal with complex stimuli, sensory signals are conveyed, via the central sensory (ascending) pathways (6), to the higher sensory structures (7). After analysis and integration within the higher sensory, association, and integrative centers (8), appropriate signals are transmitted to the higher motor centers (9) and then via the central motor (descending) pathways (10) to the lower motor centers (11). Final motor commands are sent out, via the lower motor neurons and peripheral motor (efferent) nerves (12), to the peripheral effectors—i.e., the skeletal muscles (13)—in the somatic NS, whose activation generates bodily movements. Central commands for the visceral effectors (smooth muscles and glands) are generated within special, evolutionarily older structures in the brain (limbic system, hypothalamus, and medulla oblongata) (9A) and are sent out via the autonomic (sympathetic and parasympathetic) nerve fibers (efferents) (12A) to regulate the activity of blood vessels, the heart, and digestive and other systems (13A).

CENTRAL NERVOUS SYSTEM

P.N.S.

SENSORY RECEPTOR

SENSORY CENTERS

HIGHER HIGHER HIGHER

AUTONOMIC REGULATION
9A

LOWER LOWER
ASSOCIATION CENTER

MOTOR CENTERS

P.N.S.

SOMATIC SKELETAL EFFECTOR
skeletal muscle
13

12

AUTONOMIC VISCERAL EFFECTOR
smooth muscle
12A
SYMPATHETIC & PARASYMPATHETIC NERVES
gland 13A cardiac muscle

anterior posterior

SPINAL CORD (LOWER CENTERS)

posterior
anterior

CEREBRAL CORTEX (HIGHER CENTERS)

animal cortex

Sensory, motor, and association functions are carried out by different parts of the CNS. In the spinal cord (oldest CNS area), the anterior (ventral) structures carry out motor functions while posterior (dorsal) structures perform sensory functions. The cord's middle region is concerned with association functions, connecting sensory with motor areas and the right half with the left one. In the brain cortex (the newest CNS region), sensory functions are carried out by areas located mainly in the posterior half, behind the central fissure, while motor functions are carried out by anterior (frontal) areas. In the spinal cord, the relative size of the association areas is fairly small compared to that of the sensory and motor areas. In the cortex, the size of the association and integrative areas far exceeds that of the sensory and motor areas. Note the marked increase in the size of the association/integrative areas with evolution, depicting the importance of these areas in the higher functions of the nervous system (learning, perception, language).

Housed in the skull, the *brain* consists of all the parts of the central nervous system (CNS) above the spinal cord. The brain may be divided into two major parts—a lower *brain stem* and a higher *cerebrum (forebrain)*.

BRAIN STEM REGULATES VISCERAL FUNCTIONS & BRAIN REFLEXES

The brain stem is situated directly above the spinal cord and under the brain hemispheres and is connected to these regions with fiber pathways. The brain stem is the more ancient part of the brain and consists of the medulla, pons, and midbrain. The structure and function of the brain stem are fairly similar in the lower and higher vertebrates, particularly among mammals. Brain stem structures carry out many vital somatic, autonomic, and reflexive functions. The centers for respiration, cardiovascular, and digestive functions are in the *medulla*, the "lowest" of the brain's areas. The *pons* has inhibitory control centers for respiration and interacts with the cerebellum.

Other diffusely organized areas in the reticular core of the pons and medulla constitute the *reticular formation*, which is involved in regulation of sleep, wakefulness, and attention as well as controlling the level of excitation in the higher forebrain structures (plate 106). Somatic motor centers (nuclei) in the *midbrain* are involved in regulation of walking and posture and of reflexes for head and eye movements (plate 97). The *cerebellum*, a large motor structure involved in movement coordination, is placed in the back of the brain stem (plate 97).

The anencephalic infant—In the human infant, brain stem capacities are more mature than those of the higher forebrain regions. The role of the brain stem in behavior and body functions may be shown by observing the motor and behavioral abilities of *anencephalic* ("no brain") infants, born without a forebrain. Such infants usually do not survive for long, but during their short lives, they are capable of many behaviors. They can find the nipple and suckle milk, smile, frown, cry and make other infant sounds, and move the head and limbs in a manner similar to normal newborns.

THE FOREBRAIN REGULATES HIGHER BRAIN FUNCTIONS

The human forebrain consists of two hierarchically organized regions, a lower *diencephalon* and a higher *telencephalon*.

Hypothalamus & thalamus—The diencephalon consists of the *hypothalamus* and the *thalamus*. The hypothalamus contains numerous centers (nuclei, areas) for regulating the internal environment (homeostasis), including those for controlling body temperature, blood sugar, hunger and satiety, and sexual behavior. It controls the diurnal cycles by its biological clock and regulates the activities of the endocrine system and hormones. The *thalamus* is a complex sensory-motor relay station involved in integrating sensory signals and relaying them to the *cerebral cortex*. The thalamus also participates in motor control and in regulation of cortical excitation and attention.

Cerebral hemispheres and cortex—Situated above the diencephalic structures of hypothalamus and thalamus is the telencephalon of the forebrain. It consists of two nearly symmetrical *cerebral hemispheres*. These house the *cerebral cortex*, the *basal ganglia*, and the *limbic system*. The two hemispheres are connected by a massive bundle of fibers called the *corpus*

callosum. The cerebral cortex is a network of highly organized nerve cells (gray matter) in a sheet about 5 mm thick that covers the surface of the hemispheres (cortex = bark). The neurons of the cortex are organized horizontally in six layers and vertically in functionally distinct "columns" (plates 93, 100).

The large surface area of the cortex and the need to fit this sheet within the skull produces the folds and convolutions seen in the outer brain surface (sulcus = furrow; gyrus = convolution). The cortex and the associated large mass of nerve fibers (white matter) make up the bulk of the cerebral hemispheres. In humans, the cerebral cortex is extremely well developed in both size and nerve cell organization, enabling it to be the site of the highest and most intricate analysis and integration of sensory and motor information (plate 111).

Cortical lobes—Each hemisphere (particularly its cortex) is divided into four externally visible major lobes and a large, externally hidden area, "the insula." The *frontal* lobe extends from the anterior tip of the hemisphere back to the *central sulcus* (fissure of Rolando). The posterior areas of the frontal lobe are specialized for motor functions (plate 96) and the anterior areas are involved in learning, planning, speech, and some other psychological functions (plate 111). The *occipital lobe*, located in the back of the hemisphere, carries out mainly visual functions (plate 100). The *parietal lobe* consists of the dorsal (top) and lateral areas between the frontal and occipital lobes and is specialized for somatic sensory functions (e.g., skin senses) and related association and integrative roles (plate 93). Certain areas in the parietal lobe also are very important in cognitive and intellectual processes. The *temporal lobe* comprises the hearing centers and related association areas, including some speech centers. Other areas of the temporal lobe are important in memory (plate 109). The anterior and basal areas of the temporal lobe are involved in the sense of smell and in functions related to the limbic system. A fifth major cortical area, the "insular lobe," is not visible externally and is buried deep to the lateral fissure.

The basal ganglia & the limbic system—The forebrain also houses the *basal ganglia,* a complex of mainly motor structures. In lower animals, the basal ganglia are the only higher motor structures. In humans, the basal ganglia structures work in conjunction with the motor areas of the cortex and cerebellum to plan and coordinate gross voluntary movements (plate 97). Another forebrain system is the *limbic system* or "limbic lobe." The limbic system structures—the hippocampus, the amygdala, the cingulate gyrus, and the septum—work with the hypothalamus to intimately control the expression of instinctive behavior, emotions, and drives. Also, the hippocampus and amygdala have been found to have major cognitive functions, mainly in processing of memory. Overall size and organization of the limbic system do not change significantly during the course of mammalian evolution, indicating this system's involvement with basic instinctive behaviors common to all species of mammals (plates 97, 108).

Although the motor, sensory, cognitive, and behavioral functions are fairly well localized to distinct brain areas, these regions are well connected by fiber pathways, and the brain often works as a whole. This is true for the "global functions" of the brain such as learning, memory, and consciousness.

CN: Use dark colors for B, C, E, F, and G.
1. Begin in the upper right corner by coloring the cortex of the two cerebral hemispheres (A') and the list of titles, without coloring the structures to which the titles refer. Then start with the material in the upper left corner and work your way dow[n] to the limbic system and across the basal ganglia
2. Color both views at the bottom simultaneously. The vertical broken line in the mid-sagittal view shows the location of the coronal section.

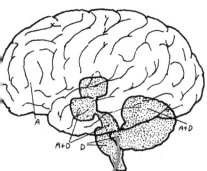

FOREBRAIN & BRAIN STEM

The brain may be divided into a forebrain (cerebrum), subserving higher nervous functions (perceptions, voluntary motor control, emotion, cognition, and language), and a brain stem, regulating internal bodily functions and involuntary reflexes, and also serving as a relay station for signal transmission to and from the forebrain.

gyrus
sulcus
central sulcus

sylvian fissure

LOBES OF THE CEREBRAL CORTEX A'

FRONTAL K
PARIETAL L
OCCIPITAL M
TEMPORAL N

The forebrain consists of two hemispheres, each divided into four major lobes—frontal, parietal, occipital, and temporal—plus the insular area (insular lobe). The lobes are covered by the cortex, a wide, thin (3–5 mm) sheet of gray matter, folded extensively to fit in the skull (hence, the convolutions—sulci and gyri). The occipital lobe performs higher visual functions; the temporal lobes house the auditory and associated language and cognitive areas; the parietal lobes perform the somatic sensory and related association functions; the frontal lobes contain the higher motor areas and those for planning and higher behavior. Deep within the lobes are the white matter (fibers), the limbic system, and basal ganglia structure of the forebrain.

BRAIN
FOREBRAIN A
CEREBRAL CORTEX A'
LIMBIC SYSTEM B
BASAL GANGLIA C
BRAIN STEM D
THALAMUS E
HYPOTHALAMUS F
MIDBRAIN G
PONS H
MEDULLA I
CEREBELLUM J

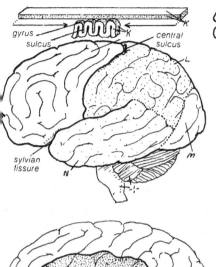

LIMBIC SYSTEM B

The forebrain's limbic system regulates and integrates expression of emotions, feelings, and drives. In lower animals, the limbic system is intimately connected with the olfactory sense. In higher animals, it is well connected with the cortex of the frontal lobes and basal ganglia. Some limbic structures (hippocampus and amygdala) are also involved in memory processing.

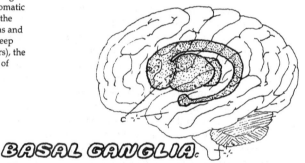

BASAL GANGLIA C

The basal ganglia structures are higher motor centers, functioning in harmony with the motor cortex. Basal ganglia lesions produce pronounced motor disorders (e.g., Parkinson's disease). In birds and lower vertebrates that lack a true cortex (neocortex), the basal ganglia are the highest motor control centers.

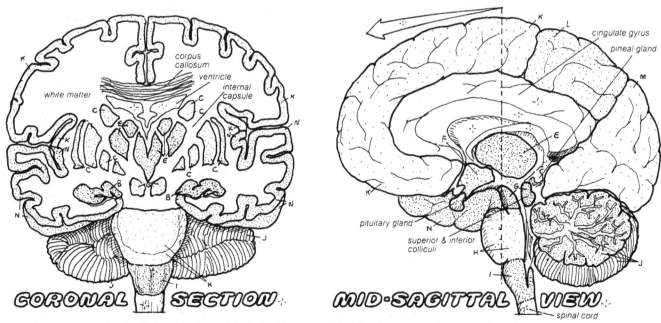

CORONAL SECTION

corpus callosum
ventricle
internal capsule
white matter

MID-SAGITTAL VIEW

cingulate gyrus
pineal gland
pituitary gland
superior & inferior colliculi
spinal cord

To view the inner structures of the brain, either a coronal (cross) section or a sagittal section can be used. The coronal section shown (left diagram) depicts the relationship between the cortex and the underlying white matter and nerve centers (basal ganglia, thalamus). Note the corpus callosum connecting the two hemispheres. The sagittal section cuts along the medial plane, exposing the hidden medial cortical structures as well as many of the brain stem structures.

The *spinal cord* (SC) is one of the two main parts of the central nervous system (CNS). The SC is a cord of nervous tissue about 40–45 cm long, extending from the neck to the loin within the inner cavity of the vertebral column. Practically all the voluntary skeletal muscles in the neck, trunk, and limbs receive their supply of motor nerves from the SC. All the sympathetic and the lumbo-sacral parts of the parasympathetic motor outputs to the skin and visceral organs also emerge from the SC. All sensory signals from the peripheral receptors of the skin, muscles, and joints in the trunk and limbs are communicated to the SC.

THE SPINAL CORD (SC) MEDIATES REFLEXES & HAS MAJOR FIBER PATHWAYS

The spinal cord performs two basic functions. First, it can act as a nerve center, integrating the incoming sensory signals and activating the motor output directly, without any brain intervention. This function is manifested in the operation of spinal reflexes, which are extremely important in defending against noxious stimuli and maintaining postural body support. Second, the SC is the intermediate nerve center (station) between the periphery and the brain. All voluntary and involuntary motor commands from the brain to the body musculature must first be communicated to the spinal motor centers, which process these signals appropriately before passing them to the muscles. Similarly, sensory signals from the peripheral receptors to the brain centers are first communicated to the SC sensory centers, where they are partly processed and integrated before delivery to the brain sensory centers. The important and massive SC fiber pathways serve in such two-way communication between the brain and the cord. Damage to the SC results in paralysis, inability to use the voluntary muscles and absence of sensation from the lower periphery, even though the brain is intact.

THE SC WHITE MATTER SURROUNDS THE INNER GRAY MATTER

The structural organization of SC can best be studied by observing a cross section of the cord. Throughout the SC length, an outer band of *white matter* uniformly surrounds an inner core of *gray matter;* the right and left halves of the SC are symmetrical. The size of the gray and white matter, however, varies between segments of the cord; it is thicker in segments related to the limbs. The white matter consists mostly of myelinated nerve fibers (axons) grouped in bundles. The cell bodies of these fibers are either in the brain (*descending fibers*) or in the SC (*ascending fibers*). The gray matter consists of nerve cells (neurons), their dendritic processes, and the numerous synapses between the nerve cells.

The SC gray matter is shaped like the letter H (or more like a butterfly, whose wings are called "horns") and is divided into three functional zones: the *dorsal (posterior) horns* are sensory in function, the *ventral (anterior) horns* are motor, and the middle zone carries out, in part, association functions between the sensory and motor zones. These zones of gray matter constitute the associative and integrative SC zones.

MOTOR, AUTONOMIC, & INTERNEURONS COMPRISE SC GRAY MATTER

The SC gray matter is populated by large and small neurons.

The large neurons comprise the motor or sensory neurons. The *motor neurons*, located in the ventral horns, are output neurons that send their motor fibers to the voluntary skeletal muscles via the ventral (motor) roots. Motor neurons are grouped in clusters, each cluster serving a different muscle. In the thoracic, lumbar, and sacral segments of the SC are separate clusters of nerve cells — the *autonomic motor neurons* — which innervate the autonomic ganglia and visceral organs (plate 85).

Sensory roles of dorsal roots and dorsal horns. The peripheral sensory input to the spinal cord arrives in the dorsal horns through the *dorsal roots* by the *primary sensory neurons*. The cell bodies of these cells are located outside the SC gray matter, in the *sensory ganglia (dorsal root ganglia)*. These primary sensory cells have a bifurcating axon: the peripheral branch brings sensory messages from sources such as the skin and joints; the central branch moves through the sensory root to enter the dorsal horn and synapse with the sensory relay cells and the interneurons of the SC. Located in the dorsal horns, the large sensory relay cells give rise to fibers that cross over to the opposite side and ascend in the SC white matter to communicate the incoming peripheral sensory signals to the higher brain centers.

Motor functions of the ventral horns. Some of the dorsal root sensory fibers continue uninterrupted to enter in the ventral horn of the same side, where they synapse directly with the motor neurons. Other sensory fibers contact small *interneurons* (association neurons) that mediate excitatory and inhibitory connections between the sensory neurons and the motor neurons in the ventral horns of the same or the opposite side. These local connections provide the nerve circuits for the operation of spinal reflexes (plate 95). Motor neurons receive input not only from the sensory neurons and the interneurons, but also from the neurons in the higher brain centers (plate 96). The spinal motor neurons are called the "final common path," because communication between the various brain neurons and the voluntary skeletal muscles occurs exclusively through these neurons.

SC HAS DESCENDING (MOTOR) & ASCENDING (SENSORY) PATHWAYS

The SC white matter is divided into bundles (columns, funiculi), each containing tens of thousands of nerve fibers (axons) that traverse between the SC and the brain. These fiber bundles form the *ascending* and *descending pathways* of the SC. The ascending pathways are sensory, taking peripherally originated messages from the spinal cord to the brain; the descending pathways are motor, bringing commands from the brain to the cord for further transmission to the muscles. The fibers of the motor and sensory pathways are segregated in distinct bundles based on function. For example, fine touch, pressure, and proprioception signals ascend in the *dorsal column* pathways while pain and temperature signals ascend in the *lateral spinothalamic* pathways . The voluntary motor signals descend in the *dorsolateral pathways*, and the involuntary motor signals descend in the *ventral pathways*.

CN: Use a dark color for F.
1. Begin in the upper left corner.
2. Color the organization of the spinal cord. Color the gray matter (D) gray and leave the white matter (E) uncolored.
3. Color the organization of gray matter, repre-

sented by the left half of the spinal cord section shown below. Note that only the borders of the three zones of gray matter are colored gray. In the right half of the illustration, color the various tracts of the white matter. The entire gray matter portion of this half is colored gray.

FUNCTIONS OF SPINAL CORD.
SENSORY SIGNALS.
PATHWAYS TO.&. FROM BRAIN.
MOTOR SIGNALS.

The spinal cord (SC) is a major CNS structure that runs through the vertebral column, from the neck to lower back. It receives sensory messages from all body parts (except the head) and sends motor fibers to voluntary muscles for movements of the limbs, trunk, and neck, as well as to involuntary muscles and glands of visceral organs. Through its multitudes of sensory and motor connections with the brain, the SC mediates communication between the body and the brain. The SC also acts as an independent integrative center for involuntary (spinal) reflexes.

ORGANIZATION OF SPINAL CORD.
GRAY MATTER.
WHITE MATTER.
SENSORY NEURON.
INTERNEURON (ASSOCIATION).
MOTOR NEURON.
ASCENDING TRACT (SENSORY RELAY N.).
DESCENDING TRACT (UPPER MOTOR N.).

DORSAL ROOT (posterior)

VENTRAL ROOT (anterior)

The SC has a basic & uniform structure throughout its length. It is arranged into an inner mass of gray matter (GM) surrounded by an outer band of white matter (WM). In cross section, GM of SC is shaped like an H or butterfly. GM consists of cell bodies of neurons, their dendrites, short axons, and synapses, making it the site of neural (synaptic) analysis, integration, and transmission. The GM is connected with the dorsal and ventral roots through which the SC communicates with the periphery. The WM consists of ascending (sensory) and descending (motor) fibers (pathways) connecting the SC with the brain. The fatty myelin sheath around the fibers gives WM its name.

ORGANIZATION OF GRAY MATTER.

The gray matter (GM) of the SC is organized into dorsal and ventral horns (DH, VH). The DH carries out sensory functions and the VH, motor functions. A middle zone is involved in association functions between DH and VH of the same and opposite sides. The DH receives sensory signals, which arrive via the dorsal roots. Sensory afferents conveying various modalities (pain, touch, etc.) travel in separate nerve bundles and terminate in different laminae of the DH. DH analyzes, integrates, and transmits these signals to association and motor neurons in the SC or to relay neurons going to the brain. The VH contains cell bodies of spinal motor neurons, the fibers of which leave the SC through the ventral (motor) roots, innervating voluntary muscles. Within each VH, motor neurons are grouped in discrete nuclei, each related to a separate muscle. The middle association zone contains inhibitory and excitatory interneurons whose short axons make specific connections between the sensory and motor elements of the DH and VH of the same and other segments. These connections underlie spinal integration and spinal reflexes.

DORSAL HORN (SENSORY).
MIDDLE ZONE (ASSOCIATION).
VENTRAL HORN (MOTOR) (NUCLEI).

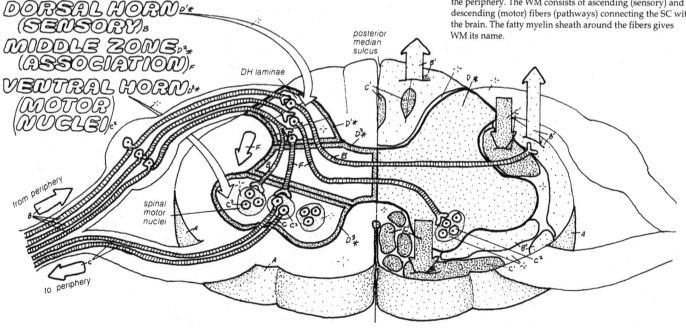

posterior median sulcus

DH laminae

from periphery

spinal motor nuclei

to periphery

ORGANIZATION OF WHITE MATTER.

The WM of SC is segregated into bundles (columns, tracts) of descending and ascending fibers (axons of large neurons). Ascending fibers are generally sensory and descending fibers are motor. Some descending fibers are for regulation of sensory input. Major ascending sensory pathways connect the SC with the medulla, brainstem reticular formation, and thalamus. Major descending tracts connect the forebrain voluntary motor areas as well as midbrain involuntary motor centers with the SC motor centers (ventral horns).

The peripheral nervous system was introduced in plate 82. Here we focus on the peripheral nerves which communicate signals between the CNS and periphery, and may be associated with the brain (*cranial nerves*) or spinal cord (*spinal nerves*). Peripheral nerves may be sensory or motor, or mixed; many of them contain *autonomic fibers*.

CRANIAL AND SPINAL NERVES

Cranial nerves are associated with the brain & perform diverse functions — The 12 pairs of cranial nerves are referred to by names or by roman numerals and emerge from different brain sites. A summary of their functions is provided here:

No.	Name	Type	Specific Function
I	Olfactory	Sensory	Afferents to olfactory bulb.
II	Optic	Sensory	Visual afferents from eye to brain.
III	Oculomotor	Somatic motor Parasympathetic	Eye movements (up, down, medial). Lens accommodation, pupil constriction.
IV	Trochlear	Motor Sensory	Eye movements (down & lateral). Afferents from muscle receptors.
V	Trigeminal	Sensory Motor	Face, teeth, nasal mucosa and mouth. Mastication (chewing).
VI	Abducens	Motor	Eye movements (lateral).
VII	Facial	Sensory Motor Parasympathetic	Pressure, proprioception from face; taste signals from anterior 2/3 of tongue. Facial expression. Stimulates salivary and tear glands.
VIII	Vestibulo-cochlear	Sensory	Hearing & balance.
IX	Glosso-pharyngeal	Motor Sensory Parasympathetic	Swallowing muscles in pharynx. Taste from posterior 1/3 of tongue; blood pressure receptors. Stimulates salivary glands.
X	Vagus	Parasympathetic motor Visceral sensory	Soft palate, pharynx, heart and digestive organs. Sensation from ear canal, diaphragm, abdominal and chest visceral organs.
XI	Accessory	Motor	Muscles of palate, pharynx, larynx, and some neck and shoulder muscles.
XII	Hypoglossal	Motor	Tongue movement.

Spinal nerves are mixed & each is associated with one vertebrae — There are 31 pairs of *spinal nerves*, each formed from the union of fibers emerging from the spinal cord *dorsal roots (sensory fibers)* and *ventral roots (motor and autonomic fibers)*. Like their corresponding vertebrae, the spinal nerves are divided into 8 cervical (neck), 12 thoracic (chest), 5 lumbar (loin, lower back), and 5 sacral (sacrum bone). There is also one coccygeal nerve. The cervical nerves innervate targets in the neck, shoulders, and arms; thoracic nerves innervate the trunk; lumbar nerves affect the legs; and sacral-coccygeal nerves supply the genitalia, pelvic, and groin areas. The largest spinal nerve, the sciatic nerve, is actually two nerves in one, and supplies the leg with both motor and sensory fibers. Some of the spinal nerves on their route to their targets form nerve plexuses: cervical plexus (C1–C5); brachial (C5–T1), and lumbosacral (T12–L4 & L4–S4).

Dermatomes & myotomes are supplied by corresponding spinal nerves — Careful nerve dissection, and examination of patients with neurological defects has revealed that each patch of the skin surface is innervated by a specific sensory nerve, with some overlap. This is best shown in the trunk where the upper trunk is supplied by thoracic nerves T2–T6 while the lower trunk by T7–T12, all in an orderly manner. These *dermatome maps* are not as orderly in the arms and legs areas due to changes in body configuration and rotation during the development of the body. In the embryo, the body is segmented into several somites, each receiving its nerve from the adjacent spinal cord segment. A dermatome is the portion of somite that becomes the skin. Similarly there are *myotomes* for muscles. The orderly correspondence of human dermatomes with spinal nerve supplies is best shown in the quadrupedal position. The dermatomes of the head are supplied in a similar fashion by the cranial nerves.

AUTONOMIC NERVES & GANGLIA

The autonomic nervous system (ANS) and the targets and functional effects of the autonomic nerves were introduced in plate 29 and in plates where specific systems and organs are controlled by the ANS. Autonomic regulation is carried out by two types of nerves: *sympathetic* and *parasympathetic*. Autonomic motor nerves regulate motility and secretion in skin, blood vessels, and visceral organs by stimulating smooth muscles and exocrine glands. This plate deals with anatomic organization, pathways and central control of autonomic nerves.

Sympathetic motor outflow via thoracic & lumbar spinal nerves — Sympathetic nerves innervate many visceral (heart, digestive organs) and peripheral (skin glands, blood vessels, and skeletal muscle arterioles) targets. Targets in the head (e.g., the iris of the eyes) receive sympathetic innervation by spinal nerves. Sympathetic nerves found within the spinal nerve trunks are usually unmyelinated, *postganglionic fibers* — their cell bodies are in the *sympathetic ganglia* chain, located on both sides of the vertebral column. The postganglionic sympathetic neurons are driven by the shorter myelinated *preganglionic sympathetic* neurons located in the spinal cord lateral horns with their axons terminating in the ganglia chain.

Sympathetic chain neurons are connected by interneurons which help in the generalized discharge characteristic of the sympathetic NS. Other sympathetic ganglia are found in the viscera, in relation to splanchnic nerves, innervating targets like the stomach and the adrenal medulla. In accord with the non-selective and diffuse function of the sympathetic NS, the sympathetic fibers innervate practically every visceral and peripheral organ in the body, particularly the blood vessels, thus controlling the blood flow in those organs.

Parasympathetic nerves emerge from the brainstem & sacral spinal cord — The parasympathetic nerves are associated with four cranial nerves: III, VII, IX, X, and with the sacral spinal nerves. A prominent parasympathetic nerve is the vagus "wanderer" (cranial nerve X), which innervates many visceral organs, including the lungs, heart, and digestive tract. The parasympathetic nerve fibers are basically preganglionic, with cell bodies in the brain stem motor nuclei and sacral spinal cord. The postganglionic neuron is short and emerges from a peripheral ganglia located near or in the target organ. Parasympathetic innervation of visceral organs is selective; profuse in heart and digestive organs but sparse in kidneys.

Hypothalamus & medulla serve in central ANS control — The ANS fibers are controlled by nerve centers in the brain stem, particularly the medulla and the hypothalamus. The medullary centers exert routine automatic control over cardiovascular, respiratory, and digestive systems. The sympathetic hypothalamic centers are involved in controlling body temperature and bodily responses to fear and excitement, fight and flight. Descending neurons from these hypothalamic and medullary centers terminate on, and stimulate the preganglionic autonomic neurons in the midbrain and spinal cord, which in turn stimulate the postganglionic neurons going to the peripheral effectors.

CN: Use dark colors for C and E.

1. Begin with the peripheral nerves. Note that the 12 cranial nerves contain various sensory, motor, and parasympathetic (autonomic) nerves. Spinal nerves contain all three, as seen in the enlarged cut nerve. Color all the peripheral nerves.

2. Color the large diagram of the spinal nerves. Begin with the anatomical portion to the left and include the directional arrows. Then color the cutaway drawing on the right side. Note that the title, autonomic efferent motor, refers to both the sympathetic and parasympathetic nerves, and is colored gray. In this large illustration, only the sympathetic system (F) is shown; the parasympathetic (G) is included in the bottom diagram.

The peripheral nervous system (PNS) consists of peripheral nervous structures and nerves that serve in the somatic and autonomic divisions. In the somatic division, sensory nerves connect the special (e.g., ear) and general (e.g., skin) sensory receptors to the spinal cord (SC) and brain, and motor nerves connect the central nervous system (CNS) to the skeletal muscles. In the autonomic division, visceral sensory fibers (inflow) and the sympathetic (S) and parasympathetic (PS) fibers (motor outflow) connect the visceral organs and effectors to the S and PS ganglia as well as to the SC and brain. Fibers within a nerve trunk vary in size and speed of conduction.

SPINAL NERVES (31 PAIR)ₐ
DORSAL ROOT & GANGLIONʙ
SOMATIC & VISCERAL AFFERENT SENSORY N.c
VENTRAL ROOT,
SOMATIC, EFFERENT MOTOR NERVE.ₑ
AUTONOMIC EFFERENT MOTOR:*
SYMPATHETIC: PREGANGLIONIC ▭ F
POSTGANGLIONIC ◇◇◇◇ F'
SYMPATHETIC CHAIN,F² GANGLION F³
PARASYMPATHETIC: PREGANG. ▭ G
POSTGANGLIONIC ◇◇◇ G' GANGLION G²

Of the total of 43 pairs of peripheral nerves, 12 are associated with the brain (cranial nerves) and 31 with the SC (spinal nerves). Cranial nerves emerge directly from the brain, but the spinal nerves form by the merger of the dorsal and ventral roots of the SC. The cranial nerves are identified by names or roman numerals; spinal nerves by name and number of corresponding vertebrae. Some cranial nerves are purely sensory, others are motor or mixed. Some contain partly parasympathetic (autonomic) fibers; others are largely parasympathetic. Spinal nerves are generally mixed, containing sensory, motor, and autonomic fibers.

PARASYMPATHETIC G
SENSORY c
MOTOR ₑ

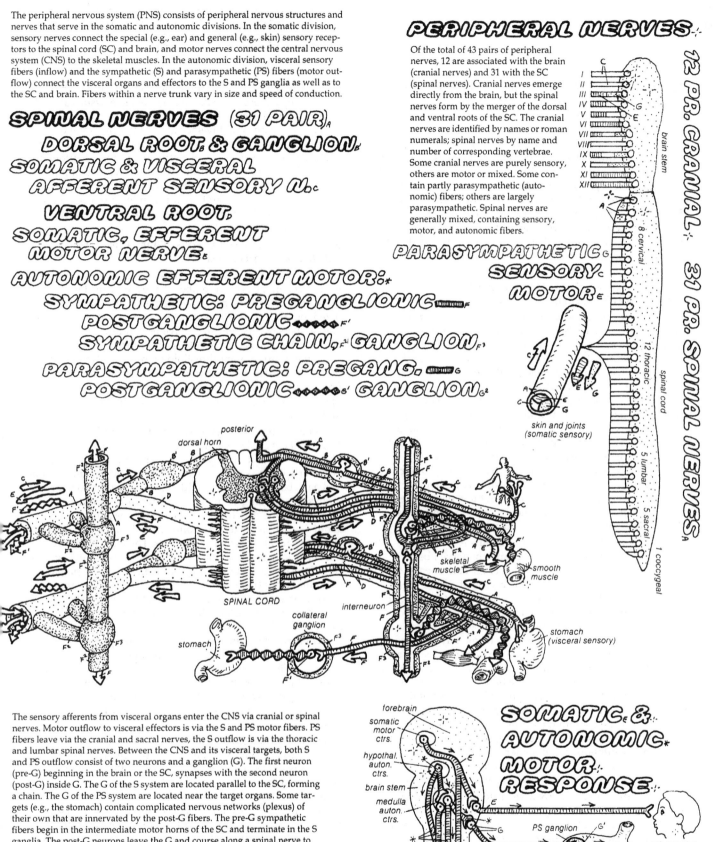

posterior
dorsal horn
SPINAL CORD
collateral ganglion
interneuron
stomach
stomach (visceral sensory)
skeletal muscle
smooth muscle

skin and joints (somatic sensory)

brain stem
8 cervical
12 thoracic
spinal cord
5 lumbar
5 sacral
1 coccygeal

The sensory afferents from visceral organs enter the CNS via cranial or spinal nerves. Motor outflow to visceral effectors is via the S and PS motor fibers. PS fibers leave via the cranial and sacral nerves, the S outflow is via the thoracic and lumbar spinal nerves. Between the CNS and its visceral targets, both S and PS outflow consist of two neurons and a ganglion (G). The first neuron (pre-G) beginning in the brain or the SC, synapses with the second neuron (post-G) inside G. The G of the S system are located parallel to the SC, forming a chain. The G of the PS system are located near the target organs. Some targets (e.g., the stomach) contain complicated nervous networks (plexus) of their own that are innervated by the post-G fibers. The pre-G sympathetic fibers begin in the intermediate motor horns of the SC and terminate in the S ganglia. The post-G neurons leave the G and course along a spinal nerve to serve blood vessels and sweat glands of the somatic area served by that spinal nerve. Other post-G fibers in the S system leave the G to reach their target via an independent visceral nerve. The neurons in the S ganglia are connected by interneurons, allowing for simultaneous and generalized discharge from several S ganglia, even when only one G is activated from the brain or periphery In contrast, the proximity of PS ganglia with their targets and lack of inter-G connections allow for discrete activation of specific targets by the PS system. Both S and PS pre-G neurons within the brain or SC are controlled by descending fibers from higher centers in the hypothalamus and medulla, enabling the hypothalamus and medullary centers to exert finite control over internal bodily functions (digestion, blood flow, etc.). Also, via its connections with the limbic system, the hypothalamus controls internal bodily responses during arousal and emotional states.

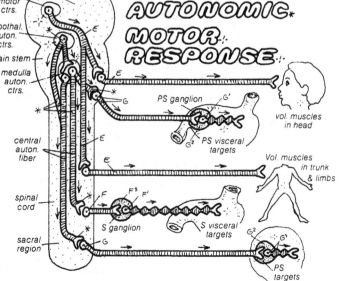

SOMATIC & AUTONOMIC ₑ
MOTOR RESPONSE *

forebrain
somatic motor ctrs.
hypothal. auton. ctrs.
brain stem
medulla auton. ctrs.
central auton. fiber
spinal cord
sacral region

PS ganglion
PS visceral targets
S ganglion
S visceral targets
PS targets

vol. muscles in head
Vol. muscles in trunk & limbs

The *peripheral nerves* (PNs) that connect the sensory receptor systems with the CNS and the CNS with the motor effectors such as muscles and glands are a critical part of the operation of the *peripheral nervous system* (plate 82). Here we describe the structure and functional properties of the PNs and their constituent nerve fibers. The targets and functional aspects of the *spinal* and *cranial nerves* as well as some aspects of the central and peripheral organization of the *autonomic nervous system* will be presented in the next plate (86). A discussion of the physiology of the autonomic nervous system and the peripheral distribution of its nerves may be found in plate 29. Detailed study of peripheral nerves and their exact distribution and targets in the body is the subject of peripheral neurology and human anatomy.

PERIPHERAL NERVES SHOW DIVERSE ANATOMIC & FUNCTIONAL PROPERTIES

The PNs form a diverse group of cord-like nerve tissue running between the CNS and the peripheral sensors and effectors, conveying various motor, sensory, and autonomic signals. They often form a distinct nerve trunk such as the sciatic nerve, which supplies the leg with motor and sensory fibers, or the optic nerve, which takes photic stimuli to the brain, or the vagus nerve, which innervates the visceral organs. The sizes of nerve trunks vary markedly; the largest is the sciatic nerve, with a diameter of nearly 1 cm (it is actually two nerves in one). Most other nerves are medium sized; still others, like some of the autonomic nerves, are nearly thread-like. Occasionally, several PN trunks merge to form a nerve plexus, such as the brachial plexus in the shoulder area or the sacral plexus in the pelvic area; in these plexi some mixing of fibers may occur among the nerve trunks and a new set of nerves may emerge from the plexi.

PNs may be sensory, motor, or mixed — A typical PN trunk consists of hundreds to thousands of nerve fibers grouped in discrete bundles called *fascicles*, each sheathed by connective tissue coats. Each of the fascicle of a PN is functionally distinct either in terms of the sensory or motor function or in terms of its target, such as the arm, hand, or particular skin zones it innervates. Since each PN consists of many fascicles, the whole of the nerve trunk may be sensory, motor, or mixed. Purely sensory or purely motor nerves are rare. For example, the *acoustic nerve* is a mainly sensory nerve that conveys auditory signals to the brain stem; the *trochlear nerve* is a mainly motor nerve responsible for some of the eye movements. Most PNs, such as the *trigeminal* or *sciatic nerves*, are mixed nerves, containing sensory, motor, and autonomic fibers. Many of the small-diameter PNs are purely autonomic, some contain autonomic motor efferents, and others have both efferents and visceral sensory afferents — an example is the vagus nerve, which contains many visceral afferents.

Internal divisions of a PN & the roles of epineurium, perineurium, and endoneurium — The structural organization of a PN is evident in its microscopic cross section. Such a view shows that each PN trunk has a tough and elastic connective tissue coat called *epineurium*, which protects it from various external forces and pressures. Within this outer coat several fascicles are found, each sheathed by a thinner and semi-tough coat called *perineurium*, which helps bundle numerous small *nerve fibers* together. The numerous nerve

fibers within each fascicle are of varying sizes but each is surrounded by a loose, soft sheath of connective tissue called the *endoneurium*. The endoneurium should not be confused with the *myelin* sheath, a highly specialized fatty membrane around the axon, made by the Schwann cells. Each nerve fiber inside a fascicle is the axon of a sensory, motor, or autonomic neuron. The three neural protective coats — epineurium, perineurium and endoneurium — are of different thickness and elastic properties, but they form an interconnected protective fabric within which the nerve fibers exist. Many PNs also contain blood vessels.

PERIPHERAL NERVE FIBERS FORM THREE TYPES BASED ON SIZE & CONDUCTION VELOCITY

The nerve fibers within each fascicle are of varying diameters. In general, large nerve fibers have a myelin sheath (myelinated fibers); some of the smaller fibers may be myelinated, but the smallest lack myelin (unmyelinated fibers). Nerve fibers of different sizes have been found to show various functional (electrophysiological) properties, expressed usually in terms of their threshold of excitation and conduction velocity of action potentials. In this regard, the nerve fibers have been classified into three general types, called *Types A, B and C*, with four subtypes for the Type A fibers (α, β, γ, and δ).

Type-A fibers are myelinated and fast conducting — Type-A fibers are large-diameter (up to 20 μm), myelinated fibers that conduct rapidly (up to 120 m/sec, nearly 250 miles per hour!). Type-A fibers show low thresholds of stimulation and have been further subdivided into α, β, γ, and δ subtypes with correspondingly smaller diameters and conduction velocities. Type Aδ fibers, the smallest of the A group, are about 5 μm in diameter and conduct at about 20 m/s. They conduct pain and thermal sensations. The largest type-A fibers (Aα) are found in the motor nerves supplying fast muscles of the hands and eyes. Type-B fibers — e.g., the preganglionic autonomic fibers — are medium-sized (1–3 μm) and conduct at about 10 m/s. Type-C fibers are the smallest fiber type (<1 μm in diameter), showing very high thresholds of excitation and slow speeds (1 m/sec). They are unmyelinated and comprise the pain and crude touch group of sensory fibers as well the postganglionic sympathetic fibers. The olfactory nerve fibers conducting afferent odor signals from the nose to the brain are the smallest (0.2 μm) and slowest (0.5 m/sec) nerve fibers known in the nervous systems.

Compound action potential of a PN reflects its various fiber types — The familiar features of single-spike action potential recorded from a single nerve fiber have been discussed before (plate 15). When action potentials are recorded from a mixed nerve like the sciatic nerve, often a complex pattern called a *compound action potential* is observed, which reflects the excitation of the various types of fibers in it. With weak stimuli, only a single peak is observed (A wave) which corresponds to the excitation of low-threshold large-diameter myelinated fibers (A type). Increasing the stimulus intensity gradually reveals correspondingly smaller peaks (B and C waves) related to the activation of smaller-diameter and higher-threshold fibers (types B and C). Under optimum recording conditions, even waves relating to the α, β, γ, and δ subtypes of the A fibers may be found (α, β, γ, and δ waves).

CN: Use red for J, blue for I.
1. Color the coats of the PN trunk at the top of the page. For purposes of coloring, the thickness of the coats has been exaggerated and far fewer vesicles and nerve fibers are shown than normally exist.

2. Color the types of nerve fibers. Note that in the lower illustration fibers representing the three size categories receive different colors whereas in the upper drawing, for simplification, all the nerve fibers receive the same color (D).

STRUCTURE OF A PERIPHERAL NERVE (PN):-

Peripheral nerves (PN) are visible cords of nerve tissue containing thousands of nerve fibers (NF) that form fascicles. PN act as communication cables of wires, carrying signals between the central nervous system and peripheral sensors and effectors. PN can be part of spinal, cranial or autonomic groups. Large mixed PN trunks contain sensory (afferent), motor (efferent) and autonomic (usually efferent) nerve fibers (NF). Individual fascicles are usually functionally distinct, containing either sensory or motor NF.

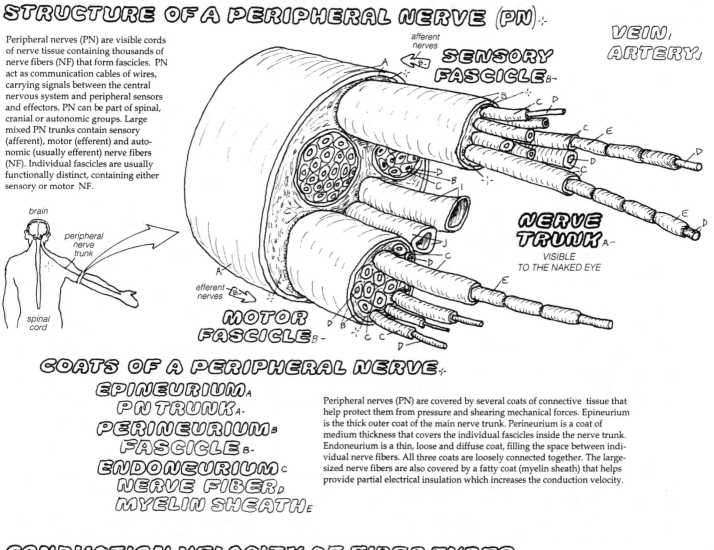

afferent nerves

SENSORY FASCICLE B-

VEIN, ARTERY.

NERVE TRUNK A-

VISIBLE TO THE NAKED EYE

efferent nerves

MOTOR FASCICLE B-

COATS OF A PERIPHERAL NERVE:-

EPINEURIUM A
PN TRUNK A-
PERINEURIUM B
FASCICLE B-
ENDONEURIUM C
NERVE FIBER,
MYELIN SHEATH E

Peripheral nerves (PN) are covered by several coats of connective tissue that help protect them from pressure and shearing mechanical forces. Epineurium is the thick outer coat of the main nerve trunk. Perineurium is a coat of medium thickness that covers the individual fascicles inside the nerve trunk. Endoneurium is a thin, loose and diffuse coat, filling the space between individual nerve fibers. All three coats are loosely connected together. The large-sized nerve fibers are also covered by a fatty coat (myelin sheath) that helps provide partial electrical insulation which increases the conduction velocity.

CONDUCTION VELOCITY OF FIBER TYPES:-

TYPES OF NERVE FIBERS D-
LARGEST (FASTEST) F
$A\alpha$ TYPE F (diameter up to 20 μm: velocity up to 120 m/sec)
$A\beta$ TYPE F' (10 μm average; up to 70 m/sec)
$A\gamma$ TYPE F² (5 μm average; up to 30 m/sec)
$A\delta$ TYPE F³ (2–4 μm; up to 20 m/sec)
MEDIUM SIZE G
TYPE B G (1–3 μm; up to 10 m/sec)
SMALLEST (SLOWEST) H
TYPE C H (< 1 μm; 0.5–2 m/sec)

The diameter of nerve fibers (NF) in a mixed, large nerve trunk occur in large sizes (type A- with subtypes $A\alpha$ >$A\beta$ >$A\gamma$ >$A\delta$), intermediate sizes (type B), and small (type C) sizes. Large NF (motor fibers to fast muscles, optic fibers) are generally myelinated and fastest in conduction velocity. Type C fibers (pain, autonomic) are the smallest and slowest. If a large mixed PN is stimulated at one point with a strong stimulus (high intensity), a compound action potential (CAP) is obtained at a distant recording site. The CAP's highest peak (peak A in drawing) occurs very rapidly, corresponding to the action potential (AP) of the largest NF. Peaks $A\alpha$-$A\delta$, B and C, occur at increasingly later times, correspond to the respective smaller sized fibers which have correspondingly higher thresholds and lower speeds. If stimulus intensity is weak, only the A-α peak of the low threshold, large fibers is seen (graph on left).

COMPOUND ACTION POTENTIAL OF A PN:-

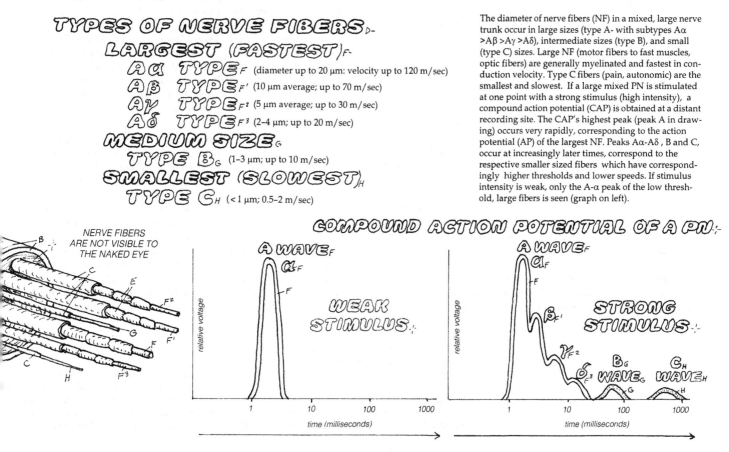

NERVE FIBERS ARE NOT VISIBLE TO THE NAKED EYE

A WAVE F
α F
F

WEAK STIMULUS.

relative voltage

time (milliseconds)

A WAVE F
α F
F
β G

STRONG STIMULUS.

γ F²
δ F³
B G **WAVE** G
C H **WAVE** H

relative voltage

time (milliseconds)

Although neurons are the excitable cells of the nervous tissue, they are not the true units of the *nervous system function*. Single neurons generate and conduct action potentials and release neurotransmitters, but by themselves are unable to carry out typical nervous system functions, such as simple reflexes or complex thought processes.

NEURAL CIRCUITS & SYNAPSES ENABLE INTEGRATION IN CNS

CNS functions are carried out by neuronal circuits and networks, the true functional units of the nervous system. A neural circuit may have two or thousands of neurons that interact with one another through *excitatory* and *inhibitory* synapses. These synapses, by providing controllable interneuronal functional connections, are responsible for the CNS *integrative* functions. Trillions of synapses are believed to exist in the human brain. Without these, reflexes would not operate, communication between the periphery and CNS as well as within the CNS would cease, and the brain's integrative operations would not take place.

CNS SYNAPSES ARE EITHER EXCITATORY OR INHIBITORY

To view synaptic interaction in a nerve circuit, consider a spinal motor neuron. To make a muscle fiber contract, this motor neuron must be excited to its threshold level; it will then discharge nerve impulses along its axon to excite its target muscle fiber. To prevent a muscle from contracting or to relax it, the motor neuron must be suppressed (inhibited).

The cell body and dendrites of this motor neuron are the site of thousands of input synapses made by the endings of sensory neurons, interneurons, or descending motor neurons originating in the brain. Some of these synapses are excitatory (E), others are inhibitory (I). The E and I synapses may occur side by side on a postsynaptic neuron. Although a neuron may receive both E and I input synapses from different presynaptic neurons, it can make only one type of output synapse, either the E or the I type, since its output terminals are all E or I.

NEURONS ARE EITHER EXCITATORY OR INHIBITORY

Neurons that provide E-type output synapses are called *excitatory* (*E-neurons*). An E-neuron excites its post-synaptic neuron or target cell, inducing it to become functional (e.g., fires action potentials). Of the E-type are all motor neurons as well as the somatic sensory neurons that connect the periphery to the CNS and the majority of the large neurons (macroneurons) that communicate between the major parts of the CNS, including the descending motor nerves from the brain. The I-type synapses, which are critical for central synaptic integration, are often made by the small *inhibitory neurons* (also called short-axon neurons, interneurons, microneurons).

Thus, if a sensory fiber from the periphery or a descending motor fiber from the brain needs to inhibit a spinal motor neuron, it must first excite the I-type interneurons, which in turn will inhibit the motor neuron via their I output synapses. In the adult CNS, all the E and I terminals to a motor neuron are permanently in place; only the pattern of nervous activity — i.e., the degree of sensory or descending motor input — determines which terminals (E or I) will be used. Of course use or disuse and learning can alter the properties of neurons and

CNS synapses by changing their efficacy, but will not change a neuron or a synapse from one type to another.

SYNAPTIC POTENTIALS DETERMINE NEURONAL FUNCTION

Activation of each synaptic terminal produces a slow, weak, and graded *synaptic potential*. These are divided into two types — *excitatory postsynaptic potential* (EPSP), which occurs in the E-type synapse, and *inhibitory postsynaptic potential* (IPSP), which occurs in the I-type synapse. (For electrical and ionic aspects of EPSP and IPSP, see captions here and plates 19, 20). If many synaptic terminals impinge on a single neuron, the EPSPs and/or IPSPs can add up algebraically, giving rise to the phenomenon of *summation*, discussed below.

Summation of synaptic potentials — In a large synapse with many synaptic vesicles — e.g., the neuromuscular junction — a single action potential causes the release of enough neurotransmitter (acetylcholine) to cause a large motor end-plate potential (a strong type of EPSP) that often results in a muscle twitch (plate 20). In the central synapses, however, the energy of a single EPSP or IPSP usually is not sufficient to activate the postsynaptic neuron. For that to occur, the excitation or inhibition level at the postsynaptic surface must increase. The algebraic accumulation of synaptic potentials at the receptive surface of a postsynaptic neuron is called *synaptic summation*. When the E-type synapses are more active than the I-type synapses, excitation will prevail; in the opposite case, inhibition will dominate in the postsynaptic neuron. If the two types of synapses are equally active, their effects cancel out. Summation of synaptic interaction is a major determinant of neuronal integration. Thus the balance of the excitatory and inhibitory input to a postsynaptic neuron determines if its axon will fire nerve impulses.

Spatial and temporal summation — There are two types of synaptic summation, *spatial* and *temporal*. Spatial summation occurs when the presynaptic input is summed across the different synaptic sites, impinging on the same postsynaptic neuron. Spatial summation can occur with both E and I types of synapses and the presynaptic terminals may belong to one or more neurons. Temporal summation — i.e., accumulation of individual synaptic potentials in time — involves a single synapse. Here an increase in the *frequency* of discharge (impulses/unit time) enhances the effectiveness of the synapse. In the E synapse, summation increases the probability of postsynaptic neuron discharge; in the I synapse, it decreases discharge probability.

Convergence, divergence, & recruitment — Opportunities for spatial summation in a neuronal circuit are created by increasing the number of synaptic terminals of a single presynaptic neuron to the same postsynaptic neuron. This is called *convergence* and an example is seen in the central terminal branching of the sensory afferents. Alternatively, the number of active presynaptic neurons firing on the receiving neuron can be increased. This called *recruitment. Convergence* of several neurons of the same type on a single postsynaptic neuron is another device to create opportunities for spatial and temporal summation. Divergence is the way a neuron uses its multiple branches to excite several postsynaptic neurons. This may not always involve synaptic summation.

CN: Use very light colors for A and B.
1. Begin with the excitatory neuron (A) and color its fiber from the brain down the spinal cord in the upper left drawing. Then color its enlargement along with an inhibitory neuron (B). The afferent (sensory) and efferent (motor) fibers are colored (A) since they are also excitatory. Color the typical excitatory and inhibitory synaptic ending of each neuron in the large illustration.
2. Color convergence and divergence next.
3. Color the spatial and temporal summation with the accompanying synaptic potential chart. In both cases, summations are shown for excitatory potentials (also occurs in inhibitory systems).

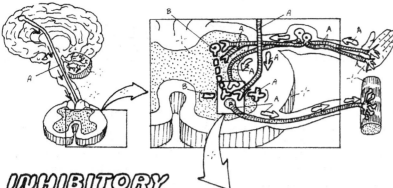

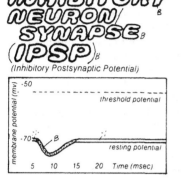

Neurons interact with other neurons through excitatory (E) and inhibitory (I) connections (synapses). E synapses activate and I synapses deactivate (inhibit) the postsynaptic neuron. The action of a neuron (exerted by its output synapses) is of either E or I type. However, each neuron can receive both E- and I-type synaptic input from other neurons. The large projection neurons are generally of the E type. The I neurons usually are small in size (microneurons) and have short axons. They often function as *interneurons* because they provide local inhibitory connections between the large E neurons.

INHIBITORY NEURON SYNAPSE (IPSP)
(Inhibitory Postsynaptic Potential)

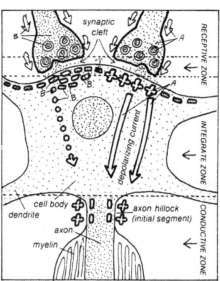

The release of the I-type neurotransmitter at the I-type synapse causes inflow of Cl^- or outflow of K^+, creating a localized *hyperpolarization* of the postsynaptic membrane (i.e., an IPSP). The increased negative charges prevent flow of depolarizing current toward the postsynaptic cell axon hillock, decreasing the probability of axon discharge.

EXCITATORY NEURON SYNAPSE (EPSP)
(Excitatory Postsynaptic Potential)

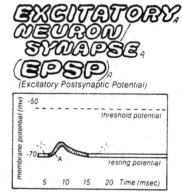

The release of E-type neurotransmitter at the E synapse causes inflow of positively charged ions (Na^+), creating localized *depolarization* (i.e., EPSP) of the postsynaptic membrane. Current flows from this area to the resting (polarized) initial segment of the axon. The strength of this current depends on the strength of EPSPs. The axon will discharge nerve impulses if the depolarized current is at or above the axon firing threshold.

CONVERGENCE

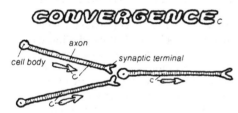

Convergence (C) occurs when a neuron receives synaptic input from several other neurons. Divergence (D) occurs when the synaptic output from one neuron is distributed to more than one neuron (by branching of the axon). C and D may occur in both E and I neurons. C and D are fundamental in the physiology of neuronal circuits because they provide the structural basis of such important synaptic phenomena as spatial summation, facilitation, and occlusion.

DIVERGENCE

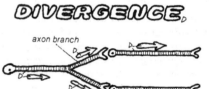

SUMMATION OF SYNAPTIC POTENTIALS
SPATIAL

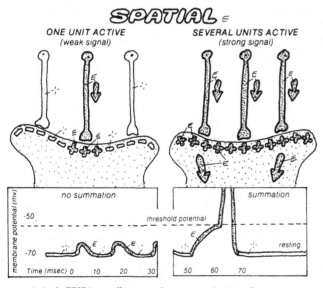

A single EPSP is usually too weak to cause activation of a neuron. One solution is to increase the number of input sites (i.e., activation of more presynaptic units). The EPSPs from all the active sites of the postsynaptic neuron now add up (spatial summation), creating the threshold current for axon discharge. Spatial summation occurs in inhibitory synapses as well, with the opposite effect.

TEMPORAL

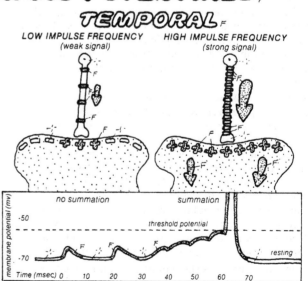

Another way to increase the strength of signals at the synapses is through summation of the synaptic potentials in "time"—i.e., temporal summation—which occurs by increasing impulse frequency at the same synapse. In the E synapse, successive input impulses cause a "rapid" buildup of positive charges, exciting postsynaptic neurons. In the I-type synapse the opposite effect is observed. Temporal and spatial summation can occur together.

Synapses are the mechanisms by which *central neurons* communicate with their peripheral targets and with one another. The importance of synapses in brain function is extremely critical and cannot be understated. We have already studied the basic structure, chemistry, and function of a typical synapse and the operation of inhibitory and excitatory synapses in previous plates (19, 87). However, the CNS synapses are anything but typical, presenting a wide diversity of structure, chemistry, and function.

NUMBERS & STRUCTURAL TYPES OF CNS SYNAPSES

The human brain has 10 billion neurons & 10 trillion synapses — The number of synapses in the brain is enormous. Assuming that the brain has 10 billion neurons and that each neuron on the average makes 1000 synapses onto other neurons, nearly 10 trillion synapses exist in the brain. The ratio of synapse number to neuron number differs widely among brain regions — 50 thousand to one in the human forebrain.

Most central synapses are of the axo-dendritic types — Most central synapses occur between axon terminals (*synaptic knobs*) of one neuron and the dendrites of another neuron (*axodendritic synapse*). Many axodendritic synapses occur on spines of dendrites. Synapses also occur between axon terminals and cell bodies (soma) of neurons (*axo-somatic*). A spinal motor neuron is the target of nearly 10 thousand synaptic knobs, of which 80% are on its dendrites and 20% on its soma. This ratio is higher in the brain pyramidal cells, where more than 95% of synapses are on dendrites.

Axo-axonal & dendro-dendritic synapses occur but are relatively rare — Rarer types of synapses occur between axon terminals of one neuron and those of other neurons (*axo-axonal synapse*); these serve in *presynaptic inhibition*. Similarly, *dendro-dendritic synapses* occur; some are *reciprocal*, associated with inhibitory neurons lacking axons; they occur in the retina and olfactory bulb and function in *local inhibitory circuits*.

CENTRAL SYNAPSES COMPRISE DIFFERENT CHEMICAL TYPES

The notion of chemical transmission between neurons and the acetylcholine and norepinephrine synapses operating in the neuromuscular junction and autonomic synapses has been introduced before (plates 19, 29). In the CNS, chemical synapses occur in a wide variety, particularly in terms of the neurotranmistter type and the postsynaptic receptor mechanism.

Acetylcholine & biogenic amine synapses regulate emotions, sleep, arousal, & higher brain functions — *Acetylcholine* synapses occur in some spinal cord and brain synapses and participate in a number of central circuits involved in regulation of sleep, wakefulness, memory, learning, and motor coordination. The biogenic amine synapses (*dopamine, norepinephrine, epinephrine, serotonin,* and *histamine*) take part in spinal cord and brain circuits. In the brain, serotonin synapses are involved in regulation of moods, appetite, pain, and pleasure. Dopamine synapses participate in diverse pathways involved in motor coordination (basal ganglia), central inhibition (olfactory bulb), cognition, and personality (frontal lobes).

Amino acid synapses show inhibitory and excitatory types — *Glutamate* and *aspartate,* the neurotransmitters in excitatory CNS synapses, are found in many brain regions, often in major projection neurons. The postsynaptic mechanisms of glutamate synapses are complex, involving calcium and other second messengers, and provide the cellular and molecular basis of some plasticity- and learning-related neuronal functions. *GABA* (*γ-amino-butyric acid*) and *glycine* operate in inhibitory CNS synapses. The majority of inhibitory brain synapses use GABA as the transmitter.

Numerous neuropeptides function as synaptic cotransmitters and neuromodulators — Over 80 *peptide neurotransmitters* are known; a few act as main transmitter (e.g., substance-P); most act as *cotransmitters* and modulators of the main neurotransmitters. *Neuromodulators* help regulate the effectiveness (efficacy) of synapses, changing their excitability levels and rapidity of functional responses. *Substance-P* and *β-endorphin* are the neurotransmitters in slow pain and central pain inhibition pathways; many brain peptide neuromodulators work as hormones elsewhere in the body — e.g., *oxytocin, TRH, gastrin, secretin, angiotensin.*

Nitric oxide & carbon monoxide are gases but have neurotransmitter function as well — Recently two gaseous compounds, *nitric oxide* (NO) and *carbon monoxide* (CO), to act as neurotransmitters in parts and functions of the brain. NO also acts as a smooth-muscle relaxing factor in the blood vessels, intestine, and penis (plate 151). NO is synthesized by the enzyme *nitric oxide synthase* from the amino acid arginine. NO and CO of the brain freely diffuse between neurons and synapses and activate the *guanylyl cyclase.*

FAST SYNAPSES ARE IDEAL FOR RELAY, AND SLOW SYNAPSES FOR INTEGRATION

The cholinergic neuromuscular synapse, where the postsynaptic receptor is also an ion channel (the nicotinic cholinergic receptor), is typical of *fast synapses.* Fast synapses are well suited for relay but not for integration of responses, an essential requirement in the CNS. Many CNS synapses are of the *slow type*, with a *long time delay* between the arrival of a presynaptic impulse and a postsynaptic response. Slow synapses involve neuromodulators and neuropeptides and act in mediation of intracellular second messengers; here the postsynaptic receptor is not an ion channel. Receptor binding activates G-proteins and enzymes (e.g., adenylate cyclase) in the postsynaptic membrane, leading to the release of second messengers (e.g., cyclic AMP), which then activate the ion channel (e.g., cAMP gated channel), allowing ion flow and electrical responses. These events take longer, accounting for the slow response. But they also allow for integration of the response and its modulation by other pre- and postsynaptic factors. Many synapses that work in arousal, attention, and neural plasticity (learning and memory) rely on slow synapses.

Many brain disorders are associated with loss or malfunction of central synapses — *Alzheimer's disease,* involving loss of memory and cognitive disorders, is associated with loss of cholinergic synapses in hippocampus and basal forebrain, areas important in cognition and memory. *Parkinson's disease* involves loss of dopamine synapses in the basal ganglia, resulting in marked motor disorders. *Schizophrenia* is associated with hyperactivity of dopamine synapses in frontal lobes, resulting in mental and cognitive dysfunction. *Epilepsy,* a brain disorder involving mild and severe brain seizures, is in part associated with decreased function of GABA synapses. Many drugs aimed to ameliorate brain disorders work on central synapses, particularly on receptor mechanisms.

CN: Use light colors for B, D, E and F.
1. Color the synapse locations. Completely color all the cell bodies, axons and dedrites.
2. Color the names of the chemical neurotransmitters.
3. Color the examples of fast and slow synapses, following the numbered sequence.

TYPES OF CNS SYNAPSES

LOCATION
1. AXO-DENDRITIC c
2. AXO-SOMATIC a
3. AXO-AXONIC b
4. DENDRO-DENDRITIC c

CNS neurons communicate with one another via numerous and diverse types of synapses which differ in structure, location, neurotransmitter chemistry and function. A CNS neuron may receive thousands of synapses from other neurons and make hundreds of synapses upon other neurons.

Synapses between axons of one neuron and dendrites of another (axo-dendritic) (1) are the most common; followed by axo-somatic synapses (2) which terminate on the cell body. Rarer types include axo-axonic synapses (between axon terminals of one neuron and the presynaptic terminal of another neuron (3), that serve in presynaptic inhibition; and dendrodendritic synapses (4) that serve in local inhibitory feedback circuits. This type may be reciprocal (i.e., both-ways).

CELL BODY a
AXON b
DENDRITE c

CHEMICAL NEUROTRANSMITTER d

NF: stands for Normal Functions
AF: stands for Abnormal Functions
(drugs used in treating abnormal functions are shown in parenthesis)

SEROTONIN d

NF: moods, sex, appetite
AF: depression; eating disorders
(anti-depressant drugs: prozac; halucinogens: LSD)

ACETYLCHOLINE d

$H_3C - C - O - CH_2 - CH_2 - \overset{+}{N} - (CH_3)_3$

NF: motor, sleep, memory
AF: Alzheimer disease;
(cholinergic drugs)

DOPAMINE d

$HO - - CH_2 - CH_2 - NH_2$ (HO)

NF: motor function; personality, thoughts
AF: Parkinson's disease (decrease in dopamine) [L-dopa]
AF: schizophrenia (increase in dopamine) [dopamine blockers]

GABA d
gamma-aminobutyric acid

$H_2N - CH_2 - CH_2 - CH_2 - COOH$

NF: central inhibition
AF: epilepsy; anxiety
(barbiturates; valium)

GLUTAMATE d

$H_2N - CH - CH_2 - CH_2 - COOH$
$ | $
$ COOH$

NF: excitatory transmission; motor & sensory relay
AF: related deficits

PEPTIDES d

over 50 known peptide neurotransmitters
examples include:
beta endorphin — NF: pain inhibition; reward, pleasure
substance-p — NF: slow pain relay
neuropeptide Y — NF: sex, appetite
angiotensin — NF: thirst, drinking behavior

Many different neurotransmitters are found in the variety of CNS synapses. Some are small organic molecules (e.g., amines and amino acids), others are large peptides. Some serve in short axon interneurons, others in long axon projection neurons. Acetylcholine synapses occur in pathways related to sleep and memory and motor coordination; ACh dysfunction is involved in Alzheimer's disease. Biogenic amines (serotonin, norepinephrine, dopamine) pathways function in arousal, sleep, emotions and moods; disorders of serotonin and dopamine occur in depression, schizophrenia and Parkinson's disease. Glutamate and aspartate synapses function in excitatory transmission and learning and memory while GABA functions in inhibitory synapses (sedation). Synapses releasing various peptide neurotransmitters occur in pain and pleasure (endorphins, substance P), in sex functions (GnRH), in feeding (neuropeptide Y) and in drinking (angiotensin) behaviors. Peptides and biogenic amines often act as neuromodulators in complex, slow adaptive synapses.

FUNCTION
PRESYNAPTIC MEMBRANE e
NEUROTRANSMITTER d
POSTSYNAPTIC MEMBRANE f
MEMBRANE RECEPTOR g
ION CHANNEL h
G PROTEIN j
EFFECTOR PROTEIN j

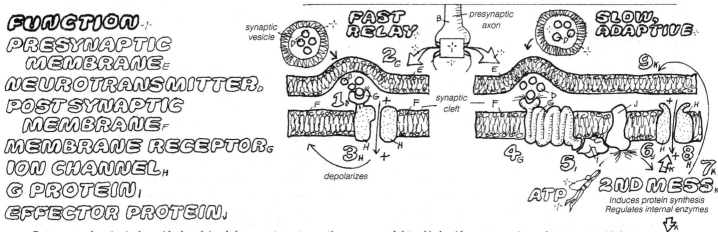

Fast synapses function in the rapid relay of signals from pre- to post-synaptic neurons. A neurotransmitter (e.g., acetylcholine in nicotinic cholinergic synapses) is released from presynaptic vesicles (1) to diffuse across the synaptic cleft and bind to a receptor protein, which is also an ion channel (2). Binding of transmitter to receptor opens the pores in ion channels; in excitatory synapses, positive ions move in, depolarizing and activating the postsynaptic neurons (3). Fast synapses may also be inhibitory. Slow adaptive synapses serve in functions requiring alteration in synaptic properties (e.g., in attention, arousal, learning, and memory). Aa neurotransmitter acting as a neuromodulator, binds with a postsynaptic membrane receptor (4) that is *not* an ion channel. Binding activates an adjacent regulatory G-protein (5), which activates an effector protein — e.g., adenylate cyclase, (6) that increases second messengers (SMs) — e.g., cAMP, calcium — in the postsynaptic cell (7). SMs perform several functions: they open (or close) ion channels (8), allowing ion movements that depolarize or hyperpolarize postsynaptic neurons. SMs also temporarily or permanently alter properties of other postsynaptic or presynaptic membrane proteins (9). These effects can be exerted directly or through changes in the synthesis of new proteins in postsynaptic cells.

The functions of the nervous system are divided into three categories—sensory, motor, and integration. Sensory receptors are a vital part of the sensory mechanisms of the NS.

SENSORY RECEPTORS TRANSDUCE EXTERNAL & INTERNAL STIMULI

The *sensory receptors* are highly specialized nerve cells, as in retinal rods and cones, or in modified dendrites, or in the skin's receptor organs, by which the nervous system detects the presence of and changes in the different *forms of energy* in the external and internal environments. The sensory receptors convert these various forms of energy into a common, unitary language of the nervous system—i.e., *nerve signals (action potentials)*, which are then communicated to the CNS. Each sensory receptor is equipped with parts that confer its ability to *detect* the stimulus and to *transduce* (convert) the physical energy into nerve signals.

For example, the skin's Pacinian corpuscle is sensitive to indentation in the skin, usually produced by a hard touch or pressure stimulus, which is detected by the corpuscle's fibrous capsule and transmitted as waves of mechanical deformation to the nerve ending segment in the corpuscle core. Nerve endings transduce pressure stimuli into waves of electrical depolarization—i.e., *graded receptor potential*—which acts as a generator potential and activates the adjoining axon segments or node of Ranvier of the sensory nerve fiber, producing action potentials for transmission to the CNS.

EACH SENSORY RECEPTOR TYPE DETECTS A SPECIFIC STIMULUS

The physical world around us contains numerous forms of energy, not all of which we are able to detect. The detectable ones are classified into the categories of *mechanical, chemical, thermal,* and *photic* (light) energies. The body may have one or more types of sensory receptors to detect any one kind of these energy forms. Some receptors, like those in the skin, are modified dendrites. In the eye's retinal photoreceptor neurons, much of the cell is modified for detection and transduction of light rays. Some receptor cells (such as the skin and smell receptors) act as independent single sensory units. In other cases, the receptors are housed as organized masses of cells within a sensory organ (such as the eye's retina). The structural integrity of the retina as a whole is essential for form and movement perception.

Types of sensory receptors—Depending on the energy form to which they respond, sensory receptors are classified as *mechanoreceptors, chemoreceptors, thermoreceptors, nociceptors,* and *photoreceptors.* Most of these are described further in this book in the appropriate plates along with a description of the sense they serve. Here a general description of the different functional categories will be given.

MECHANORECEPTORS DETECT DISPLACEMENT, STRETCH, & SOUND

The *mechanoreceptors* make up the most diverse group of sensory receptors. Found in the skin, muscles, joints, and visceral organs, they are sensitive to mechanical deformation of the tissue and cell membranes. This deformation can arise in various ways: indentation, stretch, and hair movement. *Skin receptors* include the largest variety of mechanoreceptors. Many of the sensory nerve endings encapsulated in a fibrous (connective tissue) covering are believed to be mechanoreceptors. Specialized stretch-sensitive proteins are found in the plasma membranes of mechanoreceptor elements.

Types of skin mechanoreceptors—According to some classifications, *light* (fine) *touch* is detected by the superficially located receptors, such as *Meissner's corpuscle, Merkel's disk,* and the nerve plexus found around the roots of skin hairs—*hair root plexi. Crude touch* and *pressure* are detected by the deeper receptors, such as *Krause's endbulb, Ruffini's ending,* and the *Pacinian corpuscle.*

Muscle & tendon receptors—Changes in the *muscle length and tension* are detected by the stretch receptors in the *muscle spindle;* changes in the *tendon length and tension* are detected by the *Golgi tendon organ* found in the muscle tendons. Specialized mechanoreceptors called *joint receptors* or *kinesthetic receptors* found in the joints signal changes in the *displacement* and *position* of the joint or limb in space.

Hair cells—More specialized mechanical receptors containing *hair cells* (cells with modified cilia) are found in the inner ear. Movement of the ciliary hair deforms the cell membrane, activating the hair cell. These hair cells are found in the *cochlea* (hearing organ), where they respond to mechanical waves generated by sounds, and in the *vestibular apparatus* (balance organ), where they respond to mechanical fluid waves in the vestibular apparatus caused by head movements.

Baroreceptors & stretch receptors—Walls of many visceral organs contain stretch receptors that signal distention. The *baroreceptors* in the walls of certain arteries (carotid and aorta) are well-known examples. These are sensitive to changes in the distention of the arterial wall caused by changes in blood pressure; similarly, gut wall stretch receptors respond to stomach and intestinal distention.

FREE NERVE ENDINGS DETECT HOT, COLD, OR PAINFUL STIMULI

The sensations of *warmth* and *cold* are conveyed by the *thermoreceptors,* a type of *free nerve ending* in the skin. In addition, certain neurons in the brain *hypothalamus* that take part in neural regulation of body temperature can detect changes in the *blood temperature.* Other specialized types of free nerve endings (*nociceptors*) respond to noxious stimuli that cause pain.

CHEMORECEPTORS ARE SENSITIVE TO SPECIFIC CHEMICALS

Numerous sensory stimuli of a chemical nature are detected by a variety of *chemoreceptors.* Thus, *olfactory receptor cells* in the posterior nasal cavity detect environmental odors. *Taste receptor cells* in the tongue's taste buds detect certain substances in food that may be advantageous (sweets, salts) or harmful (bitter substances) to the body. Other types of internal chemoreceptors detect changes in physiologically important blood substances. For example, sensor cells in the *carotid* and *aortic bodies* detect oxygen, and chemosensor neurons in the brain medulla detect carbon dioxide. Still other hypothalamic neurons are specialized as *osmoreceptors,* which regulate blood *osmolarity* by detecting blood *sodium* levels, and others as *glucoreceptors,* detecting blood *glucose* levels.

PHOTORECEPTORS ARE SPECIALIZED TO DETECT LIGHT ENERGY

The retina, the nervous part of the eye, contains *photoreceptor* cells (rods and cones) that detect *light* energy. The visible light rays make up a specific band in the spectrum of electromagnetic wave energy. *Rods,* being more abundant and more sensitive, serve in peripheral and night vision; *cones* work only in daylight and detect red, blue, and green colors—i.e., certain specific wavelengths of visible light.

CN: Because of the many colors needed on this plate, you may have to use the same color for different letter labels if you run short.
1. Color the three upper panels from left to right.

2. Color the various mechanoreceptors, most of which appear in the block of skin to the right.
3. Color the categories of receptors. Where the receptor is small, color the arrow pointing to it.

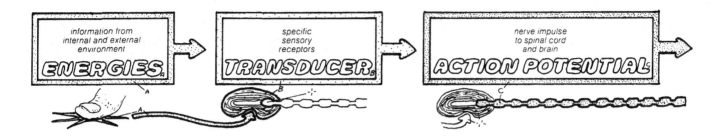

ENERGIES.	TRANSDUCER.	ACTION POTENTIAL
information from internal and external environment	specific sensory receptors	nerve impulse to spinal cord and brain

MECHANORECEPTORS.

LIGHT TOUCH:
  MEISSNER'S CORPUSCLE.
  MERKEL'S DISK.
  HAIR ROOT PLEXUS.
DEEP PRESSURE:
  PACINIAN CORPUSCLE.
CRUDE TOUCH:
  KRAUSE'S END BULB.?
  RUFFINI'S ENDING.?
MUSCLE LENGTH, TENDON & LIMB POSITION:
  MUSCLE SPINDLE.
  GOLGI TENDON ORGAN.
  JOINT/KINESTHETIC RECEP.
HEARING & BALANCE:
  HAIR CELLS.
BLOOD PRESSURE
  AORTIC & CAROTID
  BARORECEPTORS.

SKIN RECEPTORS

NOCICEPTORS.
  PAIN: FREE NERVE ENDINGS.

THERMORECEPTORS.
  WARMTH: FREE NERVE ENDINGS?
  COLD: FREE NERVE ENDINGS?
  INTERNAL TEMPERATURE:
    HYPOTHALAMIC THERMOSTAT.

INTERNAL AND SPECIAL RECEPTORS.

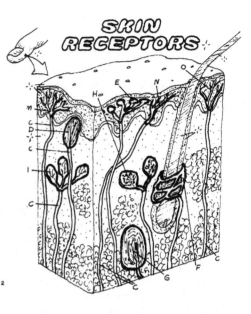

CHEMORECEPTORS.
  ODOR: OLFACTORY NEURONS.
  BLOOD O_2, CO_2, H^+:
    AORTIC & CAROTID BODIES.
    MEDULLARY CHEMORECEPTOR.
  BLOOD GLUCOSE:
    HYPOTHAL. GLUCORECEPTOR.
  OSMOLARITY LEVELS:
    HYPOTHAL. OSMORECEPTOR.
  TASTE: GUSTATORY CELLS OF TASTE BUDS.

PHOTORECEPTORS.
  LIGHT: RODS & CONES.

An important problem in sensory physiology is the mechanism by which the sensory receptors convert the energy in a stimulus (be it physical, like touch, or chemical, like odors) into the common language of nervous system communication—i.e., the nerve impulse. This process is called *sensory transduction*. In this plate, we study the functional properties of the skin sensory receptors and how they handle sensory transduction.

SENSORY NERVE ENDINGS TRANSDUCE MECHANICAL STIMULI
The Pacinian corpuscles are sensory receptor "organs" or bodies found in the skin's deep layers as well as in the visceral tissues. They are believed to be mechanical receptors sensitive to such stimuli as *pressure* and *vibration*. Each Pacinian corpuscle consists of the nerve ending of a myelinated sensory fiber wrapped in a fibrous connective tissue capsule. The nerve ending is a modified dendrite and the essential transducer of mechanical energy. As it emerges from the capsule, it is continuous with a myelinated sensory fiber (axon) and its nodes of Ranvier (plate 18).

Mechanical pressure stimuli applied to the skin surface indent the skin, resulting in stimulation of the underlying Pacinian corpuscles. Since these are deep in the skin, only strong stimuli, causing sufficiently deep indentation, as produced by pressure stimuli, can activate them. Thus, the Pacinian corpuscles are pressure sensors. The numerous layers of connective tissue fibers making up the capsule act as cushions. They protect the nerve ending against light touch stimuli and help distribute the mechanical deformation waves to all parts of the nerve ending. These waves finally reach the nerve-ending membrane and activate stretch-sensitive mechanoreceptor proteins that interact with sodium channels. As a result sodium ions flow in and depolarize the nerve-ending membrane, creating a local *receptor potential*.

RECEPTOR POTENTIALS MEDIATE STIMULUS TRANSDUCTION
The receptor potential, also called the *generator potential* because it generates the nerve impulse in the adjoining axon, is a type of *graded potential*. In contrast to *action potentials*, which occur only in axons and obey the all-or-none principle (plate 17), the receptor potentials, like all graded potentials, show varying amplitudes in response to variation in the strength of the stimulus; the stronger the stimulus, the more sodium ions enter the ending, and the higher the amplitude of the receptor potential.

As long as the receptor potential lasts, an electronic current flows between the nerve ending and the adjacent node of Ranvier, since the inside of the stimulated nerve ending acts as a positive pole while the inside of the first node, being at rest, acts as negative pole. The strength of this current is proportional to the amplitude of the receptor potential. Once the current reaches the node's excitation threshold, the node produces an action potential that is conducted along the myelinated sensory fiber.

IMPULSE RATE IN SENSORY FIBER IS PROPORTIONAL TO STIMULUS INTENSITY
The first node of Ranvier continues to fire action potentials as long as the receptor potential lasts. The amplitude of the receptor potential, being proportional to the *stimulus strength*, also determines the frequency (number/sec) of the nerve impulse. This relationship forms the basis of the frequency coding of sensory messages. That is, the stronger the stimulus, the higher the impulse frequency in the sensory nerve. This is how the CNS detects changes in stimulus intensity.

Receptor recruitment—If the stimulus intensity continues to increase, causing more extensive skin indentation while the maximal discharge frequency of the nerve fiber branch is reached, then the adjacent corpuscles are activated, first those belonging to the same sensory unit and then those belonging to the neighboring sensory neurons (units). This is called *receptor recruitment*.

SOME RECEPTORS SHOW ADAPTATION, OTHERS DO NOT
Many receptors decrease or stop firing even if the stimulus continues to be applied. These are called *rapidly adapting* receptors. In contrast, *slowly adapting* receptors continue to fire throughout the duration of the stimulus, at the same or somewhat lower rates. The receptors for fine touch and pressure (e.g., hair root plexi and Pacinian corpuscles) are rapidly adapting types; the joint and muscle mechanoreceptors, serving in proprioceptive and kinesthetic senses respectively, are of the slowly adapting type.

Pain receptors, for obvious survival-related reasons, show little adaptation. Receptor adaptation helps to increase tactile discrimination. Thanks to receptor adaptation, we are normally less aware of the presence of clothing on our skin, although we feel it when we first put it on and again briefly every time we move. Receptor adaptation may be due to the elastic properties of the connective tissue fibers engulfing the nerve ending, as in the Pacinian corpuscle capsule.

CN: Letter label A should receive the same color as was given the transducer (B) on the previous page. Action potential (C) on both plates should have the same color. Use dark colors for D and E.
1. Begin with the upper diagrams and the three stages of sensory transduction. Note that in stage 1, the symbols of positive and negative charges are colored over with the background colors, whereas in steps 2 and 3, they may or may not be colored. If they are, the positive charge receives the Na^+ permeability color (D).
2. On the lower chart, color the vertical column of short bars on the far left that represents the frequency of action potentials, and note that in the five lengthy arrows representing different receptors, light lines have been drawn to suggest the frequency of action potentials (you will have to color over these lines when you fill in the arrows).

TRANSDUCTION & RECEPTOR POTENTIAL
PACINIAN CORPUSCLE
NAKED (FREE) NERVE ENDING (DENDRITE)
MYELINATED SENSORY NERVE FIBER (AXON)

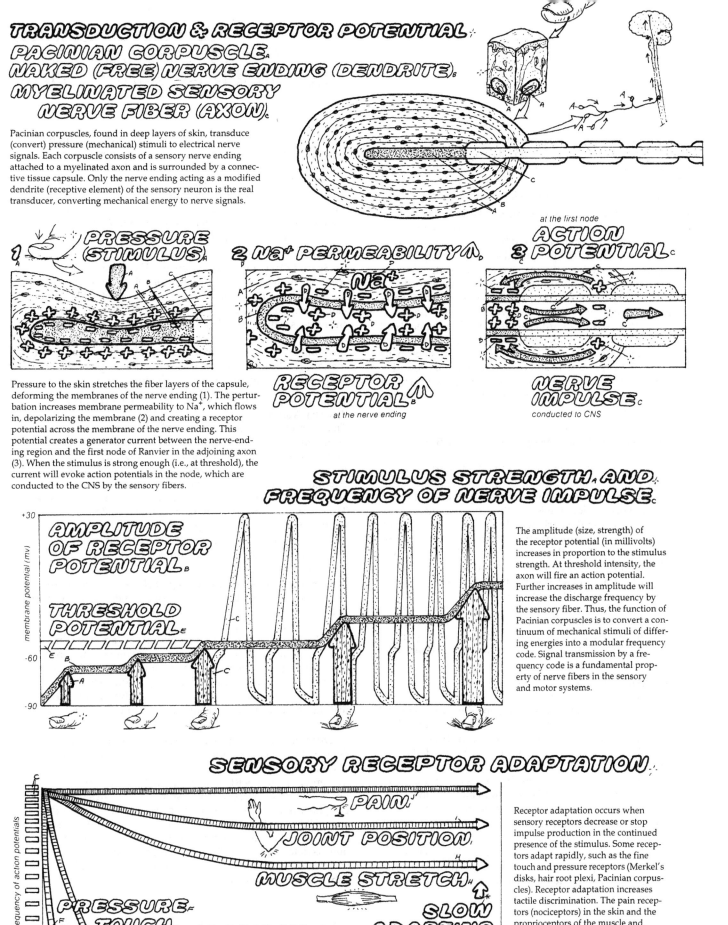

Pacinian corpuscles, found in deep layers of skin, transduce (convert) pressure (mechanical) stimuli to electrical nerve signals. Each corpuscle consists of a sensory nerve ending attached to a myelinated axon and is surrounded by a connective tissue capsule. Only the nerve ending acting as a modified dendrite (receptive element) of the sensory neuron is the real transducer, converting mechanical energy to nerve signals.

1 PRESSURE (STIMULUS)

2 Na⁺ PERMEABILITY ↑
RECEPTOR POTENTIAL ↑
at the nerve ending

3 ACTION POTENTIAL
at the first node
NERVE IMPULSE
conducted to CNS

Pressure to the skin stretches the fiber layers of the capsule, deforming the membranes of the nerve ending (1). The perturbation increases membrane permeability to Na^+, which flows in, depolarizing the membrane (2) and creating a receptor potential across the membrane of the nerve ending. This potential creates a generator current between the nerve-ending region and the first node of Ranvier in the adjoining axon (3). When the stimulus is strong enough (i.e., at threshold), the current will evoke action potentials in the node, which are conducted to the CNS by the sensory fibers.

STIMULUS STRENGTH AND FREQUENCY OF NERVE IMPULSE

AMPLITUDE OF RECEPTOR POTENTIAL

THRESHOLD POTENTIAL

membrane potential (mv)

+30
-60
-90

The amplitude (size, strength) of the receptor potential (in millivolts) increases in proportion to the stimulus strength. At threshold intensity, the axon will fire an action potential. Further increases in amplitude will increase the discharge frequency by the sensory fiber. Thus, the function of Pacinian corpuscles is to convert a continuum of mechanical stimuli of differing energies into a modular frequency code. Signal transmission by a frequency code is a fundamental property of nerve fibers in the sensory and motor systems.

SENSORY RECEPTOR ADAPTATION

frequency of action potentials

PAIN
JOINT POSITION
MUSCLE STRETCH
SLOW ADAPTING

PRESSURE-TOUCH
RAPIDLY ADAPTING

Seconds   1   2   3   4   5   6   7   8

SUSTAINED STIMULUS

Receptor adaptation occurs when sensory receptors decrease or stop impulse production in the continued presence of the stimulus. Some receptors adapt rapidly, such as the fine touch and pressure receptors (Merkel's disks, hair root plexi, Pacinian corpuscles). Receptor adaptation increases tactile discrimination. The pain receptors (nociceptors) in the skin and the proprioceptors of the muscle and joints adapt very slowly. Pain usually signals discomfort or danger and should be attended to, hence the need for its persistent awareness.

Once sensory stimuli are transduced into nerve signals, they are communicated to the CNS along the *afferent sensory nerves*—i.e., the fibers of the *primary sensory neurons*. The *cell bodies* of these neurons are in the *dorsal root ganglion*. In the spinal cord these ganglia are associated with the dorsal roots.

SENSORY NEURONS ACT AS FUNCTIONAL SENSORY UNITS

The sensory neurons are *pseudounipolar neurons;* a single process emerges from the cell body and bifurcates into peripheral and central segments. The peripheral segment, bringing sensory signals toward the cell body, was once considered as a dendrite and the centrally directed segment as an axon. In fact, the sensory neuron's true dendrite is only the short *nerve ending* of the peripheral segment, serving as the receptor transducer; the sensory neuron's remaining fiber segments all show the structural and functional properties of an axon, such as myelin and action potentials. A primary sensory neuron, its fibers and all the peripheral and central *end branchings*, and the central synaptic terminals constitute a *sensory unit*. Thus, the body periphery is served by numerous independent sensory units bringing in the various somatic sensory messages to the CNS.

RECEPTIVE FIELD (RF) OF SENSORY UNITS IN THE SKIN IS CIRCULAR

The specific area of the skin (or any other body part) served by a sensory unit is called the *receptive field* (RF) of that unit. Because the peripheral branchings of sensory fibers in the skin have a radial orientation, the RFs of the sensory units have a circular shape. When the peripheral branches of two adjacent sensory units are far apart, the stimuli impinging on the RF of one unit will not activate the adjacent unit. If the branches of two adjacent units innervate a common area of a target, the RFs will be overlapping. Overlapping RFs provide a basis for complex analysis and integration of the sensory unit's input because the stimuli impinging on the RF of one unit will also elicit impulses in the adjacent units, although to different degrees. This differential activity is detected by the CNS neurons and forms one of the bases for fine tactile discrimination, such as that in the fingertips.

TACTILE ABILITIES VARY IN DIFFERENT SKIN REGIONS

Humans have remarkable tactile abilities (touch sensitivity and discrimination). Different skin regions show differential tactile capacities. The *fingertips*, the *lips*, the *genitalia*, and the *tongue tip* are the most sensitive areas, while the back is the least sensitive. The tactile sensory skills may be divided into two categories, *intensity discrimination* and *spatial discrimination*. Intensity discrimination (sensitivity) refers to the ability to judge *stimulus strength*; spatial discrimination involves the ability to differentiate between the *locations* of *point stimuli*.

Tactile sensitivity & thresholds—To test intensity discrimination, a pointed probe is pushed against the skin surface until it is just felt. This is the *tactile threshold*, which is measured by the depth of the skin indentation formed by the probe. This depth gives a quantitative measure of tactile sensi-

tivity. The tongue tip is the most sensitive body area in this regard, followed by the fingertips, in which a mere 6-micron (μm) indentation can be detected. In the palm of the hand, this threshold is four times higher, while on the back of the hands, trunk, and legs, it is ten to twenty times higher. Note that high threshold values mean low sensitivity. Therefore, the highest tactile sensitivity is associated with the fingertips, used to investigate the environment, and the tongue tip, used to sample food type and texture.

Spatial discrimination by touch—The fingertips also show the highest *spatial discrimination* ability for touch. This is assessed by the *two-point discrimination test* in which the tips of caliper arms are placed on the skin and the distance between the arms is reduced until only *a single point* is felt. This *minimum separable distance* or "limen" is an index of spatial discrimination: the lesser this distance, the higher the discrimination. The limen value is narrowest in the fingertips (1–2 mm) and widest in the back (up to 70 mm).

UNIT NUMBER, RF SIZE, AND RF OVERLAP DETERMINE TACTILE ABILITY

The neural basis of differential tactile sensitivity and discrimination lies in part on the number of sensory units and branches per unit area of the skin. The fingertips contain many more units than the back. Therefore, stimuli of similar strength (causing the same amount of skin indentation) will activate more sensory units on the fingertip than on the back. The *convergence* (plates 87, 93) of the primary afferents from the fingertip units onto the spinal sensory relay cells is also higher in the sensitive regions. As a result, CNS cells can be activated by very weak stimuli applied to the fingertip, but not to the back.

Importance of size and overlap of RFs—The neural basis for spatial discrimination lies in the size and degree of overlap in the RFs. In the fingertips, the RFs are small and the degree of overlap high; the opposite is true for the back or legs. Therefore, in the fingertips, even two closely applied stimuli are likely to activate two different sensory units, since one point impinges on the RF of one unit and the second point on another unit's RF. As long as the CNS neuron receives messages from two separate sensory units, the two points can be distinguished. If both point stimuli fall in the RF of one unit, only one point will be perceived.

Differential unit activity & lateral inhibition—The high degree of RF overlap in the fingertips also permits other discriminative abilities. For example, if a stimulus impinges on the center of a unit's RF, it activates that unit maximally. Because of RF overlap, the same stimulus activates the RF periphery of an adjacent unit. But activation of the second unit will be weaker, since only the periphery is stimulated. This differential rate of activity between the two neighboring units is detected by the CNS neurons and forms the basis of *lateral inhibition*, which serves to sharpen contrast and enhance sensory discrimination.

CN: Use dark colors for A, G, and I.
1. Begin with the sensory unit at the top of the plate.
2. Color the illustrations of RFs. Note the many vertical bars (C') along the axon. These represent the number of impulses traveling along the axon. The cell bodies have been omitted for simplification.

3. Color the examples representing the two forms of tactile discrimination: spatial and intensity. Color the points of the calipers used to make the measurements in these demonstrations.
4. Color the explanation of lateral inhibition. Note the broken line of the inhibitory neuron (I) in the spinal cord and the dotted lines representing inhibited ascending sensory pathways.

SOMATIC (AFFERENT) SENSORY UNIT (SU)

- **SENSORY RECEPTOR** (A)
- **END BRANCHING** (B)
- **SENSORY FIBER (AXON)** (C)
- **CELL BODY** (D)
- **CENTRAL TERMINAL** (E)

Primary somatic afferent (sensory) neurons have a single long peripheral process that traverses within a peripheral nerve and branches out near or within its target organ (skin, joints, muscle spindles, tendons, teeth, tongue, etc.). The central process enters the spinal cord or brain stem to synapse with central relay neurons. A single sensory neuron with all its peripheral branching and central terminals is called a sensory unit (SU). The specific peripheral area served by a SU is called the unit's receptive field (RF).

ascending sensory pathway

spinal cord

PATHWAY (F)

RECEPTIVE FIELDS (RF)
- **STIMULUS** (G)
- **NERVE IMPULSE** (C')

Receptive fields (RF) of neighboring sensory units (SU) in the skin may be separate or overlapping. Stimulation within each separate RF evokes impulse activity only in the corresponding SU; stimuli falling within the overlapping RFs will evoke activity in all the participating units. However, as shown at right, when a tactile stimulus activates more branches of one unit than another, then impulse activity in that unit will be correspondingly higher than in the adjacent units.

firing

SEPARATE RF

quiet

OVERLAPPING RF

low activity

high activity

low activity

TACTILE DISCRIMINATION

The tactile sensitivity of a particular part of skin is directly proportional to the number of SUs innervating that area, as well as to the degree of overlap within the RFs of these SUs. Sensitivity is usually inversely related to the size of the RF. Thus, the fingertips, with many SUs and small overlapping RFs, are far more sensitive than the skin on the back, which has large, separate RFs.

Spatial discrimination: In the two-point discrimination test, the spatial discriminative ability of the skin is determined by measuring the minimum separable distance between two tactile point stimuli. The tongue tip, fingertips, and lips rank high in this ability (1–3 mm limen); the back of the hands, the back, and the legs rank low (50–100 mm).

Intensity discrimination: Sensitive areas are also better able to discriminate differences in the intensity of tactile stimuli. An indentation of 6 μm on the fingertip is sufficient to evoke a sensation. This threshold is four times higher in the palm.

SPATIAL (2-POINT) DISCRIMIN. (H)

1–2 mm

FINGER (MANY UNITS) (H)
units overlap

30–70 mm

BACK (FEW UNITS) (H)
no overlapping

INTENSITY DISCRIMIN. (G')

6μ

24μ

ONE-POINT DISCRIMINATION	TWO-POINT DISCRIMINATION
less than 1 mm	more than 1 mm

ONE-POINT DISCRIMINATION	TWO-POINT DISCRIMINATION
less than 30 mm	more than 70 mm

LATERAL INHIBITION
INHIBITORY NEURON

In the SUs with overlapping RFs, when activity in one SU is higher than in the adjacent units (e.g., by recruitment of more peripheral branches of the main SU), signal transmission from the less active neighboring units to CNS neurons is suppressed by inhibition of their synapses. This "lateral inhibition" serves to sharpen contrast and discrimination.

spinal cord

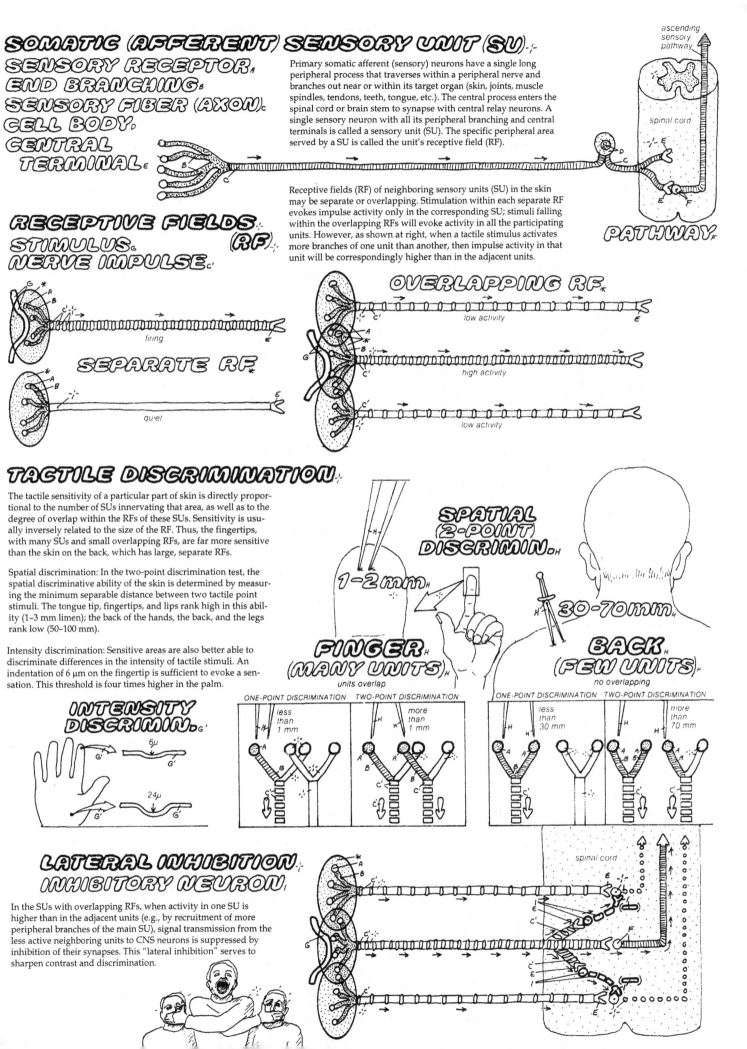

Each sensory receptor type is designed to respond mainly to one type of stimulus, which is then converted to action potentials by the receptor transducer and conveyed to the CNS (as volleys of nerve impulses) by the nerve fibers attached to the receptor. Since all sensory fibers coming from diverse receptor types use the same code for communication (action potentials), how does the CNS differentiate between these various sensory modalities? For example, how is cold sensation felt differently from heat, or pain from touch and touch from pressure? In this plate, we consider the functional segregation of somatic sensory modalities and the channeling of sensory signals along the various pathways and synaptic stations of the spinal cord and brain. Such a segregation and channeling is one of the means by which the CNS can differentiate between modalities.

MODALITIES ARE FUNCTIONALLY SEGREGATED IN SENSORY NERVES

If we dissect a single nerve fiber from the thousands found in a sensory nerve and record its activity, we find that the largest increase in its activity occurs when a particular type of stimulus is applied to its receptor ending. This *adequate stimulus* may be touch, pain, temperature, etc. If we stimulate that fiber electrically, the subject reports only the sensation associated with the adequate stimulus. The generalization that each fiber of a sensory nerve conducts signals concerning only one type of sensory modality (presumably because it is connected to only one particular type of receptor) has been named the "doctrine of specific nerve energies," which also holds true for the sensory pathways within the CNS. This principle and the specificity of sensory receptors, as well as the functional properties of the sensory cortex, to be discussed in the next plate (93), are the important bases for how sensory modalities are functionally differentiated in the nervous system.

Within a sensory nerve trunk (or bundle), different fiber types carry signals related to different sensory modalities. In general, signals related to the modalities of *crude tactile*, *pain* (nociceptive), and *temperature* (thermal) are conveyed by unmyelinated, small-diameter fibers (type C). The cell bodies of these fibers, located in the dorsal horn ganglia, are small and their central terminals in the spinal cord release peptide transmitters (e.g., substance-P by the pain fibers) (plate 94). Signals relaying fine touch and pressure as well as proprioceptive modalities (from joints and muscles) are carried by fast-conducting, large-diameter, myelinated fibers (type A) with large cell bodies.

VARIOUS MODALITIES ASCEND IN SPECIFIC SPINAL PATHWAYS

Upon entry to the spinal cord, the various sensory fibers become segregated into two categories. Thin fibers carrying pain, temperature, and crude tactile sensations, collectively referred to as the *nondiscriminative* modalities, terminate in the dorsal horn of the spinal cord where they synapse with the second-order relay cells.

Spinothalamic pathway for pain, temperature, and crude touch—The fibers of the relay cells *decussate* (cross over the midline) and enter the white matter of the opposite side to ascend in the *spinothalamic pathway* toward the brain. This pathway has two main divisions: the pain and temperature modalities are segregated in a *lateral* division and the crude tactile fibers are bundled in an *anterior* (ventral) division.

There may be further segregation of fibers within these specific pathways, so that pain and temperature fibers are further set apart. The spinothalamic fibers terminate in a specific region (nucleus) of the *thalamus*, a major subcortical sensory relay/integration center in the brain. The spinothalamic pathway and its related modalities represent a basic and ancient somatic sensory system seen in all vertebrates.

Dorsal columns for touch, pressure, and proprioception—The large myelinated sensory fibers carrying the modalities of fine touch, pressure, and proprioception (*discriminative tactile*) enter the spinal cord but do *not* terminate in the dorsal horn. Instead they ascend, without crossing, up the sensory pathways of the *dorsal (posterior) columns*, to end in the brain *medulla*, where they make their first synapse. The axons of the second-order sensory cells cross over the midline and ascend in the *medial lemniscus* pathway to end in the same area of the thalamus where the spinothalamic fibers end. The dorsal column–lemniscal system is called the *discriminative tactile* pathway because such important sensory capacities as precise localization (stimulus source), two-point discrimination, fine touch, vibrations, stereognosis (object recognition by manipulation), and limb/body position in space are all conveyed by this system. Lesions in this pathway selectively impair these sensory abilities. Phylogenetically, the dorsal column system is a more recent somatic sensory system, well developed in primates and humans and poorly developed or absent in lower species.

THALAMIC PROJECTIONS CONVEY SIGNALS TO SENSORY CORTEX

From the thalamus arise the third-order neurons forming the fibers of *somatic* (or *thalamic*) *radiation*, which project to the *sensory cortex* (primary somatic sensory cortex) located in the *postcentral gyrus*. All along the synaptic relay stations—i.e., the dorsal horn, the medulla, and the thalamus—the sensory impulses are filtered and integrated so that the messages arriving in the sensory cortex have already undergone certain processing for fine tuning. This fine-tuning process is in part controlled by the sensory cortex, which sends certain *descending sensory control* fibers to the subcortical relay stations to regulate the quality and quantity of the messages arriving in the cortex (feedback control circuits). The final analysis of sensory messages by the sensory cortex neurons is the subject of the next plate (93).

SENSORY INPUT TO SPINAL MOTOR CENTERS IS CRITICAL FOR REFLEX RESPONSE

A major function of the somatic sensory afferents from the skin, joints, and muscles is to activate the spinal reflexes (plate 95). This is accomplished by collaterals or main branches of the primary afferents as they enter the spinal cord. These branches synapse with the spinal interneurons, which will in turn synapse with the spinal motor neurons to complete the motor reflex circuits. The nociceptive fibers carrying pain signals play a critical role here because of their protective functions, but information from other modalities is also necessary for appropriate adjustment of the reflex responses.

Activation of brain stem motor centers and reticular formation—On their way to the brain, the ascending sensory fibers send *collateral branches* to the midbrain motor centers to influence involuntary motor activity and to the centers in the reticular formation to affect sleep and wakefulness (plate 106), arousal, attention, and central inhibition of pain (plate 94).

CN: Use dark colors for F, G, H, and J.
1. Before coloring the nerves and pathways, color structures A–E along with their location on the human figure. Then begin with the block of skin in the lower left corner and color the various receptors and their sensory neurons (F–H) and follow

them to the various ascending pathways (F'–H') beginning in the spinal cord.
2. After completing the pathways, color the decussation title (J) in the center, and the two arrows identifying the decussation sites.
3. Color the descending controls at the top.

SENSORY CORTEX_A
THALAMUS_B
MIDBRAIN_C
MEDULLA_D
SPINAL CORD_E

PAIN, WARMTH AND COLD:_F
LATERAL SPINOTHALAMIC PATHWAY_F'

CRUDE TACTILE:_GG TOUCH & PRESSURE
ANTERIOR SPINOTHALAMIC PATHWAY_G

Pain, thermal, and crude tactile sensations are conducted by thin unmyelinated (type C) fibers. Terminating in the dorsal horns, these fibers synapse with relay cells that cross the midline and ascend to the brain via the spinothalamic pathways (pain and temperature via lateral division; tactile via anterior division) to synapse in the thalamus. In midbrain and reticular formation, collateral synapses are made for midbrain motor reflexes and arousal, respectively. Synapses in the thalamus integrate and relay sensory signals to the somatic sensory cortex for higher perception. Sensations produced by this projection system are crude and diffuse.

DISCRIMINATIVE TACTILE:_H
FINE TOUCH & PRESSURE   2-POINT DISCRIMINATION   VIBRATION

OR PROPRIOCEPTION:_H
POSTERIOR COLUMN PATHWAY_H'

Signals from discriminative tactile senses (fine touch, pressure, vibration, two-point discrimination) as well as from kinesthetic and proprioceptive senses (limb and body position) are conducted by thick, myelinated (type A) fibers. After sending collaterals to the dorsal horn for reflexes, the main central processes ascend in the dorsal columns to synapse in the brain medulla. Relay cells cross the midline and form the medial lemniscus pathway, which ascends to thalamus. Thalamic fibers (somatic radiation) project to the somatic sensory cortex (postcentral gyrus). This system is responsible for fine tactile discrimination, for exact stimulus localization in the body and skin, and for body image and position.

SKIN_:_I

receptor _H
sensory fiber _H

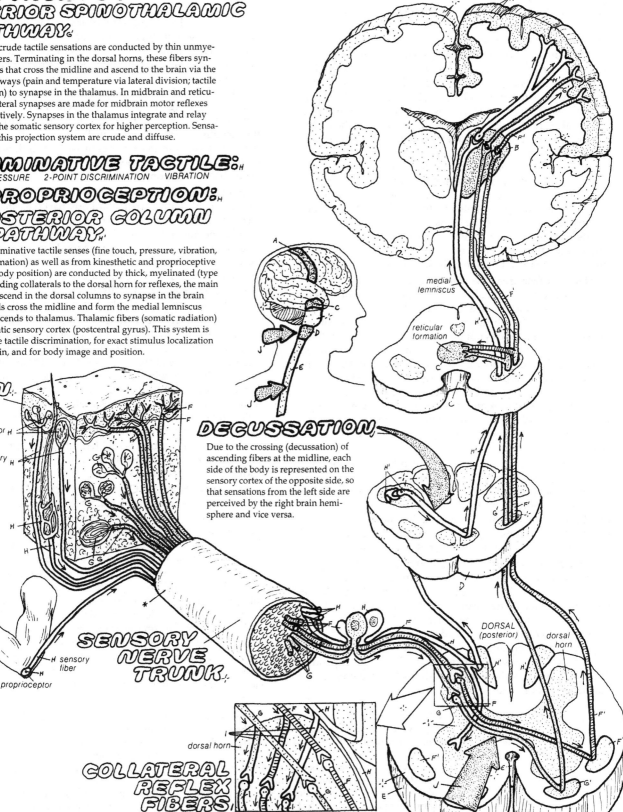

SENSORY NERVE TRUNK_J

sensory fiber _H
proprioceptor

DECUSSATION_

Due to the crossing (decussation) of ascending fibers at the midline, each side of the body is represented on the sensory cortex of the opposite side, so that sensations from the left side are perceived by the right brain hemisphere and vice versa.

COLLATERAL REFLEX FIBERS

dorsal horn

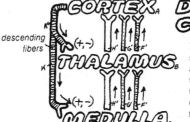

descending fibers

CORTEX_A
THALAMUS_B
MEDULLA_D

DESCENDING CONTROLS_K

Descending fibers from the sensory cortex, via excitatory and inhibitory synapses, control the ascending input from the lower relay stations (medulla, thalamus) in order to fine tune and improve quality of somatic sensory perception.

medial lemniscus

reticular formation

DORSAL (posterior)
dorsal horn

The primary sensory cortex detects the exact source and perceives the specific qualities of the various sensory stimuli. Damage to the sensory cortex impairs tactile and spatial discrimination and loss of the ability to identify objects by manipulation alone (*stereognosis*).

SENSORY CORTEX SHOWS A SOMATOTOPIC ORGANIZATION
If single hairs on an animal's skin are bent and the sensory cortical point where a potential change is evoked is mapped and the points are connected, a *somatotopic map* (body surface representation) is obtained. This map enables the brain to localize the sources of stimuli.

The human somatotopic map (sensory homunculus) — A similar sensory map for the human was recorded by the Canadian neurosurgeon Penfield about 50 years ago. Working with conscious patients undergoing brain surgery, he stimulated the sensory cortex of the postcentral gyrus with a mild electrical current. This procedure is painless since the brain has no pain receptors. The patients reported tactile sensations in different parts (e.g., the toes, the fingers, or the back) depending on the points of stimulation on the cortex. Connecting these points, Penfield noted a representation of the body on the sensory cortex, a sensory *"homunculus"* (little man). The legs are represented on the hidden medial portion of the postcentral gyrus, the trunk on the top, and the arms, hands, and head on the larger lateral surface. Due to the crossing of ascending pathways, the body's left side is represented on the right hemisphere; the right side on the left hemisphere.

Hands & face area have enlarged representation — The human sensory homunculus shows important distortions in two respects. First, the area for the hands is interposed between those of the head and the trunk. Second, the cortical area of representation is not proportional to the size of the body parts, the hands and the face having enlarged representations. Within the hand area, the fingers and the thumb have independent representation, the largest being the index finger. Lips and tongue (perioral) areas are especially large. The extent of sensory cortex area devoted to a body part is proportional to the part's sensory innervation, receptor density, tactile sensitivity, and discrimination prowess, not to its physical size.

Supplementary sensory area — In addition to the main sensory cortex, referred to as SI, a smaller supplementary somatotopic sensory map (SII) can be found in the superior wall of the sylvian fissure, between the parietal and temporal lobe, but it is less well organized functionally. The SII neurons receive their input mainly from the SI neurons. Lesions in SII result in deficits in tactile learning but not in tactile sensation.

Plasticity of sensory map — Although the sensory map (SI) normally is fixed, it can show changes in response to injury and in response to learning and use. Experiments in primates have shown that if a nerve to a finger is cut, the area of the sensory cortex devoted to the denervated finger becomes responsive to stimuli from the adjacent fingers. Also, if a finger is excessively and preferentially used in certain skilled tactile tasks, the sensory cortical area to that finger expands. These observations show that the adult sensory cortex is plastic and modifiable.

SENSORY CORTEX NEURONS FORM LAYERS & COLUMNS
The sensory cortex consists of six layers parallel to the cortical surface and numbered from top to bottom. Large and small pyramidal neurons as well as stellate and fusiform neurons are found in various layers. Layer IV neurons receive the afferent sensory input; those in layers V and VI are output neurons projecting relay and feedback control signals to other CNS areas. Smaller neurons of layers II and III serve as local association neurons connecting adjacent cortical areas.

When a microelectrode for recording the electrical potentials (responses) of single cortical neurons is inserted in the cortex perpendicular to the cortical surface, all the neurons along the electrode's path share the same receptive field and respond to the same tactile modality. When the electrode is inserted obliquely, different groups of neurons are encountered, responding to stimuli of different modalities from the same receptive field. If the electrode is moved farther away, the receptive field changes as well. Therefore the sensory cortex neurons are *functionally* organized in cylinders or *columns*, each about 3–5 mm long, less than 1 mm wide, and containing about 100,000 neurons.

Sensory columns are modality specific — Each group of columns deals with a particular body part (i.e., share similar receptive fields), and the cells of each single column respond to only one modality. Thus, in a group of columns relating to an area in a finger, one column responds to proprioception, the next to touch, and another to pressure stimuli. There are no separate columns for temperature and pain, these modalities being served by a few cells in some tactile columns; Presumably they are perceived by lower subcortical centers.

Columnar cells are feature detectors — Within each column, some cells imitate the behavior of the sensory receptors (e.g., increasing their firing rate with increasing stimulus intensity); these cells are called *simple cells*. Other cells increase their activity only when a stimulus moves across the skin in a particular direction; these cells are called *complex cells*. The response pattern of the neurons in the cortical columns is called *feature detection*. Through feature detection in the primary and secondary (association) sensory cortex, the complex world of sensory stimuli is molded into a perception pattern. The columnar organization is also found in the visual and auditory cortex (plates 100, 102).

ASSOCIATION SENSORY CORTEX CREATES BODY IMAGE
Nerve impulses from the primary sensory cortex are relayed to higher "association" sensory cortical areas in the posterior parietal lobe for further analysis, integration, and synthesis. The results of the processing of somatic sensory data, together with that from other sensory modalities (particularly visual), are critical in higher somatic perception related to the formation of body image, body orientation in space, sensory-motor integration, and behavior. For example, the association sensory cortex on one side of the brain helps to coordinate hand and eye movements on the opposite side. Individuals with lesions in the association sensory cortex may be unaware of, or fail to recognize, certain parts of their own body, such as an arm. They may comb their hair on one side, ignoring the other side, not realizing it is theirs (hemineglect syndrome).

CN: Use a dark color for W (the neurons in the cortex).
1. Begin with material at the top, noting that the two structures representing the thalamus (C) are buried deep below the thin cortex layer (A).
2. Color the representation of body parts on the sensory cortex. Also color the homunculus, which is a proportional drawing based upon the amount of cortex devoted to a particular part. Note that the last two sections of the

cortex are colored gray because they represent the pharynx and intra-abdominal structures (not shown).
3. Color the organization of the sensory cortex. Note that the left three columns deal with three specific modalities. On the right, neuron connections are within a modality.
4. Color the examples of convergence and divergence in the lower right corner, starting with the many receptor feeds (1) of convergence and working up to the cortex.

SOMATIC SENSORY CORTEX,
ASSOCIATION SENSORY CORTEX,
THALAMUS.

central sulcus (fissure)

The primary somatic sensory cortex (somatic cortex, S) is the area just behind the central fissure (postcentral gyrus). This area receives the specific somatic radiation from the thalamus and is the main cortical area for analysis and integration of sensory input from the skin and joints as well as for somatic perception and sensations related to body position and movement. Lesions of the sensory cortex do not abolish tactile sensations, but diminish tactile discrimination (localization, intensity) and sensation of body movements. The association sensory cortex is located posterior to the primary sensory cortex from which it receives input. Loss of this area results in disorders of body image and distorted complex somatic-visual sensations.

THE SENSORY HOMUNCULUS.

rat musculus

human homunculus

REPRESENTATION OF BODY PARTS ON THE CORTEX.

Stimulation of discrete areas of the somatic cortex (S) in conscious patients undergoing brain surgery evokes tingling sensations in specific body areas. Careful mapping of these projected sensations has revealed an orderly representation of the body on the S cortex (somatotopic map, sensory homunculus) with the trunk on the top, the hands at the middle, and the face at the bottom of the postcentral gyrus. The representation is not based on the size of the body part but on its sensory prowess and tactile abilities; hands, fingers, tongue, and lips have large representations but the trunk and legs have relatively small ones. A supplementary sensory area (SII) exists adjacent to the SI area, in the sylvian gyrus, with less detailed representation; it may be involved in tactile learning.

TRANSMISSION OF STIMULI TO NEURONS IN CORTEX

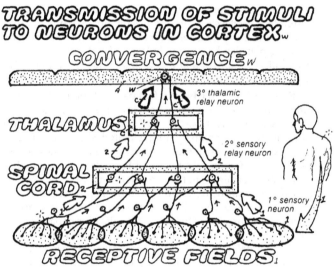

CONVERGENCE

3° thalamic relay neuron

THALAMUS

2° sensory relay neuron

SPINAL CORD

1° sensory neuron

RECEPTIVE FIELDS

The convergence from the peripheral receptors to neurons in the S cortex allows a single cortical cell to respond to sensory signals from large areas of the body (sum of the receptive fields of all the primary sensory afferents). This arrangement is observed generally in the spinothalamic projection system serving crude tactile sensation. In contrast, divergence by which a single peripheral afferent sends signals to many cortical cells allows for more refined discrimination of tactile signals. This arrangement is seen in the dorsal column-lemniscal projection system, serving discriminatory tactile sensations. The divergence pattern is so organized that with point stimulation of a skin part, some cortical cells are activated more than others, providing for contrast discrimination.

CORTEX ORGANIZATION

1 LAYERED
RECEPTIVE CELLS
PROJECTION (ASSOC.) CELLS

2 COLUMNAR:
PROPRIOCEPTIVE (JOINT)
PRESSURE
TACTILE

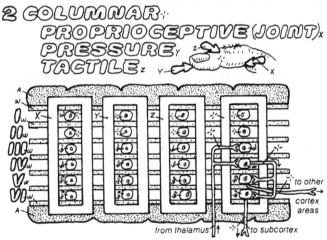

to other cortex areas

from thalamus   to subcortex

The S cortex, like other areas of the neocortex, is organized both horizontally and vertically. Horizontally, it is organized into six layers of neurons, with cells of each layer involved in a different transmission function. Layer IV cells receive the specific somatic radiation from the thalamus. Vertically the S cortex is organized into columns. Each column is about 1 mm wide and a few mm long. There are thousands of columns in the S cortex. All cells in a particular column respond to signals relating to a single modality from a distinct body area. Thus a column sensitive to tactile stimuli from a finger is located next to another column responsive to proprioceptive sensation from the joints in that finger, for example.

DIVERGENCE

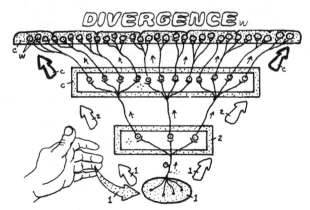

Pain or *nociception* is a complex sense, as it involves sensation as well as feelings and emotions. The neurochemistry of pain and nociception and the role of the brain in pain inhibition have much expanded the scope of pain physiology recently.

CHEMICALS STIMULATE FREE NERVE ENDINGS TO INITIATE PAIN
Pain is served by *free nerve endings* in the skin and visceral tissues. Pain can be initiated by a variety of stimuli. Strong *mechanical* stimuli (intense pressure), very hot and very cold *thermal* stimuli, and certain *chemical* stimuli (such as acidic substances) all can cause pain. Pain receptors show high thresholds, so they are usually activated by intense stimuli that often are noxious—hence the names *nociception* and *nociceptors*. Nociceptive stimuli cause *tissue damage* of varying degrees (from a pinch to a burn); this results in release of certain endogenous *nociceptive substances* such as *histamine* in the injured tissue; these then act on the nociceptive free nerve endings to initiate pain signals. Other nociceptive compounds are *serotonin*, *substance-P*, and *kinin* peptides (bradykinin, etc.). Release of K^+ ions is a main cause of pain in fatigued muscle.

PAIN IS CONVEYED BY FAST AND SLOW AFFERENT SYSTEMS
Pain transmission involves two distinct pathways, each resulting in a different pain experience. Upon stepping on a thumbtack, first a *sharp* sensation is felt (initial pain), followed after a time delay by a *dull* pain (delayed pain). The sharp or *prickling* pain can be accurately *localized* and is *short lasting*. The dull sensation, or a *throbbing* pain, is *long lasting* and *diffuse*, often ascribed to a larger body part.

Roles of type-Aδ fibers in fast pain & type-C fibers in slow pain—The sharp initial pain is conveyed by thin but myelinated, relatively *fast*, nerve fibers (*type Aδ*) and the dull, hurting, throbbing pain by unmyelinated, *slow*, type-C fibers. Conduction velocity in the Aδ fibers is about 10 times faster than in the C fibers. Both types of fibers, although somewhat segregated, terminate in the *dorsal horn* and ascend in the *spinothalamic* pathway. The fast pain fibers project directly to the thalamus and up to the *sensory cortex*. The cortical input, although minor, allows for the fine localization capacity of the sharp/fast pain system. Patients with damage to the sensory cortex are unable to localize the pain source but can still feel pain and are hurt by it. The slow pain fibers make a major input into the brain stem *reticular formation*, which mediates the central inhibition and arousal effects of this pain. These fibers terminate largely in the *thalamus*, with further input to the *limbic system*, in particular the *cingulate gyrus*, where the emotional, hurting component of pain is processed.

BRAIN CENTERS CAN INHIBIT ASCENDING PAIN SIGNALS
Electrical stimulation of the *periaqueductal gray* region of brain stem reticular formation inhibits the sensation of pain in the conscious animal. From this region *descending* fibers project to the dorsal horn, where they *suppress* the relay of afferent pain signals to the brain. Central pain inhibition helps animals and humans cope with the debilitating consequences of hurting pain arising from tissue damage and wounds during physical

stress and fighting. Central pain inhibition also permits athletes, soldiers, and presumably the yogis of India to tolerate intense bodily trauma and pain.

Endorphins mediate central pain inhibition—Descending fibers from periaqueduct region of the brain release the neurotransmitter *serotonin*, which excites certain *inhibitory interneurons* in the dorsal horn that release a peptide neurotransmitter called *enkephalin* (an *endorphin peptide*). Enkephalin suppresses relay of pain signals by the afferent type-C fibers through binding with *opiate receptor molecules* present in the synaptic nerve terminals of these fibers. *Morphine* and other opiate *analgesics* (pain killers) also bind to the same receptors. A peptide called *nociceptin*, which resembles dynorphin, has been found to have the opposite effects—i.e., it increases pain sensation when injected in the brain (*hyperalgesia*).

LARGE TACTILE AFFERENT FIBERS CAN SUPPRESS PAIN
The interneurons of the dorsal horn may also be involved in *afferent pain inhibition*. Rubbing of a skin area relieves the pain from the same or a nearby area. Rubbing activates the large, fast-conducting *tactile fibers* (type Aα), while pain is conveyed by C fibers. In the dorsal horn, central branches of touch fibers activate inhibitory interneurons, which inhibit the synaptic transmission of pain signals to relay cells. This is the basis of the *gate theory of afferent pain inhibition*. Presumably, the more powerful tactile signals dominate the transmission "gates" in the dorsal horn, suppressing and excluding access to the weaker pain signal. Afferent and central pain inhibition by endorphins may in part underlie the phenomenon of *acupuncture analgesia*.

REFERRED PAIN MAY INVOLVE SHARING OF SYNAPTIC RELAYS
If pain from a visceral organ is felt in a superficial zone, it is called a "referred pain." For example, pain originating in the heart may be felt in the inner aspects of the left arm. Pain in the ureter is felt in the testicles. Maps of referred pain are used extensively for medical diagnoses (e.g., heart conditions). The mechanism explaining referred pain may be synaptic convergence and/or facilitation of input. Afferent pain fibers originating from the same area show extensive convergence onto the dorsal horn relay cells. In certain cases, the convergence may include fibers from different body areas, usually of similar embryonic origin, causing activation of the spinal relay cell by pain stimuli originating in a different body part. Usually, one part is a visceral organ.

PHANTOM PAIN MAY BE CENTRAL IN ORIGIN
Phantom pain refers to persistent pain originating from an amputated limb. This pain was once thought to originate from irritated endings of cut nerves that project to the corresponding area of somatic sensory cortex. However, if phantom-pain patients are conditioned to "imagine" that they still possess the lost limb, the phantom pain gradually disappears. A patient with an amputated left arm is trained to constantly look at the mirror image of his moving right arm, creating the illusion that the left arm exists. The gradual disappearance of phantom pain shows that it is suppressed centrally, possibly by rearrangement of the sensory map.

CN: Use dark colors for E and H.
1. Begin on the left side of the upper panel and work your way up to the brain cortex in the upper right. Only the dorsal horn regions are colored gray.
2. Color the panel on pain inhibition and follow the numbered sequence. The process described is an elaboration of the function of the descending pathway (J) in the diagram above. Note the reduction in frequency

of nerve impulses (represented by the vertical bars that receive the color H) in the inhibited relay fiber (5). The terminal of the incoming nerve receives the color of the pain-producing substance-P (M), which is itself being inhibited (3).
3. Color the gate-theory panel.
4. Color the referred-pain panel and then the phantom-limb pain material.

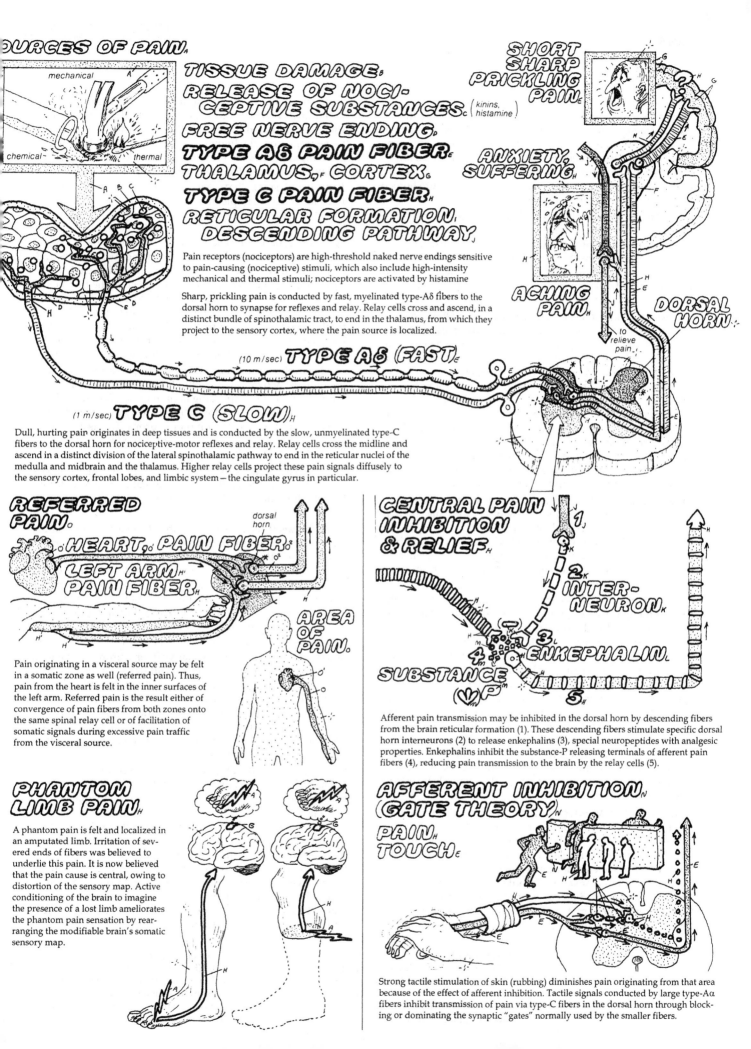

SOURCES OF PAIN

TISSUE DAMAGE
RELEASE OF NOCI-CEPTIVE SUBSTANCES (kinins, histamine)
FREE NERVE ENDING
TYPE Aδ PAIN FIBER
THALAMUS, CORTEX
TYPE C PAIN FIBER
RETICULAR FORMATION, DESCENDING PATHWAY

SHORT SHARP PRICKLING PAIN

ANXIETY, SUFFERING

ACHING PAIN

DORSAL HORN

to relieve pain

Pain receptors (nociceptors) are high-threshold naked nerve endings sensitive to pain-causing (nociceptive) stimuli, which also include high-intensity mechanical and thermal stimuli; nociceptors are activated by histamine

Sharp, prickling pain is conducted by fast, myelinated type-Aδ fibers to the dorsal horn to synapse for reflexes and relay. Relay cells cross and ascend, in a distinct bundle of spinothalamic tract, to end in the thalamus, from which they project to the sensory cortex, where the pain source is localized.

(10 m/sec) TYPE Aδ (FAST)

(1 m/sec) TYPE C (SLOW)

Dull, hurting pain originates in deep tissues and is conducted by the slow, unmyelinated type-C fibers to the dorsal horn for nociceptive-motor reflexes and relay. Relay cells cross the midline and ascend in a distinct division of the lateral spinothalamic pathway to end in the reticular nuclei of the medulla and midbrain and the thalamus. Higher relay cells project these pain signals diffusely to the sensory cortex, frontal lobes, and limbic system — the cingulate gyrus in particular.

REFERRED PAIN

HEART PAIN FIBER
LEFT ARM PAIN FIBER
AREA OF PAIN

Pain originating in a visceral source may be felt in a somatic zone as well (referred pain). Thus, pain from the heart is felt in the inner surfaces of the left arm. Referred pain is the result either of convergence of pain fibers from both zones onto the same spinal relay cell or of facilitation of somatic signals during excessive pain traffic from the visceral source.

CENTRAL PAIN INHIBITION & RELIEF

1.
2. INTER-NEURON
3. ENKEPHALIN
4.
SUBSTANCE (P)
5.

Afferent pain transmission may be inhibited in the dorsal horn by descending fibers from the brain reticular formation (1). These descending fibers stimulate specific dorsal horn interneurons (2) to release enkephalins (3), special neuropeptides with analgesic properties. Enkephalins inhibit the substance-P releasing terminals of afferent pain fibers (4), reducing pain transmission to the brain by the relay cells (5).

PHANTOM LIMB PAIN

A phantom pain is felt and localized in an amputated limb. Irritation of severed ends of fibers was believed to underlie this pain. It is now believed that the pain cause is central, owing to distortion of the sensory map. Active conditioning of the brain to imagine the presence of a lost limb ameliorates the phantom pain sensation by rearranging the modifiable brain's somatic sensory map.

AFFERENT INHIBITION (GATE THEORY)

PAIN
TOUCH

Strong tactile stimulation of skin (rubbing) diminishes pain originating from that area because of the effect of afferent inhibition. Tactile signals conducted by large type-Aα fibers inhibit transmission of pain via type-C fibers in the dorsal horn through blocking or dominating the synaptic "gates" normally used by the smaller fibers.

Reflexes are programmed, stereotyped, predictable motor responses to certain specific sensory stimuli. They are the most elementary form of nervous action and govern much of the motor behavior of simpler animals and human newborns. Some reflexes are defensive, enhancing survival—e.g., limb withdrawal in response to noxious stimuli. Other reflexes help maintain balance and posture, still others ensure homeostasis and stable internal environment.

Spinal reflexes, i.e.—those associated with *spinal cord* control of trunk and limb muscles—are the best-known reflexes. *Brain reflexes* (with reflex center in the brainstem) also exist, e.g., those for eye movements. *Somatic reflexes* involve body's skeletal muscles and motor behavior, while *autonomic reflexes* help regulate the internal environment by affecting exocrine glands, heart, and visceral smooth muscles.

THE REFLEX ARC GOVERNS THE OPERATION OF REFLEXES
The operation of any reflex requires the active participation of several components: (1) the *sensory receptor*, which detects the *stimulus*; (2) the *afferent nerve*, which conveys the sensory signal to the spinal cord or brain; (3) an integrative *synaptic center*, which analyzes and integrates the sensory input and produces motor output commands; (4) the *efferent nerve*, which conducts the motor output to the periphery; and (5) a *motor effector* (e.g., skeletal muscle, smooth muscle, glands), which carries out the response. These elements together make up a *reflex arc*. Complexity of a reflex response corresponds to the complexity of the reflex center; this in turn depends on the number of interneurons and synapses involved.

MONOSYNAPTIC STRETCH REFLEX IS THE SIMPLEST REFLEX
The *stretch reflex* is the simplest known reflex because there is only one synapse in the path of its arc (*monosynaptic reflex arc*). Large skeletal muscles involved in body support and limb movements contain spindle-shaped organs (*muscle spindles*), which act as sensory organs detecting changes in the muscle length or tension.

Role of muscle spindle stretch receptor—Muscle spindles contain the sensory receptors for the stretch reflex. Each spindle contains modified muscle fibers called spindle or *intrafusal* (= inside spindle) fibers. The middle segment of each spindle fiber acts as a mechanical *stretch receptor* that is connected to a sensory afferent nerve to spinal cord. Stretching a muscle activates the spindle stretch receptor, firing nerve signals to the spinal cord in proportion to the amount of stretch; the terminals of the spindle sensory fiber make direct excitatory synaptic contact with *alpha* (α) *motor neurons* serving the same muscle. The α motor neurons are the large neurons that excite the ordinary muscle fibers (extrafusal). Activation of α motor neurons by the spindle sensory fibers and the resultant contraction lead to shortening and return of the muscle fiber to its original length.

STRETCH REFLEX MONITORS MUSCLE LENGTH & TENSION
The stretch reflex continuously monitors the length and tension of muscle fibers and keeps these constant during rest. Since muscle tension is related to muscle length, the stretch reflex functions to enhance tonus (tone) in muscles, maintaining their readiness for action. Spindle fibers also have a *contractile segment* located on the sides of the stretch receptor. Contraction of these motor segments stretches the spindle sensory segment, activating the stretch reflex. The spindle

motor segments are driven by small spinal motor neurons called *gamma* (γ) *motor neurons* (γ *efferents*). The γ motor neurons are stimulated by sensory fibers from the periphery and by neurons from the higher brain motor centers. This higher brain input primes the muscles for postural adjustments and behaviorally controlled movements.

KNEE JERK REFLEX: BOTH MONO- & POLYSYNAPTIC PARTS
In contrast to the simple, monosynaptic stretch reflex, most spinal reflexes are *polysynaptic*, i.e.—the reflex arc and center involves one or more *interneurons* and higher numbers of synaptic connections. The *knee jerk reflex*, used in clinical diagnosis, provides an example of the operation of both mono- and polysynaptic reflex responses. One sits on a high chair with legs dangling; the patellar tendon, connecting the thigh extensor muscles to the tibia bone, is tapped just below the knee. The lower leg shows a fast reflex extension (knee jerk) due to thigh *extensor* muscle contraction. This is a stretch reflex: tapping the tendon pulls on the tendon fibers, which stretch the muscle and the spindle fibers, activating the stretch reflex.

However, execution of a proper knee jerk reflex requires not only the activation of the thigh extensor muscles but also the relaxation of the opposing *flexor* muscles. Because all motor neurons are excitatory, the only way to obtain flexor relaxation is to inhibit their motor neurons. This is achieved by *inhibitory interneurons* that are simultaneously activated by a branch of the spindle sensory fiber.

INTERNEURONS ARE CRITICAL IN COMPLEX SPINAL REFLEXES
The associative interneurons of the spinal cord, particularly the inhibitory ones, underlie the operation of all complex spinal reflexes.

Withdrawal reflex—In the limb *withdrawal reflex* (a limb *flexor reflex*), noxious (sharp or hot) stimuli activate the pain receptors and their sensory afferent fibers; the central terminals of these stimulate excitatory spinal interneurons; these, in turn, excite motor neurons going to the flexor muscles on the same side, causing ipsilateral limb withdrawal (as in touching a very hot object). Even in a simple withdrawal reflex, the ipsilateral extensors must be simultaneously relaxed.

Crossed-extensor reflex—Another example of the mediation of spinal reflexes by associative interneurons is provided by the *crossed extensor reflex*. Here, withdrawal of one leg in the standing posture throws the weight onto the other leg, exciting the contralateral leg extensors and inhibiting the flexors. The excitatory and inhibitory circuitry for activation of most of the spinal reflexes is already present at birth. The activation of any particular reflex circuit depends mainly on the stimulus type and location.

SPINAL REFLEX INDEPENDENCE & SPINAL SHOCK
Spinal reflexes can occur independently, without brain control, as seen in *spinal transected* animals and quadriplegic humans with spinal cord transection or damage. For varying durations after spinal transection, spinal reflexes do not occur. This period of *spinal shock* is short in lower animals (minutes in frogs) and long in higher animals (hours in cats, weeks to months in humans). A higher animal's large brain exerts more control over its spinal cord, compared to a lower animal, in which the spinal cord functions fairly independently. The gradual increase in motor control by the brain during evolution is referred to as *encephalization*.

CN: Use dark colors for B, D, and O.
1. Begin with the upper panel. The title for letter label O, interneuron, is found in the knee jerk reflex panel.
2. Color the stretch reflex panel, beginning with the overview in the small rectangle on the left. Then follow the numbered sequence.

3. Color the knee jerk reflex panel, noting that a muscle spindle (K & L) has been greatly enlarged for purposes of illustration. Note that the flexor muscle that is inactivated by an inhibited efferent motor nerve (D) is left uncovered. In the withdrawal reflex panel, leave the extensor muscle uncolored.

REFLEX ARC

RECEPTOR A
AFFERENT (SENSORY NERVE) B
SPINAL CORD OR BRAIN C (INTEGRATING SYNAPTIC CENTER)
EFFERENT (MOTOR) NERVE D
EFFECTOR E

SPINAL NERVE F
GANGLION G
DORSAL ROOT H
VENTRAL ROOT

Reflexes are simple, involuntary, stereotyped motor actions generated in response to specific sensory stimuli. Reflexes operate through the reflex arc.

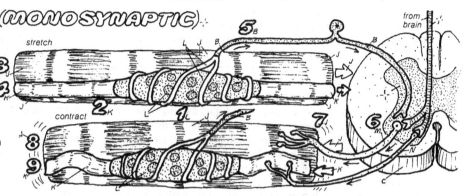

A reflex arc consists of (1) sensory receptor(s) that transduce the stimuli; (2) afferent sensory fibers that enter the spinal cord via the dorsal roots, conveying signals to the CNS; (3) an integrating center (synapses and interneurons) that analyzes the sensory input, delivering output signals to the motor neurons. The fibers of the motor neurons, forming the arc's efferent path (4), leave through the spinal ventral roots, to stimulate skeletal muscles (effectors) (5).

STRETCH REFLEX (MONOSYNAPTIC)

SKELETAL MUSCLE J
SPINDLE FIBER K
MIDDLE ZONE L
ALPHA EFF. FIB. M
GAMMA EFF. FIB. N

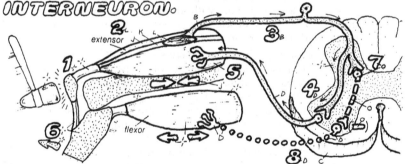

Simplest reflexes operate via only one synapse (monosynaptic reflex), as in the stretch reflex (SR), which functions to keep muscle length and tension (tonus) constant. The sensory receptor for SR (1) is in the middle segment of intrafusal fibers found in the muscle spindle (MS) (2). Stretching of the muscle (3) stretches the spindle fibers (4), activating the MS stretch receptors and the associated sensory fibers (5). These monosynaptically excite the large α motor neurons (6), which excite the ordinary muscle fibers (extrafusal) (7). Contraction of these fibers shortens the muscle (8) and relaxes the spindle fibers (9), terminating the SR and muscle contraction.

CROSS EXTENSOR REFLEX

In a standing position, painful stimulation of one foot causes flexion (withdrawal) of the ipsilateral leg, as well as the extension of the contralateral leg (crossed extensor reflex), to stabilize posture. By utilizing various inhibitory and excitatory interneurons, the ipsilateral leg flexors are activated and the extensors inhibited, and vice versa in the contralateral leg.

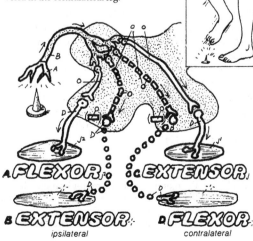

A FLEXOR   C EXTENSOR
B EXTENSOR   D FLEXOR
ipsilateral   contralateral

KNEE JERK REFLEX (EXTENSOR)
(POLYSYNAPTIC INHIBITORY SYNAPSE)
INTERNEURON

The stretch reflex is involved in the knee jerk reflex. A tap on the patellar tendon (1) stretches the extensor muscle (EM) (2) and its spindle. The spindle discharges, exciting the associated sensory fibers (3) that excite the motor neurons to the EM (4). Contraction of EM (5) extends the lower leg (6) (knee jerk). Simultaneously, ipsilateral flexors must relax for extensors to function. To do this, branches of sensory fibers from MS activate inhibitory interneurons (7), which, in turn, inhibit the motor neurons to the flexor muscle (8).

WITHDRAWAL REFLEX (FLEXOR)

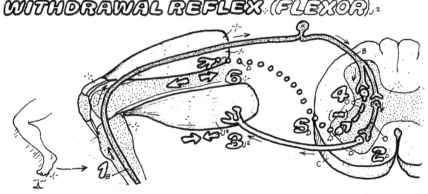

The withdrawal reflex is a defensive flexor reflex elicited in response to noxious (painful) stimulation of the foot. Sensory pain signals (1) excite motor neurons to the flexor muscles (2), eliciting flexion and withdrawal of the leg (3). Simultaneously, via inhibitory interneurons (4), motor neurons to the extensor muscles are inhibited (5) to relax the extensors of the same leg.

The centers for *voluntary motor control* are in the brain, as shown in stroke patients and those with brain damage due to gunshot wounds or accidents. In such cases marked movement deficits occur in the absence of any spinal damage.

THE MOTOR CORTEX INITIATES & EXECUTES FINE VOLUNTARY MOVEMENTS

Electrical stimulation of frontal lobe areas in animals results in muscle and limb movements. Electrical stimulation of the human frontal lobe, during neurosurgical operations, evokes movements of distinct body muscles. This area, located on the *precentral gyrus*, is called the *primary motor cortex* (PMC) or MI.

Body muscles are somatotopically mapped on the PMC— The PMC, like its sensory counterpart on the postcentral gyrus, shows a *somatotopic* organization (map of body muscles or a *motor homunculus*): the legs and trunks are on the top of the precentral gyrus, hand and finger areas on the lateral aspects, and the areas for head, tongue, and other speech muscles at the bottom of the gyrus. The representation map is contralateral for the trunk and limb muscles and bilateral for the head and speech muscles.

Similar to the sensory homunculus (plate 93), the representation in the motor homunculus is proportional to the degree of motor control and skilled movement of the part and not to the part's physical size. The small-sized, fast muscles of the hands and digits, capable of great movement versatility and fine motor control, have large cortical representations, as do the tongue and speech muscles, in contrast to small representation for the massive leg muscles.

In animals, the PMC is also somatotopically organized. Electrical stimulation of deep layers of PMC with weak currents evokes contractions of single muscles or discrete muscle groups. Surface stimulation with strong currents causes contractions of complex muscle groups, perhaps because the current spreads to wider cortical areas.

PMC neurons, columns, and layers— The PMC has six layers but is thicker than the sensory cortex and contains mainly pyramidal neurons, organized in vertical columns. Neurons in the deeper layers are the output neurons. The Betz cells, the very large pyramidal neurons once thought to be the cortical substrate of voluntary movement, comprise a small proportion of these neurons. Each column is concerned with a set of muscle fibers within each muscle. Firing pattern of these neurons (action potential frequency and volleys) determine duration and strength of contraction of their target muscles.

CORTICOSPINAL TRACT PROVIDES THE OUTPUT FOR PMC

To excite voluntary muscles, PMC output neurons must first excite the spinal motor neurons. This is done through the fibers of the *corticospinal (CS) tract* (= *pyramidal tract*). These output neurons are often referred to as the *upper motor neurons* in contrast to the *lower motor neurons* for peripheral motor efferents. The PMC and CS tract, formerly called the pyramidal system, are the chief executors of voluntary motor control, especially skilled movements. This system, present only in mammals, is well developed in monkeys and humans, where manipulation capacity is advanced and speech motor skill has evolved. Lesions of the CS tract in monkeys and humans markedly impairs initiation and execution of voluntary movements and fine motor control of distal muscles.

The corticobulbar & corticospinal tracts— Output fibers from the PMC initially form two divisions: *corticobulbar* and

corticospinal. Corticobulbar fibers descend ipsilaterally and terminate on the motor neurons of the brain stem to regulate the muscles of speech as well as tongue and head movements. The CS tract descends into the spinal cord to terminate on the spinal motor neurons; it controls trunk and limb movement.

Functional significance of segregation of CS tract— All CS fibers decussate before terminating on their spinal motor neuron targets, resulting in contralateral motor control by the brain: the PMC in the left hemisphere controls muscles on the right side and vice versa. In humans, about 80% of CS fibers decussate at the medulla level (pyramidal decussation); these fibers form the *lateral* CS tract and terminate on the spinal motor neurons that regulate movements of the *distal limb muscles* such as those of the hands and digits. This pathway is therefore critical in control of fine and skilled hand movement. The remaining 20% of CS fibers descend ipsilaterally in the *ventral (anterior) division* and cross the midline before terminating on the spinal motor neurons controlling axial and proximal muscles for gross control of trunk and limbs.

HIGHER MOTOR AREAS GENERATE COMPLEX MOVEMENT PATTERNS

Detailed stimulation of frontal lobe cortical areas anterior to the PMC has revealed two additional and major cortical motor areas: the *supplementary motor area* and the *premotor area*. These areas act as higher-order integrative "association" areas for programming of the PMC and other brain motor structures.

Supplementary motor area— This area is located just in front of the PMC, on the superior aspects of frontal lobe. Its output is mainly to the PMC and its electrical stimulation leads to complex, whole, and purposive movements. It may provide the PMC with detailed programs for complex movements; these programs are communicated to specific zones in the PMC, instructing them to initiate and execute the movements by activating their specific target muscles groups.

Premotor area— This area is also located in front of the PMC but inferior to the supplementary motor area; it communicates with the structures of basal ganglia and cerebellum (plate 97) in planning of movement and recruitment of these and other brain motor areas for initiation and execution of voluntary movements. Just before a movement is planned and initiated, the neurons of premotor and supplementary motor areas increase their firing of nerve impulses *before* the PMC neurons, in anticipation of movement. The association motor cortex receives association fibers from other cortical areas, especially from the somatic sensory association area. The illustration on the lower right corner of this plate represents diagramatically the relationship between the cortical motor and sensory areas in a cartoon of a real-life situation.

Effects of lesions— Lesions or ablation (removal) of the PMC or CS tract does not cause paralysis but produces a state of marked *paresis* (weakness or inability to initiate voluntary movement) in monkeys and humans; damage to the premotor cortex and lateral CS tract results in *poverty of skilled movements* and loss of *fine motor control* in the hands and speech muscles. When damage is restricted to areas in front of the hand motor cortex, skilled manipulation is markedly impaired; damage to the area anterior to the speech muscles area results in impaired articulation (plate 111).

CN: Use the same color for the spinal cord (C) as you used on the previous page.
1. Begin with the upper left diagram.
2. Color the representation of body muscles on the motor cortex (the motor homunculus).
3. Color the diagrams on motor and premotor cortex in the upper right region.
4. Color the material in the lower right.

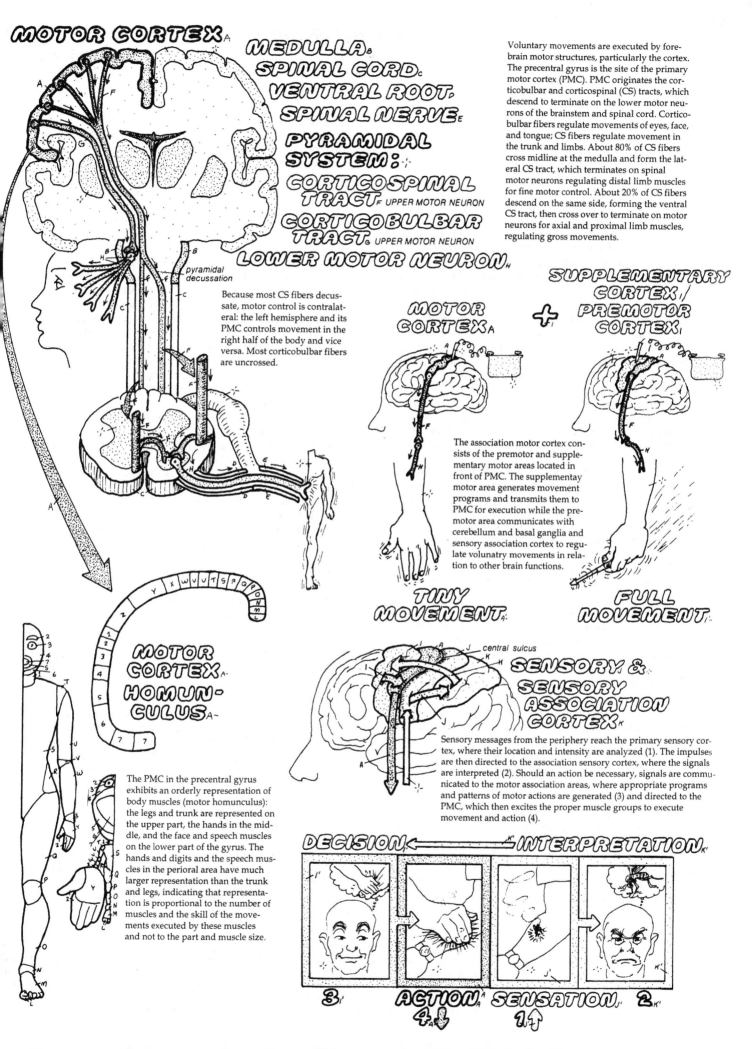

MOTOR CORTEX

MEDULLA, SPINAL CORD, VENTRAL ROOT, SPINAL NERVE, PYRAMIDAL SYSTEM: CORTICOSPINAL TRACT, UPPER MOTOR NEURON CORTICOBULBAR TRACT, UPPER MOTOR NEURON LOWER MOTOR NEURON

Voluntary movements are executed by forebrain motor structures, particularly the cortex. The precentral gyrus is the site of the primary motor cortex (PMC). PMC originates the corticobulbar and corticospinal (CS) tracts, which descend to terminate on the lower motor neurons of the brainstem and spinal cord. Corticobulbar fibers regulate movements of eyes, face, and tongue; CS fibers regulate movement in the trunk and limbs. About 80% of CS fibers cross midline at the medulla and form the lateral CS tract, which terminates on spinal motor neurons regulating distal limb muscles for fine motor control. About 20% of CS fibers descend on the same side, forming the ventral CS tract, then cross over to terminate on motor neurons for axial and proximal limb muscles, regulating gross movements.

pyramidal decussation

Because most CS fibers decussate, motor control is contralateral: the left hemisphere and its PMC controls movement in the right half of the body and vice versa. Most corticobulbar fibers are uncrossed.

MOTOR CORTEX + SUPPLEMENTARY CORTEX / PREMOTOR CORTEX

The association motor cortex consists of the premotor and supplementary motor areas located in front of PMC. The supplementay motor area generates movement programs and transmits them to PMC for execution while the premotor area communicates with cerebellum and basal ganglia and sensory association cortex to regulate volunatry movements in relation to other brain functions.

TINY MOVEMENT

FULL MOVEMENT

MOTOR CORTEX HOMUNCULUS

The PMC in the precentral gyrus exhibits an orderly representation of body muscles (motor homunculus): the legs and trunk are represented on the upper part, the hands in the middle, and the face and speech muscles on the lower part of the gyrus. The hands and digits and the speech muscles in the perioral area have much larger representation than the trunk and legs, indicating that representation is proportional to the number of muscles and the skill of the movements executed by these muscles and not to the part and muscle size.

central sulcus

SENSORY & SENSORY ASSOCIATION CORTEX

Sensory messages from the periphery reach the primary sensory cortex, where their location and intensity are analyzed (1). The impulses are then directed to the association sensory cortex, where the signals are interpreted (2). Should an action be necessary, signals are communicated to the motor association areas, where appropriate programs and patterns of motor actions are generated (3) and directed to the PMC, which then excites the proper muscle groups to execute movement and action (4).

DECISION ← INTERPRETATION

ACTION 4 — SENSATION 1 — 3 — 2

The *cerebellum* (CB) and the *basal ganglia* (BG) are *major subcortical motor systems* that help coordinate and integrate various forms of movement. These functions are performed in association with motor areas of the cerebral cortex.

CEREBELLUM IN MOTOR COORDINATION & LEARNING

The human CB is a prominent structure attached to the back of the brain stem by *cerebellar peduncles*. Its prominence is related to the disproportionate growth of the CB hemispheres, which are critical in coordination of speech and fast movements of the hands and eyes and the learning of motor tasks.

Organization and input/output systems — The CB consists of an overlying and highly folded *cerebellar cortex* and deep *cerebellar nuclei*. The CB cortex receives excitatory input from the premotor area of the cerebral cortex and the proprioceptors and muscles via the brain stem relay centers. This input is analyzed by CB cortex circuits, and the outcome is relayed to the deep CB nuclei by the *Purkinje cells*, which are inhibitory neurons, releasing the inhibitory neurotransmitter GABA (γ-amino-butyric acid). However, the neurons of CB nuclei — i.e., the true CB output — are excitatory and terminate on two targets: (1) the *red nucleus* in the midbrain, which mediates the CB output down to spinal motor systems, and (2) the *thalamus*, which mediates the CB output up to the premotor cortex.

CB FUNCTIONS ARE PERFORMED BY THREE DISTINCT ZONES

Basic knowledge of CB function comes from lesion studies in animals and effects of cerebellar damage in humans. These studies indicate that the CB is essential for proper and smooth *coordination* of balance, posture, and voluntary movements but not for their *initiation*. Motor deficits produced by CB damage occur ipsilaterally. Recent studies have implicated the CB in learning motor tasks as well. The CB shows three functional divisions: *vestibulocerebellum, spinocerebellum,* and *cerebrocerebellum*.

Balance & eye movements — The *vestibulocerebellum* is concerned with balance, eye movement, and head/eye movement coordination; this functions is executed by the *flocculonodular lobe*. The input to this lobe comes from the spinal cord and vestibular system of the inner ear and the output is to the spinal cord motor systems and the oculomotor nuclei of the brain stem.

Posture & locomotion — The *spinocerebellum* is concerned with coordination of gait, posture, and locomotion, functions executed by the *vermis* (snake), located over the midline zone. The vermis receives extensive projections from proprioceptors of joints and muscles and its output is to the midbrain and spinal cord motor systems for regulation of tension in the axial and postural muscles.

Skilled movements — The *cerebrocerebellum*, consisting of the *cerebellar hemispheres*, is concerned with coordination and smoothening of fast and skilled movements of distal muscles in the limbs and head. The input to CB hemispheres is from the *premotor cortex* and output is to cortical motor areas, via thalamus, as well as to midbrain and spinal motor systems. The cortical circuits of CB hemispheres are intimately implicated in learning functions of the CB.

CB MAY FUNCTION AS A COMPARATOR COMPUTER

The mechanisms by which the CB participates in overall coordination of movement is not well understood. One suggested mechanism is that it works as a *comparator neural device*, smoothening the movements by adjusting error signals.

The CB hemispheres have two-way communication with the motor cortex and the muscles, to coordinate motor performance. When a voluntary movement is planned (e.g., picking up a glass), the *supplementary cortex* generates the movement programs, relaying them to the *premotor* and *primary motor cortex* muscles via the corticospinal tract (plate 96). The same commands are sent by the *premotor cortex* to the CB via the *cerebellar relay nuclei* in the *pons*. The CB cortex matches these commands with the input from the joints and muscle receptors and generates appropriate error-correcting feedback signals, which it sends to the motor cortex via the thalamus. This alerts the motor cortex of the need for adjustments. At the same time, the CB, acting through the red nucleus and its descending connections with the γ motor neurons, modifies muscle tension and the stretch reflexes to bring the muscles in line with motor cortex commands.

BASAL GANGLIA AS MOTOR INTEGRATION CENTERS

The basal ganglia (BG) consist of five interconnected subcortical motor structures — three forebrain structures (*caudate, putamen, globus pallidus*) and two midbrain structures (*substantia nigra, subthalamus*); the *red nucleus* is sometimes considered as a functional part of the BG. The BG were once thought to originate the *extrapyramidal system*, regulating gross and unskilled (yet voluntary) movements — e.g., those of posture and locomotion. Recent findings of extensive two-way connections between the BG and the motor cortical areas emphasize the interaction between the two motor systems.

LESIONS OF BASAL GANGLIA PRODUCE DRAMATIC MOTOR ABNORMALITIES

Striking motor abnormalities are produced by lesions or degenerative diseases of the BG — e.g., Parkinson's and Huntington's diseases (choreas). The BG are involved in complex integrative motor functions, including gross voluntary movements — e.g., postural control or ballistic limb movements. The BG also control motor cortex output via extensive feedback connections. The output pathways of BG are consistent with these two functions.

BASAL GANGLIA OUTPUT IS MOSTLY TO MOTOR CORTEX

The major input to the BG is from the premotor cortex to the caudate nucleus and putamen. This input is integrated and relayed to the globus pallidus, which serves as an output center, projecting, by way of the thalamus, to the premotor and primary motor cortex. By this route the BG exert control on voluntary motor activity. The subthalamus and substantia nigra, the midbrain components of BG, have reciprocal connections with the caudate and putamen and also act as output centers to brain-stem motor nuclei regulating eye movements and limb and posture control. The descending *rubrospinal* and *reticulospinal* pathways may mediate axial effects by controlling the activity of γ motor neurons of the spinal cord (plate 82) and hence muscle tension, the stretch reflex, and proprioceptive and kinesthetic activities.

CN: Use the same colors as on the previous page for the premotor cortex (C) and motor cortex (D). 1. Begin in the upper left corner by coloring the titles of the various structures that make up the basal ganglia (A). Note that only the titles are present; all the structures are represented by a single structure labeled A. 2. Color the structures of the cerebellum on the left side of the page. Color the diagram of its function 3. Color the titles of the various disorders on the bottom of the page. Use the same color for the large X suggesting malfunction.

BASAL GANGLIA:
CAUDATE NUCLEUS A¹
PUTAMEN A²
GLOBUS PALLIDUS A³
SUBTHALAMUS A⁴
SUBSTANTIA NIGRA A⁵

The basal ganglia (BG) are motor structures in the forebrain (caudate, putamen, and globus pallidus) and midbrain (subthalamus, substantia nigra) that integrate voluntary and complex involuntary movements. Lesions of BG produce dramatic motor disorders (see lower panel). The BG have extensive two-way connections with cortical motor areas and the CB as well as access to descending pathways for axial and gross limb movement control.

THALAMUS B
PREMOTOR CORTEX C
MOTOR CORTEX D
RED NUCLEUS E
EXTRAPYRAMIDAL TRACTS F
MUSC. SENS. SIGNAL G
PYRAMIDAL TRACT H

The BG receive major input from cortical motor areas, especially the premotor and supplementary motor areas (1). The BG integrate these signals and feed them back to the premotor cortex (2) via the thalamus. The premotor then activates the PMC (3) to execute movement (4) and the CB (5) to coordinate it (6). Other descending pathways (7)—once called extrapyramidal—mediate control of BG over tension in axial and proximal limb muscles through changes in γ-efferents and spindle stretch function (8).

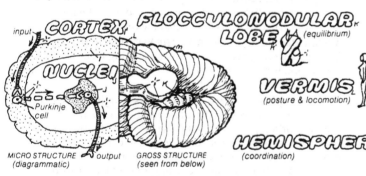

CEREBELLUM

The cerebellum (CB) is the major center for motor coordination. Three functional zones are recognized. (1) The vestibulocerebellum consists mainly of the flocculonodular lobe and coordinates balance and eye/head movement in connection with the vestibular organ in the inner ear. (2) The spinocerebellum consists of the vermis; it receives input from muscles and joints and coordinates posture, gait, and locomotion. (3) The cerebrocerebellum consists of CB hemispheres; it coordinates and smoothens fast and skilled movements by distal muscles. The CB cortex receives and integrates input from brain and periphery and relays it to CB nuclei, which serve as the CB output structure. CB nuclei are under inhibitory control of the Purkinje cells of the CB cortex.

CORTEX
FLOCCULONODULAR LOBE (equilibrium)

NUCLEI
Purkinje cell

MICRO STRUCTURE (diagrammatic) → output

GROSS STRUCTURE (seen from below)

VERMIS
(posture & locomotion)

HEMISPHERE
(coordination)

FUNCTION

The CB may function as a comparator neural device that monitors the commands of the cortical motor centers and the performance of motor effectors. Its output serves to correct error signals and improve motor performance. For example, to hold a glass steady, the supplementary MC (1) signals the appropriate motor patterns to the primary MC (2), which transmits them to arm and hand muscles (3). At the same time, the supplementary & pre-MC send similar signals to the CB (4). Muscle and joint receptors inform the CB of their performance and position (5). CB matches performance of effectors with the motor commands from the PMC and sends feedback error-correcting signals to the pre-MC (6) to adjust its motor commands appropriately. CB also signals the γ efferents to modify muscle tension and readiness tonus appropriately.

DISORDERS OF BASAL GANGLIA & CEREBELLUM

SUBSTANTIA NIGRA A⁵
PARKINSON'S DISEASE

Degeneration of dopamine-releasing neurons in the substantia nigra results in Parkinson's disease, whose frequency increases with age; it is characterized by bradykinesia (poverty of movement), rigidity and tremor (shakes), and masked face.

CAUDATE & PUTAMEN A¹ A²
CHOREA

Degeneration of neurons in the striatum (caudate-putamen) causes "choreas," diseases in which orderly progression of voluntary movements, as in walking, is replaced by rapid, involuntary dancing movements (St. Vitus dance). Huntington's disease is an example, caused by loss of GABA-releasing neurons of the striatum.

GLOBUS PALLIDUS A³
Degeneration of the globus pallidus results in "athetosis," involuntary writhing (twisting and turning) movements of the limbs.

ATHETOSIS

CEREBELLUM I
Damage to the CB results in poor posture, disrupted equilibrium, and ataxia (disorders of movement coordination). Gait is wide and appears drunken. Postural (extensor) muscles are weak (hypotonia). Speech is slurred. Voluntary movements are accompanied by "intention tremor" and dysmetria (past-pointing).

HYPOTONIA
DYSMETRIA
ATAXIA
INTENTION TREMOR

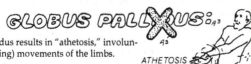

The *eye*'s anterior optical apparatus forms a sharp image on the retina, a sheet of nervous tissue located at the back of the eye on its inner surface. An image falling onto the retinal surface produces neural signals on a mosaic of photoreceptors; these convert the incident light photons to nerve signals. Thus the retina forms a two-dimensional neural image that is transmitted by the visual pathways to the subcortical and cortical integrative visual centers. These reconstruct the three-dimensional visual images that we perceive. In this plate, we study the eye's optical structures and functions.

EYE'S OPTICAL PARTS REFRACT, CONVERGE, & FOCUS LIGHT RAYS

The eye's optical functions are to refract (bend) the light rays emitted from objects and sharply focus them onto the *retina*. The objects may be as simple as a point light source (a small distant candle) or more complex, such as lines and circles, or still more complex stationary or moving forms, such as buildings or a bird in flight. The images on the retina are always smaller than the objects that produce them. In the human *fovea* (a small patch of retina in the eye's posterior pole and along its central optical axis), where spatial vision is highly developed, image size is always less than 1 mm, whether from a tree or from a leaf!

Cornea & aqueous humor are the early refracting media— Light rays pass through several *transparent* media in the eye before striking the retinal photoreceptors. These media help refract and converge the rays so that the image falling on the retina is inverted and smaller than the object. The first of these media is the *cornea*, which, because of its higher density (compared to air) and its curved surface, refracts the rays inward, like a convex lens. Defects of corneal curvature give rise to *astigmatism*. Next the rays pass through the *aqueous humor*, a viscous fluid in the eye's *anterior chamber* (between the lens and cornea). This fluid is produced in the *ciliary body* and exits through a venous channel, *canal of Schlemm*. Excessive fluid pressure in the anterior chamber (normally about 20 mm Hg) is the basis of *glaucoma*, a serious eye disorder that can cause blindness.

The iris and pupil regulate the amount of light entering the eye— The circular pupil is an aperture formed by a ring of pigmented smooth muscles called the *iris*. Contraction of the iris *sphincter* muscles constricts the pupil; that of the *dilator* muscles widens it. The pupil's well-known function is the *light reflex*. When exposed to bright light, the pupil constricts to permit less light to enter the eye; in the dark, the pupil dilates to allow more light in. Another function of the pupil is focusing for near objects (see below). The sympathetic nerves dilate the pupil and parasympathetic nerves constrict it.

The lens can change its curvature to focus an image— Light rays, after passing through the anterior chamber and the pupil, strike the eye's crystalline *lens*, which acts like a biconvex glass lens. Parallel rays (i.e., from more than 6 m) entering the lens periphery are refracted inward, converging on a *focal point* behind the lens and along its *optical axis*, a straight line passing through the lens center. In a glass lens, the *focal distance* (i.e., the distance between a lens and its focal point) is fixed. The human lens can actively change this distance (16

mm at rest) by changing its curvature (see below). Because of the lens's biconvexity, the image formed on the retina is upside down. The brain inverts this image so that our mental images of the objects are right side up. Behind the lens there is one last transparent medium, the gel-like *vitreous humor*, which helps maintain the eyeball's spherical shape.

Accommodation reflex regulates the lens-focusing mechanism— The ability of the lens to form a sharp image on the retina is best shown during focusing for a near object. The reflex that regulates this function is called the *accommodation reflex*. The lens is held in place by the *lens ligaments* attached to it and by the *ciliary muscles*. When these muscles contract, the ligaments loosen, releasing their tension on the lens. As a result, the elastic lens relaxes, assuming a spherical shape; this decreases the focal distance of the lens. Relaxation of the ciliary muscles pulls the ligaments away from the lens and increases their tension on the lens, making it flatter; this increases the lens's focal distance. In order to form sharp retinal images of distant objects, the ciliary muscles relax to flatten the lens; for near objects, they contract to increase the lens's curvature. The lens's ability to change curvature for sharp focusing is called *accommodation*. The sympathetic nerves relax the ciliary muscles for far vision and parasympathetic nerves contract it for near vision.

Accommodation also involves changes in the *pupil* size. When the eyes shift from looking at a far object to a near one, the pupil constricts. This *pupillary constriction* response increases the depth of visual field, like a pinhole, permitting sharp focus and clear vision of near objects. Because the two eyes converge during the near-vision response, the pupillary constriction occurring is also called the *convergence response*.

VISUAL PROBLEMS ARISE FROM THE EYE'S OPTICAL DEFECTS

Visual problems may be peripheral, resulting from various optical abnormalities of the eye, or they may stem from retinal or central nervous dysfunctions.

Presbyopia & cataracts are disorders of aging lenses— Lens elasticity decreases sharply during adulthood so that by the early fifties, humans totally lose the capacity for accommodation, a condition called *presbyopia*; to correct for this condition biconvex corrective lenses are required for near-vision tasks such as reading. Another aging-related lens disorder is *cataracts*, caused by pigment accumulation in the lens, which makes it opaque and causes light scattering and glare. Cataracts are now treated by replacing the lens with a plastic one.

Near- & far-sightedness are due to abnormal eye shape— Some eye abnormalities are caused by the eyeball's shape defects. Long and ellipsoid eyes cause *myopia*, in which the focal point falls in front of the retina, resulting in blurred visual images (*nearsightedness*). To be seen better, the object must be nearer to the eye. This defect is corrected by placing biconcave lenses in front of the eyes. These diverge the light rays before they enter the eye, in effect bringing the object closer. Short eyeballs cause *hyperopia*; here the focal point falls behind the retina. Hyperopic individuals see distant objects better (*farsightedness*). A biconvex corrective lens converges the light rays before they enter the eye, in effect moving the object farther away.

CN: Use your lightest colors for A, B, C, and G.
1. Begin with the structure of the eye and its comparison to a camera.
2. Color the panel on light refraction.
3. Color the defects and correction of image formation.

4. Color the three examples of near-vision adjustments, noting that presbyopia (on the right) is a problem associated with accommodation (example 1).
5. Follow the numbered sequence of steps in the control of near-vision reflexes.

STRUCTURE OF THE EYE

CORNEA_A
AQUEOUS HUMOR_B
LENS_C
CILIARY MUSCLE_D
LIGAMENTS_E
IRIS_F
VITREOUS HUMOR_G
RETINA_H
CHOROID_I
SCLERA_J
OPTIC NERVE_K

LENS_C' / DIAPHRAGM / FILM_H'

fovea

conjunctiva

The eye is like a camera. The eye's diaphragm is the iris, which changes pupil aperture, regulating the amount of light entering. The curved cornea and the biconvex lens are like a camera's compound lens, refracting and converging light rays to form inverted images on the light-sensitive retina (film). The ciliary muscles change lens curvature for sharp focusing; in a camera, the lens moves back and forth.

DEFECTS OF IMAGE FORMATION

(farsighted) HYPEROPIA_L'

DEFECT — too short

too long

MYOPIA_L² (nearsighted)

The eyeball is considered normal (emmetropic) if, at rest, it can focus parallel rays from distant objects sharply on the retina. Short eyeballs focus images behind the retina (far-sightedness; hyperopia), a defect corrected by using convex lenses (glasses). Long eyeballs focus images in front of the retina (near-sightedness; myopia), a defect corrected by using concave lenses (glasses).

convex lens — CORRECTED

concave lens

REFRACTION_L³ / LIGHT RAY_L

Light rays passing through the transparent media of different densities are bent (refraction). The denseness and curvature of the medium determine the degree of refraction. Refraction is necessary for forming a small-sized image on the retina. The eye's refractive media (cornea and lens) together act as a single convex lens system (reduced eye), enabling the formation of small inverted images of objects on the retina.

NEAR VISION ADJUSTMENTS

1. ACCOMMODATION

distant viewing

near

As a distant object moves closer, the image moves behind the retina. To keep the image sharply on retina, the lens accommodates: ciliary muscles contract, lens ligaments relax, and the lens becomes rounder. This moves the focal point closer to the lens, focusing the image. With age, the lens hardens and is less able to accommodate. After age 55, accommodation is no longer possible (presbyopia), requiring corrective lenses (glasses) for reading, etc.

PRESBYOPIA_C'

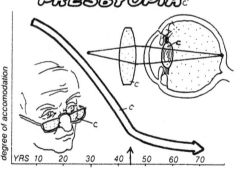

degree of accomodation

YRS 10   20   30   40   50   60   70

2. PUPIL CONSTRICTION

distant   *near*

During accommodation, the iris also constricts to narrow the pupil, permitting increased depth of focus. For very close objects, external eye muscles move the eyeballs inward (converge) to keep sharp focus.

3. CONVERGENCE

MUSCLE_M

Light rays from an approaching object (1) form a blurred image behind the retina (2). The blurred image signals (3) are sensed by the brain visual centers (4), which activate midbrain motor centers (5) to send corrective motor signals for lens accommodation. Parasympathetic fibers (6) release acetylcholine to contract the ciliary muscles (7), which relax the lens for sharp focusing; these nerves also stimulate the iris to constrict the pupil (8). Sympathetic fibers release norepinephrine to stimulate the iris to dilate the pupil (9).

CONTROL OF NEAR VISION REFLEXES

CONSTRICTS ← PARASYMPATHETIC N.

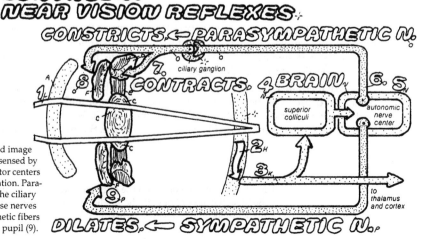

ciliary ganglion

CONTRACTS   BRAIN

superior colliculi

autonomic nerve center

to thalamus and cortex

DILATES ← SYMPATHETIC N._P

The *retina*, is a layered sheet of neural tissue that covers the eye's posterior inner surface. Except for the *blind spot* and the *fovea*, retina's structure is uniform. Retinal photoreceptors transduce the light stimuli into nerve signals and other retinal neurons transform these nerve signals into a two dimensional map that is transmitted to the brain.

RETINAL CELLS FORM A FUNCTIONAL CIRCUIT

The retina has five cell types and three cell layers. The retina's neurons include the *photoreceptor cells* (PR-cells), the *bipolar cells* (BP-cells), the *ganglion cells* (G-cells), the *horizontal cells*, and the *amacrine cells*. The outermost PR-cell layer, the middle BP-cell layer, and the innermost G-cell layer form the retina's three main layers. The horizontal cells and amacrine cells are located next to the PR-cell and G-cell layers, respectively.

Since PR-cells form the retina's outermost layer, light rays must pass through all retinal layers before striking them. The adaptive advantage of this pattern may relate to the *melanin*-containing *pigment cell layer*, found apposed to the external aspect of PR-cells. Melanin makes the interior of the eye black preventing back-reflection of light and glare. Light and dark stimuli change the excitability of PR-cells; this is sent synaptically to the BP-cells, which in turn influence the activity of the G-cells, the retina's output cells. G-cell axons, conveying nerve impulses to the brain, congregate at the *optic disk*, where they form the *optic nerve*. Light stimuli falling on the optic disc is not perceived (the blind spot) due to absence of PR-cells.

Horizontal cells modulate the activity of adjacent PR-cells and amacrine cells modulate the activity of adjacent G-cells. Since retinal neurons lack axons (except G-cells), they produce slow, graded potentials.

RODS AND CONES HAVE DIFFERENT FUNCTIONAL PROPERTIES

The retina has two kinds of PR-cells: *rods* and *cones*. Rods are much more numerous than cones, are very sensitive to light (low threshold), and function in dim light (*night vision*). The eyes of nocturnal animals contain mainly rods. Cones require more light for activation, are sensitive to colors, and function best in *day vision*. Rods are found mainly in the retinal periphery; cones are heavily concentrated in the fovea. There are also more G-cells per unit area of fovea, providing direct channels (one cone to one G-cell) between cones and brain cells. In the retinal periphery, the receptor cell–to–neuron ratio is very high (100 rods to 1 G-cell); this increases G-cell light sensitivity. The fovea is used for day vision, color vision, and that requiring great *visual acuity*, such as reading small print. When inspecting an object carefully, the eyes move so that the fovea is placed directly along the eye's optical axis. The retinal periphery is ideal for night vision, being so sensitive that candlelight can be seen 10 miles away.

MOLECULAR CASCADE AND ELECTRICAL RESPONSES IN RODS

Photoreceptor molecules in rods detect light and evoke chemical events mediated by second messengers to change the electrical properties of rods. Light hyperpolarizes the rods.

cGMP mediates the effects of light on rod PR-cells — How does light cause hyperpolarization of the rod cell membrane? In the dark, sodium ions continuously enter the rods via Na^+ channels; this maintains the cell in a depolarized state. Stimu-

lation of rod cells by light decreases the intracellular levels of *cyclic GMP* (cGMP), an intracellular second messenger. The rod Na^+ channels are *cGMP-gated* – i.e., their gates open when cGMP binds to the channel protein. Thus a decrease in cGMP results in the *closing* of Na^+ channels, hyperpolarizing the rod membrane. How does light change cGMP levels?

Rhodopsin, light-reaction, & cGMP — Rods contain, in their outer zone, numerous membranous *disks*, each containing millions of molecules of *rhodopsin*, the photoreceptor protein molecule (visual purple). Rhodopsin is a membrane-bound protein consisting of a protein, *opsin*, and a light-sensitive pigment, *retinine* (retinaldehyde, retinal). In its "11-cis" position, retinine's hydrocarbon chain is bound to opsin. Light switches the chain to the "all trans" position (*light reaction*), dissociating retinine from opsin. This separation activates an adjacent G-protein called *transducin*, which in turn activates an adjacent phosphodiesterase enzyme. This enzyme converts cGMP to 5'-GMP, thereby reducing cGMP levels. Since presence of cGMP keeps the Na^+ channels open, its absence results in closing of these channels and hyperpolarization of the rod membrane. The effect of this molecular cascade is to amplify the light signal; as a result, for each rhodopsin activated, thousands of Na^+ channels will close. Rhodopsin-like photoreceptor proteins are also found in cone cells, which are sensitive to blue, green, and red light and perform in color vision.

Rhodopsin levels are restored during dark adaptation — Staring at a highly illuminated white sheet temporarily reduces vision; closing the eyes restores vision. High illumination diminishes the supply of rhodopsin. In the dark, rhodopsin slowly reforms by recombination of opsin with *vitamin A*, the oxidized form of retinine (*dark adaptation*). Vitamin A deficiency can lead to *night-blindness* (inability to see in dim light). During dark adaptation, retinal sensitivity gradually but markedly increases (100,000 times in 20 min). A dark-adapted eye can detect a single quantum of light.

RETINAL NEURONS INTEGRATE OUTPUT OF RODS & CONES

Objects in the visual fields are a collection of light and dark points, with point image on the retina. The retina's integrative neurons utilize inhibitory and excitatory mechanisms to construct a two-dimensional map of the image and transmit it to the brain.

Synaptic excitation & inhibition play critical roles in retinal circuits — In the dark, the depolarized rods stimulate the inhibitory BP-cells, which in turn inhibit the G-cells. The inhibited G-cells do not send signals to the brain, which interprets this as a dark spot. Light stimuli hyperpolarize the rods; this removes their excitatory effect over the BP-cells, which remove their inhibitory effect over G-cells. Activated G-cells send signals to the brain, indicating the presence of light.

The retina forms a 2-dimensional map to transmit to the brain — Most objects are more complicated than a light source; these stimulate the retina's PR-cells at many different points, corresponding to the image shape and size. Signals from the retina's PR-cells are integrated by the retina's inhibitory and excitatory interneurons to construct a two-dimensional map that is transmitted to the brain by the axons of the G-cells.

CN: Use the same colors as on the previous page for A, C, and D.
1. Begin in the upper left with the small diagram of the eye. Then work from the pigment epithelium (E) upwards.

2. Color the lower panel, beginning with the enlarged rod ce (G) on the far left. Note that only the cell membrane (G') is colored, along with the disks (M') in the outer zone.

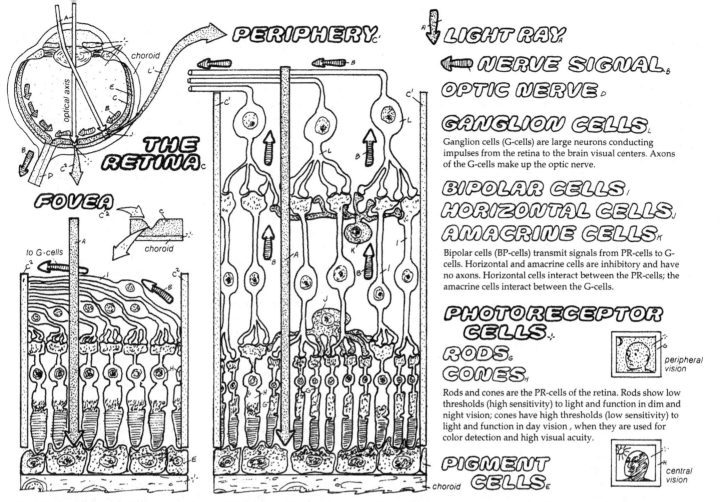

PERIPHERY

LIGHT RAY

NERVE SIGNAL

OPTIC NERVE

GANGLION CELLS

Ganglion cells (G-cells) are large neurons conducting impulses from the retina to the brain visual centers. Axons of the G-cells make up the optic nerve.

BIPOLAR CELLS, HORIZONTAL CELLS, AMACRINE CELLS

Bipolar cells (BP-cells) transmit signals from PR-cells to G-cells. Horizontal and amacrine cells are inhibitory and have no axons. Horizontal cells interact between the PR-cells; the amacrine cells interact between the G-cells.

PHOTORECEPTOR CELLS

RODS
CONES

Rods and cones are the PR-cells of the retina. Rods show low thresholds (high sensitivity) to light and function in dim and night vision; cones have high thresholds (low sensitivity) to light and function in day vision, when they are used for color detection and high visual acuity.

peripheral vision

PIGMENT CELLS

central vision

THE RETINA

FOVEA

choroid · *to G-cells*

The retinal fovea contains mainly cones. BP-cells and G-cells are pushed aside to permit light to strike the cones without interference. The nearly 1:1 ratio between the cones and G-cells (not shown) permits the fovea to be the center for high visual acuity and spatial discrimination (day vision). The retinal periphery contains mainly rods, which have a high convergence ratio to G-cells (100:1). These features make the retinal periphery highly sensitive in low illumination (night vision).

PHOTORECEPTOR CELL EXCITATION

synapse · *inner zone* · *outer zone* · *pigment cell*

PHOTORECEPTOR MEMBRANE
RHODOPSIN: *
OPSIN
11-CIS RETININE
TRANS-RETININE
DISK

(rhodopsin synthesis) dark reaction

light reaction (rhodopsin breakdown)

vitamin-A · *pigment layer*

The rods contain a light-sensitive molecule, rhodopsin (R). R consists of a protein, opsin, attached to a pigment, retinine. Light physically changes retinine from the 11-cis form to the all-trans form, triggering electrical activity in PR-cell. In the dark, R is reformed by recombination of opsin with the regenerated 11-cis retinine (dark reaction). Regeneration occurs both in the rods and in the underlying pigment layer. The latter can store 11-cis retinine as vitamin A, which it obtains from the blood.

LIGHT ABSENT

"open" Na+ channel · *HIGH c-GMP* · *Na+* · *gate* · *Na+ channel*

Rod depolarized · *BP excited* · *GC inhibited*

In the dark, R molecules are stable, signaling Na$^+$ channels to stay open. Inflow of Na$^+$ depolarizes the rods and their synapses. This activates the inhibitory BP-cells, which exert inhibition over the G-cells. Thus G-cells are quiet in the dark.

LIGHT PRESENT

"close" Na+ channel · *LOW c-GMP* · *outside* · *inside*

Rod hyperpolarized · *BP inhibited* · *GC excited*

In the light, decomposition of R molecules signals the Na$^+$ channels to close. This hyperpolarizes the rods and their synapses, leading to inhibition of BP-cells. Inhibited BP-cells allow increased activity in the G-cells and their axons, informing the brain cells of the presence of light.

From the light and dark spots impinging on its photoreceptors, the *retina* generates an orderly spatial map of the visual field to transmit to the *thalamus* and *visual cortex* of the brain via its *ganglion cells* (plate 99). These functionally specialized cells are grouped into a *P type* that encodes information on form and color and an *M type* that encodes motion and dynamic properties. Signals from these cells are sent to the *thalamus* and then to the *primary visual cortex* (PVC) and its higher-order visual areas, which synthesize a three-dimensional image of the visual field, including form, brightness, contrast, color, and motion.

Optic pathways are segregated—The axons of ganglion cells form the *optic nerve* as they leave each eye. The optic nerves of the two eyes converge at the *optic chiasm*. Here the fibers from the *nasal* segment of each retina cross over to the opposite side, and the *temporal* segments stay on the same side. After crossing, the optic nerve is called the *optic tract*. Because of the crossing, the right and left optic tracts carry signals relating to the left and right *visual fields*, respectively. Tests of visual field defects help determine possible locations of lesions in the visual pathway.

THALAMUS & MIDBRAIN SERVE AS SUBCORTICAL VISUAL CENTERS

Most of the ganglion cells axons in the optic tract, concerned with visual perception, end in the visual centers of the thalamus—the *lateral geniculate body*. The P and M ganglion cell axons make their first synapse in the *parvocellular* (P) and *magnocellular* (M) layers of the lateral geniculate. Here the different retinal signals are integrated before further transmission to the visual cortex. A smaller number of optic tract fibers enter the midbrain and terminate in the *superior colliculus* and *reticular formation*. The superior colliculus coordinates accommodation and light reflexes involving the lens and the pupil, as well as some eye and head movement reflexes; roles in visual processing also have been suggested. The reticular formation functions in cortical arousal, excitability, and sleep.

PRIMARY VISUAL CORTEX IS THE INITIAL CORTICAL SITE FOR VISUAL INTEGRATION

The neurons of the lateral geniculate body of the thalamus give rise to the fibers of the *optic radiation*, which projects onto a specific area of the posterior lobe of the cerebral cortex called the *primary visual cortex (PVC, striate cortex, area 17)*.

The retina maps onto the PVC retinotopically but the fovea has enlarged representation—The entire retina is represented on the cortex, point to point, in a precise and organized way so that each retinal half and each quadrant can be clearly demarcated (*retinotopic mapping*). Damage to circumscribed areas in the PVC (e.g., from gunshot wounds) results in blindness in specific areas of the visual field. Although retinal representation in the visual cortex is very precise, it is not equal to the shares of the *fovea* and *periphery*. The retinal fovea, less than 1 mm wide, has a very large representation, while the much larger retinal periphery has a relatively smaller representation. The reason for this lies in the higher density of the ganglion cells emerging from the fovea as well as in the essential role of the fovea in visual acuity, spatial perception, and

color vision—functions requiring more neuronal units and brain area for analysis and integration.

Visual cortex neurons are organized in layers and functional columns—The PVC neurons, like other cortical areas, are arranged in six layers. The cells of each layer have different input and output and perform different visual functions. The stellate cells in layer IV receive visual input from the thalamus. The PVC cells are also organized along functional "columns" that run perpendicular to the layers. Two general types of columnar organization have been found. The *ocular dominance columns* alternate and respond to input from the right and left eyes. The *orientation columns* contain all cells that respond to stimuli with a particular orientation. Neurons within each column have the same receptive field.

Simple & complex cells of the PVC are feature detectors & occur in the same column—In contrast to retinal ganglion cells, which have a circular receptive field and respond to a point source of light, cells of layer IV, the principal recipients of thalamic visual input, are sensitive to stimuli with forms—e.g., *lines* and *edges*. Moreover, a line stimulus must have a particular *orientation* to activate a cortical cell. These orientation-specific neurons, called *simple cells*, will respond only if a line with their preferred orientation is displaced in the receptive field. Some visual cortex neurons respond to the same stimulus in any position within the receptive field as long as the same orientation is maintained. These *complex cells* may be construed as movement detectors since they continue to respond when the stimulus is moving; complex cells are located within the same orientation column above and below the simple cells (in layers II, III, and V) and receive their input from the simple cells. Hypercomplex cells integrate input from complex cells. In addition to columns, other organized cell groups in layers 2 and 3 , called "blobs" because of their appearance in cytochrome oxidase–stained material, deal with color perception.

HIGHER-ORDER VISUAL AREAS PROCESS FORM AND MOTION PERCEPTION

Surrounding the PVC, also called V1, are the *higher-order visual areas* (numbered V2 through V6). These areas, once called the visual association areas, also have retinotopic maps and receive their input initially from the PVC. They are organized, in terms of information flow, both serially (V1 → V2 → V3 → V4 → V5 → V6) and in parallel (V1 → V2 → V4 or V1 → V5). The higher visual areas further analyze and integrate the PVC input (lines, orientations, simple color) to encode more complex and detailed visual patterns (detailed form, colored images, motion detection). Higher visual areas in monkeys and humans are extremely extensive and well developed. For higher visual perception, two parallel cortical pathways (streams) exist. A *temporal stream*, involving the higher-order association areas of the occipito-temporal lobe, deals with increasingly more complex aspects of form and color perception; information in this stream originates in the P-type retinal ganglion cells. A second *parietal stream* integrates impulses coming from M-type ganglion cells and involves spatial and motion perception. Still-higher order temporal visual areas are specialized for recognition and naming of faces. Lesions in any of these areas affect their specific visual functions.

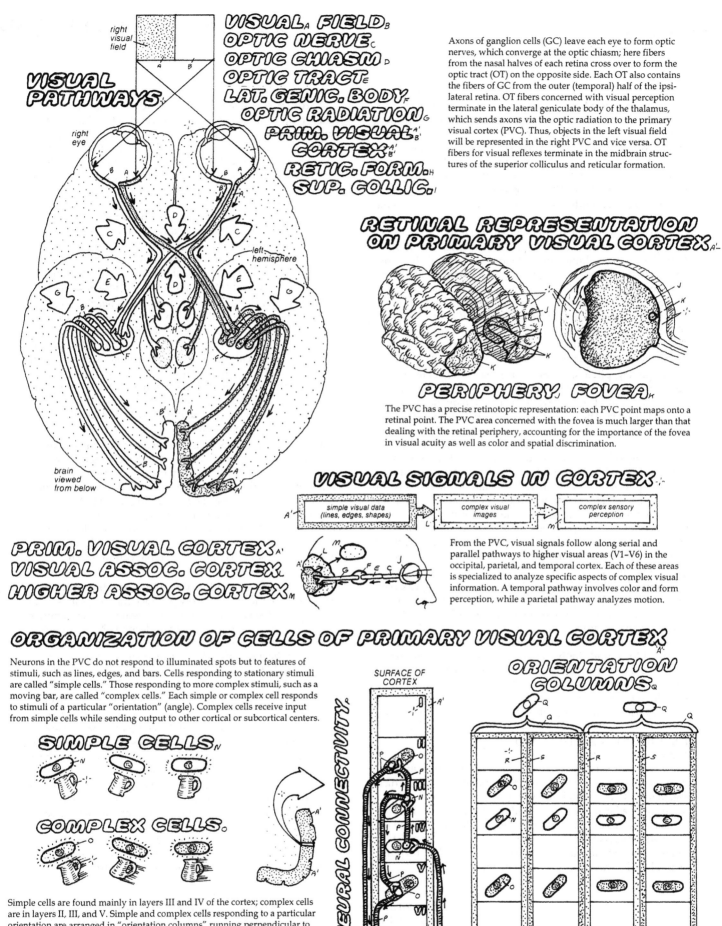

VISUAL PATHWAYS

right visual field

VISUAL FIELD.
OPTIC NERVE
OPTIC CHIASM
OPTIC TRACT
LAT. GENIC. BODY
OPTIC RADIATION
PRIM. VISUAL
CORTEX
RETIC. FORM.
SUP. COLLIC.

right eye

left hemisphere

brain viewed from below

Axons of ganglion cells (GC) leave each eye to form optic nerves, which converge at the optic chiasm; here fibers from the nasal halves of each retina cross over to form the optic tract (OT) on the opposite side. Each OT also contains the fibers of GC from the outer (temporal) half of the ipsilateral retina. OT fibers concerned with visual perception terminate in the lateral geniculate body of the thalamus, which sends axons via the optic radiation to the primary visual cortex (PVC). Thus, objects in the left visual field will be represented in the right PVC and vice versa. OT fibers for visual reflexes terminate in the midbrain structures of the superior colliculus and reticular formation.

RETINAL REPRESENTATION ON PRIMARY VISUAL CORTEX

PERIPHERY FOVEA

The PVC has a precise retinotopic representation: each PVC point maps onto a retinal point. The PVC area concerned with the fovea is much larger than that dealing with the retinal periphery, accounting for the importance of the fovea in visual acuity as well as color and spatial discrimination.

VISUAL SIGNALS IN CORTEX

| simple visual data (lines, edges, shapes) | complex visual images | complex sensory perception |

PRIM. VISUAL CORTEX
VISUAL ASSOC. CORTEX
HIGHER ASSOC. CORTEX

From the PVC, visual signals follow along serial and parallel pathways to higher visual areas (V1–V6) in the occipital, parietal, and temporal cortex. Each of these areas is specialized to analyze specific aspects of complex visual information. A temporal pathway involves color and form perception, while a parietal pathway analyzes motion.

ORGANIZATION OF CELLS OF PRIMARY VISUAL CORTEX

Neurons in the PVC do not respond to illuminated spots but to features of stimuli, such as lines, edges, and bars. Cells responding to stationary stimuli are called "simple cells." Those responding to more complex stimuli, such as a moving bar, are called "complex cells." Each simple or complex cell responds to stimuli of a particular "orientation" (angle). Complex cells receive input from simple cells while sending output to other cortical or subcortical centers.

SIMPLE CELLS

COMPLEX CELLS

NEURAL CONNECTIVITY

SURFACE OF CORTEX

SUB-CORTEX

output   input

ORIENTATION COLUMNS

OCULAR DOMINANCE COLUMNS

RIGHT EYE / LEFT EYE

Simple cells are found mainly in layers III and IV of the cortex; complex cells are in layers II, III, and V. Simple and complex cells responding to a particular orientation are arranged in "orientation columns" running perpendicular to the cortical surface. The orientation columns are placed side by side across the cortex in such a way that those receiving input from one eye alternate with those connected to the other eye (ocular dominance columns).

Sound waves are a form of mechanical energy. The various parts of the *ear* amplify the sound waves and transduce them into nerve impulses; these are relayed to the *central auditory system*, where they are perceived as meaningful sounds. Sound is produced by particles vibrating in a physical medium (air, water, or solids). Two physical parameters characterize sounds: *amplitude* and *frequency*.

Amplitude & loudness — The energy (intensity) of a sound is measured in terms of its *peak amplitude* — i.e., the maximum height of the sinusoidal oscillation. The *loudness* is the sound's perceptual equivalent and is expressed in terms of *decibels* (= 0.1 bel), a logarithmic unit. A loudness of 0 decibels corresponds to the lowest hearing threshold for humans at a standard sound pressure of 0.0002 dyne/cm^2. The loudness range of common sounds varies between 20 decibels for a whisper, 60 for normal conversation, 80 for heavy traffic, and 160 for jet planes. Each 20-decibel increase in loudness corresponds to a 10-fold increase in sound intensity. Discomfort and pain are felt at sounds of 120 and 140 decibels, respectively.

Frequency, pitch & timbre — *Frequency* refers to the number of oscillations — cycles per second, or Hertz (1 Hz = 1 cycle/sec). Audible sound waves range in frequency from 1 to over 100,000 Hz. The perceptual counterpart of sound frequency is *pitch*. A newborn's cry is perceived as high-pitched because of its high dominant frequency. After puberty, men develop larger larynxes (voice boxes) with thicker vocal cords, which produce lower sound frequency (pitch) compared to the high-pitched sound of an adult female. Average conversation pitch is 120 Hz for males and 250 for females. A third perceptual component of sound is its quality, or *timbre*, a property not easily explicable by the physical characteristics of sound waves. Identical musical tones of the same loudness emitted from two different instruments are thus perceived quite differently.

Human hearing abilities — The human ear at its peak performance — i.e., during childhood and the early teens — can detect sound waves in the frequency range of 20 to 20,000 Hz. However, the auditory sensitivity or *hearing threshold* is not the same for the various frequencies. Highest sensitivity is observed for sounds in the range of 1 to 4 KHz, which corresponds to the normal range of human speech sounds. Hearing abilities of different animals vary. Dogs can hear sounds up to 40 KHz (the dog trainer's whistle is often inaudible to the human ear). Echolocating bats emit sounds of about 60 KHz (ultrasounds) and can hear ultrasounds of up to 100 KHz.

THE MIDDLE EAR AMPLIFIES, AND INNER EAR TRANSDUCES SOUND

The ear has three parts: the outer, middle, and inner ear. The *outer ear* consists of the *pinna* (earlobe) and the *ear canal*, which together act like a funnel, collecting and channeling sound waves from a large peripheral field into the ear canal. In some animals, such as dogs and rabbits, the mobile pinna acts like a radar antenna, searching for and focusing on the source of sounds. The ear canal acts like a resonance chamber, helping to amplify the waves of specific frequencies.

Eardrum & ossicles amplify sound — The ear canal terminates in the *eardrum (tympanic membrane)*, a thin but strong round sheet of elastic connective tissue that vibrates in response to sound wave pressure. The eardrum is the first component of the *middle ear*, which also contains three distinct small bones called *ossicles*, each the size of a matchhead. They form a lever system with one end connected to the eardrum and the other end to the *inner ear's oval window*. The eardrum vibrates in unison with the first bone, the *malleus* (hammer), and the second bone, the *incus* (anvil), which is attached almost perpendicularly to the malleus, forming a lever. The incus displaces the third bone, the *stapes* (stirrup), which in turn vibrates the inner ear's oval window. The eardrum and ossicles amplify the pressure on the oval window 20 times.

Structure of the cochlea, the inner ear's hearing organ — The vibrations of the oval window set into motion various mechanical components of the *cochlea*, the hearing organ of the inner ear. The cochlea's function is to convert the sound waves into electrical signals for transmission to the brain. The cochlea ("snail") is a coiled blind tube, wide at the base and narrow at the apex. Some of its components have purely mechanical functions, others play a part in sound transduction. The coiling serves to maximize the tube's length while keeping its total size minimal.

The cochlea contains a *basilar membrane* that runs along its entire inner length and divides it into two chambers: upper (*scala vestibuli*) and lower (*scala tympani*). The two chambers communicate at the cochlea's *apex*, the *helicotrema*. A thinner membrane (*Reissner's membrane*) running through the scala vestibuli, along the basilar membrane, creates a middle chamber, the *scala media*, which is filled with a fluid called *endolymph*. Other chambers contain *perilymph*, which has a different ionic composition. These ionic variations are important in cochlear function.

Hair cells are true auditory receptors and transducers — Movements of the oval window produce *traveling waves* in the perilymph of the scala vestibuli. These waves are transmitted to the endolymph and then to the underlying basilar membrane, forcing it to vibrate. The basilar membrane supports *hair cells* of the *organ of Corti*, the true hearing receptors and the *mechanoelectrical transducers* of the auditory system. The hair cells at their apical surface have specialized ciliary structures (*stereocilia*) that are the cellular transduction elements.

The tips of the stereocilia are in contact with the *tectorial membrane* (roof membrane), and the cell bodies of the hair cells communicate via chemical synapses with the peripheral terminals of the *auditory nerve fibers*. The vibration of the basilar membrane displaces the stereocilia, producing a *receptor potential* that is synaptically transmitted to the terminals of the auditory nerve fibers. Here *nerve impulses* (action potentials) are produced and conducted by the primary auditory afferents to the brain auditory centers for further analysis and integration. The role of the various regions of the basilar membrane and its hair cells with regard to frequency and loudness discrimination is discussed in plate 102.

CN: Use dark colors for A, J, K, P, and R.
1. Begin with the upper panel, coloring all numbers and titles. Note that the three frequencies (C) illustrated are in the low amplitude range (whisper). The one example of high amplitude (B) (loud sound) is on the far right.
2. Color the structures and functions of the ear.
3. Color the structure of the cochlea (L) next, beginning with the overall view on the far left.

Note the horizontal diagram representing the entire cochlea with the spirals straightened out. The scala vestibuli (N) and the scala tympani (N^2) receive the same color because they communicate and contain the same fluids (perilymph) (P^1). The tectorial membrane (R) has been deleted for purposes of simplification. Color the sequence of events involved in the transmission of sound waves.

CHARACTERISTICS OF SOUND WAVES

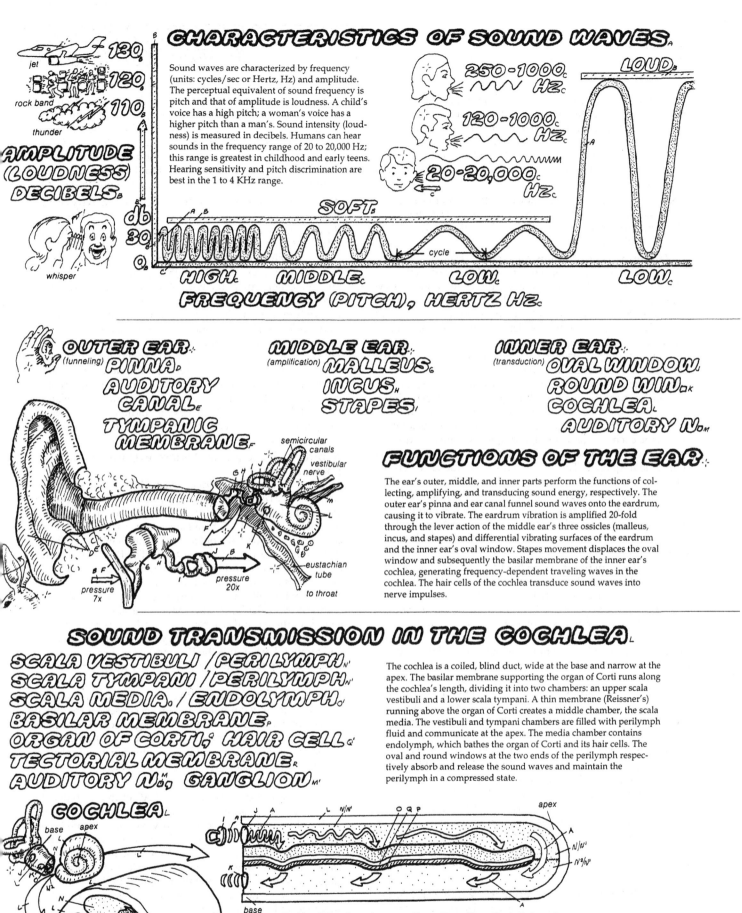

Sound waves are characterized by frequency (units: cycles/sec or Hertz, Hz) and amplitude. The perceptual equivalent of sound frequency is pitch and that of amplitude is loudness. A child's voice has a high pitch; a woman's voice has a higher pitch than a man's. Sound intensity (loudness) is measured in decibels. Humans can hear sounds in the frequency range of 20 to 20,000 Hz; this range is greatest in childhood and early teens. Hearing sensitivity and pitch discrimination are best in the 1 to 4 KHz range.

jet — 130
rock band — 120
thunder — 110

AMPLITUDE (LOUDNESS) DECIBELS

db / 30 / 0
whisper

250-1000 Hz
120-1000 Hz
20-20,000 Hz

LOUD

SOFT

cycle

HIGH   MIDDLE   LOW   LOW

FREQUENCY (PITCH), HERTZ Hz

OUTER EAR
(funneling)
PINNA
AUDITORY CANAL
TYMPANIC MEMBRANE

MIDDLE EAR
(amplification)
MALLEUS
INCUS
STAPES

INNER EAR
(transduction)
OVAL WINDOW
ROUND WIN.
COCHLEA
AUDITORY N.

semicircular canals
vestibular nerve
eustachian tube
to throat
pressure 7x
pressure 20x

FUNCTIONS OF THE EAR

The ear's outer, middle, and inner parts perform the functions of collecting, amplifying, and transducing sound energy, respectively. The outer ear's pinna and ear canal funnel sound waves onto the eardrum, causing it to vibrate. The eardrum vibration is amplified 20-fold through the lever action of the middle ear's three ossicles (malleus, incus, and stapes) and differential vibrating surfaces of the eardrum and the inner ear's oval window. Stapes movement displaces the oval window and subsequently the basilar membrane of the inner ear's cochlea, generating frequency-dependent traveling waves in the cochlea. The hair cells of the cochlea transduce sound waves into nerve impulses.

SOUND TRANSMISSION IN THE COCHLEA

SCALA VESTIBULI / PERILYMPH
SCALA TYMPANI / PERILYMPH
SCALA MEDIA / ENDOLYMPH
BASILAR MEMBRANE
ORGAN OF CORTI, HAIR CELL
TECTORIAL MEMBRANE
AUDITORY N., GANGLION

The cochlea is a coiled, blind duct, wide at the base and narrow at the apex. The basilar membrane supporting the organ of Corti runs along the cochlea's length, dividing it into two chambers: an upper scala vestibuli and a lower scala tympani. A thin membrane (Reissner's) running above the organ of Corti creates a middle chamber, the scala media. The vestibuli and tympani chambers are filled with perilymph fluid and communicate at the apex. The media chamber contains endolymph, which bathes the organ of Corti and its hair cells. The oval and round windows at the two ends of the perilymph respectively absorb and release the sound waves and maintain the perilymph in a compressed state.

COCHLEA
base   apex
apex
base

Cochlear hair cells are innervated by auditory fibers. Oval window vibration sets up traveling waves in the perilymph and endolymph, which induce the basilar membrane to vibrate. The vibration deforms the stereocilia on hair cells, resulting in receptor potentials that are is synaptically transmitted to the auditory fiber, triggering nerve impulses.

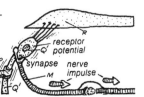

receptor potential
synapse   nerve impulse

To hear sounds, the ear and the central auditory system must decipher characteristics of the sound waves, mainly frequency, amplitude, and direction (plate 101). The discrimination of *frequency* and *loudness* is initially accomplished by the inner ear and is further refined by the auditory nerve centers. Sound is localized by the interaction of left and right ears and their respective nerve centers.

INNER EAR IN DISCRIMINATION OF FREQUENCY & LOUDNESS

In the *cochlea* of the inner ear, the fibrous *basilar membrane* (BM) and its sensory *hair cells* (HCs) play critical roles in frequency discrimination.

Various parts of BM show differential resonance — The BM, a sheet of fibrous connective tissue that supports the HCs, plays a critical role in frequency discrimination. The BM structure varies along the cochlea's length; it is narrow at its base, near the *oval window*, and wide near the apex. As a result its stiffness varies along the cochlea. Vibrations of the oval window set up *traveling waves* in the *perilymph*. Depending on the sound frequency, each wave produces a *peak amplitude* at a distinct position along the cochlea. The peak amplitude for high-frequency sounds occurs near the oval window; for low frequencies near the apex. High-frequency sounds have low wavelength and vice versa. Because waves have their highest energy at the peak amplitude, the BM shows its largest vibration at the cochlea's base for high-frequency sounds and at the apex for low-frequency sounds. The middle cochlea responds to middle frequency. This dependence of frequency discrimination on the cochlea's location is called the *place principle*. Since each BM part supports a separate set of HCs and their attached nerve fibers, the auditory nerve fibers emerging from the base of the cochlea transmit messages about high-frequency sounds while those farther away from the base do so for increasingly lower-frequency sounds.

Hair cells form two rows on the BM — The HCs are supported by the BM and its supporting cells. Each HC, at its base, is synaptically connected to an *auditory nerve* terminal. The HCs, numbering about 24,000, form two rows along the cochlea's length: a single row of 4000 *inner HCs* and three rows of *outer HCs* (20,000). The inner HCs are the true sound receptors involved in sensation; they give rise to 90% of the *auditory nerve fibers*. The outer HCs are motile and help dampen the BM vibrations, a function regulated by feedback interactions with auditory centers. The differential resonance properties of the BM regions and the distribution of inner HCs along the cochlea's length are the basic mechanisms for sound discrimination by the cochlea. The lower auditory centers discriminate sound frequency by comparing which sets of auditory fibers are active compared to the adjacent ones.

BM & auditory fibers also discriminate loudness — *High-intensity* sound waves produce larger BM displacement on the corresponding cochlear regions. This creates larger receptor potentials in the HCs of that region and consequently a higher rate of firing of nerve impulses in the associated auditory fibers. The auditory centers determine loudness by comparing the firing rate of impulses in a particular auditory nerve fiber, relative to resting levels.

Sound is localized by comparing left & right ear input — Auditory nerve centers determine a sound's source by comparing the time difference (as short as 20 μsec) between the input from the two ears. For example, if the left auditory center is activated earlier, sound is localized to the left. Since a sound from the left direction is louder to the left ear, differential loudness is another measure for sound localization. The time difference measure works best for low-frequency sounds, and loudness difference for high-frequency sounds.

AUDITORY PATHWAYS & CENTERS: COMPLEX & NUMEROUS

The central axons of the auditory neurons, with cell bodies in the cochlea's *spiral ganglia*, merge with the vestibular fibers (plate 103) to form the *vestibulo-acoustic nerve* (VIII cranial nerve), which enters the *medulla*. The auditory fibers terminate in the *cochlear nuclei*. From this synaptic station, numerous connections are made with other brain centers. (1) The medullary auditory centers serve in sound localization, auditory reflexes such as the middle ear muscle reflex, and the startle reflex. (2) The midbrain *inferior colliculus* and *reticular formation* centers are important for regulating auditory reflexes related to head and eye movements for sound localization; auditory input to the reticular formation mediates arousal, attention, and wakefulness. From the inferior colliculi, higher projections are relayed to the *medial geniculate nuclei* of the *thalamus*, which serve in auditory perception. Central auditory connections are both crossed and uncrossed with more extensive contralateral projections.

AUDITORY CORTICAL AREA PROCESSES COMPLEX SOUND FEATURES

From the thalamus, *auditory radiation fibers* project to the *primary auditory cortex (PAC)* in the *superior temporal gyrus* of the temporal lobe. The cochlear map of frequency distribution is represented on the PAC by a *tonotopic map* so that neurons in certain parts of the PAC detect higher tones while separate PAC neurons respond to lower tones. The PAC neurons discriminate the pitch and not the pure tone of a sound. The PAC neurons also form columns and are feature detectors: some respond when pitch is altered in time, others respond when loudness of particular sound frequency is altered in time. The sounds of the natural environment (bird songs, speech) show complex patterns that continuously vary in tone, loudness, and temporal order. It is the PAC and its higher integrative areas that help form meaningful auditory images of these complex sound patterns.

Effects of PAC damage — PAC damage in one hemisphere is not debilitating, but bilateral damage causes severe deficits in discrimination and temporal ordering of complex sounds and appreciation of sound qualities. Discrimination of pure tones is relatively unaffected because this function is also carried out by lower hearing centers. The finer aspects of sound localization are also affected.

Higher auditory cortical areas — PAC output is in part to *auditory association areas*, where auditory signals are further analyzed and integrated, particularly with input from higher visual cortical areas. Some of the higher-order auditory integrative areas are involved in short-term memory of sounds while still higher ones process sounds related to words and language. Wernicke's area in the temporal-parietal lobe borders is well known for its role in the sensory aspect of language and speech comprehension (plate 111).

CN: Use the same colors as on the previous plate for sound waves (A), cochlea (B), tectorial membrane (C), hair cells (D), auditory nerve (E), and basilar membrane (H). Light colors for M and N.
1. Begin with the structures of the organ of Corti at the top, moving down to the diagram of frequency analysis in the basilar membrane. Follow this to the right, where the auditory pathways (I) and the primary auditory cortex (M) are shown.
2. Color the small brain drawing.
3. Color the localization of sound drawing, noting that the arrows, identified as time and loudness differences, relate to the sound waves emanating from the tuning fork.

SOUND WAVE TRANSDUCTION & DISCRIMINATION

"ORGAN OF CORTI"

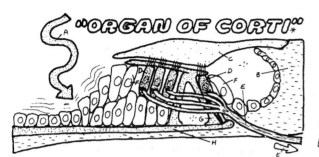

COCHLEA
TECTORIAL MEMBRANE
HAIR CELL
AUDITORY NERVE
SUPPORTING CELL
PILLAR OF CORTI
BASILAR MEMBRANE

The organ of Corti consists of the pillar of Corti and the hair cells (HCs) that are located on the supporting cells of the basilar membrane (BM). The middle-ear ossicles vibrate the cochlea's oval window. This sets up traveling waves in the perilymph of the cochlea, vibrating the BM. The peak amplitude (highest pressure) of high-Hz, narrow-wavelength sound waves occurs at the cochlea's base, vibrating the narrow, stiff BM of the basal cochlea. For low-Hz, wide-wavelength sounds, the peak amplitude shifts toward the apical cochlea, vibrating its wide, loose membrane. The BM vibrations induce receptor potentials in the HCs of the corresponding cochlea region. Of the two rows of HCs, the inner-row cells act as true auditory receptors. The HCs make synaptic contact with auditory nerve terminals. Thus, the auditory nerve fibers emerging from the cochlea's base report on high-Hz sounds and those emerging near the apex report on lower-Hz sounds. Increase in sound intensity (loudness) raises the traveling-wave amplitude, producing higher receptor potentials and ultimately higher rates of nerve impulse traffic in the auditory fibers.

Primary auditory fibers from different cochlear regions converge to form the auditory nerve, which, along with the vestibular nerve (not shown), enters the medulla as the VIII cranial nerve and synapses with the cochlear nuclei neurons. From here fibers concerned with auditory perception ascend and terminate in the midbrain's inferior colliculus and then onto the medial geniculate body of the thalamus. In this path, some fibers cross over and others ascend uncrossed. From the thalamus, higher projections (auditory radiation) project to the primary auditory cortex (PAC) in the superior temporal gyrus of the temporal lobe.

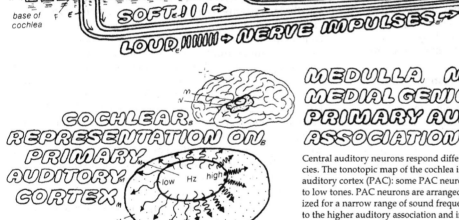

AUDITORY PATHWAYS

COCHLEAR REPRESENTATION ON PRIMARY AUDITORY CORTEX

MEDULLA, MIDBRAIN, MEDIAL GENICULATE BODY PRIMARY AUDITORY CORTEX ASSOCIATION CORTEX

Central auditory neurons respond differentially to sounds of different frequencies. The tonotopic map of the cochlea is faithfully represented on the primary auditory cortex (PAC): some PAC neurons respond best to high tones, others to low tones. PAC neurons are arranged in columns; each column is specialized for a narrow range of sound frequencies. The PAC provides major input to the higher auditory association and integrative areas located adjacent to it. In humans, these association auditory areas communicate with still higher-order cortical areas serving in language and speech functions.

LOCALIZATION OF SOUND
1. TIME DIFFERENCE
(< 3,000 Hz)
2. LOUDNESS DIFFER.
(>3,000 Hz)

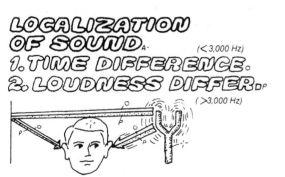

Localization of sound (identifying direction) involves interaction of the two ears. If a sound comes from a left source, it will arrive sooner and be louder to the left ear than to the right ear. By comparing the signals arriving from the two ears, the auditory centers determine the sound source. Comparison of the time difference works best for sounds below 3 KHz, that of loudness difference for sounds above 3 KHz.

AGE AND HEARING LOSS
THRESHOLD
LOSS

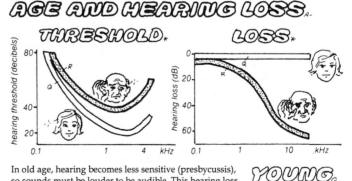

YOUNG
OLD

In old age, hearing becomes less sensitive (presbycussis), so sounds must be louder to be audible. This hearing loss is particularly severe for the high-Hz range, possibly due to selective aging damage in the basal cochlea. This selective damage interferes with audibility of consonants. With advanced age, hearing loss extends to lower-Hz ranges and eventually to 0.1–4 KHz, used in normal speech.

Vestibular system serves in the body's balance & equilibrium functions — Standing, postural control, and many normal movements often must occur against the Earth's gravity. To counter and adapt to this force, several sensory and motor mechanisms have evolved, such as the general proprioceptive sense organs in joints and the vestibular system in the head to detect changes in the body's position in relation to gravity and to ensure balance and equilibrium.

Vestibular apparatus detects changes in the head's position in space — The sensory organ for the vestibular system is the inner ear's *vestibular apparatus* (VA), which detects changes in the head's position in space. Through its connections with the brain and spinal cord, the VA sends signals to activate adaptive motor responses aimed at maintaining equilibrium. These responses involve the axial muscles of the limbs and trunk for posture and support as well as the muscles that move the head and eyes.

STRUCTURE & FUNCTION OF VESTIBULAR APPARATUS

Lesion studies reveal functions of the VA in balance — Knowledge of VA functions was largely gained by lesion studies in animals. Animals with bilateral removal of the VA are unable to stand upright, especially if blindfolded. When attempting to move forward or backward or to turn, they fall. If the VA is removed on only one side, they exhibit postural and equilibrium deficits on that side.

The VA's semicircular canals (SCC) & the utricle/saccule organs perform different functions — Selective lesions of the utricle/saccule organs (see below) alone interfere with balance in the upright position or during translational (linear) acceleration (i.e., when the body is moved forward, backward, up, or down). Lesions of the SCC interfere with balance during turning movements. Thus the utricle/saccule organs (maculae) are often activated in response to alterations in *static equilibrium*; the SCC are excited mostly in response to changes in *dynamic equilibrium*.

The macular organs detect linear acceleration — The saccule ("little sac") and utricle ("little bag") are two small swellings in the inner-ear wall, each containing one *macula* (macular organ), which is bathed by *endolymph*. Each macula is a mechanoelectrical transducing receptor organ containing *hair cells* (HCs). Each HC contains numerous apical *stereocilia* and a single *kinocilium*. Branches of the *vestibular nerve* engulf the HCs on the basal side. The cilia are embedded in an *otolithic membrane* that contains small (3–19 μm long) calcium carbonate crystals called *otoliths* (ear stones).

Macular hair cells transduce linear acceleration — When the head moves (accelerates) linearly in any direction, the maculae move along. But the otoliths, being denser than the surrounding fluid, lag behind. This distorts the position of the stereocilia, resulting in the production of a receptor potential in the HCs. This potential synaptically triggers action potentials in the vestibular nerve fiber, which are then sent to the brain. Saccule and utricle orientation are such that their maculae inform the brain about any linear movement of the head and consequently of the body. Macular activation occurs, however, mainly during the onset (acceleration) and termina-

tion (deceleration) of the motion. Thus, in a moving car or elevator, we perceive motion only during the initial and terminal phases.

Three semi-circular canals detect rotational acceleration — The SCC of the VA detect accelerations associated with rotational motion. The three fluid-filled canals are situated perpendicularly to one another. Therefore, the rotational motion of the head in any direction stimulates at least one of the three SCC. Each canal contains at each end one mechano-electrical transducing sense organ called the *ampulla*. Like the maculae, each ampulla contains HCs with similar ciliary structures. These cilia, however, are embedded in a gelatinous layer called the *cupula* ("little cup"), which extends across the canal's lumen and is attached to the canal's other wall.

Ciliated hair cells in the canal's ampulla transduce rotational acceleration — Rotational acceleration of the head moves the SCC, displacing the attached ampulla/cupula along the same direction. But the endolymph fluid in the canal lags behind, owing to its inertia. This differential movement of the fluid in relation to the cupula distorts the stereocilia, creating a receptor potential in the HCs. The receptor potentials trigger action potentials in vestibular nerve fibers. The action potentials (nerve impulses) then inform the brain's vestibular centers about the particular rotational motion.

CENTRAL VESTIBULAR SYSTEM

The vestibular fibers leave the VA, join the auditory fibers to form the vestibulo-acoustic nerve (VIII cranial nerve), which enters the *medulla*, and terminate mainly in the *vestibular nuclei*.

Vestibular nuclei are the integrating centers for the vestibular system — The vestibular nuclei motor output is chiefly directed at two targets. The first is the lower motor centers in the *spinal cord*. Here impulses activate mostly the gamma motor neurons, particularly to extensor muscles, increasing the tone and tension of the *body's muscles* (plate 95). These responses are essential for postural control and balance. The second target is the upper *midbrain motor nuclei*, particularly those regulating *eye* and *head* movements. In addition, vestibular nuclei are connected reciprocally with the vestibular part of the *cerebellum* (the flocculo-nodular lobe), which controls appropriate execution of the motor commands (plate 97). Vestibular nuclei also send fibers to the *higher somatic sensory systems* of the brain involved in the perception of head/body position (body image).

Vestibular reflexes ensure appropriate body & ocular responses to positional changes — The vestibular system's motor responses are executed through many *vestibular reflexes*. These are generally inborn, fast, and highly purposive. Thus, when we tilt to the left, our left leg extensor muscles are activated to ensure balance. Excessive tilting to one side makes us lift the opposite leg and arm in the opposite direction. If our head turns to the left while we view a stationary object to the right, *vestibulo-ocular reflexes* help maintain our gaze by moving our eyes to the right. In the absence of visual cues (as in a blind person), vestibular sensations and responses are the only cues available for orientation in space.

CN: Use dark colors for D and K.
1. Begin with the semicircular canals and detection of rotational acceleration. Color the diagram to the left of the semicircular canals. This demonstrates the three spatial planes occupied by the canals. Color the canals in the large drawing and continue up through the various blow-ups. Color the directional arrows of the cartoons.

2. Color the detection of linear acceleration by the saccule and utricle, beginning with the section of the macula (upper right), then the blow-up and the diagram on top of the heads.
3. Color the vestibular reflex by beginning in the lower left corner and following the numbered sequence. You may wish to color the relevant parts of the diagram on the right simultaneously.

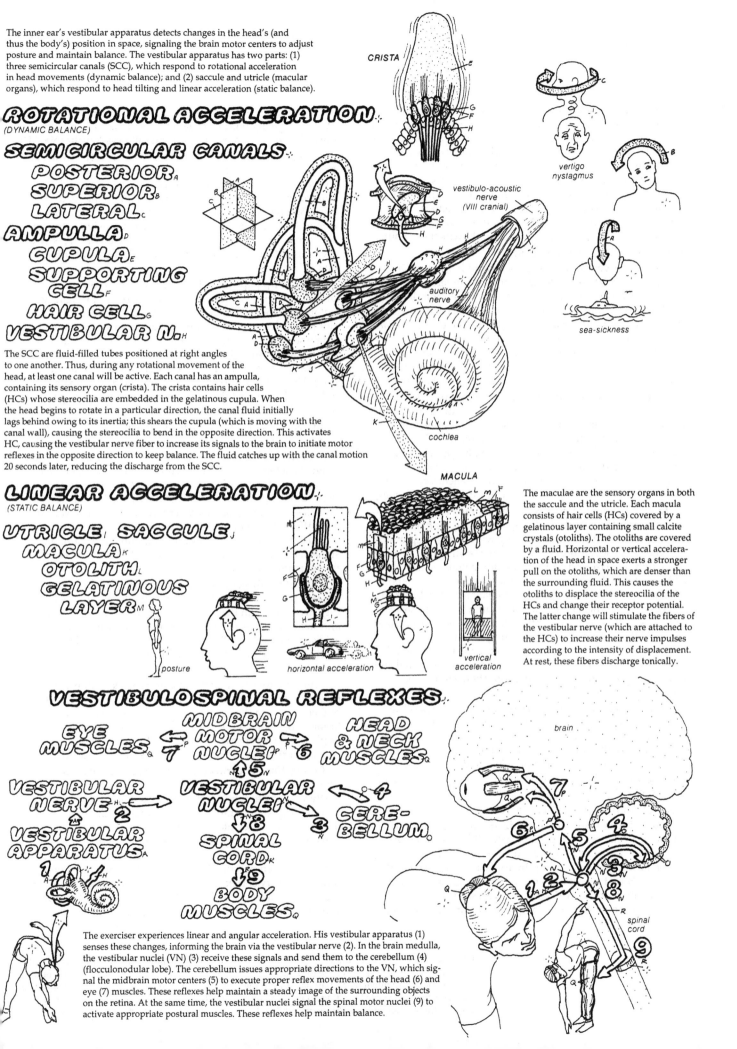

The inner ear's vestibular apparatus detects changes in the head's (and thus the body's) position in space, signaling the brain motor centers to adjust posture and maintain balance. The vestibular apparatus has two parts: (1) three semicircular canals (SCC), which respond to rotational acceleration in head movements (dynamic balance); and (2) saccule and utricle (macular organs), which respond to head tilting and linear acceleration (static balance).

CRISTA

vertigo
nystagmus

sea-sickness

vestibulo-acoustic
nerve
(VIII cranial)

auditory
nerve

cochlea

ROTATIONAL ACCELERATION
(DYNAMIC BALANCE)

SEMICIRCULAR CANALS
POSTERIOR A
SUPERIOR B
LATERAL C
AMPULLA D
CUPULA E
SUPPORTING CELL F
HAIR CELL G
VESTIBULAR N. H

The SCC are fluid-filled tubes positioned at right angles to one another. Thus, during any rotational movement of the head, at least one canal will be active. Each canal has an ampulla, containing its sensory organ (crista). The crista contains hair cells (HCs) whose stereocilia are embedded in the gelatinous cupula. When the head begins to rotate in a particular direction, the canal fluid initially lags behind owing to its inertia; this shears the cupula (which is moving with the canal wall), causing the stereocilia to bend in the opposite direction. This activates HC, causing the vestibular nerve fiber to increase its signals to the brain to initiate motor reflexes in the opposite direction to keep balance. The fluid catches up with the canal motion 20 seconds later, reducing the discharge from the SCC.

MACULA

LINEAR ACCELERATION
(STATIC BALANCE)

UTRICLE I SACCULE J
MACULA K
OTOLITH L
GELATINOUS LAYER M

posture

horizontal acceleration

vertical acceleration

The maculae are the sensory organs in both the saccule and the utricle. Each macula consists of hair cells (HCs) covered by a gelatinous layer containing small calcite crystals (otoliths). The otoliths are covered by a fluid. Horizontal or vertical acceleration of the head in space exerts a stronger pull on the otoliths, which are denser than the surrounding fluid. This causes the otoliths to displace the stereocilia of the HCs and change their receptor potential. The latter change will stimulate the fibers of the vestibular nerve (which are attached to the HCs) to increase their nerve impulses according to the intensity of displacement. At rest, these fibers discharge tonically.

VESTIBULOSPINAL REFLEXES

EYE MUSCLES Q
7
MIDBRAIN MOTOR NUCLEI P
6
HEAD & NECK MUSCLES Q
5
VESTIBULAR NERVE H
2
VESTIBULAR NUCLEI N
4
CERE-BELLUM O
VESTIBULAR APPARATUS A
1
8
SPINAL CORD R
9
BODY MUSCLES Q

brain

spinal
cord

The exerciser experiences linear and angular acceleration. His vestibular apparatus (1) senses these changes, informing the brain via the vestibular nerve (2). In the brain medulla, the vestibular nuclei (VN) (3) receive these signals and send them to the cerebellum (4) (flocculonodular lobe). The cerebellum issues appropriate directions to the VN, which signal the midbrain motor centers (5) to execute proper reflex movements of the head (6) and eye (7) muscles. These reflexes help maintain a steady image of the surrounding objects on the retina. At the same time, the vestibular nuclei signal the spinal motor nuclei (9) to activate appropriate postural muscles. These reflexes help maintain balance.

Significance & uses of taste—The sense of *taste* (gustation) serves in food choice and ingestion, nutrition, energy metabolism, and electrolyte homeostasis. Thus, the sweetness of sugars and the pleasant saltiness (in small amounts) of sodium and calcium chlorides provide a drive for intake of these essential nutrients. Indeed, salt-deprived animals and humans show preference for salty foods. Four primary taste modalities are recognized (sweet, sour, salty, bitter). The sense of taste is crucial in detecting and rejecting harmful substances in food, such as plant alkaloids, which taste bitter.

RECEPTOR MECHANISMS OF TASTE

Taste is served by special cells located in the taste buds of tongue papillae—The tongue contains numerous morphologically diverse taste *papillae*: *fungiform* (mushroom-like) papillae in the front, *vallate* (valley-like) in the back, and *foliate* (leaf-like) on the sides. Each papilla contains clusters of *taste buds*, the smallest units of taste sensation. Taste buds are round aggregates of about 50 taste cells and a lesser number of *supporting cells* and *basal cells*. Glands at the bottom and between the papillae secrete fluids that rinse the taste buds.

Taste cells are the gustatory chemoreceptors—The apical *microvilli* of the taste cells, which project into the taste pores (in the papillae canals), are the sites of taste reception. Here *tastants* (substances that produce taste) bind with the *taste receptor molecules* located in these microvilli. Basally, the taste cell is connected to the endings of the *taste nerve fiber*. Several taste cells make contact with the various branches of a single nerve fiber. The sites of contact have synapse-like properties.

Tastant binding to microvillar membrane receptors produces a receptor potential—Receptors for sugars have been isolated from the taste buds. Binding with sweet receptors activates a particular type of G-protein, resulting in activation of adenylyl cyclase and cAMP formation within the taste cells. cAMP induces the closure of the potassium channels, depolarizing the membrane and causing a *receptor potential*. Bitter substances may act through release of calcium inside the taste cells. Taste cell receptor potential spreads to the basal end and synaptically activates the nerve ending, producing *action potentials* that travel along the taste nerve fiber to the brain's gustatory centers. The amplitude of the receptor potential and frequency of action potential are proportional to the concentration of the tastant.

Taste cells continuously turn over every 10 days—The functional life of taste cells is about 10 days; after that they degenerate and die. New taste cells regenerate from the basal cells located at the bottom of the taste buds. Taste nerve fibers are important for taste cell maintenance and regeneration. Cutting a nerve fiber causes the taste cells to degenerate, and nerve regrowth causes them to regenerate.

TASTE DISCRIMINATION & PERCEPTION

Four primary tastes are recognized—Using mixtures of different substances with distinct tastes, it has been shown that various tastes and flavors can be accounted for by any one or a combination of four primary tastes: *sweet, sour, salty,* and *bitter*. Sweet sensation is produced by saccharides (sugars), but amides of aspartic acid and lead salts also taste sweet, as do some African berry proteins (with 100,000 times

sugar's sweetness). Plant alkaloids, thioureas (propyl-thiouracil), and many poisons taste bitter. Sour sensation is produced by acids (citrates, vinegar) and saltiness by salts (Na^+Cl^-).

Tongue regions show differential sensitivity to various primary tastes—Different areas of the tongue seem to be relatively specialized in this regard. The sweet sensation is evoked mainly on the tip. Areas posterior and lateral to the tip are especially sensitive to salty substances. Still more posterior and lateral areas of the tongue are responsive to the sour sensation. The bitter sensation is best detected on the back of the tongue. Considerable overlap exists, however, particularly on the tip.

Tongue tip papillae & taste cells respond to several primary tastes—While taste buds in the back of tongue are uniquely responsive to bitter taste, the taste buds in the anterior region respond to a variety of taste sensations. Indeed, single sensory fibers from the taste buds respond to all primary taste stimuli, although each fiber shows a preference to one particular primary taste (best stimulus, maximal activation). The higher taste centers presumably discriminate the individual taste qualities by extracting and comparing the information relayed simultaneously by all active taste fibers.

TASTE NERVES AND GUSTATORY CENTERS

Chordae tympani & lingual nerves convey taste signals to the medulla's gustatory nucleus—Taste sensation from the tongue's anterior two-thirds is conveyed by the *chordae tympani* nerve, a branch of the *facial nerve* (VII cranial); taste sensation from the posterior third is served by the *lingual nerve*, a branch of the *glossopharyngeal nerve* (IX cranial). The various taste afferents enter the medulla and terminate in the *gustatory nucleus*, the first major center for integrating and routing taste signals. From this center, three groups of connections are made to other brain regions.

Output to adjacent medullary centers connects taste with digestion—Short local fibers connect the gustatory nucleus to digestive centers in the medulla to regulate the *salivary gland* and *stomach* activities, causing increased salivary flow and acid secretion (plate 72). Saliva enhances gustation by helping to dissolve food substances; only dissolved substances can activate the taste buds.

Limbic connections serve in emotional & hedonic aspects of taste—The second group of connections is made with the higher centers in the *hypothalamus* and *limbic system*. The limbic projections serve the affective (hedonic—pleasant/unpleasant) responses to tastes, and the hypothalamus components contribute to hunger and satiety (plate 107).

Connections to cortical regions serve in taste discrimination & perception—The third group of connections project, partly with the tongue's tactile fibers, to the *thalamus* and *cortex* but the gustatory input is mainly uncrossed. These connections and their associated centers serve in higher gustatory perception—i.e., recognition and discrimination of taste modalities and flavors. Two separate but nearby cortical taste areas have been described. One is near the *tongue area* on the somatic sensory cortex and another is in the *insula*, adjacent to the temporal lobe cortex (plates 83, 93). Damage to these areas severely retards gustatory discrimination.

CN: Use dark colors for D and N.
1. Begin with the taste areas of the tongue.
2. Color the enlargement of the papillae (E) and the enlargement of the taste bud (G) and taste cell (H).
3. Color the section on taste pathways, while following the numbered sequence. Notice that a sagittal view of the medullary portion of the brain stem is shown in the upper square. The cerebral hemisphere is shown in a coronal view.
4. Color the two charts at the bottom.

TASTE AREAS OF THE TONGUE:
SWEET_A
SALT_B
SOUR_C
BITTER_D

Taste sensation is served by the taste buds (TB) on the tongue surface. Different areas of the tongue are selectively sensitive to different taste modalities: the tip to sweet, the back to bitter, the lateral areas to sour, and the anterolateral areas to salty substances.

PAPILLAE_E
EPITHELIUM_F
TASTE BUD_G
TASTE CELL_H
MICROVILLI_I
SUPPORTING CELL_J
NERVE FIBER_K

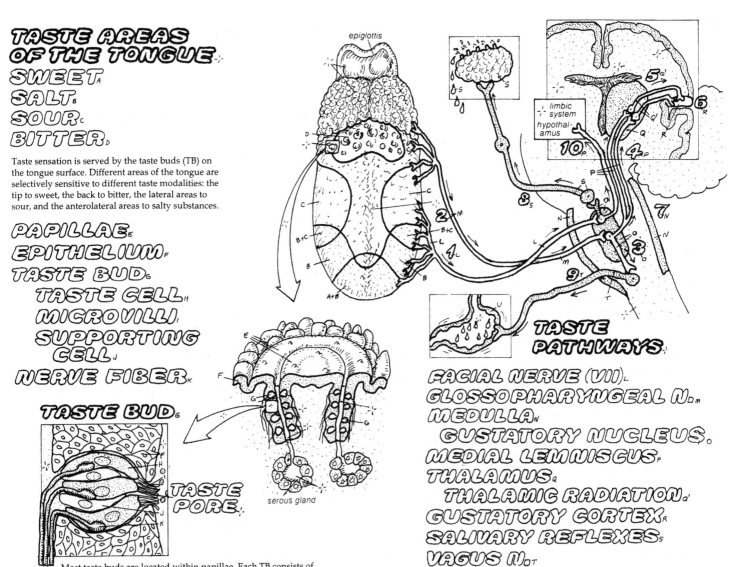

epiglottis

limbic system hypothalamus

serous gland

TASTE BUD_G

TASTE PORE_L

Most taste buds are located within papillae. Each TB consists of taste cells (TC) and supporting cells. TC have microvilli containing the receptor molecules (proteins), which bind with taste substances. Each TC is connected to the sensory endings of a taste nerve fiber.

TASTE CELL_H
RECEPTOR POTENTIAL_H'
ACTION POTENTIAL_K'

The binding of taste molecules to TC microvilli generates a receptor potential in the basal aspects of TC, which, if sufficiently strong, evokes a nerve impulse in the taste nerve fiber.

TASTE PATHWAYS:
FACIAL NERVE (VII)_L
GLOSSOPHARYNGEAL N._M
MEDULLA_N
GUSTATORY NUCLEUS_O
MEDIAL LEMNISCUS_P
THALAMUS_Q
THALAMIC RADIATION_Q'
GUSTATORY CORTEX_R
SALIVARY REFLEXES_S
VAGUS N._T
STOMACH_U

Primary taste fibers from the tongue's anterior 2/3 arrive via the facial nerve (1), those from the posterior 1/3 via the glossopharyngeal nerve (2). All signals from TC and their associated sensory fibers arrive in the medullary nerve centers for taste (gustatory nucleus) (3). Secondary fibers dealing with taste perception ascend to the thalamus via the medial lemniscus (4) and then via the thalamic radiation (5) to the cortical taste area (6) (insula, near tongue area of sensory cortex). Taste signals enhance salivary and gastric secretions. Indeed, saliva helps dissolve substances before they can stimulate taste cells. The gustatory signals for these visceral reflexes are integrated at the level of the medullary digestive centers (7). Parasympathetic motor commands are sent to the salivary glands via the facial nerve (8) and to the stomach via the vagus nerve (9). Taste signals also go to the limbic system and hypothalamus (10) for affective responses.

AFFECTIVE RESPONSE TO CONCENTRATION DIFFERENCES:

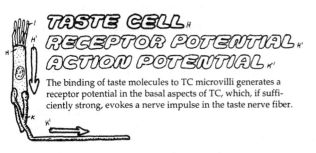

Taste sensations also provoke affective responses that vary depending on the concentration of the substance. Thus, bitter, sour, and salty substances actually taste pleasant in low concentrations.

very weak   weak   strong   full strength

CONCENTRATION →

TASTE THRESHOLDS:
SACCHARIN_A'
SUCROSE_A²
QUININE_D'
HCl_C'
NaCl_B'

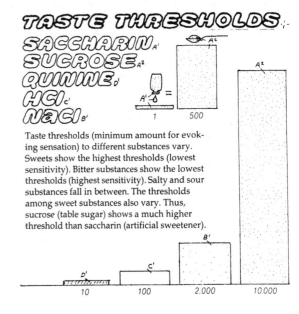

Taste thresholds (minimum amount for evoking sensation) to different substances vary. Sweets show the highest thresholds (lowest sensitivity). Bitter substances show the lowest thresholds (highest sensitivity). Salty and sour substances fall in between. The thresholds among sweet substances also vary. Thus, sucrose (table sugar) shows a much higher threshold than saccharin (artificial sweetener).

10   100   2.000   10.000

The sense of smell (olfaction) is used by humans and animals to detect odors of foods, fruits and flowers, individuals, enemies, territory, the opposite sex, etc. Olfaction evokes intensive emotional reactions resulting in strong approach/avoidance behaviors. The smell of a rose brings pleasure while that of a rotten egg causes nausea. Odor memories carry deep and emotionally rich associations. Smell is important in detecting "flavors" of foods and for regulating appetite and food intake, as shown by loss of these abilities during colds.

OLFACTORY RECEPTOR NEURONS & TRANSDUCTION

Olfactory cilia are the site of odor transduction — The *olfactory receptor neurons* (ORNs) are bipolar chemoreceptor neurons with the cell body in the olfactory mucosa (a sensory neuroepithelium) in the superior nasal sinuses. Each ORN has a dendrite that terminates in several immotile *cilia* that are immersed in a fluid mucus layer (secreted by the Bowman's glands and the supporting cells). The cilia are the site of odorant-receptor interaction and olfactory transduction. An odorant is a chemical that produces the sensation of odors.

The axons of ORNs form bundles of *olfactory nerve* that pass through the *cribriform plate* before entering the cranium and the *olfactory bulb*. In the bulb, these axons synapse with the *apical dendrites* of the secondary relay cells (*mitral cells*) at a site called the *glomerulus*.

High convergence of ORNs to mitral cells underlies high olfactory sensitivity — There are millions of ORNs, a few thousand glomeruli, and many thousands of mitral cells, providing for a high convergence ratio (1000:1) of ORNs to mitral cells. This high convergence is a basis for the great sensitivity of the olfactory system, which can detect odors at low concentration.

Odorants dissolve in mucus layer & interact with cilia of ORNs — Odorants are volatile substances that reach the nasal cavities through the nostrils. *Sniffing* (a reflex as well as a voluntary inspiratory action) increases the air flow through the upper sinuses. In the olfactory mucosa, the odorants first dissolve in the *fluid mucus layer* before activating the ORN.

Odorants bind with receptor proteins; binding is G-protein/cAMP mediated — The membranes of the immotile cilia contain *olfactory receptor proteins* to which odorants bind. The binding of the odorants to receptor proteins, activates a particular G-protein (G_{olf}), which in turn activates the membrane enzyme *adenylyl cyclase*, increasing *cyclic AMP* (intracellular messenger) levels within the ORNs. Functioning in an amplification cascade, cAMP activates gated cation (sodium, calcium) channels, depolarizing ORN membranes by increasing cation inflow. The resulting *receptor potential* induces in the ORN axon action potentials that are conducted to the olfactory bulb. The sum of receptor potentials of many ORNs produces an *electro-olfactogram* that may be recorded from the surface of the olfactory mucosa.

OLFACTORY BULB & HIGHER OLFACTORY CENTERS

Olfactory bulb is a layered structure & forms odor maps — The bulb's glomeruli are the site of the first synaptic station for the olfactory impulses. Odors appear to be mapped onto the bulb. All receptor neurons expressing a particular odorant's receptor proteins converge onto one or a small cluster of glomeruli. Odor stimulation (e.g., peppermint) enhances neuronal metabolic activity in a specific region of the bulb.

Olfactory bulb projects to the olfactory cortex & receives input from the brain — Afferent olfactory messages are processed at the level of the glomeruli and the mitral cells of the bulb and are then sent, via the *olfactory tract* (axons of mitral cells), to the higher olfactory areas of the brain (*olfactory cortex*). The olfactory bulb receives from the brain many *centrifugal fibers*, the major portion of which activate the *granule cells* of the olfactory bulb. These cells are local inhibitory interneurons that inhibit the mitral cells, as a part of a feedback loop controlling the mitral cell output.

Olfactory cortical connections mediate odor discrimination & perception — The olfactory output from the bulb to the brain mainly targets the primary olfactory cortex and the higher olfactory association areas, which process olfactory discrimination, perception, and memories. In humans, olfactory signals are further relayed to the frontal cortex, allowing for interaction with visual and auditory data.

Limbic connections mediate instinctive responses to odors and pheromones — Olfactory connections to the limbic system (plate 108), mainly the amygdala, activate smell-related emotions and instinctive behaviors. The odors that provoke instinctive, stereotyped responses are called "pheromones." The intimate connection between olfaction and the limbic system in lower animals is the basis of the term *rhinencephalon* (nose brain) for the limbic system.

Hypothalamic connections mediate feeding and hormonal responses — Other olfactory connections to the hypothalamus (plate 107) convey olfactory influences on feeding, autonomic responses, and hormonal control (particularly the reproductive hormones). Olfactory signals also activate the reticular formation, increasing arousal and attention. In humans and higher primates, the olfactory input to the limbic system, hypothalamus, and reticular formation is relayed through the olfactory cortex. In some mammals and vertebrates, a separate olfactory system working through the vomeronasal organ, the accessory olfactory bulb, and amygdala serves in pheromonal aspects of olfaction.

HIGHER NEURAL MECHANISMS OF ODOR DISCRIMINATION
Although odors may be categorized into finite primary types (i.e., floral, peppermint, musky, camphor, putrid, pungent, ethereal), human can discriminate ~10,000 odors. Both peripheral and central mechanisms share in olfactory discrimination. About 1000 genes coding for the same number of olfactory receptor protein types have been found in mammals. Each receptor type presumably interacts with different chemical types of odors (odorants). Each ORN is thought to express only one of these receptor types (but millions of molecules of it). Olfactory neurons with the same receptor types all converge into the same glomerular clusters.

These receptor proteins interact with functional chemical groups in odorants, since single ORNs can be activated by more than one odorant, although their responses are not uniform. To discriminate between the different odors, brain olfactory centers may extract the necessary information by reading the discharge pattern of many simultaneously active afferent fibers. People with a specific *anosmia* (inability to smell a particular odor) are missing the genes for the odor receptors.

CN: Use dark colors for E and F, very light for A.
1. Color the upper left illustration showing the flow of odor molecules (E) through the interior of the nasal cavity. Then color the blowup of the olfactory bulb (B) and olfactory neurons (C), starting at the bottom with the interaction of the odor molecules (E) with the olfactory neuron cilia (F). Complete the network of neural cells in the bulb.
2. Color the panel on olfactory pathways, noting that the three specific structures listed under "limbic system" (Q) are parts of the system (colored gray) but receive separate colors.

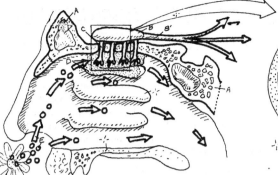

BONE_A
OLFACTORY BULB_B TRACT_{B'}
OLFACTORY NEURON_C
OLFACTORY MUCOSA_D
ODOR MOLECULES_E

CILIA_F
RECEPTOR SITE_G
SUPPORTING CELL_H
BASAL CELL_I
BOWMAN'S GLAND_{D'}
CRIBRIFORM PLATE_{A'}
GLOMERULUS_J
MITRAL CELL_K
CENTRIFUGAL FIBER_L
GRANULE CELL_M

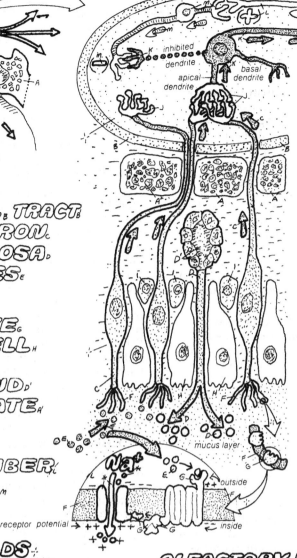

mucus layer

Na⁺

outside

inside

receptor potential

Axons of the olfactory receptor neurons (ORN) run through pores in the cribriform plate and enter the olfactory bulb (OB), where they excite the apical dendrites of mitral cells (MC) within the glomerulus. Here olfactory messages are segregated, refined, and amplified. Axons of the MC convey the excitation to higher brain centers via the olfactory tract. The granule cells of the OB inhibit MC activity by acting on the basal dendrites of MC. Centrifugal fibers from the brain suppress MC activity by exciting the granule cells. In this way the brain regulates output of the OB.

NEURON REGENERATION_C

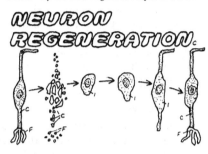

The sense of smell is served by millions of olfactory receptor neurons (ORN) located in the olfactory mucosa (OM). OM — one on each side of the head — are small areas in the roof and walls of the superior nasal sinuses. OM also contain supporting cells and Bowman's glands, which secrete the mucus coat of the OM. This mucus is essential, because odors must first dissolve in mucus before they can excite the ORN. The dendrites of ORN terminate in special cilia that are the site of olfactory transduction. Odor molecules bind with receptor proteins in the cilia, causing depolarization of ORN. ORN have a short life span (1–2 months); after degeneration, they regenerate by division and differentiation of the resting basal cells.

ODOR THRESHOLDS
(sensitivity)

The degree of sensitivity to odors, measured by odor thresholds (minimum amount needed for detection), varies depending on the odor. The threshold for mercaptans (sulfur compounds added to household natural gas) is 50,000 times lower than that for peppermint oil, whose threshold is 250 times lower than that for ether.

Hsss

MERCAPTAN_N 0.000 000 4_N
OIL OF PEPPERMINT_* 0.02_*
ETHER_* 5_* mg/L of air

SNIFFING REFLEX

Sniffing increases the speed and volume of air flow in the upper nasal cavities, promoting access of air and odors to the olfactory mucosa. Without sniffing, levels of many odors would be too low for detection.

normal breathing   sniffing

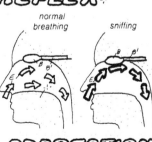

MASKING & ADAPTATION

A person can smell many different odors, but only one odor at a time, because stronger odors tend to mask the effects of weaker ones. Continuous exposure to one odor also diminishes sensitivity (adaptation). The concentration of odors must be significantly increased to overcome odor adaptation.

OLFACTORY PATHWAYS

BULB_B TRACT_{B'} → **OLFACTORY CORTEX_O** → **ASSOC. CORTEX_P**

LIMBIC SYSTEM:_{Q*}
AMYGDALA_R
SEPTUM_S
HYPOTHAL-AMUS_T

From the OB, signals are conducted along two pathways to the higher brain structures: (1) to the primary olfactory cortex and higher olfactory association areas for perception and recognition of odors and their association with other sensory data, and (2) to the limbic system structures (amygdala, septum) to activate instinctual behavior and emotions, as well as to the hypothalamus for regulation of drives and visceral activities (e.g. digestion). Olfactory signals also act on the reticular formation (not shown) to cause arousal.

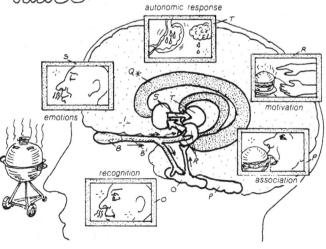

autonomic response

motivation

emotions

recognition

association

BRAIN WAVES (EEG)

Brain shows spontaneous electrical activity that can be recorded from the scalp — Brain waves, as seen on an *electro-encephalogram* (EEG), are slow and weak, originate in the cortex, and occur even at rest and in sleep. Although spontaneous, their *amplitude* and *frequency* vary depending on the state of mental (brain) activity and the recording site.

Alpha waves occur at rest, beta waves during mental activity — For a relaxed person with eyes closed, the EEG shows a moderate activity of about 8–14 Hz (c/s) called the *alpha rhythm*, which is best recorded in the back of the head over the occipital (visual) area. The regular and synchronous alpha waves disappear (*alpha block*) when the eyes open or light is flashed over the closed eyelids; they are replaced by a faster, desynchronous wave pattern called *beta waves* (>14 c/s). Beta waves, often present in the brain's frontal areas, are associated with arousal, alertness, and mental activity.

RETICULAR FORMATION IN WAKEFULNESS & AROUSAL

Reticular formation projection neurons modulate EEG waves — EEG waves originate in the cortex but are modulated in different states by subcortical neuronal systems in the medulla, pons and thalamus. These areas, collectively called the *reticular formation* (RF), form a diffuse and loosely organized region of gray matter and give rise to the *ascending reticular activating system* (ARAS), a diffuse fiber projection that fans out to innervate many areas of the forebrain. The ARAS also includes the *reticular nuclei of the thalamus*, which also project diffusely to all areas of the *cerebral cortex*.

ARAS increases cortical responsiveness to sensory stimulation, maintains wakefulness & arousal — Electrical stimulation of the midbrain RF results in beta EEG and behavioral *awakening* and *arousal*; lesions in the same RF areas result in a coma-like state with persisting synchronized *delta waves* — very slow, high-amplitude waves also seen during deep sleep. The nonspecific ARAS projections globally depolarize (excite) the cortex, in contrast to the *specific sensory* projections from the thalamus, which exert localized and functionally specific excitatory effects on the cortex. The generalized ARAS excitation facilitates cortical responsiveness to specific sensory signals from the thalamus. Normally, on their way to the cortex, the sensory signals from the *sensory afferents* stimulate the ARAS via collateral branches of their axons. The excited ARAS in turn stimulates the cortex, facilitating sensory signal reception. When afferent systems are massively stimulated (e.g., by loud noise or a cold shower), the ARAS projections trigger generalized cortical activation and arousal.

EEG CHANGES IN SLEEP STAGES

Sleep has stages characterized by distinct EEG waves — Before sleep onset, EEG shows the alpha pattern associated with rest and relaxation. As sleep progresses, the waves become slower and larger, occasionally interrupted by bursts of fast *spindle-like* activity. Human sleep consists of two states of *slow-wave* (comprising four stages) sleep and *REM* sleep. Initially the individual goes sequentially through stages 1, 2, 3, and 4 of slow-wave sleep. Each stage is characterized not only by specific EEG patterns but also by bodily or autonomic activities, such as turning over or to the sides, changes in heart and respiratory rates, bouts of sweating, and even a burst of growth hormone secretion (plate 118).

Stage 4 shows delta waves and is most characteristic of slow-wave sleep — *Stage 4*, showing synchronous large and slow *delta waves* of high amplitude (3–4 c/s), is the deepest of the four stages of slow sleep. Delta waves are so characteristic of sleep that were once called "sleep waves." Interestingly, bed-wetting in children and sleepwalking (*somnambulism*) occur mainly during stage 4. This stage is longer in early cycles of sleep and diminishes or is absent towards morning.

REM sleep shows paradoxical features and is associated with dreams — After stage 4, another distinctly different state of sleep called *paradoxical* or *REM* sleep commences. The EEG becomes beta-like (fast and desynchronized), resembling wakefulness patterns, even though the person is extremely difficult to awaken (hence the term "paradoxical sleep"). During the REM sleep, muscle tone in the neck and support muscles is suppressed — the limbs grow the floppy and the head falls — but the eyes show rapid searching movements under the closed lids (hence the designation "REM," for *rapid eye movement*). Grinding of teeth and spontaneous penile erection also occur in REM sleep. A most interesting aspect of REM sleep is its association with dreams, which may explain the beta wave (mental activation) of its EEG pattern. Humans and animals deprived of REM sleep by being awakened each time it begins get an excess amount of REM sleep when allowed to sleep. REM sleep may be involved in the brain or body's restorative functions.

Durations of sleep stages & changes with age — In adults, a complete sleep cycle (1, 2, 3, 4, REM) lasts about 90 min. REM occupies about 20% of the 8 hour total of nightly sleep; the amount of REM sleep increases during the night, with the longest REM (and dreaming) episodes occurring early in the morning. In infants, total sleep time is more than 16 hr, (50% is spent in REM). In old age, the total amount of sleep diminishes by 1–2 hr and stage 4 disappears completely.

BRAIN CENTERS REGULATING SLEEP & WAKEFULNESS

Sleep is not a passive phenomenon. While reduced sensory input facilitates sleep, it does not cause sleep. Certain regions in the hypothalamus and RF are actively involved in the generation, timing, and execution of sleep and regulation of transition between different stages.

Roles of hypothalamus — Sleep and wakefulness are parts of the *diurnal* (daily) activity cycles of humans and animals. The *suprachiasmatic nucleus* of the hypothalamus is the brain's "biological clock" and controller of diurnal cycles, and it regulates the *timing* of sleep and wakefulness. Low-frequency (8 c/s) stimulation of the posterior hypothalamus induces sleep, while high-frequency stimulation of the anterior hypothalamus awakens sleeping animals. Prostaglandins are released by the hypothalamic systems generating sleep. Certain "sleep peptides" also accumulate in the hypothalamus and cerebrospinal fluid during sleep. Their injection into awake animals causes sleep.

Cholinergic system of RF induces sleep — Three neuronal systems in the RF regulate sleep and wakefulness. The stimulation of a cholinergic system (releasing the neurotransmitter acetylcholine) actively generates sleep, while the serotonergic (releasing serotonin) and adrenergic systems (originating in the *raphe nucleus* of the pons and *locus ceruleus* of the medulla, respectively) generate and maintain wakefulness.

CN: Use a gradation of colors from light to dark for the four stages of sleep (F-I), and a dark color for J.
1. Color the afferent sensory (C) inputs from the special senses in the large head of the upper panel. Color the alpha and beta waves on the left and the remaining two diagrams.

2. Color the titles designating the stages of sleep (not the brain waves) and the amount of time spent in each stage on the chart on sleep cycles. What has not been shown in this chart is that after each cycle the sleeper quickly goes back through stages 4-1 before starting the next cycle.

WAKEFULNESS:

RETICULAR FORMATION₄
ASCENDING RETIC. ACTIVATING SYSTEM (ARAS)₄
THALAMUS RETICULAR NUCLEI₈
DIFFUSE THALAMIC PROJECTION SYSTEM₈'
SENSORY AFFERENT.   CEREBRAL CORTEX₀
ELECTROENCEPHALOGRAM (EEG)ₑ

The brain cortex shows ongoing waves of electrical activity (spontaneous EEG waves) even in the complete absence of all sensory stimuli. During rest and relaxation, the EEG is slow and synchronous (alpha waves, 8–14 cycles/sec). During alertness and concentration, the EEG shows low amplitude, faster, desynchronized patterns (beta waves).

ALERT: BETA 15-60 HZ (cycles/sec)
RELAXED: ALPHA 8-14 HZ

Many neurons in the reticular formation (RF) of the brain stem send diffuse projections to all areas of the cortex. Unlike "specific" sensory projections, these "nonspecific" projections increase general cortical excitability. The sources of these global projections are (1) RF nuclei in the lower brain stem and midbrain, and (2) RF nuclei of the thalamus and basal forebrain. Electrical stimulation of these "ascending reticular activating systems" (ARAS) causes fast EEG (beta waves) accompanied by vigilance and alertness. ARAS stimulation awakens sleeping animals. All sensory pathways send branches to the RF, helping to simulate the ARAS and enhance cortical excitability.

SLEEP.
During the 8 hr of nightly sleep, humans go through several "sleep cycles." Each cycle lasts ~1.5 hr and consists of four stages of "slow-wave" and one stage of "paradoxical" sleep. Each stage is characterized by certain bodily and EEG signs.

SLOW-WAVE SLEEP
First, the sleeper becomes relaxed and drowsy, eyes close, and the EEG is slow (alpha). As he moves to stages 2 and 3, sleep is of light to intermediate depth and the EEG becomes faster, with lower amplitudes, showing spindles. By stage 4, sleep is deep and waves are very large and slow (delta).

STAGE 1/DROWSY₆
STAGE 2/LIGHT₆
STAGE 3/INTERMEDIATE₆
STAGE 4/DEEP₁
PARADOXICAL SLEEP (RAPID EYE MOVEMENT REM)₁

Next the sleeper moves into the "paradoxical" state of sleep, during which the waking threshold is very high, the body is limp (no muscle tone, head falls), teeth grind, eyes move rapidly under closed lids (REM = rapid eye movement), and the EEG is fast (beta, like wakefulness).

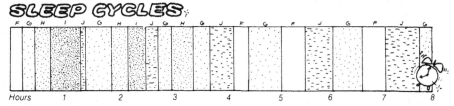

hypothalamus   pineal gland

RF cholinergic system

SLEEP GENERATORS:
1. ACETYLCHOLINE ₖ
2. PROSTAGLANDINS ₗ
3. MELATONIN ₘ

Certain areas in the brain stem RF, hypothalamus, and the pineal gland have hypnogenic (sleep-causing) activity. Cholinergic systems in the RF as well as posterior hypothalamus areas generate sleep, while the serotonergic raphe neurons and the noradrenergic neurons of the locus ceruleus (parts of the ARAS) maintain wakefulness.

During stages 1–4, heart and respiration rates are slow. Sleepwalking and bed-wetting occur during stage 4. Dreaming and increased heart and respiratory activity occur during REM. As night progresses, the duration of REM sleep increases (more dreams near morning) and that of stage 4 decreases (more sleepwalking and bed-wetting early at night). Adults spend 20% of sleep in REM, infants 50%. The elderly have less total night sleep, less REM, and no stage 4.

SLEEP CYCLES:

Hours   1   2   3   4   5   6   7   8

WAKEFULNESS VS. SLEEP:

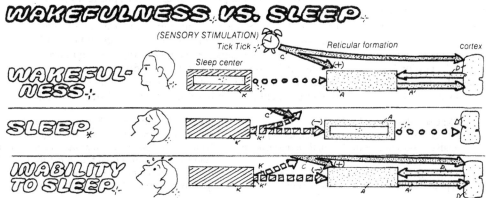

(SENSORY STIMULATION)
Tick Tick   Reticular formation   cortex
Sleep center

WAKEFUL-NESS:
SLEEP:
INABILITY TO SLEEP:

During sleep, hypnogenic centers of the RF and hypothalamus inhibit the ARAS, decreasing cortical excitability. Reduction of sensory input (darkness, silence, supine position) helps diminish activity in the ARAS. Reduced cortical excitability explains the lack of behavioral consciousness and responsiveness during sleep. Anxiety, thoughts, or sensory stimuli may counter the effects of the sleep centers on the ARAS, promoting sleeplessness. During wakefulness, the hypnogenic centers are inhibited and inactive and the ARAS can stimulate the cortex. Continuous excitatory input from the sense organs maintains vigilance.

cortical influence

brain stem

The *hypothalamus* (H) is the highest brain structure directly concerned with the body's homeostasis and integration of internal activities. Located above the brain stem and pituitary gland it receives input from other brain areas and controls the lower motor, autonomic, and endocrine systems.

STRUCTURE AND HYPOTHALAMIC INPUT/OUTPUT

Hypothalamus consists of discrete nuclei & larger "areas" — Occasionally, a particular function is ascribed to a specific nucleus (e.g., *diurnal rhythms* and the *suprachiasmatic nucleus*). In most cases, however, larger areas of the H are implicated in a function (e.g., areas in the *anterior* and *posterior* H in the regulation of *body temperature*).

Hypothalamus receives input from major sensory systems — This input may be direct, as in the case of the eye (*retino-hypothalamic tract*), or indirect, via the reticular *formation* (plate 106). The sensory input informs the H about environmental conditions (the day's length, light intensity, environmental temperature). Olfactory input mediates the effects of smell on hormones and reproduction. Sensory messages also come from the internal sensors in the mouth, digestive system, blood vessels, etc.

Emotions, drives, hormones, & blood-borne substances also provide input — Input coming from the *limbic system* (e.g., amygdala, septum) in the forebrain informs the H about the state of an individual's drives (hunger, thirst, sex) and emotions. A third category of input is provided by *hormones* and other blood-borne substances such as *sodium ions* and *glucose*, conveying messages from the body to the H about the salt, water, and energy situation.

Hypothalamus fibers project to limbic system, midbrain, lower CNS, & pituitary — Output from the H goes to the limbic system to interact with the structures controlling emotions and drives (plate 108), to the *midbrain motor centers* for controlling somatic motor responses during emotional behaviors, to the *sympathetic* and *parasympathetic* autonomic centers in the *medulla* and *spinal cord* for controlling visceral organs, and finally to the *pituitary gland* to control water, salt, metabolic, and hormonal parameters.

HYPOTHALAMIC FUNCTIONS

Hypothalamus is the "brain" of the sympathetic nervous system — A major function of the H is the control of the *autonomic nervous system*. The H has been called the "head ganglion of the sympathetic nervous system" (plates 29, 85) because of its major regulatory effects on this system. The H helps integrate autonomic responses with activities of other brain areas, with the individual animal's emotional state, and with environmental conditions conveyed via the senses.

Stimulation of the lateral H produces sympathetic responses (increase in cardiac activity, peripheral vasoconstriction, vasodilation in the skeletal muscles). The same sympathetic responses also occur during generalized physical activity (exercise, running, fight/flight response) (plate 125). In a similar manner, the H integrates the autonomic and visceral responses to emotional stress (e.g., fear).

Control of body temperature — The H contains a "thermostat" that regulates the body temperature at a "set point." Areas in the *anterior* and *posterior* H integrate the diverse sympathetic responses to cold, such as cutaneous vasoconstriction, piloerection, secretion of epinephrine and thyroxine, and the consequent increase in the metabolic rate (plate 141).

Control of food intake, salt, & water balance — The H is intimately involved in *hunger, satiety,* and *feeding behavior.* Animals with discrete lesions in the lateral H show loss of appetite, reduced food intake (anorexia), and eventual wasting. In contrast, lesions in the ventromedial H lead to excessive eating (hyperphagia) and eventual obesity. Therefore, the lateral areas of the H contain centers that enhance appetite and feeding and the ventromedial areas contain the satiety centers. Together these control *feeding behavior, energy balance,* and possibly *body weight* (plate 138). Other areas in *paraventricular nuclei* and the *dorsolateral* H are involved in the control of water intake, thirst, and salt balance (plates 66, 116, 126).

Diurnal cycles & sleep — Lesions of the suprachiasmatic *nuclei* of the H abolish or alter the *diurnal rhythms* (daily 24-hr rhythms), especially those of activity and hormones. Slices of suprachiasmatic nucleus in culture exhibit diurnal cycles of electrical activity. In the body, these endogenous rhythms are entrained by light input from the eyes. Animals show diurnal activity cycles. Humans are day active, while rodents are night active. Similarly, secretion of ACTH and cortisol (plate 127) is high in the morning and low in the evening; body temperature and BMR are low in the morning and high in the evening. The hypothalamus is also involved in active induction of sleep. Low-frequency stimulation (8 c/s) of the posterior H induces sleep, while high-frequency stimulation of the anterior areas produces arousal.

Control of sex & sexual behavior — Not only does the H control the secretion of sex *hormones* in different ways in the two sexes (plate 155), but the *preoptic areas* and the *anterior* H are intimately involved in the control and expression of sex-specific behaviors. In several mammals, a small but discrete nucleus (the *sexually dimorphic nucleus of the H*) has been found that is markedly larger in the male than in the female.

Control of endocrine system & hormones — Certain hypothalamic neurons are capable of secreting hormones. The extremely important regulatory functions of the H on the endocrine system and hormones through these neural and neurohormonal controls of the anterior and posterior pituitary glands are discussed elsewhere (plates 115–117). The hypothalamic neurons are responsible for the cyclicity and pulsatile release pattern of pituitary hormones.

Connections with the immune system, stress, & diseases — The brain is involved in control of the immune system functions and vice-versa. The H is a critical site for this interaction. Cytokines (interleukins) from white blood cells may be involved in temperature regulation and sleep changes during illness and infections. Mild stress stimulates the immune system; severe stress depresses immunity and promotes certain diseases. These effects are mediated in part by the adrenal steroids and cytokines interacting through the H.

Neuropeptides & prostaglandins in hypothalamic functions — Numerous neuropeptides participate in the functions of the H. Neurons releasing neuropeptide-Y and GnRH as their transmitters have been implicated in food intake and sexual control functions, respectively. Angiotensin neurons regulate drinking behavior. Release of different types of prostaglandins in the H is involved in many activities of the hypothalamus, including temperature control and sleep induction.

CN: Use a dark color for A.
1. Begin by coloring the hypothalamus (A) in the small cross-section of the brain, then the structures of the hypothalamus.
2. Color gray the titles representing inputs to hypothalamus and those of the outputs.

3. Color each title in the bottom panel and the structure to which it refers in the large drawing. Note that the paraventricular nucleus receives t colors to represent a dual role in H function. Als note that the lateral areas of the H are not adequately represented in this sagittal diagram.

HYPOTHALAMUS

The hypothalamus (H) is the major brain center for regulation of internal body functions. Situated above the pituitary gland, underneath the thalamus (hypo-thalamus), the H has numerous nuclei and areas, each involved in the regulation of some internal function. The H has many connections to and from the forebrain limbic structures in addition to receiving sensory input, especially smell, taste, and vision. Via its efferent (output) connections to the brainstem, spinal cord, and pituitary gland, the H controls somatic motor, autonomic motor, and hormonal secretions. The H may be divided into lateral, medial, anterior, and posterior areas.

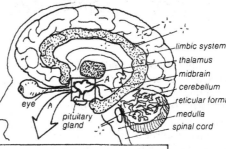

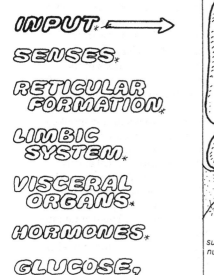

INPUT →

SENSES

RETICULAR FORMATION

LIMBIC SYSTEM

VISCERAL ORGANS

HORMONES

GLUCOSE, Na⁺

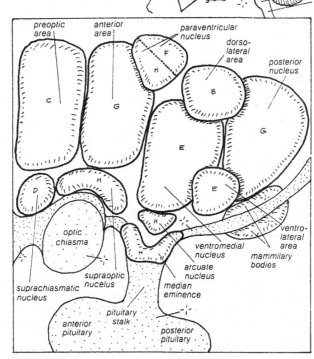

→ OUTPUT

MIDBRAIN (MOTOR)

LIMBIC SYSTEM

MEDULLA (P. SYMP.) (SYMP.)

SPINAL CORD (SYMP.)

PITUITARY (HORMONES)

HYPOTHALAMIC FUNCTIONS

Stimulation of areas in the lateral H activates generalized sympathetic responses. These areas integrate responses to physical activity and the fight/flight reactions. Stimulation of smaller areas may induce adrenal medullary release of epinephrine, or vasodilation in the skeletal muscles. The H influences the parasympathetic system also, via the parasympathetic centers in the brain medulla.

Stimulation or destruction of areas in the anterior H and preoptic areas has profound effects on the regulation of sex hormones (via the anterior pituitary) and sexual responses. In this area is a special nucleus, the sexually dimorphic nucleus, which is markedly larger (more active?) in some male mammals.

Some diurnal (daily) rhythms in bodily functions (e.g., for hormonal secretion or visceral activities) are regulated by the suprachiasmatic nuclei. Light and length of day stimulate this nucleus through direct retinal connections. This nucleus and the pineal gland (well known for its release of melatonin and mediation of light effects/diurnal functions) also interact.

Stimulation of areas in the lateral H (feeding center) increases appetite and induces eating behavior. Over time it leads to overeating and overweight. Destruction of the feeding center causes loss of appetite and wasting. Stimulation of the ventromedial H (satiety center) inhibits eating; lesions lead to overeating/obesity. Satiety and feeding centers have reciprocal inhibitory interactions. Neurons in satiety centers are sensitive to blood glucose levels (hypothalamic "glucostat") and leptin, the fat tissue hormone.

Stimulation of areas in the dorsal/lateral H induces drinking behavior. Injection of acetylcholine or angiotensin II into these areas has similar effects; lesions interfere with water regulation and electrolyte balance. Neurons in this area are sensitive to blood osmolarity and sodium levels (hypothalamic "osmostat").

Body temperature is regulated by areas in H (at 37° C). Stimulation of the anterior H activates heat loss mechanisms (cooling center); stimulation of the posterior H activates heat conservation/production (heating center). These areas have reciprocal inhibitory interactions. Some of the neurons in this hypothalamic "thermostat" are sensitive to changes in blood or skin temperature.

Parts of the H (e.g., the median eminence) act like endocrine glands, secreting hormones. Those from the posterior pituitary gland act directly on target organs (kidney, uterus, breast). Numerous H hormones regulate activities of the anterior pituitary gland, which in turn regulates activities of many organs and glands.

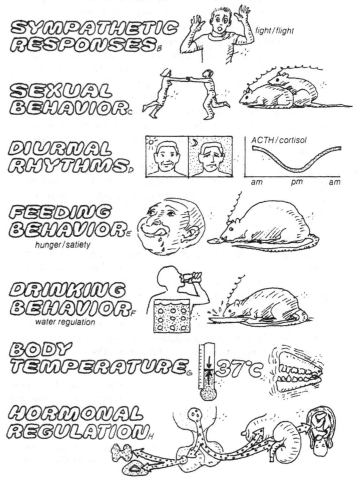

SYMPATHETIC RESPONSES — fight/flight

SEXUAL BEHAVIOR

DIURNAL RHYTHMS — ACTH/cortisol — am pm am

FEEDING BEHAVIOR — hunger/satiety

DRINKING BEHAVIOR — water regulation

BODY TEMPERATURE — 37°C

HORMONAL REGULATION

Three "brains" in one — The brain may be viewed as a hierarchy of three "separate" brains: a lower *vegetative/reflexive* brain, a higher *adaptive/skilled* brain, and an intermediate brain concerned with *emotions* and *instincts*. The vegetative brain corresponds roughly to the *brain stem* and is concerned with controlling vital bodily functions (respiration, digestion, circulation) as well as with integrating brain reflexes. The adaptive and skilled brain corresponds to the *cerebral cortex* (the neocortex). It has sensory, motor, and association/integrative areas that serve in complex perception and execution of skilled sensory and motor functions (e.g., hand movements, speech) as well as higher mental functions (e.g., learning, thoughts, introspection, planning).

Limbic system's role in emotions and instinctive behaviors — *Limbic system* (LS) structures are concerned with central (neural) control over the expression of *emotions, instinctive behaviors, drives, motivation,* and *feelings.* In lower vertebrates, the LS is called the *rhinencephalon* (smell brain) because of its intimate connection to the central *olfactory* structures. In these animals, many instinctive behaviors are guided by the sense of smell. The cerebral cortex and LS have access to brain stem motor areas, permitting them to carry out their respective adaptive and instinctive controls over behavior.

LIMBIC SYSTEM STRUCTURES AND CONNECTIVITY

Amygdala, septum, hippocampus, cingulate gyrus, anterior thalamus, & hypothalamus are the main limbic system structures — These structures are connected by complicated and occasionally reciprocal pathways, some of which constitute a loop. The *loop of Papez* (hypothalamus → anterior thalamus → cingulate gyrus → hippocampus → hypothalamus) has been considered as a possible neural circuit serving in emotions. Patients with lesions in this loop exhibit abnormalities of emotional expression. Recent research has de-emphasized the role of the hippocampus in emotion while emphasizing the role of the amygdala in these functions.

Limbic system & cerebral cortex are interconnected — The LS was once thought to be involved only in emotional and instinctive behaviors with little connection and communication with the cerebral cortex. This view is changing. The cingulate gyrus (located on the medial hemispheric surface) is a part of both the LS and the cerebral cortex and provides an important link between the adaptive cortical structures and the instinctive/emotional structures of the LS. LS structures like the amygdala and septum can also communicate, via their reciprocal connections to the cingulate gyrus, with the *higher cortical association areas.* The LS structures also receive abundant sensory input and send motor output to both voluntary and involuntary motor centers. The output to the motor cortex is via the cingulate gyrus; to the brain stem it is via the hypothalamus.

Amygdala & hippocampus are also involved in memory functions — It is now known that the hippocampus, a prominent LS structure, thanks to its numerous connections to both the older (e.g., olfactory and limbic) and the newer (e.g., the cortex) parts of the brain, is a major brain center for of functions related to processing and storage of cognitive *memory* (plate 109). Recently the amygdala has been found to play an important role in learning and memory. Presumably,

the LS allows the lower animals to form memories.

EFFECTS OF LS STIMULATION & LESIONS

Stimulation of some limbic system structures induces emotional behaviors — Electrical stimulation of certain areas in the hypothalamus and neighboring structures in animals (e.g., cat) can evoke the behavior patterns of *fear* or *aggression* similar to those observed during a cat fight (hissing, spitting, hair standing, back hunching, and slapping). Lesion studies have indicated that disconnecting the forebrain from these lower limbic areas leaves the expression of these emotions intact while removing purpose and directedness from them.

Amygdala plays critical role in brain's emotional functions — Additionally, stimulation of certain areas in the amygdala often activates aggressive responses. Bulls can be forced to charge under these conditions. Stimulation of the amygdala in a normally submissive monkey causes the animal to exhibit aggressive gestures. As a result, the monkey temporarily moves up in the group's dominance order. Violent attacks of *rage* are occasionally seen in humans with *epileptic seizure discharges* (causing excessive local electrical activity) in the amygdala; surgical removal of the amygdala eliminates the rage attacks. Removal of the amygdala in monkeys also results in timidity and passivity, similar to the effects of frontal lobotomy seen in humans. The frontal lobes are connected to the amygdala via the anterior thalamus. The hippocampus is one place where stimulation does not elicit any emotional or instinctive behaviors.

Certain limbic system areas may function as neural substrate of pleasure & reward — When rats with electrodes in the septum or associated pathways (e.g., medial forebrain bundle) are taught to electrically self-stimulate at will, they do so for a long time, preferring this stimulation of their brain to food rewards (hence the label "pleasure centers"). Human patients in preparation for neurosurgery also report pleasure when electrically stimulated in similar locations.

Responses to a rose may help explain connectivity & function of LS — Exactly how the LS circuits function in the experience of feelings and expression of emotions and instincts is not quite clear. A person's responses to the *smell* and *sight* of a rose may help explain the functions and connectivity of the LS. These stimuli bring *pleasure, fond visual associations,* a *smile,* and *autonomic responses* (e.g., changes in heartbeat). The rose scents find access to the LS via the *olfactory system,* which feeds into the amygdala and the hypothalamus (plate 105). The rose's visual sensations activate the LS via reticular formation pathways to the hypothalamus (LS) or via visual thalamic nuclei to the anterior thalamus (LS).

Activation of the LS circuits at this time presumably creates the subjective experience of pleasure/good feelings. The motor responses of smiling can be activated via the hypothalamic outflow to the brain-stem nuclei serving in control of *muscles of expression.* The hypothalamus also serves as the LS center/output for such autonomic motor responses as changes in heartbeat. To touch and pick the rose, *voluntary motor areas* of the cortex are accessed via the cingulate gyrus or hippocampus. Similar connections to *frontal lobes* and *temporal lobes* evoke the *higher aspects of feelings* (e.g., love) and fond memories, respectively.

CN: Use a dark color for C.
1. Begin in the upper left corner with the view of the limbic system (C) compared with the brain stem (A) and cerebral cortex (B). Before going on to the actual structures of the limbic system, begin coloring its connections and responses, as shown in the middle diagram. Start with the sensory inputs on the left (olfactory bulb/other senses), and follow the arrows into the darkly outlined square representing the limbic system. As you work your way through the system, simultaneously color the anatomical representation of each structure at the top of the page.
2. Color titles of limbic structures in the bottom panel.

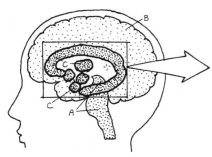

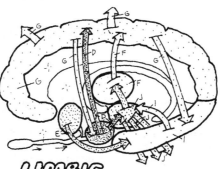

BRAIN STEM, CEREBRAL CORTEX, LIMBIC SYSTEM.

The limbic system (LS) consists of some forebrain structures and the hypothalamus. Positioned between the lower brain (serving vital functions) and the higher cerebral cortex (adaptive and skilled brain), the LS functions in motivation, emotions, and the expression of goal-directed instinctive behavior. In lower animals, the LS is intimately connected to the sense of smell.

LIMBIC STRUCTURES.

The amygdala, septum, hippocampus, cingulate gyrus, hypothalamus, and anterior thalamus and their associated fiber pathways make up the LS. The amygdala and septum help connect the primitive senses and the cortex to the LS; vision and audition find access via the thalamus. A loop within the LS (the Papez circuit) allows impulses from the hypothalamus to travel up to the anterior thalamus, on to the cingulate gyrus, and then via the hippocampus back to the hypothalamus. The cingulate gyrus and anterior thalamus provide connections between the LS and the cerebral cortex.

LIMBIC SYSTEM FUNCTIONS: EMOTIONS. INSTINCTS. DRIVES. LEARNING/MEMORY.

Some stimuli (aromas, strange sounds, a baby's smile) evoke emotions and bodily responses—e.g., "feelings" (pleasure), instinctive motor responses (smile), and visceral effects (heart rate). These responses are integrated by the LS structures, including the hypothalamus, which also provides a main output for the LS motor commands. Thus, signals for somatic motor reactions (smiling) are sent to the brain-stem motor centers. For visceral motor effects (heart rate), they are sent to the autonomic nervous centers. For neurohormonal effects, they go to the pituitary/endocrine system. Feelings are probably integrated at the higher cortical levels. The hippocampus and amygdala are also involved in learning and memory.

RESPONSES OF THE LIMBIC SYSTEM.

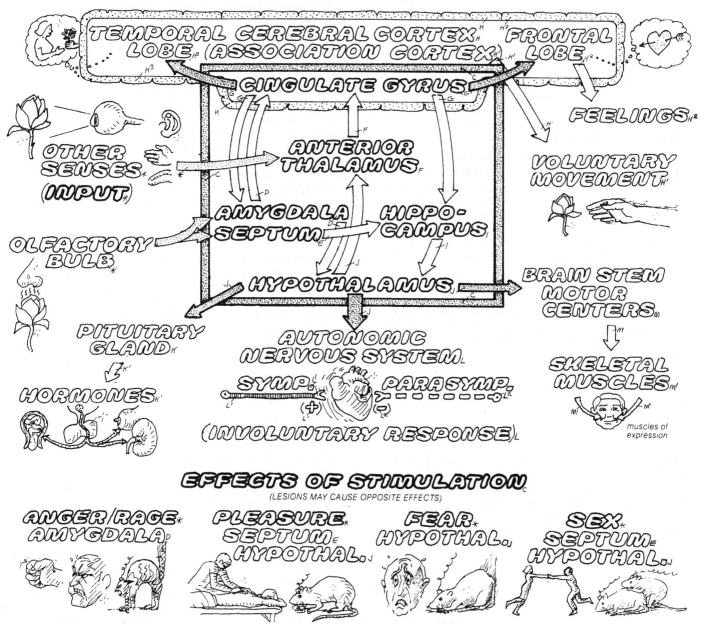

TEMPORAL CEREBRAL CORTEX LOBE (ASSOCIATION CORTEX) — FRONTAL LOBE

CINGULATE GYRUS

OTHER SENSES (INPUT)

OLFACTORY BULB

ANTERIOR THALAMUS

AMYGDALA SEPTUM — HIPPO-CAMPUS

HYPOTHALAMUS

FEELINGS

VOLUNTARY MOVEMENT

BRAIN STEM MOTOR CENTERS

PITUITARY GLAND

HORMONES

AUTONOMIC NERVOUS SYSTEM

SYMP. (+) — PARASYMP. (−)

(INVOLUNTARY RESPONSE)

SKELETAL MUSCLES

muscles of expression

EFFECTS OF STIMULATION

(LESIONS MAY CAUSE OPPOSITE EFFECTS)

ANGER/RAGE AMYGDALA

PLEASURE SEPTUM HYPOTHAL.

FEAR HYPOTHAL.

SEX SEPTUM HYPOTHAL.

Learning is a change in response to a stimulus following experience and is a major aspect of human brain function.

Habituation & sensitization are simple forms of learning—Repeated exposure to the same stimulus decreases the intensity of the response (habituation). Sensitization involves an increase in the intensity of response if the stimulus is accompanied with other positive (e.g., pleasant) or negative (e.g., unpleasant) stimuli. These non-associative forms of learning occur in all animals.

Conditioned learning involves new connections between learned stimuli & inborn responses—Here learning occurs by connecting two different stimuli. Dogs naturally salivate when exposed to the smell, sight, or taste of meat. This is an inborn reflexive response. The food (meat) is the *unconditioned stimulus* (US), the salivation the *unconditioned response* (UR). Inborn synaptic connections in the brain enable this response to occur. If the US (food) is paired with a conditioned stimulus (CS) (e.g., a bell) repeatedly (the bell must ring a few seconds before the food is presented), the animal soon shows the same UR (salivation) to the conditioned stimulus (bell) alone. The new *conditioned response* (CR) is evidence of associative learning and indicates the formation of a *new* connection in the brain between pathways for hearing and salivation.

Instrumental (operant) conditioning involves rewards & reinforcement—This form of learning (also called *trial/error* learning) is a more complex form of associative learning in which the learner takes an active part in the learning process, and reward plays a reinforcing role. Here, as in the conditioned reflexes, the ability to show and use a learned response improves with repeated exposure and practice, and the associated memories become more permanent.

STAGES OF LEARNING AND MEMORY FORMATION

Learning & memory formation occur in three stages—An initial *instantaneous stage* lasting a few seconds is followed by a *short-term* stage (minutes to hours), terminating in a *long-term* stage. For example, upon reading a new phone number, an instantaneous *working memory* is formed that will dissipate rapidly if not reinforced. Repeated exposure or active use of the phone number results in a *short-term memory* that is accessible for minutes to hours; finally it is consolidated into a long-term form that is stored permanently.

Various agents prevent consolidation of long-term memory—Long-term memory involves formation of permanent physiochemical changes in the brain, such as modified or new synapses. If animals or humans are exposed to conditions that temporarily reduce brain metabolism and protein synthesis (drugs, hypothermia) or alter the operation of brain electrical activity (electroshocks, blows) during formation of short-term memory, the learning is lost and memories cannot be recalled. However, these disturbing agents cannot erase long-term memory.

TYPES OF LEARNING & MEMORY AND BRAIN SYSTEMS

Declarative vs. procedural memory—Different types of memory are recognized. Declarative memory (explicit memory) is the more familiar conscious type and applies to memories of learned cognitive tasks (names, forms, words, symbols, events). Procedural memory (implicit, non-declarative) applies to memories formed following development of new sensory and motor skills (tracing a figure or riding bicycles).

Roles of different brain regions & pathways in learning and memory—Subcortical structures can mediate conditioned reflexes, although the presence of the cortex improves the efficacy of their formation and execution. Conditioned reflexes depend on the formation of new synaptic connections in the subcortical structures. The cortex is necessary for development and processing of working memory, which precedes formation of short-term memory. The *hippocampus* and *amygdala* of the limbic system (plate 108) are necessary for formation (consolidation) of long-term declarative (cognitive) memories although they are not the storage sites. Humans with hippocampal loss can form short-term memories but cannot consolidated them. Older memories are not lost and permanent new procedural (e.g., motor) memories also can form. The *cholinergic projections* form the *basal forebrain region* to the hippocampus and amygdala are also important in memory. These neurons show degeneration in Alzheimer's disease, which involves severe memory loss. The *adrenergic projections* from the *locus ceruleus,* which mediate arousal and attention, also facilitate learning.

CELLULAR & PHYSIOLOGICAL MECHANISMS OF LEARNING

Synaptic calcium is involved in cellular basis of learning & memory—Sensitization and habituation occur as a result of changes in the function of presynaptic sensory terminals. A decrease in the entry of *calcium ions* into the presynaptic terminal during habituation decreases the amount of transmitter released and efficacy of the synapse, while sensitization involves increased calcium entry and transmitter release. Development of short-term memory may depend on the phenomenon of *long-term potentiation* (LTP), in which postsynaptic cells, after episodes of increased presynaptic activity, maintain increased firing rates long after the stimulus has ceased. Such circuits exist in the hippocampus and involve increased calcium entry in the postsynaptic neuron. Neurons releasing glutamate and a specific type of *glutamate synaptic receptor* (NMDA receptor) are involved in these calcium-mediated responses. Specific genes and synthesis of special proteins involved in synaptic learning phenomena are now known or under investigation.

Rapid learning may involve reverberating circuits—When a stimulus is presented for the first time, the associated neuronal circuits remain active for as long as the stimulus is present. A *reverberating circuit* can prolong activity in the original circuit, even when the initial stimulus ceases. In such a circuit, the original excitatory input from, say, the sensory neurons activates parallel excitatory interneurons. These make *recurrent positive feedback* connections to the original circuit, enabling the excitation to continue. Reverberating circuits can explain *instantaneous* memories.

Modification of existing synapses & formation of new ones—Changes can occur in the properties of the *synaptic membranes* (receptors, enzymes) as well as in intracellular synaptic compartments (presynaptic and postsynaptic). These may increase the functional abilities of a synapse (*synaptic facilitation*), so that the same stimulus in a presynaptic neuron could result in a different (more or less intense) response in the postsynaptic neuron. New synapses can form in response to new experience (*synaptic growth*). Such changes occur in the cerebellar and the motor cortex following active (skilled) motor learning and involve *gene expression* and *protein synthesis*.

CN: Use dark colors for C, F, and K.
1. Color the stages of a conditioned reflex formation (top) one at a time. Note that in stage 3, neurons in the olfactory system (representing unconditioned stimulus) are not colored.

2. When coloring the material on short- and long-term memory, begin with the four neural mechanisms on the left, which are believed to be involved in learning and memory formation. Color borders of the remaining areas of this section.

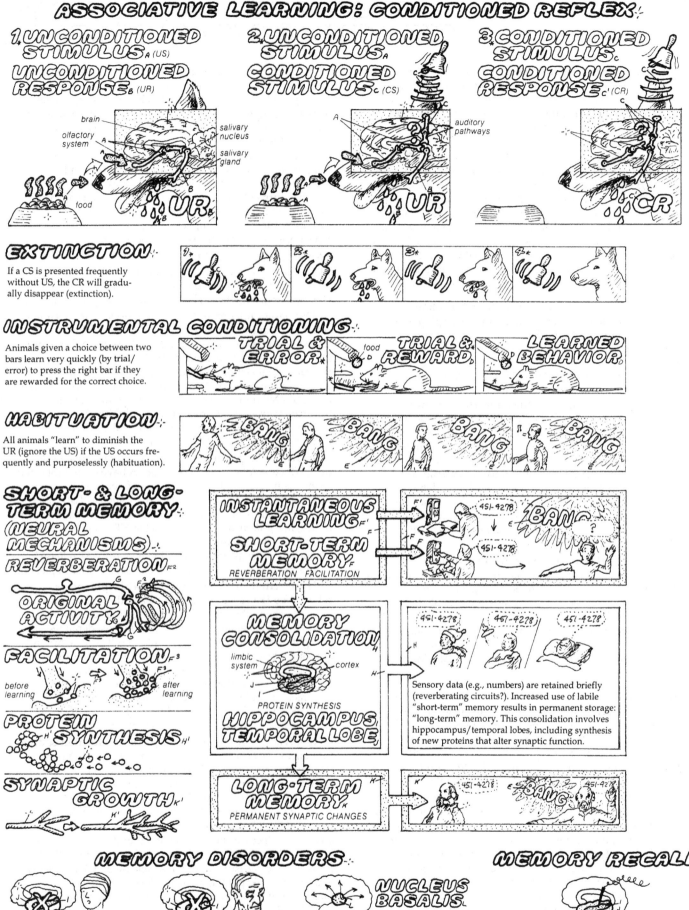

ASSOCIATIVE LEARNING: CONDITIONED REFLEX

1. UNCONDITIONED STIMULUS (US) — UNCONDITIONED RESPONSE (UR)

brain
olfactory system
salivary nucleus
salivary gland
food
UR

2. UNCONDITIONED STIMULUS — CONDITIONED STIMULUS (CS)

auditory pathways
UR

3. CONDITIONED STIMULUS — CONDITIONED RESPONSE (CR)

CR

EXTINCTION

If a CS is presented frequently without US, the CR will gradually disappear (extinction).

INSTRUMENTAL CONDITIONING

Animals given a choice between two bars learn very quickly (by trial/error) to press the right bar if they are rewarded for the correct choice.

TRIAL & ERROR — TRIAL & REWARD — LEARNED BEHAVIOR

food

HABITUATION

All animals "learn" to diminish the UR (ignore the US) if the US occurs frequently and purposelessly (habituation).

BANG BANG BANG BANG

SHORT- & LONG-TERM MEMORY (NEURAL MECHANISMS)

REVERBERATION

ORIGINAL ACTIVITY

FACILITATION

before learning
after learning

PROTEIN SYNTHESIS

SYNAPTIC GROWTH

INSTANTANEOUS LEARNING / SHORT-TERM MEMORY
REVERBERATION FACILITATION

451-4278
BANG?

MEMORY CONSOLIDATION

limbic system
cortex
PROTEIN SYNTHESIS
HIPPOCAMPUS / TEMPORAL LOBE

451-4278

Sensory data (e.g., numbers) are retained briefly (reverberating circuits?). Increased use of labile "short-term" memory results in permanent storage: "long-term" memory. This consolidation involves hippocampus/temporal lobes, including synthesis of new proteins that alter synaptic function.

LONG-TERM MEMORY
PERMANENT SYNAPTIC CHANGES

451-4278 BANG

MEMORY DISORDERS

LESIONS & INJURIES

Concussions cause retrograde amnesia possibly through damage to the hippocampus and temporal lobe structures (amygdala), which mediate consolidation.

SENILE AMNESIA

Senile memory deficits (diminished short-term memory, consolidation) may be related to aging-associated degeneration in the limbic/temporal lobe.

ALZHEIMER'S DISEASE / SENILE DEMENTIA — NUCLEUS BASALIS

Cholinergic input from the nucleus basalis to the cortex and limbic system markedly influences memory functions. Senile dementia seen in Alzheimer's disease involves, in part, loss of these projections.

MEMORY RECALL

TEMPORAL LOBE STIMULATION

Electrical stimulation of the temporal lobe in conscious patients (undergoing brain surgery) elicits vivid recall of past memories, particularly those with strong affective components.

Biogenic amines as neurotransmitters in CNS — *Norepinephrine* (NE), *dopamine* (DA), and *serotonin* (ST) belong to the group of *biogenic monoamine* neurotransmitters that serve in the central neural pathways regulating affective states (moods, motivation, feelings) and in self-awareness, consciousness, and personality.

Drug effects on CNS established roles for NE, ST, & DA in affective behavior & mental disorders — Reserpine, a plant alkaloid known to reduce hypertension, acts by decreasing storage of synaptic vesicles and availability of NE in peripheral synapses. Reserpine treatment for blood pressure also causes central "affective disorders" such as depression and loss of appetite and interest. Reserpine has been used for centuries in India to relieve mania (abnormally elevated moods) in mental patients. These observations implicate NE, ST, and DA in regulating mood and feelings (affective states).

PATHWAYS OF MONOAMINE NEURONS IN THE BRAIN

Norepinephrine & serotonin pathways originate in the midbrain & project to the forebrain — Mapping studies using fluorescent staining techniques showed that the neurons releasing these amines make up nerve groups within the reticular formation (plate 106). The cell bodies are generally located in the brain stem and the fibers ascend to the forebrain, targeting a variety of structures. There are two *NE projections*. One originates in the *locus ceruleus* of the medulla and courses up along the *medial forebrain bundle* to innervate the *cerebral cortex* and *limbic system*. NE fibers do not innervate the basal ganglia. Another NE pathway forms the lateral group innervating the hypothalamus and basal forebrain. The *ST neurons* originate mainly in the *raphe nucleus* and innervate many forebrain areas. Recent research indicates that the locus ceruleus NE system is involved mostly in arousal and attention, while the ST system regulates moods, motivation, pleasure, and well-being.

Mesocortical DA pathway is involved in behavior — The *DA pathways* also begin in the midbrain. One pathway ends in the hypothalamus, another in the basal ganglia, and a third — the *mesocortical (mesolimbic) pathway* — ends mainly in the structures of the *limbic system* and the *frontal lobes* (plate 108). This mesolimbic pathway in particular serves in complex mental functions related to the frontal lobe and limbic system (i.e., goal-directed behavior, self-awareness, thinking and planning, anxiety).

MONOAMINE SYNAPSE BIOCHEMISTRY & PHARMACOLOGY

NE & DA are synthesized from tyrosine & ST from tryptophan — Monoamine neurotransmitters are derived from amino acids: NE and DA from tyrosine, which is hydroxylated by the enzyme *tyrosine hydroxylase* to form DOPA and then DA. DA can be metabolized to NE. DA neurons lack the enzyme for this last conversion. ST comes from tryptophan and is converted by the enzyme *tryptophan hydroxylase* to ST.

Drugs influence amine synapse functions either pre- or postsynaptically — *Presynaptic effects* may include interference with: (1) transmitter synthesis by inhibiting the synthesizing *enzyme tyrosine hydroxylase*; (2) transmitter *storage* in *vesicles*; (3) transmitter *release* from vesicles; (4) *reuptake* of transmitter after release. *Postsynaptic* actions include (1) stimulation or blockage of *receptor binding* by the transmitter and (2) inhibition of the *deactivating enzyme* (plates 19, 20).

Neurotransmission drugs enhance or suppress synaptic function — The drugs that inhibit *reuptake* or the deactivating enzymes enhance synaptic function by increasing transmitter availability in the synapses. Drugs that block the *postsynaptic receptors* or inhibit transmitter synthesis or releases suppress synaptic function by reducing impulse transmission and transmitter availability, respectively. *Amphetamine* (an "upper" drug) increases release and blocks transmitter reuptake, increasing transmitter availability in the synapses, which in turn enhances synaptic function. As a result, arousal, mood, excitability, and the ability to concentrate are increased. Of course — amphetamines like other drugs — may have unpleasant side effects that appear later.

BIOGENIC AMINES AND MENTAL ILLNESS

Major depression and *schizophrenia* are two main groups of mental disorders, largely hereditary and caused by functional and chemical abnormalities in biogenic neurotransmission.

Decreased function of NE/ST synapses in depression is relieved by antidepressive drugs — Depression has been linked to *reduced activity* in the ST and possibly NE synapses. Drugs that enhance function in NE/ST synapses also improve the behavioral symptoms of the depression. For example, amphetamine, which enhances release and inhibits reuptake of the ST/NE transmitters, or drugs that inhibit the deactivating enzyme *monoamine oxidase* (MAO inhibitors) tend to relieve both neurochemical and behavioral deficiencies by increasing the NE/ST levels in the synapses.

Increasing importance of ST in affective disorders & drug effects — Alteration in ST synapses have been implicated as the main agent in the pathogenesis of depression, in contrast to NE. Many hallucinogenic drugs, such as *LSD* (*lysergic acid diethylamide*), and secondary metabolites of certain mushrooms, like *psilocin* and *mescaline*, act as ST receptor agonists (specifically stimulating 5-HT$_2$ receptors). The euphoria-producing street drug-"*ecstasy*" acts by increasing ST release in brain synapses, with major side effects caused by ST depletion.

Hyperactivity of DA synapses in schizophrenia is relieved by DA receptor blockers — Another major advance in psychochemotherapy has been in treating schizophrenia. Victims of this mental disease have delusions, deranged thoughts, and demented concepts of the self. The symptoms are sometimes accompanied by anxiety and psychosis. Some types of schizophrenia are associated with hyperactivity in the mesocortical pathway, caused by overexpression or hyperactivity of D$_2$ or D$_4$ receptors of DA. Drugs that block these receptors (*DA receptor blockers* — e.g., haloperidol) are effective in ameliorating schizoid symptoms.

Involvement of DA and endorphins in addiction & drug tolerance — *Addiction* is caused by continuous use of certain substances and drugs such as morphine and heroin (opiates), cocaine, amphetamine, nicotine, and ethanol. These drugs increase the availability of DA in synapses to act on D$_2$ receptors for DA. The pathway involved is a branch of the DA mesolimbic pathway projecting to areas such as the *nucleus accumbens*. Electrical stimulation of this and similar sites in the brain produces increased self-stimulation in animals (by bar pressing) and pleasure in humans. Release of *opioid peptides* (β endorphins, enkephalin), which reduce pain and increase pleasure and well-being, are also implicated in the phenomena of addiction and *drug tolerance*.

CN: Use dark colors for G, M, and Q and red for H.
1. At the top of the page, color the introduction and chemistry of biogenic amine neurotransmitters.
2. Color the panel on depression and its treatment, beginning with the brain diagram and related structures. Then go to the site of drug action, where an increase in levels of NE in the synapse is shown in the enlargement of the synaptic area. Conclude this section with the chemical structure of NE.
3. Do the same with the panel on dopamine.

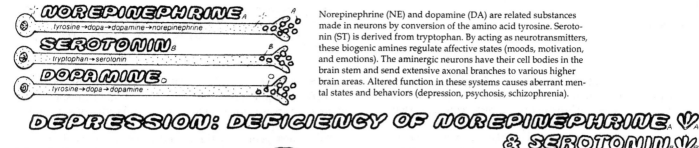

NOREPINEPHRINE A
tyrosine→dopa→dopamine→norepinephrine

SEROTONIN B
tryptophan→serotonin

DOPAMINE C
tyrosine→dopa→dopamine

Norepinephrine (NE) and dopamine (DA) are related substances made in neurons by conversion of the amino acid tyrosine. Serotonin (ST) is derived from tryptophan. By acting as neurotransmitters, these biogenic amines regulate affective states (moods, motivation, and emotions). The aminergic neurons have their cell bodies in the brain stem and send extensive axonal branches to various higher brain areas. Altered function in these systems causes aberrant mental states and behaviors (depression, psychosis, schizophrenia).

DEPRESSION: DEFICIENCY OF NOREPINEPHRINE A & SEROTONIN B

Deficiency of NE and particularly of ST in the brain may be the cause of depression. Drugs that "increase" NE/ST levels at the synapses elevate mood in normal people and alleviate depression in the mentally ill. To increase NE levels, some drugs (MAO inhibitors) reduce breakdown of NE by suppressing the deactivating enzyme in the postsynaptic neuron. Others (amphetamines) inhibit the reuptake of NE into the presynaptic neuron.

LIMBIC SYSTEM C
CEREBRAL CORTEX D
CEREBELLUM E
LOCUS CERULEUS F
NORADRENERGIC PROJECTIONS G

NE and particularly ST are important in regulating moods, pleasure, and brain excitability. NE neurons are located in the locus ceruleus in the medulla and project to the hypothalamus, thalamus, limbic system, and cortex, but not to the basal ganglia. ST neurons (not shown) begin in the raphe nucleus of the pons and innervate the same structures as well as the basal ganglia.

DRUG ACTION (ANTIDEPRESSANT)

BLOOD H

PRESYNAPTIC NEURON I
VESICLE J

POSTSYNAPTIC N. K
RECEPTOR L
NERVE IMPULSE M

synapse

DRUG 1 → REUPTAKE OF
DRUG 2 → DEACTIVATING ENZYME } NE. A ↑ SER. B ↑

SEROTONIN B
NOREPINEPHRINE A
NH₂ A
CH₂
HO-C-H A
OH
HO

SCHIZOPHRENIA: EXCESS OF DOPAMINE ↑

Drugs that reduce or block transmission at the DA synapses (DA receptor blockers—e.g., haloperidol) are the most effective agents in the treatment of schizophrenia (antipsychotic drugs). In contrast, drugs that markedly "increase" brain DA levels and transmission (amphetamine, cocaine in high doses) can cause paranoid/schizoid behavior, even in normal humans.

MIDBRAIN P
DOPAMINERGIC PATHWAY Q

In the brain, three separate DA pathways are known—one in the hypothalamus, the second in the basal ganglia, and the third originating in the midbrain, projecting to the limbic system and frontal cortex. Hyperactivity in the synapses of this mesolimbic (mesocortical) pathway, caused by increased levels of D_2 receptors for DA, may be involved in the genesis of schizoid psychosis. Schizophrenia is the most common form of mental disorder.

DRUG ACTION (DA RECEPTOR) BLOCKER (ANTIPSYCHOTIC DRUG)

DOPAMINE NH₂
CH₂
H-C-H
OH
HO

DRUG → RECEPTOR ↓ → DOPAMINE (DA)
(DA RECEPTOR BLOCKER) TRANSMISSION ↓

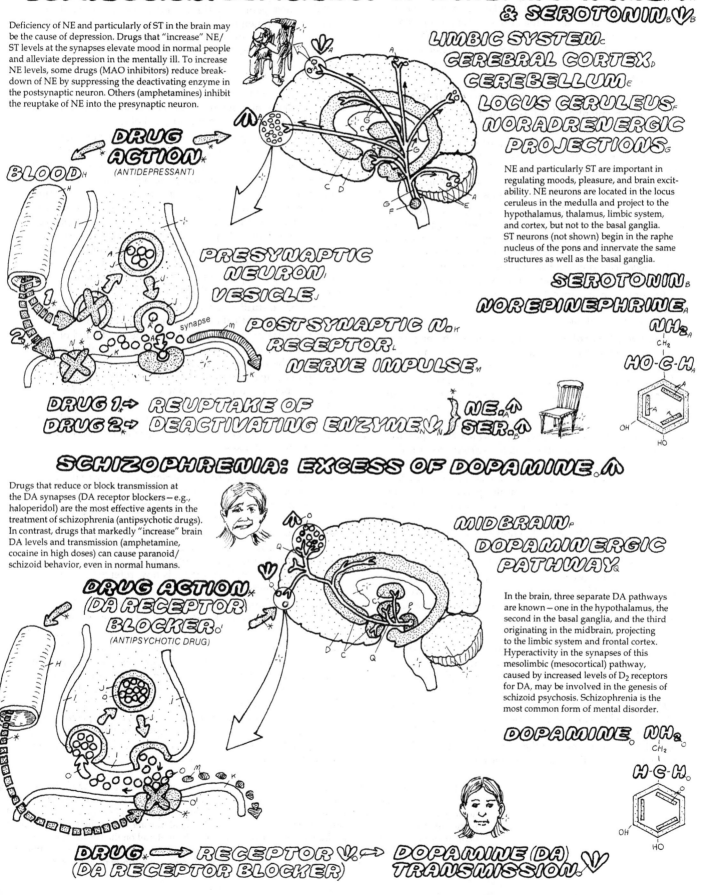

Cortical association areas are the site of higher integrative functions — In addition to areas specialized for purely *sensory* or *motor* functions, the human *cerebral cortex* contains extensive areas that are neither sensory nor motor. These "*association*" areas (*frontal, temporal,* and *parietal-temporal-occipital*) constitute the greater part of the human cerebral cortex and are involved in higher-order *integrative* functions of the brain (e.g., speech and language, planning). The equivalents of these areas are either small or not present in other mammals and primates.

ORGANIZATION OF LANGUAGE FUNCTIONS IN THE BRAIN

Only left-hemisphere lesions cause aphasia (speech disorders) — In the 19th century, French neurologist Broca noted that patients with lesions in a particular area of the *left frontal lobe* (now known as *Broca's area*), anterior to the motor speech cortex, could understand speech but had difficulty producing meaningful sentences (*motor* or *nonfluent aphasia* [aphasia = speech disorder]) and had no sign of speech paralysis. Later Wernicke, a German neurologist, noted that left-hemisphere lesions, limited to a region bordering the *parietal* and *temporal lobes* (now known as *Wernicke's area*) caused *sensory* or *fluent aphasia*, a disorder in which the patient showed poor speech "comprehension" without having any hearing problems. From these and later studies, a cerebral organization for language and speech was formulated that localized this important human faculty to certain discrete association areas of the *left hemisphere*.

A pathway for spoken language in the left hemisphere — According to this scheme, spoken words and sentences are analyzed initially by the *primary auditory areas*, then by the *secondary auditory association areas* before being relayed to the higher integrative areas (i.e., Wernicke's area of the left temporal lobe). Here the symbolic meanings of words and language are processed and understood. To speak words, signal commands are relayed from Wernicke's area via a special association fiber pathway (the *arcuate fasciculus*) to Broca's area in the frontal lobe of the same left hemisphere. Broca's area functions as the *premotor area* for speech, sending programs for the activation of appropriate speech muscles and their proper order of contraction to the *speech motor cortex* in the lower precentral gyrus. Activation of upper motor neurons in this area results in contraction of speech muscles and speech production (plate 96).

Pathways for visual language (reading, writing, sign language) — Based on observations in patients showing abnormal ability in reading (*dyslexia*) and writing (*agraphia*) of words, a similar scheme has been drawn for processing visual language (reading, writing). Thus, images of words, after processing by the *visual association areas*, are relayed via the *angular gyrus* (a higher-order visual integrative area) to the *hands premotor* area. Between the angular gyrus and hands premotor cortex, the impulses may pass through Wernicke's area. The hands premotor area communicates to the neighboring *hand motor cortex* the necessary programs for movement of the hand muscles, resulting in writing. Sign language may involve a similar scheme. The signals between the different association areas are sent via the intra-hemispheric and inter-hemispheric association tracts (see below).

HEMISPHERIC DOMINANCE VS. HEMISPHERIC SPECIALIZATION

Left hemisphere is motor dominant — Right hemisphere damage in areas equivalent to Broca's and Wernicke's areas of the left hemisphere causes few speech defects. This and the fact that most people are right-handed (i.e., motor control areas of the left hemisphere are superior or dominant to those of the right) led to the notion that the two hemispheres, though fairly symmetrical in form, are unequal in function, with the left hemisphere being *dominant* in motor and speech tasks. Recently, several minor but distinct differences between the right and left hemispheres have been found and the dominance of the right hemisphere in certain non-motor and non-verbal tasks has been recognized (see below).

The two hemispheres are connected by the corpus callosum, whose transection creates a "split brain" — This inter-hemispheric association tract, massive in humans, specifically connects the association areas of one hemisphere to the exact mirror-image areas in the opposite one, thereby transferring information between the hemispheres. Occasionally, the corpus callosum in human patients suffering from epileptic convulsions is sectioned to prevent the spread of seizures from one hemisphere to the other (*split brain* operation). Careful testing of these patients revealed that each hemisphere functions not only independently but in a different manner, as though each had functional abilities and a "mind" of its own.

Split brain studies confirm left hemisphere is specialized for verbal & analytical tasks (categorical hemisphere) — After the surgery, if a key is placed in the right hand of a blindfolded split-brain patient, the sensory signals reach the left hemisphere through crossing of sensory pathways (plate 92). Upon being asked about the nature of the object in his hand, the patient verbally responds, "A key." If the key is placed in the left hand, its sensory image is in the right hemisphere. In this cases, the patient cannot verbally describe the key although he can recognize the object (can point to the name or shape of a key).

These results imply that (1) centers for verbal expression are in the left hemisphere, (2) the right hemisphere has access to the speech centers only via the corpus callosum, and (3) the right hemisphere has full perceptive and cognitive but non-verbal motor competence. The left hemisphere, in addition to its motor and verbal superiority, appears to be specialized for logical and analytical operations; it categorizes things and reduces them to their parts in order to understand them. Accordingly, the left hemisphere is now referred to as the "categorical" hemisphere.

Right hemisphere is specialized in spatial & holistic tasks (representational hemisphere) — The right hemisphere is known to be superior in *representational* and *visuospatial* functions, in perception and discrimination of *musical tones* and *speech intonations*, in *emotional responses*, and in appreciating *humor* and *metaphor*. In broad terms, the right-hemisphere functions are holistic and spatial (hence it is labeled the "artistic" or "representational" hemisphere). Despite these functional divisions, under normal conditions, and especially with regard to global, cognitive, and adaptive functions (memory, learning), the brain functions as a whole, utilizing the capacities of its different parts in concert.

CN: Begin in the upper left corner with the list of the seven functional characteristics of the right hemisphere (A). Note that they don't refer to any particular structure. Do the same with the left hemisphere, and color the large drawing of the two hemispheres and the remaining material at the top. 1. Color the corpus callosum (C) in the large and smaller drawings below and to the left. The corpus callosum deals with communication between the two hemispheres. The other association tracts (D) handle signal transfer within the same hemisphere. 2. Color the material on the left hemisphere's role in speech and other symbolic communication functions. Starting with number 1 at the ear.

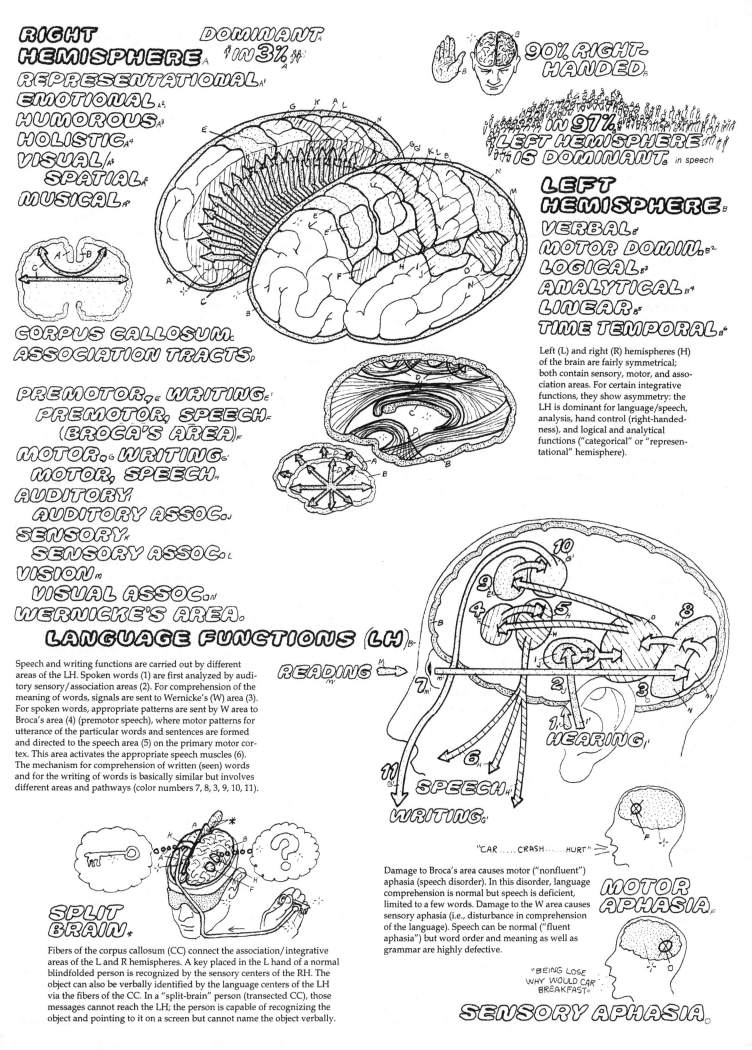

RIGHT HEMISPHERE, DOMINANT IN 3%

- REPRESENTATIONAL
- EMOTIONAL
- HUMOROUS
- HOLISTIC
- VISUAL/
- SPATIAL
- MUSICAL

CORPUS CALLOSUM.
ASSOCIATION TRACTS.

- PREMOTOR, WRITING
- PREMOTOR, SPEECH (BROCA'S AREA)
- MOTOR, WRITING
- MOTOR, SPEECH
- AUDITORY
- AUDITORY ASSOC.
- SENSORY
- SENSORY ASSOC.
- VISION
- VISUAL ASSOC.
- WERNICKE'S AREA.

90% RIGHT-HANDED.

IN 97% LEFT HEMISPHERE IS DOMINANT. in speech

LEFT HEMISPHERE

- VERBAL
- MOTOR DOMIN.
- LOGICAL
- ANALYTICAL
- LINEAR
- TIME TEMPORAL

Left (L) and right (R) hemispheres (H) of the brain are fairly symmetrical; both contain sensory, motor, and association areas. For certain integrative functions, they show asymmetry: the LH is dominant for language/speech, analysis, hand control (right-handedness), and logical and analytical functions ("categorical" or "representational" hemisphere).

LANGUAGE FUNCTIONS (LH)

Speech and writing functions are carried out by different areas of the LH. Spoken words (1) are first analyzed by auditory sensory/association areas (2). For comprehension of the meaning of words, signals are sent to Wernicke's (W) area (3). For spoken words, appropriate patterns are sent by W area to Broca's area (4) (premotor speech), where motor patterns for utterance of the particular words and sentences are formed and directed to the speech area (5) on the primary motor cortex. This area activates the appropriate speech muscles (6). The mechanism for comprehension of written (seen) words and for the writing of words is basically similar but involves different areas and pathways (color numbers 7, 8, 3, 9, 10, 11).

READING

HEARING

SPEECH

WRITING

"CAR .... CRASH ...... HURT"

SPLIT BRAIN

Fibers of the corpus callosum (CC) connect the association/integrative areas of the L and R hemispheres. A key placed in the L hand of a normal blindfolded person is recognized by the sensory centers of the RH. The object can also be verbally identified by the language centers of the LH via the fibers of the CC. In a "split-brain" person (transected CC), those messages cannot reach the LH; the person is capable of recognizing the object and pointing to it on a screen but cannot name the object verbally.

Damage to Broca's area causes motor ("nonfluent") aphasia (speech disorder). In this disorder, language comprehension is normal but speech is deficient, limited to a few words. Damage to the W area causes sensory aphasia (i.e., disturbance in comprehension of the language). Speech can be normal ("fluent aphasia") but word order and meaning as well as grammar are highly defective.

MOTOR APHASIA

"BEING LOSE WHY WOULD CAR BREAKFAST"

SENSORY APHASIA

The *brain* is active all the time, in wakefulness and sleep. Therefore, it, like the heart, is critically in need of a continuous supply of *metabolic fuel substances* (energy) and *oxygen* provided by the *blood flow*.

BRAIN'S DEPENDENCY ON GLUCOSE

Brain tissue is exclusively dependent on glucose for fuel — In contrast to other active body organs (e.g., heart, muscle), which can utilize alternative fuels such as fatty acids, the brain under normal conditions depends almost exclusively on *glucose* for its energy needs. The adult brain uses glucose at the rate of 80 mg/min and has *glycogen* reserve of 1.6 mg/g (2.2 g/brain), lasting about 2 min without glucose.

Brain's dependency on glucose is shown by effects of hypoglycemia — Marked reduction in blood glucose (*hypoglycemia*) caused by a large insulin dose or as a result of prolonged starvation may lead to fainting, convulsions, coma, or death. Normal plasma glucose range is 70–110 mg/dl. Below 60 mg, cognitive and conscious activities are disturbed. Below 50 mg, speech is slurred and movement uncoordinated; below 30 mg, unconsciousness and coma set in; at 20 mg/l, convulsions may occur, and at 10 mg permanent brain damage and loss of medullary respiratory centers cause death. The brain's critical dependency on glucose is a main reason for many neural and hormonal mechanisms to ensure high (normal) blood glucose levels (plates 131, 132).

Brain can adapt metabolically to use ketone bodies as fuel — Interestingly, after many days of starvation, the brain may develop the capacity (enzymes) to use *ketone bodies* (a product of fatty acid metabolism in the liver plate 133) as an alternative energy source. This capacity is present in the newborn brain but disappears after infancy.

BRAIN'S DEPENDENCY ON OXYGEN

Brain cells are very sensitive to oxygen deprivation — In adults, 10 seconds of *anoxia* (oxygen deprivation, extreme hypoxia) is enough to cause loss of consciousness and higher brain functions (fainting). A few minutes of hypoxia can lead to coma and severe and irreversible brain damage; at first, higher cortical centers and basal ganglia structures are damaged; death can occur because of the loss of function in vital respiratory centers of the medulla (the most resistant to hypoxia).

Brain has high rate of oxygen consumption — In adult males, the brain weighs ~1.45 kg (3 lb) (1.25 kg in females), and the brain has an *oxygen consumption rate* of about 50 ml/ per min. Thus, although the brain's weight is only 2% of the body's, it accounts for 20% of the whole body's oxygen consumption rate (*metabolic rate*).

Cellular basis of high energy needs — The brain's work depends heavily on the formation, propagation, synaptic transmission, and integration of a variety of electrochemical potentials, cellular functions requiring the maintenance of proper ionic gradients (plates 10, 11, 15). This, the brain cell membranes contain one of the largest concentrations of sodium-potassium pumps in the body. These pumps are ATP dependent and underlie the operation of the plasma membrane enzyme Na-K-ATPase, which is also present in the brain in large concentrations. The sodium pump uses most of the ATP produced in the brain (plate 10).

Brain cells have many mitochondria, mainly in synapses — To produce the large quantity of ATP required by brain cells, the citric acid cycle/oxidative phosphorylation pathway is utilized (plate 6). This accounts for the brain's large concentration of mitochondria in neuronal synapses and critical dependence on oxygen. Staining of brain tissue with cytochrome oxidase, a mitochondrial marker enzyme, shows zones of high activity in synapse-rich areas of the brain.

Neuronal synapses & synapse-rich areas are major energy consumers — In general, the synapse-rich gray matter areas have *high* metabolic rates, while synapseless white matter areas (myelinated nerve fibers) have *low* rates. Among the gray matter areas, rates vary depending on the brain region. The forebrain basal ganglia and the midbrain inferior colliculi show *very high* rates; the *cortex* of the cerebrum and cerebellum have *moderately high* rates; the thalamus and the *nuclei* of the *cerebellum* and *medulla* show medium rates; the lowest rates are associated with spinal cord white matter.

BRAIN BLOOD FLOW & CHANGES WITH FUNCTION

Brain's high blood flow rate shows regional variation — To support its high oxygen and glucose needs, the brain has an extensive vascular supply and a large, efficient blood-flow (750 ml/min, 15% of the body's). Blood flow in different brain regions is different and roughly corresponds to each area's oxygen consumption rate.

Brain's regional blood flow can be measured & is activity dependent — Applications of various modern non-invasive imaging methods such as functional MRI (fMRI, *Magnetic Resonance Imaging*) and PET (*Positron Emission Tomography*) allow blood flow changes in different brain regions to be measured under different conditions in healthy, intact humans. In general, increase in *neural activity* in any particular brain area results in a rapid increase in the flow of blood to that area, to supply increased needs for oxygen, glucose and to remove carbon dioxide.

Regional blood flow helps localize sites of higher brain functions — Frontal lobe areas show above average activity even at rest. Activity increases during contemplation, problem solving, and planning, as well as during pain and anxiety. The simple reading of words increases activity in the posterior visual areas, while thinking about the reading material spreads the activity to parietal and temporal association areas. Listening carefully to words increases activity in auditory areas of temporal-parietal cortex. Speaking words activates Broca's area and the speech cortex on the left hemisphere, and the speech sensory areas (lips, tongue, face) of both hemispheres. Thinking about words activates the large region of frontal, parietal and temporal lobes (Wernicke's area) involved in language comprehension and planning (plate 111).

Blood flow changes occur in brain diseases — Certain brain and mental diseases such as schizophrenia and depression and the senile disorders such as the dementias, e.g., Alzheimer's disease, involving reduced cognitive and memory capacities, are associated with *reduced* blood flow/ metabolic activity. Brain diseases such as epilepsy, involving convulsions, create excessive electrical activity and *increased* blood flow and metabolic activity.

CN: Use red for I, and a dark color for E.
1. Begin in the upper left corner, and go on to the diagram on the right side of the page.
2. Color the diagrams on metabolic rate, noting that the sagittal section of the brain is intended to be a composite of several views in order to show the structures involved. Note that, in the diagram on the right, only the borders are colored to indicate general regions of high or low activity.
3. In the lower panel, color only the shaded areas.

BRAIN VS. TOTAL BODY

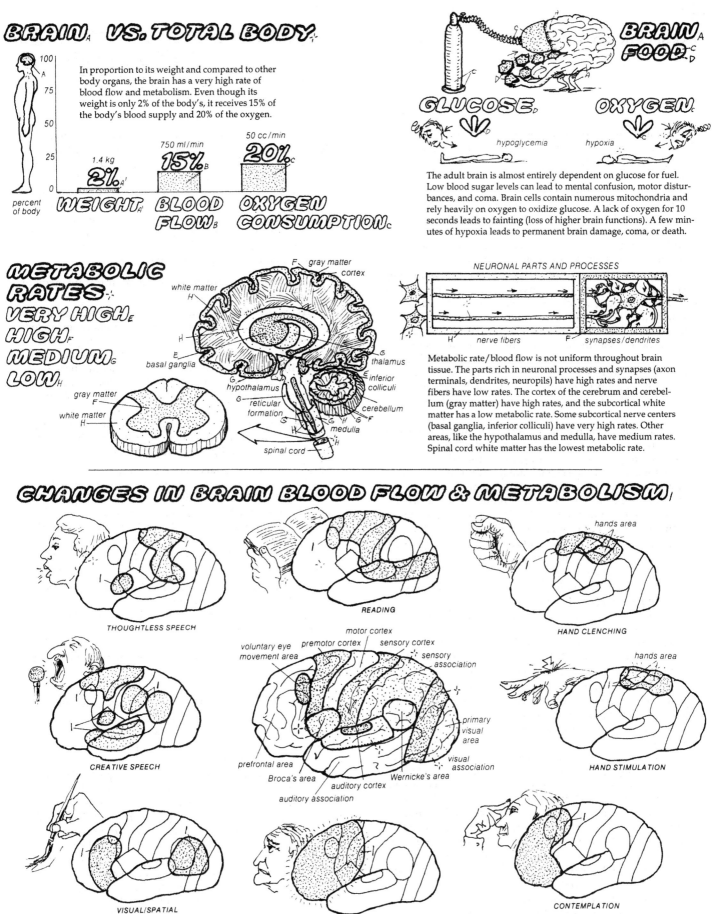

100
75
50
25
0
percent
of body

In proportion to its weight and compared to other body organs, the brain has a very high rate of blood flow and metabolism. Even though its weight is only 2% of the body's, it receives 15% of the body's blood supply and 20% of the oxygen.

1.4 kg
2%_{A'}
WEIGHT_{A'}

750 ml/min
15%_B
BLOOD FLOW_B

50 cc/min
20%_C
OXYGEN CONSUMPTION_C

BRAIN FOOD

GLUCOSE_D
hypoglycemia

OXYGEN_C
hypoxia

The adult brain is almost entirely dependent on glucose for fuel. Low blood sugar levels can lead to mental confusion, motor disturbances, and coma. Brain cells contain numerous mitochondria and rely heavily on oxygen to oxidize glucose. A lack of oxygen for 10 seconds leads to fainting (loss of higher brain functions). A few minutes of hypoxia leads to permanent brain damage, coma, or death.

METABOLIC RATES
VERY HIGH_E
HIGH_F
MEDIUM_G
LOW_H

gray matter — cortex
white matter H
basal ganglia
hypothalamus
reticular formation
thalamus
inferior colliculi
cerebellum
medulla
spinal cord

gray matter F
white matter H

NEURONAL PARTS AND PROCESSES
nerve fibers H
synapses/dendrites F

Metabolic rate/blood flow is not uniform throughout brain tissue. The parts rich in neuronal processes and synapses (axon terminals, dendrites, neuropils) have low rates. The cortex of the cerebrum and cerebellum (gray matter) have high rates, and the subcortical white matter has a low metabolic rate. Some subcortical nerve centers (basal ganglia, inferior colliculi) have very high rates. Other areas, like the hypothalamus and medulla, have medium rates. Spinal cord white matter has the lowest metabolic rate.

CHANGES IN BRAIN BLOOD FLOW & METABOLISM

THOUGHTLESS SPEECH

READING

hands area
HAND CLENCHING

CREATIVE SPEECH

voluntary eye movement area
premotor cortex
motor cortex
sensory cortex
sensory association
prefrontal area
Broca's area
auditory cortex
auditory association
Wernicke's area
primary visual area
visual association

hands area
HAND STIMULATION

VISUAL/SPATIAL

ANXIETY/PAIN

CONTEMPLATION

Different brain regions change their blood flow and neural activity in different physiological and psychological states. When a person clenches the right hand, blood flow in the premotor cortex and hand area of the motor cortex in the left hemisphere increases. Similar increases occur in the sensory cortex. During resting and contemplation, activity is higher in the frontal lobe areas than in the posterior cortical areas. During concentration, awareness, anxiety, and pain, frontal lobe activity is markedly increased, indicating the importance of frontal areas in these mental states. During silent reading, visual association areas and the area for voluntary eye movements in the premotor cortex show increased activity. Intensive talking involving expression of ideas increases activity in the auditory and speech motor cortex as well as in Wernicke's and Broca's areas.

The importance of organization in the body is implicit in the concept of the body as an "organism." To be organized, the parts of the body must be regulated to work in synchrony with one another and in harmony with the external environment. This regulation is carried out by the nervous system and *endocrine system*. The nervous system, by sending nerve signals along the peripheral nerves, functions very rapidly, adjusting the activities of the internal organs within seconds. Although rapid, these effects (e.g., changes in blood pressure, respiration, and temperature) are relatively short lasting.

Hormones of the endocrine system exert slow but long-lasting effects—Unlike the nervous system, hormones of the endocrine system, which are secreted into the blood, act slowly, their effects taking minutes to hours to days to develop; however, these effects are longer lasting than those produced by the nerves. Hormones are chemical substances secreted in minute amounts into the bloodstream by the cells of the *endocrine glands*. Traversing through the circulation, hormones bind with appropriate receptors, which are selectively present in the cells of their target organs, inducing the desired effects on growth, metabolism, or function in those organs.

Neuroendocrine regulation—The nervous and endocrine systems are capable of regulating each other's activities as well as acting in concert to bring about desired changes in body functions. The special advantage of this *neuroendocrine* system of hormonal communication is that it allows the mediation of the effects of both the environment and the brain's systems on the endocrine system. The various types of neuroendocrine communication are briefly outlined below.

Endocrine glands of the body—The *endocrine glands* are clusters of endocrine cells with distinct hormonal secretory functions. The major endocrine glands include the *pineal* (melatonin), *anterior pituitary* (growth hormone, tropins), *posterior pituitary* (ADH, oxytocin), *thyroid* (thyroxine, T3), *parathyroid* (parathormone), *adrenal cortex* (corticosteroids), *adrenal medulla* (catecholamines), *pancreatic islets* (insulin and glucagon), and *testes* (male steroids, inhibin) and *ovaries* (female steroids, inhibin). These endocrine glands and their hormonal actions are detailed in future plates.

Organs with partial endocrine function—Another category of cells with endocrine functions is made up of those found scattered individually or in small aggregates within other organs with distinctly nonendocrine functions. These organs are the *kidney* (renin, erythropoietin, calcitriol), *liver* (somatomedin), *thymus* (thymosin), *hypothalamus* (hypothalamic hormones), *heart* (natriuretic peptide), *stomach* (gastrin), and *duodenum* (secretin, CCK, GIP). The testis and ovary may also be considered in this category because they also produce the male and female gametes. The presence and location of endocrine cells within another organ are often dictated by some special functional relationship between that organ and the endocrine cells it is hosting. For example, the kidney senses decreased blood pressure and secretes renin to compensate for this deficiency.

TYPES OF HORMONAL COMMUNICATION
Endocrine–target gland interaction is the simplest type of hormonal communication—Hormones were initially con-

ceived of as substances secreted by an endocrine gland (cell) into the blood to reach a target organ, aiming to regulate or alter the activity of that organ. This form of purely hormonal communication is still applicable to many of the endocrine glands and their hormones—e.g., the pancreatic islets (insulin and glucagon). Hormonal communication may also occur between two endocrine glands. For example, the anterior pituitary secretes several tropic (trophic) hormones that stimulate other endocrine glands (*target glands*) to secrete hormones of their own (target gland hormones).

Several types of neuroendocrine communication mediate brain influences on body functions:

Direct neurosecretory control—In the simplest case, axons of certain nerve cells in the brain hypothalamus are extended into the posterior pituitary, secreting hormones (e.g., ADH) directly into the bloodstream to reach their targets (e.g., the kidney). In a more complicated case, hypothalamic nerve cells secrete regulatory hormones into a special *portal vascular system* connecting the hypothalamus with the anterior pituitary gland to control the secretion of some anterior pituitary hormones (e.g., growth hormone and prolactin) in the bloodstream. The anterior pituitary hormones then reach their own target organ(s) (e.g., adipose tissue and mammary glands).

Brain–anterior pituitary–target gland type of interaction—In a more complicated subtype of neurohormonal communication, the anterior pituitary hormones (e.g., ACTH) secreted in response to hypothalamic hormone (e.g., CRH) traverse the circulation to act as stimulating hormones (tropin) on some other endocrine gland (e.g., adrenal cortex). The latter then secretes its own target hormone (e.g., cortisol) to reach, via the blood, the final desired target organ (e.g., the liver).

Autonomic nervous system control of endocrine hormones—Another type of neurohormonal communication is the secretion of a hormone from an endocrine gland directly in response to nerve signals from *autonomic nerves*. For example, the secretion of hormones of the adrenal medulla and pineal are subject to regulation by nerve signals from the sympathetic nervous system.

PARACRINE & AUTOCRINE HORMONAL COMMUNICATION
The last major type of hormonal communication discovered during recent decades is *local* or "tissue" hormonal communication. In this type, the definition of hormone is expanded to apply to substances secreted by special paracrine cells directly into the extracellular space of a particular tissue. These hormones diffuse across short distances within the extracellular space of the same tissue to act on nearby cells (*paracrine effect*) or the same cells (*autocrine effect*). The blood, therefore, is not involved as the medium of transport for this type of local hormone unless the paracrine cells themselves are blood cells. Paracrine hormonal communication has been observed in many tissues. Prostaglandins, known to be involved in many local regulatory functions, are the best-known examples of paracrine hormones. Many growth factors exert their effects on their target cells by autocrine or paracrine means.

CN: Use red for D and dark colors for H and J.
1. Color the upper panel, starting with the endocrine glands on the left, all receiving the same color

(A). Do the same with the organ column on the right (B).
2. Color the lower panel, completing each form of communication before going on to the next one.

ENDOCRINE GLANDS.ₐ

PINEAL₁
PITUITARY₂
THYROID₃
PARATHYROID₄
PANCREAS₅
ADRENAL₆
OVARY₇
TESTIS₈

Endocrine glands secrete hormones into the blood. The classic endocrine glands with mainly endocrine functions are the pineal, pituitary, thyroid, parathyroid, and adrenal glands, pancreatic islets, and testis and ovaries. The testes and ovaries form gametes as well.

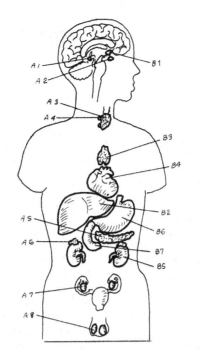

ORGANS WITH PARTLY ENDOCRINE FUNCTION.ᵦ

HYPOTHALAMUS₁
LIVER₂
THYMUS₃
HEART₄
KIDNEY₅
STOMACH₆
DUODENUM₇

Some organs contain individual cells or aggregates of endocrine cells that release hormones. These hormones often relate to the functions of the organs. Among these are the hypothalamus, liver, thymus, heart, kidney, stomach, and duodenum. The testis and ovary can also be included in this list.

FORMS OF HORMONAL COMMUNICATION.

1. ENDOCRINE.

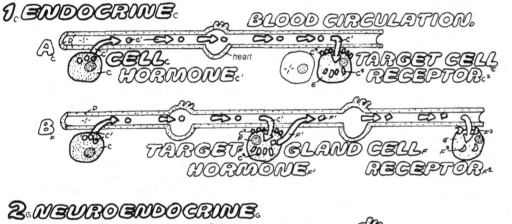

BLOOD CIRCULATION.
CELL. HORMONE.
TARGET CELL RECEPTOR.
TARGET GLAND CELL HORMONE. RECEPTOR.

Hormones are secreted into the blood to regulate the function of a distant target cell (organ). In the simplest form of hormonal communication, a hormone from an endocrine cell is transported by blood to a target cell (containing receptors for that hormone). In a more complex case, hormonal communication occurs between two endocrine glands, one serving as the target for the other. Still more complex forms involve interaction between the brain and endocrine glands. Thus a nerve cell can secrete a hormone directly into the blood. Or a nerve cell secretes a hormone to reach the pituitary via a portal circulation. The pituitary cell then secretes a tropin that acts on either a target organ or another target endocrine gland. Nerve cells are also able to stimulate endocrine cells directly.

2. NEUROENDOCRINE.

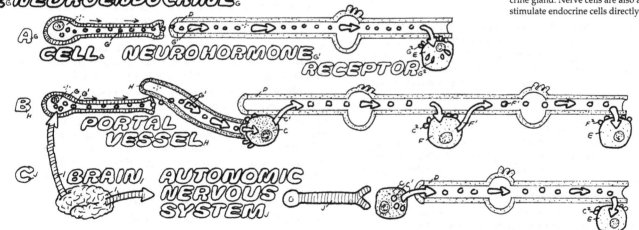

CELL. NEUROHORMONE. RECEPTOR.
PORTAL VESSEL.
BRAIN. AUTONOMIC NERVOUS SYSTEM.

3. PARACRINE.
(LOCAL TISSUE ENVIRONMENT).

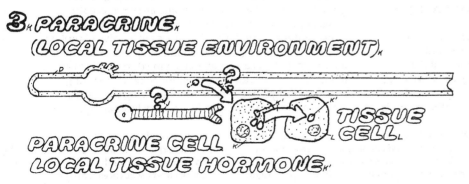

PARACRINE CELL
LOCAL TISSUE HORMONE.
TISSUE CELL.

In local hormonal communication, paracrine cells secrete local (or tissue) hormones in the extracellular fluid to reach the neighboring target cells by diffusion, bypassing the blood entirely.

AUTOCRINE.

Local hormones also may act on cells that secrete them (autocrine).

CELLULAR MECHANISMS OF HORMONE ACTION

The body has numerous hormones, exerting a wide variety of actions including growth, metabolic processes, and functional activities. These diverse hormones may be divided into two major groups, *fast acting* and *slow acting*, each with distinct mechanisms of cellular action. The individual members of each of these groups may show minor variation from the general scheme of cellular action.

SLOW-ACTING HORMONES ACT VIA NUCLEAR RECEPTORS & PROTEIN SYNTHESIS

The slow-acting hormones comprise the *steroid* hormones of the *adrenal cortex* and *gonads* (*testes and ovaries*) and the *amine* hormones of the *thyroid* gland as well as the hormones derived from *vitamin D_3 (calcitriol)*. Their actions involve binding with nuclear receptors and synthesis of new proteins. The actions of these hormones, though profound, are manifested slowly — within hours to days — but are long lasting.

Binding with plasma proteins — Less than 10% of the released steroid or thyroid hormones circulate in "free" form; the rest are bound to specific *plasma binding proteins* that are formed in the liver and act as physiological regulators of the "free" levels of hormones and prevent their loss in the kidneys during filtration.

Conversion to active hormone — In some tissues, the hormone is first converted to a more active form. Thyroxine (T4), the major thyroid hormone in plasma, is first converted in the target cell to tri-iodo-thyronine (T3), which is the cellularly active form of the hormone. Some steroid hormones undergo similar changes. Testosterone, the testis hormone, may be converted to dihydrotestosterone or even to estrogen.

Binding with nuclear receptors — Inside the target cell, the steroid or thyroid hormones move into the cell nucleus, where each binds with *a specific nuclear receptor*. These receptor proteins have a binding site for the hormone and a DNA binding domain for interaction with the nuclear DNA (genes). Hormone binding changes protein conformation, exposing the DNA binding domain.

Initiation of transcription & protein synthesis — Each receptor binds to a specific portion of the DNA (specific gene), inducing the transcription process, resulting in synthesis of a specific mRNA molecule that moves into the cytoplasm. There its code is translated into synthesizing a *specific protein* that carries out the aimed function of the hormone in the cell.

Specific proteins carry out the functions of the hormone — Hormone stimulation may create an enzyme, a receptor, or some other functional protein. Its structure and function depend on the type of the hormone and the target tissue involved. The actions of these diverse and functionally distinct gene products (proteins) within the cell are responsible for the physiological effects connected to various steroid and thyroid hormones in their particular target tissue.

FAST-ACTING HORMONES ACT VIA MEMBRANE RECEPTORS, G PROTEINS, & SECOND MESSENGERS

The fast-acting hormone group comprises the *peptide* hormones of the hypothalamus, pituitary, pancreas, and gastrointestinal tract and the *catecholamine* hormones of the adrenal medulla. These hormones bind with *plasma membrane receptors*, releasing *intracellular messengers* that activate cellular enzymes, resulting in specific hormone action within seconds to minutes. Their effects, however, are not long lasting.

Binding with membrane receptors — Peptide and catecholamine hormones bind to serpentine-type *"membrane receptors"* in their target cells. Each hormone has its own specific receptor. These receptors have sites for hormone binding and for interaction with other membrane components.

Hormone receptors interact with G proteins — Binding of the hormone with membrane receptors activates another class of regulatory membrane proteins called *G proteins*. These diverse regulatory proteins take part in coupling of the receptor proteins to other *effector proteins* in the membrane, resulting in *activation* or *inhibition* of the effector proteins. Numerous G proteins have been recognized. Many hormones and neurotransmitters and other intercellular first messengers (chemical signals) work through G proteins.

G proteins interact with membrane enzymes & release second messengers — Activation of G proteins results in activation membrane effectors such as enzymes (*adenylate cyclase*), ion channels (*calcium channels*), or other membrane receptors. Interaction of the G protein with an enzyme effector results in the formation and release of *second messengers*. The second messengers act as the intracellular chemical signals initiating the cellular actions of the hormones. Several second messengers are known, including *cyclic-AMP, cyclic-GMP, calcium*, and *inositol triphosphate* (IP$_3$) (plates 12–14) .

Role of cyclic-AMP (cAMP) as intracellular messenger — Through mediation of G proteins, binding of catecholamines and peptide hormones such as glucagon and gonadotropins results in activation of the membrane enzyme *adenylate (adenylyl) cyclase*, which converts ATP to cAMP. cAMP binds to a *protein kinase*, which in turn activates inactive enzymes by *phosphorylating* them. The phosphorylated proteins then initiate the physiologic events associated with the actions of these hormones. For example, both the pancreas hormone glucagon and epinephrine from the adrenal medulla use this mechanism to increase the release glucose from the liver. One advantage of such a cascade of events is the amplification of the signals and effects. Thus a single hormone molecule can form thousands of cAMP molecules, which in turn produce millions of phosphorylated enzymes, and they in turn form billions of glucose molecules within a few seconds.

Role of calcium ions as intracellular messengers — In some target cells, the hormone–receptor–G protein complex of events activates calcium channels, increasing the flow of extracellular calcium ions into the cells. Calcium can also be released from intracellular reserves. Calcium binds to and activates a regulatory protein called *calmodulin*. Activated calmodulin in turn activates protein kinases, which catalyze the phosphorylation of some inactive proteins into their active forms. As with cAMP, these effects also result in the amplification of the original hormonal signal.

Some rapidly acting peptide hormones (e.g., insulin & growth hormone) act via membrane enzyme receptors — Some peptide hormones such as insulin also bind to membrane receptors, but their action does *not* involve G proteins and second messengers. Instead the receptor has an intracellular domain that can act as an enzyme (tyrosine kinase); hormone binding activates this receptor enzyme, resulting in a variety of signaling systems not involving the standard second messengers. Growth hormone acts similarly.

CN: Use red for A; dark colors for C, I, and J.
1. Begin with blood circulation (A) across the top.
2. Color the steroid hormones (C) as they bind with hormone-binding proteins (A^1) in the upper-right.
3. Color the lower sequences for (J) and (J^1).

Follow the grayish numbered sequence 1–12 for steroid hormones. In the upper left corner, do the same for steps 1–4 for thyroid hormones (I).

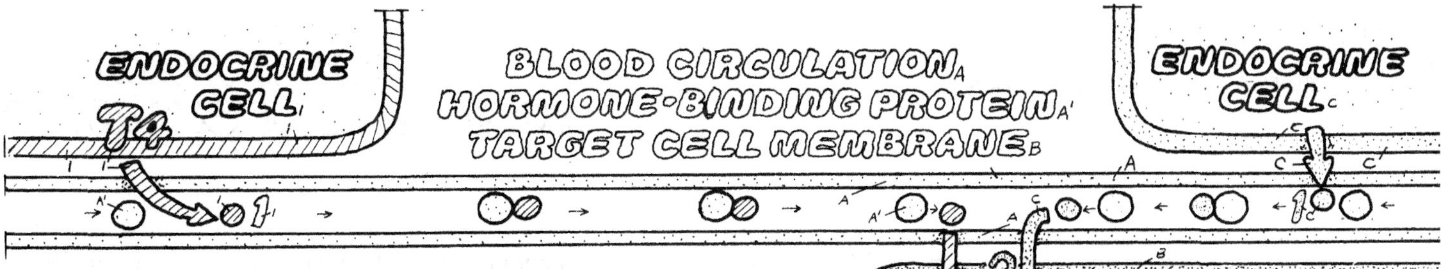

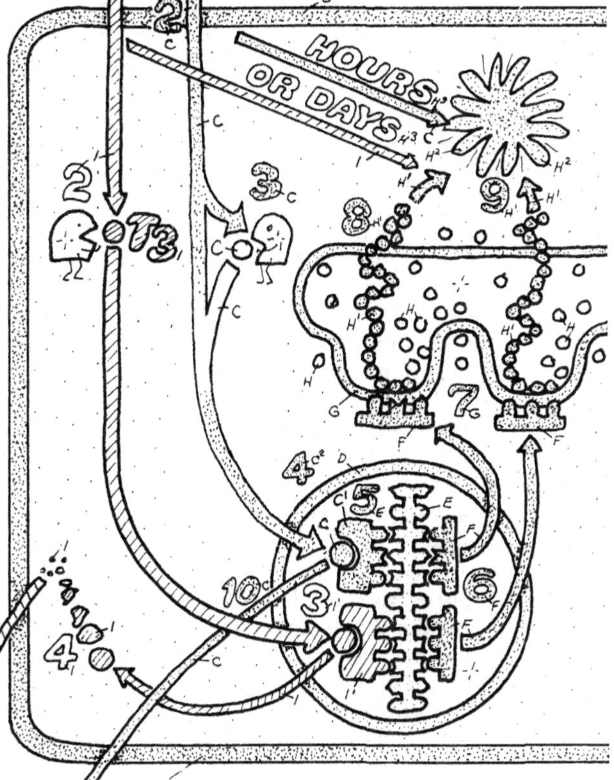

STEROID$_C$ & THYROID$_I$ HORMONES

STEROID NUCLEAR RECEPTOR$_{C'}$
CELL NUCLEUS$_D$
DNA$_E$ MESSENGER RNA$_F$
ROUGH ENDOPLASMIC RETICULUM$_G$
AMINO ACIDS$_H$ PROTEIN$_{H'}$
PROTEIN ACTION$_{,H^2}$ RESPONSE TIME $_{H^3}$

Steroid (1) and thyroid (1) hormones bind with hormone-binding proteins in plasma. Steroids enter target cells (2); some are converted to other steroids (3). The intracellular steroid (4) enters the nucleus and binds with nuclear receptors, changing their conformation; this results in transcription of DNA (5). The resulting mRNA (6) moves to the cytoplasm to initiate synthesis of a specific protein on the endoplasmic reticulum and ribosomes (7). This protein (8) then expresses the physiological action of the hormone (9). After exerting their action (10), steroids are deactivated in the liver (11) and excreted by the kidneys (12).

THYROID NUCLEAR RECEPTOR$_I$

In the target cell, thyroxine (1) is converted to T3 (2). T3 enters the nucleus and binds with a special nuclear receptor protein (3); this complex interacts with DNA, initiating synthesis of mRNA (6) and a specific protein (7– 9) as above. T3 is deactivated partly in tissues (4) and partly in the liver (11) and excreted by the kidneys (12).

HOURS OR DAYS

EXCRETION   DEGRADATION

kidney   liver

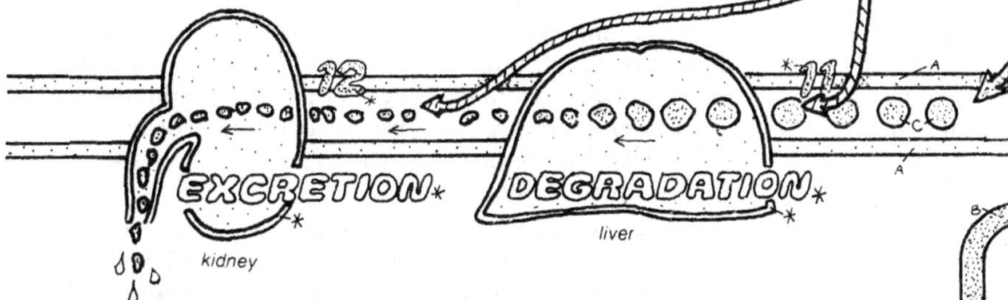

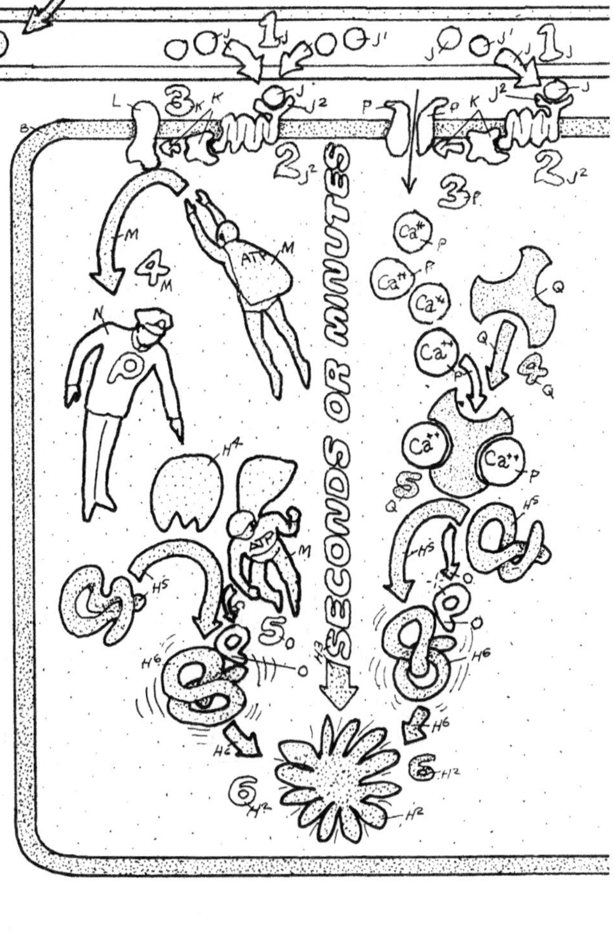

PEPTIDE$_J$ & CATECHOLAMINE$_{J'}$ HORMONES$_{-I}$

MEMBRANE HORMONE RECEPTOR$_{J^2}$
G PROTEIN$_K$
ADENYLATE CYCLASE$_L$
ATP $\rightarrow$ CYCLIC AMP$_N$ PHOSPHATE$_O$
PROTEIN KINASE$_{H^4}$
INACTIVE$_{,H^5}$ ACTIVE PROTEIN$_{H^6}$

Peptide hormones (1) and catecholamines (1) bind with special membrane-bound hormone receptor proteins (2) on the target cells. This complex interacts with regulatory G proteins that couple the receptor to adjacent membrane effectors such as the enzyme adenylate cyclase (3), which uses ATP to form cyclic-AMP (a second messenger) inside the cell (4). Cyclic-AMP activates a protein kinase, which in turn activates enzymes (5) by phosphorylation with ATP. Activated enzymes express the various physiological actions (6) of the hormones.

Ca$^{++}_O$, CHANNEL$_P$ CALMODULIN$_Q$

In some cells, peptide (1) and catecholamine (1) hormones bind with membrane receptors (2), which interact with G proteins, resulting in the opening of calcium channels and an increase in the intracellular level of Ca^{++} (3) (a second messenger). Calcium ions bind to calmodulin (4), an intracellular regulatory protein. The complex (5) promotes phosphorylation of certain enzymes to express hormone action (6).

SECONDS OR MINUTES

Hormones influence many cellular and metabolic functions. To exert their effects appropriately, the hormones should be optimally secreted and finely regulated. Many bodily diseases are caused by abnormal hormonal secretion. Regulation is basic to all endocrine functions and hormonal actions.

SELF-REGULATION AND BRAIN-MEDIATED CONTROL ARE TWO MAJOR MECHANISMS OF HORMONAL REGULATION

To regulate hormones within physiological limits or in response to physiological demands, the endocrine system uses two types of control mechanisms. In one type, hormones are controlled by a *self-regulatory* system where blood hormone level and the physiological parameter regulated by it interact automatically to maintain hormone secretion within predetermined and *set limits*. The second type of hormonal control mechanism utilizes the influence of the *nervous system* over the endocrine system to override the self-regulatory operation, initiate new hormonal responses, and/or set new baselines for hormonal secretion.

Feedback systems ensure self-regulation of hormonal levels — The operation of physical and biological systems involves an *input* and an *output*. In a self-regulated system, the output exerts control over the input (*feedback control*). When the relationship between the input and the output is inverse, so that an increase in the output leads to a decrease in the input and vice-versa, the regulation is by *negative feedback*. When the relationship is direct — i.e., an increase in output leads to a further increase in the input — the operation is by *positive feedback*.

Negative feedback ensures equilibrium — Negative feedback regulation promotes *stability* and *equilibrium*, aiming to maintain a system at a desired *set point*. All physiological *homeostatic* mechanisms, including numerous endocrine glands and their hormones, operate by negative feedback. Positive feedback regulation tends to create *disequilibrium* and a *vicious cycle*, leading to abnormal hormonal conditions and disease. Normal events occasionally depend on a positive feedback between hormones and their hypothalamic regulatory mechanisms. Examples are ovulation and parturition (plates 155, 158).

Simple negative feedback regulates many hormones & their blood effects — Regulation of hormonal secretion in the body is achieved at different levels of complexity. *Simple hormonal regulation* involves only one *endocrine gland*. Here the secretion of a hormone from an endocrine gland is controlled directly, through negative feedback, by the plasma concentration of the *physiological variable* or *parameter* the hormone is regulating. The endocrine cell usually has a receptor or a similar mechanism to detect the blood level of that parameter.

For example, decreased plasma *calcium ion* (Ca^{++}) *level* triggers an increase in *parathyroid hormone* from the *parathyroid gland*, which acts on *bone* to release calcium. Elevated plasma calcium inhibits further release of parathyroid hormone (plate 120). This negative feedback maintains optimal plasma calcium levels. Other examples are regulation of blood sugar by *insulin* and *glucagon* from the *pancreatic islets* (plate 123). These simple types of hormonal regulation and their automatic negative feedback operations are aimed at promoting *homeostasis*

and equilibrium for the physiologically important variables (e.g., blood glucose and plasma Ca^{++}) within the internal environment.

Complex hormonal regulation involves the anterior pituitary & its target glands — In *complex hormonal regulation*, the activity of one endocrine gland is controlled by hormones of another gland. Well-known examples are the control of the thyroid, adrenal cortex, and gonads by the *pituitary gland*. Removal of the pituitary leads to atrophy of these three glands and diminished levels of their hormones. Injection of pituitary extracts causes the secretory function of the atrophied glands to resume. These pituitary effects are conveyed by special *tropic hormones* that stimulate the *target glands* to grow and/or secrete their own hormones and exert negative feedback on the pituitary to inhibit the secretion of their respective tropic hormones. For this reason, the pituitary was once considered as the "master" endocrine gland, orchestrating the activities of several target glands whose hormones influence so many functions in the body. Later on it was learned that the pituitary itself is subordinate to the brain.

NEUROENDOCRINE AXIS MEDIATES BRAIN CONTROL OVER SOME ENDOCRINES AND THEIR HORMONES

Complex neurohormonal regulation involves anterior pituitary and hypothalamus — The interactions between the brain and the endocrine system are mediated by *complex neurohormonal regulation*. The pituitary is attached to the *hypothalamus*, a part of the brain involved in regulating visceral, emotional, and sexual functions. A special portal vascular system connects the hypothalamus to the anterior pituitary. Blood flowing through this portal system delivers the hormonal secretion from the nerve endings of certain hypothalamic neurosecretory cells directly to the anterior pituitary cells. These *hypothalamic hormones* regulate the release of anterior pituitary hormones.

Such a neural mechanism superimposes control of the brain over the pituitary and its target glands and mediates the effects of moods, emotions, stress, rhythmical neural activity (e.g., diurnal rhythms), and the *environment* (e.g., light, sound, temperature, and odors) on the endocrine system and hormones. Hormones of the target glands and the pituitary hormones act through *long* and *short feedback loops* to exert negative and positive feedback effects on the hypothalamic neurosecretory cells, modifying their secretion of neurohormones. Hypothalamic neurons, like the pituitary cells, contain receptors that can detect blood hormone levels.

Neuroendocrine control is also exerted via the posterior pituitary, autonomic nerves, & adrenal medulla — The nervous system and brain can control hormonal secretion via the *posterior pituitary* as well. The hypothalamus controls plasma water content, milk flow, and parturition by releasing its hormones directly into the blood at the site of the posterior pituitary. The hypothalamus also exerts rapid and direct effects on the secretion of several endocrine glands by modifying the activity of the *sympathetic* and *parasympathetic* nerves, which innervate these glands. A special case is the adrenal medulla, whose secretion of catecholamines is regulated by the sympathetic nervous system.

CN: Use dark colors for A, E, F, and G.
1. Work from the top down. Note that the size of the output arrow reflects the amount of output.

2. Color each of the three levels of hormonal regulation, working from top to bottom.

GENERAL FEEDBACK REGULATION

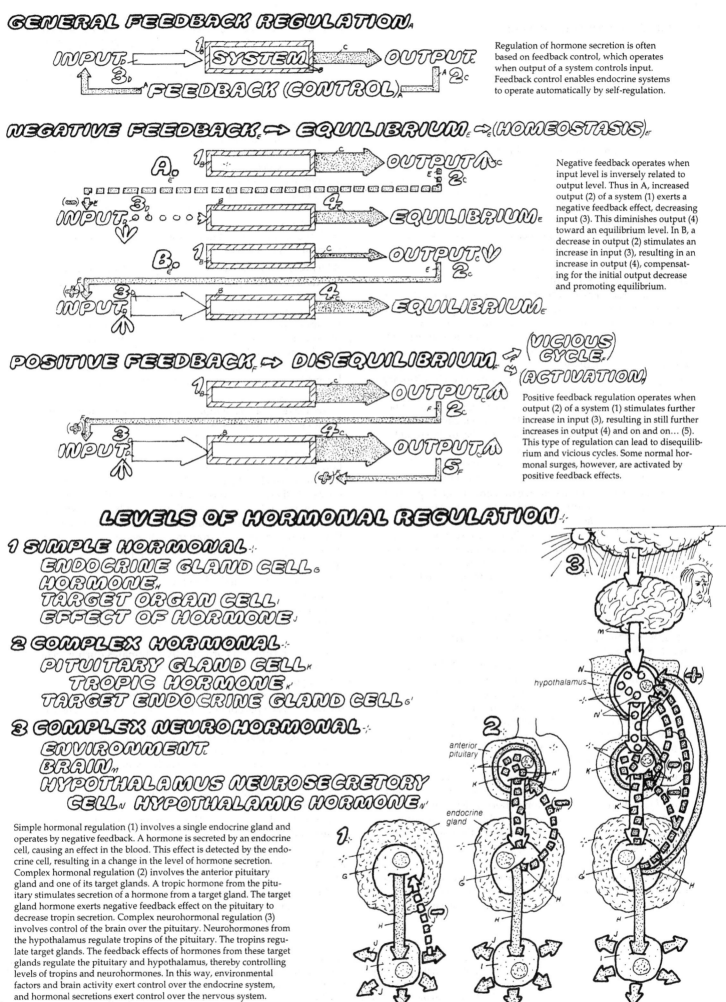

INPUT → SYSTEM → OUTPUT

FEEDBACK (CONTROL)

Regulation of hormone secretion is often based on feedback control, which operates when output of a system controls input. Feedback control enables endocrine systems to operate automatically by self-regulation.

NEGATIVE FEEDBACK → EQUILIBRIUM → (HOMEOSTASIS)

A.
OUTPUT ⋀
INPUT → EQUILIBRIUM

B.
OUTPUT ⋁
INPUT → EQUILIBRIUM

Negative feedback operates when input level is inversely related to output level. Thus in A, increased output (2) of a system (1) exerts a negative feedback effect, decreasing input (3). This diminishes output (4) toward an equilibrium level. In B, a decrease in output (2) stimulates an increase in input (3), resulting in an increase in output (4), compensating for the initial output decrease and promoting equilibrium.

POSITIVE FEEDBACK → DISEQUILIBRIUM → (VICIOUS CYCLE) ↝ (ACTIVATION)

OUTPUT ⋀
INPUT → OUTPUT ⋀

Positive feedback regulation operates when output (2) of a system (1) stimulates further increase in input (3), resulting in still further increases in output (4) and on and on… (5). This type of regulation can lead to disequilibrium and vicious cycles. Some normal hormonal surges, however, are activated by positive feedback effects.

LEVELS OF HORMONAL REGULATION

1 SIMPLE HORMONAL
ENDOCRINE GLAND CELL
HORMONE
TARGET ORGAN CELL
EFFECT OF HORMONE

2 COMPLEX HORMONAL
PITUITARY GLAND CELL
TROPIC HORMONE
TARGET ENDOCRINE GLAND CELL

3 COMPLEX NEUROHORMONAL
ENVIRONMENT
BRAIN
HYPOTHALAMUS NEUROSECRETORY
CELL HYPOTHALAMIC HORMONE

hypothalamus

anterior pituitary

endocrine gland

Simple hormonal regulation (1) involves a single endocrine gland and operates by negative feedback. A hormone is secreted by an endocrine cell, causing an effect in the blood. This effect is detected by the endocrine cell, resulting in a change in the level of hormone secretion. Complex hormonal regulation (2) involves the anterior pituitary gland and one of its target glands. A tropic hormone from the pituitary stimulates secretion of a hormone from a target gland. The target gland hormone exerts negative feedback effect on the pituitary to decrease tropin secretion. Complex neurohormonal regulation (3) involves control of the brain over the pituitary. Neurohormones from the hypothalamus regulate tropins of the pituitary. The tropins regulate target glands. The feedback effects of hormones from these target glands regulate the pituitary and hypothalamus, thereby controlling levels of tropins and neurohormones. In this way, environmental factors and brain activity exert control over the endocrine system, and hormonal secretions exert control over the nervous system.

The *pituitary* gland, located *underneath* the hypothalamus of the brain, is vital to body physiology. Some pituitary hormones—e.g., prolactin and antidiuretic hormones—exert direct actions on body organs—mammary glands and kidneys, respectively. Other pituitary hormones regulate the activity of several target endocrine glands (thyroid, adrenal, gonads). The pituitary gland is controlled by the brain and mediates the effects of the central nervous system on hormonal activity in the body, which explains its critical anatomic position underneath the brain.

PITUITARY STRUCTURE AND RELATION TO THE HYPOTHALAMUS OF THE BRAIN

The pituitary has two functional lobes (anterior & posterior) and a vestigial intermediate lobe—The pituitary (*hypophysis*) is divided into an *anterior lobe* (adenohypophysis), a *posterior lobe* (neurohypophysis), and an *intermediate lobe*. The anterior and posterior lobes are functional and secretory, but the human intermediate lobe is either absent or vestigial, consisting of a few cells with no known functions. The pituitary is connected to the hypothalamus of the brain via the *hypophyseal stalk*. The hypothalamus is critical in regulation of both the anterior and posterior lobes. This plate focuses on the structure and functions of the posterior lobe, to illustrate the concept of *neurosecretion*. Neurosecretion is also essential for understanding of anterior lobe function (plate 117) and is a cornerstone of the modern science of neuroendocrinology.

The posterior pituitary is an extension of the hypothalamus—The posterior lobe of the pituitary secretes two hormones, *antidiuretic hormone* (ADH) and *oxytocin*. The posterior pituitary is not a true endocrine gland because it does not contain true secretory cells. The gland is, in fact, an extension of the brain hypothalamus and consists mainly of nerve fibers and nerve endings of the neurons of two hypothalamic nuclei. These neurons have their cell bodies in the hypothalamus and send their axons through the *hypothalamo-hypophyseal tract* to the posterior pituitary through the hypophyseal stalk.

NEUROSECRETION: SOME BRAIN CELLS ARE MODIFIED TO SECRETE HORMONES

Hypothalamic neurosecretory cells produce hormones of the posterior pituitary—The hypothalamic nuclei that are connected to the posterior pituitary are called *supraoptic* and *paraventricular*. The neurons of these nuclei are typical examples of *neurosecretory* cells. The *cell bodies* of the neurosecretory neurons are the site of the synthesis of the hormones that are destined for the pituitary. In the case of the posterior pituitary, oxytocin and ADH, being peptides, are synthesized as larger prohormone molecules. These prohormone molecules contain the true hormone and a nonhormonal portion called *neurophysin*, which may function in hormone transport. The prohormone complexes are packed within special secretory vesicles (*Herring bodies*), which flow down the interior of the axons of the hypothalamo-hypophyseal tract, aided by rapid axoplasmic transport.

Posterior pituitary hormones are released from nerve terminals into the blood—Before reaching the nerve terminals in the posterior lobe, the hormone is split off the larger pro-

hormone and stored in the *axon terminals*, to be released into the blood capillaries and carried out to the target tissues. Neurosecretory cells maintain their electrical excitability and produce action potentials. The stimuli for hormone release are the nerve impulses arriving from the cell body in the hypothalamus down the axon membrane to the nerve terminal. Arrival of a nerve impulse triggers the flow of calcium ions into the terminal. This leads to the fusion of secretory vesicles with the terminal membrane, resulting in the release of hormone into the blood capillary.

ADH & OXYTOCIN ARE POSTERIOR PITUITARY HORMONES

ADH regulates plasma water plus blood volume & pressure—The neurons of the supraoptic nucleus make and secrete principally the antidiuretic hormone (ADH, also called vasopressin). ADH is involved mainly in regulating body water and is secreted whenever the amount of water in the blood is decreased, as in dehydration due to excessive sweating or osmotic diuresis (caused by an increase in glucose or ketone bodies or sodium loss in the urine), as well as during hemorrhage and blood loss.

Increased plasma osmolarity or decreased blood volume stimulates ADH release—The signal for ADH release is an increase in the plasma osmolarity mediated by an increase in the concentration of plasma sodium ions. The sodium elevation is sensed by specific *osmoreceptor* neurons in the hypothalamus, which in turn stimulate the supraoptic neurons to release ADH from the posterior pituitary. ADH acts principally on the collecting ducts in the kidney, by increasing their permeability to water, through increase in water channels. Water moves by osmosis from the kidney ducts to the plasma, increasing plasma water and decreasing plasma osmolarity (plate 66).

ADH is also secreted when mechanoreceptors (blood volume receptors) in the heart and pressure receptors in the vasculature are stimulated after hemorrhage and blood loss. After a hemorrhage, ADH causes vasoconstriction, leading to an increase in blood pressure (vasopressive action) (plate 47).

Oxytocin functions principally in females during lactation and parturition—Oxytocin is secreted principally by the cells of paraventricular nuclei, in response to stimulation of mechanoreceptors in the nipples of the breasts and cervix of the uterus. As a part of neurohormonal reflex arcs, sensory nerves convey the signals from the sensory receptors to the hypothalamus, leading to the secretion of oxytocin from the posterior pituitary. During labor, oxytocin acts on the myometrium of the uterus to cause massive contractions, resulting in the expulsion of the fetus (oxytocin = swift birth) (plate 158). During lactation, oxytocin acts on the myoepithelium of the mammary glands to elicit their contraction and cause the ejection of milk (plate 159). There are no known functions for oxytocin in the male.

Amino acid composition of oxytocin & ADH—Oxytocin and ADH are both polypeptides containing nine amino acids. Their structures are identical except for the substitution, in ADH, of *phenylalanine* and *arginine* in place of one of the *tyrosines* and the *leucine* found in oxytocin.

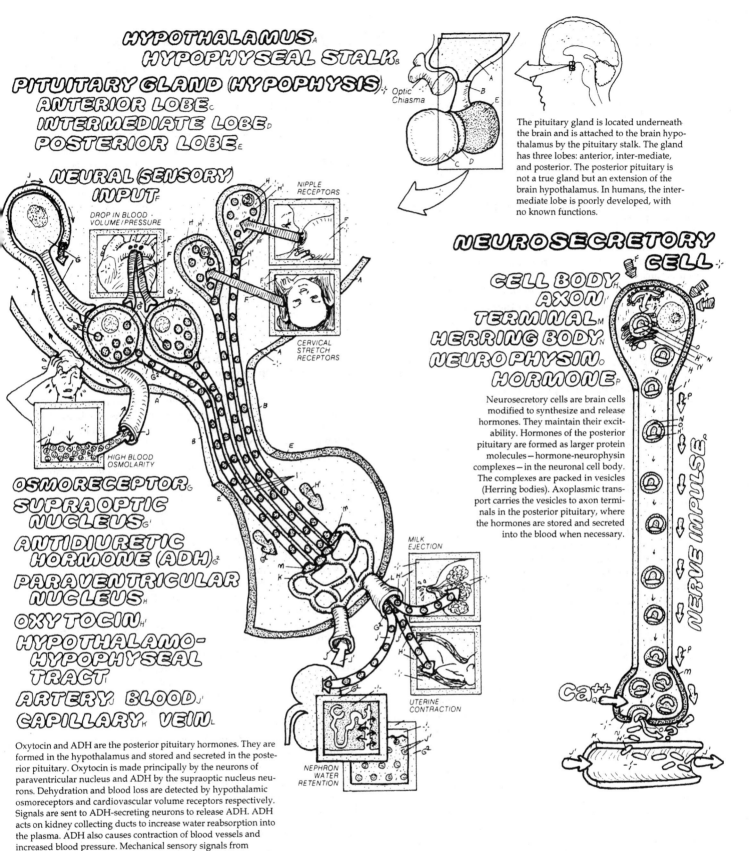

HYPOTHALAMUS ₐ
HYPOPHYSEAL STALK ᵦ
PITUITARY GLAND (HYPOPHYSIS)
ANTERIOR LOBE c
INTERMEDIATE LOBE d
POSTERIOR LOBE e

Optic Chiasma

The pituitary gland is located underneath the brain and is attached to the brain hypothalamus by the pituitary stalk. The gland has three lobes: anterior, inter-mediate, and posterior. The posterior pituitary is not a true gland but an extension of the brain hypothalamus. In humans, the intermediate lobe is poorly developed, with no known functions.

NEURAL (SENSORY) INPUT f

DROP IN BLOOD VOLUME/PRESSURE

NIPPLE RECEPTORS

CERVICAL STRETCH RECEPTORS

HIGH BLOOD OSMOLARITY

OSMORECEPTOR g
SUPRAOPTIC NUCLEUS g¹
ANTIDIURETIC HORMONE (ADH) g²
PARAVENTRICULAR NUCLEUS h
OXYTOCIN h¹
HYPOTHALAMO-HYPOPHYSEAL TRACT ᵢ
ARTERY BLOOD ⱼ
CAPILLARY k VEIN ₗ

NEUROSECRETORY CELL
CELL BODY ₕ
AXON ₗ
TERMINAL ₘ
HERRING BODY ₙ
NEUROPHYSIN ₒ
HORMONE ₚ

Neurosecretory cells are brain cells modified to synthesize and release hormones. They maintain their excitability. Hormones of the posterior pituitary are formed as larger protein molecules — hormone-neurophysin complexes — in the neuronal cell body. The complexes are packed in vesicles (Herring bodies). Axoplasmic transport carries the vesicles to axon terminals in the posterior pituitary, where the hormones are stored and secreted into the blood when necessary.

NERVE IMPULSE

Ca^{++}

MILK EJECTION

UTERINE CONTRACTION

NEPHRON WATER RETENTION

Oxytocin and ADH are the posterior pituitary hormones. They are formed in the hypothalamus and stored and secreted in the posterior pituitary. Oxytocin is made principally by the neurons of paraventricular nucleus and ADH by the supraoptic nucleus neurons. Dehydration and blood loss are detected by hypothalamic osmoreceptors and cardiovascular volume receptors respectively. Signals are sent to ADH-secreting neurons to release ADH. ADH acts on kidney collecting ducts to increase water reabsorption into the plasma. ADH also causes contraction of blood vessels and increased blood pressure. Mechanical sensory signals from nipples and cervix in females stimulate oxytocin release. Oxytocin contracts smooth muscles of mammary ducts and the uterus, aiding in milk ejection and parturition, respectively.

STRUCTURE OF POSTERIOR PITUITARY HORMONES

OXYTOCIN h¹
CYS. TYR. TYR. GLN. ASN. CYS. PRO. LEU. GLY. NH₂

ADH g²
CYS. TYR. PHE. GLN. ASN. CYS. PRO. ARG. GLY. NH₂

Oxytocin and ADH are similar peptides, each with nine amino acids. In ADH, one of the two tyrosines and the only leucine found in the oxytocin are replaced by phenylalanine and arginine, respectively.

The *anterior pituitary gland* (APG) is a true and major endocrine gland with profound actions in the body. It secretes at least six protein hormones that either regulate the hormonal secretion of other endocrine glands or directly control the activity of particular target organs. For this reason APG was once called the "master gland" but we now know that APG is in turn controlled by the brain hypothalamus.

ANTERIOR PITUITARY HORMONES HAVE DIVERSE TROPIC & TROPHIC EFFECTS

All hormones of the APG have essential growth promoting *(trophic)* effects on the cells of their *target glands;* most APG hormones also stimulate and regulate the hormonal secretion of their target endocrine glands (tropic effects). APG hormones are often collectively called the *tropin* or *trophin* hormones.

Some APG hormones regulate other endocrine glands (tropic effects) — Among the APG hormones, *thyroid-stimulating hormone* (TSH) and *adrenocortico-tropin hormone* (ACTH) regulate the hormonal secretions of the thyroid and adrenal cortex, respectively; *follicle-stimulating hormone* (FSH) and *luteinizing hormone* (LH) regulate the activities of the gonads (testes and ovaries). The tropins that regulate other endocrines generally increase the synthesis and release of the hormones of their target glands as well. Thus, TSH promotes the secretion of *thyroxine*, ACTH of *cortisol*, and FSH and LH of sex steroids (estrogen, progesterone, testosterone). Removal of the APG *(hypophysectomy)* leads to atrophy of these target glands and cessation of their hormonal secretion.

Other APG hormones promote growth & function of nonendocrine target organs (trophic effects) — Two other hormones, *prolactin* and *growth hormone* (GH) (also known as *somatotropin* [STH]), directly act on nonendocrine target organs. Prolactin acts on the exocrine mammary glands to promote milk secretion. Growth hormone promotes growth and anabolic effects in muscle and bones during development and fat breakdown and fatty acid mobilization in adult adipose tissue. Growth effects of GH are mediated by *insulin-like growth factors* (IGF, previously called *somatomedins*), hormones released by the liver and other tissues in response to GH. APG also produces other substances such as β-*lipotropin*, β-*endorphin*, and *melanocyte stimulating hormone* (MSH, intermediate lobe hormone of lower animals).

Specific cell types in the APG secrete its hormones — Based on the application of modern immunocytochemical staining methods, five major cell types in the APG have been distinguished, each secreting one or more of its hormones. Thus, *thyrotrops* secrete TSH, *corticotrops* ACTH, *somatotrops* GH, and *mammotrops* prolactin; *gonadotrops* secrete both FSH and LH. APG cells are also known as chromophilic *acidophils* or *basophils*, on the basis of their reaction to acidic or basic stains. Thyrotrops and gonadotrops are basophilic; corticotrops are lightly basophilic; somatotrops and mammotrops are acidophilic. Cells that do not stain with these dyes *(chromophobes)* are immature cells or resting corticotrops that can become active under stress.

HYPOTHALAMIC CONTROL OF ANTERIOR PITUITARY

Specific hypothalamic neurohormones control APG hormones — APG hormones are controlled by specific (mainly peptide) neurohormones *(hypophysiotropin hormones)* formed by certain hypothalamic neurons and released in extremely small amounts into a special portal circulatory system *(hypophyseal portal capillaries)* that delivers them directly to the pituitary cells, bypassing the general circulation. These neurohormones are also referred to as *hypothalamic-releasing* or *release-inhibiting hormones* (-RH, -IH), depending on whether they increase or decrease the APG hormones. For TSH, a releasing hormone (TRH) of tripeptide composition has been found; for ACTH, there is also only one releasing hormone, CRH. A single releasing hormone of decapeptide composition, GnRH, regulates release of both gonadotropins LH and FSH. For GH, there is a GRH — a large peptide — and a GIH *(somatostatin)*, a smaller peptide with 14 amino acids; for prolactin, a PRH and a PIF (dopamine) have been found. APG cells contain specific membrane receptors for the corresponding hypophysiotropins; the effects of these receptors are mediated by G-proteins and cyclic AMP.

Higher brain regions & feedback from target hormones control the hypophysiotropins — Stimuli from two sources control the release of the hypophysiotropins. One source is other brain areas mediating exogenous (environmental) stimuli and stresses as well as endogenous rhythms (see below). The other source is the feedback signals from target hormones in the plasma. For example, a decrease in plasma cortisol acts, through a negative feedback loop, on the CRH-hypothalamic neurons, increasing their secretion of CRH. This leads to an increase in the secretion of ACTH from APG, which in turn increases the secretion of cortisol from the adrenal gland. The principal site of feedback regulation for some target gland hormones (e.g., thyroid) is at the level of the APG.

Brain control over APG mediates the environmental and emotional influences on hormones — The main value of the releasing and inhibiting hormones is to permit the brain to exert a dynamic control over the endocrine system, adjusting its operation to the needs of the body. Thus, in animals, seasonal changes in light and length of day can result in the appropriate activation and inhibition of the gonads. Long-term changes in environmental temperature can result in the appropriate adjustment in basal metabolic rate and heat production by altering thyroid secretion. Similarly, the brain, in response to various stresses, can increase the secretion of antistress glucocorticoids from the adrenal cortex by increasing CRH and ACTH from the hypothalamus and APG, respectively (plate 127).

APG hormones and their hypophysiotropins are released in pulses — The secretion of most of the pituitary hormones occurs in an episodic (pulsatile) manner — i.e., there is a rhythm and a peak of secretion at regular intervals. The intervals are specific for each hormone and are in the range of one to several hours. These rhythms are believed to be caused by the episodic release of the hypothalamic-releasing hormones triggered by signals from other brain centers. The frequency and amplitude of these secretory pulses can be altered by a variety of factors, representing one way by which the brain can exert its influence over the endocrine system. Also, the diurnal pattern of ACTH — high in the morning and low in the evening — is regulated by hypothalamic mechanisms.

CN: Use red for E, purple for F, and blue for G. Use dark colors for A, B, N, and O.
1. Begin at the top. Tiny circles and squares represent hypothalamic and APG hormones in general. Names of specific hormones appear in the list of titles. Color in background colors first, and then fill in the squares and circles over those colors. Note that a sixth cell type within the anterior pituitary, chromophobe, is left blank.
2. Note at the bottom, that target glands and the bulk of their own hormones exerting feedback on the hypothalamus and pituitary are given the color of the APG hormone that stimulates them, but the arrows of feedback are colored gray for emphasis.

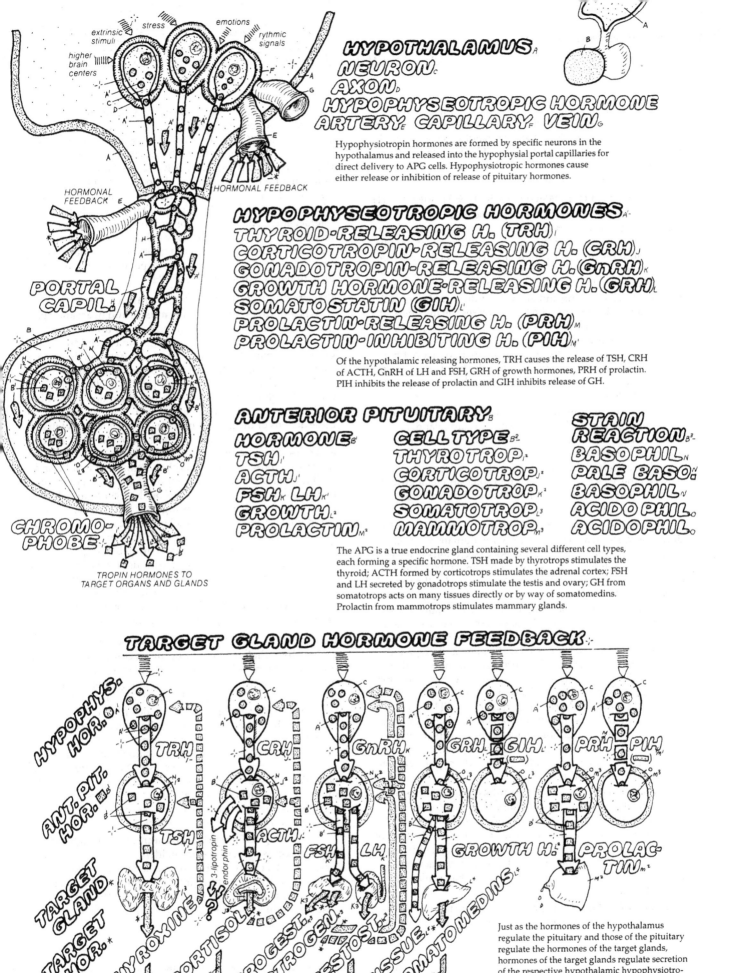

HYPOTHALAMUS
NEURON
AXON
HYPOPHYSEOTROPIC HORMONE
ARTERY. CAPILLARY. VEIN.

extrinsic stimuli · stress · emotions · rythmic signals · higher brain centers

HORMONAL FEEDBACK

HORMONAL FEEDBACK

PORTAL CAPIL.

CHROMO-PHOBE

TROPIN HORMONES TO TARGET ORGANS AND GLANDS

Hypophysiotropin hormones are formed by specific neurons in the hypothalamus and released into the hypophysial portal capillaries for direct delivery to APG cells. Hypophysiotropic hormones cause either release or inhibition of release of pituitary hormones.

HYPOPHYSEOTROPIC HORMONES
THYROID-RELEASING H. (TRH)
CORTICOTROPIN-RELEASING H. (CRH)
GONADOTROPIN-RELEASING H. (GnRH)
GROWTH HORMONE-RELEASING H. (GRH)
SOMATOSTATIN (GIH)
PROLACTIN-RELEASING H. (PRH)
PROLACTIN-INHIBITING H. (PIH)

Of the hypothalamic releasing hormones, TRH causes the release of TSH, CRH of ACTH, GnRH of LH and FSH, GRH of growth hormones, PRH of prolactin. PIH inhibits the release of prolactin and GIH inhibits release of GH.

ANTERIOR PITUITARY

HORMONE	CELL TYPE	STAIN REACTION
TSH	THYROTROP	BASOPHIL
ACTH	CORTICOTROP	PALE BASO.
FSH. LH	GONADOTROP	BASOPHIL
GROWTH	SOMATOTROP	ACIDO PHIL.
PROLACTIN	MAMMOTROP	ACIDOPHIL.

The APG is a true endocrine gland containing several different cell types, each forming a specific hormone. TSH made by thyrotrops stimulates the thyroid; ACTH formed by corticotrops stimulates the adrenal cortex; FSH and LH secreted by gonadotrops stimulate the testis and ovary; GH from somatotrops acts on many tissues directly or by way of somatomedins. Prolactin from mammotrops stimulates mammary glands.

TARGET GLAND HORMONE FEEDBACK

HYPOPHYS. HOR.

ANT. PIT. HOR.

TARGET GLAND

TARGET HOR.

TRH · CRH · GnRH · GRH · GIH · PRH · PIH

TSH · ACTH · FSH · LH · GROWTH H. · PROLAC-TIN

B-lipotropin / endorphin

THYROXINE · CORTISOL · PROGEST. · ESTROGEN · TESTOST. · TISSUE · SOMATOMEDINS

PHYSIOLOGICAL ACTIONS IN THE BODY.

Just as the hormones of the hypothalamus regulate the pituitary and those of the pituitary regulate the hormones of the target glands, hormones of the target glands regulate secretion of the respective hypothalamic hypophysiotro-pin hormones and the pituitary tropins. This is due to negative and positive feedback effects on hypothalamus neurons and pituitary cells.

GROWTH HORMONE: GROWTH & METABOLIC EFFECTS

Human *growth hormone* (GH) is a protein (a single-chain polypeptide of 191 amino acids) secreted by the somatotrop cells of the *anterior pituitary gland*. Somatotrops constitute the majority of the cells in the pituitary. GH actions fall into two categories: those that promote *tissue* and *body growth* and those that influence *metabolism*.

EFFECTS OF GROWTH HORMONE ON BODY & TISSUE GROWTH

GH stimulates growth of bone, muscle, & visceral tissues but not that of the brain & gonads — Removal of the anterior pituitary in a growing animal stops growth, while injections of GH cause resumed growth. A clear effect of GH on tissue growth is seen in bones. The *epiphyseal plate*, a band of proliferating and developing cells in the *epiphysis* (head) of long bones, is thick in young and growing animals, a sign of active bone growth. Treatment with GH increases the thickness of the epiphyseal plate (the *tibia test*), accompanied by proliferation of bone cells and increase in bone formation and bone length (plate 121). GH also promotes growth of many types of soft tissue, especially muscle tissue, the heart, and visceral organs, although some tissues such as the brain and gonads are *not* affected.

GH effects on growth occur after birth — The observation that anencephalic newborns, with no pituitary gland, are of normal size at term indicates that GH does not influence embryonic and fetal growth. The effects of GH on human growth are exerted during the postnatal period, particularly between the ages of 2–16 years. Embryonic and fetal growth are regulated by other hormones, such as the *insulin-like growth factors* (IGF-1 & IGF-2).

Increased secretion of GH leads to gigantism, while its decrease causes dwarfism — In growing humans with pituitary or hypothalamic tumors, excessive secretion of GH leads to *gigantism*. Pituitary giants are more than 8 ft tall. Absence or reduced levels of growth hormone during childhood lead to *dwarfism*. Adult dwarfs of pituitary origin have a small body but normal head size and usually do not show mental retardation. In individuals who have a selective lack of GH because they lack its gene but have an otherwise normal pituitary, stature is short, but sexual maturation and pregnancy may still occur, with normal offspring. Some individuals may have circulating GH but do not have GH receptors at the target cell membranes; these will also have short stature. Children with unusually short stature due to low GH levels can now be treated with synthetic human GH protein, available through modern biotechnology methods. Dwarfism and gigantism can also be produced in growing animals by removing the pituitary gland (hypophysectomy) or treatment with excess GH, respectively.

Excessive secretion of GH in adults causes acromegaly — In adults with excessive growth hormone secretion, bones that can no longer grow in length due to fusion (closure) of the epiphyseal plates grow in width. This leads to the typical picture of *acromegaly*, where the abnormal growth of bones in the digits, toes, mandible, and back lead to a characteristic bodily deformity. Acromegalic people also have enlarged visceral organs.

Effects of GH on tissue growth are mediated by IGFs — The effects of GH on tissue growth are in part direct and in part mediated by certain tissue growth factors that were previously called *somatomedins* but are now known as *Insulin-like Growth Factors* (IGFs). The IGFs are secreted by the *liver* and also locally produced in target tissues in response to stimulation by GH. Two IGFs, IGF-1 & IGF-2, are known; their protein structure resembles that of the pancreatic hormone *insulin*. IGFs have receptors of their own and promote *cell proliferation* and *protein synthesis* in their target cells. IGFs and GH interact at the epiphyseal plate and through the liver to promote bone growth. The specific mechanisms by which IGFs and GH stimulate growth of other tissues are under investigation. As mentioned above, IGFs independently regulate fetal growth with no involvement of GH.

EFFECTS OF GROWTH HORMONE ON METABOLISM

GH mobilizes fatty acids for muscle and heart and spares glucose for brain — In addition to growth-promoting (anabolic) actions that are normally manifested in children and growing animals, GH exerts important effects on the metabolism of fats and carbohydrates, particularly in adults. GH acts on the *fat cells* of *adipose tissue*, stimulating *lipolysis* (breakdown of triglyceride fat stores) and mobilization of the liberated *fatty acids*. The adrenal hormone cortisol is necessary for these actions of the growth hormone. The mobilized fatty acids are released into the blood and are oxidized by the heart and muscle for energy, in preference to glucose. GH also acts directly on muscle cells, promoting amino acid uptake and inhibiting glucose uptake, by opposing the action of insulin (anti-insulin action). GH also acts on the liver to mobilize its glucose reserves. These effects together lead to sparing of blood glucose and its increased levels. These actions of GH are important during stress or sustained exercise and also during fasting and starvation. By sparing blood glucose, GH provides this critical energy source for use by the brain, which depends exclusively on glucose and cannot use fatty acids to obtain energy. Possibly the increase in nightly GH secretion (see below) helps to provide glucose for the brain, since the body is in a mild fasting state during the night sleep.

REGULATION OF GROWTH HORMONE SECRETION

Two hypothalamic neurohormones control secretion of growth hormone — The secretion of GH is regulated by two neurohormones from the *hypothalamus*: a *growth hormone-releasing hormone* (GRH), which stimulates the secretion of GH, and *somatostatin* (growth hormone–inhibiting hormone, GIH), which inhibits the release of GH. Somatostatin is a 14 amino acid peptide with disulfide bonds and is secreted tonically, while GHR is a much larger peptide that is secreted in pulses preceding the GH release pulse. The secretion of GH is *pulsatile* – i.e., shows *episodic bursts* (peaks) — with about 4-hr intervals. Early during the night sleep, a big burst of GH secretion occurs, presumably to increase plasma glucose for brain use during the fasting state of sleep.

Circulating IGFs regulate GH secretion through the hypothalamus — There is some evidence that circulating IGF-1 regulates GH secretion by exerting a negative feedback effect on the hypothalamus, inhibiting GRH and stimulating somatostatin release. The plasma levels of fatty acids and glucose also influence GH secretion by acting on the hypothalamic neurons.

CN: Use the dark colors from the previous page for the hypothalamus (A) and anterior pituitary (D).
1. Begin at the top, and color down to adipose tissue (G) and liver (H).
2. Color the section in the upper right corner.
3. Color the GH effect on cell metabolism.
4. Color the secretions of somatomedins (I) from the liver and the effects on growth.

↑GH, ↑FFA, ↑GIU...

HYPOTHALAMUS A
NEUROSECRETORY CELL B
GROWTH HORMONE- RELEASING HOR. (GRH) A'
SOMATOSTATIN (GIH) C

ANTERIOR PITUITARY D
SOMATOTROP CELL E
GROWTH HORMONE (GH) F

ADIPOSE TISSUE G
LIVER H
SOMATOMEDINS I
MUSCLE TISSUE J
CELL METABOLISM J'

GH secretion is elevated during stress (e.g., sustained exercise or fasting). GH acts on the fat cells of adipose tissue, promoting conversion of stored fat (triglycerides) to fatty acids and glycerol (fatty acid mobilization).

METABOLIC EFFECTS OF GH J'

Mobilized fatty acids are released in the blood and used by the heart and skeletal muscles as fuel for cellular oxidation and ATP formation. Such a shift to fat metabolism spares glucose for use by the brain, ensuring normal brain function during stress or starvation.

Secretion of GH is stimulated by a hypothalamic-releasing hormone (GRH) and inhibited by a release-inhibiting hormone (GIH; somatostatin). Increased blood levels of IGFs, fatty acids, and glucose exert feedback effects on the hypothalamus to reduce GH secretion.

SOURCES OF DEFECTS INVOLVING GH

Abnormalities of GH hypersecretion may be due to tumors in the hypothalamus (increased GRH) (1) or in the pituitary (2). Congenital hypoplasia or absence of the pituitary (2, 3) lead to diminished growth and stature. Specific genetic defects involving GH (3) (absence of GH gene, absence of GH or IGF receptors) can also lead to decreased growth.

GROWTH EFFECTS OF GH VIA SOMATOMEDINS

PROTEIN SYNTHESIS M'
AMINO ACID UPTAKE M

GH, through release of somatomedins (IGFs) from the liver, increases amino acid uptake and promotes protein synthesis at the ribosomal level, leading to cell growth.

CELL PROLIFERATION N

GH, acting together with IGFs (somatomedins), increases mitosis of cartilage cells in the epiphyseal plates of long bones, leading to proliferation of cartilage and their transformation to bone cells.

SKELETAL GROWTH O

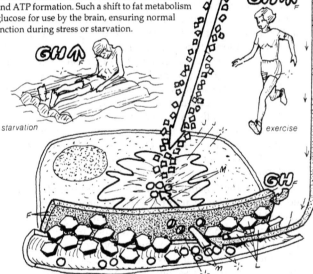

starvation *exercise*

AMINO ACIDS n GLUCOSE l

GH inhibits uptake of glucose and promotes uptake of amino acids in muscle tissue, leading to hyperglycemia and glucose sparing.

longitudinal growth *thickening*

IN CHILDREN
EPIPHYSEAL PLATES P
SECRETION LEVELS

IN ADULTS ACROMEG-ALY

EPISODIC GH SECRETION F

IN SLEEP F

In man, GH secretion shows episodic (pulsatile) bursts in 4-hr intervals. During early sleep, a big burst occurs, but secretion is inhibited during the REM stage.

Hours
0   12   18   24

12 yrs *12* *12*

DWARFISM *normal* GIGANTISM

In childhood, GH deficiency leads to stunted bone growth and stature (dwarfism). Hypersecretion causes gigantism due to continued proliferative activity in the epiphyseal plates and growth of long bones.

In adults, if the epiphyseal plates are fused, GH hypersecretion leads to continued growth in bone width, causing deformities in the bones of the back, digits, toes, face, etc. (acromegaly).

The *thyroid* is a butterfly-shaped endocrine gland located in the neck, anterior and lateral to the larynx. It receives a rich blood supply and secretes two closely related hormones, *thyroxine* (T4, tetra-iodothyronine) and *tri-iodothyronine* (T3).

ACTIONS & REGULATION OF THYROID HORMONES

Regulation of the metabolic rate is the main action of THs in adults—THs increase the metabolic rate by increasing the rate of *oxygen consumption* and *heat production* in many body tissues including the heart, muscles, and visceral tissues, but not in the brain, lymphatics, and testes. This *calorigenic action* of THs is critical for adaptation of animals and human infants to external cold and heat. Calorigenic actions of THs may take hours to days to develop fully but are long lasting.

THs also influence cardiovascular functions by increasing heart rate and contractility and vascular responsiveness to catecholamines, resulting in increased blood pressure. Brain function, excitability, and behavior also are affected by THs, through enhancing the action of catecholamines.

THs regulate growth and development—THs influence differentiation and growth of numerous tissues, including soft (muscle) and hard (bone) tissues and visceral organs; testicular growth and development and sperm production in animals are controlled by THs. The most critical and well-known developmental effects are on the brain. Absence or deficient THs lead to underdevelopment of the brain and mental retardation (see below). THs act synergistically with growth hormones and are necessary for their synthesis.

The pituitary tropin TSH is the primary regulator of thyroid function—The synthesis and release of THs are controlled by the pituitary tropin hormone *thyrotropin* (TSH). TSH increases synthesis and secretion of THs (T4 and T3); excess TSH leads to increased cell number (hyperplasia) and size (hypertrophy) of the gland (*goiter*). Goiters can occur in disease or usually in response to iodine deficiency. Decreased TSH levels lead to thyroid atrophy and reduced secretion. TSH is regulated by negative-feedback effects of circulating THs on the anterior pituitary. Increased plasma T4 acts directly on the pituitary, diminishing TSH release, and vice versa.

Brain controls thyroid through hypothalamic TRH—The brain influences the secretion of THs as well. Hypothalamus neurons produce thyrotropin-releasing hormone (TRH), which regulates the release of TSH from the anterior pituitary. The brain senses changes in environmental temperature (heat and cold) through its peripheral thermoreceptors and makes appropriate adjustments in hypothalamic TRH release. A marked and prolonged decline in environmental temperature triggers a rise in TRH, which causes an increase in TSH levels; this in turn leads to greater secretion of THs and increased heat production. TRH, a tripeptide, was the first hypothalamic hypophysiotropin whose chemical identify was determined. There is no known inhibitory neurohormone from the hypothalamus for regulation of TSH.

THYROID HISTOPHYSIOLOGY

The thyroid gland consists of numerous *follicles* with many blood capillaries between the follicles. Each follicle has a single row of *follicular cells* (thyroid epithelial cells) surrounding a cavity (lumen) filled with a colloidal substance, the *colloid*. The colloid is the storehouse of a large protein, *thyroglobulin*, synthesized by thyroid cells and secreted into the lumen to help in the synthesis of THs. Thyroglobulin has many tyrosine residues, the precursor amino acid for synthesis of THs.

Thyroid cells & colloid collaborate in synthesis & release of THs—*Iodide* is actively transported into the thyroid cells by transporter proteins and then moved into the colloid, where it is oxidized into *iodine*. Enzymes attach iodine to the *tyrosine* residues in the thyroglobulin. The iodinated tyrosines are then converted to mono-iodo-thyronine and di-iodothyronine and finally into thyroxine and T3. Bits of the colloid, containing the hormone, are taken in by pinocytosis; lysosomal enzymes release the THs from the thyroglobulin. The free hormones diffuse into the blood.

Thyroxine (T4) may be a prohormone and is converted to T3 in target cells—The thyroid produces 10 times more T4 than T3. T4 may be a prohormone, since after entry into target cells, it is mostly converted to T3. Also, the nuclear receptors for THs have much higher affinity for T3 than T4. THs are slow-acting hormones exerting most of their effects through nuclear receptors (plate 115). In the blood, THs bind with special blood proteins (*thyroid-binding globulins*, TBG), which carry them to their target tissues. There the THs are freed from these carriers, entering target cells to exert their actions.

ABNORMALITIES OF THYROID FUNCTION

Hyperthyroidism involves increased metabolic rate, weight loss, irritability, and cardiovascular changes—Excessive secretion of THs (*hyperthyroidism*) is often associated with *Graves' disease*, an autoimmune disease caused by abnormal antibodies against TSH receptors, resulting in excessive thyroid stimulation. Hyperthyroid individuals have a high BMR (up to +100%). The enhanced heat production depletes energy reserves (liver glycogen and body fat), leading to weight loss and thinness. These individuals are also irritable and nervous and show increased cardiovascular and respiratory activities. Protruding eyeballs (exophthalmus) is one of the signs of hyperthyroidism. Some individuals develop goiters. The follicular cells become enlarged, and the colloid appears depleted. Hyperthyroidism may also be caused by tumors of the thyroid, pituitary, or hypothalamus.

Adult hypothyroidism involves decreased BMR, myxedema, and reduced activity—Hypothyroidism may be caused by disorders of the thyroid or pituitary or hypothalamic failures. In adults, hypothyroidism results in diminished BMR (down to –40%) and the syndrome of *myxedema*. Myxedemic individuals have thick skin, a puffy face (edema), husky voice, and coarse hair. They are slow in physical and mental activities and may act deranged.

Developmental hypothyroidism is associated with dwarfism and mental retardation—In infants and children, thyroid deficiency results in the syndrome of *cretinism*. Cretins are dwarfed and mentally retarded due to abnormal brain development. Cretinism can result from maternal iodine deficiency or congenital thyroid problems. It can be reversed by thyroxine replacement therapy treatment, begining at birth.

CN: Use red for J. Use the same colors for the hypothalamus (A) and anterior pituitary (C) as were used on the previous pages.
1. Begin with the upper panel. Include the chemical structure (the points of iodine attachment) of the two hormones.
2. Color the middle panel beginning with the thyroid gland (E). Follow the numbered sequence in the large illustration. Note that at step 5, iodide (K becomes iodine (F3) and it and the attached tyrosine molecule receive the new color. The arrows of movement reflect the color of the molecules involved.
3. Color the arrows in the lower illustrations.

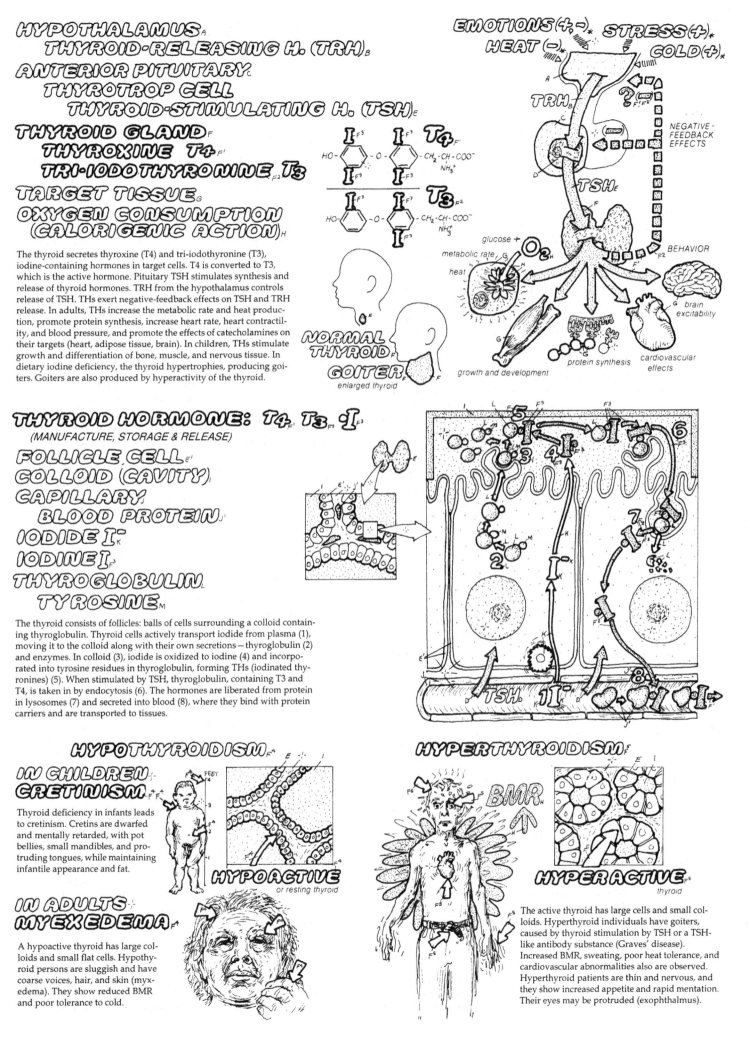

HYPOTHALAMUS (A)
THYROID-RELEASING H. (TRH) (B)
ANTERIOR PITUITARY
THYROTROP CELL
THYROID-STIMULATING H. (TSH) (E)
THYROID GLAND (F)
THYROXINE T4 (F')
TRI-IODOTHYRONINE T3 (F2)
TARGET TISSUE (G)
OXYGEN CONSUMPTION (CALORIGENIC ACTION) (H)

The thyroid secretes thyroxine (T4) and tri-iodothyronine (T3), iodine-containing hormones in target cells. T4 is converted to T3, which is the active hormone. Pituitary TSH stimulates synthesis and release of thyroid hormones. TRH from the hypothalamus controls release of TSH. THs exert negative-feedback effects on TSH and TRH release. In adults, THs increase the metabolic rate and heat production, promote protein synthesis, increase heart rate, heart contractility, and blood pressure, and promote the effects of catecholamines on their targets (heart, adipose tissue, brain). In children, THs stimulate growth and differentiation of bone, muscle, and nervous tissue. In dietary iodine deficiency, the thyroid hypertrophies, producing goiters. Goiters are also produced by hyperactivity of the thyroid.

EMOTIONS (+, −), STRESS (+), HEAT (−), COLD (+)

NEGATIVE-FEEDBACK EFFECTS

BEHAVIOR

glucose +
metabolic rate
heat

brain excitability

protein synthesis

cardiovascular effects

growth and development

NORMAL THYROID
GOITER
enlarged thyroid

THYROID HORMONE: T4, T3, I
(MANUFACTURE, STORAGE & RELEASE)
FOLLICLE CELL (E')
COLLOID (CAVITY) (I)
CAPILLARY
BLOOD PROTEIN (J)
IODIDE I⁻ (K)
IODINE I (F3)
THYROGLOBULIN (L)
TYROSINE (M)

The thyroid consists of follicles: balls of cells surrounding a colloid containing thyroglobulin. Thyroid cells actively transport iodide from plasma (1), moving it to the colloid along with their own secretions—thyroglobulin (2) and enzymes. In colloid (3), iodide is oxidized to iodine (4) and incorporated into tyrosine residues in thyroglobulin, forming THs (iodinated thyronines) (5). When stimulated by TSH, thyroglobulin, containing T3 and T4, is taken in by endocytosis (6). The hormones are liberated from protein in lysosomes (7) and secreted into blood (8), where they bind with protein carriers and are transported to tissues.

HYPOTHYROIDISM (F)
IN CHILDREN CRETINISM

Thyroid deficiency in infants leads to cretinism. Cretins are dwarfed and mentally retarded, with pot bellies, small mandibles, and protruding tongues, while maintaining infantile appearance and fat.

HYPOACTIVE
or resting thyroid

IN ADULTS MYXEDEMA

A hypoactive thyroid has large colloids and small flat cells. Hypothyroid persons are sluggish and have coarse voices, hair, and skin (myxedema). They show reduced BMR and poor tolerance to cold.

HYPERTHYROIDISM (F)

BMR

HYPERACTIVE
thyroid

The active thyroid has large cells and small colloids. Hyperthyroid individuals have goiters, caused by thyroid stimulation by TSH or a TSH-like antibody substance (Graves' disease). Increased BMR, sweating, poor heat tolerance, and cardiovascular abnormalities also are observed. Hyperthyroid patients are thin and nervous, and they show increased appetite and rapid mentation. Their eyes may be protruded (exophthalmus).

In humans, the *parathyroid* glands are four small bodies, each the size of a lentil, embedded in the superior and inferior poles of *thyroid* tissue, although there are no anatomic or physiologic connections between the thyroid and parathyroid glands. Two types of cells are found in the parathyroid gland: *chief* and *oxyphil*. In response to a decrease in the level of plasma *calcium* ions (Ca^{++}), the chief cells secrete the *parathyroid hormone*, which acts on bone and the kidneys to elevate the plasma calcium level. The function of oxyphil cells is not well understood. They may be degenerated chief cells.

IMPORTANCE OF CALCIUM IONS

Plasma calcium is critical for normal nerve and muscle function — Plasma calcium regulates the electrical activity of excitable cells (nerve and muscle), heart contraction, and blood clotting. Calcium level in plasma is therefore intricately regulated by complex hormonal mechanisms. The normal plasma calcium level is 10 mg/100 ml.

Hypocalcemia may be fatal — A marked reduction below critical limits (*hypocalcemia*) increases nerve and muscle excitability but decreases neurotransmitter release at synapses and neuromuscular junctions. The net result of hypocalcemia is *spasmic contractions* in muscles (*tetany*). A characteristic clinical sign seen with *hypocalcemic tetany* is *Trousseau's sign* (flexion of the wrist and thumb with extension of the fingers). Spasms of respiratory muscles interfere with respiration and can be fatal. Such problems are usually produced after surgical removal of the thyroid tissue, if the parathyroids have been inadvertently removed. In experimental animals, removal of the parathyroid gland leads, within four hours, to a marked reduction in plasma calcium levels and eventually to death unless calcium is given by infusion.

THREE HORMONES REGULATE PLASMA CALCIUM LEVELS

Three hormones participate in the regulation of plasma calcium: parathyroid hormone (*parathormone*, *PTH*) from the parathyroid glands, *calcitonin* from the thyroid, and *calcitriol* from the kidney. The role of PTH is central and vital.

Parathyroid cells release PTH in response to low plasma calcium — The plasma calcium level is monitored by specific membrane *calcium receptors* present in the parathyroid chief cells that are coupled to intracellular IP_3 and calcium via G-proteins. A decrease in plasma calcium below a set limit signals the release of PTH into the circulation; PTH acts directly on bone and the kidneys and indirectly on intestinal mucosa to raise plasma calcium. This rise in turn reduces PTH secretion by negative feedback action on the parathyroid.

In chronic cases of decreased plasma calcium (as in *rickets* or kidney diseases) or in certain physiological conditions when calcium utilization is intensive, such as pregnancy or lactation, parathyroid glands increase in size (hypertrophy). The enlarged gland is more sensitive to reductions in calcium levels and secretes PTH more efficiently, so that a 1% decrease in calcium results in a 100% increase in plasma PTH concentration.

PTH mobilizes calcium from bone to increase plasma calcium — PTH acts on bone tissue, increasing resorption of calcium from the bone matrix and elevating plasma calcium

levels. PTH action on bone tissue occurs at two levels. The principal effect is to stimulate the *osteoclasts*. These bone cells digest the bone matrix, increasing the level of calcium ions in the matrix fluid (*bone fluid*). This calcium is then exchanged with plasma. There may even be an increase in the number of osteoclasts. A more rapid action that elevates calcium levels in minutes is the pumping of calcium from the bone fluid to the plasma. This transport occurs across an extensive membrane system separating the bone fluid from the extracellular fluid (plasma). These membranes are formed by the processes of *osteocytes* and *osteoblasts* (plate 121).

PTH also acts on kidneys to retain calcium — PTH also increases plasma calcium by increasing the renal tubular reabsorption of this ion. A more effective action of PTH in this connection is the increase in the renal excretion of *phosphate* (HPO_4^-) ions (phosphaturia). Usually, the product of calcium and phosphate ion concentrations in plasma is constant. Thus, a decrease in phosphate concentration would lead to an increase in calcium concentration. Parathyroid hormone also increases calcium absorption in the small intestine by stimulating the formation of the active form of vitamin D in the kidneys (see below).

Vitamin D & its metabolites increase intestinal absorption of calcium — *Vitamin D_3* (*cholecalciferol*) can be obtained in the diet or produced from cholesterol derivatives in the skin in the presence of ultraviolet radiation in sunlight. To become active, vitamin D_3 is first converted to *calcidiol* in the liver; further conversion of calcidiol in the kidney proximal tubules produces *calcitriol*, the most potent form of vitamin D. PTH stimulates the formation of calcitriol in the kidney. Calcitriol is considered a kidney hormone since it is produced by the kidney and secreted in the blood to elevate plasma calcium.

The main function of calcitriol is to stimulate the intestinal epithelial cells (enterocytes) to increase their synthesis of *calbindin-D* proteins, which function as *calcium transporter proteins*, absorbing calcium from the intestinal lumen (plate 79). This effect is mediated via nuclear receptors similar to steroid hormone receptors. Increased intestinal absorption increases plasma calcium levels. Calcitriol also enhances PTH action on bone cells. Vitamin D_3 and its derivatives are called *secosteroids*.

Calcitonin decreases plasma calcium by increasing its absorption by bone — Calcitonin is a hormone secreted by the *parafollicular C cells* of the thyroid. It is secreted in response to an *increase* in the plasma calcium. Calcitonin's function is to decrease plasma calcium levels, an effect opposite to that of PTH. Calcitonin affects bone on two levels: it stimulates the osteoblasts to increase their absorption of calcium and its deposition into bone; and it inhibits the osteoclasts, thereby decreasing bone resorption of calcium and its loss to blood.

The actions of calcitonin are important in growing children because bone growth requires inhibition of bone resorption and stimulation of bone deposition. Calcitonin is also important during pregnancy and lactation, when it helps protect maternal bones from the excess calcium loss initiated by parathormone. In normal adults, however, its action is less important than that of PTH, which is sufficient for minute-to-minute regulation of plasma calcium.

CN: Use red for E, yellow for G, a very light color for bone (P), and a dark color for B.
1. Begin at the top by coloring the thyroid (A) and parathyroid (B) and plasma calcium (E).
2. Color the PTH title (B') and follow its influence

on bone (P) and then color the C cells (O) of the thyroid on the far left, and follow the influence of calcitonin (O^1) (and PTH) down to the bottom.
3. Go back to the PTH title and follow its actions in the intestine; color the material below it.

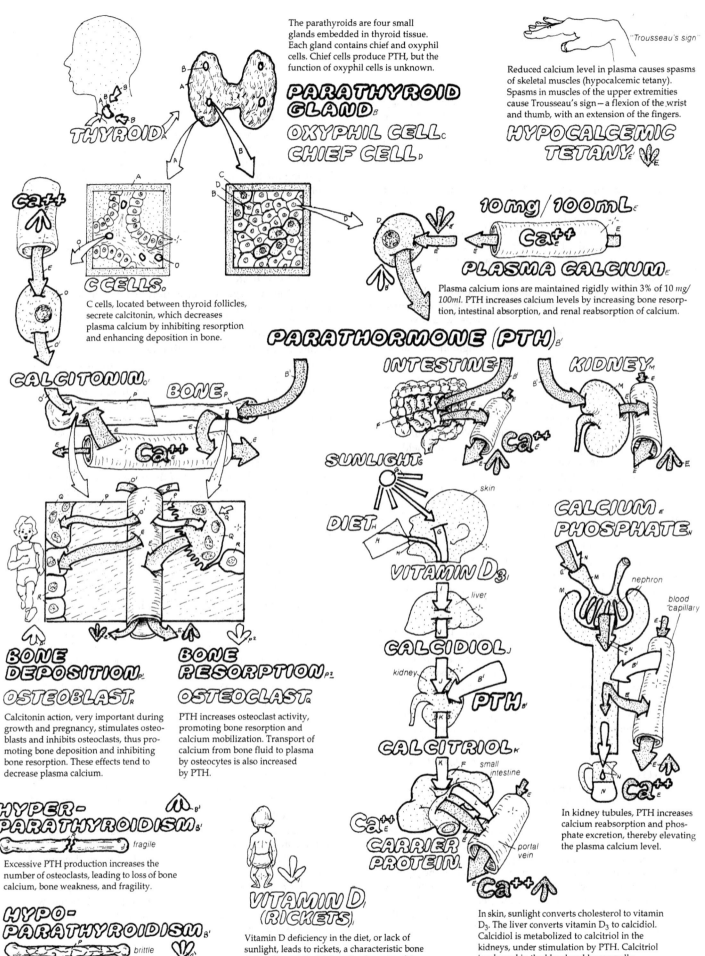

The parathyroids are four small glands embedded in thyroid tissue. Each gland contains chief and oxyphil cells. Chief cells produce PTH, but the function of oxyphil cells is unknown.

THYROID

PARATHYROID GLAND

OXYPHIL CELL

CHIEF CELL

"Trousseau's sign"

Reduced calcium level in plasma causes spasms of skeletal muscles (hypocalcemic tetany). Spasms in muscles of the upper extremities cause Trousseau's sign — a flexion of the wrist and thumb, with an extension of the fingers.

HYPOCALCEMIC TETANY

Ca++

C CELLS

C cells, located between thyroid follicles, secrete calcitonin, which decreases plasma calcium by inhibiting resorption and enhancing deposition in bone.

10 mg / 100 mL

Ca++

PLASMA CALCIUM

Plasma calcium ions are maintained rigidly within 3% of 10 mg/100ml. PTH increases calcium levels by increasing bone resorption, intestinal absorption, and renal reabsorption of calcium.

PARATHORMONE (PTH)

INTESTINE

KIDNEY

CALCITONIN

BONE

Ca++

SUNLIGHT

DIET

Ca++

VITAMIN D₃

skin

CALCIUM PHOSPHATE

liver

nephron

blood capillary

CALCIDIOL

kidney

PTH

BONE DEPOSITION

OSTEOBLAST

Calcitonin action, very important during growth and pregnancy, stimulates osteoblasts and inhibits osteoclasts, thus promoting bone deposition and inhibiting bone resorption. These effects tend to decrease plasma calcium.

BONE RESORPTION

OSTEOCLAST

PTH increases osteoclast activity, promoting bone resorption and calcium mobilization. Transport of calcium from bone fluid to plasma by osteocytes is also increased by PTH.

CALCITRIOL

small intestine

Ca++

CARRIER PROTEIN

portal vein

Ca++

In kidney tubules, PTH increases calcium reabsorption and phosphate excretion, thereby elevating the plasma calcium level.

HYPER-PARATHYROIDISM

fragile

Excessive PTH production increases the number of osteoclasts, leading to loss of bone calcium, bone weakness, and fragility.

HYPO-PARATHYROIDISM

brittle

In hypoparathyroidism, reduced secretion of PTH inactivates the osteoclasts, decreasing bone plasticity and increasing hardness and brittleness of bones.

VITAMIN D (RICKETS)

Vitamin D deficiency in the diet, or lack of sunlight, leads to rickets, a characteristic bone deformity seen in children. Inadequate calcium absorption in the intestine and deposition in bone weakens the growing bones. Long bones of the lower extremities bend under stress.

In skin, sunlight converts cholesterol to vitamin D₃. The liver converts vitamin D₃ to calcidiol. Calcidiol is metabolized to calcitriol in the kidneys, under stimulation by PTH. Calcitriol is released in the blood and hormonally stimulates intestinal absorption of calcium.

Bones support the body and provide leverage for muscles and movement. Bones also harbor the brain, spinal cord, and bone marrow and provide a storehouse for calcium. Hormones help release bone calcium into the plasma, preventing reduction from the normal level of this important ion in the blood (plate 120).

Bone is a living tissue with active cells and blood supply—Although bone appears hard and inert, it is in fact an active tissue, supplied by nerves and blood vessels. Various bone cells (see below) are continuously active, even in the adult, building and rebuilding, repairing and remodeling the bone in response to strains, stresses, and fractures.

BONE STRUCTURE & CELL TYPES

A long bone has two heads (epiphyses) & a shaft (diaphysis) & consists of compact & spongy bone types—A cross-section of the long bone reveals dense and cavernous areas. Dense areas contain *compact bone*; cavernous areas consist of *spongy bone*. The diaphysis of a long bone contains mainly compact bone; the epiphysis contains both compact and spongy bone.

Compact bone consists of numerous repeating lamellar units (Haversian systems)—Microscopic examination of the compact bone in the diaphysis reveals many cylindrical units, called *Haversian systems (osteon)*. These cylindrical units run along the bone length and are packed tightly and held together by a special cement. Each Haversian system consists of concentric plates (*lamellae*) surrounding a *central canal* through which blood vessels and nerves run. The central canal communicates with numerous smaller *lacunae* located throughout the Haversian system. The many lacunae in turn communicate via smaller passageways (*canaliculi*), which permit blood and nerves to reach bone cells.

Bone cells & bone matrix are the functional parts of bone—Physiologically, bone tissue consists of two compartments: first, a metabolically active cellular compartment made up of *bone cells* and second, a metabolically inert extracellular compartment, the *bone matrix*, consisting of a mixture of *organic* and *inorganic* materials. The organic part is made of *collagen* fibers—extremely tough fibrous proteins—and the *ground substance* (*glycoproteins* and *mucopolysaccharides*). The inorganic component of the bone matrix consists of a mineral of calcium and phosphate called *hydroxyapatite crystals* – (Ca_{10} [PO_4]6 [OH]2). To make the bone matrix, bone cells deposit the hydroxyapatite crystals on a mesh of collagen fibers and glycoproteins, a process called *calcification*. The calcified matrix gives the bone its remarkable hardness and strength.

THREE TYPES OF BONE CELLS: OSTEOBLASTS, OSTEOCYTES, AND OSTEOCLASTS

Osteoblasts make bone & osteocytes maintain bone—Osteoblasts, usually found near the bone surfaces, are the young bone cells that secrete the organic substances of the matrix—i.e., the collagen fibers and ground substance. Once totally surrounded by the secreted matrix, osteoblasts markedly diminish their bone-making activity, turn into mature bone cells, and continue function as osteocytes, helping in calcium exchange and daily maintenance of bone tissue. Osteocytes are found in or near the lacunae. They develop extensive processes (*filopodia*) that run through the canaliculi, connecting them with the other osteocytes. These membra-

nous processes facilitate the exchange of nutrients, especially calcium, between the bone and blood.

Osteoclasts remodel, repair, and dissolve bone—The third type of bone cell is the osteoclast, which resembles a blood macrophage. Osteoclasts have important functions in repairing fractures and remodeling new bone. To accomplish their tasks, osteoclasts secrete *lysosomal enzymes* (e.g., the protease *collagenase*) into the bone matrix. These enzymes digest the matrix proteins, liberating calcium and phosphate. Because of their ability to digest bone, osteoclasts are targets for hormones, such as parathormone, that promote bone resorption and calcium mobilization.

BONE GROWTH AND ITS HORMONAL REGULATION

Hyaline cartilage cells of epiphyseal plates initiate bone formation—The development of bone is usually preceded by the formation of hyaline cartilage. Most of the fetal skeleton consists of cartilage. In long bone, the growth and elongation begins after birth, continuing through adolescence. Elongation is achieved by the activity of two hyaline cartilage plates (*epiphyseal plates*), located between the shaft and the head on each end of the bone. Germinal cells in these plates continuously produce new cartilage cells, which migrate toward the shaft, forming a template. Next, young bone cells (osteoblasts) move into these areas, constructing new bone over these templates.

In this manner, the length of the bone shaft increases at both ends, and the two heads of the bone move progressively apart, increasing bone length. The width of the epiphyseal plates is proportional to the rate of growth. In the growing child, the plates are wide and active, with proliferating cells. As growth proceeds, growth rate diminishes and plate thickness gradually decreases, becoming narrow at puberty and disappearing by adulthood (*epiphyseal closure*). Longitudinal bone growth is not possible after this stage, which occurs at different ages for different bones.

Growth hormones, IGFs, thyroid hormones, & androgens stimulate bone growth—During childhood, *growth hormones* stimulate plate growth. The effect of growth hormone is mediated by insulin-like growth factor (IGF-1). Thyroid hormones are required for the action of growth hormone as well. Thyroid hormones also promote bone differentiation. *Androgens* stimulate bone growth in puberty and are important in the *adolescent growth spurt*. In males, the testes provide the androgens; in growing females, the adrenal cortex is the source of androgens. However, in late adolescence, androgens enhance the closure of epiphyseal plates, thus terminating growth in height. The action of androgens on epiphyseal plate growth and closure may be preceded by intracellular conversion of androgen to estrogen. In adults, excess GH promotes bone growth only in width, leading to the thick bones characteristic of *acromegalic* individuals (plate 118).

Repair of bone fracture involves the hyaline cartilage and aid of osteoblasts & osteoclasts—During repair of a fracture, a special type of connective tissue, the *hyaline cartilage*, develops at the fracture site, forming a *callus*. The callus serves as a model for new bone growth and protects the healing bone against the deforming stress forces acting on it. When new bone replaces the callus, osteoclasts remodel the bone into its original shape by digesting the extra bone.

CN: Use red for A, a pale yellow or tan for B, and very light colors for C and D.
1. Begin with the bone structures at the top. Notice that only one group of Haversian systems (d1) has been selected for coloring.

2. Color the three types of bone cells and two illustrations underneath demonstrating their functions.
3. Color the three steps in fracture repair.
4. Color the hormonal regulation of bone growth shown at the bottom.

BONE STRUCTURE

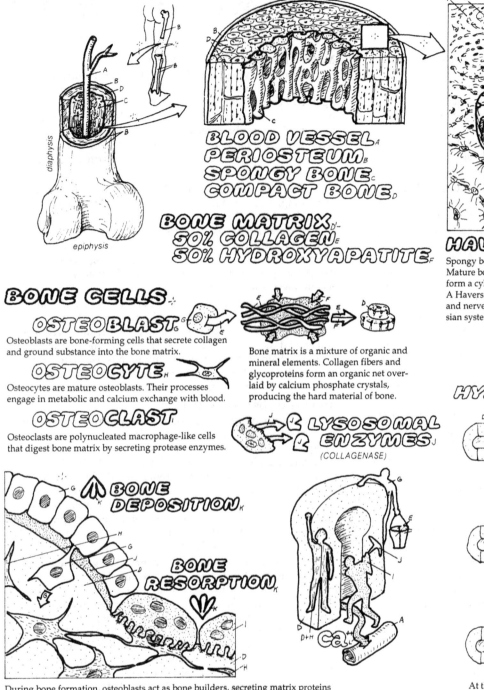

BLOOD VESSEL A
PERIOSTEUM B
SPONGY BONE C
COMPACT BONE D

BONE MATRIX D'
50% COLLAGEN E
50% HYDROXYAPATITE F

diaphysis

epiphysis

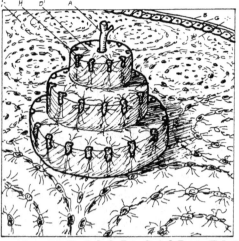

HAVERSIAN SYSTEMS D'

Spongy bone contains cavities; compact bone does not. Mature bone is lamellar. Several concentric lamellae form a cylindrical unit, the Haversian system (osteon). A Haversian canal at its middle contains blood vessels and nerves. A long bone consists of numerous Haversian systems running parallel to the length.

BONE CELLS

OSTEOBLAST G

Osteoblasts are bone-forming cells that secrete collagen and ground substance into the bone matrix.

OSTEOCYTE H

Osteocytes are mature osteoblasts. Their processes engage in metabolic and calcium exchange with blood.

OSTEOCLAST I

Osteoclasts are polynucleated macrophage-like cells that digest bone matrix by secreting protease enzymes.

Bone matrix is a mixture of organic and mineral elements. Collagen fibers and glycoproteins form an organic net overlaid by calcium phosphate crystals, producing the hard material of bone.

LYSOSOMAL ENZYMES J
(COLLAGENASE)

BONE DEPOSITION K

BONE RESORPTION K'

During bone formation, osteoblasts act as bone builders, secreting matrix proteins (collagen and ground substance). Calcification of the matrix isolates osteoblasts, which mature into osteocytes with extensive processes. Osteocytes participate in exchange of nutrients and calcium with blood. Osteoclasts help shape bone by digesting extra bone pieces. When stimulated by parathyroid hormone, osteoclasts release calcium to compensate for plasma calcium deficiency.

FRACTURE REPAIR *
HYALINE CARTILAGE M

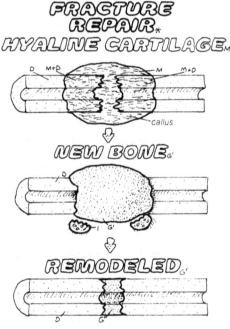

callus

NEW BONE G'

REMODELED G'

At the edges of bone fractures, hyaline cartilage proliferates, forming a callus. This helps support the fracture and serves as a model for bone formation. Infiltration by bone cells transforms the callus to bone, which is then remodeled and shaped by digestive actions of osteoclasts.

GROWTH

EARLY YOUTH

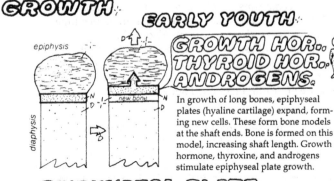

GROWTH HOR..
THYROID HOR..
ANDROGENS

epiphysis

diaphysis

new bone

In growth of long bones, epiphyseal plates (hyaline cartilage) expand, forming new cells. These form bone models at the shaft ends. Bone is formed on this model, increasing shaft length. Growth hormone, thyroxine, and androgens stimulate epiphyseal plate growth.

EPIPHYSEAL PLATE N

MATURITY
"epiphyseal closure"

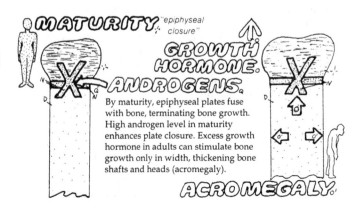

GROWTH HORMONE.
ANDROGENS Q

By maturity, epiphyseal plates fuse with bone, terminating bone growth. High androgen level in maturity enhances plate closure. Excess growth hormone in adults can stimulate bone growth only in width, thickening bone shafts and heads (acromegaly).

ACROMEGALY

The pancreas is a major *mixed* (both *exocrine* and *endocrine*) gland located in the abdominal cavity, under the stomach. By bulk, most of the gland (98%) deals with the exocrine functions—i.e., secretion of digestive enzymes and bicarbonate by the pancreatic acini and ducts, respectively (plate 76). Endocrine cells of the pancreas secrete peptide hormones that are vital to the regulation of carbohydrate metabolism and blood sugar. Pancreas access to hepatic portal vein blood allows it to detect levels of nutrients absorbed from the intestine.

The endocrine cells of the pancreas are located in the islets of Langerhans—The *islets of Langerhans* consist of 1–2 million round clusters (islets) of cells, scattered throughout the pancreas tissue between the exocrine acini. A rich bed of special blood capillaries with large pores and access to *hepatic portal vein* blood surrounds the islets. Each islet is a collection of several different types of cells, each secreting one of the pancreatic hormones. By using specific immunocytochemical staining methods, researchers have identified three types of cells: A, B, and D cells (also known as α, β, and δ cells). *A cells*, less numerous and located peripherally, secrete the hormone *glucagon*. *B cells*, located centrally, are more numerous and secrete the hormone *insulin*. *D cells* are sparse and secrete the local hormone *somatostatin*. Recently an F cell type has also been found, forming a *pancreatic polypeptide* with unknown hormonal function.

INSULIN & GLUCAGON ARE THE MAIN PANCREAS HORMONES

Insulin and glucagon regulate carbohydrate metabolism in tissues and ensure the maintenance of optimal plasma *glucose* (blood sugar) *levels*. Blood sugar usually increases after eating a meal, particularly meals rich in carbohydrates.

A rise in plasma glucose releases insulin, which increases glucose uptake by cells—By increasing the transport of glucose across cell membranes, insulin enhances the availability of glucose in cells and promotes its utilization. As a consequence, blood sugar is decreased. In this regard, insulin functions as a *hypoglycemic hormone*—i.e., one that decreases blood sugar. Insulin's specific target tissues are muscle, fat, and liver.

A decline in plasma glucose releases glucagon, which increases glucose release from the liver—Glucagon also enhances carbohydrate utilization for body cells but does so between meals, when blood sugar and insulin levels are low. Glucagon's function is to mobilize glucose from its major storage source, *liver glycogen*, into the blood. In doing so, glucagon functions as a *hyperglycemic hormone*—i.e., one that increases blood sugar. Low blood sugar levels are encountered also during fasting, necessitating glucagon action.

Somatostatin regulates insulin and glucagon secretion locally—Somatostatin, a local tissue hormone, is the third peptide secreted from the pancreas; it tends to dampen the rise in both insulin and glucagon, preventing a sudden rise in their secretory pattern. Somatostatin is also released to the circulation and may inhibit growth hormone secretion from the pituitary, since growth hormone has anti-insulin effects. Insulin and glucagon also regulate each other's secretion directly: glucagon stimulates insulin secretion while insulin inhibits glucagon secretion.

MECHANISMS OF INSULIN SYNTHESIS

Insulin, a polypeptide with two chains, is formed from proinsulin—Insulin is a protein (polypeptide) hormone made up of two peptide chains (*A chain* and *B chain*) connected at two amino acid loci by disulfide bridges (bonds). This is the form in which insulin is released into the blood and acts on its receptors on the target cells. Insulin is synthesized within the B cells on the endoplasmic reticulum in the form of a larger single peptide chain called *proinsulin*.

To form insulin, the C chain is removed—During later processing, proinsulin folds because of the formation of disulfide bridges. During packaging in the vesicles of the Golgi apparatus, *protease enzymes* convert the folded proinsulin to insulin by hydrolyzing the peptide bonds of the long chain at two locations, splitting the original single chain into two pieces—insulin (the connected A and B chains) and a second piece, the *C chain*.

Insulin and the C chain are transported together within secretory vesicles to the B-cell membrane, where they are released into the blood by vesicle exocytosis. Insulin circulates and exert its action on target cells. There is no known function for the C chain, but its blood concentration is used clinically as an indication of endogenous insulin, a useful measure for diabetic people receiving exogenous insulin injections.

REGULATION OF INSULIN RELEASE

Plasma glucose levels regulate insulin secretion—Insulin release by B cells is regulated by the plasma glucose through a negative feedback system. An increase in plasma glucose, occurring usually after a meal, is detected by the B cells, resulting in increased insulin secretion into the blood. Insulin acts on tissues by increasing glucose uptake and utilization. As a result, blood sugar levels are decreased. Release of insulin is thought to occur in two phases—an immediate phase, involving a sharp peak that lasts for several minutes, followed by a slow rising response lasting for two hours.

Insulin secretion is triggered by increased glucose entry and a rise in B-cell calcium—The B-cell mechanism involved in glucose-insulin feedback involves interaction between glucose entry and calcium release; calcium rise is required for exocytosis and insulin release. Glucose is transported into the B cells by specific glucose transporter proteins (GLUT-2). Entering glucose molecules are quickly oxidized by glucokinase, forming ATP; ATP acts on nearby ATP-sensitive K^+ channels, which close. This depolarizes the B-cell membrane, which in turn opens voltage-sensitive Ca^{++} channels, allowing influx of calcium. This rise in intracellular calcium triggers an initial rapid release of insulin from B cells by promoting fusion of insulin-containing vesicles and exocytosis.

Intracellular rise in cyclic GMP enhances insulin secretion and synthesis—Persistent glucose entry activates additional mechanisms, involving a rise in intracellular cyclic GMP; this rise releases more calcium from intracellular reserves (Ryanodine receptors). Also, cGMP triggers increased synthesis of insulin, ensuring availability of insulin for prolonged secretion (hours), until the hyperglycemia is terminated. These mechanisms also explain the biphasic release of insulin mentioned above.

CN: Use a yellowish color for A, red for H, purple for I, and another bright color for E.
1. Begin at the top and color down to and including the two test tubes symbolizing blood glucose levels.
2. Color the formation of insulin (E^1) from proinsulin (M) shown at the bottom left. This process

occurs at step (7) in the B-cell illustration.
3. Color the steps of insulin synthesis shown in the diagrammatic B cell. Note that insulin (E^1) is represented by two small parallel bars, which stand for the A chain (E^2) and B chain (E^3) in the previous illustration. Don't color the interior of the capillary (I).

PANCREAS.A

EXOCRINE: 98%.B
PANCREATIC ACINI.A'
PANCREATIC DUCT.B'

ENDOCRINE: 2%.c
ISLETS OF LANGERHANS.c
A CELLS - GLUCAGON.D'
B CELLS - INSULIN.E'
D CELLS - SOMATOSTATIN.F'

SPLENIC VEIN.G
GLUCOSE LEVEL (BLOOD SUGAR).H

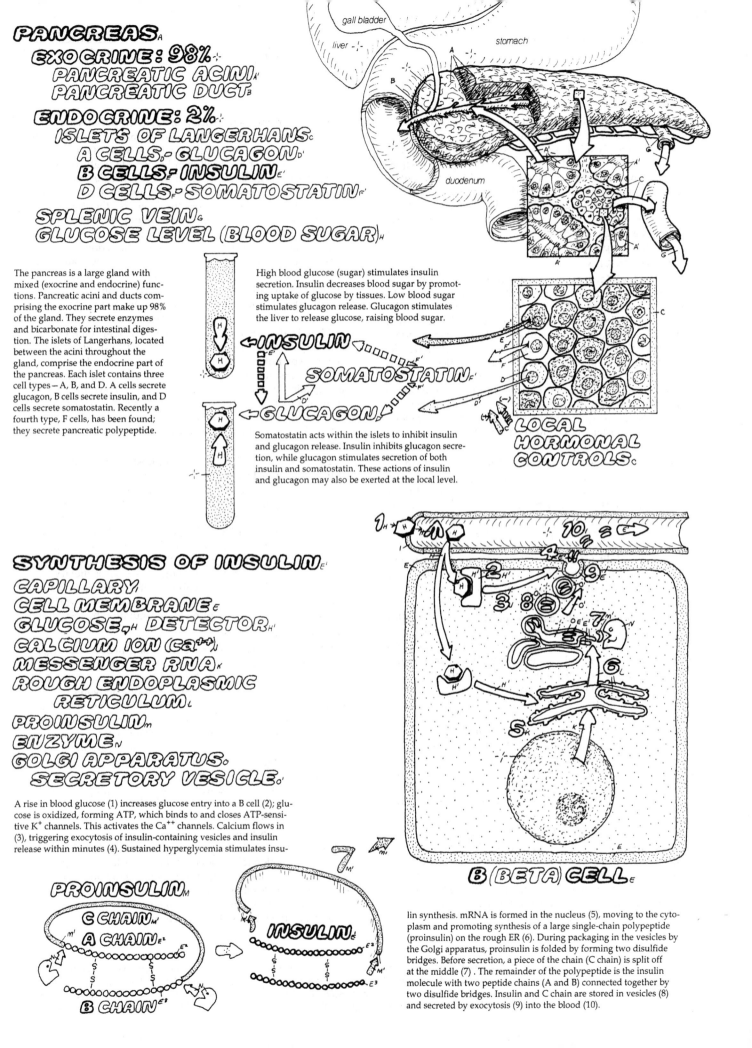

gall bladder

stomach

liver

duodenum

The pancreas is a large gland with mixed (exocrine and endocrine) functions. Pancreatic acini and ducts comprising the exocrine part make up 98% of the gland. They secrete enzymes and bicarbonate for intestinal digestion. The islets of Langerhans, located between the acini throughout the gland, comprise the endocrine part of the pancreas. Each islet contains three cell types—A, B, and D. A cells secrete glucagon, B cells secrete insulin, and D cells secrete somatostatin. Recently a fourth type, F cells, has been found; they secrete pancreatic polypeptide.

High blood glucose (sugar) stimulates insulin secretion. Insulin decreases blood sugar by promoting uptake of glucose by tissues. Low blood sugar stimulates glucagon release. Glucagon stimulates the liver to release glucose, raising blood sugar.

←INSULIN.E'
SOMATOSTATIN.F'
←GLUCAGON.D'

LOCAL HORMONAL CONTROLS.c

Somatostatin acts within the islets to inhibit insulin and glucagon release. Insulin inhibits glucagon secretion, while glucagon stimulates secretion of both insulin and somatostatin. These actions of insulin and glucagon may also be exerted at the local level.

SYNTHESIS OF INSULIN.E'
CAPILLARY.D
CELL MEMBRANE.E
GLUCOSE.H DETECTOR.H'
CALCIUM ION (Ca^{++}).J
MESSENGER RNA.K
ROUGH ENDOPLASMIC RETICULUM.L
PROINSULIN.M
ENZYME.N
GOLGI APPARATUS.O
SECRETORY VESICLE.O'

A rise in blood glucose (1) increases glucose entry into a B cell (2); glucose is oxidized, forming ATP, which binds to and closes ATP-sensitive K$^+$ channels. This activates the Ca^{++} channels. Calcium flows in (3), triggering exocytosis of insulin-containing vesicles and insulin release within minutes (4). Sustained hyperglycemia stimulates insulin synthesis. mRNA is formed in the nucleus (5), moving to the cytoplasm and promoting synthesis of a large single-chain polypeptide (proinsulin) on the rough ER (6). During packaging in the vesicles by the Golgi apparatus, proinsulin is folded by forming two disulfide bridges. Before secretion, a piece of the chain (C chain) is split off at the middle (7). The remainder of the polypeptide is the insulin molecule with two peptide chains (A and B) connected together by two disulfide bridges. Insulin and C chain are stored in vesicles (8) and secreted by exocytosis (9) into the blood (10).

B (BETA) CELL.E

PROINSULIN.M
C CHAIN.M'
A CHAIN.E'
B CHAIN.E'

INSULIN.E

Normal fasting levels of blood glucose are in the range of 70-110 mg/100 ml of plasma, a range that remains constant throughout life. Two hormones of the *pancreatic islets, insulin* and *glucagon,* act to maintain normal blood sugar levels (plate 122). Insulin, released when blood sugar is high (after meals), decreases blood sugar levels by acting on muscle, fat, and liver cells. Glucagon is released when blood sugar is low and increases it by acting on the liver. Although insulin is hypoglycemic and glucagon hyperglycemic, both have a common aim of providing ample glucose (energy) to body cells.

ACTIONS OF INSULIN

Insulin facilitates membrane transport of glucose — The primary action of *insulin* is to facilitate membrane transport of *glucose* molecules from blood plasma into certain target cells such as *muscle* (heart, skeletal, and smooth) and adipose (*fat*) cells, and to increase glucose utilization by other cells (liver and fat). In the absence of insulin, the membranes of muscle and fat cells are impermeable to glucose, regardless of blood glucose level. Plasma insulin binds with *insulin receptors* in the plasma membranes of the target cells. This binding stimulates a chain of events resulting in the increase in the number of specific *glucose transporter proteins* in target cell membranes. Glucose transporters move glucose from blood plasma into the cell cytosol, making it available for use by the target cells.

Insulin increases the number of glucose receptors in membrane — Muscle and fat cells contain a particular type of insulin-sensitive glucose transporter proteins called GLUT-4. Normally a few of these transporters reside in the cell membranes. Binding of insulin with its receptors stimulates the rapid incorporation of many more GLUT-4 transporters from a cytoplasmic pool into the membrane. This markedly enhances the ability of muscle cells to take up glucose from plasma. When the insulin level is low (between meals), these transporters return back to the cytoplasm until next time glucose and insulin levels increase.

Exercise mobilizes glucose transporters of muscle cells — Increased muscle activity during exercise promotes new GLUT-4 glucose transporters and their incorporation into the plasma membrane, thereby increasing the ability of muscle cells to take up plasma glucose. This effect is independent of insulin's action, but when coupled with insulin, it has a pronounced effect on glucose transport and use in muscle and the body as a whole.

Insulin receptors are large proteins with both extracellular and intracellular domains — Binding of insulin results in conformation changes in the insulin receptor protein, activating its intracellular domain, which has *tyrosine kinase* activity. This activity results in *phosphorylation* of other proteins, ending in the various actions of insulin in cells, including mobilization of GLUT-4 transporters.

Fate of glucose in muscle cell — Muscle cells normally prefer to use glucose for oxidation and *cellular energy.* Once inside a muscle cell, glucose either is directly oxidized to provide ATP for muscle contraction or is stored by being incorporated into *glycogen,* a polymer of glucose (see below). Glycogen formation occurs during rest. During muscle activity, glycogen is broken down into glucose.

Insulin promotes formation of fat in adipose tissue cells — Insulin also promotes glucose entrance into the fat cells of adipose tissue by mobilizing insulin-sensitive GLUT-4 glucose transporters into fat cell membranes. Here, glucose is not utilized for energy but is metabolized to form *glycerol.* Fat cells utilize this glycerol and *fatty acids* obtained from plasma and liver to form *triacylglycerols (triglycerides),* the storage form of body fat. In addition, insulin, acting through intracellular signaling pathways, stimulates the action of enzymes for *lipogenesis* (fat formation) and inhibits the action of specific *lipases* (the *hormone-sensitive lipase*) that break down fats (*lipolysis*). All these actions promote deposition and storage of fat in cytoplasmic fatty granules of fat cells, resulting in their hypertrophy. Lipogenetic and hypertrophic action of insulin on adipose tissue is one of the underlying causes of obesity.

Liver is also a main target of insulin — Liver cell membranes have non-insulin-sensitive glucose transporters and therefore do not require insulin for glucose entry. However, they have insulin receptors. Insulin, acting through intracellular signaling systems, stimulates the liver cell enzymes that promote utilization of glucose for synthesis of glycogen, amino acids, proteins, and fats, particularly fatty acids. These fatty acids are used in part by the adipose tissue to form triglycerides (plates 133, 134).

Brain cells, kidney tubules, and intestinal mucosa do not require insulin — Free permeability of these tissues to glucose is adaptive. Brain cells rely solely on glucose for their energy needs and require a steady glucose supply. So major alterations in insulin secretion indirectly impair brain function, by changing blood sugar levels. High insulin levels (e.g., by injection) cause *hypoglycemia,* depriving the brain of energy and ATP, resulting in confusion, cognitive dysfunction, and even convulsions and death. Intestinal mucosa absorbs dietary glucose, and kidney tubules reabsorb filtered glucose back into the blood, in mechanisms unrelated to glucose use for cellular energy, thus exempting their regulation by insulin.

ACTIONS OF GLUCAGON

Glucagon increases liver secretion of glucose — Glucagon is released between meals when blood sugar drops below 70 mg/100 ml. Glucagon binds with specific *glucagon receptors* in the membranes of liver cells. This binding activates the enzyme *adenylate cyclase,* increasing the concentration of cyclic AMP within the liver cells. Cyclic AMP, acting as a second messenger, initiates a cascade of chemical reactions involving activation (phosphorylation) of the enzymes of *glycogenolysis* (breakdown of glycogen to glucose). Through an amplification mechanism, billions of enzyme molecules are mobilized within seconds to break down glycogen, the highly branched polymer of glucose (*glycogen tree*) and release its monomers, the glucose molecules.

Glucagon also stimulates synthesis of new glucose molecules from amino acids in the liver, a process called *gluconeogenesis.* This action takes longer and is important during nighttime sleep and in adaptation to fasting or starvation. Glucose molecules mobilized by the action of glucagon leak into the blood, providing plasma sugar for use by the steady consumers: the brain, kidney, and heart.

ACTIONS OF INSULIN

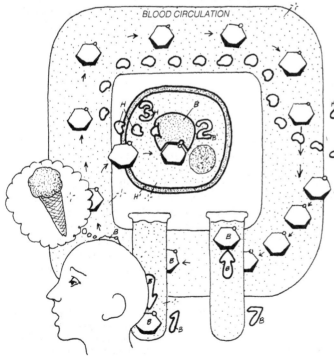
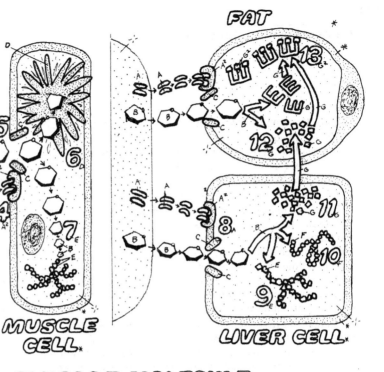

BLOOD CIRCULATION

MUSCLE CELL*

FAT

LIVER CELL*

Increased blood sugar levels after meals (1) are detected by special mechanisms (2) in B cells, triggering insulin release (3). Insulin binds with its receptors (4) in tissue. In muscles, binding increases entry of glucose (5), which is either oxidized for energy (6) or stored as glycogen (7). Glucose enters liver cells without insulin help, but insulin binding (8) stimulates formation of glycogen (9), proteins (10), and fatty acids (11) from glucose. The fatty acids are used in the liver and sent to fat cells (12). In fat cells, insulin promotes glucose entry, enhancing its conversion to glycerol and fatty acids. These esterify to form triglycerides (13), which are stored. As a result of the effects of insulin on muscle, liver, and fat cells, the blood glucose level decreases (14).

GLUCOSE MOLECULE b
B CELL a'
GLUCOSE DETECTOR b'
INSULIN a
INSULIN RECEPTOR a2
GLUCOSE GATE c
CELL METABOLISM d
GLYCOGEN SYNTHESIS e
PROTEIN SYNTHESIS f
FATTY ACID SYNTHESIS g
GLYCEROL → TRIGLYCERIDES g2

ACTIONS OF GLUCAGON h

A CELL h'
GLUCAGON h
GLUCAGON RECEPTOR h2
ENZYME REACTIONS i
GLYCOGEN e'

BLOOD CIRCULATION

LIVER CELL*

GLYCOGEN TREE
(highly branched polymer of glucose)

Between meals, the blood glucose level drops (1). This is detected by the A cells (2), stimulating the release of glucagon (3). Glucagon binds with its receptors (4) in membranes of liver cells, increasing the level of cyclic AMP within hepatocytes. cAMP activates a cascade of enzymes (5) to degrade glycogen into glucose (6). The liver releases glucose into the blood, elevating levels of blood glucose (7) and its supply to tissues.

INSULIN DEFICIENCY HAS WIDESPREAD DELETERIOUS EFFECTS

Insulin is a major regulator of *metabolism*, as evidenced by the widespread deleterious and sometimes catastrophic consequences that follow its deficiency. Insulin deficiency occurs following surgical removal of the pancreas accidental toxic damage to B cells, or as a consequence of *diabetes mellitus*.

Excessive & prolonged hyperglycemia—Insulin-deficient individuals have reduced uptake of *glucose* into the muscle and fat tissues and increased release of glucose into the blood, leading to high blood sugar levels (*hyperglycemia*) ranging from two times the normal (before a meal) to four times (after a meal). Also it also takes longer (6–8 hours) for glucose levels after meals to return to pre-meal levels, compared to 1–2 hours in normal individuals.

Increased breakdown of fat & protein reserves—In insulin and glucose deficiency, muscle cells utilize alternative energy sources such as their *fat* and *protein* reserves, resulting in muscle wasting, weakness, and weight loss. Weight loss is further worsened by events in the fat cells of the adipose tissue. Glucose cannot enter these cells, the absence of insulin removes the inhibition of lipases, resulting in increased breakdown of stored triglycerides and mobilization of fatty acids.

Starvation in the midst of plenty—Loss of body fat contributes to the thinness of the young diabetic patient or insulin-deficient individuals. The malnourished state of the tissues and the individual, in the presence of high blood sugar, is why diabetes is called the disease of "starvation in the midst of plenty" and insulin is called the "hormone of abundance."

Increased blood ketones and ketoacidosis may cause coma—The mobilization of fatty acids provides a ready source of fuel for the energy-starved heart and muscle tissue. However, excessive production of fatty acids results in formation of *keto acids* (*ketone bodies*), particularly in the liver. The ketone bodies enter the blood, causing *ketosis* and *ketoacidosis*. In addition, the ketone bodies are excreted in the urine, worsening the *osmotic diuresis* caused by glucose (see below). If untreated, metabolic and ketoacidosis suppress the higher nervous functions, causing coma. Ultimately, the depression of the brain respiratory centers leads to death.

Glycosuria is a consequence of hyperglycemia—The kidney reabsorbs filtered glucose completely, so that the urine is normally free of sugar. In hyperglycemia, if glucose levels exceed 170 mg glucose/100 ml of plasma, the kidney's reabsorptive capacity is exceeded. Extra glucose spills into the urine. One of the well-known signs of diabetes mellitus and insulin deficiency is the presence of sugar in urine (*glycosuria*).

Polyuria & polydipsia are consequences of glycosuria—The excess glucose in the urine causes *osmotic diuresis* (excess water in urine) and *polyuria* (excess urine production). Polyuria results in decreased plasma volume and increased plasma osmolarity, leading to activation of hypothalamic thirst centers and excessive drinking of water (*polydipsia*). Diabetic individuals are characterized by frequent urination and drinking during the night. Excessive water loss may cause severe *dehydration* and *osmotic shock*, which can lead to irreversible brain damage, coma, and death.

TWO TYPES OF DIABETES

Diabetes is a common metabolic disease, afflicting 15 million people in the United States (6% of the population). Two types of diabetes are now recognized: the *juvenile* type (Type I) and the *maturity-onset* type (Type II).

Type I diabetes is characterized by insulin deficiency—Type I (*insulin-dependent diabetes mellitus*, IDDM) occurs mostly in juveniles and results from autoimmune destruction of insulin-producing pancreas B cells. It represents 10% of all diabetes and shows low familial association. All signs and symptoms of insulin deficiency occur in juvenile diabetes. If untreated, it can be fatal because of *ketoacidosis* and *dehydration shock*. Treatment involves regular injection of insulin before meals, proper diet and meal planning, and exercise.

Type II diabetes is characterized by insulin resistance—Type II diabetes (non-insulin-dependent diabetes mellitus, NIDDM) is 10 times more frequent than Type I and occurs mostly in adults over 40 years of age. It is partly related to increased body fat, since in susceptible people it is often associated with prolonged obesity. Type II shows a strong familial association, in particular with familial obesity. Indeed, Type II is thought to be preventable or delayed if body fat content is kept low. Type II is characterized by *insulin resistance*, as blood insulin levels may be even higher than normal. Insulin resistance may stem from the decreased number of insulin receptors in muscle and fat cells in response to excessive and long-term insulin production. Membrane glucose transporters are also markedly reduced. This condition may be limited to increased carbohydrate intake, increased insulin secretion, and hypertrophy of fat cells over a long time period (years).

Type II diabetes involves hyperglycemia and glycosuria but not ketoacidosis—Since plasma insulin is ineffective, signs similar to complete insulin deficiency develop in Type II diabetes, such as hyperglycemia, glycosuria, polydipsia, and weight loss. Only ketosis and ketoacidosis do not occur. Although Type II diabetic individuals can be treated with extra insulin, simple weight reduction (decrease in body fat content) frequently ameliorates the symptoms in early stages. In later stages, however, use of oral hypoglycemic drugs and insulin may be necessary.

Nerve and vascular damage in untreated diabetes—Untreated Type II diabetes causes nerve and vascular damage, leading to neuropathy, blindness, atherosclerosis, heart attacks, kidney disease, and gangrene. The mechanisms for these pathologies are not known; increased *glycation* – i.e., non-enzymatic binding of glucose to various body proteins— and increased formation of tissue *sorbitol* may be involved.

Insulin therapy is essential for Type I and helpful for later stages of Type II diabetes—The chain of abnormal events seen in a diabetic patient can be halted and reversed to some extent by regular treatment with exogenous insulin. Human insulin, synthesized by bacteria through bioengineering technology, is now available. Proper diet and meal planning are important. Exercise, by mobilizing glucose transporters, helps enhance glucose utilization, which in turn reduces the need for insulin and decreases fat storage.

CN: Use purple of A, red for B, yellow for F. Use a dark color for I.
1. Begin by coloring the title "Starvation..." and the membrane of the shrinking body cell in the upper left corner. Then begin with number 1 to the right of the glucose level at the top of the page and follow the numbered sequence. It would be useful to color all the glucose (B), ketone bodies (I), and water molecules (J) in the illustrations.
2. Color the glucose tolerance test down below.

DIABETES:
STARVATION IN THE MIDST OF PLENTY

pancreas — no insulin

BODY CELL

HYPER-PHAGIA

acetone breath

BLOOD ACIDITY
KETOSIS
COMA
DEATH

SPILL OVER

KETONURIA
POLYURIA
GLYCOSURIA

POLYDIPSIA
(excessive thirst)

CAPILLARY A
GLUCOSE B
INSULIN RECEPTOR C
GLUCOSE GATE D
GLYCOGEN STORES E
FAT STORES F
PROTEIN STORES G
METABOLISM H
KETONE BODIES I
WATER J
KIDNEY NEPHRON K

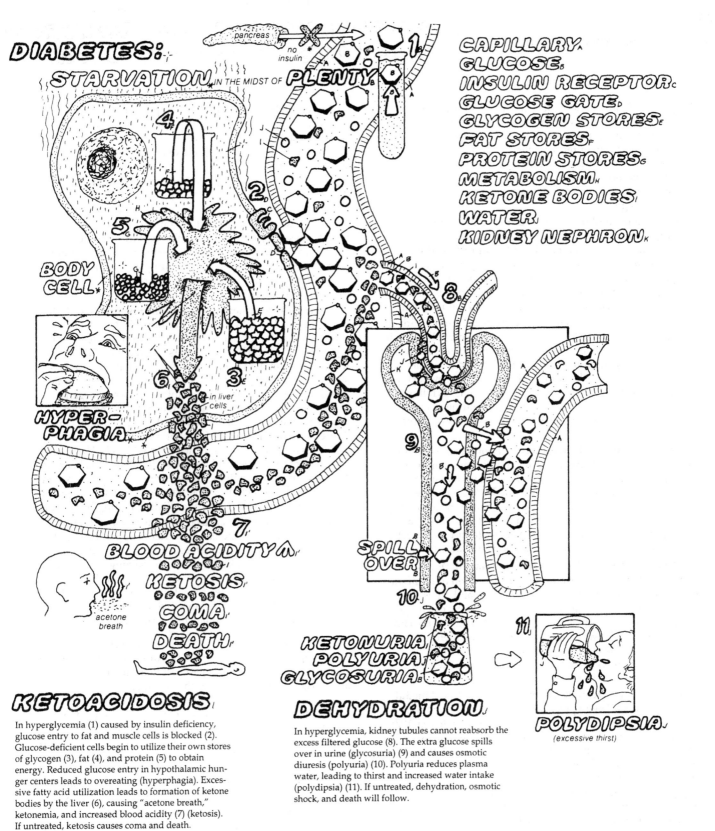

KETOACIDOSIS

In hyperglycemia (1) caused by insulin deficiency, glucose entry to fat and muscle cells is blocked (2). Glucose-deficient cells begin to utilize their own stores of glycogen (3), fat (4), and protein (5) to obtain energy. Reduced glucose entry in hypothalamic hunger centers leads to overeating (hyperphagia). Excessive fatty acid utilization leads to formation of ketone bodies by the liver (6), causing "acetone breath," ketonemia, and increased blood acidity (7) (ketosis). If untreated, ketosis causes coma and death.

DEHYDRATION

In hyperglycemia, kidney tubules cannot reabsorb the excess filtered glucose (8). The extra glucose spills over in urine (glycosuria) (9) and causes osmotic diuresis (polyuria) (10). Polyuria reduces plasma water, leading to thirst and increased water intake (polydipsia) (11). If untreated, dehydration, osmotic shock, and death will follow.

GLUCOSE-TOLERANCE TEST

Insulin-deficient or diabetic people have higher fasting blood sugar levels. When patients are given a load of glucose after fasting (glucose tolerance test), blood glucose increases to a much higher level and decreases over a much longer duration than in normal individuals.

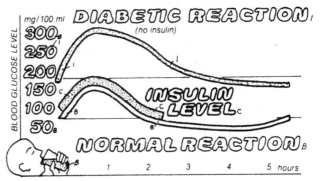

DIABETIC REACTION
(no insulin)

INSULIN LEVEL

NORMAL REACTION

mg/100 ml — BLOOD GLUCOSE LEVEL

300
250
200
150
100
50

1   2   3   4   5 hours

Adrenal glands are paired endocrine organs located above the kidneys and concerned with preparing and protecting the body to fight stress. Each adrenal consists of two separate *endocrine glands*—an outer *adrenal cortex* and an inner *adrenal medulla*. The two glands have different embryological origin, structure, and hormonal secretions, but with respect to responses to stress, their functions are synergistic and aimed at a common goal.

The adrenal medulla is a part of the sympathetic nervous system—The adrenal medulla is essentially a modified *sympathetic ganglion*. The secretory cells of the adrenal medulla, called the *chromaffin* cells, are equivalent to *postganglionic* sympathetic neurons that have lost their axons (plate 29). The regulation of adrenal medullary hormones is intimately linked to the activation of the sympathetic nervous system.

Catecholamines are synthesized from the amino acid tyrosine—The chromaffin cells of the adrenal medulla contain secretory vesicles filled with *epinephrine* (E) and *norepinephrine* (NE). These biogenic amines, collectively called *catecholamines*, are synthesized from the amino acid *tyrosine* via several enzymatic chemical reactions in the chromaffin cells, as follows: tyrosine → dopa → dopamine → norepinephrine → epinephrine. Tyrosine can be of dietary origin or synthesized from phenylalanine.

Two chromaffin cell types secrete epinephrine & norepinephrine—E and NE are released from different chromaffin cell types, one secreting E and another NE. In humans, 80% of the total catecholamine output is E and 20% is NE, reflecting the larger population of E cells. Dopamine and the opiate peptide *endorphin* are also secreted from the adrenal medulla. Endorphin may have anti-stress analgesic (anti-pain) effects but the function of dopamine secretion is not known.

SYMPATHETIC SYSTEM REGULATES ADRENAL MEDULLA

Sympathetic stimulation increases secretion of epinephrine & norepinephrine—The secretion of E and NE is under sympathetic control. During intense sympathetic discharge, preganglionic nerves from the spinal cord stimulate the chromaffin cells to release E and NE. Intense sympathetic activation occurs during emotional conditions such as fear and excitement or stressful and strenuous muscular exercise (running, physical exertion, and struggle).

Hypothalamus regulates sympathetic & adrenal medullary responses—The hypothalamus is, among other things, the highest center for sympathetic regulation (plates 85, 107). During excitement and stress, various brain areas activate the *hypothalamus*. Excitatory fibers from the hypothalamus descend in the spinal cord, stimulating the *preganglionic* sympathetic neurons. These neurons release *acetylcholine* in the sympathetic ganglia, stimulating the *postganglionic* neurons. The fibers of these neurons innervate the visceral organs and skin, releasing NE at their targets. In the case of the adrenal medulla, a long preganglionic fiber (via a splanchnic nerve) stimulates the chromaffin cells (the fiberless postganglionic neurons) to release catecholamines directly in to the blood.

ROLES OF CATECHOLAMINES IN FIGHT-FLIGHT RESPONSES

The E and NE hormones help prepare the body for stressful situations, such as fight-flight response, involving physical exertions; exercise can bring about similar responses. Lets consider a person who is running fast.

Increased heart activity & blood pressure—The need for increased oxygen and nutrients to fuel muscles demands increased delivery of blood by the heart. Thus, cardiac output (heart rate and contractility) must be increased (plate 44) to meet this demand.

Vasodilation, vasoconstriction, & bronchiolar dilation—To increase blood flow to the heart and muscles, the blood vessels (arterioles) to them must be dilated, while those to the skin and visceral organs must be constricted to reduce blood flow to these organs and shunt the blood to where it is most needed (muscles and heart). At the same time, the respiratory activity must be increased and the lung bronchioles dilated to supply more oxygen to the tissues and remove more carbon dioxide from them.

NE & E cause systemic & metabolic responses—All these responses are brought about by various effects of E and NE acting on the target organs. E acts mainly on the heart, causing increases in both its rate and its contractility. NE acts on the arterioles in the visceral organs to cause their vasoconstriction. This increases peripheral resistance and systemic blood pressure. Decreasing flow into the visceral organs causes the blood to be shunted to the muscles and heart. This differential response occurs because the heart contains mainly β receptors, which bind preferentially with E, and visceral arterioles have α receptors, which bind with NE. The smooth muscles of bronchioles and those of arterioles of the heart and muscle contain β receptors. These receptors, when activated by E, relax the bronchiole smooth muscles, causing vasodilation in the heart and muscles and dilation in lung bronchioles.

Metabolic , pupil, & arousal responses—Metabolically, the body demands an increased nutrient supply during physical stress. E increases glycogen breakdown in the liver and lipolysis of fat in the adipose tissue, mobilizing ample fuel substances (glucose and fatty acids). Lastly, catecholamines act on the brain to increase arousal, alertness, and excitability. They also act on the iris of the eye to dilate the pupil, allowing more light into the eyes, enhancing peripheral vision.

CELLULAR MECHANISM OF ACTION OF CATECHOLAMINES

Epinephrine & norepinephrine bind with α & β adrenergic receptors—Catecholamines exert their diverse effects on their target cells by binding with specific *adrenergic receptors* on the membranes of their target cells. Two major types of adrenergic receptors are known: α and β . NE binds mainly with the α receptors, while E can bind with both types. The particular responses of the target organs to each of these two catecholamines depend on the kind and number of receptors present in the cells of the organ. Also, because sympathetic nerve fibers release only norepinephrine, they tend mainly to activate the α receptors. Adrenal medullary secretion, being a mixture of both catecholamines, tends to activate both types of receptors. Many drugs that modify the functions of cardiovascular and respiratory systems exert their effect on the α and β receptors. Recently additional receptor subtypes have been found for β receptors.

Mediation of effects by intracellular second messengers—Activation of these receptors is coupled to intracellular second messengers, such as cyclic-AMP and calcium, and phosphorylation of specific functional proteins within the target cells.

CN: Use red for K and bright colors for C & D.
1. Begin with the material in the upper panel.
2. Color the middle section. Note the four inputs from the brain in the upper left side.

3. In the lower panel, color each response to catecholamines and its respective number or letter in the illustrations, using the two colors denoting norepinephrine (D) and epinephrine (C).

ADRENAL GLANDS.

ADRENAL MEDULLA.

adrenal cortex

CATECHOLAMINES.

EPINEPHRINE (E) (ADRENALINE).

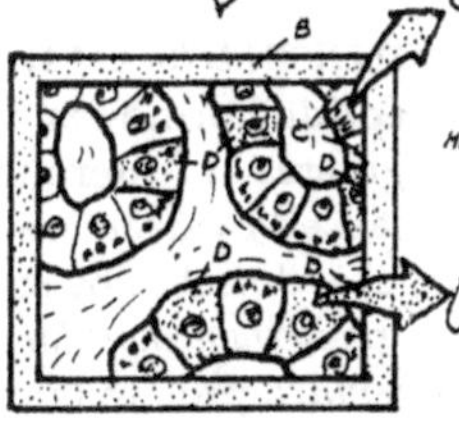

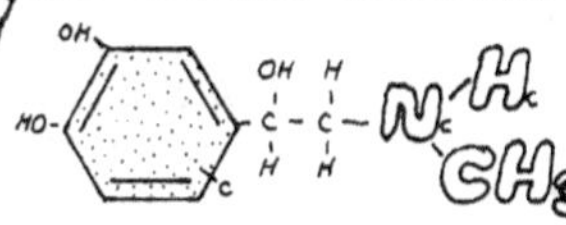

80%

NOREPINEPHRINE. (NE).

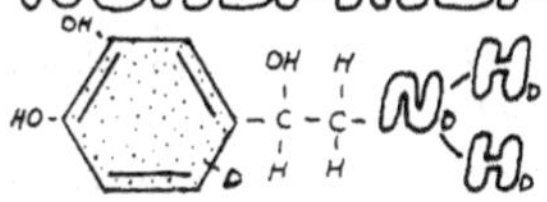

20%

The adrenal medulla, the inner part of the adrenal gland, is part of the sympathetic NS. Chromaffin cells of the adrenal medulla secrete NE and E (catecholamines). These hormones are derived from the amino acid tyrosine, which can be synthesized from phenylalanine. E and NE are stored in vesicles in E and NE cell types and secreted in response to stresses that activate the sympathetic system. E consitutes 80% of the secretion.

STRESS * EMOTIONS * EXERCISE *

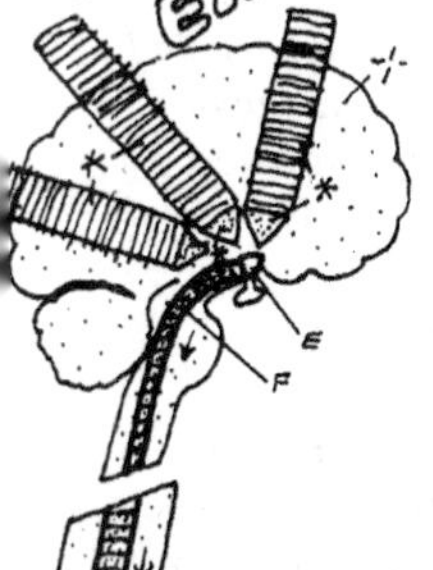

Chromaffin cells of the adrenal medulla are modified postganglionic sympathetic neurons. Sympathetic fibers secrete NE only, while chromaffin cells secrete both E and NE . Stimuli activating the sympathetic NS also activate the adrenal medulla. In target cells, E and NE bind with specific α and β adrenergic receptors. Some tissues have the α type, some β, and some both. E binds more avidly with the β type, while NE binds mostly with the α type. The differential distribution of receptors in tissues underlies the different actions of catecholamines.

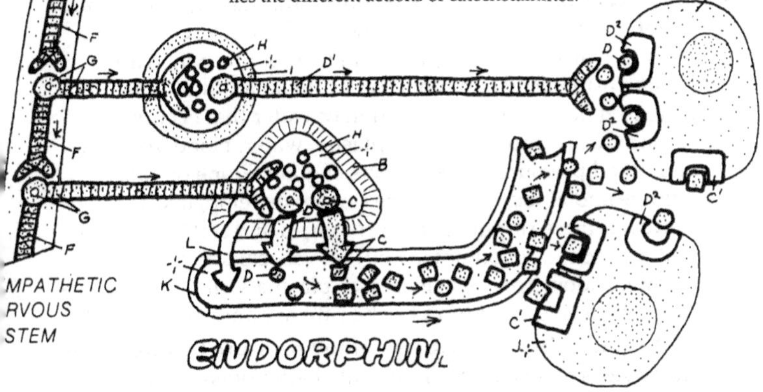

SYMPATHETIC NERVOUS SYSTEM

ENDORPHIN.

BRAIN./ HYPOTHALAMUS.
SPINAL CORD.
PREGANGLIONIC NEURON.
ACETYLCHOLINE.
SYMPATHETIC GANGLION.
POSTGANGLIONIC N..
NOREPINEPHRINE.
TARGET CELL.
ALPHA RECEPTOR.
BETA RECEPTOR.

ADRENAL MEDULLA.
EPINEPHRINE.
NOREPINEPHRINE.
BLOOD VESSEL.

FIGHT, FLIGHT OR EXERCISE.

NOREPINEPHRINE CAUSES.

A VASOCONSTRICTION IN SKIN, KIDNEY, DIGESTIVE TRACT & SPLEEN.
B DECREASES DIGESTIVE ACTIVITY.
C GLYCOGENOLYSIS.
D LYPOLYSIS (FATTY ACID MOBILIZATION)
E INCREASED HEART ACTIVITY.
F BRAIN AROUSAL.
G HAIR ERECTION.
H BLOOD PRESSURE RISE.

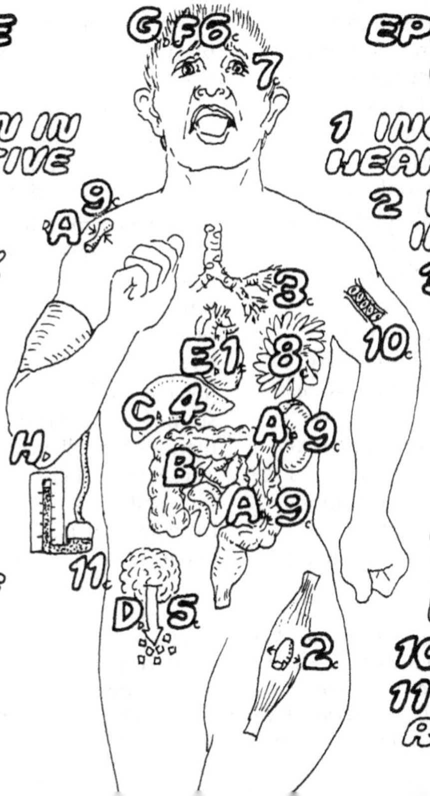

EPINEPHRINE CAUSES.

1 INCREASED HEART ACTIVITY.
2 VASODILATION IN MUSCLE.
3 BRONCHIOLE DILATION.
4 GLYCOGENOLYSIS.
5 LIPOLYSIS.
6 BRAIN AROUSAL.
7 PUPIL DILATION.
8 INCREASED BMR.
9 VASOCONSTRICTION IN SKIN, KIDNEY, ETC.
10 BLOOD CLOTTING.
11 BLOOD PRESSURE RISE.

The adrenal glands are two endocrine glands in one. The inner gland, the adrenal medulla, was considered in plate 125. Here we introduce the *adrenal cortex*, the source of *corticosteroid* hormones that are essential for life, and then focus on the regulation and functions of *aldosterone* hormone.

The adrenal cortex has three zones, each secreting a different type of steroid hormone — The adrenal cortex is divided into three distinct zones, each secreting a specific type of corticosteroid with distinct functions. The outermost zone (*zona glomerulosa*) secretes the hormone *aldosterone*. Aldosterone is a *mineralocorticoid* involved in the regulation of plasma salts (*sodium* and *potassium*), *blood pressure*, and *blood volume*. The middle zone (*zona fasciculata*) secretes *glucocorticoid* hormones, chiefly *cortisol*, which regulates the metabolism of glucose, especially in times of stress (plate 127). The innermost zone of the cortex (*zona reticularis*) secretes *sex steroids*, chiefly androgens (plate 128). In this plate, we focus on aldosterone and its actions in salt balance and blood pressure regulation.

SODIUM AND POTASSIUM ARE ESSENTIAL TO LIFE

Sodium is the chief electrolyte of the plasma and extracellular fluid — Sodium is extremely important for the excitability functions of cell membranes, especially those of the nerve and muscle tissue (plates 10, 11, 15–18). Sodium levels are also critical for regulating plasma and extracellular water volume and for blood pressure. Since reductions in sodium levels are hazardous to bodily functions, numerous factors, including the hormone aldosterone, regulate the plasma sodium level, particularly if it falls below normal levels (~140 mmoles/L).

Potassium is the chief *intracellular* electrolyte — Changes in extracellular potassium concentration markedly influence the resting membrane potential of all body cells and thereby the intracellular concentration of potassium which by itself is critical for cellular enzymes and protein synthesis. Abnormal rise in plasma potassium disturbs cardiac and brain functions and may be fatal. The potassium level in plasma is therefore kept low, within appropriate limits (~4 mmoles/l).

ACTIONS AND REGULATION OF ALDOSTERONE

Aldosterone regulates plasma sodium and potassium by acting on kidney tubules — Indeed, the absence of aldosterone, as occurs with removal of the adrenals (adrenalectomy), is fatal unless followed by proper treatment with hormone or salt therapy. Aldosterone acts mainly on the cells of kidney tubules, stimulating them to synthesize new protein molecules. By acting as enzymes or transporters, these proteins enhance the tubular transport (reabsorption) of sodium from the kidney tubule lumen into the plasma. Aldosterone action on sodium indirectly decreases plasma potassium levels by promoting potassium *secretion* into the kidney tubules, thereby decreasing its levels (plates 65, 69).

A decrease in sodium & an increase in potassium in plasma trigger aldosterone release — These conditions may arise through alterations in the dietary or intestinal intake of these electrolytes. Also, loss of blood, leading to decreased blood volume and blood pressure (as occurs during a hemorrhage), is a strong stimulus for aldosterone release. Increased levels of potassium have a marked and rapid effect on aldosterone secretion, as they act directly on the cells of the zona glomerulosa. In contrast, the mechanism by which a decrease in sodium stimulates aldosterone release is slow because it involves several steps, as described below.

RENIN & ANGIOTENSIN SYSTEM REGULATE ALDOSTERONE

A chain of peptide hormones, involving renin & angiotensin I & II, regulates aldosterone secretion — A decrease in blood pressure caused by decreased sodium intake or blood loss (hemorrhage) is detected by sensors in the kidney arterioles adjacent to the *juxtaglomerular apparatus*, stimulating them to release *renin*, a protein hormone. Renin acts as an enzyme, splitting a large polypeptide called *angiotensinogen*, which is secreted by the liver and normally circulates in the blood. The resultant smaller polypeptide, called *angiotensin I*, is rapidly converted to a still smaller peptide called *angiotensin II* as the blood circulates through the lungs and some other tissues. The *angiotensin converting enzyme* (ACE) in the capillaries of these tissues is responsible for the final conversion. Control of this enzyme is now of great interest in cardiovascular diseases.

Angiotensin II releases aldosterone, which increases sodium reabsorption in kidney tubules — Angiotensin II acts on the cells of the zona glomerulosa, stimulating aldosterone secretion. Aldosterone acts on the renal tubules, enhancing sodium reabsorption. Increased plasma sodium increases plasma osmolarity and blood pressure. In addition, the increased obligatory reabsorption of water that occurs after sodium reabsorption restores plasma water, blood volume, and blood pressure. These effects of aldosterone on blood volume and blood pressure are slow to develop, taking hours; however, these responses are prolonged and stable compared to direct actions of angiotensin II on vasoconstriction (see below). Aldosterone also increases plasma sodium by increasing reabsorbing of sodium by the salivary and sweat glands.

Angiotensin also increases blood pressure by direct arteriolar vasoconstriction — To elevate blood pressure directly, angiotensin II binds with *angiotensin receptors* on the smooth muscle of the arterioles, causing vasoconstriction, which in turn causes increased peripheral resistance. These conditions rapidly increase the blood pressure (plate 47).

Potassium stimulates aldosterone release — Elevated plasma potassium levels, as mentioned above, directly stimulate release of aldosterone by the cells of the zona glomerulosa. Aldosterone decreases potassium levels in the plasma by increasing secretion of this ion in the urine. In the renal tubules, potassium is secreted in exchange for sodium, which is then reabsorbed (plate 65).

Absence of aldosterone may lead to death — The function of aldosterone function in sodium and potassium balance is so essential that loss of the adrenals results in death unless replacement therapy and increased dietary salt and water are provided. Adrenal loss eliminates aldosterone. Loss of aldosterone increases sodium loss in urine, which in turn decreases plasma levels of sodium and increases plasma levels of potassium. These conditions lead to serious heart and brain abnormalities as well as to dehydration and shock. Rats with adrenal insufficiency are known to spontaneously increase their salt intake. Presumably, increased salt appetite results from an increase in the animal's salt taste threshold. Aldosterone is one reason the adrenal cortex is so essential to life.

CN: Use red for E, very light colors for A, B, and C.
1. Begin with the zones of the adrenal cortex.
2. Proceed to "Aldosterone: Sodium Control," and follow the stages from the initial drop in blood volume and pressure (and loss of sodium) to the reabsorption of sodium at the bottom of the page.
3. Color "Aldosterone: Potassium Control," and the diagram of the kidney nephron in the lower right corner, summarizing the effects of aldosterone on both sodium and potassium.

ADRENAL CORTEX
CORTICOSTEROIDS

ADRENAL MEDULLA

ZONA GLOMERULOSA
MINERALOCORTICOID:
ALDOSTERONE

ZONA FASCICULATA
GLUCOCORTICOID:
CORTISOL

ZONA RETICULARIS
SEX STEROIDS:
ANDROGEN
ESTROGEN

ALDOSTERONE: SODIUM CONTROL

Aldosterone increases reabsorption of sodium in renal tubules, elevating plasma sodium and leading to increased blood water, volume, and pressure. A decrease in sodium or blood pressure is detected by the juxtaglomerular apparatus, stimulating the release of renin in the blood. Renin acts as an enzyme, converting angiotensinogen — secreted by the liver — to a shorter polypeptide, angiotensin I. In the lung, angiotensin I is further converted to a shorter and highly potent peptide, angiotensin II.

JUXTA-GLOMERULAR APPARATUS

ANGIOTENSINOGEN

Angiotensin II causes vasoconstriction, elevating blood pressure. Angiotensin II also stimulates the adrenal cortex to release aldosterone. Aldosterone acts on the kidneys to increase sodium reabsorption. Water follows by osmosis. Increased plasma sodium and water elevate blood pressure and volume as well as compensating for decreased sodium.

VASOCON-STRICTION
(RAPID RESPONSE)

BLOOD VOL/PRESSURE

hemorrhage

excessive perspiration

upright posture

afferent arteriole

distal tubule

Na$^+$

RENIN

blood

ANGIOTENSIN I

ENZYME ACTION

lungs

ANGIOTENSIN II

adrenal

kidney

ALDOS-TERONE

Na$^+$ X

Na$^+$ H$_2$O

REABSORPTION

(SLOW RESPONSE)

blood

arteriole

ALDOSTERONE: POTASSIUM CONTROL

Aldosterone acts to decrease the potassium level in plasma. An increase in potassium is detected by the adrenal cortex, stimulating the release of aldosterone. Aldosterone acts on the kidneys to increase the urinary excretion of potassium by exchanging potassium with sodium, which is reabsorbed in the tubules.

EXCRETION

K$^+$

KIDNEY   NEPHRON

afferent arteriole

capillary

collecting duct

proximal tubule

distal tubule

K$^+$

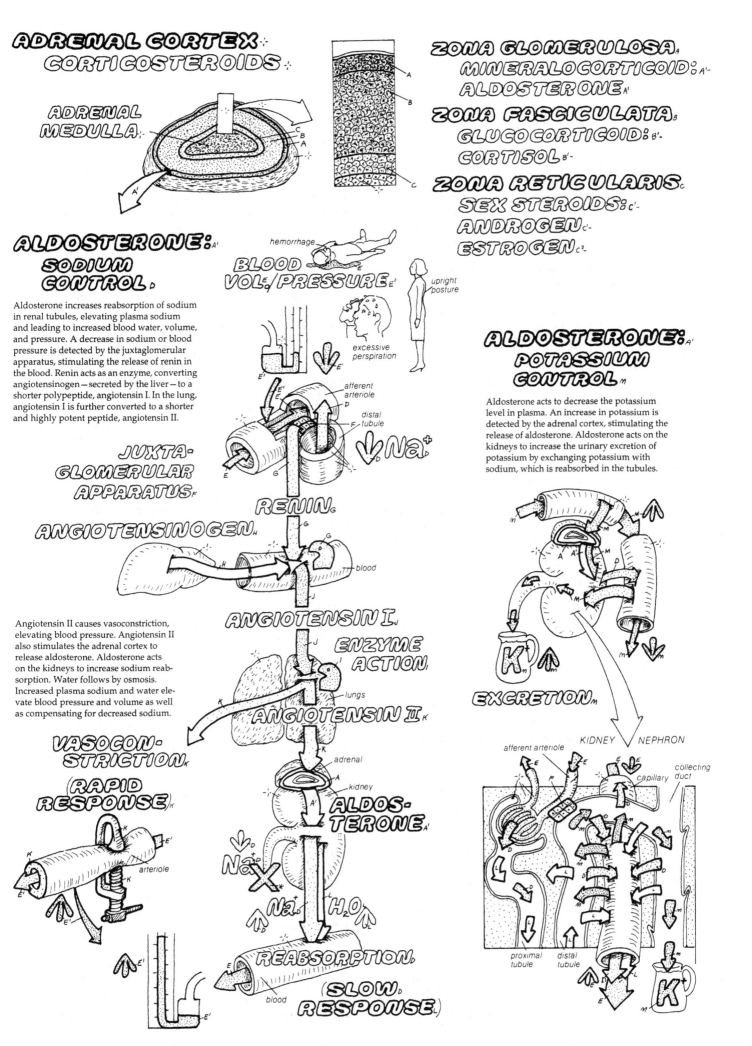

Cortisol is the chief steroid hormone secreted by the cells of the middle part of the Human *adrenal cortex* (*zona fasciculata*), diverse life-saving effects in the body.

Cortisol promotes gluconeogenesis & increases blood glucose — Cortisol's best-known action is to increase blood *glucose* supply for tissues, mainly the brain and heart. It acts by promoting catabolism of *proteins* and by stimulating conversion of the resultant *amino acids* to glucose (*gluconeogenesis*). Gluconeogenesis occurs principally in the *liver*. It is for this role in carbohydrate metabolism that cortisol and similar steroids are called "gluco"-corticoids.

Cortisol, secreted in response to various stresses, is essential for life — Cortisol has numerous other effects in the body. Along with gluconeogenic action, many of these are intimately related to body responses in various "stress" conditions. Some of these responses are short term, exerted in conjunction with the actions of catecholamines from the adrenal medulla. Actions of cortisol in the stress response are exerted independently from the adrenal medulla and are longer lasting. Because cortisol is so important in defending the body against noxious and traumatic stresses, it is considered essential for life. Adrenalectomized animals and humans may die if exposed to sudden unexpected stresses.

REGULATION OF CORTISOL SECRETION

Stress induces CRH & ACTH release from hypothalamus & pituitary — Many of stressful conditions (cold, fasting, starvation, loss of blood pressure [hypotension], hemorrhage, surgery, infections, pain from wounds and fractures, inflammations, severe exercise, and even emotional traumas) can act on the brain to release *CRH* (corticotropin-releasing hormone) from the *hypothalamus* into the hypophysial portal blood. CRH stimulates the release of *ACTH* (corticotropin), a polypeptide hormone, from the *corticotrop cells* of the *anterior pituitary*.

ACTH stimulates cortisol secretion from adrenal cortex — ACTH circulates in the blood and acts on the zona fasciculata of the adrenal cortex, stimulating the synthesis and release of cortisol. Once the cortisol level is sufficiently elevated, CRH and ACTH secretion is decreased through the negative-feedback effect of cortisol on the hypothalamus. This reduces the cortisol level back to the normal baseline condition. When stress is chronic, the brain overrides this control. Continued stimulation of the zona fasciculata by ACTH leads to hypertrophy (excess growth) of this area and enlargement of the adrenal cortex. Other zones remain unaffected.

Effects of cortisol and catecholamines are synergistic — In many instances of *short-term* responses to stress, both cortisol and catecholamines are secreted from the adrenal gland. The increased release of cortisol occurs rapidly, within a few minutes. Although the effects of catecholamines in these instances are well known, those of cortisol are not. Cortisol may promote the effects of catecholamines. For example, the vasoconstriction and fatty acid–mobilizing effects of catecholamines are markedly reduced in the absence of cortisol.

ACTIONS OF CORTISOL IN RESPONSE TO STRESS

Release of cortisol is a physiologic adaptation to stress — The effects of cortisol in promoting *long-term* metabolic adaptation are better known. This adaptation is necessary to improve defenses, promote tissue repair and wound-healing;

providing nutrients by way of glucose and amino acids.

During stress, cortisol helps provide endogenous glucose for the brain & other glucose-utilizing organs — Consider, for example, an injured animal, immobilized with broken bones, or a human stranded in the sea, overcome by starvation, fatigue, sunburn, anxiety, and despair (stress conditions). Food intake being nil, liver and muscle glycogen stores are soon exhausted, threatening the supply of glucose to the brain and the heart. This may have disastrous consequences, because under normal conditions, the brain relies practically entirely on glucose for its energy needs. Adequate supplies of amino acids are also needed for injured tissues undergoing regeneration and repair. The amino acid–mobilizing and gluconeogenic actions of cortisol are essential in combating these stress-related deficiencies.

Cortisol promotes protein catabolism and conversion of amino acids to glucose — Cortisol acts on muscle, bone, and lymphatic tissue, stimulating the catabolism of their labile protein reserves. The "mobilized" amino acids pour into the bloodstream and are taken to the liver, where they are deaminated (the amine group is removed) and converted to glucose (gluconeogenesis). Cortisol also stimulates the synthesis of gluconeogenic enzymes in the liver to stimulate gluconeogenesis. The newly formed glucose enters the blood and ensures an adequate fuel supply for the brain and heart. Cortisol also reduces glucose uptake by muscle tissue, sparing it for the brain and heart.

Amino acids are also used for tissue repair — Not all of the amino acids liberated by tissue catabolism are utilized for gluconeogenesis; some are shunted to tissues that need them for repair and regeneration. Others are used in the liver for synthesis of blood proteins necessary for survival. Under the influence of cortisol and catecholamines, the triglycerides of the fat cells are broken down and fatty acids are mobilized. The latter can be used by the muscle, heart, and liver (but not the brain) for energy.

Excessive cortisol can cause certain stress diseases — In *chronic stress*, excess cortisol may have detrimental and harmful effects. Thus, atrophy of lymphatic nodes, reduction in white blood cells (decreased immunity), hypertension and vascular disorders, and possibly stomach ulcers occur after severe and prolonged stresses.

High doses of cortisol have therapeutic effects — Large doses of cortisol (*pharmacologic doses*) have therapeutic effects against tissue *inflammations* produced by wounds, allergies, or *rheumatoid arthritis* (joint disease). It is not known how these pharmacologic effects of cortisol are exerted or whether they occur during "physiological" defenses.

Cortisol shows a diurnal secretion cycle — Normally, the secretion of cortisol shows a "diurnal" (daily) cycle, the secretion rate being highest in the morning and lowest in the evening. This diurnal cycle is regulated by centers in the hypothalamus and is independent of stress (plate 107).

Permissive actions of cortisol — Several actions of cortisol are "permissive." Thus, cortisol must be present for glucagon and growth hormone to exert their actions on the liver (glycogenolysis) and adipose tissue (lipolysis) and for catecholamines to cause vasoconstriction.

CN: Use red for C, yellow for I, and the same color as the previous page for the zona fasciculata (A).
1. Begin in the upper left corner by coloring the area of the adrenal cortex that secretes cortisol (A[1]). Then begin the numbered sequence of events in which stress causes a drop in blood pressure (C).
2. Color the numbered sequence) of the metabolic response to cortisol secretion.
3. Below, color the titles, tablet, capsule, and hypodermic needle (vehicles for delivery of cortisol).

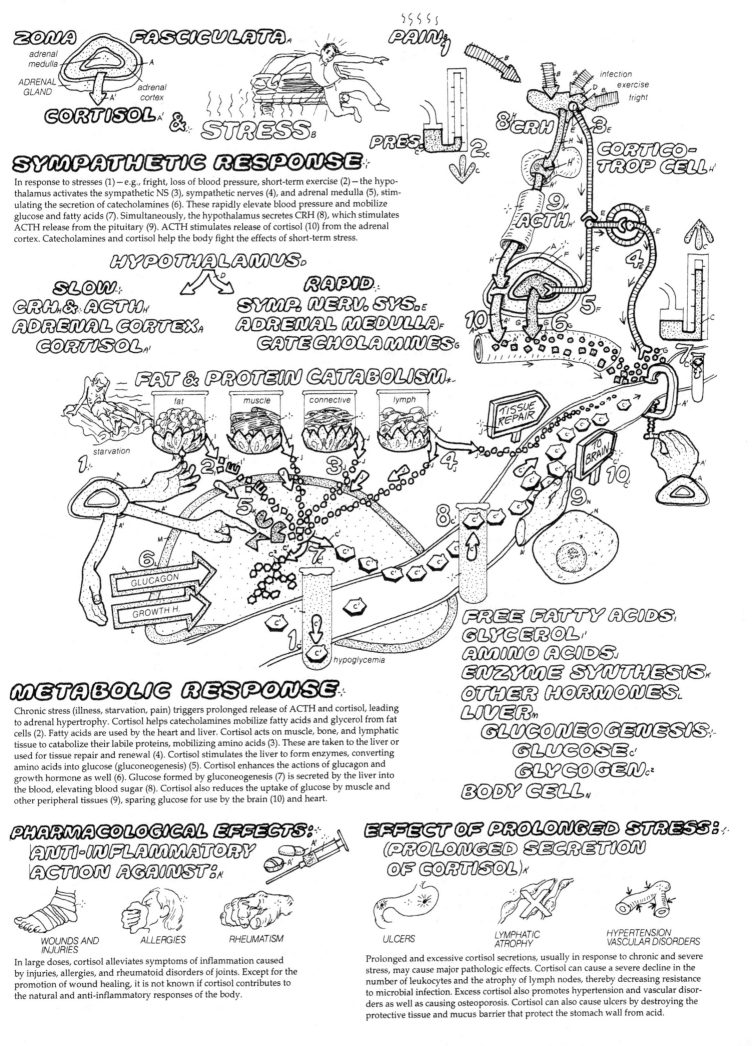

ZONA FASCICULATA

adrenal medulla
ADRENAL GLAND
adrenal cortex

CORTISOL & STRESS

PAIN
PRES.
CRH
ACTH
CORTICO-TROP CELL
infection
exercise
fright

SYMPATHETIC RESPONSE

In response to stresses (1) – e.g., fright, loss of blood pressure, short-term exercise (2) – the hypothalamus activates the sympathetic NS (3), sympathetic nerves (4), and adrenal medulla (5), stimulating the secretion of catecholamines (6). These rapidly elevate blood pressure and mobilize glucose and fatty acids (7). Simultaneously, the hypothalamus secretes CRH (8), which stimulates ACTH release from the pituitary (9). ACTH stimulates release of cortisol (10) from the adrenal cortex. Catecholamines and cortisol help the body fight the effects of short-term stress.

HYPOTHALAMUS

SLOW
CRH & ACTH
ADRENAL CORTEX
CORTISOL

RAPID
SYMP. NERV. SYS.
ADRENAL MEDULLA
CATECHOLAMINES

FAT & PROTEIN CATABOLISM

fat   muscle   connective   lymph

starvation

TISSUE REPAIR

TO BRAIN

GLUCAGON
GROWTH H.

hypoglycemia

FREE FATTY ACIDS
GLYCEROL
AMINO ACIDS
ENZYME SYNTHESIS
OTHER HORMONES
LIVER
GLUCONEOGENESIS
GLUCOSE
GLYCOGEN
BODY CELL

METABOLIC RESPONSE

Chronic stress (illness, starvation, pain) triggers prolonged release of ACTH and cortisol, leading to adrenal hypertrophy. Cortisol helps catecholamines mobilize fatty acids and glycerol from fat cells (2). Fatty acids are used by the heart and liver. Cortisol acts on muscle, bone, and lymphatic tissue to catabolize their labile proteins, mobilizing amino acids (3). These are taken to the liver or used for tissue repair and renewal (4). Cortisol stimulates the liver to form enzymes, converting amino acids into glucose (gluconeogenesis) (5). Cortisol enhances the actions of glucagon and growth hormone as well (6). Glucose formed by gluconeogenesis (7) is secreted by the liver into the blood, elevating blood sugar (8). Cortisol also reduces the uptake of glucose by muscle and other peripheral tissues (9), sparing glucose for use by the brain (10) and heart.

PHARMACOLOGICAL EFFECTS:
ANTI-INFLAMMATORY ACTION AGAINST:

WOUNDS AND INJURIES   ALLERGIES   RHEUMATISM

In large doses, cortisol alleviates symptoms of inflammation caused by injuries, allergies, and rheumatoid disorders of joints. Except for the promotion of wound healing, it is not known if cortisol contributes to the natural and anti-inflammatory responses of the body.

EFFECT OF PROLONGED STRESS:
(PROLONGED SECRETION OF CORTISOL)

ULCERS   LYMPHATIC ATROPHY   HYPERTENSION VASCULAR DISORDERS

Prolonged and excessive cortisol secretions, usually in response to chronic and severe stress, may cause major pathologic effects. Cortisol can cause a severe decline in the number of leukocytes and the atrophy of lymph nodes, thereby decreasing resistance to microbial infection. Excess cortisol also promotes hypertension and vascular disorders as well as causing osteoporosis. Cortisol can also cause ulcers by destroying the protective tissue and mucus barrier that protect the stomach wall from acid.

ADRENAL SEX STEROIDS

Adrenal cortex also produces sex steroids — Cells of the inner part of the adrenal cortex (*zona reticularis*) secrete *sex steroids*, principally *androgens* and small amounts of *estrogen* and *progesterone*. The major adrenal androgen secreted is de-hydro-epi-androsterone (DHEA), a 17-ketosteroid. DHEA is a testosterone precursor and can be converted to testosterone in certain peripheral target tissue. Adrenal androgens have five times less potency than *testosterone*, the major male sex steroid secreted by the *testes*. In adults, the secretion of adrenal sex steroids is stimulated by ACTH and not by the pituitary gonadotropins.

Actions of adrenal androgens in the female — The adrenal androgens are the main source of the male sex steroids in females. Adrenal androgens may contribute to libido (sex drive) in women. They stimulate growth and maintenance of pubic and axial hair. Under normal conditions, adrenal sex androgens exert mainly anabolic effects in females. These androgens stimulate formation of red blood cells and help the epiphyseal closure of long bones, terminating a girl's growth.

Adrenal sex steroids in males — The function of adrenal androgens in the adult male is probably unimportant, because of the presence of large amounts of the testicular androgen, testosterone (plate 152). The secretion of estrogen, the female sex steroid, from the adrenal cortex is minor, but some of the adrenal androgen is converted to estrogen in the blood or peripheral tissues, accounting for the estrogen in male blood.

Adrenal androgens in children (adrenal adrenarche) — In children, a surge in the secretion of adrenal androgens (DHEA) occurs, with an onset (*adrenarche*) between 8 to 10 years and a peak by about 20 years, with no sex difference in the pattern of rise or age. This pubertal surge in adrenal androgens is due either to enzymatic changes in the zona reticularis cells or to the secretion of a possible special tropic hormone from the anterior pituitary (*adrenal androgen stimulating hormone*). This surge may exert significant effects on puberty; in girls it may stimulate bone and muscle growth and help terminate bone growth by causing epiphyseal closure. The adrenal androgens and their peripheral conversion to estrogens may contribute to the fat accumulation and distribution in pubescent children.

DISORDERS OF THE ADRENAL CORTEX

Disturbances in the secretion of adrenal steroid hormones, caused by atrophy, tumors, or enzymatic abnormalities in the cells of the adrenal cortex, bring about dramatic changes in the individual. These changes provide some of the classic demonstrations of pathological effects due to absence or excesses of the hormones.

Adrenogenital syndrome is caused by abnormalities in steroid synthetic enzymes — Normally, adrenal androgens have little masculinizing effect, as evident by the fact that eunuchs (males without testes) have female-like appearance, even though they still have adrenal androgens. Occasionally, however, owing to the growth of tumors or cellular (enzymatic) disorders, the adrenal cortex begins to secrete large amounts of androgens. For example, enzymes that normally convert androgens to cortisol in the adrenal cortex may become deficient. As a result, instead of cortisol, adrenal cells secrete androgens. However, absence of cortisol in the blood triggers secretion of ACTH by negative feedback; stimulating the adrenal to secrete more androgen. Soon a vicious cycle is set up, flooding the body with adrenal androgens.

Adrenogenital syndrome involves the development of male appearance in females — Increased circulating androgens in mature women results in the appearance of secondary male sexual characteristics such as body and facial hair, muscular growth, male body configuration (due to differential fat distribution), and voice and genital changes, creating the striking clinical picture of *adrenogenital syndrome*. Similar effects may be seen in young girls, in which case a precocious male-type *pseudopuberty* is observed (*virilism*). In young boys, this condition causes precocious development of external male characteristics in the absence of testicular development. The accelerated growth of bones and muscle in these boys often leads to stunted stature because of premature fusion of epiphyseal plates (plate 121).

Cushing's syndrome: effects of cortisol excess — Excessive secretion of cortisol, caused either by adrenal tumors or by ACTH-secreting pituitary tumors, leads to the development of *Cushing's syndrome* (disease). The excess secretion of cortisol causes catabolism of proteins, wasting of muscles, and fatigue. Decreased protein synthesis and increased protein breakdown in bones lead to weakening of the bone matrix (osteoporosis). Loss of connective tissue in the skin leads to bruises and poor wound healing. Blood pressure and blood sugar levels are markedly increased.

Body fat is redistributed from lower to upper parts, including the abdomen, back, neck, and face, giving rise to the "buffalo torso" appearance. In the face, loss of subcutaneous connective tissue causes edema. Along with deposited fat, this condition, produces the characteristic "moon face." The illness is often accompanied by behavioral and mental disorders ranging from simple euphoria to full-blown psychosis.

Addison's disease: Effects of adrenocortical insufficiency — Sometimes as a result of cancer or infectious diseases (tuberculosis) or in some autoimmune diseases, the adrenal cortex atrophies, resulting in diminished secretions of adrenal steroid hormones. This condition, called *Addison's disease*, is a very serious clinical disorder that, if untreated, can lead to death. Decreased secretion of aldosterone results in loss of sodium and water, leading to loss of blood pressure, dehydration, and cardiovascular and neurologic abnormalities.

Decreased secretion of cortisol diminishes the gluconeogenic ability of the liver. Hence, blood sugar cannot be raised in fasting. Decreased cortisol diminishes resistance to stress both because of the lack of direct protective actions of cortisol in the body (e.g., reduced gluconeogenesis) and by reduced response to catecholamines. As a result, during stress the body becomes almost helpless, succumbing to shock and death in response to such simple stresses as cold or hunger. But, most patients, if untreated, die because of inability to fight stresses caused by infectious agents (e.g., bacteria).

The decreased cortisol levels in Addison's victims lead to increased secretion of ACTH as well as MSH (melanocyte-stimulating hormone), which is co-produced by pituitary corticotrops. These hormones increase skin pigmentation, one of the classic signs of Addison's disease.

CN: Use red for G, yellow for F, and light brown for H. Use the same colors for the zona reticularis (A), cortisol (D) and aldosterone (I) as on the previous two plates.
1. Color the upper panel dealing with sex steroids, noting that the ovary (C) and testis (B¹) receive the same color as their principal hormone.

2. Color the three examples of disorders of the adrenal cortex, beginning with adrenogenital syndrome. For the woman on the left, color each symbol representing an increase or decrease. In the case of Addison's disease, begin with the total adrenocortical shutdown in the upper left corner.

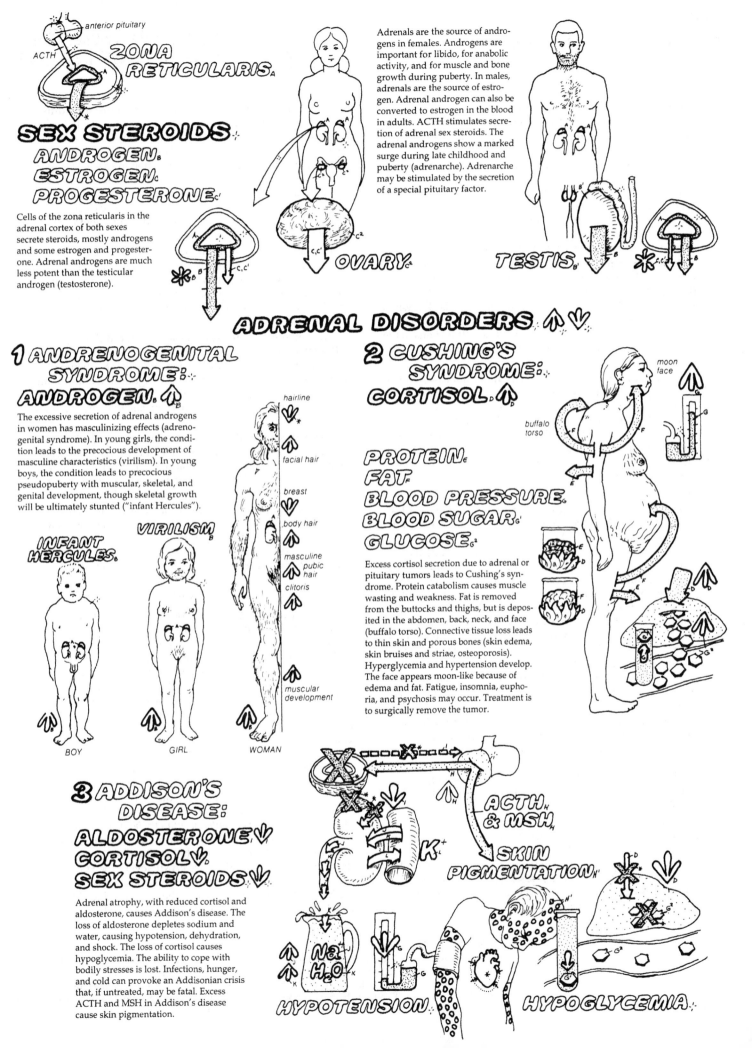

anterior pituitary

ACTH

ZONA RETICULARIS.

SEX STEROIDS.
ANDROGEN.
ESTROGEN.
PROGESTERONE.

Cells of the zona reticularis in the adrenal cortex of both sexes secrete steroids, mostly androgens and some estrogen and progesterone. Adrenal androgens are much less potent than the testicular androgen (testosterone).

Adrenals are the source of androgens in females. Androgens are important for libido, for anabolic activity, and for muscle and bone growth during puberty. In males, adrenals are the source of estrogen. Adrenal androgen can also be converted to estrogen in the blood in adults. ACTH stimulates secretion of adrenal sex steroids. The adrenal androgens show a marked surge during late childhood and puberty (adrenarche). Adrenarche may be stimulated by the secretion of a special pituitary factor.

OVARY.

TESTIS.

ADRENAL DISORDERS ⇑ ⇓

1 ANDRENOGENITAL SYNDROME:
ANDROGEN. ⇑

The excessive secretion of adrenal androgens in women has masculinizing effects (adrenogenital syndrome). In young girls, the condition leads to the precocious development of masculine characteristics (virilism). In young boys, the condition leads to precocious pseudopuberty with muscular, skeletal, and genital development, though skeletal growth will be ultimately stunted ("infant Hercules").

INFANT HERCULES.

VIRILISM.

hairline ⇓

facial hair ⇑

breast ⇓

body hair ⇑

masculine pubic hair ⇑

clitoris ⇑

muscular development ⇑

BOY

GIRL

WOMAN

2 CUSHING'S SYNDROME:
CORTISOL. ⇑

moon face

buffalo torso

PROTEIN.
FAT.
BLOOD PRESSURE.
BLOOD SUGAR.
GLUCOSE.

Excess cortisol secretion due to adrenal or pituitary tumors leads to Cushing's syndrome. Protein catabolism causes muscle wasting and weakness. Fat is removed from the buttocks and thighs, but is deposited in the abdomen, back, neck, and face (buffalo torso). Connective tissue loss leads to thin skin and porous bones (skin edema, skin bruises and striae, osteoporosis). Hyperglycemia and hypertension develop. The face appears moon-like because of edema and fat. Fatigue, insomnia, euphoria, and psychosis may occur. Treatment is to surgically remove the tumor.

3 ADDISON'S DISEASE:
ALDOSTERONE ⇓
CORTISOL ⇓
SEX STEROIDS ⇓

Adrenal atrophy, with reduced cortisol and aldosterone, causes Addison's disease. The loss of aldosterone depletes sodium and water, causing hypotension, dehydration, and shock. The loss of cortisol causes hypoglycemia. The ability to cope with bodily stresses is lost. Infections, hunger, and cold can provoke an Addisonian crisis that, if untreated, may be fatal. Excess ACTH and MSH in Addison's disease cause skin pigmentation.

K^+

ACTH & MSH.

SKIN PIGMENTATION.

Na^+ H_2O

HYPOTENSION.

HYPOGLYCEMIA.

THE CONCEPT OF LOCAL HORMONES

Local hormones are released into the tissue environment to act as autocrine or paracrine chemical agents—Local or tissue hormones are specific, highly active, usually short-lived chemical messengers released by cells into their tissue environment (extracellular fluid) in order to act on the same or other cells in the immediate vicinity. The local hormones that act on the same cells from which they are released are called *autocrines* or *autacoids*; those acting on other cells are called *paracrines* (plate 113). Local hormones may act independently or mediate actions of systemic hormones.

Prostaglandins are typical local hormones—Among the substances known to act as local hormones are *prostaglandins* (PGs) and related substances (*thromboxane* and *leukotriens*). Substances such as serotonin and histamine are also known to act occasionally as local hormones (e.g., in the blood or stomach mucosa). Some disorders are linked to malfunctions in the local hormones. Several major drugs (e.g., aspirin) interfere with the action of local hormones.

Growth factors may also act as local hormones—Actions of some growth factors (nerve growth factor, epidermal growth factor, insulin-like growth factors) also have been described in terms of local hormones. Some factors are released in response to stimulation by a systemic hormone to exert their action as a local hormone: the action of growth hormone on some tissues is mediated by release of insulin-like growth factor (IGF), which exerts a local hormone effect.

PROSTAGLANDINS: STRUCTURE, FORMATION, & FUNCTIONS

Prostaglandins are synthesized from arachidonic acid—Prostaglandins (PGs) are closely related substances released by certain body cells. The PGs are 20-carbon complex fatty acids with a hydrocarbon ring, derived from the enzymatic modification of *arachidonic acid*, a 20-carbon unsaturated fatty acid. Arachidonic acid is a component of plasma membrane phospholipids. A membrane-bound *lipase* enzyme (phospholipase A) hydrolyzes the membrane phospholipids to release arachidonic acid. Different body cells, employing various enzymes, then use arachidonic acid to form the different PGs. Major PGs are PG-Es and PG-Fs, but other PGs (PG-A to PG-I) are also known. Arachidonic acid itself is an essential fatty acid and must be provided in the diet, as it not made in the body; dietary deficiency of this *essential fatty acid nutrient* can lead to illness, partly because of prostaglandin deficiency.

Prostaglandins induce contractions of uterine & intestinal smooth muscle—This action may be important in *sperm transport* in the female reproductive tract. PGs released intrinsically by the uterine tissue are also important in uterine contractions during *parturition*; in fact, certain PGs acting as drugs induce *miscarriage*. PGs also stimulate contraction of smooth muscles in other tissues, such as those of the *gut wall*.

Prostaglandins induce vasodilation & bronchiole dilation—In the lung bronchioles, certain PGs induce dilation by causing relaxation of the bronchiole smooth muscle, an effect with therapeutic value in the respiratory disorders of *asthma*. Certain other PGs cause *vasodilation* in the blood vessels; this has proven important in the treatment of the vascular disorder *hypertension*.

Regulation of reproductive function by prostaglandins—In addition to the actions involving smooth muscle, PGs play numerous roles in other body tissues. Some of these actions may be in conjunction with the *endocrine hormones*; others may be independent of them. Thus, in addition to sperm transport and parturition, PGs are important in several other aspects of reproductive functions such as *follicular growth, ovulation, embryonic implantation*, and *corpus luteum atrophy* (plates 153, 157). PGs are also known to be involved in the generation of premenstrual tension syndrome (PMS).

Prostaglandins play a role in hypothalamic temperature regulation—Release of PGs in the hypothalamus raises *body temperature*. When excessive, this response leads to *fever* (plate 141). The antipyrogenic (fever-reducing) action of aspirin involves inhibition of the PG-forming enzyme in the hypothalamus.

Prostaglandins & leukotriens are produced during inflammatory responses—A well-known case is the release of prostaglandins in arthritic disorders of the joints (rheumatic and osteo-arthritis) (plate 146). The analgesic and anti-inflammatory effects of aspirin and related compounds against these painful disorders are caused in part by *inhibiting* the enzymes that form the PGs.

Some prostaglandins stimulate platelet aggregation, while others inhibit it—Aggregation of platelets is important in the process of blood clot formation. Certain PGs prevent *clot formation* by inhibiting platelet aggregation, while other PGs (thromboxanes) promote *clotting* (plate 145).

Prostaglandins inhibit stomach acid secretion—This action has important implications for the treatment of stomach ulcers. Aspirin and related drugs are known to cause or worsen stomach ulcers. An effect thought to be exerted by aspirin inhibiting certain PG-forming enzymes which increases acid formation by the parietal cells, leading to ulcers in the stomach and duodenal walls (plates 73, 81).

PROSTAGLANDINS & CYCLIC-AMP MAY INTERACT IN ENDOCRINE HORMONE ACTIONS

Some PG actions are exerted in conjunction with those of the endocrine hormones. Also, an intimate interaction seems to exist between the PGs (released extracellularly) and the second messengers, like cAMP and cGMP (released intracellularly). Thus, certain hormones (first messengers), on reaching their target, activate receptor mechanisms, causing the release of PGs in the extracellular medium. These PGs in turn activate, in the same or nearby cells, the membrane enzyme *adenyl cyclase*. This increases cAMP levels, bringing about the action of the hormones (first messengers). In this way, PGs can *amplify* or *antagonize* the action of a systemic hormone in the tissue environment, depending on the type of PGs and the second messenger (cAMP or cGMP) involved. Thus, some PGs mimic the effects of *anterior pituitary hormones*, particularly those that increase cAMP levels in the target cell (e.g., TSH, ACTH, and prolactin) (plate 114). In other cases, where the pituitary hormones decrease cAMP levels, the PGs *antagonize* the action of these hormones. Thus, PGs cause *diuresis* in the kidney tubule, the opposite of the effect produced by ADH, the *posterior pituitary hormone* (plates 66, 116).

CN: Use red for C, a bright color for E^2, very light colors for A, D, and G, and a dark color for B.
1. Begin in the upper left corner and complete the diagram down to the large central arrow (E^2). Note that local hormones (E) refer to both paracrine (E^1) and autocrine (F). Prostaglandins (E^2) are given a paracrine color on this page.
2. Complete the various actions of PGs and the related local hormones (thromboxanes [E^3] and leukotriens [E^4]). Do not color the illustrations, but do color the PG arrows and the symbols of increase and decrease.

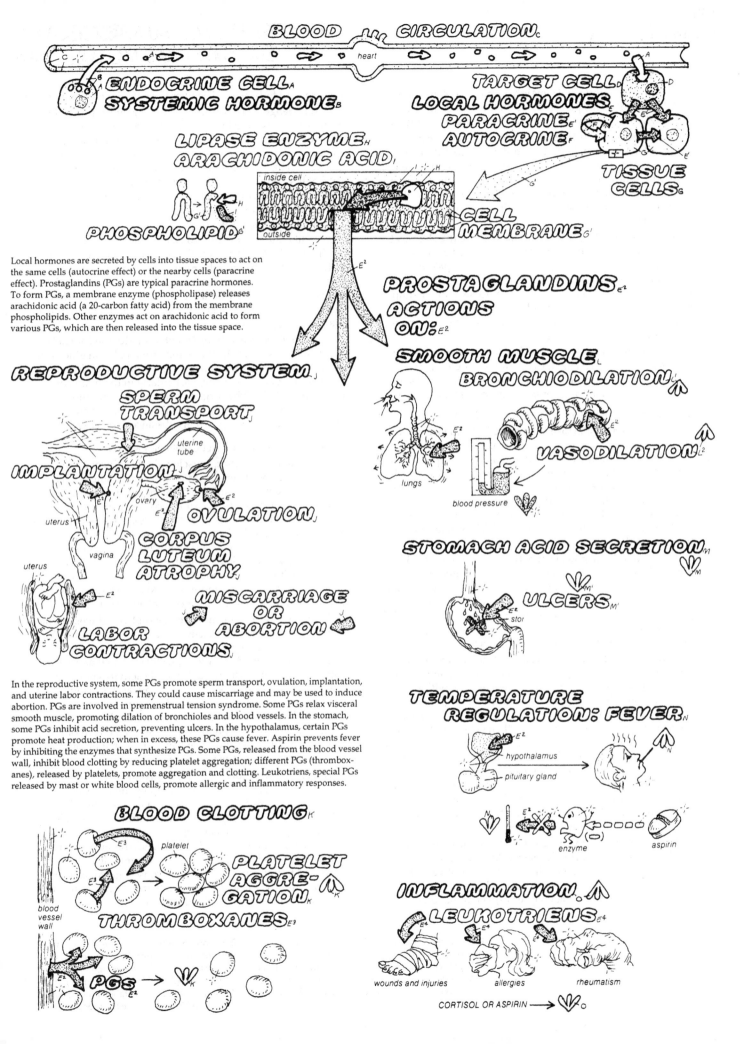

BLOOD CIRCULATION

heart

ENDOCRINE CELL
SYSTEMIC HORMONE

TARGET CELL
LOCAL HORMONES
PARACRINE
AUTOCRINE

TISSUE CELLS

LIPASE ENZYME
ARACHIDONIC ACID

inside cell
outside

CELL MEMBRANE

PHOSPHOLIPID

Local hormones are secreted by cells into tissue spaces to act on the same cells (autocrine effect) or the nearby cells (paracrine effect). Prostaglandins (PGs) are typical paracrine hormones. To form PGs, a membrane enzyme (phospholipase) releases arachidonic acid (a 20-carbon fatty acid) from the membrane phospholipids. Other enzymes act on arachidonic acid to form various PGs, which are then released into the tissue space.

PROSTAGLANDINS
ACTIONS
ON:

REPRODUCTIVE SYSTEM

SPERM TRANSPORT

uterine tube

IMPLANTATION

ovary

OVULATION

uterus

vagina

CORPUS LUTEUM ATROPHY

MISCARRIAGE OR ABORTION

LABOR CONTRACTIONS

SMOOTH MUSCLE
BRONCHIODILATION

lungs

VASODILATION

blood pressure

STOMACH ACID SECRETION

ULCERS

stom

In the reproductive system, some PGs promote sperm transport, ovulation, implantation, and uterine labor contractions. They could cause miscarriage and may be used to induce abortion. PGs are involved in premenstrual tension syndrome. Some PGs relax visceral smooth muscle, promoting dilation of bronchioles and blood vessels. In the stomach, some PGs inhibit acid secretion, preventing ulcers. In the hypothalamus, certain PGs promote heat production; when in excess, these PGs cause fever. Aspirin prevents fever by inhibiting the enzymes that synthesize PGs. Some PGs, released from the blood vessel wall, inhibit blood clotting by reducing platelet aggregation; different PGs (thromboxanes), released by platelets, promote aggregation and clotting. Leukotriens, special PGs released by mast or white blood cells, promote allergic and inflammatory responses.

TEMPERATURE
REGULATION: FEVER

hypothalamus
pituitary gland

enzyme

aspirin

BLOOD CLOTTING

platelet

PLATELET AGGRE-GATION

blood vessel wall

THROMBOXANES

PGS

INFLAMMATION
LEUKOTRIENS

wounds and injuries   allergies   rheumatism

CORTISOL OR ASPIRIN ⟶

The body uses *carbohydrates* mainly as fuel to obtain energy (ATP and heat). This plate focuses on the metabolic physiology of *glucose* and how it is handled by the liver and muscle.

Starches, fruits, milk, & table sugar as dietary carbohydrates — Dietary carbohydrates are starches, found in foods such as bread, rice, pasta, and potatoes. In Western societies, nearly half the daily caloric intake is from carbohydrates; in developing countries, carbohydrates provide most if not all of the calories. Fruits, beans, and milk are also sources of carbohydrates.

Poly-, oligo-, di-, & mono-saccharides are the chemical forms of carbohydrates — Carbohydrates may be *simple sugars* (six-carbon *monosaccharides*, principally *glucose, galactose,* and *fructose*), *oligosaccharides* (chains of two to ten simple sugars), or *polysaccharides*, larger polymers of glucose or other simple sugars. Polysaccharides occur in starches; *disaccharides* are found in milk (lactose) and table sugar (sucrose). The monosaccharide *fructose* is the sugar in fruits.

Only simple sugars are absorbed — All carbohydrates are digested by intestinal enzymes into three simple sugars: glucose, galactose, and fructose. These are absorbed across the *intestinal mucosa* and transported via the *portal vein* to the liver.

The Liver Is the Body's Glucostatic Organ

Liver converts galactose & fructose into glucose — Simple sugars freely enter the liver cells, where galactose and fructose are enzymatically converted to glucose. This process is very efficient; the only sugar normally found in blood is glucose. During the absorptive phase, the absorbed glucose can enter the blood directly. At other times, the liver is the only source of blood glucose. The glucose pool in the liver can be easily exchanged with that in the blood. The tissues obtain the glucose they need from the blood glucose pool.

Liver can release glucose into the blood — When the blood glucose level is low, the liver releases glucose into the blood; when the glucose level is high, the liver cells take up and store glucose. Thus, through its various storage and conversion processes involving glucose, and through its exceptional ability to release glucose, the liver acts as a *glucostat*, helping to maintain the blood sugar within normal levels.

Liver stores glucose as glycogen — After a meal rich in carbohydrates, blood sugar is elevated, resulting in increased glucose uptake by the liver cells. Excess glucose within liver cells promotes the incorporation of glucose into *glycogen*, a polymer of glucose, via a process called *glycogenesis*. This is how animals store excess glucose in their cells, chiefly those of the liver and muscle. Glucose residues in glycogen are bound together along branched chains, forming a treelike structure, the "glycogen tree." Excess glycogen can precipitate in the cytoplasm to form glycogen granules, found abundantly in liver and muscle cells. When the pool of free glucose in the liver cells diminishes, glycogen is partly broken down, by a process called *glycogenolysis*, to release free glucose.

Liver can convert excess glucose to proteins and fats — The liver's ability to form glycogen is limited. As a result, the extra glucose entering the liver in the absorptive and early postabsorptive phase is converted to amino acids and proteins as well as to fats (triacylglycerols) via formation of

glycerol and fatty acids. The liver is an efficient fat maker.

Gluconeogenesis: forming glucose from protein & fat — To make glucose, the liver degrades *proteins* into *amino acids,* some of which (e.g., alanine) undergo *deamination* to form pyruvic acid, which can be converted to glucose by *reverse glycolysis*. This process of *gluconeogenesis* is carried out by special liver enzymes and is a major source of new and endogenous glucose for the liver and blood, particularly during fasting and starvation (plate 127).

Another source for the formation of new glucose is the *glycerol* liberated by the breakdown of *triglycerides* (*lipolysis*) in the liver and fat cells. Glycerol molecules can be recombined to form glucose through the reverse steps of glycolysis. The liver is not able to convert fatty acids to glucose, as it lacks the necessary enzymes. *Lactic acid* is another source for endogenous glucose, being converted to *pyruvate* and then to glucose via reverse glycolysis.

Tissues & organs rely differently on glucose for fuel — Some tissues, such as the brain, rely principally on glucose for their energy needs. Depriving the brain of glucose leads to serious irreversible damage, particularly in the brain cortex tissue (plate 112). Other organs, such as the heart and skeletal muscle, prefer to use glucose for this purpose but are also equipped to use alternative fuel substances, such as *fatty acids*.

How Muscle Utilizes Glucose for Energy

In oxygen abundance, muscle oxidizes glucose completely to CO_2 & water — In an actively exercising muscle, glucose is taken up rapidly from the blood and converted to *glucose-6-phosphate* (G-6-P). G-6-P is converted to pyruvate by the enzymes of glycolysis (*aerobic glycolysis*) and, when oxygen is available, to CO_2 and water by the enzymes of the Krebs cycle in mitochondria. The glycolytic breakdown of glucose to pyruvate yields a small amount of ATP (2 ATP/glucose). Mitochondrial oxidation of pyruvate to CO_2 and water yields a great deal more ATP (38 ATP/Glu) (plate 5), which the muscles use for their contraction work (plate 27).

In oxygen deficiency, muscle metabolizes glucose to lactate — When oxygen is deficient, muscle uses pyruvate to form lactic acid (lactate), a process called *anaerobic glycolysis*. This will yield two more ATP, although still far less than that obtained in the mitochondria. If muscle activity continues, lactic acid builds up, leaks into the blood, and is taken up by the liver, where lactate is converted to pyruvate and then to glucose. The production of lactic acid in muscle, its transport to the liver, its conversion to glucose, and the return of the glucose to the muscle and the eventual reformation of muscle lactic acid constitutes the *Cori cycle*.

At rest, muscle stores glucose as glycogen — When the muscle is resting, the glucose taken up by muscle cells is converted to G-6-P. Because the muscle is not utilizing ATP, the G-6-P is used to form glycogen, thus storing the available glucose. During activity, this glycogen is converted back to G-6-P, which is shunted directly for glycolysis. Because the appropriate enzyme is lacking, the G-6-P of the muscle cannot be converted to free glucose. Therefore muscle glycogen can be used only for the muscle's own needs and cannot contribute directly to homeostasis of blood glucose.

CN: Use red for A, light blue for D, blue for G, purple for K, and a bright color for H.
1. Begin with the three types of carbohydrate molecules at the top. Color the glycogen molecule (E). that it receives a different color from the individual glucose (A) molecules.

2. Follow the numbered sequence beginning with the entry of three monosaccharides (A, B, C) into the portal vein in the upper right. Note conversion of galactose (B) and fructose (C) into glucose (A). All stages of glucose metabolism in the active muscle cell below receive the muscle color (L).

MONOSACCHARIDE (SIMPLE SUGAR)

Glucose, galactose, and fructose are the principal monosaccharides. They are six-carbon sugars and are metabolically interconvertible.

OLIGOSACCHARIDE (2-10 MONOSAC.)

Oligosaccharides are sugars containing two or more monosaccharides. Important disaccharides are lactose (glu-gal), sucrose (fru-glu), and maltose (glu-glu).

POLYSACCHARIDE

Polysaccharides are polymers of monosaccharides. Polysaccharides in starches are the main dietary source of carbohydrates for humans. Glycogen, a polymer of glucose with long branched chains, serves as storage for glucose in animal cells.

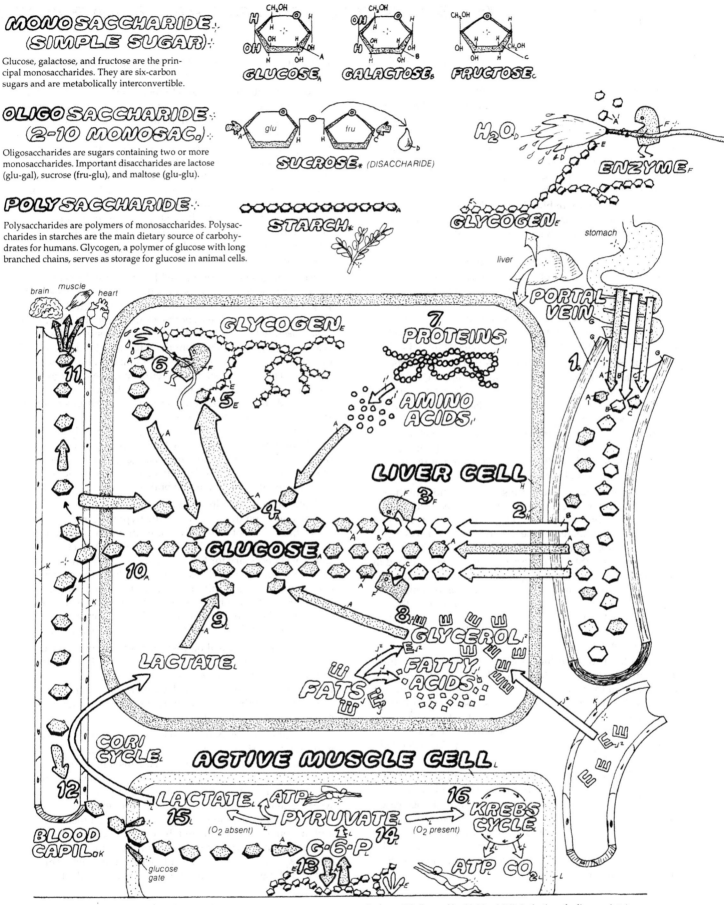

GLUCOSE · **GALACTOSE** · **FRUCTOSE**

SUCROSE (DISACCHARIDE)

H_2O · **ENZYME** · **GLYCOGEN**

STARCH

stomach · liver · **PORTAL VEIN**

GLYCOGEN · **PROTEINS** · **AMINO ACIDS** · **LIVER CELL**

brain · muscle · heart

GLUCOSE

LACTATE · **GLYCEROL** · **FATTY ACIDS** · **FATS**

CORI CYCLE · **ACTIVE MUSCLE CELL**

BLOOD CAPIL · glucose gate

LACTATE · **ATP** · **PYRUVATE** (O_2 absent) · **G-6-P** · **KREBS CYCLE** (O_2 present) · **ATP** CO_2

Carbohydrates are digested in the intestines to form monosaccharides—glucose (Glu), galactose, and fructose (1). These enter the liver via the portal vein (2). In the liver, enzymes (3) convert fructose and galactose to Glu (4). Excess Glu is stored in the liver as glycogen (glycogenesis) (5). Glycogen can be broken down into Glu (glycogenolysis) (6). Glu may also be formed by the conversion of amino acids (gluconeogenesis) (7). Glycerol from the breakdown of triglycerides (lipolysis) can also be a source of new Glu (8). Lactate formed in anaerobic glycolysis is also convertible to Glu (9). The Glu pool in the liver freely exchanges with the pool in the blood (10). In fasting, the liver maintains a constant blood sugar level, ensuring Glu supply to tissues (11). In muscle, Glu is taken up (12) and phosphorylated to Glu-6-P. Glu-6-P is either converted to glycogen (13) or oxidized to pyruvate (glycolysis) (14) to generate ATP. In the absence of oxygen, pyruvate is converted to lactate (anaerobic glycolysis) (15) for additional ATP. Excess lactate moves into the blood and liver to form new Glu (Cori cycle). In the presence of oxygen, pyruvate is utilized by the mitochondrial Krebs cycle (16) for more efficient ATP generation.

For most tissues, *glucose* is the ideal fuel substance for cellular energy production. It is the preferred fuel for the heart and skeletal muscle, and normally is the brain's exclusive fuel.

Blood glucose level is kept constant around 100 mg/dl. Given the central role of the brain and heart in body function and survival, an ample supply of glucose must be provided to these organs at all times. This is accomplished by regulating blood glucose concentration around a presumably optimal level of 1 g/l (80 to 110 mg/dl) of plasma at all ages (5 g/whole blood in adults).

Brain helps regulate blood sugar through behavioral & neurohormonal mechanisms — Along with the purely hormonal mechanisms involving insulin and glucagon (plate 132), the brain, in particular its hypothalamic centers, provides for a complex behavioral and neurohormonal homeostatic system aimed to restore the optimal glucose level whenever it deviates critically from the normal range. Hypoglycemia (low blood sugar) has serious consequences for brain and heart function (see below). Here we focus on the neurobehavioral and neurohormonal mechanisms that elevate the blood sugar level whenever it falls below set limits.

ROLES OF HYPOTHALAMIC FEEDING & SATIETY CENTERS

Hypothalamus has a glucostat mechanism — Certain neurons in the *hypothalamus* that constitute a *glucostatic center* can detect changes in blood sugar levels. These neurons have a high metabolic rate (oxygen and glucose consumption), which permits them to detect changes in the glucose level within their cytoplasm and consequently in the blood (plate 138). These neurons are the only ones in the brain that require insulin for glucose uptake and entry.

Low blood sugar induces hunger — Reductions in the glucose level a few hours after a meal are detected by glucostat neurons, which activate the *hypothalamic feeding (hunger) center*. This center increases the appetite and food-seeking behavior, ultimately leading to increased *food intake*, or ingestion (plates 107, 138). The dietary carbohydrates absorbed in the intestine are converted to glucose in the liver, which releases glucose into the blood. This condition of temporary hyperglycemia stimulates the release of insulin from the pancreatic islets; insulin in turn promotes the entry of glucose into tissues, including the neurons of the hypothalamic glucostatic center.

High blood sugar induces satiety — Hypothalamic glucostat neurons detect high blood sugar levels, and their output signals inhibit the feeding control center and activate the satiety control center of the hypothalamus. As a result, appetite is reduced and a state of satiety prevails, at least for a few hours. Satiety and appetite suppression can also be produced by increased activity of the sensory nerves from the distended stomach after food ingestion.

Neural & hormonal signals from gut wall also control satiety — Food ingestion also triggers release of hormones from the gut wall that act on the hypothalamus to decrease food intake. The duodenal hormone CCK (*cholecystokinin*), among some other gut peptides, is known to exert such short-term feedback effects. Long-term suppression of food intake in the hypothalamus is also provided by *leptin*, a fat tissue hormone (plates 134, 139).

HYPOTHALAMUS COORDINATES RELEASE OF HORMONES THAT ELEVATE BLOOD SUGAR

Roles of catecholamines — In response to a relative fall in blood sugar, which may occur between meals, the glucostatic center initiates a series of actions to counteract this decline and elevate the blood sugar. Thus, the center initially activates the hypothalamic center for control of the *sympathetic nervous center*, leading to the release of *norepinephrine* from sympathetic nerves and *epinephrine* from the adrenal medulla. The catecholamines increase *glycogenolysis* in the liver and *lipolysis* in the adipose tissues. Glycogenolysis directly increases the glucose pool in the liver. Lipolysis provides *glycerol* for conversion to glucose in the liver. Also, such tissues as muscle use the *fatty acids* mobilized from the adipose tissue, sparing glucose for the brain and heart.

Roles of growth hormone — When food intake is delayed for a long time or reduced as a result of fasting or sustained physical exercise, blood sugar falls to its lower limit. These conditions stimulate the hypothalamus to liberate *growth-hormone releasing hormone* (GRH), which in turn stimulates the release of *growth hormone* from the pituitary gland (plate 118). Growth hormone acts on fat cells, mobilizing fatty acids and glycerol. The mechanism for this effect may be through increasing the sensitivly of fat cells to catecholamines. As mentioned above, utilization of fatty acids by tissues results in sparing of glucose; also, glycerol contributes to gluconeogenesis in the liver. As a result, the blood glucose supply increases. In addition, growth hormone acts on muscle tissues to decrease glucose utilization in exchange for an increase in the uptake of *amino acids*. This effect also spares glucose for more essential uses (e.g., in the brain).

Roles of cortisol — In addition to growth hormone, during fasting and hypoglycemic stress, the hypothalamus also stimulates the release of *cortisol* from the adrenal cortex by activating the *corticotropin-releasing hormone* (CRH)–ACTH axis. Cortisol is necessary for the action of growth hormone on fat cells. Cortisol also mobilizes amino acids from the muscle and connective tissue and stimulates their utilization for gluconeogenesis in the liver. Like growth hormone, cortisol inhibits glucose intake by nonessential user tissues, such as skeletal muscle, to spare this sugar for the brain and heart (plate 127).

Minor neural control on insulin & glucagon secretion — Except for the stimulatory role of the vagus in insulin release, the nervous system does not play a major role in the release of the pancreatic hormones, which by themselves carry out much of the routine compensatory mechanism to keep blood glucose constant (plates 123, 132).

Hypoglycemia may have severe consequences in heart and brain function — During prolonged starvation, all the above-mentioned restorative mechanisms fail; blood glucose levels inevitably fall below critical limits as consumption by the heart and brain continues. Below 60 mg/dl, the heart weakens and nervous, cognitive, and conscious activities become disturbed. Below 50 mg/dl, speech becomes slurred and movement uncoordinated. Blood glucose below 30 mg/dl may cause unconsciousness and coma; at 20 mg/l, convulsions may occur, and at 10 mg/dl, permanent brain damage and loss of medullary respiratory centers causes death.

CN: Use red for A and the same colors as on the previous page for glycerol (C), liver (O), and glycogen (P). Use a dark color for B.
1. Begin with number 1 in the upper left corner, indicating a drop in the blood glucose level (A^1). Color this portion of the circulatory system (A, A^1) moving clockwise and down to the middle of the page.
2. Color the hypothalamus (B) and glucostat (A^2).

Color number 2 and follow its path into the liver at the bottom, but do not color the reactions within the liver just yet.
3. Do the same with 3, 4, and 5.
4. Color the conversion of substances in the liver to glucose, and the release of glucose into the blood circulation; follow this compensatory response to low blood sugar.

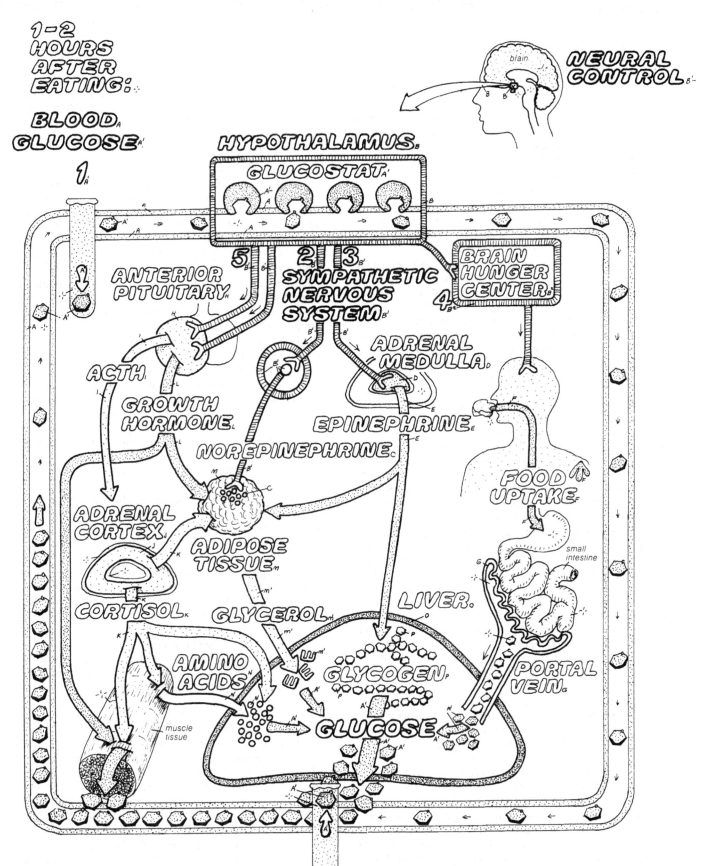

HYPOTHALAMUS.

GLUCOSTAT

1

5 2 3

ANTERIOR PITUITARY

SYMPATHETIC NERVOUS SYSTEM

BRAIN HUNGER CENTER.

4

ADRENAL MEDULLA.

ACTH

GROWTH HORMONE.

NOREPINEPHRINE.

EPINEPHRINE.

FOOD UPTAKE

small intestine

ADRENAL CORTEX

ADIPOSE TISSUE.

LIVER.

CORTISOL

GLYCEROL

AMINO ACIDS

muscle tissue

GLYCOGEN.

PORTAL VEIN.

GLUCOSE

One to two hours after a meal, the continuous use of Glu by tissues, especially the brain, reduces blood Glu (1) levels. This is detected by "glucostat" neurons in the hypothalamus, which initiate a series of compensatory responses. Sympathetic activation liberates norepinephrine from nerve fibers (2) and epinephrine from the adrenal medulla (3). Epinephrine stimulates glycogenolysis in the liver. Both catecholamines promote lipolysis in adipose tissue, providing fatty acids and glycerol. Mobilized fatty acids are used by the heart and muscles, sparing Glu for the brain and raising blood Glu. Glycerol is converted to Glu in the liver. Later, activation of hypothalamic feeding centers (4) leads to

food ingestion, increasing Glu intake and the hepatic Glu pool. If food intake is further delayed, the hypothalamus stimulates release of GH and ACTH from the pituitary (5). GH promotes lipolysis in adipose tissue and decreases Glu uptake in muscle. ACTH stimulates cortisol release. Cortisol facilitates GH action on fat cells, decreases Glu uptake by muscle, and stimulates gluconeogenesis in the liver. Cortisol also increases protein catabolism in muscle to provide amino acids for gluconeogenesis in the liver. Increased Glu output by the liver and decreased Glu uptake by muscle increase blood Glu levels, ensuring Glu availability to vital organs such as the brain and heart.

This plate describes the integration of all the hormones involved in the regulation of blood sugar (glucose). Because hypoglycemia is a potentially life-threatening condition (plate 131), many hormones act to increase the blood sugar level. Only one hormone, insulin, is specifically involved in lowering the blood sugar, and that this action is not the hormone's primary goal but a consequence of its action. Being a center for storage and production of glucose, the liver serves as a target for almost all the hormones involved in blood sugar regulation and carbohydrate metabolism.

HORMONES THAT RAISE BLOOD SUGAR

Hormones that act to raise the blood sugar level are glucagon from the pancreatic islets, epinephrine and norepinephrine from the adrenal medulla, growth hormone from the pituitary, and cortisol from the adrenal cortex.

Epinephrine & glucagon act rapidly to raise blood sugar level—The stimulus for secretion of these hormones is a drop in plasma glucose level, a condition that normally occurs in the interval between meals. Epinephrine and glucagon have a common mechanism of action to increase the blood sugar level. They both stimulate liver cells to increase glycogenolysis, thereby mobilizing liver glucose (plates 123, 125). These two hormones, though different chemically, both act on membrane receptors in the liver that are coupled to G-proteins. Binding activates adenylate cyclase, which increases the concentration of cAMP in the liver cells. Acting as the "second messenger," cAMP, through an amplifying cascade of effects, activates the liver enzyme phosphohydrolase, which acts on the liver glycogen, liberating glucose molecules. Once the pool of free glucose in the liver increases, the excess glucose is secreted into the blood, compensating for the hypoglycemia. The action of glucagon is regulated by a purely hormonal feedback, while that of epinephrine involves mediation of the brain and sympathetic nervous system.

Growth hormone & cortisol act slowly to raise blood sugar level—The need for these hormones arises during times of metabolic stress, such as fasting, strenuous and prolonged exercise, and immobility, when food intake is considerably delayed. Prolonged (but not dangerous) hypoglycemia and stress signals from the brain provide the stimulus for secretion of growth hormone and cortisol. These hormones increase blood sugar indirectly, either by increasing the levels of substrates for gluconeogenesis (e.g., amino acids and glycerol) or by reducing the entry or utilization of glucose in certain tissues (e.g., muscle), thereby sparing blood glucose and raising its level in the blood.

Cortisol increases blood sugar by catabolizing tissue proteins & enhancing liver conversion of amino acids to glucose—Cortisol promotes protein catabolism in peripheral tissues such as the skeletal muscle, liberating amino acids. In addition, cortisol stimulates the synthesis of certain liver enzymes—i.e., those of deamination and gluconeogenesis—which convert the liberated amino acids into glucose. Cortisol also decreases glucose uptake by muscle (plate 127).

Growth hormone increases blood sugar by helping to mobilize fatty acids & sparing glucose use—Growth hormone acts on fat cells of adipose tissue to increase lipolysis of triglycerides (triacylglycerols), mobilizing glycerol and fatty acids (plate 118). Glycerol is converted to glucose in the liver by reverse glycolysis, raising glucose levels in the liver and blood. Meanwhile, the use of fatty acids as fuel by muscle, heart, and liver tissues spares the glucose for consumption by those tissues that are more critically dependent on it (especially the brain). Catecholamines have actions similar to those of growth hormone in adipose tissue and on carbohydrate metabolism (plate 125). However, the action of growth hormone takes longer to develop and lasts longer, being in the long run more effective for survival.

HORMONES THAT LOWER BLOOD SUGAR

Insulin lowers blood sugar by increasing tissue uptake of glucose—The principal hypoglycemic hormone is insulin, produced in the pancreas's islets of Langerhans. Secreted in response to an increase in blood glucose level shortly after meals, insulin increases glucose entry into muscle and fat tissue and promotes glycogen synthesis and storage in the liver; causing a decrease in blood sugar levels (see plate 123).

High levels of thyroid hormones can also lower blood sugar by increasing metabolic rate—Elevated levels of thyroid hormones, as in hyperthyroidism or during long-term adaptation to cold, can also cause hypoglycemia by increasing the metabolic rate, but regulation of blood sugar is not the primary effect (plate 119).

LIVER IN HORMONAL GLUCOSE HOMEOSTASIS

Liver is target of all glucose-regulating hormones — All hormones acting to regulate blood sugar (insulin, glucagon, growth hormone, cortisol, epinephrine) do so in part by acting on the liver. The liver has special membrane and nuclear receptors for all these hormones as well as a variety of intracellular second messengers and signaling systems, such as cyclic AMP, to mediate the effects of the hormones and receptors.

Liver is first to receive absorbed glucose—The liver, via its special connection to the small intestine (portal vein), has direct access to the carbohydrates absorbed from the intestine, making it at once a center for the synthesis, delivery, storage, and production of glucose.

Liver's glucose transporters are not insulin dependent—The liver takes up a large portion of the abundant blood glucose after meals, a task performed by the liver's special glucose transporters. These transporters are not dependent on insulin (insulin-insensitive GluT2).

Liver is storehouse of glycogen—The liver is the major organ regulating the homeostasis of carbohydrate metabolism generally and blood sugar specifically. The liver has enzymes that convert glycogen to glucose and glucose to glycogen. Approximately 500 g of glycogen is stored in the liver. Assuming this to be equal to 500 g of glucose, then the liver has 100 times more glucose than the whole blood.

Only liver can secrete glucose—Because the liver contains a special enzyme, glucose-6-phosphatase, that hydrolyzes the glucose-6-phosphate to free glucose, it is the only organ in the body that can secrete glucose into the blood when the level of glucose in the liver exceeds that in the blood. This gives the liver the unique roles of glucose exchanger and glucostat.

Liver can make glucose from amino acids, glycerol, & lactate but not from fatty acids—The liver also converts glycerol to glucose and the reverse, and amino acids to glucose and the reverse. But the liver cannot synthesize glucose from fatty acids, a characteristic shared by all animal cells.

CN: All the titles to be colored appear on the preceding page and should receive the same colors, though the letter labels may differ.
1. Begin in the upper left corner with the long arrows representing various hormones, glycerol (F), and amino acids (G).
2. Color the three primary users of blood glucose in the upper right.
3. Color the two hormones (thyroid and insulin) that lower glucose levels.
4. Color the hormonal influences on the liver, in the lower panel.

HORMONES THAT RAISE BLOOD GLUCOSE LEVELS

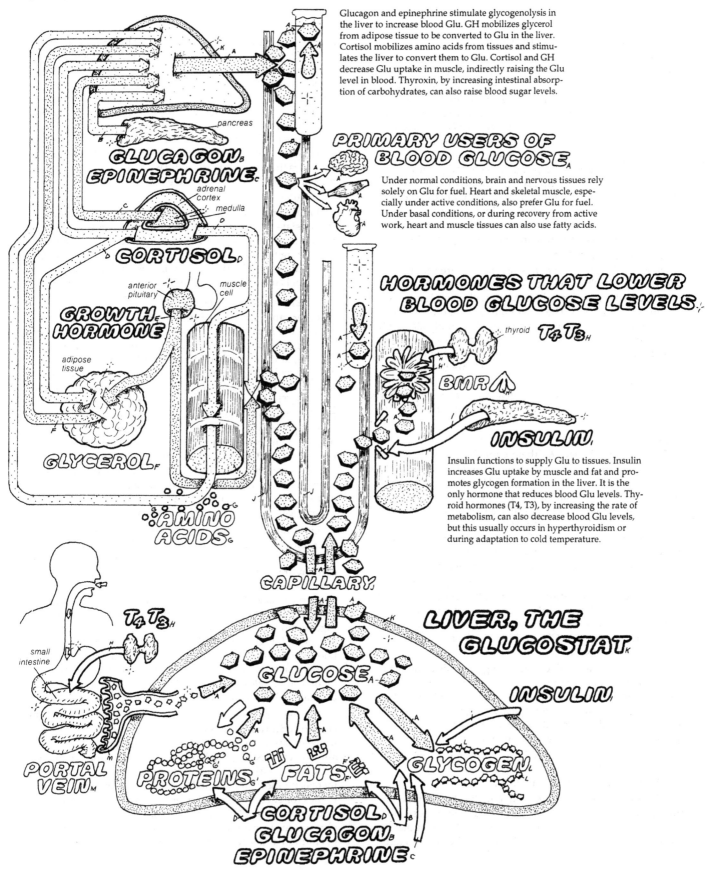

Glucagon and epinephrine stimulate glycogenolysis in the liver to increase blood Glu. GH mobilizes glycerol from adipose tissue to be converted to Glu in the liver. Cortisol mobilizes amino acids from tissues and stimulates the liver to convert them to Glu. Cortisol and GH decrease Glu uptake in muscle, indirectly raising the Glu level in blood. Thyroxin, by increasing intestinal absorption of carbohydrates, can also raise blood sugar levels.

PRIMARY USERS OF BLOOD GLUCOSE

Under normal conditions, brain and nervous tissues rely solely on Glu for fuel. Heart and skeletal muscle, especially under active conditions, also prefer Glu for fuel. Under basal conditions, or during recovery from active work, heart and muscle tissues can also use fatty acids.

HORMONES THAT LOWER BLOOD GLUCOSE LEVELS

Insulin functions to supply Glu to tissues. Insulin increases Glu uptake by muscle and fat and promotes glycogen formation in the liver. It is the only hormone that reduces blood Glu levels. Thyroid hormones (T4, T3), by increasing the rate of metabolism, can also decrease blood Glu levels, but this usually occurs in hyperthyroidism or during adaptation to cold temperature.

pancreas

GLUCAGON
EPINEPHRINE

adrenal
cortex
medulla

CORTISOL

anterior
pituitary   muscle
cell

GROWTH
HORMONE

adipose
tissue

GLYCEROL

AMINO ACIDS

CAPILLARY

thyroid $T_4 T_3$

BMR

INSULIN

LIVER, THE GLUCOSTAT

small
intestine

$T_4 T_3$

PORTAL VEIN

GLUCOSE

PROTEINS   FATS   GLYCOGEN

INSULIN

CORTISOL
GLUCAGON
EPINEPHRINE

The liver, acting as a glucostat, is the main organ for Glu homeostasis. The liver converts other monosaccharides to Glu. In the liver, glycerol from fat catabolism, amino acids from protein catabolism, and lactate from anaerobic glycolysis can be converted to Glu. The liver stores Glu as glycogen and mobilizes Glu by glycogenolysis. It is the only organ that can secrete Glu into the blood, as it contains the enzyme Glu-6-phosphatase. Numerous hormones, including insulin, glucagon, cortisol, epinephrine, and thyroid hormones, act on the liver to influence these metabolic processes.

TYPES OF ADIPOSE TISSUE & USES OF FATS

Body fats are divided into fuel & structural types — Fuel fats are stored in fat depots of the *adipose tissue*, which consists of *fat cells* with large cytoplasmic stores of fat. Adipose tissue is active, continuously forming and degrading fats. It occurs in the abdominal cavity, within or around organs (muscle, heart), and under the skin. The *subcutaneous fat* helps in thermal insulation. *Brown fat* is a form of subcutaneous fat with many mitochondria that liberate mainly heat (not ATP) upon oxidation, to protect the body against cold temperatures. Brown fat location, structure, and physiology are discussed in plates 140 and 141. *Structural fats* (phospholipids, cholesterol) are not utilized for energy; phospholipids occur in cell membranes, and cholesterol functions in synthesis of steroid hormones, vitamin D and neural myelin tissue (plate 135).

BASIC CHEMISTRY OF FATS

Triglycerides (TG, *triacylglycerols, neutral fats*) are the storage fats (fat depots). They are esters of *glycerol* and three *fatty acids* (FA). FA are long hydrocarbon chains with a single carboxylic acid group at one end. The longer the chain and the smaller the number of double bonds, the lower the fluidity state of the FA and the associated TG. The most commonly occurring body FA are palmitic, stearic, and oleic acids with chains between 14 and 16 carbon atoms long. TG are broken down either completely to glycerol and FA or incompletely to FA and mono- or diglycerides. The breakdown of TG (*lipolysis*) is catalyzed by various *lipase* enzymes in the intestine, liver, and adipose tissue.

FATS AS ENERGY SOURCE

Fats take small space & liberate much energy — Fats are ideal for fuel storage because per unit weight they occupy less volume and produce more energy (ATP) than carbohydrates or proteins. When oxidized, 1 g of fat produces 9.3 kcalories — 2.3 times more than 1 g of carbohydrate or protein. Some tissues easily utilize FA for energy; 60% of the heart's basal energy requirement is derived from fats, chiefly FA. Skeletal muscle also utilizes FA to obtain energy — especially during recovery from strenuous exercise — to replenish the exhausted supply of ATP, creatine phosphate, and glycogen (plate 27).

FA undergo β-oxidation to form ATP or for conversion to amino acids — To liberate their energy, the FA are degraded to acetate (acetyl CoA) by a process called *β-oxidation*. The acetyl CoA is then oxidized to CO_2 and H_2O in the mitochondria to produce ATP (plate 6).

Glycerol can be oxidized through glycolysis or used to form glucose — Both of the products of triglyceride lipolysis — glycerol and FA — can be utilized for energy production. Glycerol can be converted to intermediates of glycolysis and then to pyruvate, which then enters the *Krebs cycle* to form ATP (plate 6). Alternatively, glycerol can be converted to *glucose* in the liver (gluconeogenesis); glucose is used by tissues such as the brain for fuel.

FAT METABOLISM IN ADIPOSE TISSUE

Glycerol and FA are esterified to make storage fats (lipogenesis) — After a carbohydrate meal, the adipose tissue fat cells, stimulated by insulin, take up the abundant plasma glucose and convert it to glycerol and FA. The glycerol (an alcohol) and FA (*acids*) are then *esterified* to form TG (*lipogenesis*). Fatty meals increase the blood *chylomicrons* — very large size lipoprotein particles transporting the absorbed TG and cholesterol in the blood (plate 79). Within the capillaries of adipose tissue and liver, an enzyme called *lipoprotein lipase* hydrolyzes the glycerides, freeing glycerol and FA. These are taken up by fat cells and re-esterified to form storage TG. TG with sufficiently long chains tend to solidify and are therefore easily stored. Increased storage of solid fats in the cytoplasm increases the size of fat cells, which accumulate in the thick *fat pads* of the adipose tissue. If excessive, this condition leads to *obesity* (plate 139).

Lipase enzymes degrade stored fats to glycerol and FA (lipolysis) — When stimulated by catecholamines and other hormones (plate 134), the TG are lipolyzed by lipase enzymes, mobilizing the glycerol and FA into the blood. The mobilized FA are then used by the heart, muscles, and liver for energy. Glycerol is usually taken up by the liver to make new glucose.

FAT METABOLISM IN THE LIVER

Liver can convert fats into proteins & glucose and vice versa — The *liver*, like the adipose tissue, is capable of forming, degrading, and storing fats, although the fat granules in the liver hepatocytes are not intended for long-term storage. The particular importance of the liver lies in its capability for metabolic interconversion between fats, carbohydrates, and proteins. The liver hepatocytes contain all the enzymes required for these chemical transformations. For example, excess glucose can be metabolized into fatty acids, which are then either incorporated into TG or mobilized for consumption by tissues. Glycerol can be converted to glucose by *reverse glycolysis* and then to *glycogen*. FA can be converted to some *amino acids* and vice versa. These amino acids can then make *proteins*. The only reaction that the liver, and animal cells in general, cannot do in this regard is convert FA into glucose.

Liver can form cholesterol & ketone bodies — A major liver role in fat metabolism is to make *cholesterol* and *ketone bodies*. Cholesterol metabolism is detailed in plate 135. When carbohydrates are low in the diet or in cells (as in diabetes), the liver degrades FA to acetate (acetyl CoA). When the available pool of acetyl CoA exceeds the loading capacity of the mitochondria, the acetate molecules are instead *condensed* together to form compounds such as *acetoacetic acid, acetone*, and other *keto acids*, collectively called ketone bodies. Ketone bodies leak out from the liver into the blood, where they are excreted in the *kidneys*. Excessive amounts of ketone bodies in blood lead to *ketosis* and *metabolic acidosis*, conditions that may be fatal, such as untreated insulin-deficiency diabetes (Type I).

Ketones are normally excreted but may be used as fuel in certain conditions — In normal adults, ketone bodies are little utilized for energy. However, in newborns, pregnant women, and individuals subjected to prolonged starvation, many tissues, particularly the brain, undergo metabolic adaptation, increasing their rate of uptake and utilization of the ketone bodies for energy. This ability accounts not only for the continued function of the brain (an organ that usually uses only glucose) in starvation, but also for the lack of ketone toxicity in children and starved individuals.

CN: Use red for I and yellow for C. Use light colors for glycerol (A), fatty acids (B) and light blue for D.
1. Color the upper materials on the structure of fat. Note that in the cube representing TG in solid form, all letters receive the triglyceride color (C).
2. Color the metabolism of fat following the numbered sequence, starting with the consumption of the hamburger. Note that step (7) ends with the formation of glucose in the liver. Then proceed to (8) in the lower cell.

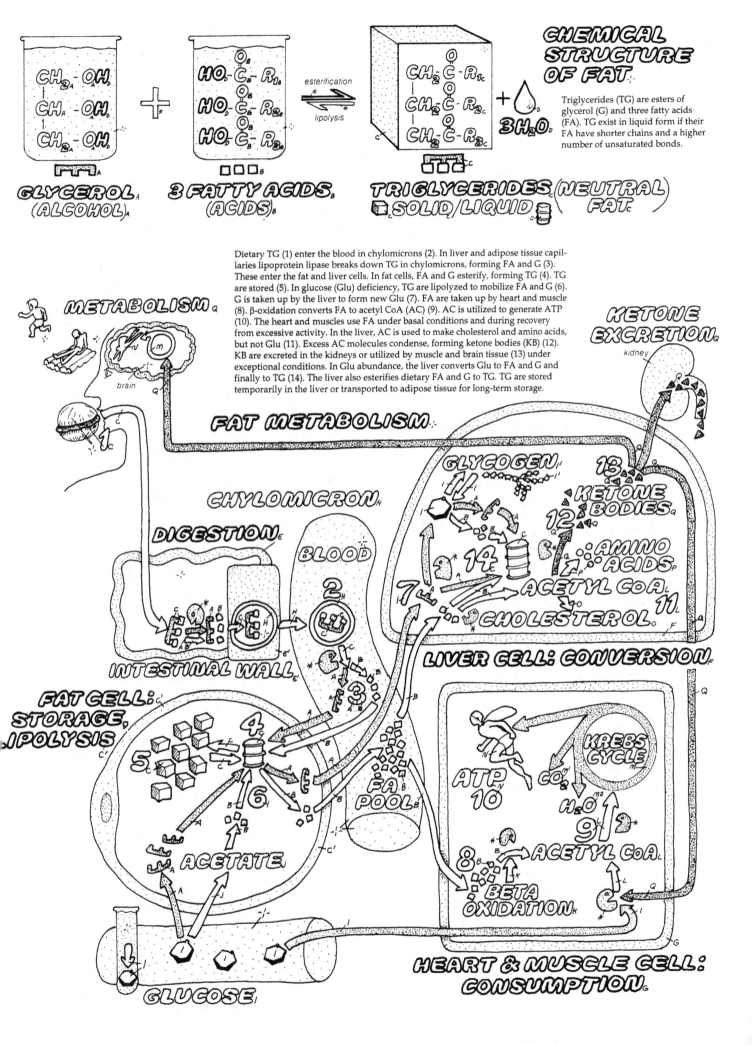

CHEMICAL STRUCTURE OF FAT

GLYCEROL (ALCOHOL) + 3 FATTY ACIDS (ACIDS) ⇌ (esterification / lipolysis) TRIGLYCERIDES (NEUTRAL FAT) SOLID/LIQUID + $3H_2O$

Triglycerides (TG) are esters of glycerol (G) and three fatty acids (FA). TG exist in liquid form if their FA have shorter chains and a higher number of unsaturated bonds.

FAT METABOLISM

Dietary TG (1) enter the blood in chylomicrons (2). In liver and adipose tissue capillaries lipoprotein lipase breaks down TG in chylomicrons, forming FA and G (3). These enter the fat and liver cells. In fat cells, FA and G esterify, forming TG (4). TG are stored (5). In glucose (Glu) deficiency, TG are lipolyzed to mobilize FA and G (6). G is taken up by the liver to form new Glu (7). FA are taken up by heart and muscle (8). β-oxidation converts FA to acetyl CoA (AC) (9). AC is utilized to generate ATP (10). The heart and muscles use FA under basal conditions and during recovery from excessive activity. In the liver, AC is used to make cholesterol and amino acids, but not Glu (11). Excess AC molecules condense, forming ketone bodies (KB) (12). KB are excreted in the kidneys or utilized by muscle and brain tissue (13) under exceptional conditions. In Glu abundance, the liver converts Glu to FA and G and finally to TG (14). The liver also esterifies dietary FA and G to TG. TG are stored temporarily in the liver or transported to adipose tissue for long-term storage.

METABOLISM
brain

KETONE EXCRETION
kidney

CHYLOMICRON
DIGESTION
BLOOD
INTESTINAL WALL

GLYCOGEN
KETONE BODIES
AMINO ACIDS
ACETYL CoA
CHOLESTEROL
LIVER CELL: CONVERSION

FAT CELL: STORAGE, LIPOLYSIS
ACETATE

FA POOL

KREBS CYCLE
ATP
CO_2
H_2O
ACETYL CoA
BETA OXIDATION
HEART & MUSCLE CELL: CONSUMPTION

GLUCOSE

REGULATION OF FAT METABOLISM

Like *carbohydrates*, fats are used for fuels; fat and carbohydrate metabolism are intimately linked, and many of the neural and hormonal factors regulating carbohydrate metabolism also take part in regulation of fat metabolism.

HORMONES THAT INCREASE FAT DEPOSITION

Insulin is the main hormone promoting fat formation (lipogenesis) — Insulin is secreted after a carbohydrate meal. Its action on fat cells involves three effects. Initially, insulin increases glucose uptake; fat cells use glucose for lipogenesis, forming FA and glycerol that are further esterified to triglycerides. The second action of insulin occurs if hyperglycemia is marked and prolonged; in this condition insulin also promotes the activity and synthesis of lipogenetic enzymes, increasing the efficiency of fat cells to make fat from carbohydrates. This action of insulin also occurs in the liver. The third action of insulin on fat cells is to inhibit the hormone-sensitive lipase, resulting in decreased lipolysis. In short, insulin reduces FA mobilization and increases fat formation and storage in adipose tissue and the liver. This response is adaptive for the body energy economy during times of food availability, since the extra food is stored as fat and these fat stores can be mobilized in times of need to provide fuel for the body.

Estrogens stimulate fat formation in the female — Female bodies normally have 5% more body fat than those of males. Estrogens, the female sex hormones, promote the extra fat reserves and the distribution of subcutaneous fat in the female body that begins at puberty and occurs during pregnancy.

FACTORS THAT PROMOTE FAT BREAKDOWN (LIPOLYSIS)

Sympathetic system & catecholamines act rapidly to stimulate lipolysis & mobilize fatty acids — During strenuous physical activity or in starvation, food intake is delayed, and the blood sugar level is threatened. A drop in blood sugar (glucose) triggers responses of the *hypothalamic glucostat* (plates 131, 138). The *sympathetic nervous system* becomes activated, causing the sympathetic nerves and adrenal medulla to release the *catecholamines (epinephrine, norepinephrine)* that stimulate *lipolysis* in the fat cells of adipose tissue (plate 125).

Catecholamines cause release of cAMP & activation of hormone-sensitive lipase in fat cells — Catecholamines bind with their receptors, which are coupled to G-proteins and adenylate cyclase activation, increasing the level of cyclic-AMP (second messenger) inside the fat cells. The action of cyclic-AMP results in stimulation of a special enzyme in the fat cell, called *hormone-sensitive lipase*. This lipase catalyzes the conversion of stored *triglycerides* to *fatty acids* (FA) and *glycerol*, but its activation is subject to the stimulation of catecholamines from plasma or sympathetic nerves.

Fatty acids act as fuels for heart and muscle — The mobilized FA enter the blood, furnishing fresh fuel for the heart and muscle and sparing glucose. Glycerol is taken up by the *liver* and converted to glucose (gluconeogenesis). These events will reverse hypoglycemia and increase the glucose supply to the brain.

Growth hormone sensitizes fat cells to catecholamines, increasing fatty acid mobilization — In long-term strenuous physical activity or starvation, the hypothalamic glucostat induces the hypothalamic mechanisms for regulation of *growth hormone* (GH) to release it from the pituitary. GH has a marked lipolytic action on the fat and liver cells, mobilizing FA and glycerol (plate 118). These effects take longer to appear but are longer lasting than those of the catecholamines. To exert these effects, GH increases the sensitivity of fat and liver cells to catecholamines, possibly by increasing their receptors or intracellular mediators, so that catecholamines will induce more intensive lipolysis, resulting in greater FA mobilization. GH, by inhibiting the use of carbohydrates in muscles, increases their use of FA for energy, thereby sparing glucose.

Cortisol has diverse stimulatory effects on lipolysis & fat mobilization — Another hormone that enhances lipolysis is *cortisol*, released by the adrenal cortex. Cortisol is released quite rapidly in response to stimulation by the stress-induced release of pituitary hormone, ACTH, which is in turn released in response to the release of the CRH from the hypothalamus (plates 117, 127). Cortisol's lipolytic actions are exerted in several ways. A very important way is "permissive" action. Cortisol must be present in order for the lipolytic actions of both catecholamines and GH on the fat cells to be expressed. The nature of the permissive action of cortisol is not known. Cortisol also imparts direct effects, increasing fat catabolism in adipose tissue and FA oxidation and *ketone body* formation in the liver. These actions occur during the body's responses to prolonged fasting and starvation (plate 127).

ROLE OF LEPTIN IN LONG-TERM REGULATION OF BODY FAT

Leptin is a fat tissue hormone — Recently a protein hormone named leptin (from Greek word for "thin") has been found in the blood that actively takes part in long-term fat metabolism. Leptin is secreted from fat tissue; its secretion rate and plasma content are proportional to body fat content.

Leptin exerts feedback effect on hypothalamus to regulate body fat content — Increase in fat content is associated with increased leptin secretion, which acts on the hypothalamus to decrease appetite and food intake and possibly to increase release of fat-mobilizing hormones (GH, catecholamines?) to use the extra fat for energy. Mice lacking leptin or its receptors are obese. Leptin's effect on the hypothalamus is overridden in special circumstances — e.g., animals preparing for hibernation, pregnancy, and lactation. The role of leptin in human obesity is detailed in plate 139.

ABNORMALITIES OF FAT METABOLISM

Normal fat content for adult human males and females is about 15% and 20% of body weight, respectively. Increased body fat may occur in normal conditions, as in pregnancy, or in abnormal conditions, as in *obesity*. Causes of obesity may be genetic or environmental or both (plate 139). Poor dietary habits, including excessive intake of fats and carbohydrates, and reduced physical activity are among the controllable factors. Abnormally low body fat may lead to (or be caused by) *anorexia nervosa*. Obesity and anorexia may be associated with endocrine changes, resulting in altered lipolysis and lipogenesis. Insulin and leptin secretory changes are believed to contribute to obesity. Altered fat metabolism occurs in Type II diabetes and in vascular diseases (e.g., coronary heart disease). Hypothyroidism is often associated with increased fat and hyperthyroidism with decreased fat.

CN: Use red for F, purple for E, and the same colors for fatty acids (B), glycerol (C), triglycerides (D), and liver (L) as used on the previous page.
1. Begin at the top and follow numbers 1-5 to both the fat (D^1) and liver (L) cells without coloring the actions in those cells. Then color 6 and 7 and follow the elements floating in the capillary to the liver cell (L) on the right.
2. Color the bottom panel, beginning with number 1 in the lower left corner.

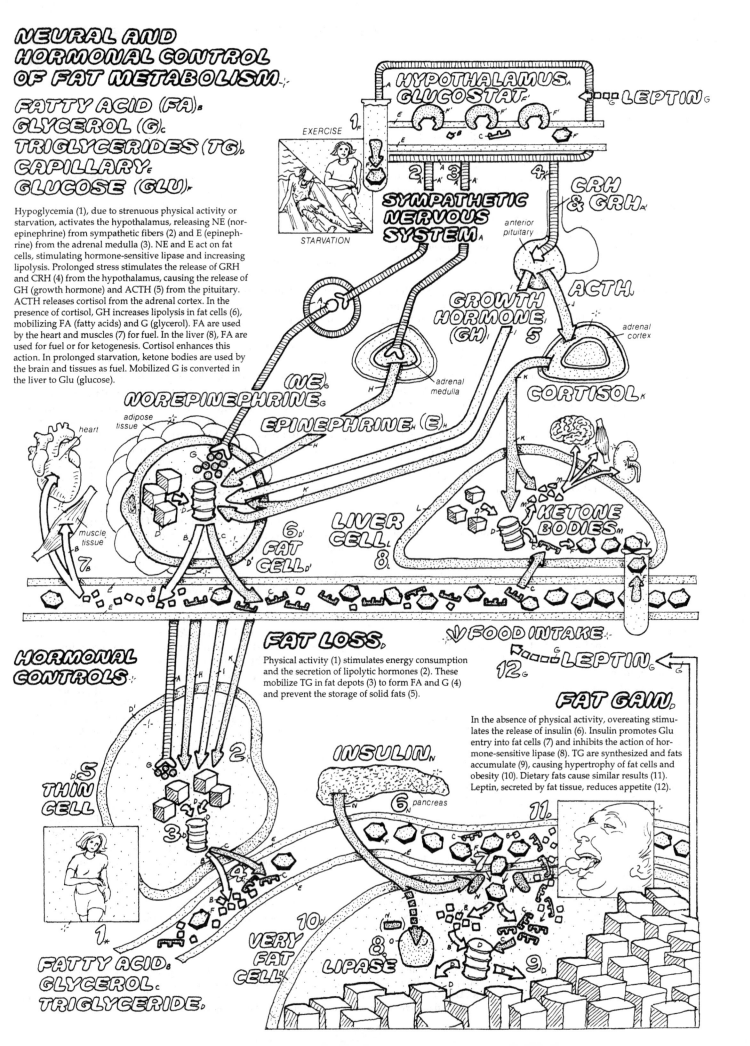

NEURAL AND HORMONAL CONTROL OF FAT METABOLISM

FATTY ACID (FA)b
GLYCEROL (G)c
TRIGLYCERIDES (TG)d
CAPILLARYe
GLUCOSE (GLU)f

Hypoglycemia (1), due to strenuous physical activity or starvation, activates the hypothalamus, releasing NE (norepinephrine) from sympathetic fibers (2) and E (epinephrine) from the adrenal medulla (3). NE and E act on fat cells, stimulating hormone-sensitive lipase and increasing lipolysis. Prolonged stress stimulates the release of GRH and CRH (4) from the hypothalamus, causing the release of GH (growth hormone) and ACTH (5) from the pituitary. ACTH releases cortisol from the adrenal cortex. In the presence of cortisol, GH increases lipolysis in fat cells (6), mobilizing FA (fatty acids) and G (glycerol). FA are used by the heart and muscles (7) for fuel. In the liver (8), FA are used for fuel or for ketogenesis. Cortisol enhances this action. In prolonged starvation, ketone bodies are used by the brain and tissues as fuel. Mobilized G is converted in the liver to Glu (glucose).

HYPOTHALAMUS GLUCOSTAT — LEPTIN

EXERCISE

STARVATION

SYMPATHETIC NERVOUS SYSTEM

CRH & GRH

anterior pituitary

ACTH

GROWTH HORMONE (GH) 5

adrenal cortex

adrenal medulla

NOREPINEPHRINE (NE)g

EPINEPHRINE (E)h

CORTISOL

heart

adipose tissue

LIVER CELL 8

KETONE BODIES

muscle tissue

6 FAT CELL

FOOD INTAKE

LEPTIN 12

HORMONAL CONTROLS

FAT LOSS
Physical activity (1) stimulates energy consumption and the secretion of lipolytic hormones (2). These mobilize TG in fat depots (3) to form FA and G (4) and prevent the storage of solid fats (5).

FAT GAIN
In the absence of physical activity, overeating stimulates the release of insulin (6). Insulin promotes Glu entry into fat cells (7) and inhibits the action of hormone-sensitive lipase (8). TG are synthesized and fats accumulate (9), causing hypertrophy of fat cells and obesity (10). Dietary fats cause similar results (11). Leptin, secreted by fat tissue, reduces appetite (12).

THIN CELL

INSULIN

pancreas

VERY FAT CELL

LIPASE

FATTY ACIDb
GLYCEROLc
TRIGLYCERIDEd

Cholesterol has many functions but is not a fuel—
Cholesterol is a structural fat of animal origin. It is a *sterol* synthesized from *acetate*, mainly in the *liver*; the adrenal cortex and gonads also synthesize cholesterol. In the liver it is used to make *bile salts* to facilitate intestinal fat digestion. It is the precursor for gonadal and adrenocortical steroids and vitamin D_3 formation in the skin. Cholesterol is a component of the neural-tissue *myelin sheath* and in the *corneum* of skin (the outer keratinized layer), which helps make the skin waterproof and prevents water evaporation. In the *membranes* of cells and some cellular organelles, cholesterol helps stabilize the movement of *phospholipid* chains.

Dietary vs. endogenous cholesterol—In the body, cholesterol may be exogenous (dietary) or endogenous—i.e., synthesized in the tissues, chiefly the liver. Dietary cholesterol comes solely from foods of animal origin (egg yolk, liver, fatty meats, cheese). Cholesterol is absorbed in the intestine with other fats inside *chylomicrons*, large *lipoprotein particles* (LPPs) (plate 79). Chylomicrons are digested by the enzyme *lipoprotein lipase* in the capillaries of liver and adipose tissue. *Triglycerides* are delivered to the adipose tissue, and remaining cholesterol and phospholipids are delivered to the liver (plate 133).

Cholesterol can be made in the liver from acetate—To make cholesterol, liver acetate (acetyl-CoA) undergoes several reactions to form *mevalonic acid*, which is converted first to *squalane* and then to cholesterol. Cholesterol regulates its own synthesis by substrate inhibition of the enzyme that forms mevalonic acid. Dietary cholesterol inhibits the liver's synthesis of cholesterol in the same manner. Most of the liver cholesterol is used to form bile salts (e.g., cholate) that help digest fat by *emulsification* (plate 77).

Liver cholesterol is exported to tissues packed in lipoprotein particles—The liver supplies cholesterol to most tissues through the blood. Cholesterol is packed in LPPs that resemble chylomicrons. LPPs have different fat and protein composition and are of different size and density. The higher the fat content of the LPPs, the lower their density. Each LPP has a core of hydrophobic fats (triglycerides and cholesterol as *cholesteryl esters*) engulfed in a coat of hydrophilic proteins and phospholipids. The protein in the coat is called *apoprotein* (*apolipoprotein*), a very important protein since it binds with the tissue receptors that bind LDL.

Lipoprotein particles vary in size, density, & lipid content—LPPs vary in size between 10 to 80 μm. Cholesterol for export is transported in the largest of plasma LPPs—*very low density lipoproteins* (VLDL). In the plasma, these are transformed to smaller lipoproteins—*low density lipoproteins* (LDL) and *intermediate density lipoproteins* (IDL)—by the actions of enzymes. Cholesterol delivered directly to tissues is in LDL particles. Once inside the tissue cells, cholesterol is utilized for the variety of functions named above. The excess cholesterol is packed in the smallest of LPP—*high density lipoproteins* (HDL)—and transported back to the liver for processing.

Thyroid & sex hormones influence cholesterol levels—Thyroid hormones decrease plasma cholesterol by increasing its uptake by the liver and tissues. *Estrogen*, the female sex steroid hormone, decreases cholesterol level, while *androgen*, the male sex steroid hormone, increases it. These sex steroid effects may relate to the higher incidence of atherosclerosis (see below) in men.

CHOLESTEROL, ATHEROSCLEROSIS, & HEART DISEASE

Cholesterol contributes to atherosclerosis & heart disease—This disease is responsible for nearly half of all deaths, mostly in men and in the elderly. When the inner wall of an artery is damaged, *platelets* adhere to the site of damage, stimulating *fibrosis*. Plasma cholesterol is deposited on these lesions, along with *calcium ions*, forming hard, calcified *cholesterol plaques*. Atherosclerosis is involved in many arterial diseases, such as *arteriosclerosis* (hardening of the arteries).

Buildup of plaques in the lumen of coronary arteries (coronary occlusion disease) reduces blood flow to various cardiac regions, causing coronary *ischemia*. Plaques also facilitate *blood clot* (thrombus) formation, which blocks blood flow to a heart region (thrombosis), leading to heart attacks. Similar events can occur in brain arteries, leading to strokes (brain attacks).

LDL gives up cholesterol to the vascular plaques—It is now believed that high levels of plasma cholesterol, particularly the cholesterol in LDL ("bad cholesterol"), favors plaque formation. Presumably, LDL particles, coming from the liver to the tissues in high numbers and with specific receptors for tissue binding, are more likely to attach to damaged arterial walls and deposit their cholesterol at these sites. In contrast, cholesterol in HDL particles ("good cholesterol") traveling from tissues to the liver is not deposited in the lesions.

Low plasma cholesterol & high HDL:LDL ratios reduce plaques & heart disease—The normal range of plasma cholesterol is 120–220 mg/dl (average 170 mg/dl). High dietary intake of cholesterol contributes to the disease by increasing plasma cholesterol. Lowering of plasma cholesterol by diet and drugs (which inhibit cholesterol synthesis) reduces plaque formation and may even reverse it. The recommended maximum plasma cholesterol level for men with family history of heart disease is 180 mg/dl, since incidence of arterial plaques and heart attacks increases above this limit; for women at risk, this value is 200 mg/dl.

In addition to decreased total plasma cholesterol, a high ratio of HDL to LDL cholesterol appears to ward off plaque formation and heart disease. Ideal values for HDL cholesterol ("good cholesterol") are > 35 mg/dl and for LDL cholesterol ("bad cholesterol") are < 130 mg/dl . Women have lower LDL and higher HDL levels than men, probably because of their higher estrogen levels, which may account for their lower incidence of heart disease in their forties.

Role of dietary fatty acids—Dietary fatty acids have been shown to play a role in plasma cholesterol and atherosclerosis through influencing the cholesterol content of LPP. *Monounsaturated fats* present in olives, almonds, avocados, and canola are highly recommended since they increase the HDL and decrease the LDL cholesterol (i.e., increase their ratios); *polyunsaturated fats* (vegetable oils) are also good, but they mainly increase HDL cholesterol. *Saturated fats* (butter, animal fats, and hydrogenated vegetable oils) should be kept to minimum, since they increase the LDL cholesterol.

CN: Use the same colors as were used on the previous page for B and K. Use purple for D and red for M. Use a bright color for A.
1. Color the chemical structure of cholesterol at the top of the page and its bodily functions in the upper right corner.
2. Follow the numbered sequence, starting at the ingestion of dietary fat and cholesterol.

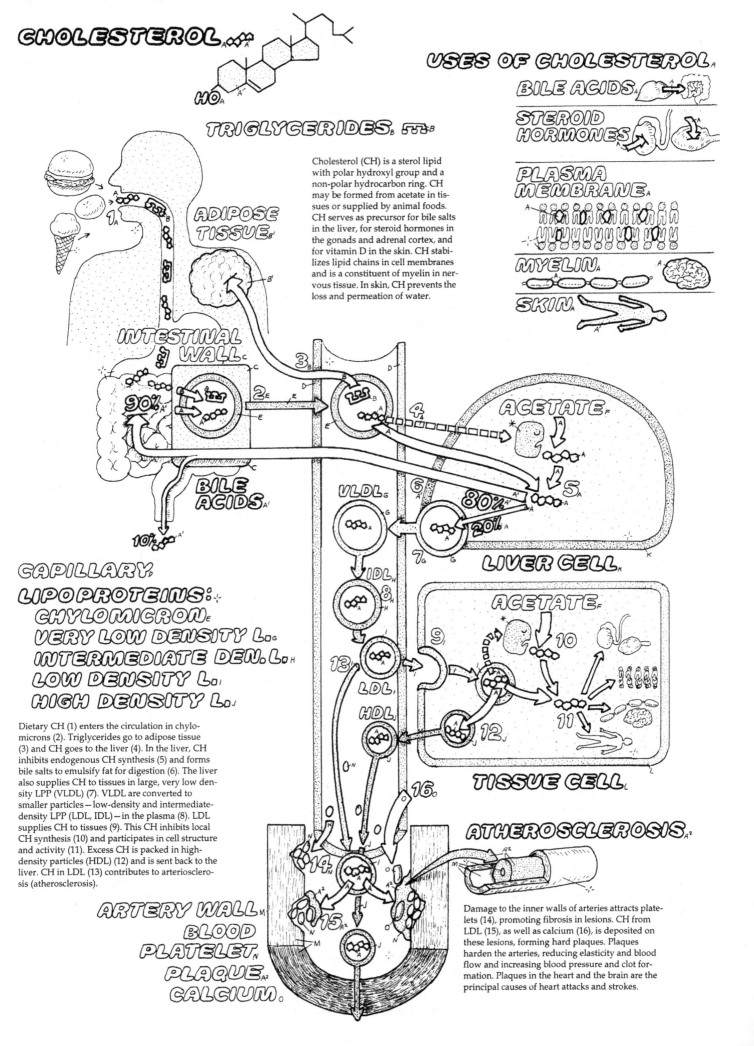

CHOLESTEROL

TRIGLYCERIDES

USES OF CHOLESTEROL

BILE ACIDS

STEROID HORMONES

PLASMA MEMBRANE

MYELIN

SKIN

Cholesterol (CH) is a sterol lipid with polar hydroxyl group and a non-polar hydrocarbon ring. CH may be formed from acetate in tissues or supplied by animal foods. CH serves as precursor for bile salts in the liver, for steroid hormones in the gonads and adrenal cortex, and for vitamin D in the skin. CH stabilizes lipid chains in cell membranes and is a constituent of myelin in nervous tissue. In skin, CH prevents the loss and permeation of water.

ADIPOSE TISSUE

INTESTINAL WALL

BILE ACIDS

ACETATE

VLDL

LIVER CELL

IDL

ACETATE

LDL

HDL

TISSUE CELL

CAPILLARY
LIPOPROTEINS:
CHYLOMICRON
VERY LOW DENSITY L.
INTERMEDIATE DEN. L.
LOW DENSITY L.
HIGH DENSITY L.

Dietary CH (1) enters the circulation in chylomicrons (2). Triglycerides go to adipose tissue (3) and CH goes to the liver (4). In the liver, CH inhibits endogenous CH synthesis (5) and forms bile salts to emulsify fat for digestion (6). The liver also supplies CH to tissues in large, very low density LPP (VLDL) (7). VLDL are converted to smaller particles—low-density and intermediate-density LPP (LDL, IDL)—in the plasma (8). LDL supplies CH to tissues (9). This CH inhibits local CH synthesis (10) and participates in cell structure and activity (11). Excess CH is packed in high-density particles (HDL) (12) and is sent back to the liver. CH in LDL (13) contributes to arteriosclerosis (atherosclerosis).

ATHEROSCLEROSIS

ARTERY WALL
BLOOD
PLATELET
PLAQUE
CALCIUM

Damage to the inner walls of arteries attracts platelets (14), promoting fibrosis in lesions. CH from LDL (15), as well as calcium (16), is deposited on these lesions, forming hard plaques. Plaques harden the arteries, reducing elasticity and blood flow and increasing blood pressure and clot formation. Plaques in the heart and the brain are the principal causes of heart attacks and strokes.

STRUCTURE, VARIETY, AND SIGNIFICANCE OF PROTEINS

Proteins are of primary importance in cellular, tissue, and bodily structures and functions. Over 100,000 different proteins are thought to exist in the body, on the basis of 100,000 genes in the human genome and the one gene–one protein relation. Not all proteins occur in all cells. Proteins are made of different combinations of about twenty naturally occurring *amino acids*, which vary in structure but share a common feature: the presence of a *carboxyl acid* and *amino* group. Amino acids can be joined by *peptide bonds*, forming *peptide chains* – hence *dipeptides, tripeptides, oligopeptides,* and *polypeptides*. Proteins are basically large polypeptides of one or more chains. Proteins are synthesized within cells from amino acids (plate 4). Of the twenty amino acids making up the proteins, the body can only synthesize twelve, starting with glucose or fatty acids; the other eight must come from the diet (the dietary *essential amino acids* – e.g., leucine, tryptophan). Proteins are plentiful animal and plant foods.

ROLES OF LIVER IN PROTEIN METABOLISM

The liver is a center for catabolism & anabolism of amino acids & proteins — Amino acids form a labile pool in the liver, utilized to make the liver and blood proteins as well as glucose, fats, and energy (ATP). Liver amino acids can be exchanged with a second pool in the *blood,* which in turn exchanges with a third pool within the *tissue cells*. The liver is capable of making all the non-essential amino acids as well as numerous proteins and degrading them, forming purines and pyrimidine bases of the nucleic acids from amino acids.

Blood proteins are made mainly in the liver — The liver forms and secretes most of the *blood proteins: albumins* (transpor hormones and fatty acids, regulate plasma osmotic pressure), *globulins* (enzymes; transport hormones), and *fibrinogen* (needed for blood clotting). The human liver forms up to 50 g of these proteins every day.

The liver can oxidize amino acids to form ATP or convert them to glucose and fats — For this purpose, amino acids are first *deaminated*, forming various *keto acids* such as *pyruvic* and *alpha-keto-glutaric* acids. These are then oxidized in the *Krebs cycle* to form ATP. Amino acids are equal to carbohydrates in their capacity to release metabolic energy and form ATP. Formation of *keto acids* from amino acids allows conversion to glucose *(gluconeogenesis)* or to *fatty acids* and *glycerol (lipogenesis),* from which glycogen and triglycerides, respectively, can be formed for storage (plates 127, 133).

The liver forms urea as a result of amino acid deamination — Deamination of amino acids produces *ammonia* (NH_3), a gas that is toxic for the liver and other tissues. The liver detoxifies ammonia by converting it into *urea,* a far less toxic, water-soluble substance. To form urea, two molecules of NH_3 react with one molecule of CO_2. The actual formation of urea occurs through a chain of enzyme-catalyzed reactions called the *urea cycle,* involving the amino acids *ornithine, citruline,* and *arginine,* which act as intermediates. Urea diffuses into the blood and is excreted into the urine by the kidney.

METABOLISM OF PROTEINS IN TISSUES

Tissues require amino acids for growth, repair, and normal turnover of cellular proteins — These are obtained from the pool of amino acids in the blood, which in turn is in equi-librium with that in the liver. The cells of each tissue make their own specific proteins (i.e., those characteristic of each cell). These and other general cellular proteins are continuously formed on *ribosomes* and broken down in *lysosomes*. The *turnover rate* of proteins depends on the type of proteins and tissue. The liver enzymes show a fast turnover rate (a few hours); structural proteins show a slow rate (e.g., a few months for bone collagen).

Tissues form both general and specific proteins — Besides the general metabolic enzymes common to all cells, different tissues contain special proteins that perform unique functions. Thus, *antibodies,* the defense proteins, are secreted by the white blood cells (leukocytes). *Hemoglobin,* the oxygen-transporting protein, is found in the red cells. *Collagen,* the most abundant protein in the body, is secreted by the bone and cartilage cells and fibroblasts. *Actin* and *myosin* are the contractile protein of muscle tissue.

Tissue proteins are normally spared from oxidation but are catabolized during starvation — Consistent with their function as the building blocks of proteins, amino acids, particularly those of tissue proteins, are normally spared from metabolic oxidation, carbohydrates and fats being used preferentially. During prolonged fasting and starvation the amino acids of the liver and tissues are catabolized as fuels, but even then, proteins of the heart and brain are spared.

HORMONAL REGULATION OF PROTEIN METABOLISM

Growth hormone, insulin, & insulin-like growth factors (IGFs) promote protein anabolism and growth — Hormones profoundly influence protein metabolism. Thus, *growth hormone* and *insulin* increase the uptake of amino acids and protein synthesis in certain tissues such as muscle and bone. The effects of growth hormone are mediated by insulin-like growth factors (IGFs), usually secreted locally by tissues. IGFs are responsible for fetal growth. Absence of growth hormone or IGFs leads to cessation of growth and dwarfism. Bone and muscle growth is particularly severely affected while growth of heart, nerve, and brain tissue is spared.

Thyroid hormones promote synthesis of specific enzymes & proteins — In the heart and muscles *thyroxine* stimulates protein synthesis by increasing ribosomes and forming many specific, functionally significant proteins. These effects promote functional differentiation of these tissues, such as development of contractile proteins. In the liver and kidney, thyroid hormones induce the formation of specific proteins (e.g., Na-K-ATPase of the membrane pump). These hormones are synergistic in their anabolic effects on protein synthesis in many tissues, especially bone and muscle.

Effects of sex & adrenal steroids — *Estrogens* and *androgens* promote protein synthesis and cell proliferation in target reproductive tissues. Androgens from the testes and adrenal cortex are also known for their anabolic effects on muscle and bone growth. This action is important for the growth spurt at puberty (plate 152). *Cortisol* from the adrenal cortex promotes protein catabolism in many tissues, such as muscles and lymphatics, during stress and starvation, but in the liver, cortisol increases amino acid uptake and synthesis of enzymes for gluconeogenesis (plate 127), an important adaptive action during metabolic stress.

CN: Use a bright color for A, red for H, and very light colors for I and J.
1. Color the chemical formulas in the upper panel.
2. Color the four degrees of protein structure.
3. In the protein metabolism panel, follow the numbered sequence after coloring the liver (I) and tissue (J) cells, and the blood vessel (H).
4. In hormonal controls, note that growth hormone, somatomedins, and insulin operate together and are given a single color (N).

CHEMICAL STRUCTURE

CARBON ATOM.
AMINO GROUP.
CARBOXYL GROUP.
RADICAL GROUP.
PEPTIDE BOND.

SYNTHESIS
(ANABOLISM)

BREAKDOWN
(CATABOLISM)

>50 AA = PROTEIN
POLYPEPTIDE

HYDROLYSIS

AMINO ACID. AMINO ACID.

DIPEPTIDE

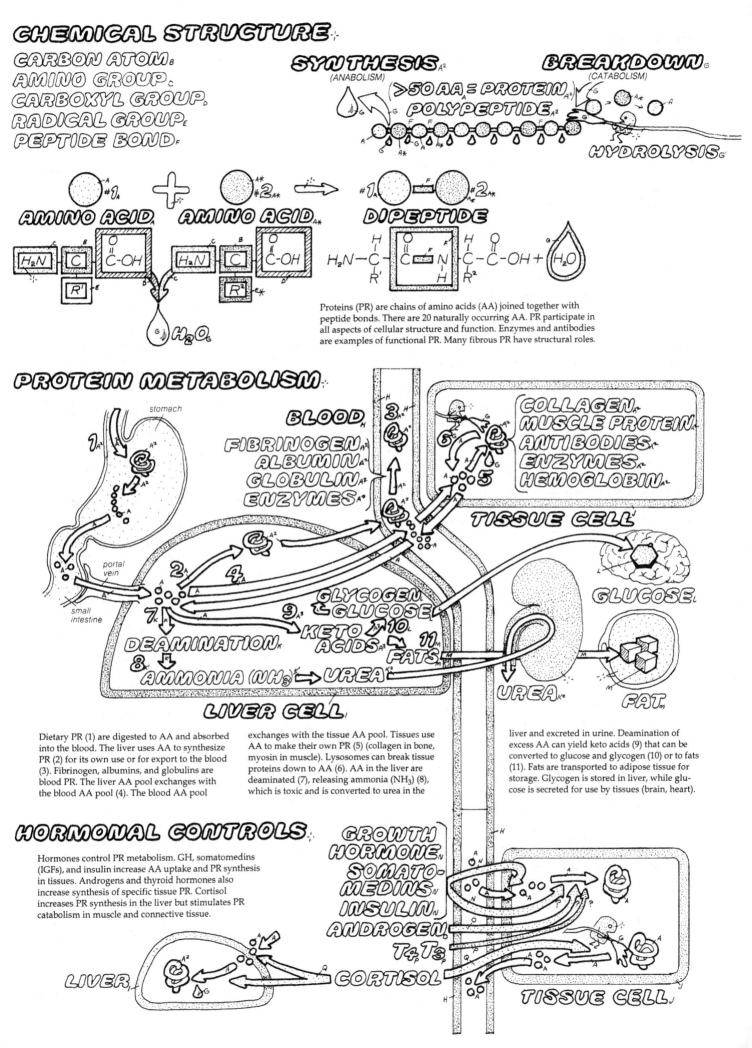

Proteins (PR) are chains of amino acids (AA) joined together with peptide bonds. There are 20 naturally occurring AA. PR participate in all aspects of cellular structure and function. Enzymes and antibodies are examples of functional PR. Many fibrous PR have structural roles.

PROTEIN METABOLISM

BLOOD

FIBRINOGEN
ALBUMIN
GLOBULIN
ENZYMES

COLLAGEN
MUSCLE PROTEIN
ANTIBODIES
ENZYMES
HEMOGLOBIN

TISSUE CELL

GLUCOSE

GLYCOGEN & GLUCOSE

KETO ACIDS

FATS

DEAMINATION

AMMONIA (NH₃) → UREA

UREA

FAT

LIVER CELL

Dietary PR (1) are digested to AA and absorbed into the blood. The liver uses AA to synthesize PR (2) for its own use or for export to the blood (3). Fibrinogen, albumins, and globulins are blood PR. The liver AA pool exchanges with the blood AA pool (4). The blood AA pool exchanges with the tissue AA pool. Tissues use AA to make their own PR (5) (collagen in bone, myosin in muscle). Lysosomes can break tissue proteins down to AA (6). AA in the liver are deaminated (7), releasing ammonia (NH₃) (8), which is toxic and is converted to urea in the liver and excreted in urine. Deamination of excess AA can yield keto acids (9) that can be converted to glucose and glycogen (10) or to fats (11). Fats are transported to adipose tissue for storage. Glycogen is stored in liver, while glucose is secreted for use by tissues (brain, heart).

HORMONAL CONTROLS

Hormones control PR metabolism. GH, somatomedins (IGFs), and insulin increase AA uptake and PR synthesis in tissues. Androgens and thyroid hormones also increase synthesis of specific tissue PR. Cortisol increases PR synthesis in the liver but stimulates PR catabolism in muscle and connective tissue.

GROWTH HORMONE
SOMATO-MEDINS
INSULIN
ANDROGEN
T₄ T₃
CORTISOL

LIVER

TISSUE CELL

OXIDATION OF NUTRIENTS, METABOLIC HEAT & METABOLIC RATE

Fuels burned outside the body release heat but may or may not result in work — The burning of most organic fuel substances in the air (i.e., combining them with oxygen: oxidation, oxygenation) produces carbon dioxide and water. However, in this process all the energy stored in the chemical bonds of the fuel substance is released as heat, with no work produced. Thus, the efficiency (i.e., ability to generate work from a form of energy) of this process is zero.

In a power plant, a fuel (e.g., coal) is burned to generate heat; heat drives the generator turbines; turbines produce electricity, which is put to the service of homes and shops to be utilized for a variety of work. Here heat, as a form of energy, is converted into usable energy (i.e., work).

Cellular oxidation also releases heat but some energy is captured as ATP for cellular functions — The human body is also a machine. To carry out its vital functions, it has to perform work, for which the body requires energy. The body obtains this energy by consuming and burning fuels (e.g., carbohydrates and fats). The oxidation of foodstuffs releases heat; however, in contrast to the power plant in the above example, the body is not able to convert the liberated heat directly into work. Instead, body cells couple the oxidation of foodstuffs with the generation of ATP, the high-energy chemical intermediate (plates 5, 6). ATP is then used for the variety of chemical (e.g., synthesis), mechanical (e.g., muscle contraction), electrical (e.g., nerve function) activities that the body cells need to carry out to survive and thrive.

Efficiency of the body machine — Some of the energy liberated during the oxidation of fuel substances is released as heat (metabolic heat), which is not wasted entirely because it can be utilized to keep the body warm. This is very useful in cold-blooded animals (poikilotherms, ectotherms) and an absolutely essential requirement in warm-blooded species (homeotherms, endotherms) like humans.

Heat is the ultimate form of energy — Even the energy utilized to do cellular work is ultimately converted to heat. Not only does the hydrolysis of ATP occurring during work generate heat, but some of the energy used to do the actual work is also converted to heat; for example, muscle contraction creates friction; friction creates heat.

MEASURING BODY ENERGY NEEDS IN TERMS OF HEAT (CALORIES)

Calorie — One Calorie (with a capital C = 1 kilocalorie) is defined as the amount of heat required to raise the temperature of 1 g of water by one degree $^\circ$ C).

Activities & foodstuffs can be expressed in heat units (calories) — Based on the universal utility of heat, it is a common practice to measure all bodily energy processes in terms of heat units (i.e., calories). The energy values of food substances (their usefulness for the body energy needs) are also best measured in terms of their fuel heat value (caloric values of foodstuffs). Thus 1g of carbohydrate or protein produces 4.1 kcalories, and 1 g of fat produces 9.3 kcalories.

Direct calorimetry measures total body heat produced — An accurate method for measuring body energy needs is by *direct calorimetry* (i.e., measuring the exact amount of heat the body produces). A sitting subject is placed in a *room calorimeter*, an insulated chamber to minimize heat exchange with the outside. The heat released from the body is used to warm a

stream of water running within a tube in the room. The increase in water temperature (outflow minus inflow), after conversion to calories, is equivalent to the amount of calories generated (or required) by the body.

Indirect calorimetry (spirometry) measures total oxygen consumed — In this method, the total amount of *oxygen* consumed during a particular time period is measured using a *spirometer*. Oxygen is inhaled from the spirometer tank through a mouthpiece. The decline in oxygen content of the tank is registered by a device (e.g., a kymograph). The CO_2 produced is absorbed by a soda lime tank. One liter of oxygen gas utilized during the burning of any foodstuff, inside or outside the body, generates 4.82 Cal. By knowing the total volume of oxygen utilized per unit time, we can determine the subject's total caloric production (or requirement).

METABOLIC RATE AND FACTORS INFLUENCING IT

Metabolic rate & what it takes to sustain it — Using the above calorimetric methods, the *metabolic rate* (MR) of the body under different conditions can be determined. The *basal metabolic rate* (BMR) is the amount of energy required to sustain the body at rest in a supine position. In the average adult human male, weighing about 70 kg (154 lb), BMR is about 2000 Cal/day. Therefore, based on the caloric values of foods, it takes 480 g of carbohydrates or proteins, or 215 g of fats, to sustain this person at rest for one day. This translates to about 30 apples, or 900 g (2 lb) of bread, or 800 g (1.8 lb) of meat, or 9 cups of cooked beans.

Physical activity is a major factor influencing metabolic rate — Metabolic rate decreases during sleep and increases during activity. During walking it is twice that during sitting, and running involves a rate three times higher than walking; climbing stairs produces a rate twice that of running.

Surface area to mass ratio is a major determinant of BMR — When calculated per unit mass (weight), BMR is higher in small animals, since they have a high surface area to mass ratio. The small body mass does not permit heat storage within the core, and the relatively large surface area facilitates heat loss. Thus, a mouse, which consumes less than 4 Cal/day compared to 5000 Cal/day for a horse, has a BMR of 200 Cal/day/kg, twenty times that of the horse. This is also why small animals (e.g., birds) seem to be constantly eating and why children have a higher BMR than adults.

Hormones & autonomic nervous system regulate metabolic rate — Thyroid hormones, growth hormone, and catecholamines from the adrenal medulla increase the metabolic rate, as do the androgen and progesterone sex hormones. Hypothyroid individuals may have a 40% decrease and hyperthyroid individuals a 100% increase in their BMR. Testosterone may be one reason for the higher BMR of human males. Increased thyroid hormones and progesterone are in part responsible for the normal increase in BMR in pregnant women. Increased progesterone accounts for higher BMR in women in the postovulatory period of their monthly cycle. Lowered sympathetic activity and catecholamine secretion may be the cause of lower BMR in the elderly. Ingestion of foods also increases the metabolic rate; the absorption of foods (independent of food utilization) has an even greater effect. These effects, called the "specific dynamic action of foods," are most marked for proteins (30% increase in MR).

CN: Use red for B, a bright color for C, and light blue for F.
1. Begin with the ingestion of food and follow the process throughout the diagram. Do the same with the test-tube reaction to the right.
2. Color the two ways of measuring metabolism.
3. Color the factors influencing metabolic rate. Each title receives a different color.

METABOLISM & HEAT PRODUCTION

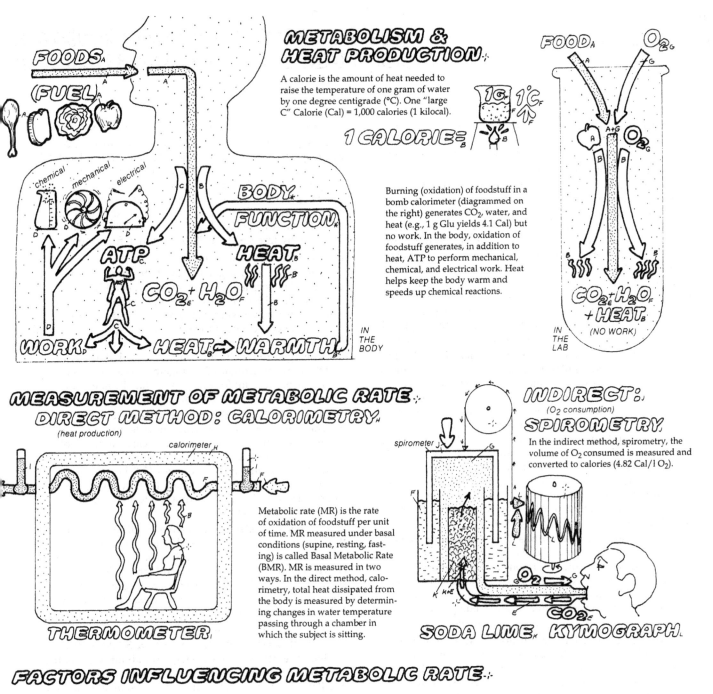

A calorie is the amount of heat needed to raise the temperature of one gram of water by one degree centigrade (°C). One "large C" Calorie (Cal) = 1,000 calories (1 kilocal).

1 CALORIE =

Burning (oxidation) of foodstuff in a bomb calorimeter (diagrammed on the right) generates CO_2, water, and heat (e.g., 1 g Glu yields 4.1 Cal) but no work. In the body, oxidation of foodstuff generates, in addition to heat, ATP to perform mechanical, chemical, and electrical work. Heat helps keep the body warm and speeds up chemical reactions.

FOODS (FUEL)

chemical mechanical electrical

BODY FUNCTION

ATP

HEAT

$CO_2 + H_2O$

WORK   HEAT → WARMTH

IN THE BODY

FOOD O_2

$CO_2 + H_2O + HEAT$

(NO WORK)

IN THE LAB

MEASUREMENT OF METABOLIC RATE
DIRECT METHOD: CALORIMETRY
(heat production)

calorimeter

Metabolic rate (MR) is the rate of oxidation of foodstuff per unit of time. MR measured under basal conditions (supine, resting, fasting) is called Basal Metabolic Rate (BMR). MR is measured in two ways. In the direct method, calorimetry, total heat dissipated from the body is measured by determining changes in water temperature passing through a chamber in which the subject is sitting.

THERMOMETER

INDIRECT:
(O₂ consumption)
SPIROMETRY

spirometer

In the indirect method, spirometry, the volume of O_2 consumed is measured and converted to calories (4.82 Cal/l O_2).

O_2 CO_2

SODA LIME   KYMOGRAPH

FACTORS INFLUENCING METABOLIC RATE

MASS VS. SURFACE AREA

	Cal/day	Cal/day/kg
	4	200
	750	50
	2000	30
	5000	10

Absolute MR (Cal/day) increases with body mass. In terms of per unit mass (Cal/day/kg), MR is higher in smaller animals because of a larger surface area to mass ratios (i.e., lose more heat).

ACTIVITY

Cal/hr   65   100   200   600   1200

Muscular activity increases MR, since muscles are the biggest consumers of energy. The lowest MR occurs during sleep; the highest MR from heavy exercise.

SEX & AGE

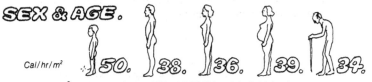

Cal/hr/m²   50   38   36   39   34

MR (Cal/hr/m²) declines with age. Although BMR in adults is higher in men (androgen effect), it is highest in pregnant women (thyroid/progesterone effect).

SYMPATHETIC NERVES & HORMONES

Sympathetic activity releases epinephrine and norepinephrine from the adrenal medulla and norepinephrine from nerve fibers. These increase MR by stimulating cellular metabolism. Thyroid hormones have the greatest such effects on MR. Sex hormones (androgens) and growth hormone also increase MR.

EATING

Food ingestion increases MR (specific dynamic action of food). Mere ingestion increases MR by 510%. Absorption of proteins (AA) has a longer-lasting and greater effect on MR (+ 30%). Malnutrition and starvation decrease MR by 30%.

BODY TEMPERATURE

MR increases by 15% for each °C increase in body temperature. This may become a serious problem in the case of fever.

ENVIRONMENTAL TEMP. CLIMATIC ADAPTATION

Increase and decrease in room temperature above and below optimal level (20°C) increase MR. In long-term adaptation, BMR is lower in tropical climates and higher in cold climates.

A healthy body operates on a *balance* between *energy input* (fuel – i.e., food intake) and *energy output* (energy expenditure – e.g., organ function, physical activity, heat production).

REGULATION OF FOOD INTAKE

Hypothalamic feeding & satiety centers control food intake – Increased activity in the *feeding center* promotes appetite, feeding behavior, and food intake. The feeding center is normally inhibited by the *satiety center*. Lesions of the satiety center cause rats to overeat and become obese. Lesions of the feeding center cause loss of appetite and extreme weight loss (anorexia). These centers may also control the set point for body weight. A type of obesity (*hypothalamic obesity*) and the loss of appetite and extreme thinness seen in *anorexia nervosa* may be the result of disturbances of these brain regulators of food intake (plates 107, 131, and 134). Central and peripheral factors influence activities of these centers and food intake.

Blood glucose, hormones, & neural signals from the gut wall are short-term regulators of food intake – According to the *glucostatic theory* of food-intake regulation, the satiety center neurons are sensitive to *blood glucose* levels: high levels increase and low levels decrease their activity. Low blood sugar reduces the inhibition of the feeding center by the satiety center, resulting in hunger and feeding. After absorption, increased blood sugar activates the satiety center, which will inhibit the feeding center. During ingestion, the stimulation of gustatory and mechanical sensors in the mouth and distention of the stomach inhibit the feeding center. Food in the gut also induces release of gut peptide hormones (e.g., cholecystokinin [CCK]) that stimulate the satiety center.

Leptin from fat tissue inhibits long-term food intake – Food intake can also be regulated on a long-term basis. A recently discovered protein hormone from the fat tissue – leptin – is thought to exert this effect. Increased fat is associated with an increased amount of leptin, which acts on the hypothalamus to suppress appetite and feeding (plate 140).

COMPOSITION & CALORIC VALUES OF BODY FUEL STORES

Caloric values of foods are different – Most *fats* (except cholesterol and phospholipids), most *proteins,* and all pure *carbohydrates* can be utilized as *fuels*, but not all fuels have similar *calorigenic* value. Caloric value of fuels is the amount of calories released upon complete oxidation to CO_2 and water (4.1 kcal/g of carbohydrate or protein, 9.3 kcal/g of fat).

Fats are ideal for fuel storage – In addition to providing more calories per unit weight, fats occupy less space, making them ideal substances for long-term fuel energy storage. Carbohydrates are very efficiently utilized for fuel by all cells, but they require much water and space for storage.

Fuel stores account for about 25% of body weight – In an average male weighing ~70 kg (154 lb), about 25% of the body weight is potentially available as fuel. The carbohydrate fuel sources, totaling about 520 g, are glucose (20 g in blood and liver) and glycogen (400 g in muscles and 100 g in liver). Proteins (~10 kg) make up 14% of the adult male body; of this, only 6 kg can be utilized as fuel (usually during fasting or starvation and mostly from liver proteins and muscles). Neutral fats in the *fat depots* (~10.5 kg) make up about 15% of the body weight, all of which can be utilized as fuel. These

fuel stores account in total for ~18 kg or 25% of body weight.

Fats, proteins, & carbohydrates account for 78%, 20%, & 2%, respectively, of total caloric values of fuels – The total fuel value of these substances is the product of the total weight of each fuel store and the respective caloric value per unit weight. In the example of an average-sized male (70 kg), a total of ~125,000 kcal is available in the body. Fats comprise ~78% of this total (98,000 kcal), proteins ~20% (25,000 kcal), and carbohydrates ~2% (2000 kcal) of total fuel value.

REGULATION OF ENERGY OUTPUT

Working of body organs accounts for basal caloric expenditure – Many organs (brain, heart, liver, and kidneys) are constantly working, utilizing the body's caloric intake or stores. Other organs, such as the muscles and digestive organs, work only part of the time. Some require a great deal of energy; the brain, liver, and muscles each use up 20% of body energies routinely, the heart 12%, the kidneys 8%, and the remaining organs 20% together. These basic energy requirements account for the basal metabolic rate of the average-sized human (2000 kcal/day).

Physical activity is the major regulator of energy expenditure – Arm, leg, and trunk muscles use energy during ordinary activities. More intense physical activities (uphill walking, climbing stairs, running) increase energy utilization up to 10-fold (climbing stairs utilizes 1200 kcal/hr of energy, six times more than walking) (plate 137). By utilizing food calories, increased physical activity reduces the need for storage of energy as fat.

Influence of sedentary vs. active conditions – Among those humans for whom physical activity is an integral part of routine daily life and work, obesity is less frequent. Modern conveniences (car, TV remote control) and sedentary office work reduce physical activity, encouraging fat storage and obesity (plate 139). Regular, voluntary exercise is a useful way of inducing fuel utilization for those with sedentary life styles; it also prevents fat storage and obesity.

FUEL MOBILIZATION DURING FASTING & STARVATION

Body fuel stores provide energy for about two months of survival – Prolonged fasting or starvation reduces activity and BMR, but vital organs still work to sustain life. Fuel stores are mobilized to provide energy and ensure survival. With a daily requirement of 2000 kcal for basic body needs for a 70 kg man, a total of ~125,000 kcal is available from all fuel stores (2000 from 500 g of carbohydrates, 25,000 from 6 kg of fuel proteins, and 98,000 from 11 kg of fats). Dividing the total of 125,000 kcal by 2000 kcal/day, we obtain ~62 days as the maximum period one can survive without food (with a lack of water, vitamins, or minerals, death occurs earlier).

Carbohydrates are the first to be utilized & tissue proteins the last – Liver and muscle glycogen and blood and liver glucose are the first to be utilized, providing energy for about one day. During the next days and weeks, all fat depots and the remaining labile protein reserves of the liver are mobilized. Finally, tissue proteins of muscle and bone are catabolized to use their amino acids for *gluconeogenesis*. Ketone formation is promoted. The brain and other tissues adapt to use ketones for energy (plate 127).

CN: Use yellow for A and red for C. Use dark colors for F, G, and J.
1. Color the upper two panels. Continue down the right side of the page through starvation.

2. Color the food-intake regulation diagram. Note the dotted line of step 2, which indicates that the normal inhibiting action (step 10) of the satiety center on the feeding center is turned off.

FUEL VALUES OF FOODSTUFF

FATS, PROTEINS, CARBOHYDRATES.

The caloric value (Cal/g) of fats is 9.3, compared to 4.1 for proteins (PR) and carbohydrates (CH). In a 70 kg male, fats constitute 15%, PR 14%, and CH <1% of body weight. All of the fat but only 6 kg of the protein can be used for fuel. Hence the total fuel value of fat is 78%, compared to 20% for PR and 2% for CH.

9.3 Cal/g.
4.1   4.1

RELATIVE WEIGHT

TOTAL FUEL VALUE.

78%

H_2O

15%
14%
0.7%

20%
2%

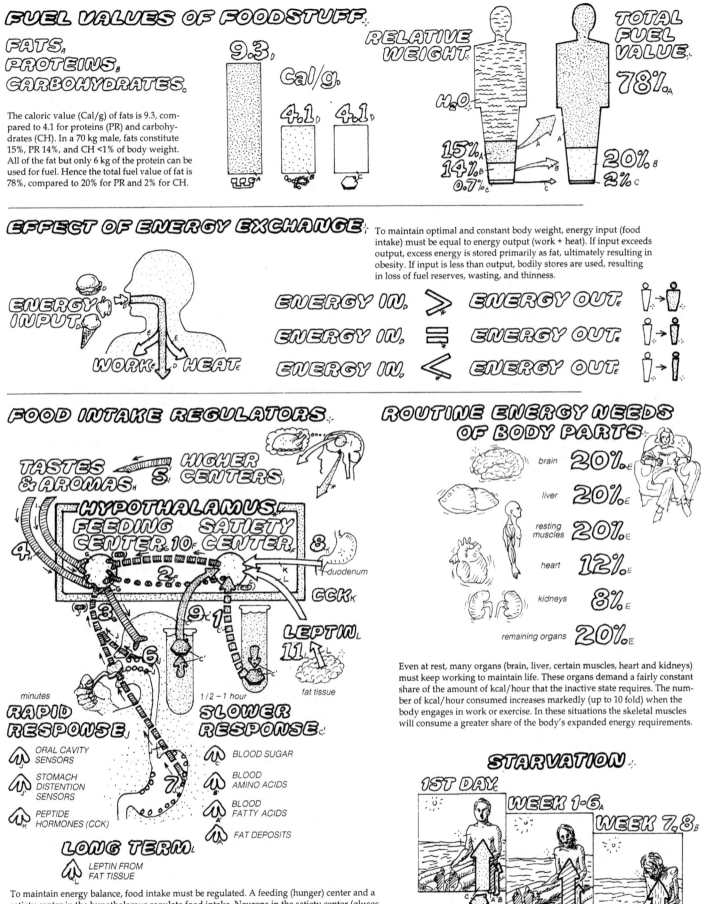

EFFECT OF ENERGY EXCHANGE:

To maintain optimal and constant body weight, energy input (food intake) must be equal to energy output (work + heat). If input exceeds output, excess energy is stored primarily as fat, ultimately resulting in obesity. If input is less than output, bodily stores are used, resulting in loss of fuel reserves, wasting, and thinness.

ENERGY INPUT.
WORK   HEAT.

ENERGY IN. > ENERGY OUT.
ENERGY IN. = ENERGY OUT.
ENERGY IN. < ENERGY OUT.

FOOD INTAKE REGULATORS.

TASTES & AROMAS.

HIGHER CENTERS.

HYPOTHALAMUS
FEEDING CENTER.10   SATIETY CENTER.

duodenum

CCK

LEPTIN

fat tissue

minutes

RAPID RESPONSE.

- ORAL CAVITY SENSORS
- STOMACH DISTENTION SENSORS
- PEPTIDE HORMONES (CCK)

1/2 – 1 hour

SLOWER RESPONSE.

- BLOOD SUGAR
- BLOOD AMINO ACIDS
- BLOOD FATTY ACIDS
- FAT DEPOSITS

LONG TERM.

- LEPTIN FROM FAT TISSUE

To maintain energy balance, food intake must be regulated. A feeding (hunger) center and a satiety center in the hypothalamus regulate food intake. Neurons in the satiety center (glucostat) respond to changes in blood Glu (1). Reduction in blood Glu decreases neuron activity, releasing the feeding center from inhibition (2). This stimulates the appetite and food seeking (3). Odors, tastes (4), and thoughts of food also stimulate the feeding center (5). Ingestion of food activates sensory nerves in the oral cavity (6) and causes distention of stomach (7), inhibiting the feeding center. Food entering duodenum induces the release of CCK which stimulates satiety center (8). Absorption of food increases blood Glu (9), increasing the activity of the satiety center, which in turn inhibits the feeding center (10). Increase in the size of fat deposits releases a hormone, leptin that has a long term stimulating effect on satiety center (11).

ROUTINE ENERGY NEEDS OF BODY PARTS

brain	20%
liver	20%
resting muscles	20%
heart	12%
kidneys	8%
remaining organs	20%

Even at rest, many organs (brain, liver, certain muscles, heart and kidneys) must keep working to maintain life. These organs demand a fairly constant share of the amount of kcal/hour that the inactive state requires. The number of kcal/hour consumed increases markedly (up to 10 fold) when the body engages in work or exercise. In these situations the skeletal muscles will consume a greater share of the body's expanded energy requirements.

STARVATION.

1ST DAY.   WEEK 1-6.   WEEK 7,8.

Fasting and starvation lead to the loss of fuel reserves. The meager CH reserves are the first to be used. Next, fats and labile PR are consumed. The structural-tissue PR are the last to be used. The brain and heart are usually spared. Ketones are mobilized from fatty acids, and the brain adapts by relying on ketones for energy.

Obesity is measured in terms of % body fat—Obesity is related to excess body fat and is measured as *percent body fat* relative to ideal body weight. Normal fat content in mature healthy males is between 12 and 18% (average, ~15%) of ideal body weight; 18 to 23% (average ~20%) for females. Men with >25% and women with >30% body fat are considered obese.

Body mass index (BMI) expresses weight relative to height & is proportional to fat content—BMI is defined as the ratio of body weight (in kg) divided by height (in m^2) or 704 × weight in lb/height (in inches^2). BMI does not directly measure body fat but normalizes weight by height; however, BMI increases in proportion with body fat. Most people with normal fat content have a BMI of 20 to 25; those below this range have less than normal and those above 25 have higher than normal BMI. Females with normal fat content often fall in the lower range of normal BMI (20 to 23), while males fall in the higher segment (23 to 25). BMI is less useful for individuals for large muscle mass (body builders, athletes), since they have high BMI with low body fat (<10%).

Overweight vs. obesity—According to 1998 guidelines of the National Institutes of Health, individuals with BMI between 25 and 29.9 are considered *overweight*; those over a BMI of 30 are *obese*. Three grades of obesity are recognized: Grade 1, with BMI of 30 to 34.9; Grade 2, with BMI of 35 to 39.9; and Grade 3, with BMI of 40 or more. Obese individuals in the last group show a high mortality rate (*morbid obesity*).

TYPES OF OBESITY

Hypertrophic vs. hyperplastic-hypertrophic obesity—Increased fat deposition is associated with increased size (*hypertrophy*) of fat cells. If fat cells increase in number (*hyperplasia*), the potential for excess fat gain is even more. Thus, two types of obesity are recognized. *Hypertrophic obesity* is the usual type occurring in adults and can be more easily managed by caloric restriction and exercise. The second type, *hyperplastic-hypertrophic obesity*, occurs mainly in children, whose bodies are in the state of hyperplastic growth, and also during pregnancy. This type of may be more difficult to reverse since it is not possible to eliminate the fat cells by diet or exercise, although it is possible to shrink them by these means.

Abdominal vs. lower-body obesity—Abdominal obesity occurs when excess body fat accumulates in the abdomen, around the waist (*pot belly, beer belly, apple-shape obesity*). This type occurs more often in men and shows a higher association with diseases that are related to obesity (diabetes and heart disease). Lower-body obesity results from fat accumulation in the buttocks and thighs. Such bodies resemble the shape of a pear (*pear-shape obesity*). This type is seen more often in women and during pregnancy. The *waist-to-hip ratio* (WHR) is based on differential fat distribution; it should be <1 in normal men and <0.8 in normal women. Higher values indicate increased abdominal obesity and health risk.

GENETIC & ENVIRONMENTAL ASPECTS OF OBESITY

Genetics play a major role in obesity—Both genetic and environmental factors contribute to obesity. Evidence for genetic factors comes from the fact that some individuals maintain stable body weight for much of their lives, without accumulating fat, while others easily gain fat after a few heavy meals or periods of low activity. Also, obesity runs in families. A child with two obese parents is very likely to follow suit.

Recent rise in obesity incidence shows environmental factors in obesity—Environmental determinants are best evidenced by obesity-prone people in whom a voluntary decrease in activity or increase in food intake leads to excess fat. Based on the BMI index, currently 50% of the U.S. population is overweight (BMI of 25 to 30) and 35% is obese (BMI >30), nearly twice the 1960s values. The causes of this recent rise in obesity lie in *decreased physical activity* at home, school, and workplace as well as in *increased consumption of food* (total caloric intake), fast foods and restaurant foods, and increased dietary fat in certain groups (not all).

Mere reduction of fat in the diet does not prevent weight gain unless it is accompanied by decreased caloric intake. Excess carbohydrates (pasta, french fries, sugar) are easily converted to body fat, particularly if one is inactive. Another problem is the rise in the incidence of childhood obesity, which, in view of the threat of hyperplastic obesity, may have further detrimental consequences.

Alterations in leptin, brown fat, insulin, hypothalamic food intake centers, & physical activity all may cause obesity—The physiological bases of differential propensity to obesity are not well understood. Differences in *brown fat* content, concentration of and receptors for the *leptin* hormone, *hypothalamic mechanisms* regulating food intake (*appetite* and *satiety*), and *hypersensitivity to insulin* may be involved. Leptin is a protein hormone released by the fat tissue in proportion to the body fat content; it acts via its hypothalamic receptors to depress food intake and increase activity. Animal experiments implicate the genes for leptin and its receptors in obesity. Mice missing the genes for leptin or its receptors show increased food intake and obesity. Treatment of these mice with leptin reverses the obesity. In humans, leptin levels are higher in women than in men and in obese individuals than in normal ones. But, leptin injection does not reduce obesity.

Insulin promotes fat storage following excess food intake (carbohydrates in particular); over a long time obesity may result (plate 123). The critical importance of physical activity and muscular exercise in weight control and weight loss as well as lack of activity in obesity were described in plate 138.

OBESITY MAY BE NORMAL OR ABNORMAL

Normal fat gain—Increased body fat is adaptive in humans living in cold regions with low food supply. Animals preparing for hibernation are markedly obese. Many humans increase food intake and body fat in winter and lose it in the summer. Pregnant women actively produce additional body fat to support fetal growth and lactation. Pubertal females deposit fat in their breasts and hips; without this extra fat, menarche is delayed and pregnancy does not occur.

Several major diseases are associated with obesity—Diabetes, coronary heart disease, hypertension, gallstones, ulcers, and kidney diseases occur more frequently in obese people. Among people with Type-II diabetes, 85% were obese prior to disease onset. Fat loss markedly improves Type-II diabetes and may be preventive. Similarly, low fat diet and weight loss decrease heart disease and high blood pressure; one mechanism for these effects may be through reduction of total plasma and LDL cholesterol.

CN: Use yellow for F and light colors for A–E.
1. Color the chart panels and corresponding numbers.

Color the little people on each side of the caption
2. Color the material under "Types of Obesity."

BODY MASS INDEX (BMI):

NORMAL WEIGHT VS. OBESITY:

NORMAL WEIGHT A
OVERWEIGHT B
OBESITY:
GRADE 1 C
GRADE 2 D
GRADE 3 E

$$BMI = \frac{WEIGHT\ (kg)}{HEIGHT\ (m^2)}$$

$$BMI = 704 \times \frac{WEIGHT\ (lb)}{HEIGHT\ (in^2)}$$

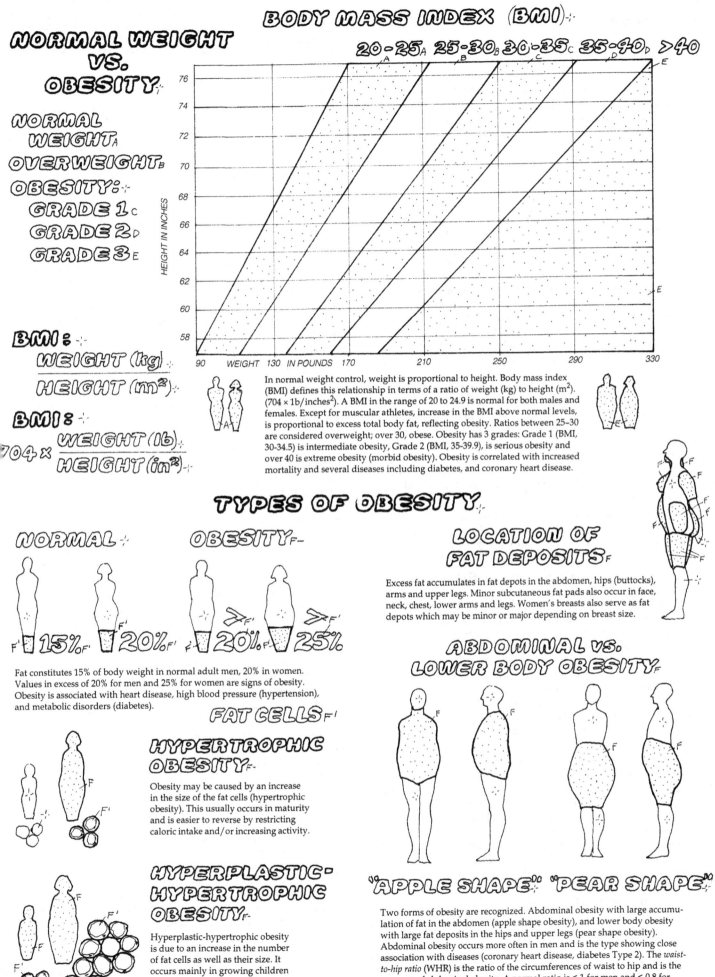

20-25 A   25-30 B   30-35 C   35-40 D   >40 E

HEIGHT IN INCHES

76 74 72 70 68 66 64 62 60 58

90   WEIGHT 130 IN POUNDS 170   210   250   290   330

In normal weight control, weight is proportional to height. Body mass index (BMI) defines this relationship in terms of a ratio of weight (kg) to height (m²). (704 × 1b/inches²). A BMI in the range of 20 to 24.9 is normal for both males and females. Except for muscular athletes, increase in the BMI above normal levels, is proportional to excess total body fat, reflecting obesity. Ratios between 25–30 are considered overweight; over 30, obese. Obesity has 3 grades: Grade 1 (BMI, 30-34.5) is intermediate obesity, Grade 2 (BMI, 35-39.9), is serious obesity and over 40 is extreme obesity (morbid obesity). Obesity is correlated with increased mortality and several diseases including diabetes, and coronary heart disease.

TYPES OF OBESITY

NORMAL   OBESITY F

15% F'   20% F'   20% F'   25%

Fat constitutes 15% of body weight in normal adult men, 20% in women. Values in excess of 20% for men and 25% for women are signs of obesity. Obesity is associated with heart disease, high blood pressure (hypertension), and metabolic disorders (diabetes).

FAT CELLS F'

HYPERTROPHIC OBESITY F-

Obesity may be caused by an increase in the size of the fat cells (hypertrophic obesity). This usually occurs in maturity and is easier to reverse by restricting caloric intake and/or increasing activity.

HYPERPLASTIC-HYPERTROPHIC OBESITY F-

Hyperplastic-hypertrophic obesity is due to an increase in the number of fat cells as well as their size. It occurs mainly in growing children and pregnant women and is more difficult to reverse.

LOCATION OF FAT DEPOSITS F

Excess fat accumulates in fat depots in the abdomen, hips (buttocks), arms and upper legs. Minor subcutaneous fat pads also occur in face, neck, chest, lower arms and legs. Women's breasts also serve as fat depots which may be minor or major depending on breast size.

ABDOMINAL VS. LOWER BODY OBESITY F

"APPLE SHAPE" "PEAR SHAPE"

Two forms of obesity are recognized. Abdominal obesity with large accumulation of fat in the abdomen (apple shape obesity), and lower body obesity with large fat deposits in the hips and upper legs (pear shape obesity). Abdominal obesity occurs more often in men and is the type showing close association with diseases (coronary heart disease, diabetes Type 2). The waist-to-hip ratio (WHR) is the ratio of the circumferences of waist to hip and is the measure of abdominal obesity. A normal ratio is < 1 for men and < 0.8 for women). For an adult to have a larger waist than hip size is a sign of obesity.

CONSTANCY & VARIATION IN BODY TEMPERATURE

Body temperature is kept constant by balancing heat gain & loss — Mammals and birds are warm-blooded animals (*homeotherms, endotherms*) since they can maintain a constant body temperature (around 37°C for humans and other mammals). To maintain the temperature of any system (or body) at a constant point, a balance must exist between *heat gain* and *heat loss*. The *metabolic heat* generated by oxidation of foodstuffs in the visceral organs and tissues (*body core*) is a constant source of heat (plates 5, 6, 137). This heat can be increased by *muscular activity* (shivering, running); by nervous and hormonal factors such as *sympathetic nervous activity, catecholamines,* and *thyroid hormones. Food ingestion,* especially protein foods, can also increases metabolic heat.

Body temperature varies in different regions — Although it would be ideal for all parts of the body to operate at 37°C, only the body core (i.e., the brain and visceral organs and tissues in the trunk) operate at this optimum temperature. The tissues of the extremities and the skin, being far from the core heat source and in direct contact with the outside, have much lower temperatures. For example, in a room temperature of about 21°C, hand and foot skin temperatures are about 28°C and 21°C, respectively. These values are about 34 to 35°C in a room temperature of about 35°C, because in the absence of limb movement, the only heat source is the arterial blood flowing from the viscera. Presumably, this heat is not sufficient to keep the limb tissues warm enough. *Frostbite* and *gangrene* (tissue death) in feet and hands that occurs at extremely cold temperatures is due to failure of an adequate blood and heat supply to these regions.

Diurnal variation in core temperature — Even the core temperature is not constant at all times. A circadian (diurnal, daily) cycle exists, core temperature being lowest in morning (36.7°C) and highest in the evening (37.2°C) (plate 107).

PHYSICAL & PHYSIOLOGICAL ASPECTS OF HEAT EXCHANGE & BODY TEMPERATURE

Physical mechanisms of heat exchange include radiation, convection, & conduction — The body can gain heat from external sources (sun rays, heaters). This kind of passive heat exchange is achieved by *radiation* (from the sun), by *convection* from near sources (e.g., heat from a heater in a room), and by *conduction* involving direct contact of warm objects (a heating blanket with the skin). Conduction, convection, and radiation can also work in reverse: to increase heat loss from the body. Heat loss occurs passively if the body is exposed to outside temperatures below its temperature; sitting on a cold chair warms the seat (conduction); the presence of a crowd in a cold room increases the room temperature (convection).

Skin plays a major role in heat exchange & thermal regulation — The body is equipped with physiological mechanisms that actively increase or decrease heat loss. The *skin* is the organ that plays a central role here. Heat can be lost through the skin in two ways: by direct exchange of heat between the blood circulating in the skin and the outside or by *evaporation* of water from the skin surface. The skin contains special thoroughfare-like blood *capillaries* that do not exchange nutrients with skin cells but function solely in exchanging heat with the outside.

Blood flows in these vessels whenever they are open. In cold weather, these vessels close up (cutaneous *vasoconstriction*), markedly reducing blood circulation in the skin and minimizing heat loss. In a hot environment, these special thermoregulatory vessels are open (cutaneous *vasodilation*), allowing through blood flow and increasing heat loss to the outside. Because of these mechanisms, blood flow in the skin, which at its height is nearly 10% of the total cardiac output, can vary more than a hundredfold, making heat exchange in the skin via blood circulation a very efficient and effective mechanism for heat loss and heat conservation.

Evaporation of water helps heat loss & occurs by insensible perspiration & sweating — The second way heat can be lost from the skin or other exposed surfaces such as the respiratory ducts is by *evaporation* of water. Water has a very high heat capacity (0.6 Cal/g), which means that the loss of 1 g of water from the body is accompanied by the loss of 600 calories of heat. This water loss occurs in two ways — *insensible perspiration* and active *sweating*.

Insensible perspiration — Loss of water from the skin at lower temperatures is called insensible perspiration because water diffuses through skin cells and pores and quickly evaporates; no sweat drops are formed. Similarly, considerable water and heat are lost in the respiratory passageways every day. Insensible perspiration accounts for the loss of more than 0.5 l of water per day (360 Cal of heat — about 20% of daily basal caloric production)!

Sweating — When internal body temperatures increase to above 37°C, active secretion of *sweat* containing water and salt by the *sweat glands* begins, markedly raising the rate of water evaporation and heat loss. Sweat glands are exocrine (eccrine) glands located abundantly in many parts of the skin (e.g., the forehead, palms, and soles). Animals lacking sweat glands, like the dog, *pant*, thereby markedly increasing air flow in the respiratory passages, resulting in similar increases in evaporation and heat loss rates.

Body hair (fur) decreases heat loss — A third way by which skin is able to decrease heat loss is by the use of *body hair (fur)*, a mechanism of little value in humans but of great value in the furry animals (bear, sheep, etc.), particularly those living in cold climates. In cold weather, skin hairs stand up (*piloerection*), which causes *entrapment* of the air in the hair web. The trapped air forms an insulating layer because blood now exchanges heat not with the flowing cold air on the outside, but with a stationary air layer in the hair web. Warm clothes in humans, particularly wool, perform the same role.

Subcutaneous fat pads prevent heat loss — The skin by itself is only a weak insulation. However, in many animals, including humans, the fat under the skin (*subcutaneous fat pads*) has the dual function of acting as both a very effective insulation and a source of metabolic energy.

Brown fat produces much heat during oxidation — In fetuses, newborns, and infants of humans as well as in many other animals, a special type of fatty tissue, *brown fat*, is present. The numerous *mitochondria* of these fat cells oxidize the fat in such a way as to produce a great deal more heat than ATP. This heat acts as a furnace, generating heat to protect against the cold. This heat may be one reason why newborns do not shiver when exposed to cold. Most adult humans lose their brown fat, but some may keep it (plate 133).

CN: Use light blue for F, brown for J, and red for L.
1. Color titles A–F and their corresponding structures in the upper right diagram. Then color the one below it. Complete the material on perspiration (F¹) to the left.
2. Color the panels on piloerection and brown fat.
3. Color the bottom panel, beginning at the center, then do the outer temperature extremes.

CORE TEMPERATURE (T)

The continuous operation and oxidative metabolism of visceral organs generates heat in the body core. By balancing heat loss and gain, core temperature (T) is kept constant around 37°C (98.6°F).

SKIN AS INSULATOR

Skin mediates the heat exchange between body core and environment. Special thermoregulatory mechanisms enable the skin to change its insulating capacity.

RADIATION

The body can exchange heat with distant objects of different T via radiation. Heat from the sun or from heaters is conveyed by radiation.

CONDUCTION

Direct contact with hot or cold objects allows a heat exchange by conduction. A cold chair will thus be warmed by a seated individual.

CONVECTION

Movement of air in a room (convection) increases the heat exchange between body and environment. A fan blowing air will help the body to cool off.

EVAPORATION

The evaporation of water from body surfaces is an efficient way of losing heat because of the great deal of heat needed to evaporate water (0.6 Cal/g of water).

INSENSIBLE PERSPIRATION

.6L/DAY

SWEATING

SWEAT GLAND

A major source of heat loss is evaporation from body surfaces (skin, respiratory tract). When outside T is cool (below 20 ° C), water evaporates from exposed skin and lungs by insensible perspiration. When environmental T approaches body T, sweating (water and salt secretion by sweat glands) increases. Evaporation of sweat increases heat loss. Dogs, which have no sweat glands, lose water and heat by panting, which increases perspiration through respiratory surfaces.

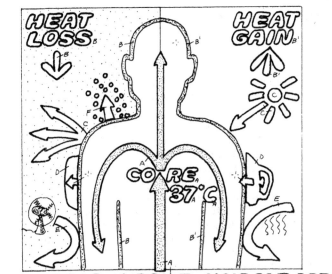

HEAT LOSS

HEAT GAIN

CORE 37°C

BMR, FOOD INTAKE, MUSCLE ACT.

Heat gain is increased by food intake and by muscular activity as well as by radiation from sunlight, etc. Heat may be gained or lost by contact with warm or cold objects. Hair and clothing reduce heat loss while sweating and evaporation increase heat loss.

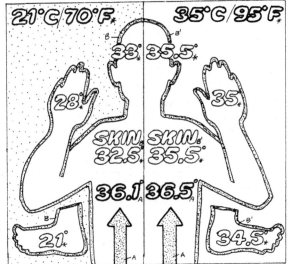

21°C/70°F   35°C/95°F

33   35.5

28   35

SKIN 32.5   SKIN 35.5

36.1   36.5

21°   34.5°

BODY TEMP.

Body T is not uniform throughout. While body core is homeothermic, the extremities vary in T. In a cool room (21°C), core T is near 37°C, while hand and foot T are 28 and 21°C respectively. In a hot room (35°C), the T of the extremities approaches room T.

PILOERECTION
HAIR SHAFT
MUSCLE
TRAPPED AIR

moving air

stationary air

In furry animals, cold T causes hair to stand up (piloerection), increasing the thickness of a trapped layer of air and minimizing heat loss. In humans, clothing performs similar functions.

BROWN FAT
FAT CELL
MITOCHONDRION
FAT GRANULE

A special type of fatty tissue (brown fat), rich mitochondria, can generate much heat (but less ATP) through lipolysis. Brown fat is found primarily in infants and animals and is located in the back and around the scapulae.

ENVIRONMENTAL TEMPERATURE

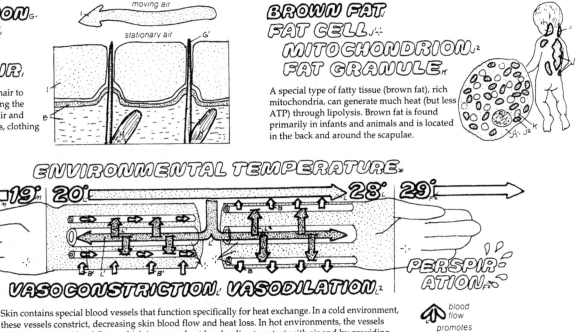

19°   20°   28°   29°

HEAT PROD.

VASOCONSTRICTION   VASODILATION

PERSPIRATION

blood flow prevents heat loss

blood flow promotes heat loss

Skin contains special blood vessels that function specifically for heat exchange. In a cold environment, these vessels constrict, decreasing skin blood flow and heat loss. In hot environments, the vessels dilate, increasing blood flow, which increases heat loss by direct contact with air and by providing fluids for sweat glands. These vessels are supplied by sympathetic nerve fibers.

To keep its core temperature constant, the body uses physiological mechanisms that change the extent of *heat loss* and *heat gain* in the body. A decrease in temperature increases heat gain and decreases heat loss and vice versa.

Hypothalamus has a "thermostat" — The thermoregulatory adjustments of the body are controlled by a "hypothalamic thermostat-like center" (HTC), whose neurons have a normal *setpoint* at 37 °C and respond to changes in skin and blood temperature. Deviation of hypothalamic temperature away from this setpoint activates bodily responses in the opposite direction, to return core temperature to the normal level. The HTC works in conjunction with other hypothalamic, autonomic, and higher nervous centers (plate 107). Some of these responses are involuntary, through the autonomic nervous system, some are neurohormonal, and others are semi-voluntary or voluntary behaviors.

PHYSIOLOGICAL RESPONSES TO COLD

Cold is communicated to brain centers by blood & skin cold receptors — When exposed to cold, one's skin temperature quickly drops, stimulating the *skin cold receptors*; and *blood* flowing in the skin is cooled. The impulse activity of cold receptor nerves increases with decreasing degrees of skin temperature. Their signals are received by both the HTC and the higher *cortical centers*. The HTC is also activated by the change in *blood temperature*. The HTC initiates responses that promote heat gain while inhibiting brain centers that promote heat loss.

Sympathetic centers initiate heat gain responses & inhibit heat loss responses — Activation of *sympathetic centers* results in several responses: (1) *cutaneous vasoconstriction*, or constriction of skin blood vessels (because of norepinephrine released from sympathetic fibers), decreased cutaneous blood flow, and decreased heat loss; (2) increase in *metabolic rate*, causing thermogenesis due to increased adrenal medullary *epinephrine* secretion; (3) contraction of body hair muscles, resulting in *piloerection* (hair standing up, also called *horripilation*), which traps air in a thick layer next to the skin, decreasing heat loss (piloerection is particularly effective in furry animals but of little value in humans); and (4) increase in *brown fat* oxidation, causing thermogenesis (a response that important mainly in infants, some animals and few adults).

In addition to the above responses mediated by the sympathetic system, a *shivering center* in the hypothalamus is activated, which in turn activates *brain stem motor centers* to initiate involuntary contraction of skeletal muscles. Shivering generates a lot of heat.

Voluntary & involuntary behaviors help increase heat gain & decrease heat loss — Cold also activates some compensatory *behavioral responses* directed at increasing heat production or decreasing heat loss. For example, *curling up* decreases exposed body surface area and therefore heat loss. *Huddling* and *cuddling* seen in animals and humans, *voluntary physical activity* (rubbing the hands, pacing), and sheltering next to a heat source and wearing warm clothing are other examples of voluntary cold-fighting responses. Voluntary or semi-voluntary behaviors are activated by the responses of the higher brain centers (cortex and limbic system) to the uncomfortable sensation of cold. In many animals and in children, prolonged exposure to a cold climate increases the basal secretory rate

of *thyroid hormones*, which, by their potent calorigenic actions, increase heat production (plate 119). As a result of these compensatory responses, the body gets warmer. The HTC detects the increasing warmth and diminishes the heat-producing and heat-loss-prevention responses.

PHYSIOLOGICAL RESPONSES TO HEAT

Heat is reported to brain centers by blood & skin warmth receptors — When the body is exposed to heat (e.g., from the sun, fire, or excessive clothing), body temperature rises. Here too, both skin *warmth receptors* and blood convey the changes to the HTC. But warmth receptors are less effective than blood, because there are fewer of them than there are cold receptors and because blood volume and flow in the skin are high during exposure to heat (vasodilation). The HTC initiates compensatory responses, some of which increase heat loss and others decrease heat production.

Cutaneous vasodilation & sweating greatly increase heat loss — In response to heat, the adrenergic activity of the sympathetic nervous system, controlling cutaneous vasoconstriction and metabolic rate, is inhibited, resulting in *cutaneous vasodilation* and reduced metabolic rate. These responses increase heat loss from the skin and decrease heat production in the core. If heat is sufficiently intense, a particular division of the autonomic nervous system (*cholinergic sympathetic fibers, releasing acetylcholine*) that innervates the *sweat glands* is activated, causing sweating. Sweating markedly increases heat loss from the skin and is an effective involuntary heat-fighting response in humans (600 Cal heat lost per liter of sweat).

Behavioral responses to heat help increase heat loss & decrease heat gain — A hot person becomes lethargic and tends to rest or lie down with limbs spread out. These behavioral states decrease heat production and increase heat loss. Heat loss is also enhanced by wearing loose and light clothing, fanning, drinking cold drinks, and swimming.

CAUSES AND MECHANISMS OF FEVER

Fever is caused by release of cytokines from white blood cells in response to infection — *Fever*, defined as an increase in core body temperature of one to several degrees, occurs during illness when infectious microbial agents enter the body. These microbes release *toxins* (exogenous pyrogens = fever producers) that stimulate the *white cells* (monocytes, macrophages) to release their *cytokine* hormones (e.g., *interleukins*). These cytokines act as *endogenous pyrogens* and stimulate the HTC neurons, raising their setpoint (e.g., to 40 °C). To reach this new setpoint, one shivers to increase heat gain; skin vasoconstriction decreases heat loss.

Fever is a natural defense response — The hot state of the body is detrimental to bacterial growth and deactivates toxins. When the infection is cured, pyrogen secretion decreases and the HTC is set back to its original 37 °C. Now the body attempts to cool down by skin vasodilation and sweating.

High fevers may have deleterious consequences. *Heat shock* may occur with very high temperatures (above 42 °C) and may be fatal if not treated. Treatment consists of physical measures to cool the body or certain drugs. Aspirin reduces fever by inhibiting the formation of *prostaglandins* (plate 129). These local hormones are produced in the hypothalamus as a result of the action of cytokines such as interleukins.

CN: Use red for D and dark colors for A, B, and E.
1. At the very top of the page color the titles "heat loss" and "heat production" and their respective "increase and decrease" symbols. These symbols will appear throughout, summarizing the result of each factor in the thermal adaptive process.

2. Color the events that follow a drop in environmental temperature. Use the order of letter labels C-J as your guide.
3. Go to the bottom panel and color the adaptation to a rise in environmental temperature.
4. Color the elements of fever.

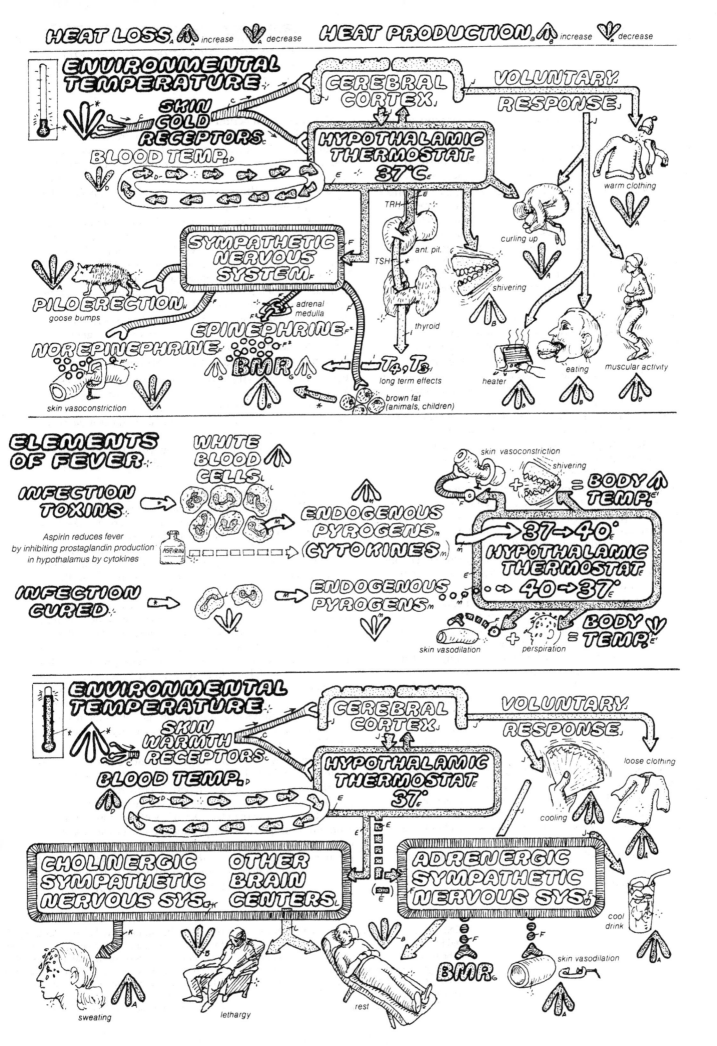

HEAT LOSS ⋏ increase ⋎ decrease HEAT PRODUCTION ⋏ increase ⋎ decrease

ENVIRONMENTAL TEMPERATURE

SKIN COLD RECEPTORS

BLOOD TEMP.

CEREBRAL CORTEX

VOLUNTARY RESPONSE

HYPOTHALAMIC THERMOSTAT 37°C

warm clothing

curling up

SYMPATHETIC NERVOUS SYSTEM

TRH

ant. pit.

TSH

shivering

PILOERECTION
goose bumps

adrenal medulla

EPINEPHRINE

thyroid

eating

muscular activity

NOREPINEPHRINE

BMR

T_4, T_3
long term effects

heater

skin vasoconstriction

brown fat (animals, children)

ELEMENTS OF FEVER

WHITE BLOOD CELLS ⋏

skin vasoconstriction

shivering

INFECTION TOXINS

Aspirin reduces fever by inhibiting prostaglandin production in hypothalamus by cytokines

ENDOGENOUS PYROGENS (CYTOKINES)

+ = BODY TEMP. ⋏

37→40° HYPOTHALAMIC THERMOSTAT 40→37°

INFECTION CURED

ENDOGENOUS PYROGENS ⋎

skin vasodilation

+ = BODY TEMP. ⋎

perspiration

ENVIRONMENTAL TEMPERATURE

SKIN WARMTH RECEPTORS

BLOOD TEMP.

CEREBRAL CORTEX

VOLUNTARY RESPONSE

HYPOTHALAMIC THERMOSTAT 37°

cooling

loose clothing

CHOLINERGIC SYMPATHETIC NERVOUS SYS.

OTHER BRAIN CENTERS

ADRENERGIC SYMPATHETIC NERVOUS SYS.

cool drink

BMR

skin vasodilation

sweating

lethargy

rest

Blood serves as the body's principal extracellular fluid and is critical for its maintenance and survival. Blood is involved in several physiological regulation systems, ensuring homeostasis and defense mechanisms of the body. Blood transport, its fluid and flow properties are critical for its functions.

BLOOD TRANSPORTS OXYGEN, NUTRIENTS, HORMONES, HEAT
The flow of blood through the tissues permits its numerous *transport functions*, ensuring nutrition and respiration for body cells. Blood obtains *nutrients* such as glucose and vitamins from the small intestine and *oxygen* from the lungs and delivers these to the body cells as it passes through the tissue capillaries. Blood also removes the toxic waste products of cellular metabolism (*metabolites*), such as *urea* and *carbon dioxide*, from the tissue environment and eliminates them as it circulates through the kidneys and lungs, respectively.

Roles in red & white cell transport and hemostasis — The *red blood cells* (RBCs) contain the oxygen-binding protein *hemoglobin*. They help transport oxygen from the lungs to the tissues and carbon dioxide from the tissues to the lungs. Blood also transports the *white blood cells* (WBCs) to injury sites, where they defend the body by destroying invading microorganisms and their toxins. Blood flow also permits transport of the antibodies, body's major defense molecules, to diverse targets. Platelets and proteins of the blood also participate in hemostasis, the process by which injured blood vessels are plugged to prevent blood loss.

Roles in hormone transport & heat exchange — In addition, blood carries the *hormones* from the endocrine glands to their target organs in other locations. Blood is also important for temperature regulation, transporting heat from the warm body core to the colder extremities (limbs). Blood flow through the skin is critical for heat exchange, increasing in hot environments to allow heat loss and decreasing in cold environments to preserve heat.

BLOOD CONSISTS OF PLASMA & HEMATOCRIT (BLOOD CELLS)
Two components, a cellular one called the *hematocrit* and a fluid medium called the *plasma*, make up blood tissue. The blood cells float freely within the plasma. Separation of blood into these two components is achieved by spinning (centrifuging) it in a small capillary tube (hematocrit tube) which separates the blood into a colorless fluid supernatant on the top and a red precipitate on the bottom. The supernatant, amounting to about 55% of the blood volume, is the *plasma*, consisting mainly of water (91%), which helps dissolve the blood proteins (e.g., *fibrinogen, albumins*, and *globulins*) as well as nutrients, hormones, and electrolytes.

Hematocrit consists of red and white blood cells and platelets — The hematocrit forms the remaining 45% of the blood volume and consists mainly of red blood cells (*erythrocytes*), the most abundant of the blood cells. Blood cells are often referred to as *"formed elements."* The white blood cells (*leukocytes*) and the platelets (*thrombocytes*), being smaller in number, constitute only a small fraction of the hematocrit, forming a very thin yellowish band between the red hematocrit and the colorless plasma supernatant.

Comparison of serum and plasma — Another way to separate the blood fluid and cells is to allow a drop of blood to stand for a while. It will separate into a dense red core called the *clot*, surrounded by a colorless fluid called the *serum*. The clot has a composition similar to that of the hematocrit; the serum resembles the plasma. However, the serum lacks the plasma protein fibrinogen, which relates to the clot part.

Males have higher hematocrit — Blood accounts for about 8% of the body weight. On the average, human males have more blood (5.6 L) than females (4.5 L), although blood volume in females increases during pregnancy. Males' blood is also more cellular (higher proportion of red cells in particular), having a hematocrit of about 47%, compared to 42% in females and children. The higher blood volume of males reflects their larger size, and the higher hematocrit reflects a higher density of red cells. This is a response to higher metabolic rate and increased oxygen needs in males, factors that are compatible with higher muscle mass and work load.

BONE MARROW IS THE MAJOR SOURCE OF BLOOD CELLS
Most plasma proteins are made by the liver; various sources in the body contribute other dissolved plasma constituents. Blood cells, however, are formed mainly in the *bone marrow*. The mass of bone marrow in a single bone may appear insignificant, but the total mass of bone marrow tissue in the body is very large; the bone marrow, the skin, and the liver are the three largest body organs.

Red & yellow marrow as the active & resting source of blood cells — In the adult, active bone marrow is the *red marrow* found in bones of the trunk and head (*sternum, ribs, vertebrae*, and *skull*). The red marrow in these bones is the *primary source* for blood cells. In growing children, red marrow is also found in the long bones of the lower extremities (*femur* and *tibia*). In the adult, the latter bones do not entirely lose their ability to make blood cells, but provide possible secondary sources for blood cell formation. This source is activated when the primary sources are unable to keep up with the demand. Under such conditions, the *liver* and *spleen* can also make blood cells. Indeed, the liver is the principal source of RBCs in embryonic and early fetal periods; the spleen produces RBCs slightly later in fetal life. In extreme emergencies, such as massive blood loss resulting from hemorrhage or destruction of generative cells of marrow following exposure to ionizing radiation, the adult's liver and spleen, as well as the resting *yellow marrow* in the secondary sources, can again produce new blood cells.

Bone marrow stem cells form the blood cells — Blood cells are formed in the red marrow from the proliferation and differentiation of *stem cells*, which permanently reside there. One line of stem cells forms the red cells, another the white cells, and yet another the platelets. A variety of hormonal and humoral controls adjust the production rate of various blood cells in response to physiological needs. For example, the kidney hormone *erythropoietin* stimulates red cell production, and a hormone called *thrombopoietin* stimulates platelet production. Several humoral factors regulate white cell production.

CN: Use a very light or straw color for A. Use red for D.
1. Begin with the illustrations in the upper panel. Color gray the titles of all elements that make up 2% of plasma. These substances are given individual colors in the lower left diagram.
2. Color the formed elements. For all practical purposes, the number of red blood cells is equal to the hematocrit, and both receive the color red.
3. Color the sources of formed elements in red bone marrow. Note that the adolescent and the fetus have different primary sources and are colored accordingly.
4. Color the transport function of blood.

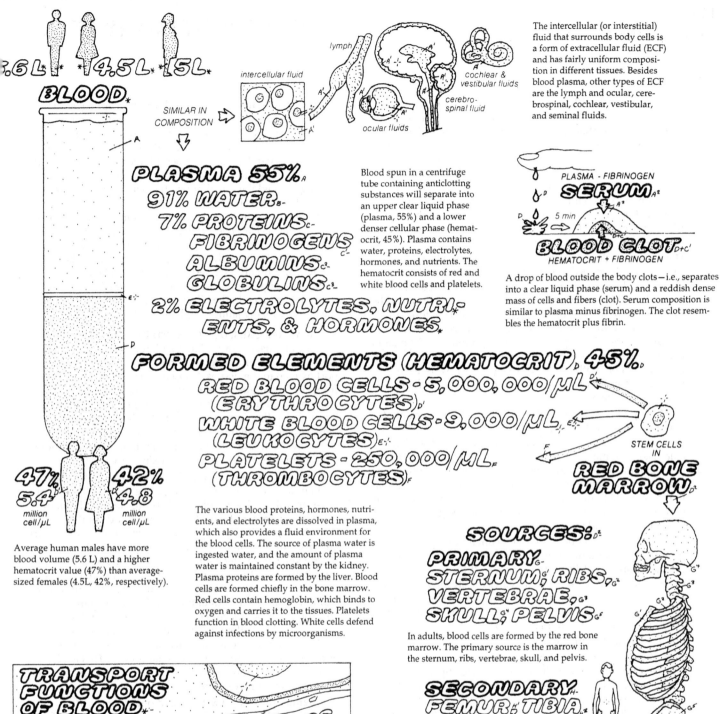

5.6 L **4.5 L** **5 L**

BLOOD

The intercellular (or interstitial) fluid that surrounds body cells is a form of extracellular fluid (ECF) and has fairly uniform composition in different tissues. Besides blood plasma, other types of ECF are the lymph and ocular, cerebrospinal, cochlear, vestibular, and seminal fluids.

SIMILAR IN COMPOSITION

intercellular fluid

lymph

cochlear & vestibular fluids

cerebro-spinal fluid

ocular fluids

PLASMA 55%
91% WATER
7% PROTEINS
FIBRINOGENS
ALBUMINS
GLOBULINS
2% ELECTROLYTES, NUTRIENTS, & HORMONES

Blood spun in a centrifuge tube containing anticlotting substances will separate into an upper clear liquid phase (plasma, 55%) and a lower denser cellular phase (hematocrit, 45%). Plasma contains water, proteins, electrolytes, hormones, and nutrients. The hematocrit consists of red and white blood cells and platelets.

PLASMA - FIBRINOGEN

SERUM

5 min

BLOOD CLOT
HEMATOCRIT + FIBRINOGEN

A drop of blood outside the body clots—i.e., separates into a clear liquid phase (serum) and a reddish dense mass of cells and fibers (clot). Serum composition is similar to plasma minus fibrinogen. The clot resembles the hematocrit plus fibrin.

FORMED ELEMENTS (HEMATOCRIT), 45%
RED BLOOD CELLS - 5,000,000/µL
(ERYTHROCYTES)
WHITE BLOOD CELLS - 9,000/µL
(LEUKOCYTES)
PLATELETS - 250,000/µL
(THROMBOCYTES)

STEM CELLS IN

RED BONE MARROW

47%
5.4
million cell/µL

42%
4.8
million cell/µL

Average human males have more blood volume (5.6 L) and a higher hematocrit value (47%) than average-sized females (4.5L, 42%, respectively).

The various blood proteins, hormones, nutrients, and electrolytes are dissolved in plasma, which also provides a fluid environment for the blood cells. The source of plasma water is ingested water, and the amount of plasma water is maintained constant by the kidney. Plasma proteins are formed by the liver. Blood cells are formed chiefly in the bone marrow. Red cells contain hemoglobin, which binds to oxygen and carries it to the tissues. Platelets function in blood clotting. White cells defend against infections by microorganisms.

SOURCES:
PRIMARY
STERNUM, RIBS, VERTEBRAE, SKULL, PELVIS

In adults, blood cells are formed by the red bone marrow. The primary source is the marrow in the sternum, ribs, vertebrae, skull, and pelvis.

SECONDARY
FEMUR, TIBIA

If necessary, blood cells can be formed by the marrow in the femur and tibia (secondary source). During adolescence, the femur and tibia are additional primary sources.

primary in adolescence

TERTIARY
LIVER & SPLEEN

In emergencies (excessive blood loss or degeneration of bone marrow), blood cells can be formed in liver and spleen. In the embryo, these organs are the primary source of blood cells.

primary in fetus

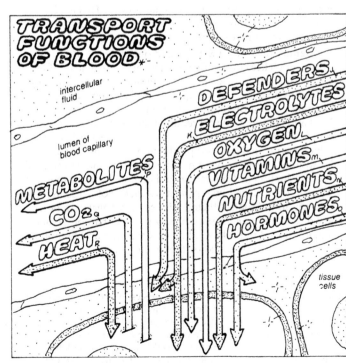

TRANSPORT FUNCTIONS OF BLOOD

intercellular fluid

lumen of blood capillary

DEFENDERS
ELECTROLYTES
OXYGEN
VITAMINS
NUTRIENTS
HORMONES

METABOLITES
CO₂
HEAT

tissue cells

A main function of blood is that of a fluid medium for transport of dissolved substances and blood cells. Blood transports oxygen, nutrients, and hormones to tissues and removes products of cellular metabolism (metabolites)—e.g., CO_2 and urea—away from the tissues. Blood also helps to exchange heat between the body core and periphery, as well as transporting white blood cells (defender cells) to sites of injury and infection.

The *red blood cells* (RBCs, *erythrocytes*) are the most abundant type of blood cells. At a density of about 5×10^6 per µL of blood, they total about 30×10^{12} cells in circulation. The RBCs' main function is in oxygen transport, and their shape is highly adapted to this function. Circulating RBCs resemble biconcave discs, having average dimensions of 7.5 µm by 2 µm (1 µm at the center). The biconcave shape maximizes the rate of diffusion of oxygen into and out of the RBC and its binding with the *hemoglobin* molecules packed within. The shape of RBCs can change as they move through various blood vessels; in veins they inflate, and in capillaries they fold.

HEMOGLOBIN TRANSPORTS O₂ IN RED BLOOD CELLS

Mature circulating RBCs have no a *nucleus* or cytoplasmic organelles. Instead, all the available space is packed with hemoglobin, the blood's reddish oxygen-binding protein. The protein part of the hemoglobin (*globin*) consists of four subunits. Each subunit is attached to a *heme* molecule, an iron-containing organic porphyrin ring compound. One iron atom is bound to each of the hemes. In the ferrous state (Fe^{++}), iron binds reversibly with molecular oxygen (O_2). Thus, each hemoglobin molecule can bind and transport up to four O_2 molecules (plate 53).

Hemoglobin content of RBC is critical—The amount of hemoglobin in the blood determines the blood's capacity to carry oxygen. This capacity diminishes in blood disorders (*anemias*). Anemias can occur becaause of either reduced hemoglobin content in the RBCs or reduced RBC production in the bone marrow (see below). Concentration of hemoglobin in the normal blood is about 160 g/L in males and 140 g/L in females. This amounts to a total of 900 and 700 g per whole blood in males and females, respectively.

In addition to hemoglobin, RBCs contain the cytoskeletal protein *tubulin* and the contractile protein *actin*, which aid in the mechanisms of RBC shape changes. RBCs do not have mitochondria but possess the enzymes for glycolysis (anaerobic oxidation of glucose). They also contain 2,3-diphosphoglycerate (DPG), a compound that regulates the binding of oxygen with hemoglobin (plates 53, 54).

LIFE CYCLE OF RED BLOOD CELLS

Red cells are formed in bone marrow—Formation of red cells (*erythropoiesis*) occurs in the *bone marrow*. Special *stem cells* residing in the marrow proliferate to give rise to all types of blood cells. Progenitor cells of RBCs (*erythroblasts*) differentiate into RBCs within a few days. During this process the developing RBCs synthesize and pack hemoglobin within their cytoplasm and then ultimately lose their nuclei. The enucleated red cells are mature and ready to function. At this time they leave the bone marrow and enter the bloodstream, where they begin to transport oxygen and carbon dioxide.

Erythropoietin stimulates RBC production—Several factors regulate RBC production, the most important being the arterial *oxygen pressure* (pO_2). In low pO_2 conditions, such as at high altitude, the low oxygen pressure in the arterial blood stimulates the kidneys to release a hormone, *erythropoietin*, into the blood. Erythropoietin stimulates the bone marrow to increase formation of new RBCs. The increased RBC number compensates for decreased O_2 transport per RBC.

Aged RBCs are destroyed by liver macrophages—The life span of circulating RBCs is about four months, after which they age. Aged cells are recognized by the tissue *macrophages* (*Kupffer cells*), large white blood cells residing in the *liver* and *spleen* blood capillaries. The macrophages phagocytize and destroy the aged RBCs and remove them from circulation. During hemoglobin breakdown, the heme is metabolized to *iron* and *bilirubin*. The iron is recycled by bone marrow for hemoglobin synthesis and bilirubin is removed by the liver into the bile as *bile pigments*, which are excreted in the intestine with the feces. These pigments give feces their light brown color. Some of the bile pigments are reabsorbed, recirculated, and finally excreted in the kidney. The bilirubin metabolite pigments cause the yellow color of urine. If bilirubin is not excreted in the bile, it backflows and accumulates in the blood, causing jaundice (plate 77).

ANEMIAS ARE CAUSED BY REDUCED RBCS OR HEMOGLOBIN

Anemias are diseases associated with reduced content of blood hemoglobin. Anemias decrease the blood's capacity to transport oxygen to the tissues. Consequences of anemias range from simple fatigue to death. Various conditions can cause anemia, including *direct blood loss* due to severe menstruation, internal bleeding from a gastrointestinal ulcer, accidental hemorrhage, or *failure of bone marrow* to produce new RBCs. This may occur as a result of exposure to high doses of ionizing radiation or to certain drugs, toxins, or viruses.

Pernicious anemia results from vitamin B₁₂ deficiency—Certain dietary deficiencies or digestive disorders also may cause anemias. Absence of *vitamin* B_{12} (cyanocobalamine), a substance necessary for erythropoiesis, cou¹d result in severe RBC shortage. Vitamin B_{12} is plentiful in foods of animal origin (liver, meats, milk) but is absent in plant foods, so a strictly vegetarian diet could result in serious deficiency in this vitamin, leading to *pernicious anemia*. More frequently, pernicious anemia is caused by a diminished absorption of vitamin B_{12} in the intestinal tract. To facilitate vitamin B_{12} absorption, the stomach secretes a protein called the *intrinsic factor* (plate 79). Following diseases of the stomach (gastritis) or after its surgical removal, intrinsic-factor production declines, reducing absorption of vitamin B_{12}. This leads to diminished hemoglobin synthesis and RBC production and ultimately to pernicious anemia.

Anemia due to iron & folic acid deficiency, kidney failure, or sickle cell condition—Dietary deficiencies of *folic acid* and *iron* may cause anemia. In women, iron is routinely lost during menstruation, requiring replacement. During pregnancy and growth, the increased need for production of RBCs and hemoglobin will require increased dietary intake of iron, folic acid, and vitamin B_{12}. Kidney disease or loss of a kidney results in decreased production of erythropoietin. In the absence of erythropoietin, bone marrow is not stimulated, RBC production is diminished, and anemia results. Anemias can also be produced by *increased destruction* of RBCs, which occurs in individuals afflicted with *sickle cell anemia*, a hereditary disease of red cells particularly prevalent among black people. Sickled red cells tend to stick together and hemolyze and are rapidly destroyed by the macrophages.

CN: Use red for A structures and dark colors for B, D, and J.
1. Color the upper panel, noting the four hemes (B) that carry the oxygen molecule (H), here represented by an arrow, with the oxygen title being found in the section below it.
2. Color the process of erythropoiesis, and follow the numbered sequence.
3. Color regulation of red blood cell production.
4. Color the various anemias due to reduced red blood cells, noting the letter labels of the titles carefully. Not all are colored red.

RED BLOOD CELL (ERYTHROCYTE)

Red blood cells (RBCs) have no cellular organelles; instead they are packed with hemoglobin (Hb), which carries oxygen. The biconcave disc shape of RBCs allow for rapid diffusion of oxygen. This shape changes as RBCs squeeze through the narrow capillaries.

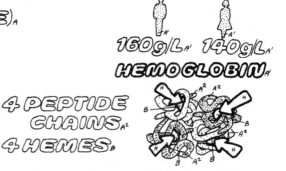

160 g/L 140 g/L
HEMOGLOBIN

4 PEPTIDE CHAINS
4 HEMES

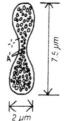

Hemoglobin (Hb) has a protein part, globin, with four subunits (two alpha chains and two beta chains). Each subunit has a heme. Each heme has one iron, which in the ferrous state (Fe^{++}) binds with one O_2.

ERYTHROPOIESIS

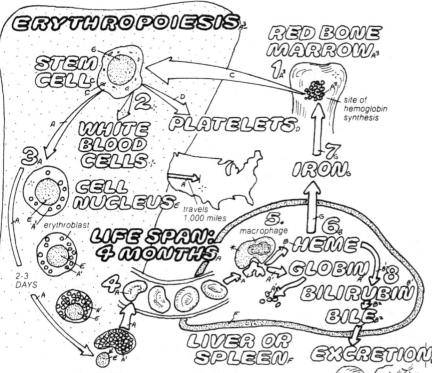

Formation of red cells or erythrocytes (erythropoiesis) in adults occurs in red bone marrow (1). The stem cells that form blood cells (2) divide, forming progenitors of red cells (erythroblasts), which possess nuclei (3). Within 2–3 days, these cells develop, fill up with Hb, lose their nuclei, and enter circulation (4). After four months, RBCs age and are destroyed by macrophages in liver and spleen (5). During catabolism of Hb, iron is released from heme (6) and recycled for heme and Hb synthesis in bone marrow (7); heme is metabolized to bilirubin (8) and excreted via bile or the kidneys.

REGULATION OF RBCs

RBC formation is regulated by the kidney hormone erythropoietin. A reduction in blood O_2 levels (e.g., after a hemorrhage or at high altitude) stimulates the kidney to increase erythropoietin release; erythropoietin stimulates the red bone marrow to increase red cell formation. An increased number of RBCs enhances O_2 transport and tissue levels.

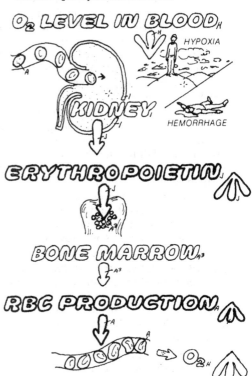

O_2 LEVEL IN BLOOD
KIDNEY
ERYTHROPOIETIN
BONE MARROW
RBC PRODUCTION

CAUSES OF ANEMIA

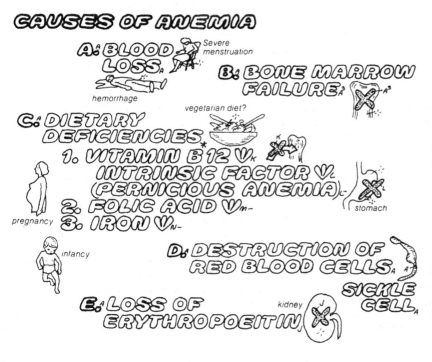

A. BLOOD LOSS — Severe menstruation, hemorrhage

B. BONE MARROW FAILURE

C. DIETARY DEFICIENCIES — vegetarian diet?
1. VITAMIN B12, INTRINSIC FACTOR (PERNICIOUS ANEMIA)
2. FOLIC ACID
3. IRON — pregnancy, infancy, stomach

D. DESTRUCTION OF RED BLOOD CELLS — SICKLE CELL

E. LOSS OF ERYTHROPOEITIN — kidney

Anemias are characterized by reduced hemoglobin (Hb) and RBC levels and are associated with many problems of reduced oxygen supply to tissues. Anemias are caused by different factors and conditions: **A.** Blood loss due to hemorrhage, internal bleeding, or, more frequently, severe menstruation. **B.** Diseases of bone marrow (aplasia) caused by endogenous abnormalities or by exposure to ionizing radiation or toxic chemicals. **C.** Dietary deficiencies. Dietary intake of iron, folic acid, and vitamin B_{12} must be increased in pregnancy and during growth. Since iron, folic acid, and vitamin B_{12} are essential for erythropoiesis, their reduced intake causes anemia. Pernicious anemia occurs in cases of severe vitamin B_{12} deficiency, which may occur in a strictly vegetarian diet lacking in vitamin B_{12}. A more frequent cause is the absence of the intrinsic factor, a protein required for the intestinal absorption of vitamin B_{12}; intrinsic factor is secreted by the stomach gland cells and is reduced following gastritis or stomach loss by surgery or cancer. **D.** Following the increased destruction of abnormal red cells, as in sickle cell anemia. **E.** Diseases of the kidney resulting in the reduced secretion of erythropoietin.

When blood from two different individuals is mixed outside the body, the *red blood cells* (RBCs) may clump together, separate from the plasma, and precipitate as solid masses, a process called *agglutination*. Consequences of agglutination are severe if it occurs inside the body — e.g., during blood transfusion. The mechanisms of agglutination are of critical importance in pathology and clinical medicine and have been well researched in relation to blood composition and physiology, genetics, and other aspects of blood compatibility.

Humans form different blood groups — On the basis of agglutination reactions, humans are classified into genetically determined blood groups. Blood from certain groups can be mixed without any undesirable consequences (agglutination), but blood of members of certain other groups cannot be mixed. The basis of these differences is the genetically determined immunological disparity between the blood types.

Cellular mechanisms of agglutination — The external surface of RBC plasma membrane contains several different *complex oligosaccharides*, *glycolipids*, and *glycoprotein* substances called *agglutinogens*, which have *antigenic* properties. They tend to react with specific antibody-like protein molecules called *agglutinins* present in the blood plasma of other individuals. The reaction is similar to the antigen-antibody reactions that occur in the immunological defense system of the body (plates 147 & 148). The types of agglutinogens are unique to individuals from a common gene pool. Identical twins have the same sets of agglutinogens, but fraternal twins may have different sets.

If the blood of an individual with a specific agglutinogen is mixed with blood containing the agglutinin against that particular agglutinogen, the agglutinins will bind with agglutinogens of several different RBCs. As a result, the affected RBCs clump together, or agglutinate. Agglutination of blood may result in anemia and other serious blood and vascular disorders. Antigenic substances similar to the agglutinogens of the RBC surface are also found in some other tissues but agglutination usually occurs in the blood, owing to the presence of both plasma agglutinins and RBC agglutinogens and also because, unlike the red cells, other tissue cells are stationary.

THE ABO & RH SYSTEMS: THE TWO MAJOR BLOOD GROUPS
Based on the various agglutinogens and agglutinins present in the blood of different individuals and the miscibility of blood between them, several blood groups have been identified. The *ABO system* and the *Rhesus (Rh) system* are the best-known groupings.

Four blood types within the ABO group — In the ABO system, humans form four different blood groups — A, B, AB, and O, on the basis of two agglutinogens, A and B, and their corresponding agglutinins. Members of the A blood group carry the agglutinogen A on their red cells and the agglutinin B in their plasma. Subtypes of A1 and A2 have been described based on different amounts of antigens on the red cell surfaces. Members of the B group carry agglutinogen B and agglutinin A. Members of the AB group have both agglutinogens A and B but none of the corresponding agglutinins. Members of the O group have both agglutinins in their plasma but neither of the two agglutinogens.

The agglutinogens A and B of the RBCs are glycolipids, while those in other tissues are glycoproteins. Among white people, A and B types comprise 41% and 10% of the population, while the O and AB constitute 45% and 4%, respectively.

Type A blood can be mixed with A and B with B, but not A with B — Blood from type A individuals can be mixed with that of other type A people. Similarly, members of the B blood group can share their blood with other members of that group. However, type A blood, with agglutinogen A, should not be mixed with type B blood, since type B blood contains agglutinin A, which reacts with agglutinogen A, resulting in agglutination; this has possibly lethal consequences.

Type O people are universal donors & type AB universal recipients — Individuals belonging to the O group are called *universal donors* because they lack agglutinogens A and B on their red cells, therefore eliminating the chances of agglutination in the recipient. Members of the AB group are called *universal recipients* as they lack agglutinins A and B in their plasma; allowing them to accept blood from the other three types.

The Rh blood grouping is based on the Rh antigen — Another important blood group system is the Rh system, which is based on the presence of the *Rh factor (antigen-D, agglutinogen-D)* on the surface of RBCs. Individuals with this factor are called *Rh positive, or Rh(+)*; those lacking it are called *Rh negative, or Rh(−)*. Rh(+) people outnumber the Rh(−) ones by about 6 to 1. In contrast to the ABO system, agglutinin-D reacting against the Rh factor is not normally circulating, but is present within several weeks of exposure to the agglutinogen (the Rh factor). Upon second exposure to Rh(+) blood, the Rh(−) recipient experiences a severe agglutination reaction.

Hemolytic disease of the newborn, as an example of Rh incompatibility — The most serious cases of agglutination due to Rh incompatibility are observed in fetuses and newborns. The offspring of an Rh(+) father and an Rh(−) mother will usually (but not always) be Rh(+). The Rh factor (antibody) in the fetus will be antigenic to the mother's blood. During delivery of the fetus, some fetal blood is mixed with the Rh(−) maternal blood. Within a few weeks, the mother produces an antibody against the Rh agglutinin. During a second pregnancy with an Rh(+) fetus, these antibodies may enter the fetal blood and react with the fetal RBCs, causing their agglutination and lysis, a disorder called *erythroblastosis fetalis*, or *hemolytic disease of the newborn*. These fetuses and newborns are at risk of severe anemia.

Treatment of erythroblastosis fetalis — The incidence of this disorder increases with each subsequent Rh(+) pregnancy. To prevent the consequences of erythroblastosis fetalis, the newborn's blood can be replaced with Rh(−) blood, enabling the infant to survive for a few months. By the time the infant's own Rh(+) red blood cells are produced, all traces of the maternal Rh agglutinin will have disappeared. To prevent erythroblastosis fetalis from occurring, the Rh(−) mother can be injected, after the first Rh(+) pregnancy, with some Rh-agglutinin, like a vaccination. In time, the treated mother produces high titers of antibodies against the Rh-agglutinin (itself an antibody). These anti-antibodies deactivate all maternal Rh-agglutinins, preventing their transfer to the next fetus.

CN: Use red for D.
1. Color the four blood groups at the top, noting that the plasma protein (E) in the test tube diagrams remains uncolored. The arrows at the bottom of the tubes represent acceptable transfusion possibilities.

2. Color the blood typing process on the right. Note that the cluster of clumping red cells in the upper right corner is to be colored.
3. Color the samples of blood group percentages in the box on the left. Note that the individual's head (circle) remains uncolored.

BLOOD AGGLUTINATION

RED BLOOD CELL ANTIGENS
- AGGLUTINOGEN A
- AGGLUTINOGEN B

BLOOD PLASMA ANTIBODIES
- AGGLUTININ A
- AGGLUTININ B

Agglutination, or clumping of many red cells together, is caused by the reaction between the agglutinogens (complex glycolipid substances) on the surface of red cells from one person and specific antibody-like agglutinin proteins in the plasma of another individual. When plasma agglutinins simultaneously react with the agglutinogens on different RBCs, clumping occurs. Agglutination is similar to immune reactions, agglutinogens being the antigens and agglutinins the antibodies.

ABO BLOOD GROUPS

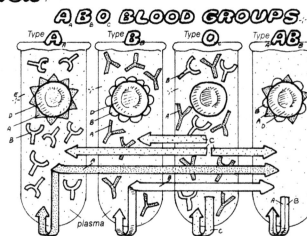

Type A Type B Type O Type AB

plasma

On the basis of two different agglutinogens, A and B, on the surface of red cells, humans are divided into four major blood groups. Individuals having agglutinogen A are type A, those with agglutinogen B are type B, and those having both agglutinogens are type AB. Type O people have neither A nor B agglutinogens. Type A blood plasma contains agglutinin B, and type B blood has agglutinin A. Plasma of type AB has neither agglutinin; type O plasma has both agglutinins. Therefore, type A blood must not be mixed with type B and vice versa, or agglutination will result. However, those with type O blood can donate blood to all other types (universal donors) but can receive only from another O type. Type AB individuals can receive blood from all types (universal recipients) but can give blood only to another AB type.

Blood types are inherited. Identical twins have the same blood type. Incidence of each blood type varies markedly within major human races.

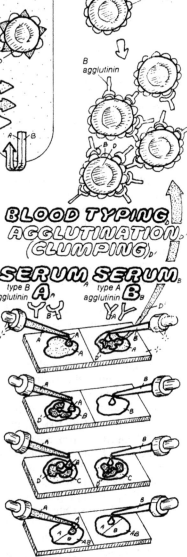

BLOOD TYPING, AGGLUTINATION (CLUMPING)

SERUM type B agglutinin A   SERUM type A agglutinin B

Blood types of individuals can be determined by mixing samples of their blood (each of the 4 groups is shown here on a separate slide) with a drop of a known serum (plasma) containing either anti-A or anti-B agglutinin (antibody), to see which combination causes agglutination.

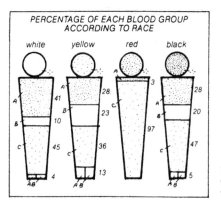

PERCENTAGE OF EACH BLOOD GROUP ACCORDING TO RACE

white   yellow   red   black

	white	yellow	red	black
A	41	28		28
B	10	23	3	20
C	45	36	97	47
AB	4	13		5

RHESUS BLOOD GROUP SYSTEM

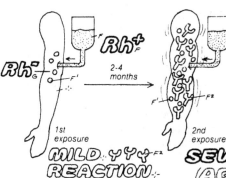

Rh⁺ ANTIGENS
ANTI Rh⁺ ANTIBODY

The red cells in most humans also contain another agglutinogen, the Rh factor. The blood of an Rh(–) person receiving Rh(+) blood will not agglutinate immediately. However, the recipient's body will soon respond by producing an anti-Rh agglutinin (antibody) against the foreign protein, the Rh factor in the donor's RBC. Upon a second transfusion with Rh(+) blood, agglutination occurs.

Rh⁻   Rh⁺

2–4 months

1st exposure   2nd exposure

MILD REACTION   SEVERE REACTION (AGGLUTINATION)

HEMOLYTIC DISEASE OF THE NEWBORN

1. 1ST PREGNANCY
2. AFTER DELIVERY
3. 2ND PREGNANCY
4. EXCHANGE OF BLOOD
5. AVOIDING FUTURE PROBLEMS IN STEP 2.

(1) Rh(+) fathers and Rh(–) mothers may have an Rh(+) fetus. Fetal cells containing the Rh factor may enter the maternal blood during pregnancy or at birth. (2) Within months, the mother produces anti-Rh antibodies (agglutinins). (3) Upon a second pregnancy, these antibodies may enter the fetal blood, causing agglutination (erythroblastosis fetalis; hemolytic disease of the newborn). (4) The newborn's blood has to be replaced by Rh(–) blood to eliminate the maternal anti-Rh. When the infant's own Rh(+) blood is produced again, no maternal anti-Rh will be present to cause agglutination. (5) One preventive approach is to inject the postpartum mother once with anti-Rh agglutinin. She will produce antibodies that will deactivate all anti-Rh agglutinins, eliminating any future effect.

To prevent blood loss after tissue injury, blood vessels constrict and form barriers (*platelet plug* and *blood clot*) to seal the site of injury. These events constitute *hemostasis*.

Serotonin release from platelets causes vasoconstriction in injured vessels — Tissue injury often severs the connective tissue of the vascular wall, exposing *collagen fibers*. The fragile blood *platelets* flowing by these rough surfaces adhere and rupture, releasing their *serotonin,* a potent local *vasoconstrictor* agent that immediately stimulates contraction of *smooth muscle cells* in the walls of injured arterioles and even the smaller arteries. This constriction effectively but temporarily reduces or blocks blood flow in these vessels.

FORMATION OF HEMOSTATIC (PLATELET) PLUG

Platelets form a hemostatic plug , temporarily sealing the injury site — After vasoconstriction, a longer-lasting measure occurs, consisting of formation of a *hemostatic plug (platelet plug)*. To form this barrier, platelets aggregate and adhere to each other and to the site of injury, creating a loose and temporary seal against blood loss.

Thromboxane A$_2$ released from injured platelets induces platelet aggregation — *Thromboxane A$_2$* is an *eicosanoid* (related to prostaglandins) and causes the nearby platelets to aggregate and adhere to those already bound to the injured wall, forming a clump. This clump quickly grows by continued platelet aggregation and finally forms a temporary hemostatic plug (platelet plug) against blood loss. To ensure that platelet plug growth is restricted to the wound area, the endothelial cells adjacent to the wound area release another eicosanoid, a prostaglandin compound called *prostacyclin* (PGI$_2$), which strongly inhibits platelet aggregation.

FORMATION & BIOCHEMISTRY OF BLOOD CLOTS

Fibrin fibers form a net over the platelet plug, trapping red cells & forming a blood clot — The rapidly forming and transient platelet plug is next reinforced by deposition of a meshwork of a fibrous protein called *fibrin* that forms a net over the platelet plug. The net traps red blood cells, forming a fairly rigid and strong barrier. Initially loose, the fibrin net gradually tightens, forming a true *blood clot* and sealing the wound until tissue regeneration repairs the wall.

Fibrin is formed from fibrinogen by action of thrombin — The *liver* produces a large protein called *fibrinogen* (profibrin), circulateing in the blood. Vessel wall injury activates a blood protease enzyme called *thrombin,* hydrolyzing fibrinogen to *fibrin,* a smaller fibrous protein. Thrombin normally circulates in the blood in its inactive form, *prothrombin.*

Thrombin is activated by factor X, via the intrinsic or extrinsic pathways — The activation of prothrombin is the key step in the clotting mechanism and requires the presence of *calcium ions* and a protein factor called *factor X* (ten). Activation of factor X can occur by either of two pathways: the *intrinsic* (blood) pathway involves the activation of factor XII, which originates from blood-related sources; the *extrinsic* (tissue) pathway involves the production from the injured tissue (endothelial cells) of another enzyme called *thromboplastin* (factor III). Thromboplastin can directly activate factor X, but factor XII must activate several other factors, which in turn activate factor X. The precipitated fibrin is initially loose; in the presence of another blood factor (factor XIII), it becomes tight, rigidifying the clot.

Antithrombin & heparin prevent clot formation in absence of injury — Blood clots pose a major danger if they form in healthy blood vessels, as they block blood flow in small vessels and are responsible for major health problems such as heart attacks and strokes. To prevent unnecessary formation of blood clots, circulating *anticlotting* factors such as *antithrombin-III* and heparin inhibit the activation of thrombin and thereby clot formation. Heparin, an endothelial cell surface protein, acts as a cofactor for activation of antithrombin-III.

Low doses of aspirin have anticlotting actions — By inhibiting the cyclo-oxygenase enzyme that forms thromboxane A$_2$, aspirin reduces aggregation of platelets, which precedes clot formation (see above). Aspirin use has been a major factor in reducing strokes and heart attacks in patients with coronary heart disease.

CLOT CONTRACTION AND DISSOLUTION

Clot contraction improves hemostasis & facilitates vessel wall healing — Once a clot forms, it begins to *contract.* Contraction is an active process requiring ATP and contraction of *actin* filaments in the platelet *pseudopods.* Clot contraction causes extrusion of the plasma trapped within the clot and shortening of the pseudopods. Because the edges of the clot are attached to the edges of the injured tissue, clot contraction is believed to bring the injured edges closer together, improving hemostasis and facilitating wound closure and repair.

Digestion of fibrin by plasmin helps lyse blood clot — Once an injured vessel wall has healed, the clot dissolves and is removed. Plasmin (fibrinolytin) digests the fibrin net, resulting in clot breakdown. Plasmin is formed from a precursor protein called *plasminogen.* Activation of plasminogen is brought about by a protein called *tissue plasminogen activator* (tPA), produced by the endothelial cells of blood vessels and incorporated in the clot.

Clinical use of clot lytic agent — Since blood clots are involved in heart attacks and brain strokes, their destruction (lysis) is of major clinical interest. Plasminogen activators such as tPA are injected immediately after the onset of a heart attack or stroke to lyse the clots in coronary and brain vessels. The improved circulation reduces permanent damage to heart and brain tissue.

ABNORMALITIES OF CLOT FORMATION

Hemophilia is caused by deficiency of blood clotting factors — Several diseases and nutritional deficiencies interfere with proper clotting. *Hemophilia* (bleeding sickness) is a group of hereditary diseases characterized by deficient hemostasis and continued blood loss after injury. It is caused by the lack of one of the blood clotting factors. Deficiency in factor VIII occurs in *type A,* the most common form of hemophilia (75%), which affects males (e.g., male descendants of Queen Victoria of England). Preventive treatment consists of injections of the missing clotting protein.

Reduced platelets & vitamin K deficiency also diminish blood clotting — Reduced platelet production (*thrombocytopenia*) by the bone marrow caused by ionizing radiation damage, disease, or toxic exposure of the bone marrow to drugs is another cause of deficient clotting. A third cause is dietary deficiency of *vitamin* K, required for the synthesis of *prothrombin* in the liver.

CN: Use red for A and dark colors for I and L.
1. Completely color the illustrations 1–5 depicting the formation of a blood clot. Do not color the material including the two pathways (under illustrations 3 and 4) until you have completed illustration 5. Note that in 1 and 2 the blood (A) is colored as a solid band, whereas in 3–5 only the red cells (A^1) and the platelets (E) are colored.
2. Color the diagram of the two pathways leading to the dissolution of a blood clot (under 5).

FORMATION OF THE BLOOD CLOT

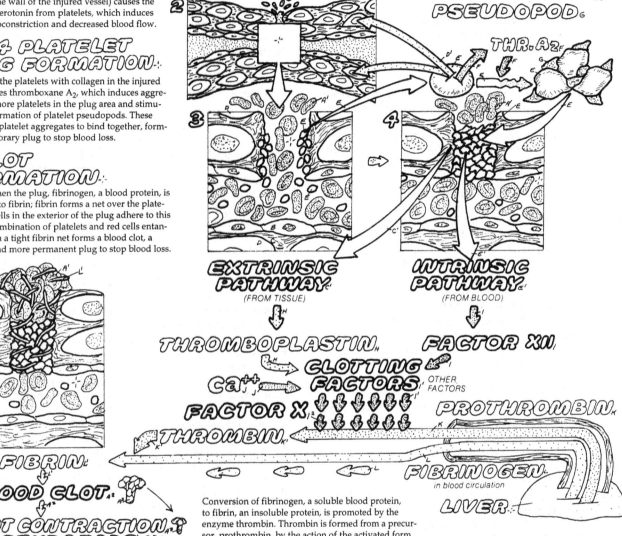

1 INJURY TO WALL OF BLOOD VESSEL

Injury (e.g., a cut) to a blood vessel is followed by a series of reactions that result in the formation of a blood clot, which seals the injured opening and prevents the loss of blood (hemostasis).

2 VASOCONSTRICTION

Adhesion of blood platelets to the exposed collagen fibers (in the wall of the injured vessel) causes the release of serotonin from platelets, which induces strong vasoconstriction and decreased blood flow.

3 & 4 PLATELET PLUG FORMATION

Contact of the platelets with collagen in the injured wall releases thromboxane A_2, which induces aggregation of more platelets in the plug area and stimulates the formation of platelet pseudopods. These enable the platelet aggregates to bind together, forming a temporary plug to stop blood loss.

5 CLOT FORMATION

To strengthen the plug, fibrinogen, a blood protein, is converted to fibrin; fibrin forms a net over the platelets. Red cells in the exterior of the plug adhere to this net. The combination of platelets and red cells entangled within a tight fibrin net forms a blood clot, a stronger and more permanent plug to stop blood loss.

BLOOD_A_ RED CELLS_A'_
ENDOTHELIAL CELL_B_
COLLAGEN FIBER_C_
SMOOTH MUSCLE_D_
PLATELETS_E_
VASOCONSTRICTOR_D'_
THROMBOXANE A_2_F_
PSEUDOPOD_G_

THR. A_2_F_

EXTRINSIC PATHWAY
(FROM TISSUE)

INTRINSIC PATHWAY
(FROM BLOOD)

THROMBOPLASTIN_H_

FACTOR XII_I_

Ca^{++}_J_ → CLOTTING FACTORS

OTHER FACTORS

FACTOR X_I_

PROTHROMBIN_K_

THROMBIN_K'_

PROTHROMBIN_K_

FIBRIN_I_

FIBRINOGEN_L_
in blood circulation

LIVER

Conversion of fibrinogen, a soluble blood protein, to fibrin, an insoluble protein, is promoted by the enzyme thrombin. Thrombin is formed from a precursor, prothrombin, by the action of the activated form of factor X. Activation of factor X depends on a cascade of reactions involving the formation of several other factors via two routes: extrinsic (from injured tissue) or intrinsic (from blood). Injured tissue releases thromboplastin; the other factors (particularly calcium ions) are provided by the blood. During tissue repair, the clot contracts by withdrawal of the pseudopods. To dissolve the clot, the enzyme plasmin lyses (breaks up) the fibrin net; plasmin is formed from an inactive precursor, plasminogen.

BLOOD CLOT_A''_

CLOT CONTRACTION (PSEUDOPODS)_G_

PLASMINOGEN_m_

PLASMIN_m_ → CLOT DISSOLVES

CONDITIONS PREVENTING CLOTTING

HEMOPHILIA*

Hemophilia is a hereditary disorder in which one or more of the blood clotting factors are absent. As a result, blood clots slowly.

CLOTTING FACTORS
FIBRIN

THROMBOCYTOPENIA*

In the disease thrombocytopenia, platelet formation by the red bone marrow is defective. Platelet deficiency prevents clotting.

PLATELET PRODUCTION
RED MARROW_A4_

VITAMIN K DEFICIENCY*

Vitamin K, provided in the diet or by intestinal bacteria, is one of the factors necessary for clotting, as it is needed for prothrombin synthesis in the liver.

PROTHROMBIN
CLOTTING FACTORS

TYPES AND GENERAL FUNCTIONS OF WHITE BLOOD CELLS

Both forms of white cells (granulocytes & agranulocytes) function in body's defense against infections—Based on their morphology, *white blood cells* (WBCs, leukocytes) are classified as *granulocytes*—i.e., those with cytoplasmic granules (*neutrophils, eosinophils,* and *basophils*)—and *agranulocytes*—i.e., those without granules (*monocytes, macrophages,* and *lymphocytes*). There are more granulocytes than agranulocytes. Among granulocytes, neutrophils are the most abundant; among agranulocytes, lymphocytes outnumber the others. Regardless of their differing morphology, all WBCs share a common function—i.e., helping to defend the body against foreign cells and infections, although each type performs a specific function (see below and plates 147, 148).

White blood cells originate mostly in the bone marrow—Granulocytes and agranulocytes originate in the *bone marrow*, where they form from the proliferative division of committed *stem cells*. Upon entry into the circulation, most of these WBCs participate in the inborn and nonspecific defensive reactions to invading infectious agents as well as in response to tissue injury and inflammation.

Most lymphocytes originate from lymphatic organs—The less numerous lymphocytes originate from another line of stem cells that reside either in the bone marrow or in parts of the *lymphatic system*. Upon formation, the immature lymphocytes temporarily migrate into certain lymphatic organs (*lymph nodes, thymus*), where they differentiate and mature, becoming specialized to carry out their major function: defending the body against invading microorganisms through "acquired" immune reactions.

Functionally, white blood cells are divided into two broad categories: (1) those that participate in *nonspecific inborn immune responses* to infections and *inflammations* caused by tissue injury; and (2) those that take part in *acquired immune responses*. Lymphocytes participate mainly in the second category; other white cells take part in the first.

NATURAL (NON-SPECIFIC) RESPONSES TO INJURY & INFECTIONS

Three lines of defense in response to injury—The granulocytes and phagocytic agranulocytes provide an early phase as well as three separate and consecutive lines of defense against microbial infections in an injury site. Each phase involves active participation of one type of WBCs.

Release of heparin & histamine by basophils and tissue mast cells begins the early response to injury—Upon injury to the protective epithelial tissue covering the body, *microbes* (e.g., bacteria) enter the body, release their toxins, and create local infection. This condition stimulates the *mast cells* (which resemble the basophils but reside in tissues) to release their granules containing *heparin* and *histamine* within the tissue spaces. Nearby *basophils* may do the same in the blood. Heparin prevents blood coagulation; histamine causes *vasodilation* and *increased permeability* of the local blood vessels to *blood proteins* and blood cells. Blood proteins and fluids leak into the injured site, causing *edema* (swelling). Gradually, the fluid in the swelling clots, trapping the bacteria and preventing their further penetration into the body.

Tissue macrophages provide the first line of defense by phagocytizing the microbes—At this time, the *tissue macro-phages* that permanently reside in many tissues, such as skin and lungs, attack the microbes and destroy them by *phagocytosis*. For this reason, the tissue macrophages are called the *first line of defense*. Phagocytosis consists of engulfing the microbes via the formation of *pseudopods*, followed by *endocytosis* of the phagocytic vesicle. Next, the *endocytotic vesicle* is incorporated into the *lysosomes* of the phagocytes, where the microbe is digested by *lysosomal enzymes*.

Migration of neutrophils to the injury site creates the second line of defense—A few hours after injury, if the infection persists, the number of neutrophils increases severalfold in the blood, particularly near the infection site. The neutrophils squeeze through the spaces between the *capillary endothelial cells* by forming *filopodia* and displacing themselves (*diapedesis*). Once inside the injured site, the neutrophils begin to phagocytize the microbes in the same manner as the tissue macrophages. Neutrophils make up the *second line of defense*.

Monocytes provide the third line of defense—If the tissue macrophages and neutrophils do not adequately counter the infection, then the *agranular monocytes* move into the injury site in the same manner as the neutrophils. Monocytes are initially small and incapable of phagocytosis. Within an hour after leaving the blood, they enlarge, attaining a shape like that of the tissue macrophages. Then they begin to phagocytize the microbes and the dead neutrophils. Monocytes may in fact be the source of new tissue macrophages, which die after phagocytosis. The monocytes are called the *third line of defense*. Usually, these three lines of defense are sufficient to eliminate the source of infection.

Release of cytokines & chemokines stimulates proliferation & migration of WBCs—The increase in number (proliferation) and migratory responses (*chemotaxis*) of the phagocytic WBCs to the injury site are controlled by humoral factors, mainly *cytokines* and *chemokines*, released from the injured tissue and/or certain white cells. The phagocytes find their way to the site of injury by *chemotaxis* or similar guiding mechanisms. Chemokines are special types of cytokines. *Interleukin-3* stimulates the migration and chemotaxis of eosinophils, while *interleukin-8* induces these responses in the neutrophils. Other humoral factors increase the permeability of blood *sinusoids* in the bone marrow, releasing fresh neutrophils and monocytes into the blood. A *adhesion molecules* on the surface of cells regulate the attachment of WBCs to the tissue cells and infectious agents.

Fibroblasts proliferate to seal the injured site & begin wound repair—The terminal phases of the natural defense involves wound repair. Gradually, the *fibroblast* cells of the connective tissue proliferate, sealing off the injured tissue to begin repair. A *pus sac* forms, containing fluid, dead cells, and dead microbes. This pus is either extruded or gradually cleared off by the macrophages.

Fever reaction & acquired immune responses will be activated against persisting infections—If these nonspecific rapid natural defense reactions are not sufficient to eliminate the infection, the toxin intrusion in the blood activates other defensive responses such as the fever reaction and, more effectively, lymphocyte reactions, which lead to acquired immune responses (see plates 147, 148).

CN: Use red for A, purple for J, lightest colors for structures C–H, and dark colors for I, K, and N.
1. Color the various white blood cells at the top of the page, beginning with their origin in red bone marrow (A).
2. Color the nonspecific response to a microbe invasion, following the numbered titles. When coloring the second and third boxes, color in the background or larger structures before coloring the smaller ones, such as proteins (K) or microbes (1). For number 3, color the tiny histamine molecules as well as the mast cell (E^1) Color the numeral 6, but not the arrow representing the movement of fluid into the tissues.
3. Color the enlargement of phagocytosis and the ma rophage action below it.

WHITE BLOOD CELLS.

White blood cells (WBCs, leukocytes) defend the body against foreign infections (bacteria, viruses). Most WBCs originate in bone marrow from undifferentiated stem cells. Lymphocytes proliferate primarily in the lymphatic organs (thymus, spleen, lymph nodes).

GRANULOCYTES.*
NEUTROPHILS. 60–70%c
EOSINOPHILS. 2–4%d
BASOPHILS. .5–1%e
MAST CELL.e'

The **granulocytes** (neutrophils, eosinophils, and basophils) constitute the majority of WBCs; their cytoplasm contains granules and their nuclei are polymorphic. Granulocytes and the monocytes take part in the natural immune responses (inflammation, phagocytosis) against invading microbes. **Neutrophils, which** constitute the bulk of granulocytes, are phagocytes and are capable of diapedesis. They recognize bacteria, adhere to them, expand, and produce pseudopods to engulf and digest them. The number of neutrophils in the blood increases markedly after infection. **Eosinophils** constitute 2–4% of WBCs; they are weakly phagocytic but exhibit strong chemotaxis (attraction to injury and infection sites). They may be specialized for digesting complex products of antigen-antibody reactions. **Basophils** constitute the smallest population of WBCs. They may release histamine (a vasodilator), heparin (an anticoagulant), and possibly serotonin and bradykinin (vasoconstrictors) into the blood. **Mast** cells are similar to basophilic WBCs but are found only in tissues; they release histamine and heparin from their granules.

RED BONE MARROW.a
STEM CELL.b
LYMPHATIC SYSTEM

AGRANULOCYTES.*
MONOCYTES. 3–8%f
TISSUE MACROPHAGE.f'
LYMPHOCYTES. 20–25%g
B-LYMPHOCYTE.g'   ANTIBODIES
T-LYMPHOCYTES.g"

KILLER T-CELLS
HELPER T-CELLS
SUPPRESSOR T-CELLS

Monocytes have a single large nucleus and few granules in the cytoplasm. They enter injured tissue and transform into **tissue macrophages**, phagocytizing bacteria and tissue debris. Monocytes may be the source of all permanent tissue macrophages—e.g., those in the liver and lung. **Lymphocytes** participate in acquired immune responses of the body against specific viruses and bacteria. Lymphocytes form two types. **B-cells** deactivate bacteria and viruses by producing specific antibodies (humoral immunity); the antibodies are actually secreted by a differentiated form of B-cells, the plasma cells. **T-cells** attack all foreign cells (cell-mediated immunity). T-cells originate in the thymus and are divided into killer (cytotoxic), helper, and suppressor subtypes.

NATURAL IMMUNITY / NONSPECIFIC RESPONSE: INFLAMMATION & PHAGOCYTOSIS.

1. TISSUE DAMAGE.h
2. MICROBES ENTER BODY.
3. MAST CELLS RELEASE HISTAMINE.e'
4. VASODILATION.
5. PROTEIN PERMEABILITY.⋀
6. FLUID SWELLS TISSUE.h
7. DIAPEDESIS OF NEUTROPHILS.c
8. PHAGOCYTOSIS OF MICROBES.c
9. MONOCYTES FOLLOW.f
10. DEATH OF MICROBES.
11. PUS SAC DEVELOPS.L
12. TISSUE REPAIR.h

After an injury (1), bacteria invade tissue space (2); local mast cells liberate histamine (3), promoting vasodilation (4) and vascular permeability; plasma proteins and fluids flow in (5), causing local edema (6); fibrin formation clots this fluid, trapping bacteria. Now stationary tissue macrophages (1st line of defense) begin to phagocytize bacteria. Next, blood neutrophils leave blood vessels by diapedesis (7) and begin massive phagocytosis (8) (2nd line of defense). In more extensive injuries, blood monocytes migrate to the site (9), transform into macrophages, and help neutrophils eliminate microbes (10) (3rd line of defense). A pus sac containing dead cells and debris develops (11) and is either extruded or gradually cleared away during tissue repair (12) by epithelial and fibroblast cells.

PHAGOCYTOSIS.c
LYSOSOME.m

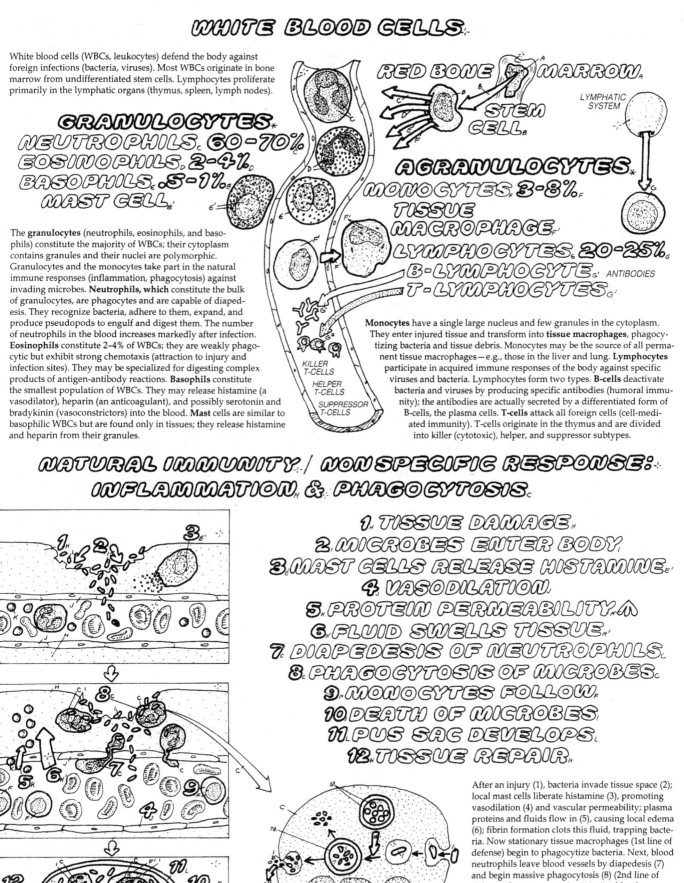

Phagocytes engulf bacteria and digest them within their lysosomes.

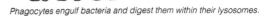

Two types of acquired immunity — The *lymphocytes* partici-pate in immune responses that develop *slowly* and *specifically* against particular foreign substances (*antigens*). This response is genetically programmed but occurs only *after* exposure to the antigen (*acquired immunity*). Two types of acquired immune responses are: *humoral-* or *antibody-mediated responses*, carried out by the *B-lymphocytes* (this plate) and *cell-mediated responses*, carried out by the *T-lymphocytes* (next plate, 148).

ACTIVE ANTIBODY-MEDIATED IMMUNITY

Antigens are usually foreign proteins (e.g., toxins), either free-floating or on the surface of infectious organisms — The presence of antigens is sensed by special receptor molecules on the surface of *B-lymphocytes* in the *lymph nodes*. The early precursor B-cells are in the bone marrow but migrate into bursa-like organs — e.g., lymph nodes — to mature. The numerous B-cells are specialized to detect a certain type of antigen. The B-cells are genetically programmed to express the receptors that recognize specific antigens. These surface receptors are a type of antibodies (see below).

Antibodies are produced by clones of plasma cells — The detection of an antigen sensitizes the B-cells; they transform into larger secretory cells called *plasma cells*. These proliferate, forming a *clone*, which synthesizes and secretes, for export to the plasma and tissues, a large amount of a specific protein called an *antibody*, whose function is to bind and neutralize the antigens (see below) of mainly bacterial origin. The forma-tion and proliferation of plasma cells are controlled by the release of *cytokines* from the *helper T-lymphocytes* (plate 148). Antibody production against foreign antigens is a form of *active immunity* and takes from *days* to *weeks* to fully develop.

Several classes of antibodies — All the varieties of antibod-ies produced against the many different antigens are protein molecules (*immunoglobulins, Ig*). IgG (γ-*globulins*) and IgM are the most abundant types and function against bacterial and viral infections. The IgD type exists on the surface of B-cells for recognizing antigens, and the IgE-type antibodies partici-pate in allergic reactions. The IgA-type antibodies are *secretory immunoglobulins* produced by a type of resident plasma cells and released into the secretions of the gastrointestinal and respiratory mucosa and milk.

Antibody molecules have variable & constant seg-ments — Each antibody is roughly Y-shaped, consisting of *heavy chains* (peptide chains) and two *light chains*. The heavy chains provide the *constant* part of the antibody molecule, which is the same in all antibodies; the light chains, located in the arms of the Y (attached to the heavy chains), constitute the *variable* and functionally significant part of the molecule. Thus, each antibody has two sites, one on each of the variable arms, for interaction with the antigen. The extreme diversity of antibodies is largely based on structural variation (amino acid composition) in the variable protein chains.

Antigen-antibody binding deactivates microbes & their toxins — Upon encountering an antigen in the blood or tissue fluids, the antibodies bind with the antigen molecules in order to neutralize and deactivate them. Deactivation occurs by *direct* combination, causing *precipitation* (agglutination) or by *masking* the active sites of the antigens. If antigen is a free toxin molecule, the antigen-antibody complexes tie together, forming clumps to be engulfed by phagocytes. Binding of antibodies to surface antigens of microbes causes the microbes to be recognized, attacked, and destroyed by the phagocytes.

The complement system helps antibodies destroy microbes — Antibodies can also achieve the same goals *indi-rectly* by activating the *complement system*, which consists of a series of enzymes arranged to catalyze a cascade of chemical events. The combination of a single antibody molecule with the antigen activates this cascade, which rapidly mobilizes millions of enzymes that quickly lyse the microorganism to which the antigen is attached or cause agglutination and similar defensive reactions.

Memory cells learn to make antibodies for future encounters with antigens — After antigens are removed, the antibodies diminish in number. Upon second exposure to the same antigen, a large amount of the same appropriate anti-body is produced. This enhanced response is due to a type of plasma cell called the *memory cell*. B-cells produce memory cells upon their first exposure to the antigen. Memory cells "learn" how to produce the antibody but rest until the second exposure to the same antigen, when they are activated and form clones that produce large amounts of the antibody. This is the basis of long-term immunity against infections.

PASSIVE IMMUNITY, VACCINATION, & AUTOIMMUNE REACTIONS

Passive immunity refers to antibody transfer across placenta & via milk — Embryos and younger fetuses are essentially devoid of antibodies, as they live in a protected environment. Some maternal antibodies (IgG antibodies) can transfer across the placenta. Maternal antibodies are also provided after birth in the form of milk IgA immunoglobu-lins. The newborn intestinal mucosa can engulf and absorb whole IgA antibody proteins intact. This ability lasts for the first few weeks of life and is one reason breast feeding is encouraged even if for a short duration. The colostrum (first milk) is specially enriched in antibodies (plate 159).

Vaccination involves artificial activation of memory cells — The memory cells are involved in the phenomenon of immunization by *vaccination*, in which the body is inten-tionally exposed to a small amount of dead or transformed antigen (e.g., dead smallpox virus) in order to sensitize the immune system and form memory cells. When the body is exposed to the same antigen later (e.g., during a real smallpox infection), antibody production will be quick and intense and usually effective.

Therapeutic use of antibodies — Monoclonal antibodies produced in the laboratory by culturing B cells can be given to patients suffering from specific diseases. The results are effec-tive but short lasting. The use of antibodies to find and kill tumor and cancer cells is now under investigation.

Autoantibodies are the cause of autoimmune diseases — Occasionally, antibodies are mistakenly produced against certain normal surface proteins of the body's own cells. Also, antibodies produced against a foreign protein may mistak-enly attack a normal cell surface protein that resembles the antigen (cross-reaction). These attacks of *autoimmune anti-bodies* cause damage to or death of the invaded cell and pro-duce a wide variety of diseases (*autoimmune diseases*), such as Grave's disease (hyperthyroidism), Type-I diabetes, myas-thenia gravis, rheumatoid arthritis, and multiple sclerosis.

CN: Use the same colors as on previous page for antigen (A) and blood circulation (F).
1. Follow the numbered sequence above, beginning in the upper left rectangle.
2. Color the lymph node sites in the body.
3. Complete antibodies and the complement system.
4. Give a separate color to each class of antibodies.
5. Color acquired and passive immunity material below.

ACQUIRED IMMUNITY:
SPECIFIC ANTIBODY RESPONSE

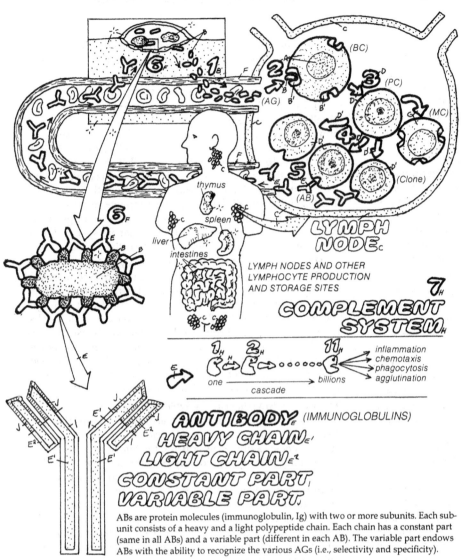

thymus
spleen
liver
intestines

LYMPH NODES AND OTHER
LYMPHOCYTE PRODUCTION
AND STORAGE SITES

(BC)
(PC)
(MC)
(Clone)
(AG)
(AB)

LYMPH
NODE

RECOGNIZES ANTIGENS
B-LYMPHOCYTE

FORMS CLONES, SECRETES ANTIBODIES
PLASMA CELL

MAKES ANTIBODIES FOR FUTURE USE
MEMORY CELL

Antigens (AGs) are foreign protein or polysaccharide substances on the surface of microbes entering the body (1). In the lymph nodes, AGs are detected by receptors on B-lymphocytes (BCs) (2). Each BC is genetically programmed to respond to one particular AG. Sensitized BCs transform into plasma cells (PCs) (3). PCs divide, forming a clone (4); the clone produces antibodies (ABs) (5) rapidly and profusely. Each AB is specific for an AG. The ABs circulate in the lymph and blood, binding to and deactivating AG (6). ABs can inactivate AG either directly or indirectly by activating the complement system (7), a cascade of enzyme reactions in the plasma that facilitates direct actions of ABs and promotes chemotaxis and inflammatory responses, causing lysis or phagocytosis of AG cells.

COMPLEMENT SYSTEM

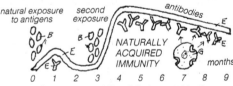

one — *cascade* → *billions*

inflammation
chemotaxis
phagocytosis
agglutination

MICROBE/ANTIGEN
B-LYMPHOCYTE
AG-RECEPTOR
PLASMA CELL
CLONE
ANTIBODY
BLOOD CIRCULATION
MEMORY CELL

ANTIBODY (IMMUNOGLOBULINS)
HEAVY CHAIN
LIGHT CHAIN
CONSTANT PART
VARIABLE PART

ABs are protein molecules (immunoglobulin, Ig) with two or more subunits. Each subunit consists of a heavy and a light polypeptide chain. Each chain has a constant part (same in all ABs) and a variable part (different in each AB). The variable part endows ABs with the ability to recognize the various AGs (i.e., selectivity and specificity).

CLASSES OF ANTIBODIES

IgA
(*secretory immunoglobulins*) in milk, gastric and respiratory mucosal fluids

IgD
B-cells surface, recognize antigens

IgE
participate in allergic reactions

IgG
(*g-globulins*) main types — fight bacterial and viruses

IgM

main types — fight bacterial and viruses

MEMORY CELL FUNCTION

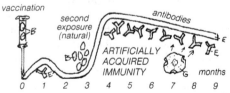

natural exposure to antigens   *second exposure*   *antibodies*

NATURALLY ACQUIRED IMMUNITY

months

After exposure to AG and during sensitization, some PCs transform to memory cells (MCs). These remain dormant in the lymph nodes for long periods. Upon further exposures to the AG, MCs will evoke a pronounced, exaggerated response (AB production) that will rapidly deactivate AGs. The MC response is the basis of natural and long-term immunity against bacteria and some viruses.

vaccination   *second exposure (natural)*   *antibodies*

ARTIFICIALLY ACQUIRED IMMUNITY

months

The MCs response is also the basis of the practice of immunization and vaccination practices. Small amounts of AG (dead or live) is given by injection. B-cells detect the AG and form PCs to make ABs and also MCs. These remain dormant until a natural invasion of the same antigen or microbe when MCs will evoke a pronounced, exaggerated response (specific AB production) that will rapidly deactivate the natural AGs.

PASSIVE IMMUNITY

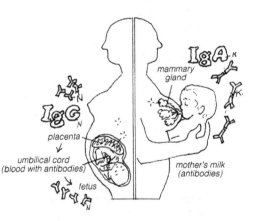

mammary gland
placenta
umbilical cord (blood with antibodies)
fetus
mother's milk (antibodies)

Some ABs (IgG type) transfer from mother to fetus across placenta to supply the fetus with natural immunity. Antibodies (type IgA) also are secreted in the milk for transfer to newborn via suckling. Newborn's intestinal mucosa can absorb whole IgA antibody proteins intact. This ability lasts for the first few weeks of life, showing the value of breast feeding in early weeks of life. The colostrum (first milk) is specially enriched in antibodies.

T-lymphocytes (T-cells) function in *cell-mediated* acquired immunity. They move around in the blood, lymph, and tissue fluids, where they seek, invade, and destroy virus-infected body cells, cells of infectious organisms (fungi, slow-acting bacteria), foreign tissue cells (grafts, transplanted organs), and abnormal body cells (cancer and tumor cells). T-cells bind to antigens on their target cells and kill them directly without the use of antibodies.

"Cytotoxic" & "helper" are the two main types of T-cells — The total lymphocyte population in adult humans is about 2 trillion (nearly 1 kg), most of which are T-cells. Four types of T-cells are recognized. The *cytotoxic* and *helper* types are the main and best-known ones. The *suppressor* T-cells, which inhibit activity of other T-cells and B-cells, and *memory* T-cells, which "learn" to react against future cellular infections, are also recognized but less is known about them.

ATTACK & DESTROY FUNCTIONS OF CYTOTOXIC T-CELLS

The cytotoxic T-cells carry out the attack and destroy functions that characterize T-cell lymphocytes, even though they comprise ~20% of the T-cell population. To learn how cytotoxic T-cells attack and destroy targets, we examine the example of a virus-infected cell.

Cytotoxic T-cells destroy their target cells by adding perforin channel molecules — Viruses infect body cells by entering and transforming them into abnormal "host" cells. Host cells synthesize viral protein, a fragment of which (antigen) is combined with some of host's own specific protein molecules and inserted in the host cell membrane ("antigen presentation"; see below). Each type of cytotoxic T-cell has special receptors on its surface to recognize a particular antigenic complex (see below); once the target is recognized as "bad" or "non-self," cytotoxic T-cells bind to it and release cytoplasmic granules that contain *perforin*. Perforin molecules form large pores in the membrane of host cells, allowing influx of water and various ions, resulting in cell swelling and death. Viruses are released to the outside and phagocytized by macrophages.

Abnormal body cells (tumor cells, cancer cells) produce *endogenous antigens* that are also complexed with the cell's own protein molecules and presented in the membranes for recognition and attack by cytotoxic T-cells. Invasion and killing mechanisms are the same as for virus-infected cells.

Antigen presentation involves combination with MHC proteins — In order for the cytotoxic T-cells to recognize an infected cell, a cancer cell, or a graft cell, the target must "present" its "foreign" or "abnormal" antigen. A fragment of the antigen (small peptide) is combined with a particular class of the cell's own specific proteins — i.e., the MHC (major histocompatibility complex) proteins — and inserted in the target cell membrane. Class I MHC proteins, which occur on the surface of all nucleated body cells, interact with cytotoxic T-cells, enabling these cells to recognize and attack (or avoid) all types of body cells.

MHC proteins serve as "self" signals — These proteins are unique to every individual (like fingerprints); only identical twins have the same MHC proteins. Their presence on cell surfaces without antigens, signals the T-cells to consider them as "self" and avoid them. The ability to differentiate the self from "non-self" is acquired by T-cells early in life while they inhabit the thymus. Transplant tissues express non-self MHC

proteins and are therefore attacked (rejection). The closer the genetic match between donor and recipient, the less is the chance of T-cell attack and tissue rejection.

REGULATORY & SECRETORY FUNCTIONS OF HELPER T-CELLS

Helper T-cells regulate many immune responses — Helper T-cell constitute the majority of T-cells (~75%); they bind to and interact with the B-cells and macrophages, releasing cytokines to regulate their functions as well as those of the T-cells. The helper T-cells are therefore called the "master switch" of the immune system. The AIDS virus (HIV) exerts its devastating action in the body by invading the helper T-cells. Helper T-cells bind with macrophages and B-cells by combining with their surface antigens; these antigens are complexed with class II MHC proteins that are specific for the immune cells and lacking in other body cells.

Helper T-cells release hormone-like cytokines — The cytokines (*lymphokines*, *interleukins*) regulate proliferation and activity of other lymphocytes. About fifteen different cytokines are now known. For example, without stimulation by the cytokine interleukin-2 from helper cells, cytotoxic T-cells do not proliferate and conduct their attack-and-destroy function. Cytokines are also released as chemical signals between the helper cells and antibody-producing B-cells and the macrophages. In the absence of helper-cell cytokines, antibody production by B-cells is markedly reduced.

THYMUS: IN MATURATION & DIFFERENTIATION OF T-CELLS

Thymus is the site of "early learning" & maturation of T-cells — The thymus, a primary *lymphatic organ* located in the anterior chest cavity, grows during the fetal and postnatal stages; after puberty it gradually becomes fatty and atrophic. The T-cell ancestors migrate from the bone marrow to the thymus (hence their "T" designation) during fetal and early postnatal life; here they proliferate, differentiate, and mature — i.e., develop the ability to recognize antigens, and differentiate into various types (helper, cytotoxic). Thymus removal in early life, but not in adulthood, causes severe T-cell-mediated immune deficiency.

Mature T-cells leave the thymus and circulate in the blood and the secondary lymphatic organs (e.g., lymph nodes), where they seek and destroy infected and abnormal body cells. T-cells have a long life (years) compared to B-cells, which are short-lasting. Since T-cells can also proliferate by forming clones in the blood and lymphatics, not all adult T-cells are truly thymus derived.

Thymus secretes the hormone thymosin to stimulate T-cell development — This protein hormone is present during development and maturity but declines after middle age. Its decline is thought to underlie the decline in the ability of the body to eliminate tumor and cancer cells during aging.

T-cells develop their diverse receptors against antigens in the thymus — These receptors, which occur on the plasma membranes of T-cells and function like antibodies, are two chained proteins with specific variable regions for binding to diverse antigens; these regions vary among the individual T-cells, enabling each T-cell to interact with a certain type of antigens. Both cytotoxic and helper T-cells have these receptors. Receptor diversity and distribution among different members of T-cells develop during their maturation phase in the thymus.

CN: Use the same colors for A, B, C, and H as were used for those items on the previous plate. 1. Begin with number 1 in the lymph node of the upper drawing. Note that in the blow-up of step 5, cell membranes are not colored 2. Complete the remaining material.

ACQUIRED IMMUNITY: CELL-MEDIATED RESPONSE.

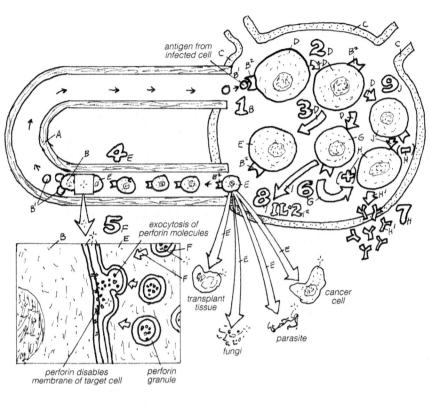

antigen from
infected cell

exocytosis of
perforin molecules

transplant
tissue

fungi

parasite

cancer
cell

perforin disables
membrane of target cell

perforin
granule

Labels
BLOOD CIRCULATION_A
INFECTED (TARGET) CELL_B
VIRAL ANTIGEN_B'
LYMPH NODE_C
T-LYMPHOCYTE_D
CYTOTOXIC T-CELL_E
(KILLER T-CELL)
AG RECEPTOR_B²
PERFORIN GRANULE_F
HELPER T-CELL_G
B-LYMPHOCYTE_H
ANTIBODIES_H'
CYTOKINES_I
SUPPRESSOR T-CELL_J

The AG on slow-acting bacteria (tuberculosis), fungi, cancer cells, and cells of transplanted tissue (1) sensitize another type of lymphocyte—i.e., T-cells (TC) (2). Sensitized TCs proliferate (3), forming several subtypes. Cytotoxic (killer) TCs (4) contain AB-like receptor molecules, enabling them to bind with AG on infected or foreign cells. After attachment, TCs release granules containing perforins, which form large pores on membrane of the antigenic cell (5), causing swelling and death. Another TC type, the helper TC (6), enhances AB production by BCs (7) and activated cytotoxic T-cells. Helper TCs produce cytokines (lymphokines), hormone-like substances (8) that regulate functions of other TCs and BCs. The suppressor TC (9) opposes the action of the helper TC, homeostatically regulating immune responses.

HELPER T-CELLS_G & CYTOKINE FUNCTIONS_I

IL-2_I²
MACROPHAGE_K
IL-1_I'

Helper T-cells release hormone-like cytokines (interleukins, e.g. IL2) to regulate many functions including activation and proliferation of cytotoxic cells and promote their attack on infected cells; cytokines from helper also stimulate B-cells to secrete antibodies and certain viruses and stimulate macrophages and NK (natural killer) cells to perform their phagocytosis of microbes. Cytokines secreted from certain macrophages (IL-1) stimulates helper cells to begin their function.

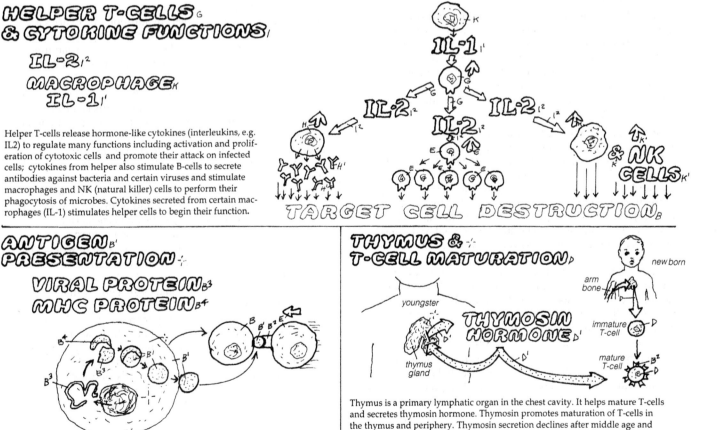

IL-1_I'
IL-2_I²
IL-2_I²
IL-2_I²
& NK CELLS_K'
TARGET CELL DESTRUCTION_B

ANTIGEN_B' PRESENTATION

VIRAL PROTEIN_B³
MHC PROTEIN_B⁴

All infected, abnormal, foreign cells must "present" their antigens on their surface in order for cytotoxic T-cell to recognize, attack and kill them. Virus infected cells synthesize viral proteins. These are combined with cells own MHC proteins and inserted in the cell membrane for recognition by cytotoxic T-cells which bind to these antigen complexes and launch their attack by their perforin molecules which cause target cell death.

THYMUS & T-CELL MATURATION_D

new born
arm bone
immature T-cell_D
mature T-cell_B²
youngster
THYMOSIN HORMONE_D'
thymus gland

Thymus is a primary lymphatic organ in the chest cavity. It helps mature T-cells and secretes thymosin hormone. Thymosin promotes maturation of T-cells in the thymus and periphery. Thymosin secretion declines after middle age and may cause reduced cell-mediated immunity in the aged. Removal of thymus in neonates (but not adults) results in marked immune deficiency against viruses, tumors, cancer and foreign cells.

T-cell ancestors migrate from bone marrow to thymus in the fetus and neonate. Here they differentiate and mature, i.e., develop specific receptors to detect antigens. Once mature, they leave the thymus to circulate in blood or lymph or lymph organs where they attack antigen bearing (abnormal, infected, foreign) cells.

OVERVIEW OF THE HUMAN REPRODUCTIVE SYSTEM

The physiologic systems that we have studied in the previous sections function to ensure the proper maintenance of the individual and are fairly similar in males and females. In this last section, we will focus on the *reproductive system*, whose function is aimed at ensuring the continuation and survival of the species and whose parts and organs are *sexually dimorphic* – i.e., they are structurally and functionally different in males and females.

Reproductive system performs diverse sexual & reproductive functions – Reproductive organs are a mixtures of ducts, organs, and glands. Some of the glands are endocrine, secreting sex hormones; others are exocrine, secreting mucus or various fluids for maintenance of the germ cells, or *gametes*. Some reproductive organs function in development of the gamete and the embryo, while others are important in copulation and in transmission and transport of the gametes.

Sex hormones stimulate growth and function of reproductive organs – The organs of the reproductive system grow and function in response to the stimulation provided by the male and female *sex steroid hormones*, secreted by the *gonads*. The gonads are in turn stimulated by the *gonadotropin hormones* released by the *anterior pituitary* gland. In the absence of these hormonal stimuli, the target glands and organs will cease to function and will atrophy.

Reproductive functions begin at puberty and show aging – Though the various organs of the reproductive system are formed during the embryonic period, the normal functions of this system begin during puberty and last for about forty years in women, terminating in "menopause," cessation of ovarian function, which typically occurs in the early fifties. In men, reproductive functions decline slowly with advancing age.

OVERVIEW OF THE MALE REPRODUCTIVE SYSTEM

The reproductive system of the human male consists of the *penis* and *scrotum*, the *testes, prostate,* and *seminal vesicles,* the *epididymis* and *vas deferens,* and the *bulbourethral glands.* The penis and scrotum, which contains the testes, are externally visible; the remaining organs are internal. The two testes (testicles) are the only organs with *endocrine* functions. They secrete the hormone *testosterone,* the most potent of the androgenic (= male-producing) sex steroids. The testes also produce the male gametes, *spermatozoa* (sperm), through a process called *spermatogenesis.* Testicular functions are controlled by gonadotropin hormones from the anterior pituitary gland. The epididymis consists of convoluted tubules that help store and mature the sperm. The vas deferens is a conduit for sperm delivery during *emission* and *ejaculation,* events occurring in the male during sexual excitation and intercourse.

The prostate and seminal vesicles are *exocrine glands* producing the plasma of the *semen,* a fluid essential for activity and survival of the sperm within the female reproductive system. The penis, with its inflatable tissue, acts as the organ of *intromission,* delivering sperm through its *urethral canal* and depositing them in the *vagina* of the female, near the *uterine cervix.* The scrotum is a sac containing the testicles, and, through extension and retraction, it maintains the temperature of the testes a few degrees *below* body temperature, to ensure proper spermatogenesis.

Secondary sex characteristics in the human male – The secondary sexual characteristics of the human male are large body size, enhanced muscular and skeletal growth, wide shoulders and narrow pelvis, enlarged larynx and vocal cords (producing a low-pitched voice), facial and body hair, pubic and axillary (armpit) hair, and receding scalp hairlines and baldness (if genetically susceptible). These characteristics appear after puberty in response to increasing levels of the testosterone hormone and may include psychological changes such as active and aggressive attitudes and independence, although these latter may occur in females as well.

OVERVIEW OF THE FEMALE REPRODUCTIVE SYSTEM

In the female, the main sexual and reproductive organs are the *ovary, uterus, uterine tube (Fallopian tube, oviduct),* and *vagina,* which constitute the internal sex organs. The *labia majora, labia minora,* and *clitoris* constitute the external *genitalia* (the *vulva*). The two ovaries act as the main endocrine glands of the female system, secreting the female sex steroid hormones, *estrogen* and *progesterone.* In addition, the ovaries are the site of formation and release of the female gametes – the *ova,* or *eggs* – by a process called *oogenesis.* The ovarian functions are controlled by gonadotropin hormones from the anterior pituitary gland.

The uterine tubes are the site of *fertilization* and transport of the egg as well as the young *embryo.* The uterus is the organ of *pregnancy,* providing a nest for *implantation* and growth of the young embryo. The uterine muscles generate the contractions necessary for birth (*parturition*). The clitoris is densely innervated by tactile receptors and functions in female sexual excitation. The vagina is adapted to receive the penis and sperm during intromission and ejaculation. As the *birth canal,* the vagina also participates in delivery and birth of the newborn. The female *breasts* contain fatty tissue and *mammary glands,* which secrete milk for nourishment of the newborn.

Secondary sex characteristics in the human female – The human female secondary sexual characteristics include a wide pelvis, narrow shoulders, high-pitched voice, non-receding scalp hairlines, and soft skin. Females are shorter than males on the average and have less muscle and bone mass. Females have larger subcutaneous and deep fat deposits, which underlie the shape of breasts, buttocks, and thighs in women. Mature human females, like males, possess axillary and pubic hair; pubic hair has the form of an inverted triangle, the opposite of its form in the male. Facial and body hair are absent or very soft and sparse in women. All these characteristics are promoted by the female sex hormone estrogen.

CN: Use dark colors for C and H. Begin with the male system, coloring the same structure in both the side view and the smaller frontal view above, before going on to the next structure.

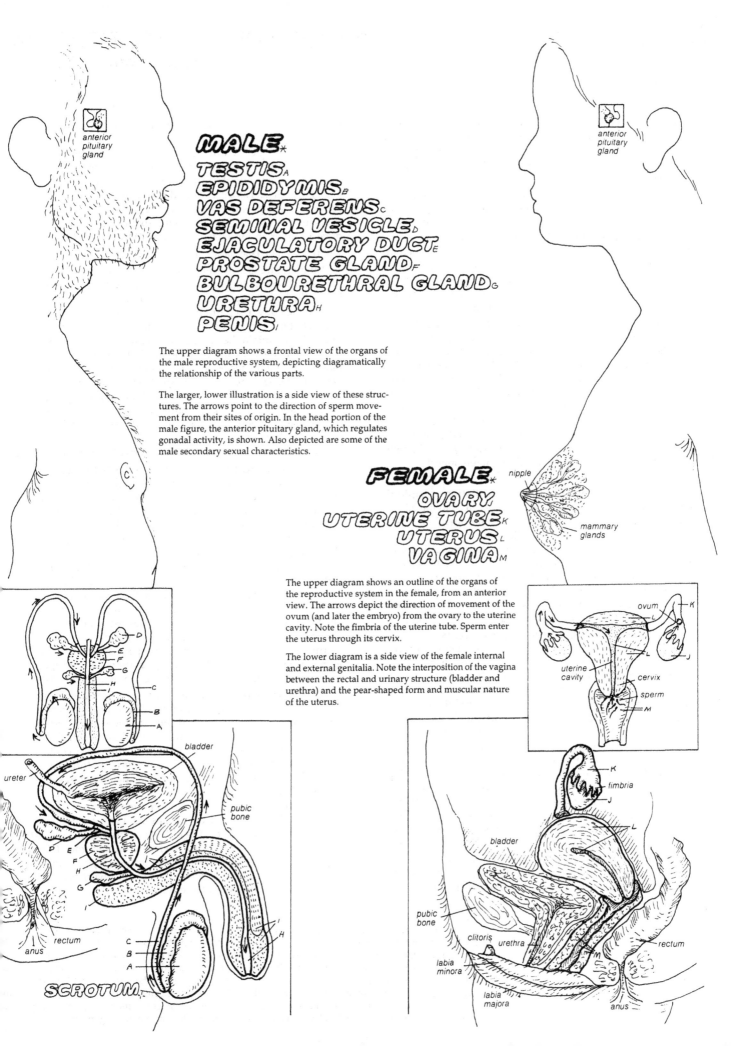

MALE *

TESTIS A
EPIDIDYMIS B
VAS DEFERENS C
SEMINAL VESICLE D
EJACULATORY DUCT E
PROSTATE GLAND F
BULBOURETHRAL GLAND G
URETHRA H
PENIS I

anterior
pituitary
gland

The upper diagram shows a frontal view of the organs of
the male reproductive system, depicting diagramatically
the relationship of the various parts.

The larger, lower illustration is a side view of these structures. The arrows point to the direction of sperm movement from their sites of origin. In the head portion of the
male figure, the anterior pituitary gland, which regulates
gonadal activity, is shown. Also depicted are some of the
male secondary sexual characteristics.

FEMALE *

OVARY J
UTERINE TUBE K
UTERUS L
VAGINA M

anterior
pituitary
gland

nipple

mammary
glands

The upper diagram shows an outline of the organs of
the reproductive system in the female, from an anterior
view. The arrows depict the direction of movement of the
ovum (and later the embryo) from the ovary to the uterine
cavity. Note the fimbria of the uterine tube. Sperm enter
the uterus through its cervix.

The lower diagram is a side view of the female internal
and external genitalia. Note the interposition of the vagina
between the rectal and urinary structure (bladder and
urethra) and the pear-shaped form and muscular nature
of the uterus.

ovum

uterine
cavity

cervix

sperm

bladder

ureter

pubic
bone

rectum

anus

SCROTUM

fimbria

bladder

pubic
bone

clitoris

urethra

labia
minora

labia
majora

rectum

anus

The testes—the male gonads—are located within the scrotum, a modified cutaneous sac, and perform two major functions: (1) *spermatogenesis*, the formation of the male *gametes* (sperm, spermatozoa), and (2) secretion of the male sex hormone, *testosterone*.

Testis functional histology—Each testis is divided into lobules, each containing one to four long, convoluted *seminiferous tubules* (STs); the STs, each 0.2 mm wide and 70 cm long, are the site of sperm formation. A *basement membrane* supports the *germinal epithelium*. Two major cell types are attached to the basement membrane: the *spermatogonia*, or the *primordial germinal cells*, and the *Sertoli cells*, non-germinal, supportive epithelial cells. Testosterone is produced by the *Leydig cells*, located in the spaces between the STs (plate 151).

SPERM ARE FORMED BY MITOSIS & MEIOSIS OF GERM CELLS
Spermatogenesis is a complex process involving repeated mitotic and meiotic divisions of the germ cells and requires the support functions of the Sertoli cells. The spermatogonia are diploid, with 46 chromosomes (22 pairs of somatic and one pair of sex chromosomes, XY). They divide by *mitosis* into daughter cells; one adheres to the basement membrane and maintains the *germinal line*, the other moves into the epithelial matrix, forming a *primary spermatocyte*, which divides by *meiosis* (plate 150) to form *secondary spermatocytes* and finally *spermatids*. The spermatogenic cells are interconnected by *cytoplasmic bridges*, enabling them to divide in synchrony.

Sperm cells are formed from differentiation of haploid spermatids—The spermatids are *haploid* cells, having 22 somatic chromosomes and one sex chromosome, X or Y. The final stage of spermatogenesis, called *spermiogenesis*, involves morphological differentiation of spermatids into unique, specialized cells called *spermatozoa (sperm, sperm cells)*, each of which possesses a tail for flagellar motility. Sperm cells are released into the ST lumen by a process called *spermeation*. The entire duration of spermatogenesis takes about 10 weeks; over 500 sperm cells are produced from each spermatogonium.

THE SERTOLI CELLS SUPPORT SPERM FORMATION
The Sertoli cells (SC) participate in spermatogenesis in several ways. (1) They support the germ cells, moving them along their inward migration within the epithelium. (2) The SC play an important role in spermiogenesis by engulfing and digesting the remaining pieces of cytoplasm and cellular debris *(residual bodies)* left over from spermatid transformation into sperm. The SC also help release sperm into the lumen (spermeation). (3) The SC secrete a fluid into the ST lumen that assists in sperm transport out of the testis into the epididymis. (4) The SC provide nutrients and metabolites to the developing germ cells and provide a testis–blood barrier that isolates the germ cells from the blood and protects the sperm from antibodies and T-cell attack; sperm cells carry several antigens on their surface that are foreign to the body's immune system. The barrier also stops leakage of these antigens to the blood.

Secretions of the Sertoli cells—To perform their functions, the SCs require the male sex steroid hormone, *testosterone*. Testosterone is also required for the development of

germinal cells in the STs and for the final maturation of the sperm in the *epididymis*. Testosterone is provided to the SCs from the Leydig cells, through the basement membrane. To maintain a high local concentration of testosterone, the SCs form and secrete into the lumen the *androgen-binding protein (ABP)*, which acts as a carrier and reservoir for testosterone. Adult SCs also produce the hormone inhibin for regulation of testis function by the pituitary gland (plate 152); fetal SCs produce a Mullerian Inhibiting Substance (MIP) (plate 160).

FACTORS THAT INFLUENCE SPERM FORMATION
Temperature is a critical environmental factor in sperm formation. Spermatogenesis is optimal at 32 °C, five degrees below the body temperature. If the testes are strapped tightly against the body, the germinal epithelium regresses, with no effects to the Leydig cells and hormone production. The scrotum helps maintain the optimal temperature for the testis by retracting in the cold and relaxing in a hot environment. Blood entering the testes is cooler than normal due to special vascular mechanisms. Other physical factors: as X-ray and ionizing radiation, are also detrimental for spermatogenesis. Malnutrition, alcoholism, cadmium salts, and some drugs reduce spermatogenesis. Gossypol, a cottonseed oil taken orally, attacks spermatids and can act as a specific male contraceptive. Vitamin E is essential for spermatogenesis in rats.

SPERM MATURATION TAKES PLACE IN THE EPIDIDYMIS
The sperm released into STs are not functionally mature; they are not motile and cannot fertilize an ovum. Functional maturity occurs mainly in the *epididymis* and partly in the female tract *(capacitation)*. Sperm move to the epididymis through the *rete testis*, a network of anastomosing conduits. Transport is aided by a fluid produced by the SCs. The *efferent ducts* connect the rete testis to the epididymis, a long convoluted tubule that has three parts: head, middle segment, and tail. In the course of two weeks, sperm move from the head part to the middle segment and finally into the tail segment, where the functionally mature sperm are stored. Testosterone and special proteins secreted from epididymis wall cells stimulate sperm maturation. Mature sperm are expelled through the *vas deferens* upon sexual excitation and ejaculation. Unused sperm undergo aging and death and are removed by phagocytosis.

Number & characteristics of sperm cells—Mature sperm cells are 50 μm long and highly motile. They comprise a head, a mid-piece, and a tail. The head contains the *nucleus* and the *acrosome*; the *acrosomal enzymes* are essential for penetration of the egg during fertilization (plate 156). The mid-piece contains the mitochondria that provide ATP for sperm motility. The sperm tail functions as a motile *flagellum* enabling it to swim at a rate of about 1 mm/min. Sperm production begins in boys at puberty (14 years) and continues to old age, at a rate of about 200×10^6 per day; there is a seasonal increase in the winter and a steady decline during old age. Sperm number is critical for fertility. About 100×10^6 sperm per ml of semen (300×10^6 per ejaculate) is normal for proper fertility. Sperm counts below 20% of normal are infertile. The count reversibly diminishes following repeated ejaculations.

CN: Use the same colors as on the previous page for (A), (D), and (E). Use red for G.
1. Begin with the diagram in the upper right corner, and then color the large illustration to its left.
2. Color the enlargement of a tubule section, beginning at the bottom of the square. Interstitial cells (F) are shown outside the tubule, secreting testosterone (F^1) into the Sertoli cells (O) as well as

into an adjacent blood capillary (G). The various cells within the tubule have their identifying labels placed within them, and both their cytoplasm and nucleus should be colored. Note the huge nucleus and cytoplasm of the Sertoli cells (O), which form a backdrop to the much smaller cells adjoining them.
3. Color the stages of spermatogenesis, beginning with the dividing spermatogonium (K).

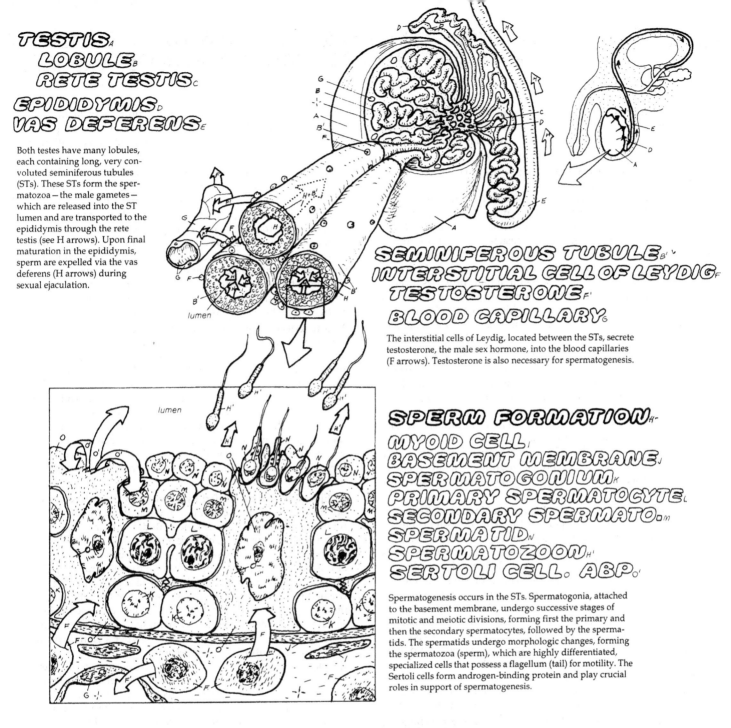

TESTIS A
LOBULE B
RETE TESTIS C
EPIDIDYMIS D
VAS DEFERENS E

Both testes have many lobules, each containing long, very convoluted seminiferous tubules (STs). These STs form the spermatozoa — the male gametes — which are released into the ST lumen and are transported to the epididymis through the rete testis (see H arrows). Upon final maturation in the epididymis, sperm are expelled via the vas deferens (H arrows) during sexual ejaculation.

lumen

SEMINIFEROUS TUBULE B'
INTERSTITIAL CELL OF LEYDIG F
TESTOSTERONE F'
BLOOD CAPILLARY G

The interstitial cells of Leydig, located between the STs, secrete testosterone, the male sex hormone, into the blood capillaries (F arrows). Testosterone is also necessary for spermatogenesis.

lumen

SPERM FORMATION H
MYOID CELL I
BASEMENT MEMBRANE J
SPERMATOGONIUM K
PRIMARY SPERMATOCYTE L
SECONDARY SPERMATO- M
SPERMATID N
SPERMATOZOON H'
SERTOLI CELL O ABP O'

Spermatogenesis occurs in the STs. Spermatogonia, attached to the basement membrane, undergo successive stages of mitotic and meiotic divisions, forming first the primary and then the secondary spermatocytes, followed by the spermatids. The spermatids undergo morphologic changes, forming the spermatozoa (sperm), which are highly differentiated, specialized cells that possess a flagellum (tail) for motility. The Sertoli cells form androgen-binding protein and play crucial roles in support of spermatogenesis.

STAGES OF SPERMATOGENESIS H

46 K' CHROMOSOMES K'
23 N' CHROMOSOMES N'

MITOSIS K'

1ST MEIOSIS L'

2ND MEIOSIS M'

CYTOPLASMIC BRIDGE P
RESIDUAL BODY P'

Spermatozoa contain only one half the number of chromosomes found in spermatogonia. This reduction is accomplished through division by meiosis; spermatogonia (diploid) divide first by mitosis to produce primary spermatocytes and preserve their own line. Each diploid (2n) spermatocyte divides by meiosis, forming tetraploid spermatocytes (4n chromosomes), which then go through two meiotic divisions to form four haploid (n chromosomes) spermatids/spermatozoa. Cytoplasmic bridges between several germ cells (a clone) allow their synchronous divisions. Spermatozoa are released into the lumen by the shedding their extra cytoplasmic remnants (residual bodies).

Upon sexual excitation, mature sperm are mobilized out of the epididymis and mixed with the secretions of the *accessory sex glands* to form *semen,* which is expelled through the urethra and urethral orifice in the penis head.

SEMEN CONSISTS OF SPERM CELLS & SECRETIONS OF MALE ACCESSORY GLANDS

Semen is a milky fluid containing sperm from the testes plus *seminal plasma,* the secretions of the male accessory glands— the *seminal vesicles, prostate,* and the *Cowper's* or *bulbourethral* glands. Secretions of the seminal vesicles and prostate constitute 60% and 20% of semen volume, respectively; the sperm account for 10%, and alkalines and mucus from the Cowper's glands make up another 10%. Seminal fluid provides an adequate environment for sperm nourishment and survival in the female tract. Seminal vesicles provide nutrients—chiefly *fructose,* but *lipids,* some *amino acids,* and *vitamins B* and *C* are also found. All are needed for sperm activity and survival. Prostaglandins, also from the seminal vesicles, may aid in sperm transport by stimulating smooth muscles in the female tract.

Like blood, semen clots when outside the body; this clot will then liquefy. The proteins and enzymes (fibrinogen, phosphatase, fibrinolysin) needed for clotting and lysing are secreted by the prostate. Semen contains zinc and electrolytes (K^+, Na^+, Ca^{++}, Mg^{++}, HCO_3^-, and Cl^-) and shows a somewhat alkaline pH of about 7.4, due to bicarbonate buffers from the Cowper's glands. Bicarbonate neutralizes acidity related to vaginal secretions and urine passage in the urethra; acidity is detrimental to the sperm.

PENILE ERECTION INVOLVES NEURAL & VASCULAR RESPONSE

The main function of the penis is to ensure deposit of sperm deep in the vagina, near the uterine cervix. Penile insertion and penetration of the vagina is called *intromission.* The penis is normally short and flaccid. To enable intromission, the penis develops a state of *erection,* which transforms it into a hardened, lengthened organ capable of penetrating the vagina. Erection occurs following sexual excitation and involves dilation of penile arterioles, allowing substantial inflow of blood to penile *erectile tissue.* Two *cavernous bodies* run along the dorsal and lateral aspects of the penis and one *spongy body* lies along its ventral aspect; the spongy body surrounds the urethra and fills the penis head *(glans).* The erectile tissue consists of numerous small elastic chambers of modified vascular and connective tissue that can fill with blood.

Erection is controlled by spinal reflexes and psychogenic stimuli from the brain. During sexual excitation, penile arterioles dilate and blood fills the chambers of the erectile bodies, leading to their turgidity and inflation. The resultant pressure closes the elastic venous outlets, trapping the blood within the erectile tissue, leading to penile hardening and erection. Erection is brought about by a somatic-autonomic spinal reflex. The glans penis contains many tactile receptors whose stimulation initiates a *parasympathetic* reflex response. The neural center for control of the erection reflex is in the sacral segments of the spinal cord. Efferent parasympathetic fibers in the *pelvic splanchnic nerves* release the neurotransmitter acetylcholine to induce vasodilation of penile arterioles. Other neurotransmitters that cause vasodilation of penile arterioles are the polypeptide *VIP* and the gaseous substance *nitric oxide* (NO). In humans, erection response can be induced by descending influences from the brain over the spinal centers.

These responses are induced by sight, sound, and smell as well as psychic influences of imagination and dreaming. Anxiety and fear responses can easily inhibit the activation of the erection reflex or interrupt it.

Erectile dysfunction (impotence) may be treated with drugs. Reduced ability for penile erection is called *erectile dysfunction* or *impotence* and occurs in some adults and many elderly males. In the past, psychological influences were thought to be the main cause of impotence, but recently attention has focused on the physiological and vascular abnormalities of the penile tissue. The drug Viagra, now used widely to correct erectile dysfunction or improve performance, enhances the vasodilatory function of the neurotransmitter nitric oxide. Autonomic nerve damage due to advanced untreated diabetes may also cause impotence.

EJACULATION IS THE REFLEX EXPULSION OF SEMEN

The expulsion of semen from the penis occurs in two stages, *emission* and *ejaculation proper.* Emission refers to the movement of sperm from the epididymis up along the *vas deferens* into the *ejaculatory duct.* This movement is accomplished by rhythmic contractions of the smooth muscles in the wall of the vas deferens. These muscles are controlled by *sympathetic nerves* from the lumbar spinal cord centers. Similar sympathetic signals cause contraction of the prostate and seminal vesicles simultaneously with the arrival of sperm into the ejaculatory duct, so that the contents of the prostate and seminal vesicles are added to them.

Ejaculation is controlled by nerves and smooth & skeletal muscles. Once the semen is in the ejaculatory duct, a new reflex for ejaculation proper is activated, involving somatic motor fibers in the *pudendal nerve* and the skeletal muscle *bulbospongiosus,* at the base of the penis. Repeated contractions of this muscle expel the semen through the urethra in a pulsatile manner. During emission, the urethra is prelubricated and washed by mucoid and alkaline secretions of the bulbourethral (Cowper's) glands that facilitate semen passage during ejaculation and neutralize acid remaining from past urination. Abnormalities in the functions of the Cowper's glands result in painful ejaculation. Sensory receptors for emission and ejaculation reflexes are located mainly at the *glans penis;* their centers are in the lumbar spinal cord. These centers are less influenced by the brain than the erection centers are. As a result, the ejaculation reflex, unlike the erection reflex, cannot be interrupted by inhibitory stimuli from the brain.

HUMAN SEXUAL RESPONSE HAS FOUR PHASES

The bodies of human males and females show a general pattern of stereotypic responses during sexual activity. Four phases are recognized that occur consecutively. We review the male pattern here. In the *excitement phase,* erotic stimuli from the penis, genital region, and other erogenous zones (lips, armpits, earlobes, groin) and/or psychic influences activate the penile erection reflex. In the second or *plateau phase,* erection intensity increases and the ejaculation reflex is facilitated. The *orgasmic phase* involves a climax—completion of ejaculation—accompanied by intense muscular contractions of the face and pelvic, chest, and leg areas. This phase is accompanied by an intense sensation of pleasure as well as marked cardiovascular and respiratory responses. During the final *resolution phase,* the entire body relaxes and blood leaves the penis, returning it to its normal flaccid state .

CN: Use dark colors for A, C, and light blue for L.
1. Color the titles in order as you follow the numbered sequence, beginning with a rise in external temperature (A') lowering the

scrotum (A) and testes (B).
2. Color the neural regulation diagrams with the title "inputs."
3. Color the actions of Viagra in the lower left corner.

SCROTUM A
 TESTIS B
  SEMINIFEROUS TUBULE B'
 SPERM C
EPIDIDYMIS D
VAS DEFERENS E
SEMINAL VESICLE F
PROSTATE GLAND G
BULBOURETHRAL GLAND H
URETHRA I

ERECTION OF PENIS J
 SENSORY NERVE J
 PARASYMPATHETIC NERVE K
 ERECTILE VASCULAR TISSUE L

EJACULATION OF SEMEN M
 SYMPATHETIC NERVE M
 BULBOSPONGIOSUS M. N

(1) The optimal temperature for sperm formation is 32°C, five degrees below body temperature. The scrotum maintains the testes at an appropriate distance from the body. (2) Sperm formation occurs in the testes and takes two and a half months. (3) Newly formed sperm are continuously released into the lumen of the seminiferous tubules and transported to the epididymis, where they stay for two weeks to mature. (4) Contractions of the vas deferens transport sperm into the ejaculatory ducts, where they are joined by the secretions of the prostate and seminal vesicles and are expelled out through the urethra. (5) The seminal vesicles secrete nutrients. (6) The prostate secretes proteins and enzymes for sperm survival. (7) The bulbourethral glands secrete alkalines and lubricants to facilitate sperm transport. (8) Erection of the penis is a vasocongestive response of the penile erectile tissues—the corpus cavernosum and the corpus spongiosum. (9) The bulbospongiosus muscle expels semen through the urethra.

NEURAL CONTROLS:
Parasympathetic signals from the sacral spinal cord initiate erection by dilating the penile arterioles. Sympathetic signals from the lumbar spinal cord initiate ejaculation through the contraction of the vas deferens, prostate, and seminal vesicles. Signals from the brain can excite or inhibit these spinal centers.

Acetylcholine, VIP (vasoactive intestinal peptide), and nitric oxide (NO) are neurotransmitters that induce vasodilation of penile arterioles, which produces the erection response. The drug Viagra reduces erectile dysfunction (impotence) by enhancing the action of the NO neurotransmission system.

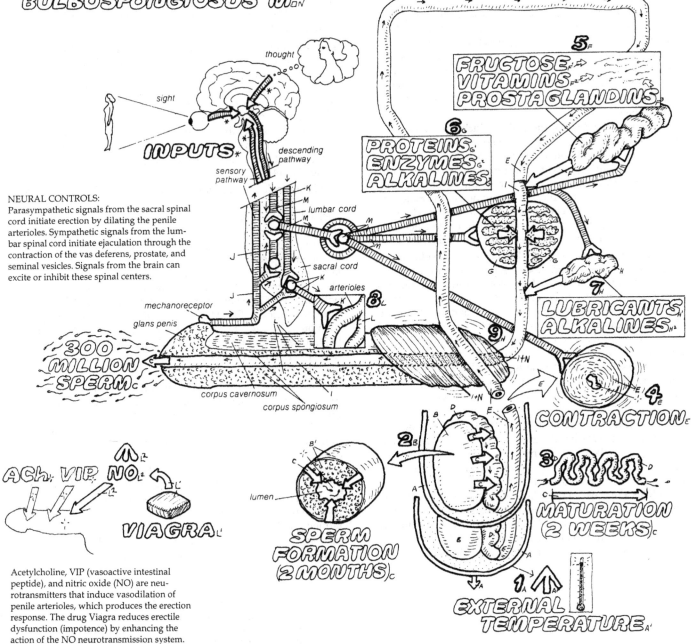

Testosterone (T) is the principal testicular hormone, secreted from the *interstitial cells of Leydig* at a rate of 10 mg per day. T is a steroid made from cholesterol and is the principal circulating androgen ("male maker" hormone). Other androgenic steroids are *di-hydroxy-testosterone* (DHT) and *de-hydro-epi-androsterone* (DHEA). DHEA is a precursor of T synthesis and is the main adrenal gland androgen. DHT is formed by conversion of T by the enzyme α-*reductase* and is present in plasma and in some body cells. Androgenic potency of T is less than DHT but much higher than DHEA.

TESTOSTERONE EXERTS THREE MAJOR TYPES OF ACTIONS
T has widespread effects in the body, which may be divided into three groups: (1) effects in adult male sexuality and reproduction; (2) actions on the development of the reproductive system and brain of the fetal male, as well orchestrating male puberty and body growth and behavior changes; and (3) non-reproductive, anabolic effects in the adult.

Stimulation and maintenance of the adult male reproductive system—In adult males, the steady secretion of testosterone (1) maintains spermatogenesis and the secretory functions of the accessory sex organs and glands—epididymis, prostate, and seminal vesicles; (2) maintains male secondary sex characteristics, including muscle and bone mass; and (3) promotes sex drive (libido) and other brain and mental effects.

Actions on the developing male & during puberty—The testes of the embryo, fetus, and neonate secrete T during these stages. In childhood, the testes remain inactive, only to start up again during puberty. The reproductive organs of the embryo initially are sexually indifferent and bipotential. In the male embryo, T promotes differentiation of the male-type genitalia. During fetal development, T promotes development of male-type hypothalamic systems, which regulate neural control of reproductive hormones and male sexual behavior.

During puberty, secretion of T in boys rises steadily from 10 years of age through adolescence, peaking in the early twenties. In adolescent boys, T promotes growth and maturation of the *primary sex organs* (e.g., testes, penis) and *accessory sex glands* (e.g., prostate, seminal vesicles), and development of *secondary sex characteristics* (low-pitched voice, dense facial and body hair, enhanced muscular and skeletal growth). T also acts on the brain to promote final maturation of the brain centers involved in regulating sexual activity and sexual behavior. Thus, immature boys transform into young men with fertile sperm and interest in the opposite sex, sexual activity, and procreation.

Anabolic and non-reproductive effects—T has widespread general anabolic effects on body cells and tissues that may or may not be related to maleness. Androgens enhance anabolism in many tissues by increasing synthesis of proteins and stimulating tissue growth. Increasing levels of T in adolescent boys increase bone growth and calcium deposition and enhance muscle mass by increasing protein synthesis. However, peak levels of T in post-adolescent boys induce the closure of epiphyseal plates of the bone, thereby terminating bone growth. Other non-reproductive effects

include increasing the size of the kidneys and formation of red blood cells in the bone marrow. Large doses of T are used to stimulate tissue growth in emaciated patients and to enhance muscle mass in athletes. However, negative side effects of increased libido and decreased fertility (sperm production) discourage such uses.

Cellular mechanisms of T actions in target tissues—The cellular mechanism of action of T in its targets follows the general pathway for steroid hormones (plate 114). In the adult male reproductive tissue, T diffuses into a target cell nucleus to bind with nuclear androgen receptors possessing binding sites for T and DNA, initiating gene action and synthesis of mRNA and proteins that mediate T actions. In the developing brain, T is first converted to estrogen by neuronal aromatase before receptor binding. In certain body tissues during sexual maturation and puberty, T is first converted to DHT by the target cell α-reductase; DHT then binds to the androgen receptor. The affinity of DHT for androgen receptors is higher than that of T.

TESTICULAR FUNCTIONS REGULATED BY PITUITARY LH & FSH
The testes' functions are controlled by LH and FSH, two gonadotropin glycoprotein hormones from the anterior pituitary gland. LH controls T release by Leydig cells and FSH acts on Sertoli cells to control spermatogenesis. LH and FSH actions follow these steps: binding with plasma membrane receptors → activation of membrane G-proteins → activation of membrane adenylate cyclase → formation of cyclic AMP, which brings about the cellular effects of LH/FSH on target cells (plate 12, 114).

LH controls Leydig cells & T production—Steady plasma levels of T in mature males are achieved by the negative feedback effect of T on the *hypothalamus* and *anterior pituitary*. A decrease in T level stimulates the hypothalamus to release more *gonadotropin-releasing hormone* (GnRH), which stimulates the anterior pituitary to release LH into the blood. LH stimulates the Leydig cells to increase T release. If T levels increase above the normal set point, the same feedback mechanism will diminish GnRH and LH levels and restore the T level to normal. Release of GnRH occurs in *pulses* every 1–2 hours, each pulse lasting a few minutes. Changes in T levels change the frequency and intensity of GnRH pulses. Pulsatile secretion of LH is critical, since continuous secretion of GnRH desensitizes the pituitary, reducing plasma LH levels.

FSH controls Sertoli cells & spermatogenesis—Sperm formation in the testes is regulated mainly by the gonadotropin FSH from the anterior pituitary. FSH exerts tropin and trophic actions on the Sertoli cells, stimulating their various functions—chiefly, support of spermatogenesis and secretion of *androgen binding protein* (ABP). The Sertoli cells in turn secrete a peptide hormone, *inhibin*, which acts on the anterior pituitary to regulate FSH release by negative feedback. In fact, inhibin has the potential for use as a male contraceptive, because high doses of it cause reduced sperm production by reducing FSH secretion. LH is also important for spermatogenesis, but its effect is mediated by release of T, which in turn stimulates Sertoli cell function (plate 151).

CN: Use red for B and a dark color for A.
1. Begin with testosterone (A) functions as shown by the three arrows from an interstitial cell (G) in the right central portion of the page.
2. Go to the titles at the top of the page.

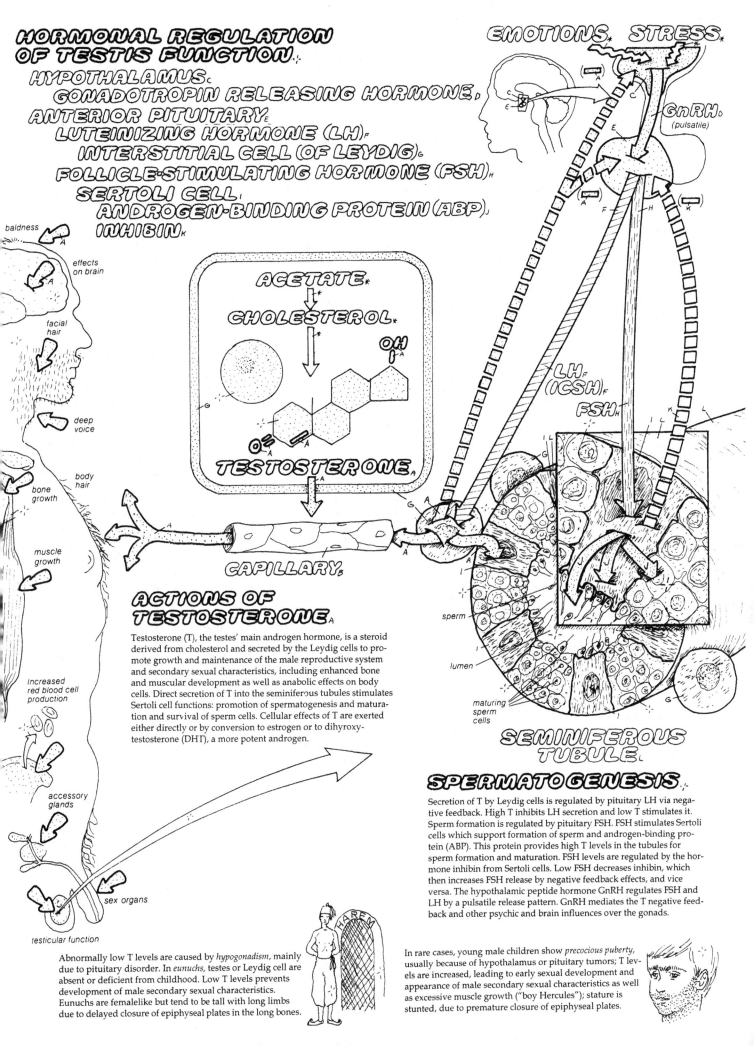

HORMONAL REGULATION OF TESTIS FUNCTION.

HYPOTHALAMUS.
GONADOTROPIN RELEASING HORMONE.
ANTERIOR PITUITARY.
LUTEINIZING HORMONE (LH).
INTERSTITIAL CELL (OF LEYDIG).
FOLLICLE-STIMULATING HORMONE (FSH).
SERTOLI CELL.
ANDROGEN-BINDING PROTEIN (ABP).
INHIBIN.

EMOTIONS. STRESS.

GnRH. (pulsatile)

baldness

effects on brain

facial hair

deep voice

bone growth

body hair

muscle growth

Increased red blood cell production

accessory glands

sex organs

testicular function

ACETATE.

CHOLESTEROL.

OH

TESTOSTERONE.

CAPILLARY.

LH (ICSH).

FSH.

sperm

lumen

maturing sperm cells

SEMINIFEROUS TUBULE.

ACTIONS OF TESTOSTERONE.

Testosterone (T), the testes' main androgen hormone, is a steroid derived from cholesterol and secreted by the Leydig cells to promote growth and maintenance of the male reproductive system and secondary sexual characteristics, including enhanced bone and muscular development as well as anabolic effects on body cells. Direct secretion of T into the seminiferous tubules stimulates Sertoli cell functions: promotion of spermatogenesis and maturation and survival of sperm cells. Cellular effects of T are exerted either directly or by conversion to estrogen or to dihydroxytestosterone (DHT), a more potent androgen.

SPERMATOGENESIS.

Secretion of T by Leydig cells is regulated by pituitary LH via negative feedback. High T inhibits LH secretion and low T stimulates it. Sperm formation is regulated by pituitary FSH. FSH stimulates Sertoli cells which support formation of sperm and androgen-binding protein (ABP). This protein provides high T levels in the tubules for sperm formation and maturation. FSH levels are regulated by the hormone inhibin from Sertoli cells. Low FSH decreases inhibin, which then increases FSH release by negative feedback effects, and vice versa. The hypothalamic peptide hormone GnRH regulates FSH and LH by a pulsatile release pattern. GnRH mediates the T negative feedback and other psychic and brain influences over the gonads.

Abnormally low T levels are caused by *hypogonadism*, mainly due to pituitary disorder. In *eunuchs*, testes or Leydig cell are absent or deficient from childhood. Low T levels prevents development of male secondary sexual characteristics. Eunuchs are femalelike but tend to be tall with long limbs due to delayed closure of epiphyseal plates in the long bones.

In rare cases, young male children show *precocious puberty*, usually because of hypothalamus or pituitary tumors; T levels are increased, leading to early sexual development and appearance of male secondary sexual characteristics as well as excessive muscle growth ("boy Hercules"); stature is stunted, due to premature closure of epiphyseal plates.

The *ovary* performs two functions: (1) formation, development, and release of the egg (ovum); and (2) secretion of the female sex hormones, *estrogen* and *progesterone*.

THE OVARY HELPS FORM, MATURE, & RELEASE OVA

Oocytes form only prenatally & occur within follicles—In the ovaries of the developing female embryo, the *oogonia* proliferate by mitosis to form millions of *primary oocytes* but cease to divide thereafter. This is in contrast to males, where spermatogonia divide after puberty and through old age. The fetal primary oocytes begin meiosis but stop at the prophase stage and remain in this arrested state through birth and childhood until puberty, when division is resumed. Each primary oocyte is surrounded by a thin layer of *granulosa cells*, forming the *primary* or *primordial follicle*. The follicular cells are important in nourishment and development of the oocytes and their preparation for fertilization.

Most follicles and their oocytes are lost before puberty—The initial number of primary follicles, 3 million per fetal ovary, diminishes to about 1 million at birth and 100,000 by puberty, a loss of more than 95%. The loss is called *atresia* and may be a form of programmed cell death (*apoptosis*). Atresia continues throughout maturity so that by 40 years, less than 1000 and by 50 years no primary follicles and oocytes are found in the ovaries. The loss of primary follicles is the main cause of *menopause*, the cessation of menstrual cycles and fertility in women over fifty. Since less than 500 eggs in all will be released in a woman during her reproductive period (15 to 50 years), the atretic losses of eggs may not be so significant. But aging of ova and follicles may underlie developmental abnormalities such as Down's syndrome.

FOLLICULAR & LUTEAL PHASES OF OVARIAN CYCLE

Development and release of ova occur in cycles—In contrast to sperm production by the testes, which occurs continuously, maturation and release of ova in the ovaries of mature women occur cyclically at 28-day intervals. During each cycle, usually one oocyte undergoes development and the resulting ovum is released at midcycle (day 14). About 1% of ovarian cycles involve development of multiple oocytes and ovulation, resulting in the birth of fraternal twins, triplets, etc. The ovarian cycle is the basis of the menstrual cycle (plate 154).

In the follicular phase, a follicle matures, forming theca and granulosa cells and a fluid-filled antrum—At the beginning of each 28-day cycle a few primary follicles begin to grow. A week later, only one—the *dominant follicle*—continues development while the others regress. In the dominant follicle, follicular cells proliferate, forming several layers of the *granulosa cells* surrounding the oocyte. Later another layer, *theca interna cells*, forms around the granulosa cells, with a *basement membrane* in between. The granulosa cells form a cavity around the oocytes called the *antrum*, filled with *antral fluid*; this fluid is rich in some proteins, hormones, and *hyaluronic acid*, a sticky substance. Theca interna and granulosa cells also produce *estrogen*, the female sex hormone, for release into the blood and the antral cavity, respectively. A fully mature Graffian follicle reaches a size of about 2 cm. The development of a follicle during the follicular phase is regulated by the gonadotropins FSH and LH. FSH is required for proliferation of granulosa and theca cells; LH preferentially stimulates estrogen secretion. Theca interna cells have many receptors for LH. Estrogen also aids follicular development.

Zona pellucida & cumulus oophorus surround the oocyte—In the Graffian follicle, the oocyte is surrounded by a zone of transparent jellylike substance, the *zona pellucida*. This zone is in turn surrounded by a thin layer of granulosa cells, forming the *cumulus oophorus* ("egg cloud"), which is continuous with the main mass of granulosa cells. The oocyte and its membranes are exposed to the antral fluid, which helps nourish and mature the developing egg. By days 12–13 of the cycle, the oocyte along with its surrounding cell layers is often found floating in the antrum.

A burst in LH and internal follicular changes lead to ovulation—By day 14 (midcycle), the *Graffian follicle*, is seen protruding from the weak ovarian surface, ruptures, releasing the oocyte, with its appendages of follicular cells and antral fluid, into the peritoneal cavity near the *fimbria* of the *uterine tube*. An event called *ovulation*. Ovulation is caused by increased pressure of the antral fluid and lysis of the follicular wall and is associated with increased release of histamine, prostaglandins, and proteolytic enzymes as well as bleeding in the follicle. A burst of LH lasting for 2–3 days stimulates ovulation. This burst and the cyclical changes in LH and FSH during the cycle are controlled by the gonadotropin-releasing hormone GnRH from the hypothalamus along with the feedback effects of estrogen and progesterone (plate 155).

Formation & secretions of the corpus luteum constitute the luteal phase—After ovulation, the remaining follicular cells form the *corpus luteum* (yellow body), which secretes mainly progesterone and some estrogen and grows for at least a week (*mature corpus luteum*). Formation of the corpus luteum and its growth are stimulated chiefly by the pituitary hormone LH, but FSH is also needed. If the egg is fertilized and an embryo forms, an LH-like hormonal signal from the young embryo (hCG, human chorionic gonadotropin, plate 157) stimulates the corpus luteum to grow further and secrete larger amounts of progesterone and estrogen to maintain pregnancy. In the absence of fertilization and hCG, the corpus luteum degenerates into the *corpus albicans* (white body).

GROWTH, MATURATION, & FINAL DIVISIONS OF THE OVUM
Early in the follicular phase, the primary oocyte resumes its meiotic division and grows in size. High concentrations of hormones and growth factors in the antral fluid may aid the ovum's development. The first meiotic division is completed before ovulation, forming the secondary oocyte and one *polar body* (cell). The secondary oocyte receives half the chromosomes and all the cytoplasm, while the polar body receives little cytoplasm but an equal share of chromosomes. Before ovulation, the secondary oocyte begins its second meiotic division but stops at metaphase and the ovum undergoes ovulation. After fertilization, the ovum completes its second meiotic division, forming the mature and haploid *female pronucleus* and the second polar body. The first polar body may also divide, forming three polar bodies altogether.

CN: Use yellow for G. Use light colors throughout.
1. Begin with the diagram in the upper right corner, coloring the 2 million "eggs."
2. Color the follicle-stimulating hormone (B') chartline and then the development of a single follicle, starting with a primary follicle (B'). When you reach the 14th day of the sequence, color the luteinizing hormone (G[3]) and the corpus luteum (G) development.
3. Color the stages of oogenesis along the bottom.

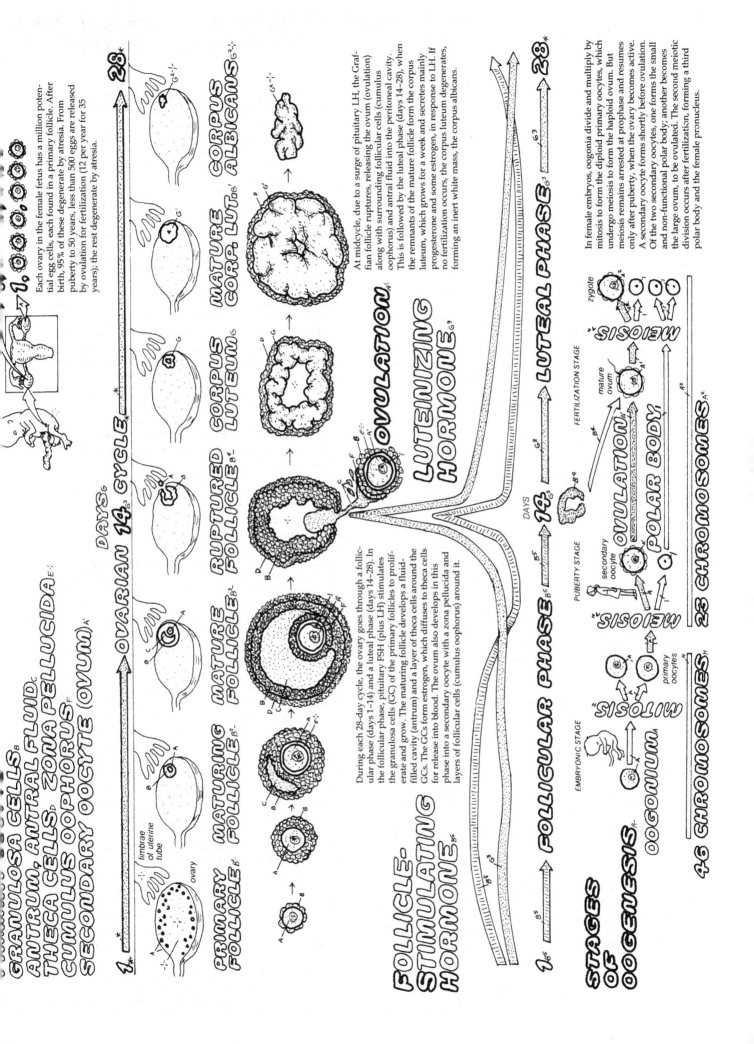

GRANULOSA CELLS_B
ANTRUM, ANTRAL FLUID_C
THECA CELLS_D ZONA PELLUCIDA_E
CUMULUS OOPHORUS_F
SECONDARY OOCYTE (OVUM)_A'

Each ovary in the female fetus has a million potential egg cells, each found in a primary follicle. After birth, 95% of these degenerate by atresia. From puberty to 50 years, less than 500 eggs are released by ovulation for fertilization (12 per year for 35 years); the rest degenerate by atresia.

OVARIAN 14 CYCLE 28

DAYS_G

PRIMARY FOLLICLE_B'   MATURING FOLLICLE_B²   MATURE FOLLICLE_B³   RUPTURED FOLLICLE_B⁴   CORPUS LUTEUM_G   MATURE CORP. LUT._G⁶   CORPUS ALBICANS_G⁷

During each 28-day cycle, the ovary goes through a follicular phase (days 1-14) and a luteal phase (days 14-28). In the follicular phase, pituitary FSH (plus LH) stimulates the granulosa cells (GC) of the primary follicles to proliferate and grow. The maturing follicle develops a fluid-filled cavity (antrum) and a layer of theca cells around the GCs. The GCs form estrogen, which diffuses to theca cells for release into blood. The ovum also develops in this phase into a secondary oocyte with a zona pellucida and layers of follicular cells (cumulus oophorus) around it.

FOLLICLE-STIMULATING HORMONE_B⁵

At midcycle, due to a surge of pituitary LH, the Graffian follicle ruptures, releasing the ovum (ovulation) along with surrounding follicular cells (cumulus oophorus) and antral fluid into the peritoneal cavity. This is followed by the luteal phase (days 14-28), when the remnants of the mature follicle form the corpus luteum, which grows for a week and secretes mainly progesterone and some estrogen, in response to LH. If no fertilization occurs, the corpus luteum degenerates, forming an inert white mass, the corpus albicans.

OVULATION_A'

LUTEINIZING HORMONE_G³

In female embryos, oogonia divide and multiply by mitosis to form the diploid primary oocytes, which undergo meiosis to form the haploid ovum. But meiosis remains arrested at prophase and resumes only after puberty, when the ovary becomes active. A secondary oocyte forms shortly before ovulation. Of the two secondary oocytes, one forms the small and non-functional polar body; another becomes the large ovum, to be ovulated. The second meiotic division occurs after fertilization, forming a third polar body and the female pronucleus.

FERTILIZATION STAGE   zygote

mature ovum

MEIOSIS_A²

OVULATION_A

POLAR BODY_I

PUBERTY STAGE   secondary oocyte

MEIOSIS_A²

EMBRYONIC STAGE

STAGES OF OOGENESIS_A

OOGONIUM_A

MITOSIS

primary oocytes

46 CHROMOSOMES_H

23 CHROMOSOMES_A²

FOLLICULAR PHASE_B⁵ 14   LUTEAL PHASE_G³ 28

DAYS 1 14

Estrogen and *progesterone* are the hormones of the ovary. They are steroid compounds derived ultimately from cholesterol. *Estradiol*, the most potent and the main estrogen secreted, has two hydroxyl groups; progesterone has two ketone groups. As female sex hormones, they regulate many aspects of female reproduction, sexuality, and secondary sex characteristics.

Granulosa and theca cells participate in estrogen secretion—In primates, estrogen can be formed by both *granulosa* and *theca interna* cells of ovarian follicles. The theca cell layer is vascularized, allowing access to plasma cholesterol used for synthesis of estradiol, which is released into the plasma. The granulosa cell layer is avascular; these cells lack access to plasma cholesterol and synthesize estradiol by converting androgen precursors, which diffuse from theca cells. Estrogen from granulosa cells is released into the follicle antrum to stimulate ovum growth. Estrogen secretion by theca and granulosa cells is stimulated by pituitary LH and FSH.

Cyclical changes in estrogen and progesterone secretion—During the *follicular phase*, estrogen secretion increases as follicle cells grow and proliferate; peak levels are reached by days 12–13 of the ovarian cycle. After ovulation, estrogen output diminishes due to the transformation of the follicle into a corpus luteum, but secretion continues into the third and fourth weeks. Progesterone secretion increases after ovulation when LH stimulates formation of the corpus luteum. *Luteal cells* of the corpus luteum are the source of progesterone and have receptors for the gonadotropins LH and FSH, both of which are necessary for optimal secretion of female sex steroids. Progesterone secretion peaks by the middle of the luteal phase (days 20–22) and declines thereafter. The lowest levels of both estrogen and progesterone occur in the absence of fertilization. Pregnancy promotes survival of the corpus luteum and marked increases in estrogen and progesterone secretion.

UTERINE ENDOMETRIUM SHOWS A MONTHLY CYCLE
The principal actions of estrogen and progesterone in the female reproductive system are on the *uterine endometrium*. This uterine mucosal lining is the site of *implantation* of the young embryo. To prepare for implantation, the endometrium undergoes cyclical changes, building up its wall to receive the embryo and destroying it in the absence of fertilization. The endometrial cyclical changes occur as a result of changes in the plasma levels of ovarian estrogen and progesterone and therefore follows the pattern of the ovarian cycle.

Estrogen promotes endometrial proliferation & thickening—Estrogen stimulates the epithelial cells of the *basal layer* of the endometrium to proliferate, forming a thick mucosa and numerous *endometrial (uterine) glands* with extensive blood vessels (*spiral arteries* and *veins*). These events constitute the *proliferative* phase of the endometrial cycle (days 6–14). At ovulation, the endometrium is fully grown (about 5 mm thick). The *myometrium*, the smooth muscle layer under the endometrium, is less affected.

Progesterone promotes secretion of endometrial glands—After ovulation, increasing levels of progesterone from the corpus luteum stimulate the endometrial gland to secrete a juice rich in proteins and glycogen that is important for survival and maintenance of the preimplantation and implanting embryo and for adherence of the implanted embryo. This part of the endometrial cycle, promoted by the action of progesterone, is termed the *secretory phase* and lasts through days 14–28 of the cycle. Progesterone is needed to sustain pregnancies.

Menstruation is caused by shedding and bleeding of endometrial tissue—In the absence of fertilization, the hormonal signals from the embryo for survival of the corpus luteum—i.e., human chorionic gonadotropin (hCG)—will not occur. The corpus luteum regresses, decreasing estrogen and progesterone secretion in the later part of the secretory phase. This weakens the endometrium, reducing blood flow and causing local oxygen deficiency (*ischemic phase*). By day 28, the endometrium begins to collapse and shed. Endometrial debris, along with some blood, constitutes the *menstrual flow* (menstruation, menses). This *menstrual phase* lasts about five days. The growth of follicles and increasing estrogen output during the next follicular phase terminate the menstrual phase and begin the next proliferative phase. Although the menstrual phase is the last phase of the endometrial cycle, in keeping with the events of the ovarian cycle it is customarily represented as the first phase (days 1–5).

Menarche and menopause—Menstrual cycles commence at puberty (*menarche*), usually at 12 to 13 years of age. Early cycles usually lack ovulation. In the early fifties, menstrual cycles cease (*menopause*). This event is a result of exhaustion of ovarian follicles and signals the end of reproductive functions, but not sexual activity. Menstrual cycles do not occur during pregnancy and in many lactating women.

OTHER EFFECTS OF ESTROGEN & PROGESTERONE

Effects on oviduct, myometrium, lactation, and feedback regulation—In the oviduct, estrogen stimulates the development of extensive mucosal folds and cilia, which function in transport of the ovum and young embryo. During pregnancy, estrogen stimulates the growth of uterine smooth muscle mass (myometrium), which functions in parturition and birth contractions (plate 158). Estrogen and progesterone stimulate mammary gland growth and support lactation (plate 159). Estrogen is mainly responsible for negative and positive feedback effects on the hypothalamus involved in regulation of its secretion (plate 155). In the brain and certain other tissues that are targets of male androgen hormones, estrogen is the true intracellular hormone mediating the androgenic effects, since androgens are converted to estrogens by aromatase.

Estrogen promotes puberty and secondary sex characteristics in females—During puberty, estrogen (along with adrenal androgens) enhances bone calcium deposition and growth. It also promotes growth of the uterus, vagina, and oviducts, as well as the mammary glands. Estrogen is responsible for development of secondary sex characteristics in adolescent females and their maintenance during maturity. These include soft skin and increased subcutaneous fat, particularly in breasts and buttocks, leading to the mature female shape. Estrogen promotes the growth of a wide pelvis and closure of epiphyseal plates in long bones. Some female secondary sex characteristics, such as a high-pitched voice, narrow shoulders, smaller bone and body mass, and lack of facial and body hair, are due to the absence of male androgens.

Estrogen may protect against aging diseases—Heart attacks due to coronary occlusion and abnormal cholesterol metabolism are rare in premenopausal women but increase sharply after menopause when plasma estrogen is deficient. Estrogen in the brain may diminish the effects of Alzheimers disease. Estrogen deficiency underlies the marked increase in osteoporosis and bone fractures in elderly women. Estrogen replacement therapy can ameliorate these aging disorders.

CN: Use the same colors for FSH (A) and LH (C) as on the preceding page. Use red for E, blue for H.
1. Begin with the bottom panel, and follow the FSH contribution to the ovarian cycle and the growth of the follicular cells into the sex hormone cycle panel. Then do the LH and luteal phase portion of the ovarian cycle.
2. Go to the upper left corner and color the diagram of the uterus. Color the enlargement of the uterine wall section.

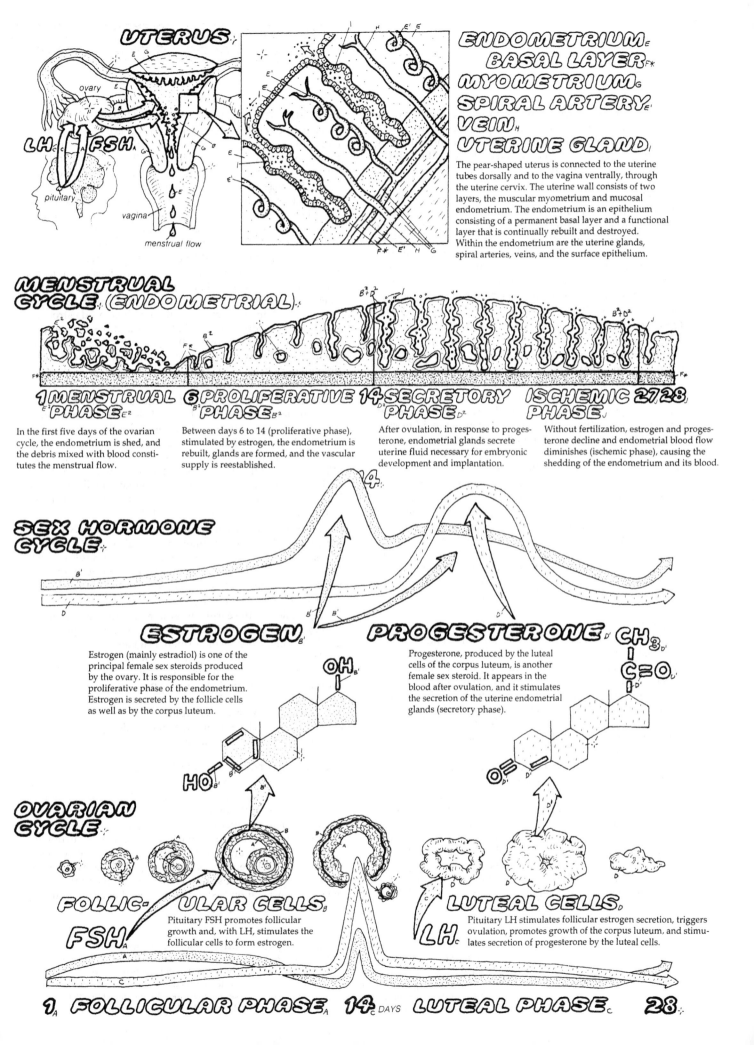

UTERUS

ovary

LH FSH

pituitary

vagina

menstrual flow

ENDOMETRIUM
BASAL LAYER
MYOMETRIUM
SPIRAL ARTERY
VEIN
UTERINE GLAND

The pear-shaped uterus is connected to the uterine tubes dorsally and to the vagina ventrally, through the uterine cervix. The uterine wall consists of two layers, the muscular myometrium and mucosal endometrium. The endometrium is an epithelium consisting of a permanent basal layer and a functional layer that is continually rebuilt and destroyed. Within the endometrium are the uterine glands, spiral arteries, veins, and the surface epithelium.

MENSTRUAL CYCLE (ENDOMETRIAL)

1 MENSTRUAL PHASE

In the first five days of the ovarian cycle, the endometrium is shed, and the debris mixed with blood constitutes the menstrual flow.

6 PROLIFERATIVE PHASE

Between days 6 to 14 (proliferative phase), stimulated by estrogen, the endometrium is rebuilt, glands are formed, and the vascular supply is reestablished.

14 SECRETORY PHASE

After ovulation, in response to progesterone, endometrial glands secrete uterine fluid necessary for embryonic development and implantation.

ISCHEMIC PHASE 27 28

Without fertilization, estrogen and progesterone decline and endometrial blood flow diminishes (ischemic phase), causing the shedding of the endometrium and its blood.

SEX HORMONE CYCLE

ESTROGEN

Estrogen (mainly estradiol) is one of the principal female sex steroids produced by the ovary. It is responsible for the proliferative phase of the endometrium. Estrogen is secreted by the follicle cells as well as by the corpus luteum.

OH

HO

PROGESTERONE CH₃

Progesterone, produced by the luteal cells of the corpus luteum, is another female sex steroid. It appears in the blood after ovulation, and it stimulates the secretion of the uterine endometrial glands (secretory phase).

C=O

O=

OVARIAN CYCLE

FOLLIC-ULAR CELLS

FSH

Pituitary FSH promotes follicular growth and, with LH, stimulates the follicular cells to form estrogen.

LUTEAL CELLS

LH

Pituitary LH stimulates follicular estrogen secretion, triggers ovulation, promotes growth of the corpus luteum, and stimulates secretion of progesterone by the luteal cells.

1 FOLLICULAR PHASE 14 DAYS LUTEAL PHASE 28

In the testes spermatogenesis and testosterone secretion occur continuously at a steady rate. The *ovary*, however, shows a cyclical pattern of activity. Thus, follicle formation (including ovum growth) and ovulation, as well as the formation and regression of the corpus luteum, all occur in sequence within a single cycle that is then repeated. Similarly, the secretion of ovarian hormones *estrogen* and *progesterone* follow a cyclical pattern, estrogen appearing in the follicular phase followed by progesterone in the luteal phase (plate 154). Average duration of the ovarian cycle in the mature human female is 28 days. Cycles begin at puberty and are interrupted only during pregnancy and lactation and by illness, and they cease after the age of fifty. Here, we study how the ovarian hormones, the anterior pituitary gonadotropins, and the hypothalamus interact to ensure the orderly operation of the ovarian cycle.

HYPOTHALAMUS & PITUITARY REGULATE THE OVARIES

Gonadotropins LH & FSH directly regulate follicular and luteal functions — The anterior pituitary secretes two gonadotropin hormones that regulate the activity of the ovary — the *follicle-stimulating hormone* (FSH) and the *luteinizing hormone* (LH). Gonadotropins are glycoprotein hormones secreted from the basophilic gonadotrope. Both LH and FSH are necessary for ovarian activity, although each may act in different phases of the cycle. FSH is essential in proliferation and growth of granulosa cells early in the follicular phase and later for increase of the theca interna cells.

LH, however, stimulates estrogen production and release by follicle cells as well as inducing ovulation and growth of the corpus luteum and its secretion of progesterone and estrogen. Gonadotropins exert their actions on granulosa and theca interna cells by binding to their own receptors on plasma membranes, acting via the G-protein → adenylate cyclase → cyclic AMP pathway (plate 12,114). The young granulosa cells have mainly FSH receptors but mature ones also carry LH receptors. The later-forming theca interna cells have both LH and FSH receptors.

Pulsatile release of GnRH from the hypothalamus controls pituitary LH & FSH — Secretion of LH and FSH are controlled by the *gonadotropin-releasing hormone* (GnRH), a peptide neurohormone released from the hypothalamus. GnRH is synthesized by GnRH-containing hypothalamic neurons, which release GnRH by their axon terminals into the *portal hypophyseal capillaries*, for rapid and direct delivery to the anterior pituitary gonadotrope cells. Receptors for GnRH are found on the plasma membrane of the gonadotropes. GnRH action is cAMP mediated.

GnRH release is not continuous but occurs in approximately hourly pulses. To increase gonadotropin secretion, GnRH amount per pulse (pulse amplitude) or the number of pulses (pulse frequency) increases, and vice versa. LH release is also known to take place in pulses that occur shortly after each GnRH pulse, but FSH release is less pulsatile and occurs more slowly. The frequency and amplitude of the GnRH pulses are under the control of two mechanisms — a hypothalamic "clock" that sets the overall duration of the cycle and the timing of major events within the ovarian cycle and the negative feedback control of estrogen on the hypothalamus.

ESTROGEN CONTROLS GnRH RELEASE THROUGH FEEDBACK

Negative feedback at onset — Low levels of estrogen at the end of the ovarian cycle, acting via *negative feedback*, stimulates the hypothalamus to increase its pulsatile output of GnRH. This leads to increased output of FSH and LH from the pituitary. FSH rises sharply in the first days and remains high for most of the follicular phase; LH shows a steady increase. FSH and LH stimulate follicular growth and secretion of estrogen. By day 13 of the cycle, estrogen level peaks while FSH and LH diminish, due to the negative feedback inhibition by estrogen.

Positive feedback at midcycle — At this point a new *positive feedback* mechanism comes into play: high estrogen levels cause a marked rise in LH (an LH burst) and a moderate one in FSH levels. Exactly how negative feedback switches to positive feedback in midcycle is not known. Increased estrogen increases the frequency of GnRH pulses and possibly augments the GnRH receptors on the gonadotrope cells, enhancing their sensitivity to GnRH pulses. These events produce the preovulatory burst of LH secretion that triggers the process of ovulation within several hours.

Return of negative feedback in the luteal phase — The postovulatory high levels of LH (and also of FSH) promote corpus luteum growth and progesterone (with some estrogen) release. By midluteal days the negative feedback effect returns; high estrogen and progesterone act to lower LH and FSH for most of the luteal phase. If the estrogen level is kept high from the beginning of the cycle, ovulation will not occur. This observation is the basis for the use of estrogen-like compounds in contraceptive pills (plate 161).

Inhibin from ovarian granulosa cells inhibits FSH secretion — The protein hormone *inhibin* also plays a role in ovarian regulation. Inhibin is secreted by the granulosa cells and exerts a negative feedback effect on the pituitary to inhibit FSH secretion. Inhibin levels are low in the follicular phase and high in the luteal phase.

Regression of the corpus luteum marks the end of an ovarian cycle — The corpus luteum begins to regress around day 25 of the cycle. Absence of hormonal signals from the implanted embryo (hCG) and the reduced LH and FSH levels signal the regression. A number of local hormones, such as prostaglandins and proteolytic enzymes, promote lysis of the corpus luteum. Regression of the corpus luteum reduces progesterone and estrogen output. This event marks the end of the ovarian cycle and promotes the shedding of the endometrium and menstruation.

FACTORS INFLUENCING OVARIAN FUNCTION

Illness, malnutrition, severe stress, and emotional crises interfere with the operation of the ovarian cycle. Stress and emotional crises act on the higher brain centers and, from there, on the hypothalamus, interfering with the pattern of GnRH release. Often the release is inhibited, leading to reduction in FSH and LH levels and consequently diminished secretion of sex hormones. Depending on the timing of the stress, diminished estrogen may cause undue menstruation (spotting) or delayed menstruation (secondary amenorrhea) that occurs in the absence of endometrial proliferation.

CN: Use the same colors as on the preceding page for FSH (D), LH (E), estrogen (G), and progesterone (H). Use light colors for A and C.
1. Color the large control illustration in the center.
2. Color the three bottom panels. Color only the bold portions of the hormone levels. Color gray that portion of the endometrium involved during the period described. The dotted ascending line in the left panel represents reduced levels of estrogen. Note that bold dotted lines in the right panel represent a cessation of FSH and LH secretion.
3. Color the diagram in the upper right.

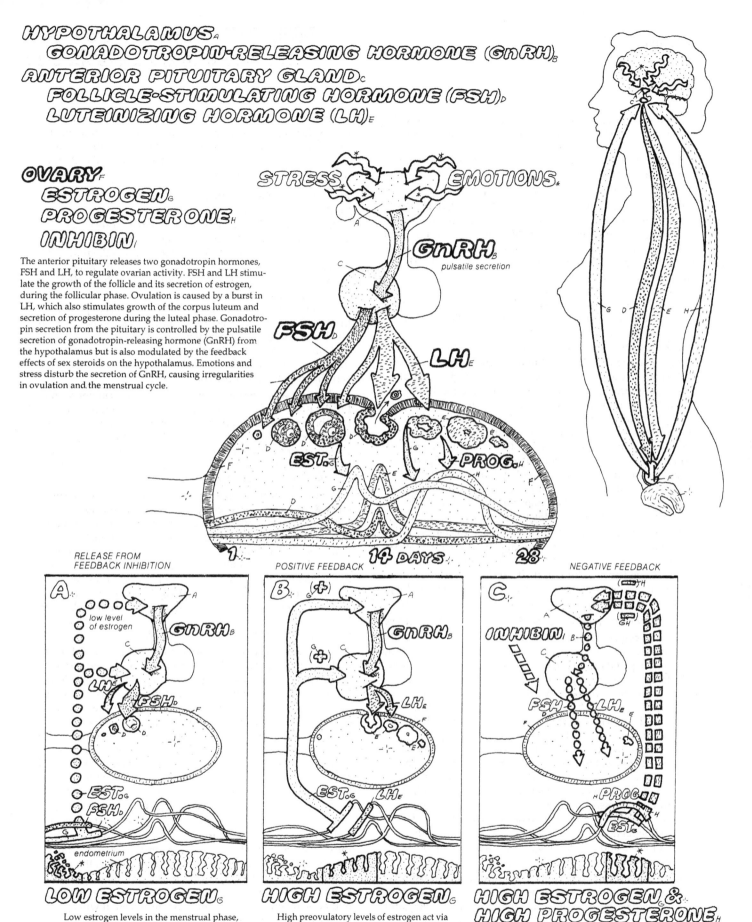

HYPOTHALAMUS_A

*HYPOTHALAMUS*_A
*GONADOTROPIN-RELEASING HORMONE (GnRH)*_B
*ANTERIOR PITUITARY GLAND*_C
*FOLLICLE-STIMULATING HORMONE (FSH)*_D
*LUTEINIZING HORMONE (LH)*_E

*OVARY*_F
*ESTROGEN*_G
*PROGESTERONE*_H
*INHIBIN*_I

The anterior pituitary releases two gonadotropin hormones, FSH and LH, to regulate ovarian activity. FSH and LH stimulate the growth of the follicle and its secretion of estrogen, during the follicular phase. Ovulation is caused by a burst in LH, which also stimulates growth of the corpus luteum and secretion of progesterone during the luteal phase. Gonadotropin secretion from the pituitary is controlled by the pulsatile secretion of gonadotropin-releasing hormone (GnRH) from the hypothalamus but is also modulated by the feedback effects of sex steroids on the hypothalamus. Emotions and stress disturb the secretion of GnRH, causing irregularities in ovulation and the menstrual cycle.

STRESS* EMOTIONS*

GnRH_B
pulsatile secretion

FSH_D LH_E

EST._G PROG._H

1   14 DAYS   28

A
RELEASE FROM FEEDBACK INHIBITION

low level of estrogen

GnRH_B
LH_E FSH_D

EST._G
FSH_D
endometrium

LOW ESTROGEN_G

Low estrogen levels in the menstrual phase, acting via a negative feedback mechanism, increase release of GnRH, which in turn increases FSH and LH release (see also panel C). These stimulate follicular growth and elevate estrogen levels to their peak by day 13 of the ovarian cycle.

B
POSITIVE FEEDBACK

(+) (+)
GnRH_B
LH_E

EST._G LH_E

HIGH ESTROGEN_G

High preovulatory levels of estrogen act via positive feedback to increase GnRH pulse frequency which triggers a burst of LH release by day 14. High LH leads to ovulation, growth of the corpus luteum, and its secretion of progesterone, which peaks at day 22; estrogen secretion continues at lower levels. Inhibin from granulosa cells inhibits FSH secretion in the luteal phase.

C
NEGATIVE FEEDBACK

INHIBIN
FSH_D LH_E

PROG._H
EST._G

HIGH ESTROGEN & HIGH PROGESTERONE_H

In the absence of fertilization and an implanted embryo, high estrogen and progesterone levels reduce secretion of LH and FSH by negative feedback. Reduced gonadotropin levels lead to regression of the corpus luteum and decreased output of sex steroids, then menstruation. The cycle continues with panel A.

The sperm and egg, the male and female gametes, are highly specialized for their respective functions in fertilization and the first stage of development. Fertilization occurs in the ampulla of the uterine duct and involves release of calcium and activation of the egg, fusion of male and female pronuclei, and formation of the zygote.

SPERM: SPECIALIZED FOR MOTILITY & OVUM FERTILIZATION
The fully mature human *spermatozoon* (sperm, sperm cell) is highly specialized for its functions of motility and fertilization of the egg. A sperm is 60 μm long and has three parts—*head, middle piece,* and *tail.* The *head* contains the nucleus, bearing the genetic material—condensed chromatin (DNA). The sperm acrosome is a large lysosome with hydrolytic enzymes (e.g., *hyalouronidase, acrosin*) to lyse the membranes around the egg and facilitate sperm penetration. The middle piece has many mitochondria. The sperm tail, the body's only known flagellum, enables the sperm's swimming movements.

Sperm are able swimmers but move in random directions in the female tract—When deposited in the vagina, sperm enter the cervical canal and swim through it to reach the uterus. Sperm remaining in the vagina die from to exposure to vaginal acids. Cervical mucus is alkaline and therefore optimal for sperm survival. In the uterus, sperm swim in various directions; some reach the uterine tube but many terminate in different loci where they age and die. Sperm swim at a rate of 3 mm/min, allowing them to reach the uterine tube within an hour. Since some sperm reach the oviduct within minutes, their transport may be facilitated by uterine contractions, possibly induced by prostaglandins.

Fewer than one in a million sperm reach the egg—Of the 300 million sperm deposited in the vagina during intercourse, only 0.1% reach the uterine tube and less than a hundred reach the egg (*ovum*). High mortality and random sperm movement are some reasons so many sperm are required for fertility, even though only one is sufficient for fertilization.

EGG: SPECIALIZED FOR NUTRIENT STORAGE & FERTILIZATION
Human egg contains large nutrient reserves and several membranes—Compared to the sperm, the human egg (ovum) is a very large cell (up to 200 μm in diameter), mainly because of its large stores of *cytoplasmic granules* (*yolk*) that contain nutrients for the young embryo. The ovulated egg is surrounded by a layer of *follicular cells* (cumulus oophorus, corona radiata), which help support the egg metabolically and nutritionally. The follicular cells are glued together by *hyalouronic acid,* a sticky mucopolysaccharide ("intercellular cement"). Between the layer of follicular cells and the egg plasma membrane is the *zona pellucida,* a 5-μm-thick membrane made of a transparent, jellylike substance. The follicular cells and the egg send microvillar projections through the zona pellucida, possibly for interchange of substances. The zona pellucida helps provide mechanical support for the young embryo and protects it against maternal antibodies and macrophages.

Cilia and uterine tube contractions aid in egg transport—After ovulation, the sweeping movements of the uterine tube and its *fimbriae* create suction, drawing the immobile egg (and cumulus oophorus) into the uterine tube. Uterine tube contractions and the constant oarlike beating of the cilia on the epithelial cells of the *mucosal folds* propel the egg toward the uterus. Estrogen is necessary for uterine tube contraction and for ciliary formation and beating. Within hours after ovulation, the egg reaches the ampulla of the uterine tube. At this time, it is fully ripe for fertilization.

FERTILIZATION OCCURS IN OVIDUCT & INVOLVES MULTIPLE SPERM-EGG INTERACTIONS
Capacitation & acrosome reaction ensure release of acrosomal enzyme—In order to penetrate the egg, the sperm must first undergo *capacitation,* which involves the removal of the acrosomal glycoprotein coat that prevents premature release of the acrosomal enzymes. Substances that induce sperm capacitation may come from the oviduct or from follicular cells. As a capacitated sperm prepares to penetrate the egg, the acrosome releases its enzymes (*acrosome reaction*). Hyalouronidase lyses the hyalouronic acid, dispersing the follicular cells and allowing the sperm to make its way through these cells. Other lysosomal enzymes, such as acrosin, digest parts of the zona pellucida.

Binding with sperm receptors ensures sperm entry and fertilization—Contact of the sperm with the egg plasma membrane is enhanced by binding of the sperm to a specific *sperm receptor* (ZP-3) on the zona pellucida. Next, aided by a sperm surface protein called *fertilin,* the egg plasma membrane engulfs the sperm, which is taken in, head and tail. This is the main stage of *fertilization.*

The zona reaction forms a barrier against more sperm entry & prevents polyspermy—Entry of the first sperm is followed immediately by the *zona reaction,* a rapid chemical modification of the zona pellucida that blocks penetration by more sperm. The cause of the zona reaction is the outflow of some substances originating from granules in the *cytoplasm* of the egg. Failure of the zona reaction leads to *polyspermy,* which is not compatible with normal development. Before this permanent barrier, an immediate and temporary barrier is formed by the change in egg's membrane potential and release of calcium within the egg cytoplasm.

Calcium release triggers activation of the egg, fusion of male & female pronuclei, and formation of the zygote—The increased calcium release together with sperm penetration results in *activation* of the egg, including its metabolic awakening; also the last meiotic division of the egg nucleus occurs, expelling the last polar body and forming the *female pro-nucleus.* Meanwhile the sperm tail degenerates, and the sperm nucleus swells and enlarges, forming the *male pronucleus.* The last stage of fertilization is the *fusion* of male and female pronuclei, resulting in the recombination of chromosomes of the male and female gametes and the formation of the *zygote nucleus.*

Sperm and egg longevity & survival—Within the female reproductive tract, sperm can survive up to four days, especially those stored in the cervical mucosa and nourished by the cervical mucus. However, when appropriately frozen, sperm can be kept several years and still maintain the ability to fertilize an ovum. Long-term storage of human eggs has recently become possible. The egg has a shorter life span after ovulation (about one day), and if not fertilized, it will age and degenerate. The optimum time for fertilization is within the first twelve hours after ovulation.

CN: Use light colors for D, G, H, and M-R.
1. Begin with the uterine tube.
2. Color the material on the spermatozoon (F)
3. Color the ovum and the five stages of fertilization. Note that the dotted line in the last one represents the degenerating tail of the spermatozoon.

Sperm deposited in the vagina swim through the cervix into the uterus and up into the uterine tube. Since sperm move in all directions, fertilization requires a very large number of sperm. Of the 300 million deposited, 0.1% reach the oviduct and about 100 reach the uterine tube ampulla to meet the ovum. Uterine contractions triggered by prostaglandins may also facilitate sperm transport.

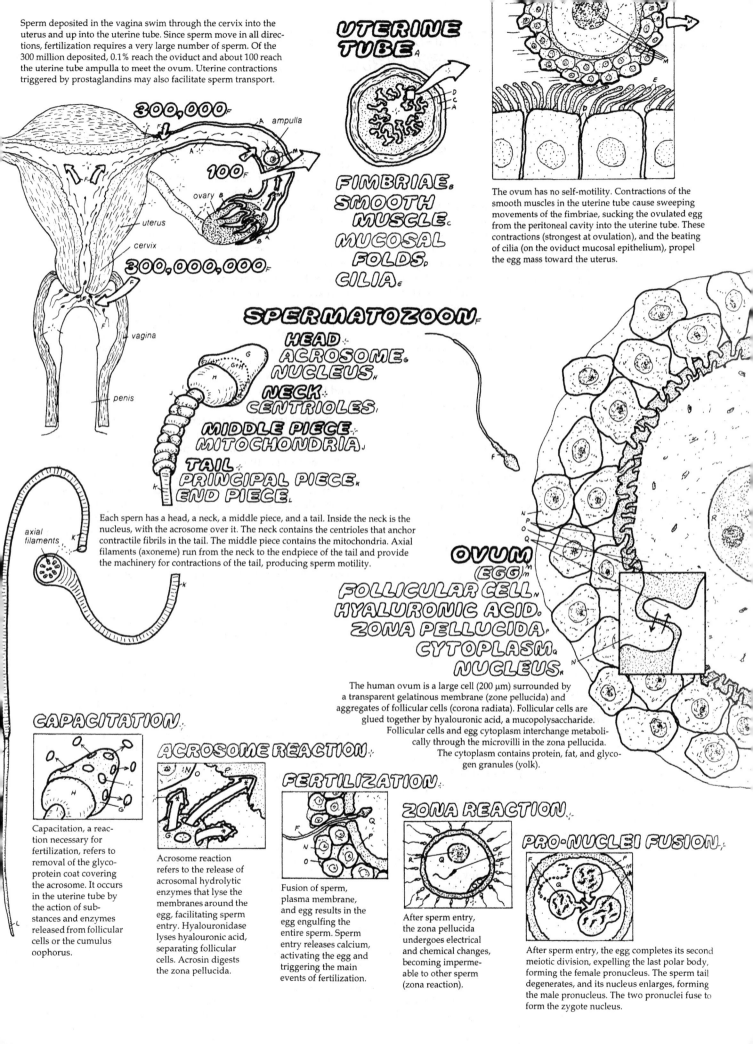

300,000

100

300,000,000

ampulla

ovary

uterus

cervix

penis

vagina

UTERINE TUBE

FIMBRIAE.
SMOOTH MUSCLE.
MUCOSAL FOLDS.
CILIA.

The ovum has no self-motility. Contractions of the smooth muscles in the uterine tube cause sweeping movements of the fimbriae, sucking the ovulated egg from the peritoneal cavity into the uterine tube. These contractions (strongest at ovulation), and the beating of cilia (on the oviduct mucosal epithelium), propel the egg mass toward the uterus.

SPERMATOZOON

HEAD
ACROSOME.
NUCLEUS.
NECK.
CENTRIOLES.
MIDDLE PIECE.
MITOCHONDRIA.
TAIL.
PRINCIPAL PIECE.
END PIECE.

axial filaments

Each sperm has a head, a neck, a middle piece, and a tail. Inside the neck is the nucleus, with the acrosome over it. The neck contains the centrioles that anchor contractile fibrils in the tail. The middle piece contains the mitochondria. Axial filaments (axoneme) run from the neck to the endpiece of the tail and provide the machinery for contractions of the tail, producing sperm motility.

OVUM
(EGG)
FOLLICULAR CELL.
HYALURONIC ACID.
ZONA PELLUCIDA.
CYTOPLASM.
NUCLEUS.

The human ovum is a large cell (200 μm) surrounded by a transparent gelatinous membrane (zone pellucida) and aggregates of follicular cells (corona radiata). Follicular cells are glued together by hyaluronic acid, a mucopolysaccharide. Follicular cells and egg cytoplasm interchange metabolically through the microvilli in the zona pellucida. The cytoplasm contains protein, fat, and glycogen granules (yolk).

CAPACITATION

Capacitation, a reaction necessary for fertilization, refers to removal of the glycoprotein coat covering the acrosome. It occurs in the uterine tube by the action of substances and enzymes released from follicular cells or the cumulus oophorus.

ACROSOME REACTION

Acrosome reaction refers to the release of acrosomal hydrolytic enzymes that lyse the membranes around the egg, facilitating sperm entry. Hyalouronidase lyses hyaluronic acid, separating follicular cells. Acrosin digests the zona pellucida.

FERTILIZATION

Fusion of sperm, plasma membrane, and egg results in the egg engulfing the entire sperm. Sperm entry releases calcium, activating the egg and triggering the main events of fertilization.

ZONA REACTION

After sperm entry, the zona pellucida undergoes electrical and chemical changes, becoming impermeable to other sperm (zona reaction).

PRO-NUCLEI FUSION

After sperm entry, the egg completes its second meiotic division, expelling the last polar body, forming the female pronucleus. The sperm tail degenerates, and its nucleus enlarges, forming the male pronucleus. The two pronuclei fuse to form the zygote nucleus.

The individual, with diverse cells, tissues, and organs, originates from a single cell—the *zygote*—and goes through the stages of embryo, fetus, infant, child, and adolescent before reaching maturity.

DEVELOPMENT OF THE YOUNG EMBRYO

Following fertilization, the zygote undergoes cell *proliferation* to increase the number of its cells and form the rudiments of the young embryo. In later stages, embryonic cells undergo *differentiation* to form the body's diverse cell types and tissues.

Cleavage divisions form the young embryo—Cell proliferation is accomplished by several *mitotic* divisions (cleavage), resulting in embryos with 2, 4, 16, and 32 daughter cells, called the *blastomeres*. As a result, the zygote transforms into a *young embryo* consisting of a ball of uniform cells (*morula*, mulberry). The blastomeres become increasingly smaller in size, since they the utilize zygote's cytoplasmic stores, and are surrounded by its *zona pellucida*, which still persists. Thus the morula is the same size as the zygote.

Young embryo is not motile and must be transported to the uterus—During cleavage, the young embryo is propelled in the uterine tube toward the uterus by the action of the cilia of the mucosal lining and by the contractions of the oviduct. It takes about four days to traverse the uterine tube; by this time, the embryo is in the morula stage.

Formation of inner cell mass and trophoblast signal differentiation—Upon entering the uterus, the embryonic cells segregate, forming an internal group of cells (*inner cell mass*) and a sheet of cells (*trophoblast*) that surround a *cavity*. At this stage (day 5), the embryo is called a *blastocyst*. The early *blastocyst* still has the zone pellucida, which soon degenerates allowing it to obtain nutrients and oxygen directly from the uterine fluid. This results in the embryo's growth and formation of the *late blastocyst*, in which the trophoblast cells become flattened and active. A diffuse external layer of syncytiotrophoblast cells with poorly defined cellular forms surrounds an orderly internal layer of cytotrophoblast cells. The various parts of the trophoblast will later form the *placenta* and embryonic membranes (e.g., the amniotic sac), while the inner cell mass cells forms the *embryo* proper.

Implantation involves burrowing of the blastocyst embryo into the endometrium—By days 6–7, the growing embryo begins to attach itself to the uterus to obtain nutrients and oxygen directly from the maternal blood. The syncytiotrophoblast releases *lysosomal enzymes*, which lyse the *endometrium*, allowing the blastocyst to burrow into it. This event is called *implantation*. After implantation, the endometrium heals and covers the embryo. As a result the human embryo grows within the uterine endometrium and not in the uterine cavity.

Ectopic pregnancies are not viable—Implantation usually occurs in the dorsal wall of the uterus, but it also may occur in various *ectopic sites* in the uterine tube, the cervix, or even the peritoneal cavity. Ectopic pregnancies usually are not viable. Tubal pregnancies create medical emergencies because the growing embryo and placenta cause blood vessel rupture and internal hemorrhage.

Embryonic tissues arise from three germinal layers—One week after implantation, the inner cell mass begins to differentiate and forms three germinal layers—ectoderm, mesoderm, and endoderm. The cells of these layers proliferate, migrate, and differentiate to produce the tissues and organs of the developing embryo. Ectoderm produces the cells of the nervous system and skin, muscle and bone tissue originate from mesoderm, and the inner lining of visceral organs develops from endoderm.

Trophoblast forms the placenta, the organ for nutrient & gas exchange—After implantation, the trophoblast proliferates to form the *chorionic villi*, which exchange nutrients, respiratory gases, and metabolites with the *maternal blood vessels* through special blood sinuses. The chorionic villi and maternal vessels later form the *placenta*, a separate, anatomically distinct and physiologically critical organ that serves in essential nutritive and respiratory functions of the embryo and fetus; the placenta also has endocrine cells that secrete hormones for the duration of pregnancy (plate 158).

Trophoblast cells of the embryo secrete the hCG hormone to stimulate survival and growth of corpus luteum—After implantation, syncytiotrophoblast cells in the chorionic villi secrete a peptide hormone called *human chorionic gonadotropin* (hCG) into the maternal blood. hCG is a glycoprotein hormone and resembles LH in structure and physiological properties. hCG binds to LH receptors on the corpus luteum and promotes its survival and growth and its secretion of progesterone and estrogen. These hormones in turn maintain the endometrium in optimal condition for gestation. The detection of hCG in maternal blood is possible during the second week of implantation and in urine three weeks after implantation; this ability is the basis of most modern immunochemical pregnancy tests.

TWINNING: DIZYGOTIC VS. MONOZYGOTIC DEVELOPMENT

Dizygotic development forms fraternal twins—Ovarian cycles normally involve development of a single follicle and release of one egg at ovulation. Growth of more than one follicle results in multiple ovulation and formation of two or more zygotes. Each of these implants separately, forming *fraternal twins or triplets*, or more, not necessarily of the same sex.

Monozygotic development forms identical twins—If the two blastomeres from a single zygote separate at the first cleavage division, or if the single inner cell mass divides into two separate masses, each blastomere or cell mass will proceed to form an independent embryo. Because these embryos share a common *genotype*—i.e., an entire set of genes (*genome*)—they will be alike in sex and with respect to physical characteristics (*phenotype*). *Identical twins* are the result of such common development; they may share a placenta or may have their own. Dizygotic twinning may be hereditary but the occurrence of monozygotic twinning appears to be accidental. Normal incidence of twinning is 1% for fraternal twins and 0.3% (one in about 300 pregnancies) for identical twins.

CN: Use red for P. Use light colors throughout, except for structures L, N, and O, which receive dark colors. Use your lightest colors for H and M.
1. Begin with the entrance of the ovum into the uterine tube. Note that the zona pellucida (K) and polar body (L) titles are in the upper right corner. Color the day numbers gray. Color the ectopic implantation (N) sites (marked by large asterisks).

2. Continue with the three-dimensional drawing of the later blastocyst at day 6 and the large drawing of day 12. Color in the uterine endometrium (M) in the large drawings before dotting in the lysosomal enzymes (O) in day 6.
3. Color the three hormonal influences in the lower right corner.
4. Color the diagrams illustrating twin formation.

EARLY STAGES OF DEVELOPMENT

OVUM_A
FERTILIZATION_B
ZYGOTE_C
2-CELL STAGE_D
4-CELL STAGE_E
8-CELL STAGE_F

MORULA_G
EARLY BLASTOCYST (5)*
INNER CELL MASS_H
BLASTOCYST CAVITY
TROPHOBLAST_J
LATE BLASTOCYST (6)*

ZONA PELLUCIDA_K
POLAR BODIES_L
UTERINE ENDO-
  METRIUM_M
ECTOPIC IMPLAN-
  TATION SITES_N

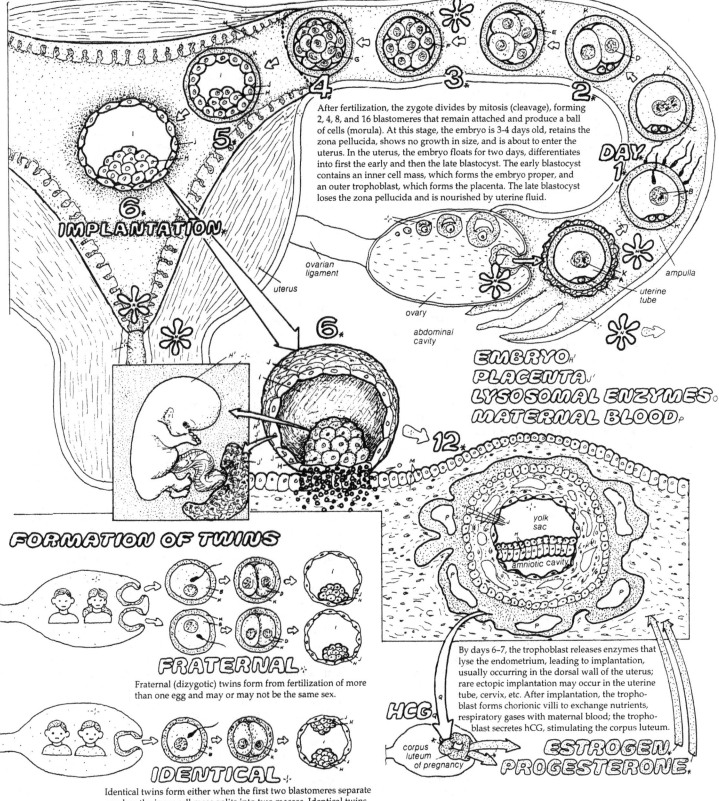

After fertilization, the zygote divides by mitosis (cleavage), forming 2, 4, 8, and 16 blastomeres that remain attached and produce a ball of cells (morula). At this stage, the embryo is 3-4 days old, retains the zona pellucida, shows no growth in size, and is about to enter the uterus. In the uterus, the embryo floats for two days, differentiates into first the early and then the late blastocyst. The early blastocyst contains an inner cell mass, which forms the embryo proper, and an outer trophoblast, which forms the placenta. The late blastocyst loses the zona pellucida and is nourished by uterine fluid.

IMPLANTATION*

ovarian ligament

uterus

ovary

abdominal cavity

ampulla

uterine tube

DAY 1*

EMBRYO_H'
PLACENTA_J'
LYSOSOMAL ENZYMES_O
MATERNAL BLOOD_P

yolk sac

amniotic cavity

FORMATION OF TWINS

FRATERNAL

Fraternal (dizygotic) twins form from fertilization of more than one egg and may or may not be the same sex.

IDENTICAL

Identical twins form either when the first two blastomeres separate or when the inner cell mass splits into two masses. Identical twins have the same sex and genotype and very similar phenotypes.

By days 6-7, the trophoblast releases enzymes that lyse the endometrium, leading to implantation, usually occurring in the dorsal wall of the uterus; rare ectopic implantation may occur in the uterine tube, cervix, etc. After implantation, the tropho-blast forms chorionic villi to exchange nutrients, respiratory gases with maternal blood; the tropho-blast secretes hCG, stimulating the corpus luteum.

HCG_Q

corpus luteum of pregnancy

ESTROGEN_R
PROGESTERONE*

Duration of human *pregnancy* or *gestation* is about 270 days (~38 weeks) from conception. Pregnancy is divided into three trimesters (3-month-long periods). The first covers the development of the *embryo*, the second and third entail the growth and development of the *fetus*. Pregnancy involves major hormonal and metabolic changes in the mother. Maternal and fetal hormonal mechanisms end pregnancy by inducing birth (*parturition*), which occurs about 284 days after the first day of the last menstruation before pregnancy.

PLACENTAL HORMONES STIMULATE PREGNANCY CHANGES

Placental hCG stimulates formation of corpus luteum of pregnancy — After implantation, placental syncytiotrophoblast cells secrete an LH-like gonadotropin (hCG, *human chorionic gonadotropin*) in maternal blood. hCG in maternal blood is the basis of most modern pregnancy tests (plate 157). hCG helps form the *corpus luteum of pregnancy* that secretes large amounts of progesterone and estrogen in the maternal blood. These steroids cause menstruation to cease, a familiar sign of pregnancy, and, through negative feedback, inhibit the pituitary gonadotropins, preventing further follicular development and ovulation. Maternal estrogen and progesterone stimulate the growth and secretions of the endometrium to ensure support of the developing embryo and promote growth and proliferation of the *myometrium* (uterine smooth muscle wall) as well as the *mammary glands* and breast tissue.

Placenta also secretes sex steroids — By the end of the first trimester, other placental endocrine cells secrete increasing amounts of estrogen and progesterone into the maternal blood, augmenting the corpus luteum. Ovarian removal during the second and third trimesters will not terminate pregnancy, since the placenta can replenish ovarian steroids. The hCG level peaks during the first trimester and falls off gradually thereafter. But it continues to stimulate the ovary and placenta to produce female sex steroids. These hormones continue to exert their effects on the uterine endometrium and myometrium as well as stimulating some of the bodily and metabolic changes in the pregnant woman — e.g., increased subcutaneous fat, fluid retention, gain in body fat and weight.

Placental hCS initiates fat utilization in the mother and spares glucose for the fetus — Another protein hormone, *human chorionic somatomammotropin* (hCS), with properties similar to growth hormone and prolactin, is secreted from the placenta into the maternal blood throughout gestation. hCS antagonizes the action of maternal insulin, sparing glucose and amino acids for the fetus; hCS also mobilizes fatty acids for maternal tissues. Fetal growth is reduced in cases of hCS deficiency because of the decreased nutrient supply to the fetus. hCS also stimulates growth of maternal mammary glands (plate 159).

Major events of embryonic and fetal development & control of fetal growth — During the embryonic period (weeks 1–8), development consists largely of proliferation and differentiation of cells and tissues and *organogenesis*, the formation of the organs and systems. Major organs form during weeks 4–8, making this period a significant and critical one in terms of the effects of drugs and other teratological agents on embryonic development. By the third month, the embryo is called a *fetus*. The fetal period (3–9 months) is characterized chiefly by growth of the tissues, organs, and body of the fetus, but

differentiation in several tissues and systems, such as the nervous system, continues to occur. Fetal growth is regulated by fetal *insulin* and *insulin-like growth factors* (IGF 1, IGF 2) but not by fetal growth hormone. Anencephalic fetuses, which have no pituitary, are of normal body size.

Functional development of fetus & newborn — Fetal movements can be felt during the second trimester. Later on fetal startle reflexes can be induced in response to sudden loud noises. Third-trimester fetuses have open eyes and occasionally suck their thumbs. Maturation of lungs, blood, and immune system and subcutaneous fat formation continue during the third trimester. Fetuses born before term are called *premature*. Eight-month-old preterm fetuses are often viable but the viability of six-month-old fetuses requires intensive medical care. Immediately after birth, reduced oxygen and increased CO_2 in the plasma stimulates breathing and activates the newborn lungs. *Umbilical arteries* and veins close, as well as the connections between the left and right atria (*foramen ovale*) and between the pulmonary artery and aorta (*ductus arteriosus*), forming the mature pattern of circulation.

SEVERAL HORMONES REGULATE PARTURITION

Throughout pregnancy, estrogen stimulates growth of uterine myometrium to support the fetus during pregnancy and expel it at birth. Mild uterine contractions begin with the fourth month. Contractions become strong and rhythmic hours before birth, resulting in fetal expulsion through the cervix and vagina (birth canal). Birth or *parturition* is regulated by hormonal signals from the fetus and mother, including estrogen, *oxytocin, prostaglandins,* and *relaxin*.

Regulation of labor onset may involve fetal and maternal signals — Estrogen increases uterine smooth muscle excitability and progesterone decreases it. In some species, a drop in progesterone before birth allows estrogen to stimulate uterine contractions, initiating labor. In sheep and possibly in humans cortisol from fetal adrenal glands increases before labor, is converted to estrogen by placenta, and induces uterine contractions. Prostaglandins from uterine glands also induce myometrium contractions in the early birth stage.

A neurohormonal reflex involving oxytocin release promotes fetal expulsion — One of the functions of the posterior pituitary hormone *oxytocin* is to stimulate uterine contraction. During late pregnancy estrogen induces a 100-fold increase in oxytocin *receptors*, enabling it to exert powerful contractile effects on the uterus. During the first stage of labor, pressure from the fetal head dilates the cervix, stimulating the *cervical/ stretch receptors* and activating a *neurohormonal reflex*. Sensory nerves from the cervical stretch receptors stimulate the *hypothalamus* and *posterior pituitary* to release oxytocin. Pulsatile release of oxytocin increases during labor due to a positive feedback loop and is terminated after fetal expulsion and relaxation of the cervix. Oxytocin and prostaglandins also help expel the placenta (*afterbirth*), which occurs shortly after the birth of the baby and constitutes the third phase of labor. Parturition can be induced by injections of oxytocin, which are often given during labor to aid in delivery. To facilitate birth, another peptide hormone, relaxin, is secreted during gestation by the corpus luteum of pregnancy and by the placenta. Relaxin softens the cervix as well as the ligaments and joints of the pelvic bones.

CN: Use same colors for the first four hormones (A–D) used in the earlier plates of this chapter. Use a dark color for F and N and a light color for O.
1. Begin with the titles of the upper half, and color all the material in the large rectangle. Color the pregnant woman, beginning with the secretion of hCG (or HCG). Note the large word PLACENTA

with an HCG arrow indicating secretion by and stimulating the placenta to produce greater amounts of the hormones represented by a solid line of arrows.
2. Color the parturition panel. Complete the left illustration before going on to the next. Note that the wall of the cervix is uncolored.

PREGNANCY

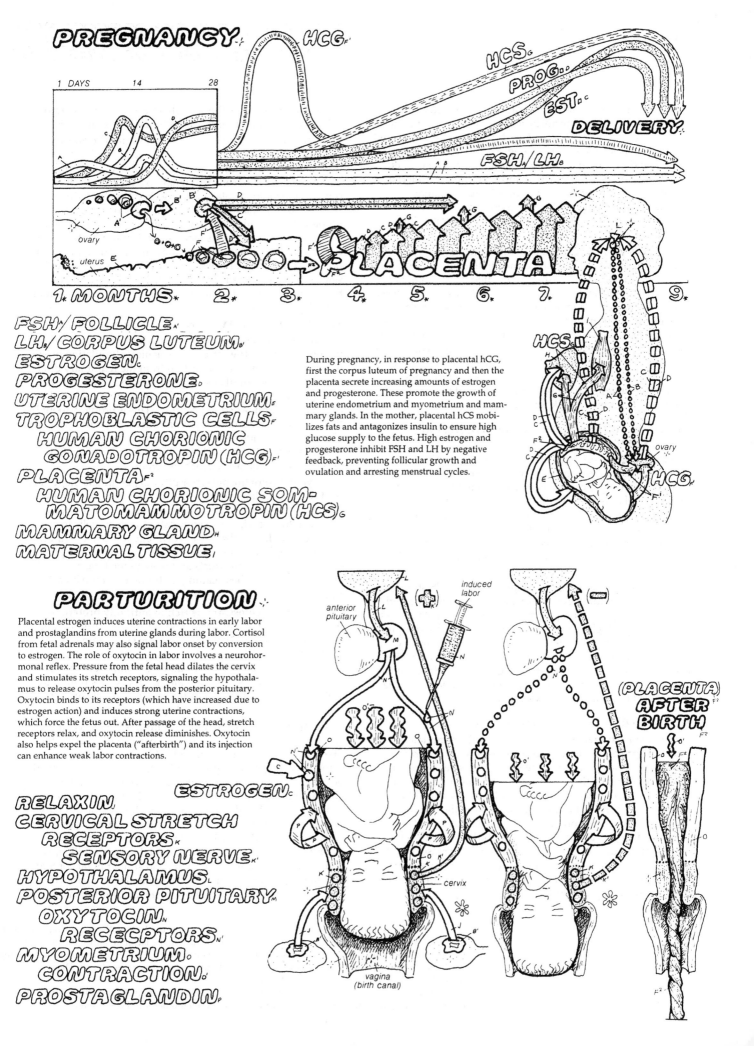

1 DAYS 14 28

ovary

uterus

HCG

HCS

PROG.

EST.

DELIVERY

FSH/LH

PLACENTA

1. MONTHS   2   3   4   5   6   7   9

HCS

ovary

HCG

FSH/FOLLICLE
LH/CORPUS LUTEUM
ESTROGEN
PROGESTERONE
UTERINE ENDOMETRIUM
TROPHOBLASTIC CELLS
HUMAN CHORIONIC GONADOTROPIN (HCG)
PLACENTA
HUMAN CHORIONIC SOM-MATOMAMMOTROPIN (HCS)
MAMMARY GLAND
MATERNAL TISSUE

During pregnancy, in response to placental hCG, first the corpus luteum of pregnancy and then the placenta secrete increasing amounts of estrogen and progesterone. These promote the growth of uterine endometrium and myometrium and mammary glands. In the mother, placental hCS mobilizes fats and antagonizes insulin to ensure high glucose supply to the fetus. High estrogen and progesterone inhibit FSH and LH by negative feedback, preventing follicular growth and ovulation and arresting menstrual cycles.

PARTURITION

Placental estrogen induces uterine contractions in early labor and prostaglandins from uterine glands during labor. Cortisol from fetal adrenals may also signal labor onset by conversion to estrogen. The role of oxytocin in labor involves a neurohormonal reflex. Pressure from the fetal head dilates the cervix and stimulates its stretch receptors, signaling the hypothalamus to release oxytocin pulses from the posterior pituitary. Oxytocin binds to its receptors (which have increased due to estrogen action) and induces strong uterine contractions, which force the fetus out. After passage of the head, stretch receptors relax, and oxytocin release diminishes. Oxytocin also helps expel the placenta ("afterbirth") and its injection can enhance weak labor contractions.

anterior pituitary

induced labor

(+)

(−)

(PLACENTA) AFTER BIRTH

ESTROGEN

cervix

vagina (birth canal)

RELAXIN
CERVICAL STRETCH RECEPTORS
SENSORY NERVE
HYPOTHALAMUS
POSTERIOR PITUITARY
OXYTOCIN
RECECPTORS
MYOMETRIUM
CONTRACTION
PROSTAGLANDIN

Mammals, as their name implies, are characterized by nourishing their newborn directly by *milk*, secreted by the mother's *mammary glands*, located in the *breast*. The size of the breast in human females depends on the degree of mammary gland growth and the amount of *fatty tissue* interspersed between the gland's lobules. Mammary glands and breasts undergo initial growth during *puberty* but prodigious mammary growth occurs mainly during *pregnancy*. Lactation refers to the formation of milk by the mammary glands following childbirth.

Mammary glands consist of alveoli and ducts—The mammary glands are exocrine glands with extensive *alveoli* and *ducts*. Alveolar cells extract raw materials (glucose, fatty acids, and amino acids), synthesize the milk proteins, lactose, and other nutrients, and secrete the milk into the *alveolar sacs*. The milk flows initially through small ducts that converge to form larger ducts, which connect with their outlets in the nipples. Specific smooth muscle *myoepithelial cells* form contractile rings around the mammary ducts. Contractions of these ducts force the milk out. Nipples also contain tactile receptors whose stimulation during suckling are important for *milk ejection* and continued milk formation.

HORMONES CONTROL MAMMARY GROWTH DURING VARIOUS STAGES

Puberty stage: estrogen stimulates duct growth, progesterone alveolar growth—During adolescence, in response to rising levels of sex steroids from the ovaries, the mammary glands begin to develop. *Estrogen* enhances duct growth and *progesterone* stimulates alveolar development. Alveoli are sparse in adolescents. Several other hormones (*insulin, growth hormone, prolactin* and *adrenal glucocorticoids*) also are necessary for the successful actions of sex steroids on mammary growth at this stage.

Pregnancy stage: sex steroids, prolactin, & hCS stimulate prodigious mammary growth—During pregnancy, the high levels of estrogen and progesterone from the placenta, as well as increasingly high levels of prolactin from the anterior pituitary, stimulate prodigious development of the mammary glands in preparation for milk production. The placental hormone *chorionic somatomammotropin (hCS)* (plate 158), as well as cortisol, insulin, and thyroid and growth hormones, synergize the effects of prolactin and sex steroids.

PROLACTIN REGULATES MILK FORMATION

Reduced estrogen after birth initiates the stimulatory effects of prolactin on milk formation—Prolactin hormone from the anterior pituitary is the major hormone stimulating milk production by the alveolar cells, but this effect is blocked by high levels of placental estrogen. Maternal sex steroid levels decrease sharply at birth following loss of the placenta. This condition allows prolactin to freely stimulate milk secretion, which begins about 1–3 days after birth. Prolactin level is highest at parturition; in lactating women, it decreases by 50% during the first postpartum week and reaches pregestation levels by 6 months postpartum. How does prolactin continue to stimulate milk production?

Suckling serves as the stimulus for pulsatile prolactin release & continued milk production—Prolactin levels increase sharply after each suckling episode. The effective stimulus for continued secretion of prolactin and milk production is the suckling-induced stimulation of the nipple tactile receptors. These sensory signals excite the hypothalamic centers controlling prolactin release, resulting in reduced secretion of *hypothalamic release-inhibiting hormone* for prolactin (dopamine) as well as increase in *hypothalamic release hormone* for prolactin. These effects increase prolactin pulsatile release from the anterior pituitary, resulting in continued milk formation. Regular artificial massages of the nipples will have the same effect. Prolonged absence of such regular stimulation of the nipples leads to decline in prolactin release and cessation of milk production.

MILK EJECTION INVOLVES A NEUROHORMONAL REFLEX

Nipple stimulation & afferents to hypothalamus form the neural part of the reflex—Mechanical stimulation of the nipple also enhances milk ejection from the mammary ducts. Secreted milk accumulates in the alveoli and ducts but will not flow out unless the myoepithelial smooth muscle cells around the mammary ducts contract. This is brought about by the action of the hormone *oxytocin* from the posterior pituitary gland. A neuroendocrine reflex controls this process. Sensory impulses generated by sucking stimuli travel up the afferent nerves from the breast to reach the brain.

Oxytocin from the posterior pituitary contracts mammary ducts and stimulates milk outflow—This activates the hypothalamus–posterior pituitary system and promotes oxytocin release in the blood, which stimulates the ducts cells to cause milk ejection. In the absence of such regular sensory stimuli from the nipples, the secreted milk accumulates in the glands, causing inflation of the ducts and pain, in the long run, milk production diminishes, leading to drying up of the breast.

MILK IS A RICH SOURCE OF INFANT NUTRITION

Initial milk (colostrum) is rich in antibody proteins—Around the time of childbirth, and before the onset of milk formation, the mammary glands secrete small amounts of a thick substance called *colostrum*, which contains no fat and little water but is rich in proteins and other milk constituents. Colostrum is a rich source of maternal antibodies (immunoglobulin A). The infant's intestine is capable of absorbing immunoglobulins, which provide passive immunity.

Composition of normal human milk—Milk is a complete source of nutrition for the newborn, particularly during the first year, although human infants often suckle for two years before weaning. Milk is produced initially at a rate of 500 mL per day; this rate doubles by the sixth month of lactation. Human milk contains 88% water, carbohydrates (lactose), protein (casein, lactalbumins), and fat (cholesterol and fatty acids such as linoleic acid) as well as many vitamins and minerals. Initially, milk is richer in protein but later the fat and lactose proportions increase. Compared to cow's milk, it is higher in lactose, lower in protein, and similar in fat content. Except for iron and vitamin D, the mineral and vitamin content of human milk make it a complete food for the infant.

CN: Use same colors as on the previous page for estrogen (F) and progesterone (G).
1. Color the stages of breast development, completing each before going on to the next. Note the increased size of the prolactin (J) arrow to reflect the greater amount of flow. Note, too, that in the pregnancy panel, part of the estrogen (F) output has the effect of blocking the prolactin (J) effect on the breast. In the lactation panel, the blowup of a portion of breast development shows the secretion of milk globules from the alveoli. These are left uncolored.
2. Color the chart illustrating the comparison between mother's and cow's milk.

ADOLESCENCE *

Early in puberty, the mammary glands are poorly developed and breasts contain little subcutaneous fat. During adolescence, estrogen promotes the development of the mammary ducts and the deposition of fatty tissue, while progesterone induces alveolar development. Growth hormones, glucocorticoids, and insulin are also necessary. Prolactin secretion from the anterior pituitary is low due to the strong inhibitory effect of hypothalamic-inhibiting hormone.

BREAST /:
DUCT A
FATTY TISSUE B
ALVEOLI c
NIPPLE D
TOUCH RECEPTOR D'
MYOEPITHELIAL CELL E

HORMONES /:
ESTROGEN F
PROGESTERONE G
PROLACTIN-INHIBITING HORMONE H
PROLACTIN-RELEASING HORMONE I
PROLACTIN J
HUMAN CHORIONIC SOMATOMAMMOTROPIN K
GLUCOCORTICOID L
INSULIN M
GROWTH HORMONE N

YOUNG ADULT *

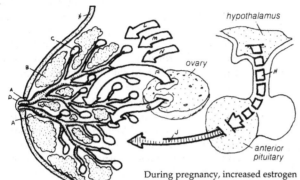

PREGNANCY *

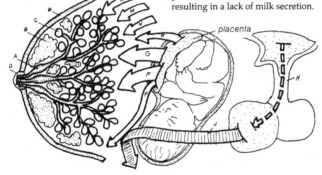

During pregnancy, increased estrogen and progesterone markedly promote mammary growth. Placental glucocorticoids and somatomammotropin and insulin are also needed for mammary and breast growth. Prolactin, the hormone that stimulates milk production, increases during pregnancy, but high estrogen and progesterone levels block prolactin's stimulation of alveoli, resulting in a lack of milk secretion.

COMPARISON OF MILK /:
MOTHER'S P
COW'S Q

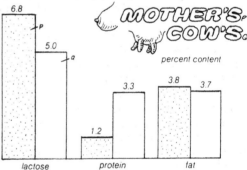

Milk contains all the nutrients needed for infant growth. Human milk contains carbohydrates (lactose), protein (casein, lactalbumins), fat, minerals, and vitamins. Cow's milk contains the same nutrients, but in different proportions.

MILK FORMATION PROLACTIN J
MILK EJECTION OXYTOCIN O

LACTATION *

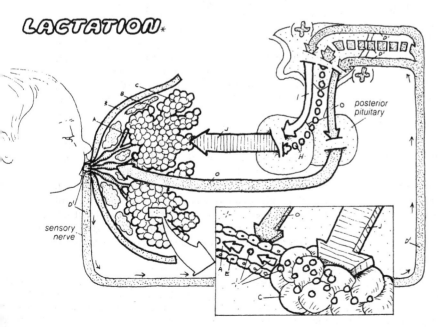

After childbirth, estrogen and progesterone levels decline sharply because of loss of the placenta. Prolactin, no longer inhibited by sex steroids, stimulates the alveoli and milk is produced. The infant's sucking activity stimulates nipple tactile receptors. Sensory impulses activate the hypothalamus and cause increased secretion of prolactin-releasing hormone. This stimulates surges of prolactin release, ensuring continued milk formation. Infant sucking stimuli also cause release of oxytocin from the posterior pituitary. Oxytocin then stimulates the contractions of myoepithelial cells of the mammary ducts, forcing milk out of the nipple.

SEX DETERMINATION & SEXUAL DEVELOPMENT

The true gender of an individual is based primarily on the type of sex chromosomes (X and Y) the zygote has at conception. Somatic cells of normal males have 22 pairs of somatic chromosomes plus one X and one Y chromosome (XY). Normal females have 22 + XX combination. The expression of the specific sex-related genes during the various phases of development produces the biological and behavioral phenotype of sexuality. Thus, as explained below, human genetic sex is determined at fertilization, the phenotypes of gonads and genitalia at embryonic weeks 8 and 12 respectively, the brain hypothalamus sexual phenotype in the late fetal period, and final maturation of the reproductive system and the secondary sex characteristics during puberty and adolescence. The male hormone testosterone (T) plays critical roles in of sexual development. The actions of sex hormones on sexual maturation during puberty are discussed in plates 128 and 152.

X & Y chromosomes determine the genetic sex at fertilization — In males, meiosis of spermatocytes results in two types of sperms, one bearing the X chromosome and another with the Y chromosome. Meiosis of primary oocytes in the female produces only X-bearing eggs. Fertilization of the egg by a sperm carrying an X chromosome produces a XX zygote — i.e., female — while a Y-bearing sperm produces an XY zygote (male). The genetic sex is determined by the father.

X & Y sperms show functional differences — The X- and Y-bearing sperms show functional differences as well; Y sperms are lighter and may swim faster, explaining why more male zygotes are formed even if the numbers of X and Y sperm number are equal. Indeed, the majority of spontaneously aborted embryos are male, yet the male-to-female sex ratio at birth is 107 to 100, implying higher rates of male conceptions. Differences between X and Y sperms are the basis of efforts to separate the two sperm types to predetermine the sex of the conceptus.

Embryonic gonads initially are sexually indifferent and bipotential — Up to the sixth week, the embryo does not show signs of sexual differentiation. The primordial gonads appear identical in both sexes and are sexually bipotential. Each gonad has a medulla and a cortex. In genetic males, by the 8th embryonic week, the cortex regresses and the medulla forms the fetal testis. The Leydig and Sertoli cells differentiate in the fetal testis and secrete testosterone (T) and Mullerian Inhibitory Substance (MIS), respectively. In the female embryo, the medullary part regresses and cortex develops into an ovary. The fetal ovary does not secrete any hormones.

Y-chromosome genes induce testis formation — The regression of the cortex and formation of the fetal testis are the result of the action of a single *testis-determining gene*, called the *SRY gene* (Sex-determining Region of Y chromosome), located on the short arm of the Y chromosome. The expression of this gene in the male embryo produces SRY protein, which acts as a transcription factor and promotes testis formation.

Ovary development occurs autonomously in the absence of SRY gene — Female embryos are missing the Y chromosome and hence the SRY gene, so the SRY protein will not be expressed. In the absence of these influences, the medulla part of the indifferent gonad degenerates and the cortex part autonomously develops into the female ovary around the eighth week of development.

Testosterone & MIS from the embryonic testis determine sexual development of the genitalia — Sex organs of the 7-week embryo are undifferentiated and the potential exists for development in the male or female direction. By the 12th week — i.e., the early fetus stage — the appropriate sex organs have differentiated in each sex from their bipotential primordia. This development depends on testis secretions. T from Leydig cells is secreted in the fetal blood and acts on the *Wolffian ducts* to induce the development of male *internal genitalia* (epididymis, seminal vesicles, vas deferens) on both sides. T also promotes the development of male *external genitalia* (penis and scrotum); for this action T is converted to another androgen, DHT (dihyrotestosterone), by the target tissue enzyme 5-α-reductase.

Sertoli cells release *MIS* (*Mullerian Inhibiting Substance*) into their immediate tissue environment. MIS from each testis induces regression of *Mullerian ducts* (primordia of female genitalia). In the absence of T and MIS — i.e., in a female embryo — the female internal and external genitalia develop autonomously from differentiation of the Mullerian ducts and other primordia.

Chromosomal, enzymic, and hormonal anomalies lead to abnormal genitalia — Embryos with no X chromosomes do not survive, but X-chromosome trisomy ("superfemale") is not abnormal. In the absence of a Y chromosome (XO; Turner syndrome), gonads will not differentiate but female genitalia will; puberty will not occur, owing to sex hormone deficiency. In the XXY pattern (Kleinfelter's syndrome), testes, male genitalia, and secondary sexual characteristics may develop, but not the seminiferous tubules. Absence of 5-α-reductase enzymes, converting T to DHT, creates female external genitalia in males (*male pseudohermaphrodites*). Female embryos exposed to high androgen levels from a fetal or maternal source (e.g., adrenal tumors) develop male external genitalia and deranged internal genitalia (*female pseudohermaphrodites*)

Testosterone regulates sexual differentiation of the hypothalamus — The hypothalamic mechanisms underlying sexual behavior and neuroendocrine control are undifferentiated and bipotential in the developing fetus. T induces differentiation of the *male-type hypothalamus*, promoting a *continuous* (non-cyclical) secretory pattern of GnRH pulses and gonadotropins FSH and LH. In rodents, this effect occurs in the neonate. A specific hypothalamic region, the Sexually Dimorphic Nucleus, is larger and well-developed in the male. T determines this effect and male-type behavior. Injections of T in newborn female rats results in a male-type GnRH pattern and sexual behavior. T effects on the developing brain are mediated by neuronal estrogen produced in the brain tissue from T by the action of aromatase enzyme.

In the absence of T (as in normal females), the hypothalamus spontaneously develops the *female-type* cyclical and pulsatile regulatory mechanisms of GnRH and of FSH and LH secretions as well as female sexual behavior. Similar T effects on hypothalamic sexual development occur in fetal monkeys; behavior is affected more than GnRH cyclicity. Structural differences in the hypothalamus of human males and females are known, but not the exact functional and behavioral correlates.

CN: Use very light colors for D, E, J, and K. Use colors previously used for estrogen (O), progesterone (G), FSH (N), and LH (O).
1. Begin at the bottom, coloring in the mature testis and ovary, and color the long arrows up the sides to the top panel, where those colors become primary spermatocytes and oocytes.
2. Color the embryonic stage, first doing the developing testis. Begin with the testis closest to the titles, and then do the outer one. Note that in the outer one, except for the Leydig cells (G), the entire structure receives the medulla color (E).

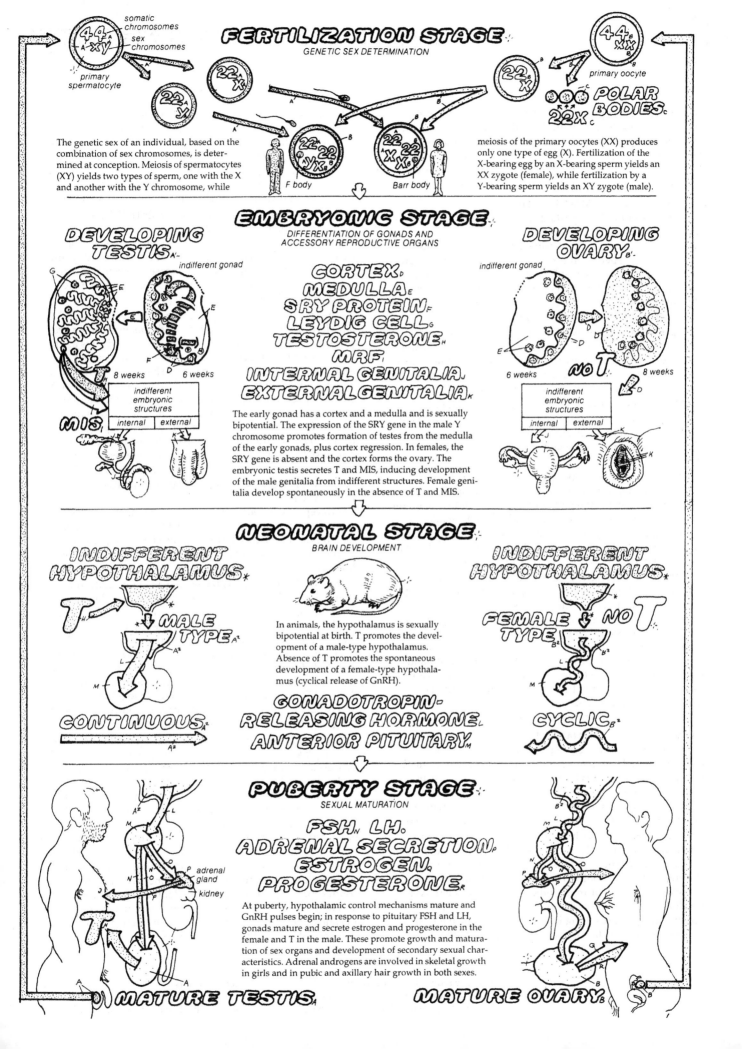

FERTILIZATION STAGE
GENETIC SEX DETERMINATION

somatic chromosomes
sex chromosomes
primary spermatocyte

primary oocyte

POLAR BODIES

F body
Barr body

The genetic sex of an individual, based on the combination of sex chromosomes, is determined at conception. Meiosis of spermatocytes (XY) yields two types of sperm, one with the X and another with the Y chromosome, while

meiosis of the primary oocytes (XX) produces only one type of egg (X). Fertilization of the X-bearing egg by an X-bearing sperm yields an XX zygote (female), while fertilization by a Y-bearing sperm yields an XY zygote (male).

EMBRYONIC STAGE
DIFFERENTIATION OF GONADS AND ACCESSORY REPRODUCTIVE ORGANS

DEVELOPING TESTIS
indifferent gonad
8 weeks   6 weeks
MIS
indifferent embryonic structures
internal | external

CORTEX
MEDULLA
SRY PROTEIN
LEYDIG CELL
TESTOSTERONE
MRF
INTERNAL GENITALIA
EXTERNAL GENITALIA

The early gonad has a cortex and a medulla and is sexually bipotential. The expression of the SRY gene in the male Y chromosome promotes formation of testes from the medulla of the early gonads, plus cortex regression. In females, the SRY gene is absent and the cortex forms the ovary. The embryonic testis secretes T and MIS, inducing development of the male genitalia from indifferent structures. Female genitalia develop spontaneously in the absence of T and MIS.

DEVELOPING OVARY
indifferent gonad
6 weeks   NO T   8 weeks
indifferent embryonic structures
internal | external

NEONATAL STAGE
BRAIN DEVELOPMENT

INDIFFERENT HYPOTHALAMUS
T
MALE TYPE
CONTINUOUS

In animals, the hypothalamus is sexually bipotential at birth. T promotes the development of a male-type hypothalamus. Absence of T promotes the spontaneous development of a female-type hypothalamus (cyclical release of GnRH).

GONADOTROPIN-RELEASING HORMONE
ANTERIOR PITUITARY

INDIFFERENT HYPOTHALAMUS
FEMALE TYPE   NO T
CYCLIC

PUBERTY STAGE
SEXUAL MATURATION

FSH   LH
ADRENAL SECRETION
ESTROGEN
PROGESTERONE

adrenal gland
kidney

At puberty, hypothalamic control mechanisms mature and GnRH pulses begin; in response to pituitary FSH and LH, gonads mature and secrete estrogen and progesterone in the female and T in the male. These promote growth and maturation of sex organs and development of secondary sexual characteristics. Adrenal androgens are involved in skeletal growth in girls and in pubic and axillary hair growth in both sexes.

T

MATURE TESTIS

MATURE OVARY

Normal fertility depends on the proper functioning of the reproductive systems in both the male and the female. One of every six couples has an infertility-related problem that prevents normal pregnancy. Causes of infertility include problems with sperm, eggs, and ovulation. Hormonal treatment and *in vitro* fertilization have reduced infertility.

FACTORS AFFECTING MALE FERTILITY

Sperm number is important for male fertility — In males, low sperm count and/or a high proportion of abnormal sperms is a major cause of sterility. Normal male ejaculate has about 300×10^6 sperms (100×10^6/ml of semen), even though only one sperm is sufficient for fertilization. Men with sperm counts below 20% of normal are sterile; sperm count between 20% and 40% of normal increases fertility to 50%. The basis for high sperm number is detailed in plate 156.

Ejaculation frequency — Since the sperm production rate is constant at about 200×10^6/day, frequent ejaculation leads to a low sperm number in ejaculate and reduced fertility. About 3–4 ejaculations per week are in accord with adequate sperm delivery from the epididymis and normal fertility.

Abnormal sperm — Abnormal sperm cells with no tail, two tails, coiled tails, no heads, two heads, or small heads make up about 20% of the sperm population in normal fertile men; higher proportions are associated with increasingly higher degrees of infertility.

High temperature — Sperm formation proceeds optimally at 32°C , five degrees below core body temperature. If the testes are kept inside the body or too close to it, the seminiferous tubules reversibly degenerate and sperm formation ceases. Tight clothing worn by athletes may result in a decline in sperm counts diminishing fertility; 30 minutes in a hot bath (43–45°C) may lead to a 90% decline in sperm number.

Other factors affecting male fertility — Excessive consumption of alcohol, major stresses, malnutrition, some infections (mumps), and cadmium salts as well as some natural compounds and drugs reduce sperm count and fertility. Gossypol, a cottonseed oil compound, inhibits spermatogenesis reversibly by inactivating the spermatids. Gossypol and the hormone inhibin — known to inhibit FSH, Sertoli cells, and spermatogenesis — are therefore potential male contraceptives. X-irradiation and other forms of ionizing radiation decrease male fertility. Sperm production is higher in winter regardless of scrotum temperature.

FACTORS AFFECTING FEMALE FERTILITY

Aging of sperm & egg — Sperms in the female tract undergo aging but retain motility and the ability to fertilize for up to four days; survival is best in the cervical mucus. Freshly ovulated eggs are mature; the best time for fertilization is about 12 hours after ovulation. Thereafter, eggs gradually age and become overripe, unable to be fertilized.

Fertility declines with age in females — Since oogonia divide in the embryonic period only, the ovarian oocytes are as old the female herself. Most oocytes die by atresia during childhood. Decline continues throughout maturity; by age 50, the ovaries have no primary follicles and oocytes left. As a result of ovarian losses, menstrual cycles and ovulation become irregular, resulting in a gradual decline in fertility. Pregnancy rates decline from the early forties to the late forties; by fifty years, nearly all women are sterile. This is the *menopause* phase, characterized by cessation of menstrual cycles, ovulation, fertility, and pregnancy. Although males show a gradual reduction in fertility during old age, the testes do not show aging changes similar to those of the ovaries and there is no male equivalent of the female menopause. Men have been known to father children even in their eighties.

Hormonal treatments enhance female fertility — Injections of gonadotropins LH and FSH or hCG enhance the number of follicles developing in the ovaries as well the chances of their ovulation and corpus luteal growth; as a result, fertility and the likelihood of pregnancy are increased. Recently, purified GnRH or its analogs (clomiphene) has been used to increase endogenous LH and FSH.

In vitro fertilization — If *in vivo* fertility treatments fail, *in vitro* fertilization methods may be undertaken. Women are primed with hormones, as described above, and eggs are harvested from the oviduct. Sperms are collected, washed, and added to the eggs in glass tubes or dishes. After fertilization, zygotes with pronuclei are allowed to develop to the 4- to 8-cell stage; several such embryos are placed in the uterine cavity of a progesterone-treated woman and allowed to implant. The *in vitro* methods increase the probability of pregnancy from 0 to about 20% — a remarkable feat, given that the normal rate of successful pregnancy in fertile couples is 40% at best. Fertility treatments occasionally result in multiple births.

"CONTRACEPTION" REFERS TO MEASURES TO PREVENT PREGNANCY

Contraception or "birth control" may be achieved by a variety of mechanical and physiological methods. Contraceptive methods aim to reduce fertility or the chance of pregnancy by preventing ovulation, the encounter of sperm and egg, fertilization, or implantation.

The rhythm method is based on time of ovulation & sperm survival — In the *rhythm method*, coitus is abstained from during the period of ovulation, when women are most fertile (from 4 days before to 3 days after ovulation, in consideration of the 4-day survival time of sperms in the female tract. The time of ovulation can be estimated by measuring *basal body temperature* every morning before leaving bed. One to two days after ovulation, body temperature rises about 0.4°C (1°F), from a low of 36.4°C at day 13 to a high at 36.8°C at day 22; the rise is caused by the postovulatory increase in progesterone from the corpus luteum.

Contraceptive pills work by inhibiting ovulation — *Oral contraceptive pills* contain synthetic estrogen or estrogen and progesterone; they prevent ovulation by feedback inhibition of the rise of LH in the cycle and its burst at ovulation. One pill is taken each day for 21 days, from fifth day of menstruation. Menstruation resumes a day or two after the last pill. Women desiring pregnancy regain normal cycles one to several months after discontinuing the pill. Pregnancy within the first 1–3 months is not encouraged, because of the possibility of multiple ovulation and pregnancy owing to the rebound of pituitary gonadotropins. Other hormones with potential as contraceptive agents are GnRH and inhibin. Continuous and high levels of GnRH desensitize the pituitary, reducing LH and FSH secretion and preventing ovulation. Inhibin reduces FSH and inhibits follicular growth.

CN: Use red for C and a dark color for E.
1. Begin with the four factors affecting male fertility.
2. Note the presence of sperm (A) among the

female factors.
3. Color the methods of contraception.
4. Color the two most common sterilization sites.

FERTILITY

FACTORS AFFECTING MALES:

SPERM NUMBER

NORMAL:
100,000,000
per mL/semen

STERILE:
<20,000,000
per mL/semen

A normal sperm count (~100 million/ml of semen) is necessary for male fertility. Below 40% of normal, fertility is reduced by 50%. Below 20%, fertility ceases.

EJACULATIONS

% of sperm in ejaculate

number of ejaculations

Since the sperm production rate is limited (~200 million/day), repeated ejaculation gradually decreases the number of sperm in the ejaculate.

BAD SPERM

About 20% of sperm are abnormal: they may have no tails, two tails, twisted tails, no heads, two heads, or shrunken heads. More abnormalities decreases fertility.

TEMPERATURE

body temperature

32°C   37°C

Sperm form normally at 32°C (5°C below body temperature). If the testes are subject to heat, seminiferous tubules reversibly degenerate, interrupting sperm production and causing sterility.

FACTORS AFFECTING FEMALES:

AGE OF SPERM, EGG

(viable)   ovulated egg

optimal fertility

1-4 DAYS   1 DAY

A woman's optimal fertility period in each monthly cycle is within 2 days of ovulation. Sperm usually survive ~1 day, but some up to 4 days. Most eggs live ~1 day, few up to 2 days.

FEMALE AGE

eggs per ovary   *fertility rate*   *menopausal*

pubescent   30   40   50

Fertility is high in females in their 20s and early 30s, declining beyond the late 30s to very low levels in the late 40s and ceasing around 50 (menopause).

HORMONES / IN VITRO FERTILIZATION

EGG   SPERM

LH   FSH   hCG   1

uterine cavity

2   3   4   5

For *in vitro* fertilization, a woman is first treated with hormones (FSH, LH, hCG, GnRH) to increase follicular growth and ovulation rates (1). Eggs are collected (2) and mixed with male sperm (3) in a test tube (dish) for fertilization (4). Young embryos are then transferred to the uterus of the progesterone-treated mother and allowed to implant (5).

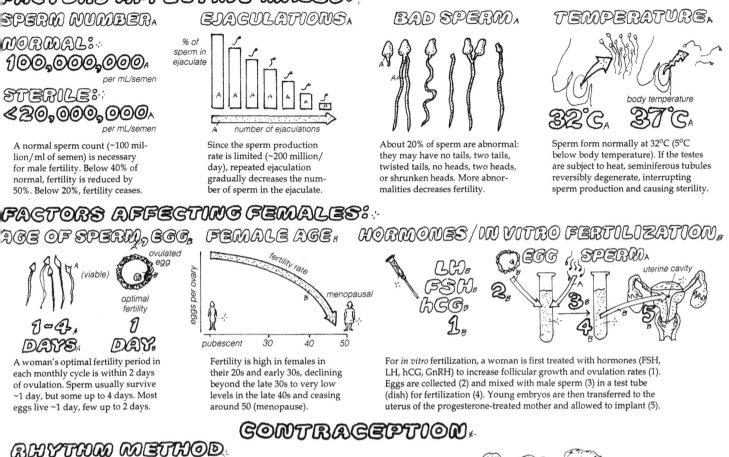

CONTRACEPTION

RHYTHM METHOD
MENSTRUATION
SPERM VIABILITY
OVULATION
EGG VIABILITY
SAFE DAYS

In the rhythm method, coitus is avoided during the week when the likihood of pregnancy is high. This period (4 days before ovulation to 3 days after) is based on the maximum survival time of sperm (4 days) and egg (2 days) in the female reproductive tract, allowing for ovulation variability.

C°   37.2   37 (98.6)   36.7

1 2 3 4 5 6 7 8 9 10 11 12 13 14 15 16 17 18 19 20 21 22 23 24 25 26 27 28
DAY

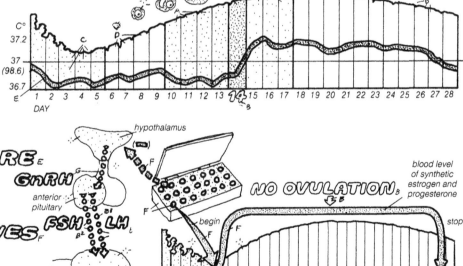

BASAL TEMPERATURE

Basal body temperature, taken early in the morning, before leaving bed, shows a rise of ~0.4°C (~1°F) after ovulation. This rise indicates ovulation and is caused by progesterone secretion; it lasts until the next menstruation.

ORAL CONTRACEPTIVES

Oral contraceptives (pills containing synthetic estrogen and progesterone) are taken by women for 21 days following menstruation. The rapid increase of these estrogen-like substances in the blood inhibits the release of FSH and LH, preventing follicular growth and ovulation.

hypothalamus
GnRH
anterior pituitary
FSH LH
ovary (no follicle growth)

NO OVULATION
blood level of synthetic estrogen and progesterone
begin   stop

STERILIZATION

TUBAL LIGATION
UTERINE TUBE

Tubal ligation (cutting and tying or cauterization of the uterine tubes) results in permanent sterility, since sperm can no longer reach the ovulated egg. Tubal ligation and vasectomy (illustration to the right) have a 50% chance of reversibility.

VASECTOMY,
VAS DEFERENS

Cutting and tying the two vas deferens ducts is a simple operation that permanently obstructs the delivery of sperm through the vas deferens during ejaculation, thereby causing male sterility.